Modeling and Design of Secure Internet of Things

IEEE Press
445 Hoes Lane
Piscataway, NJ 08854

Modeling and Design of Secure Internet of Things

Edited by

Charles A. Kamhoua

Laurent L. Njilla

Alexander Kott

Sachin Shetty

IEEE PRESS

WILEY

Published by John Wiley & Sons, Inc., Hoboken, New Jersey.
Published simultaneously in Canada.

Library of Congress Cataloging-in-Publication data applied for

ISBN: 9781119593362

Set in 9.5/12.5pt STIXTwoText by SPi Global, Pondicherry, India

Cover Design: Wiley
Cover Image: © Photographer is my life./Getty Images

10 9 8 7 6 5 4 3 2 1

Contents

About the Editors *ix*
List of Contributors *xiii*
Foreword *xix*
Preface *xxiii*

1 **Introduction** *1*
Charles A. Kamhoua, Laurent L. Njilla, Alexander Kott, and Sachin Shetty

Part I Game Theory for Cyber Deception *27*

2 **Game-Theoretic Analysis of Cyber Deception: Evidence-Based Strategies and Dynamic Risk Mitigation** *29*
Tao Zhang, Linan Huang, Jeffrey Pawlick, and Quanyan Zhu

3 **A Hypergame-Based Defense Strategy Toward Cyber Deception in Internet of Battlefield Things (IoBT)** *59*
Bowei Xi and Charles A. Kamhoua

4 **Cooperative Spectrum Sharing and Trust Management in IoT Networks** *79*
Fatemeh Afghah, Alireza Shamsoshoara, Laurent L. Njilla, and Charles A. Kamhoua

5 **Adaptation and Deception in Adversarial Cyber Operations** *111*
George Cybenko

6 **On Development of a Game-Theoretic Model for Deception-Based Security** *123*
Satyaki Nan, Swastik Brahma, Charles A. Kamhoua, and Laurent L. Njilla

7 **Deception for Cyber Adversaries: Status, Challenges, and Perspectives** *141*
Abdullah Alshammari, Danda B. Rawat, Moses Garuba, Charles A. Kamhoua, and Laurent L. Njilla

Part II IoT Security Modeling and Analysis *161*

8 **Cyber-Physical Vulnerability Analysis of IoT Applications Using Multi-Modeling** *163*
Ted Bapty, Abhishek Dubey, and Janos Sztipanovits

9 **Securing Smart Cities: Implications and Challenges** *185*
Ioannis Agadakos, Prashant Anantharaman, Gabriela F. Ciocarlie, Bogdan Copos, Michael Emmi, Tancrède Lepoint, Ulf Lindqvist, Michael Locasto, and Liwei Song

10 **Modeling and Analysis of Integrated Proactive Defense Mechanisms for Internet of Things** *217*
Mengmeng Ge, Jin-Hee Cho, Bilal Ishfaq, and Dong Seong Kim

11 **Addressing Polymorphic Advanced Threats in Internet of Things Networks by Cross-Layer Profiling** *249*
Hisham Alasmary, Afsah Anwar, Laurent L. Njilla, Charles A. Kamhoua, and Aziz Mohaisen

12 **Analysis of Stepping-Stone Attacks in Internet of Things Using Dynamic Vulnerability Graphs** *273*
Marco Gamarra, Sachin Shetty, Oscar Gonzalez, David M. Nicol, Charles A. Kamhoua, and Laurent L. Njilla

13 **Anomaly Behavior Analysis of IoT Protocols** *295*
Pratik Satam, Shalaka Satam, Salim Hariri, and Amany Alshawi

14 **Dynamic Cyber Deception Using Partially Observable Monte-Carlo Planning Framework** *331*
Md Ali Reza Al Amin, Sachin Shetty, Laurent L. Njilla, Deepak K. Tosh, and Charles A. Kamhoua

15 **A Coding Theoretic View of Secure State Reconstruction** *357*
Suhas Diggavi and Paulo Tabuada

16 **Governance for the Internet of Things: Striving Toward Resilience** *371*
S. E. Galaitsi, Benjamin D. Trump, and Igor Linkov

Part III IoT Security Design *383*

17 **Secure and Resilient Control of IoT-Based 3D Printers** *385*
Zhiheng Xu and Quanyan Zhu

18 **Proactive Defense Against Security Threats on IoT Hardware** *407*
Qiaoyan Yu, Zhiming Zhang, and Jaya Dofe

19 **IoT Device Attestation: From a Cross-Layer Perspective** *435*
Orlando Arias, Fahim Rahman, Mark Tehranipoor, and Yier Jin

20 Software-Defined Networking for Cyber Resilience in Industrial Internet of Things (IIoT) *453*
Kamrul Hasan, Sachin Shetty, Amin Hassanzadeh, Malek Ben Salem, and Jay Chen

21 Leverage SDN for Cyber-Security Deception in Internet of Things *479*
Yaoqing Liu, Garegin Grigoryan, Charles A. Kamhoua, and Laurent L. Njilla

22 Decentralized Access Control for IoT Based on Blockchain and Smart Contract *505*
Ronghua Xu, Yu Chen, and Erik Blasch

23 Intent as a Secure Design Primitive *529*
Prashant Anantharaman, J. Peter Brady, Ira Ray Jenkins, Vijay H. Kothari, Michael C. Millian, Kartik Palani, Kirti V. Rathore, Jason Reeves, Rebecca Shapiro, Syed H. Tanveer, Sergey Bratus, and Sean W. Smith

24 A Review of Moving Target Defense Mechanisms for Internet of Things Applications *563*
Nico Saputro, Samet Tonyali, Abdullah Aydeger, Kemal Akkaya, Mohammad A. Rahman, and Selcuk Uluagac

25 Toward Robust Outlier Detector for Internet of Things Applications *615*
Raj Mani Shukla and Shamik Sengupta

26 Summary and Future Work *635*
Charles A. Kamhoua, Laurent L. Njilla, Alexander Kott, and Sachin Shetty

Index *647*

About the Editors

Charles A. Kamhoua is a Senior Electronics Engineer at the Network Security Branch of the US Army Research Laboratory (ARL) in Adelphi, MD, where he is responsible for conducting and directing basic research in the area of game theory applied to cyber security. Prior to joining the Army Research Laboratory, he was a researcher at the US Air Force Research Laboratory (AFRL), Rome, New York, for 6 years and an educator in different academic institutions for more than 10 years. He has held visiting research positions at the University of Oxford and Harvard University. He has coauthored more than 200 peer-reviewed journal and conference papers that include 5 best paper awards. He is a coinventor of 3 patents and 4 patent applications. He has been at the forefront of several new technologies, coediting three books at Wiley-IEEE Press entitled *Assured Cloud Computing*, *Blockchain for Distributed System Security*, and *Modeling and Design of Secure Internet of Things*. He has presented over 60 invited keynote and distinguished speeches and has co-organized over 10 conferences and workshops. He has mentored more than 60 young scholars, including students, postdocs, and Summer Faculty Fellow. He has been recognized for his scholarship and leadership with numerous prestigious awards, including the 2019 US Army Civilian Service Commendation Medal, the 2019 Federal 100-FCW annual awards for individuals that have had an exceptional impact on federal IT, the 2019 IEEE ComSoc Technical Committee on Big Data (TCBD) Best Journal Paper Award, the 2018 ARL Achievement Award for leadership and outstanding contribution to the ARL Cyber Camo (cyber deception) project, the 2018 Fulbright Senior Specialist Fellowship, the 2017 AFRL Information Directorate Basic Research Award "For Outstanding Achievements in Basic Research," the 2017 Fred I. Diamond Award for the best paper published at AFRL's

Information Directorate, 40 Air Force Notable Achievement Awards, the 2016 FIU Charles E. Perry Young Alumni Visionary Award, the 2015 Black Engineer of the Year Award (BEYA), the 2015 NSBE Golden Torch Award – Pioneer of the Year, and selection to the 2015 Heidelberg Laureate Forum, to name a few. He has been congratulated by the White House, the US Congress, and the Pentagon for those achievements. He received a B.S. in electronics from the University of Douala (ENSET), Cameroon, in 1999, an MS in Telecommunication and Networking from Florida International University (FIU) in 2008, and a PhD in Electrical Engineering from FIU in 2011. He is currently an advisor for the National Research Council postdoc program, a member of the FIU alumni association and Sigma Xi, and a senior member of ACM and IEEE.

Laurent L. Njilla joined the Cyber Assurance Branch of the US Air Force Research Laboratory (AFRL), Rome, NY, as a Research Electronics Engineer in 2015. As a researcher, he is responsible for conducting and directing basic research in the area of cyber defense, cyber physical system, cyber resiliency, hardware security, and the application of game theory, category theory, and Blockchain technology. He is the Program Manager of the Center of Excellence (CoE) in Cyber Security for the Historically Black Colleges and Universities & Minorities Institutions (HBCU/MI), and the Program Manager of the Disruptive Information Technology Program at AFRL/RI. He has coauthored over 70 peer-reviewed journal and conference papers with a best paper award. He is a coinventor of 2 patents and 3 patent applications. Coediting of two books at Wiley-IEEE Press entitled *Blockchain for Distributed System Security* and *Modeling and Design of Secure Internet of Things*. His mentorship of young students and scholars is recognized with multiple awards including Air Force Notable Achievement awards, FIU Distinguished Alumni in Government Service award, and the 2015 FIU World Ahead Graduate award. Prior to joining the AFRL, he was a Senior Systems Analyst in the industry sector for more than 10 years. He is a reviewer of multiple journals and serves on the technical program committees of several international conferences. He received his BS in Computer Science from the University of Yaoundé-1 in Yaoundé, Cameroon, an MS in Computer Engineering from the University of Central Florida (UCF) in 2005, and a PhD in Electrical Engineering from Florida International University (FIU) in 2015. He is a member of the National Society of Black Engineer (NSBE).

Dr. Alexander Kott serves as the ARL's Chief Scientist. In this role, he provides leadership in development of ARL technical strategy, maintaining technical quality of ARL research, and representing ARL to external technical community.

Between 2009 and 2016, he was the Chief, Network Science Division, Computational and Information Sciences Directorate, US Army Research Laboratory headquartered in Adelphi, MD.

He was responsible for a diverse portfolio of fundamental research and applied development in network science and science for cyber defense.

In particular, he played a key role in initiating the Network Science Collaborative Technology Alliance, among the world-largest efforts to study interactions between networks of different types. His efforts helped start Cyber Security Collaborative Research Alliance, a unique program of creating basic science of cyber warfare.

In 2013, Dr. Kott served as the Acting Associate Director for Science and Technology of the ARL's Computational and Information Sciences Directorate; in 2015, he also served as the Acting Director of the Computational and Information Sciences Directorate.

Beginning his Government career, between 2003 and 2008, Dr. Kott served as a Defense Advanced Research Programs Agency (DARPA) Program Manager responsible for a number of large-scale advanced technology research programs. Technologies developed in programs under his management ranged from adversarial reasoning, to prediction of social and security phenomena, to command and control of robotic forces.

His earlier positions included Director of R&D at Carnegie Group, Pittsburgh, PA, and Information Technology Research Department Manager at AlliedSignal, Inc., Morristown, NJ. There, his work focused on novel information technology approaches, such as Artificial Intelligence, to complex problems in engineering design, and planning and control in manufacturing, telecommunications, and aviation industries.

Dr. Kott received the Secretary of Defense Exceptional Public Service Award and accompanying Exceptional Public Service Medal, in October 2008.

He earned his PhD from the University of Pittsburgh, Pittsburgh, PA in 1989, where his research proposed AI approaches to innovative design of complex systems.

He has published over 80 technical papers and served as the initiator, coauthor, and primary editor of over 10 books, including *Advanced Technology Concepts for Command and Control*, 2004; *Information Warfare and Organizational Decision Process*, 2006; *Adversarial Reasoning: Computational Approaches to Reading the Opponent's Mind*, 2006; *The Battle of Cognition: The Future Information-Rich Warfare and the Mind of the Commander*, 2007; *Estimating Impact: A Handbook of Computational Methods and Models for Anticipating Economic, Social, Political and Security Effects in International Interventions*, 2010; *Cyber Defense and Situational Awareness*, 2015; *Cyber Security of SCADA and other Industrial Control Systems*, 2016; and *Cyber Resilience* (2019).

Sachin Shetty is an Associate Director in the Virginia Modeling, Analysis and Simulation Center at Old Dominion University. He holds a joint appointment as an Associate Professor with the Department of Computational, Modeling and Simulation Engineering. Sachin Shetty received his PhD in Modeling and Simulation from the Old Dominion University in 2007. Prior to joining Old Dominion University, he was an Associate Professor with the Electrical and Computer Engineering Department at Tennessee State University. He was also the associate director of the Tennessee Interdisciplinary Graduate Engineering Research Institute and directed the Cyber Security laboratory at Tennessee State University. He also holds a dual appointment as an Engineer at the Naval Surface Warfare Center, Crane, IN. His research interests lie at the intersection of computer networking, network security, and machine learning. He has published over 200 research articles in journals and conference proceedings. He has also edited four books in the areas of blockchain, Internet of Things, moving target defense, and dynamic spectrum access. Two of his research papers have been selected at the top 50 Blockchain academic papers in 2018. His laboratory conducts cloud and mobile security research and has received over $12 million in funding from National Science Foundation, Air Office of Scientific Research, Air Force Research Lab, Office of Naval Research, Department of Homeland Security, and Boeing. He is the site lead on the DoD Cyber Security Center of Excellence, the Department of Homeland Security National Center of Excellence, the Critical Infrastructure Resilience Institute (CIRI), and Department of Energy, Cyber Resilient Energy Delivery Consortium (CREDC). He is the recipient of Fulbright Specialist award, EPRI Cybersecurity Research Challenge award, DHS Scientific Leadership Award, and has been inducted in Tennessee State University's million dollar club. He has served on the technical program committee for ACM CCS, IEEE INFOCOM, IEEE ICDCN, and IEEE ICCCN. He is a Senior Member of IEEE.

List of Contributors

Fatemeh Afghah
School of Informatics
Computing and Cyber Systems
Northern Arizona University
Flagstaff, AZ, USA

Ioannis Agadakos
SRI International
New York, NY, USA

Kemal Akkaya
Department of Electrical and
Computer Engineering
Florida International University
Miami, FL, USA

Hisham Alasmary
Department of Computer Science
University of Central Florida
Orlando, FL, USA

Abdullah Alshammari
Data Science and Cybersecurity
Center (DSC2)
Department of Electrical Engineering
and Computer Science
Howard University
Washington, DC, USA

Amany Alshawi
National Center for Cyber Security
Technology
King Abdulaziz City for Science and
Technology
Riyadh, Saudi Arabia

Md Ali Reza Al Amin
Computational Modeling and
Simulation Engineering
Old Dominion University
Norfolk, VA
USA

Prashant Anantharaman
Department of Computer Science
Dartmouth College
Hanover, NH
USA

Afsah Anwar
Department of Computer Science
University of Central Florida
Orlando, FL, USA

Orlando Arias
University of Central Florida
Orlando, FL, USA

Abdullah Aydeger
Department of Electrical and Computer Engineering
Florida International University
Miami, FL, USA

Ted Bapty
Institute for Software Integrated Systems
Vanderbilt University
Nashville, TN, USA

Erik Blasch
US Air Force Research Laboratory
Rome, NY, USA

J. Peter Brady
Department of Computer Science
Dartmouth College
Hanover, NH, USA

Swastik Brahma
Department of Computer Science
Tennessee State University
Nashville, TN, USA

Sergey Bratus
Department of Computer Science
Dartmouth College
Hanover, NH, USA

Jay Chen
Accenture Technology Lab
Arlington, VA
USA

Yu Chen
Department of Electrical and Computer Engineering
Binghamton University
SUNY, Binghamton
NY, USA

Jin-Hee Cho
Department of Computer science
Virginia Tech
Falls Church, VA, USA

Gabriela F. Ciocarlie
SRI International
New York, NY, USA

Bogdan Copos
Google Inc.
Mountain View, CA, USA

George Cybenko
Dorothy and Walter Gramm Professor of Engineering
Dartmouth College
Hanover, NH, USA

Suhas Diggavi
University of California
Los Angeles, Los Angeles, CA, USA

Jaya Dofe
Department of Computer Engineering
California State University
Fullerton, CA, USA

Abhishek Dubey
Institute for Software Integrated Systems
Vanderbilt University
Nashville, TN, USA

Michael Emmi
Amazon Inc.
New York, NY, USA

S. E. Galaitsi
US Army Engineer Research and Development Center
Vicksburg, MS, USA

Marco Gamarra
College of Engineering
Old Dominion University
Norfolk, VA, USA

Moses Garuba
Data Science and Cybersecurity Center (DSC2)
Department of Electrical Engineering and Computer Science
Howard University
Washington, DC
USA

Mengmeng Ge
School of Information Technology
Deakin University
Geelong, Victoria, Australia

Oscar Gonzalez
College of Engineering
Old Dominion University
Norfolk, VA, USA

Garegin Grigoryan
Computing and Information Sciences
Rochester Institute of Technology
Rochester, NY, USA

Salim Hariri
Department of Electrical and Computer Engineering
University of Arizona
Tucson, AZ, USA

Kamrul Hasan
Virginia Modeling Analysis and Simulation Center
Old Dominion University
Norfolk, VA, USA

Amin Hassanzadeh
Accenture Technology Lab
Arlington, VA, USA

Linan Huang
Department of Electrical and Computer Engineering
Tandon School of Engineering
New York University
Brooklyn, NY, USA

Bilal Ishfaq
Department of Computer Science and Software Engineering
University of Canterbury
Christchurch, New Zealand

Ira Ray Jenkins
Department of Computer Science
Dartmouth College
Hanover, NH, USA

Yier Jin
University of Florida
Gainesville, FL, USA

Charles A. Kamhoua
US Army Research Laboratory
Adelphi, MD, USA

Dong Seong Kim
School of Information Technology and Electrical Engineering
University of Queensland
Brisbane, Queensland, Australia

Vijay H. Kothari
Department of Computer Science
Dartmouth College
Hanover, NH, USA

Alexander Kott
US Army Research Laboratory
Adelphi, MD, USA

Tancrède Lepoint
Google Inc.
New York, NY, USA

Ulf Lindqvist
SRI International
San Luis Obispo, CA, USA

Igor Linkov
US Army Engineer Research and Development Center
Vicksburg, MS, USA

Yaoqing Liu
Computer Science
Fairleigh Dickinson University
Teaneck, NJ, USA

Michael Locasto
SRI International
New York, NY, USA

Michael C. Millian
Department of Computer Science
Dartmouth College
Hanover, NH, USA

Aziz Mohaisen
Department of Computer Science
University of Central Florida
Orlando, FL, USA

Satyaki Nan
Department of Computer Science
Tennessee State University
Nashville, TN, USA

David M. Nicol
Department of Electrical and Computer Engineering
University of Illinois at Urbana-Champaign
Champaign, IL, USA

Laurent L. Njilla
Cyber Assurance Branch
US Air Force Research Laboratory
Rome, NY, USA

Kartik Palani
Dartmouth College
Hanover, NH, USA

Jeffrey Pawlick
Department of Electrical and Computer Engineering
Tandon School of Engineering
New York University
Brooklyn, NY, USA

Fahim Rahman
University of Florida
Gainesville, FL, USA

Mohammad A. Rahman
Department of Electrical and Computer Engineering
Florida International University
Miami, FL, USA

Kirti V. Rathore
Department of Electrical and Computer Engineering
University of Illinois at Urbana-Champaign
Champaign, IL
USA

Danda B. Rawat
Data Science and Cybersecurity Center (DSC2)
Department of Electrical Engineering and Computer Science
Howard University
Washington, DC, USA

Jason Reeves
VMWare, Inc.
Palo Alto, CA, USA

Malek Ben Salem
Accenture Technology Lab
Arlington, VA, USA

Nico Saputro
Department of Electrical and Computer Engineering
Florida International University
Miami, FL, USA
and
Department of Electrical Engineering
Parahyangan Catholic University
Bandung, Indonesia

Pratik Satam
Department of Electrical and Computer Engineering
University of Arizona
Tucson, AZ, USA

Shalaka Satam
Department of Electrical and Computer Engineering
University of Arizona
Tucson, AZ, USA

Shamik Sengupta
Department of Computer Science and Engineering
University of Nevada
Reno, Reno, NV, USA

Alireza Shamsoshoara
School of Informatics
Computing and Cyber Systems
Northern Arizona University
Flagstaff, AZ, USA

Rebecca Shapiro
Champlain College
Burlington, VT, USA

Sachin Shetty
Virginia Modeling Analysis and Simulation Center
Old Dominion University
Norfolk, VA, USA

Raj Mani Shukla
Department of Computer Science and Engineering
University of Nevada
Reno, Reno, NV
USA

Sean W. Smith
Department of Computer Science
Dartmouth College
Hanover, NH
USA

Liwei Song
Princeton University
Princeton, NJ, USA

Janos Sztipanovits
Institute for Software Integrated Systems
Vanderbilt University
Nashville, TN, USA

Paulo Tabuada
University of California
Los Angeles, Los Angeles, CA, USA

Syed H. Tanveer
Department of Computer Science
Dartmouth College
Hanover, NH, USA

Mark Tehranipoor
University of Florida
Gainesville, FL, USA

Samet Tonyali
Department of Electrical and
Computer Engineering
Abdullah Gul University
Kayseri, Turkey

Deepak K. Tosh
Department of Computer Science
University of Texas at El Paso
El Paso, TX, USA

Benjamin D. Trump
US Army Engineer Research and
Development Center
Vicksburg, MS, USA

Selcuk Uluagac
Department of Electrical and
Computer Engineering
Florida International University
Miami, FL, USA

Bowei Xi
Department of Statistics
Purdue University
West Lafayette, IN, USA

Ronghua Xu
Department of Electrical and
Computer Engineering
Binghamton University
SUNY, Binghamton
NY, USA

Zhiheng Xu
Department of Electrical and
Computer Engineering
Tandon School of Engineering
New York University
Brooklyn, NY, USA

Qiaoyan Yu
Department of Electrical and
Computer Engineering
University of New Hampshire
Durham, NH, USA

Tao Zhang
Department of Electrical and
Computer Engineering
Tandon School of Engineering
New York University
Brooklyn, NY, USA

Zhiming Zhang
Department of Electrical and
Computer Engineering
University of New Hampshire
Durham, NH, USA

Quanyan Zhu
Department of Electrical and
Computer Engineering
Tandon School of Engineering
New York University
Brooklyn, NY, USA

Foreword

I am pleased to offer this Foreword to *Modeling and Design of Secure Internet of Things*.

Cybersecurity theorists and practitioners alike face the challenge of securing a new, global information technology ecosystem. Already, over half the people alive today are connected to the Internet, making moot the question of whether cyberspace is, indeed, its own "domain." 5G Internet, Internet Protocol Version 6 (IPv6), Artificial Intelligence (AI), and ubiquitous connectivity are converging to create new ways of managing infrastructures, businesses, government services, and other aspects of daily life. IPv6 is providing virtually unlimited (to be more precise, 2128 or approximately 3.4 × 1038) Internet Protocol addresses, which will allow the connectivity of any device to which an IP address can be assigned. 5G Internet offers high-speed, direct connectivity between and among "traditional" information technology devices and the Internet of Things (IoT) devices that will encompass our world. Indeed, the authors note that by 2020, perhaps 50 billion devices will be connected to the Internet and that, on average, each person will possess seven connected devices. Advanced, cloud-based analytics will provide us the means to find patterns and meaning in the behavior of the devices and networks that will comprise this new ecosystem; AI will allow us to direct the behavior of these devices, and the businesses and infrastructures they populate. Ubiquitous connectivity provided by today's carriers and tomorrow's low-earth orbit constellation-based carriers will provide the means by which people, networks, and devices will be connected constantly – everywhere, on land and sea, and in the air. The technologies needed to create "smart cities" will be combined, commoditized, productized, and taken to market as tech giants such as Google, Alibaba, and Amazon compete to build connected communities throughout the world. In fact, the convergence of these technologies is likely to create a new x-as-a-service environment one might call "6G," in which business services, including analytics and AI are offered as-a-service within and through global networks.

Where will this new ecosystem-of-things make its mark? My answer: everywhere! Critical and business infrastructures will rely increasingly on data from connected devices to optimize performance, from transportation and energy infrastructures, to complex global chains, and to adjusting the behavior of implanted medical devices. Manufacturers will modulate production on the fly, even as manufacturing becomes more distributed and 3D manufacturing devices expand in their presence.

National security systems will also depend increasingly on this new IoT-enabled ecosystem. Weapons systems will be comprised of IP-enabled devices designed to optimize performance, maintenance, and integration into the battlefield. Weapons systems and sensor with IoT devices will be connected via 5G battlefield networks to each other, to warfighters, and to commanders as the battlefield of the (very near) future becomes, in the authors' words, the "Internet of Battlefield Things." Autonomous and semi-autonomous platforms will collect data and may, depending on the rules of combat, carry and deploy weapons of their own.

Securing this new ecosystem will be hard, and the authors of *Modeling and Design of Secure Internet of Things* are making a powerful contribution to those seeking to tackle this challenge. The cybersecurity of these new networks, comprised of ever-more-numerous IoT devices, connected via 5G technology, and mediated by AI, will depend on new ways of understanding how these networks behave, including how they should behave and how they really behave. Such networks are more complex; they change constantly and in complex ways and are, therefore, more dynamic than the networks to which we are accustomed. *Modeling and Design of Secure Internet of Things* describes the techniques by which we can gain the understanding we need to secure them. The book's organization reflects a multimodal approach to securing IoT networks. "Game Theory and Deception" allows us to explore adversary behavior, efforts to deceive our adversaries, and ways adversaries might detect and counter that deception. In effect, "Game Theory and Deception" helps us understand the human threat to the security of our networks, and how human design can confront this threat.

"Modeling" takes us further, giving us the opportunity to study network behavior in the face of a broad range of attacks (e.g. stepping-stone attacks, polymorphic advanced persistent threats), and the effects of the defenses we might employ and manage. "Design" completes our exploration by applying what we have learned about effective cybersecurity technologies and architectures, overlaying them against the architectures of the advanced IoT networks we seek to defend. Overall, *Modeling and Design of Secure Internet of Things* is a comprehensive exploration of how best to secure the evolving IT ecosystems from which we intend to profit, and that our adversaries seek to exploit and attack.

The authors have assembled an impressive group of contributors to this volume, many of whom have worked at or with the Army Research Laboratory and with our NATO partners. Dr. Alexander Kott (ARL's Chief Scientist), Dr. Charles A. Kamhoua (ARL electronics engineer and Fulbright Fellow), Dr. Laurent L. Njilla (a cybersecurity leader at the Air Force Research Laboratory), and Dr. Sachin Shetty (Associate Professor in the Virginia Modeling, Analysis, and Simulation Center at Old Dominion University) are an impressive quartet guiding this exploration of advanced cybersecurity for complex networks and the Internet of Things.

I am confident that cybersecurity theorists and practitioners alike will profit from the discussions offered in this volume, and the world will be made safer as they do.

Samuel Sanders Visner
Director, National Cybersecurity Federally Funded Research and Development Center, The MITRE Corporation
Adjunct Professor, Cybersecurity Policy, Operations, and Technology, Georgetown University
Program in Science and Technology in International Affairs

Preface

The ubiquitous adoption of Internet of Things (IoT) technologies in commercial and military sectors has resulted in the widespread availability of various IoT solutions. However, the massive scale and distributed nature of such devices may introduce security and privacy challenges. IoT device manufacturers have not implemented security mechanisms, making IoT devices vulnerable when connected to the Internet. In addition, IoT devices and networks do not have resources typically available in traditional IT networks to host sophisticated security solutions; thus, it is challenging to port any of the existing security solutions to IoT domains. These challenges necessitate the need to comprehensively and accurately characterize the attack surface in IoT, conduct systematic modeling and analysis of the threats and potential solutions, and propose secure design solutions that balance the trade-off between cost and security risk.

This book examines issues in modeling and designing secure IoT to provide a flexible, low-cost infrastructure; reduce the risks of exploitable attack surfaces; and improve survivability of physical processes. The contributions address design issues in developing secure IoT, such as secure software-defined network-based network orchestration, networked device identity management, tactical battlefield settings, and smart cities. The book has encompassing themes that drive the individual contributions, including modeling techniques to secure IoT, game-theoretic models, cyber deception models, moving target defense (MTD) models, adversarial machine learning models in military and commercial domains, and empirical validation of IoT platforms. It synthesizes a mix of earlier work (on topics including MTD and cyber agility) as well as newer, cutting-edge research findings that promise to attract strong interest (on topics including Internet of Battlefield Things, advanced persistent threats, and cyber deception).

The editors would like to acknowledge the contributions of the following individuals (in alphabetical order): Fatemeh Afghah, Ioannis Agadakos, Kemal

Akkaya, Hisham Alasmary, Ehab Al-Shaer, Abdullah Alshammari, Amany Alshawi, Md Ali Reza Al Amin, Prashant Anantharaman, Afsah Anwar, Zahid Anwar, Orlando Arias, Abdullah Aydeger, Ted Bapty, Erik Blasch, J. Peter Brady, Swastik Brahma, Sergey Bratus, Gabriela F. Ciocarlie, Jay Chen, Yu Chen, Jin-Hee Cho, Bogdan Copos, George Cybenko, Suhas Diggavi, Jaya Dofe, Qi Duan, Abhishek Dubey, Michael Emmi, S. E. Galaitsi, Marco Gamarra, Moses Garuba, Mengmeng Ge, Oscar Gonzalez, Garegin Grigoryan, Salim Hariri, Kamrul Hasan, Amin Hassanzadeh, Linan Huang, Bilal Ishfaq, Ira Ray Jenkins, Yier Jin, Dong Seong Kim, Vijay H. Kothari, Tancrède Lepoint, Ulf Lindqvist, Igor Linkov, Yaoqing Liu, Michael Locasto, Michael C. Millian, Aziz Mohaisen, Mujahid Mohsin, Satyaki Nan, David M. Nicol, Kartik Palani, Jeffrey Pawlick, Fahim Rahman, Mohammad A. Rahman, Kirti V. Rathore, Danda B. Rawat, Jason Reeves, Malek Ben Salem, Nico Saputro, Pratik Satam, Shalaka Satam, Shamik Sengupta, Alireza Shamsoshoara, Rebecca Shapiro, Raj Mani Shukla, Sean W. Smith, Liwei Song, Janos Sztipanovits, Paulo Tabuada, Syed H. Tanveer, Mark Tehranipoor, Samet Tonyali, Deepak K. Tosh, Benjamin D. Trump, Selcuk Uluagac, Bowei Xi, Ronghua Xu, Zhiheng Xu, Qiaoyan Yu, Tao Zhang, Zhiming Zhang, and Quanyan Zhu.

We would like to thank Michael De Lucia, Paul Ratazzi, Robert Reschly, Sidney Smith, and Michael Weisman for technical review support. We would also like to extend thanks and acknowledgment to the US Army Research Laboratory technical editors Amber Bennett, Sandra Fletcher, Mark A. Gatlin, Carol Johnson, Martin W. Kufus, Sandy Montoya, Jessica Schultheis, and Nancy J. Simini, who helped edit and collect the text into its final form, and to Victoria Bradshaw, Mary Hatcher, and Louis Vasanth Manoharan of Wiley for their kind assistance in guiding this book through the publication process.

1

Introduction

Charles A. Kamhoua[1], Laurent L. Njilla[2], Alexander Kott[1], and Sachin Shetty[3]

[1] *US Army Research Laboratory, Adelphi, MD, USA*
[2] *Cyber Assurance Branch, Air Force Research Laboratory, Rome, NY, USA*
[3] *Virginia Modeling Analysis and Simulation Center, Old Dominion University, Norfolk, VA, USA*

1.1 Introduction

1.1.1 IoT Overview

Wireless technologies such as Wi-Fi, Bluetooth, Mesh networks, Zigbee, and RFID are ubiquitous in supporting mobile devices and applications. According to the Cisco Visual Networking Index, the number of mobile-connected devices exceeded the world population in 2014, with over half a billion devices introduced each year [1].

It is expected that there will be a steady transition to smarter mobile devices and an exponential increase in machine-to-machine connections. Global mobile data traffic may experience a sevenfold increase between 2016 and 2021. The explosion of mobile devices and traffic will lead to a more connected world, where by 2020 each person will own an average of seven connected devices, with over 93% of adults using smart phones for online services. It is anticipated that 2.7% of all things in the world (over 50 billion) will be connected. Also, the adoption of cloud computing and big data analytics paves the way for a smarter world, with smart energy, smart cities, smart health, smart transport, smart agriculture, smart industry, and smart living.

The Internet of Things (IoT) is the inter-networking of physical devices, vehicles, buildings, and other items embedded with electronics, software, sensors,

Modeling and Design of Secure Internet of Things, First Edition. Edited by Charles A. Kamhoua, Laurent L. Njilla, Alexander Kott and Sachin Shetty.

Published 2020 by John Wiley & Sons, Inc.

actuators, and network connectivity that enable these objects to collect and exchange data [1].

Figure 1.1 depicts the IoT system architecture comprising four key components: Things, Networking, Cloud, and Information Processing. The Things component is responsible for integrating physical devices with sensors and actuators; collecting and processing data; and serving as the interface to the physical world. The Things component is also responsible for addressing device diversity and capability with high-end devices such as drones, smart homes, cameras, laptops, smartphones, and tablets, and low-end devices such as sensors, actuators, and passive entities such as barcodes, QR-codes, and RFID. The Networking component handles network connectivity and heterogeneous communication links such as Ethernet, Wi-Fi, cellular, Bluetooth, Zigbee, Long Range Wide Area Network (LoRAWAN), Narrow Band IoT (NB-IoT), IPv6 over Low-Power Wireless Personal Area Networks (6LoWPAN), and so on. It is also responsible for providing interconnectivity of various wireless access technologies and interference management. Further, the network can be empowered by blockchain technology to securely handle transactions among untrusted IoT devices. Our prior book fully describes blockchain for distributed system security [2]. The Cloud component is responsible for supporting domain-centric applications through its infrastructure in data center, storage, and service providers. Although we do not address cloud security in this book, we refer any interested reader to our prior book describing cloud computing details [3]. Finally, the Information Processing component is responsible for providing the intelligence to realize smart IoT by using big data analytics and machine learning to transform data to information and eventually aid in game-theoretic modeling, optimization, and decision-making.

1.1.2 IoT Security and Privacy Challenges and Opportunities

Though IoT has greatly impacted systems in commercial and military domains, there is a growing concern about the security risks introduced by IoT devices. IoT vendors and manufacturers are not typical security experts and do not emphasize security aspects in the design and implementation of IoT devices. IoT device manufacturers have not implemented security mechanisms, making IoT devices vulnerable when connected to the Internet. For example, in December 2013, the first IoT botnet was discovered by Proofpoint [4], which included IoT devices such as smart TVs, baby monitors, and other smart devices found in modern homes. In late 2016, there were reported distributed denial of service (DoS) attacks on popular websites, such as Netflix, Twitter, Spotify, Airbnb, and Reddit, from a network of consumer IoT devices [5]. Researchers have also demonstrated vulnerabilities in medical devices, such as pacemakers and implantable cardiac defibrillators [6]. An insulin pump was hacked, which resulted in a fatal dosage delivery

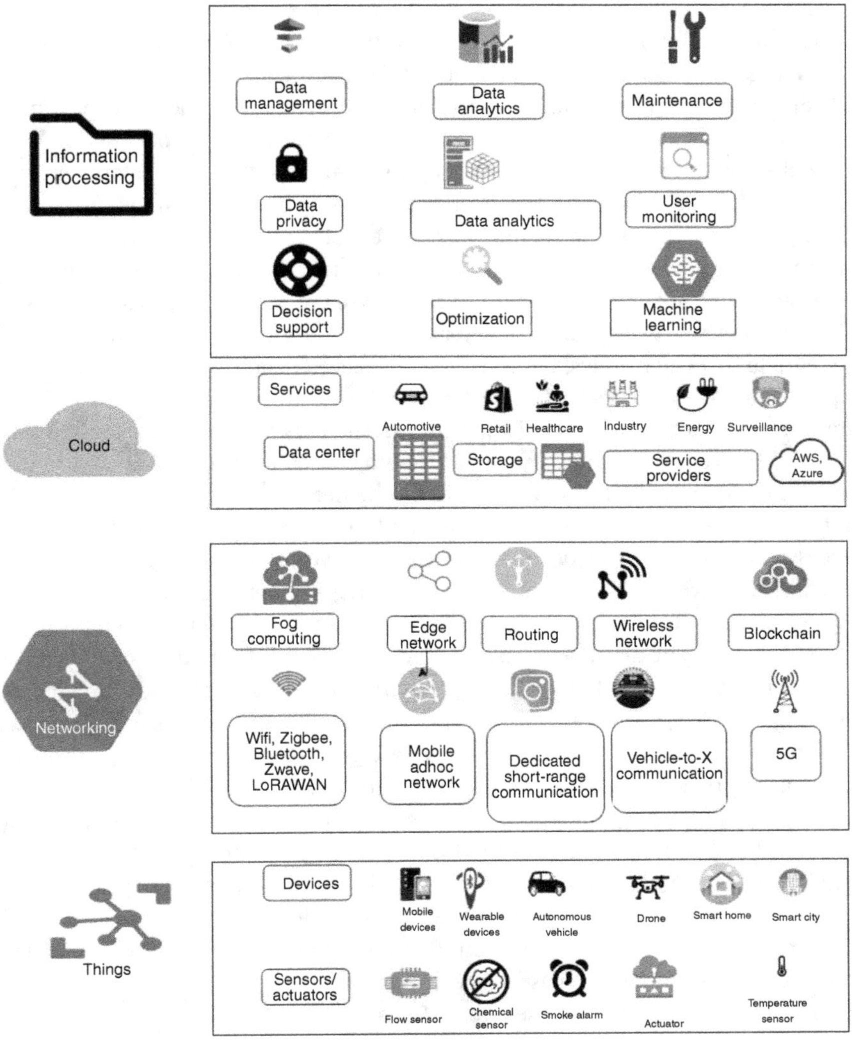

Figure 1.1 IoT system architecture.

over the air. Smart vehicles are also susceptible to attack. In 2015, Miller and Valsasek demonstrated the takeover of a jeep on a highway [7]. The pervasive networked computing capabilities provided by IoT increase their security risks as attack enablers rather than attack targets. The IoT devices can be unwitting participants in a botnet, which could lead to secondary attacks. In a 2008 attack on a Turkish oil refinery, security analysts found that vulnerabilities in surveillance camera communication software enabled the attackers to gain entry and penetrate deeper into the internal network [8]. Thus, the cameras were used as stepping stones to gain access to the network housing the critical assets.

There are several reasons for the need for IoT security. IoT devices are mass produced rapidly to be low-cost commodity items without security protection in their original design. These devices are highly dynamic, mobile, and heterogeneous without common standards. As a result, those systems are the frequent targets of cyberattacks that aim to disrupt mission effectiveness. Particularly, the low-end devices are not capable of supporting sophisticated security solutions due to constrained computing, storage, and networking requirements. Typically, most IoT devices and networks are characterized by weak security configurations and inadequate security policies, making them subject to data loss, theft, and reverse-engineering. Unprotected IoT devices can be used as “stepping stones” by attackers to launch more sophisticated attacks, such as advanced persistent threats (APTs). These challenges and the high risk and consequence of IoT attacks in the battlefield and on commercial systems drive the need to accelerate basic research on IoT security. It is imperative to understand the natural world and the physical process(es) under IoT control, and how these real-world processes can be compromised before recommending any relevant security countermeasure.

In addition to security risks, the collection and analysis of personal identifiable information, geolocation, and health and financial information can lead to increased privacy risks. Exploitation of smartphone sensors that are capable of inferring mood, stress level, personality type, smoking habits, sleep patterns, and physical movement can compromise safety. Researchers have reported the potential of privacy leakage through gesture-control devices [9].

There is a large body of security research activities in mobile and wireless networks that can be leveraged from the IoT networking technologies [10]. Conventional cryptography-based network security techniques have been proposed to secure wireless networks. However, these crypto solutions may not be feasible for many low-end devices. In addition, there is sufficient opportunity for insider attack to circumvent crypto-based security checks. This book presents a comprehensive suite of solutions to secure IoT from the devices to the overall IoT infrastructure.

1.2 Overview

The focus of this book is to provide modeling and design techniques for securing IoT in both commercial and military environments. The contributions address design issues in developing secure IoT, such as secure software-defined network (SDN)-based network orchestration, networked device identity management, tactical battlefield settings, and smart cities. The book has encompassing themes that drive the individual contributions, including modeling techniques to secure IoT, game theory, cyber deception, moving target defense (MTD) in military and commercial domains, and empirical validation of IoT platforms. It synthesizes a mix of earlier work (on topics including MTD and cyber agility) as well as newer, cutting-edge research findings that promise to attract strong interest (on topics including Internet of Battlefield Things [IoBT] [11], APTs, and cyber deception).

We describe properties underlying the formal foundations of modeling secure IoT techniques and practical design issues for deployment in commercial and military domains. We also present several attack surfaces in IoT and secure solutions that need to be developed to reach full potential. To the best of our knowledge, this is the first book to provide formal modeling techniques to analyze the attack surfaces in IoT in a comprehensive fashion. The content is accessible and readable to a broader security audience. The topics provide graduate students, cybersecurity researchers, engineers, and military officers detailed insights into the effectiveness of secure design solutions in IoT for industrial and military domains. In addition, the security analysis will aid in understanding and quantifying the impact of new attack surfaces introduced by IoT deployments. Our vision for modeling and design of secure IoT is presented herein through synergistic individual research contributions.

1.3 Roadmap

This book was organized to represent a defense-in-depth approach to layered security, with each layer representing cyber deception, modeling, and design (Figure 1.2). With three layers, each chapter was organized into three parts, with contributions from over 100 authors. Within each part we first present techniques to secure the IoT devices. Next, we describe the methodologies to realize a secure interconnection between the IoT devices. Finally, we provide the necessary distributed information processing techniques required to achieve scalable and robust protection.

Part 1: Game Theory for Cyber Deception

The focus of Part 1 is to provide the reader a deep-dive into game-theoretic techniques used to realize cyber-deception schemes for diverse IoT settings. We have organized the content in six chapters, where each chapter provides insight into how game-theoretic techniques can be used as a tool to design cyber-deception schemes to protect IoT devices. The focus is on adapting in adversarial cyber settings, ensuring trust management, understanding the interactions between

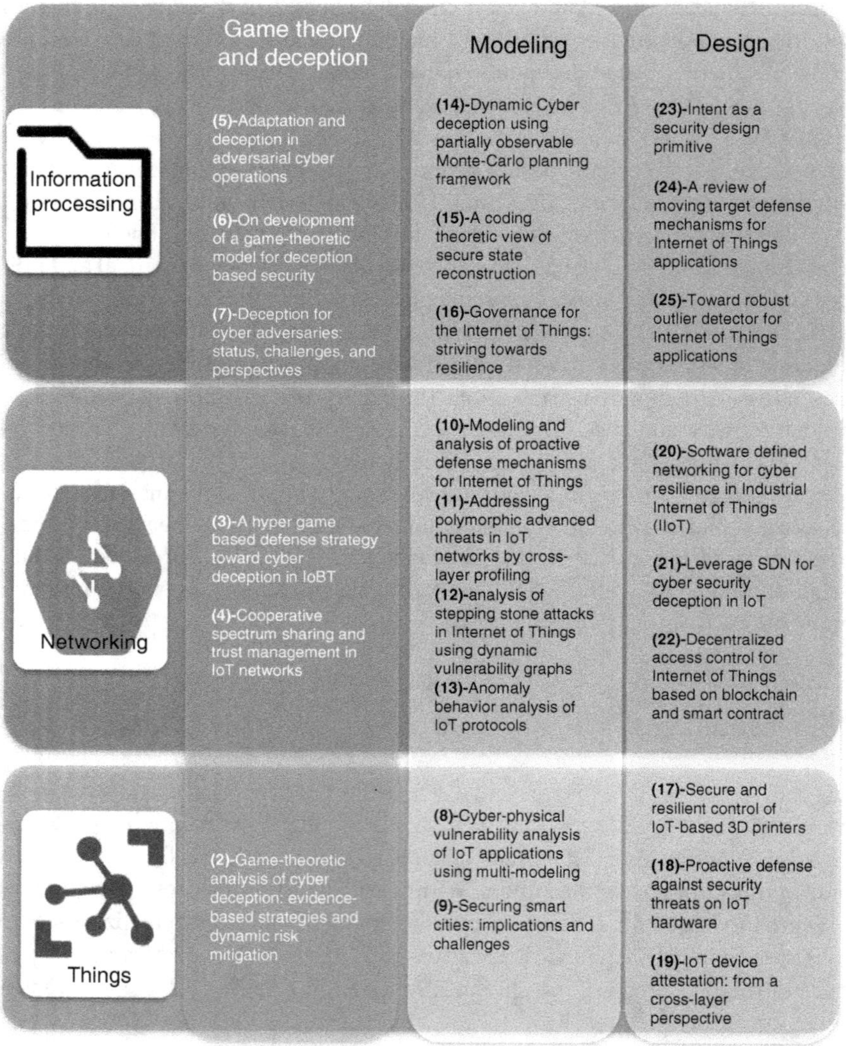

Figure 1.2 Book roadmap.

defender and attacker by analyzing the impact of configuring honeynets, and characterizing the information that the defender can use to understand adversaries strategies, tactics, and impact of attack vectors. Part 1 will introduce the reader to deception techniques in the context of cyber security that will aid the defender in convincing the adversary that the fake information and systems are legitimate.

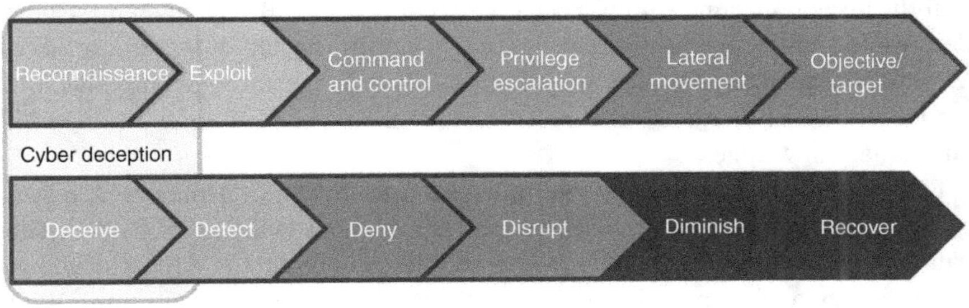

Figure 1.3 Cyber kill chain.

It also explores mechanisms used in cyber deception to trap the adversary and useful information that can be gained by the defender to understand adversaries' strategies, tactics, and the severity of the attack. Cyber deception is a proactive defense that allows the defender to act very early in the cyber kill chain (Figure 1.3) and prevent the more dangerous subsequent stage. Specifically, Part 1 of this volume is divided into the following chapters:

Chapter 2: Game-Theoretic Analysis of Cyber Deception: Evidence-Based Strategies and Dynamic Risk Mitigation, *Tao Zhang, Linan Huang, Jeffrey Pawlick, and Quanyan Zhu*

In this chapter, we provide an overview of deception games in three different environments. The chapter introduces the baseline signaling game models to capture the information asymmetry, and dynamic and strategic behaviors of deceptions. To enrich the game models with exogenous factors, we extend the baseline framework to include evidence through side-channel knowledge acquisition.

The chapter first presents a leaky deception over a binary information space model based on signaling game framework with a detector that gives off probabilistic evidence of the deception when the sender acts deceptively. The results on pooling and partially pooling equilibria of the game show that high-quality detectors eliminate some pure-strategy equilibria, and detectors with high true-positive rates encourage more honest signaling than the low false-positive rates counterpart. The analysis of the equilibrium also supports the finding that receivers obtain optimal outcomes for equal-error-rate detectors, and deceptive senders sometimes benefit from highly accurate deception detectors.

The chapter later focuses on a class of continuous one-dimensional information space and uses a signaling game model to characterize the deception by taking into account the cost of deception. Analysis of the perfect Bayesian Nash equilibrium (PBNE) reveals the deceivability of the deception. The equilibrium results show that the deception game admits a class of partial-pooling PBNE that identifies the deceivable and the undeceivable regions of a one-dimensional information space. We allow the receiver to acquire additional knowledge through investigation. Under certain conditions, the receiver can maintain the equilibrium with additional knowledge.

Finally, the chapter explores the multistage incomplete-information Bayesian game model for defensive deception for advanced persistent threats (APTs). The game model captures the multistage and multiphase structure of APTs and allows the defender to dynamically form belief based on observable footprints. With conjugate priors, the dynamic game problem can be solved using extended-state dynamic programming that admits an equilibrium with forward belief update and backward induction.

Chapter 3 : A Hyper-Game-Based Defense Strategy Toward Cyber Deception in Internet of Battlefield Things (IoBT), *Bowei Xi and Charles A. Kamhoua*

Having established the need for game-theoretic techniques for cyber deception in Chapter 1, in this chapter we shift gears by studying the cyber-deception strategies in diverse networked configurations, such as Internet of Battlefield Things (IoBT). IoBT is an Internet of Things (IoT) of connected devices on the military network with the ability to collect operational field data, to compute on the data, and to upload its information to the network. Adversaries could exploit vulnerabilities in IoBT devices and attempt to compromise the critical nodes on the network. This chapter considers a targeted attack scenario, where an adversary compromises an edge device and, through exploitation of vulnerabilities of connected IoBT devices, launches a targeted attack on a critical device on the network. A Bayesian attack graph demonstrates the potential attack paths from a compromised edge device to a target-critical device and the associated probabilities. In this chapter, we developed a defense strategy based on a two-player hypergame against a targeted attack on critical IoBT devices. Unlike other game-theoretic frameworks where both players share a common perception of the game, a hyper game allows the players in the game to have their own perceptions of the game. Because of this property, a hyper game has been used to model cyber conflicts. The hyper game provides an important theoretic framework for us to model the interaction between adversary and defender. It also provides useful insights on how to create misperceptions so that an adversary cannot obtain the correct network topology. The hypergames are played repeatedly. At each newly compromised node, the adversary updates its perception of the network topology and chooses the next node to attack, whereas the defender feeds the adversary deceptive information. Meanwhile, we use a time-varying attack graph to model how an attack spreads on an IoBT network. Our approach provides quantitative metrics such as a recommended time to disable all the links to a target device, a recommended time to reboot the network devices to clean up all potentially compromised devices, and a recommended time to reset the network topology.

Chapter 4: Cooperative Spectrum Sharing and Trust Management in IoT Networks, *Fatemeh Afghah, Alireza Shamsoshoara, Laurent L. Njilla, and Charles A. Kamhoua*

In Chapter 3, the game-theoretic approach characterized in interplay between adversary and defender in Internet of Battlefield Things (IoBT) settings. The networking environments were static in nature. In this chapter, the cognitive radio network is chosen to understand the cyber-deception strategies in a dynamic spectrum access context. Here, we provide a practical physical-layer secrecy solution to protect the communication of licensed users from eavesdroppers. In this model, the licensed users can take advantage of cooperative jamming provided by unlicensed Internet of Things (IoT) users available in their proximity to hide their message from the eavesdroppers. The unlicensed users can also assist the spectrum owners with cooperative relaying service to enhance the quality of their communication. In return, the licensed users will provide the unlicensed users with the chance of using their radio spectrum to compensate their services. This model can provide a practical solution for dynamic spectrum sharing in future IoT networks where a dedicated spectrum is not available to some new IoT technologies. The global mobile data traffic of IoT with potentially billions of wireless devices is expected to exceed 30 exabytes per month in the near future. However, the dedicated IoT spectrum and conventional spectrum-sharing techniques are not able to provide the IoT spectrum demand. The common spectrum sensing or geolocation-based spectrum access methods are too conservative in protecting the incumbent users' rights that do not allow flexible spectrum-sharing mechanisms to facilitate the needs of fast-growing IoT technologies. The developed spectrum-leasing mechanism in this chapter can provide a realistic solution by offering a win-win for both parties.

A main concern with enabling shared access techniques to the radio spectrum, particularly the federal bands, is security. In addition to traditional security threats in wireless networks, the networks with dynamic spectrum allocation mechanisms are also vulnerable to various exogenous, malicious, and selfish attacks, where unlicensed users take advantage of the ad hoc nature of such networks. While cooperative jamming provided by the unlicensed users can potentially enhance the secrecy rate of the licensed users, it involves the assumption of having unconditionally cooperative and trustable secondary users (SUs) as considered in most previously reported studies. However, this assumption is far too optimistic in communication networks, noting the non-altruistic nature of cognitive SUs (e.g. IoT devices). IoT devices often have limited energy; therefore, the natural tendency of such non-altruistic users can lead to selfish attacks where they violate their commitments to the primary users after they are granted the spectrum access. In this chapter, a decentralized game-theoretic trust-management mechanism is developed to enable cooperative spectrum leasing to potentially selfish IoT devices.

Chapter 5: Adaptation and Deception in Adversarial Cyber Operations, *George Cybenko*

Chapters 3 and 4 focused on game-theoretic models to address security issues in Internet of Battlefield Things (IoBT) and Internet of Things (IoT) spectrum environments. In this chapter, we take a step back to focus on the interplay between adaption and deception in IoT systems. Recent progress in reinforcement learning has resulted in remarkable advances in game playing technology, to the extent that machines are now demonstrating superior play against human experts in games such as Go and some versions of Texas Hold'em poker. Those advances use a variety of combinations of computational game theory and deep reinforcement learning. However, there are several fundamental reasons that such successes do not translate immediately to more complex adversarial interactions such as those that arise in cyber operations. This chapter reviews some of those reasons and how they relate to deceptions that are possible by any player in adaptive, adversarial systems.

Chapter 6: On Development of a Game-theoretic Model for Deception-based Security, *Satyaki Nan, Swastik Brahma, Charles A. Kamhoua, and Laurent L. Njilla*

Having established the nature of deception in adversarial IoT environments, in Chapter 6 we look into a formal game-theoretical model for characterizing deception in Internet of Things (IoT) environments. The game-theoretic techniques to achieve cyber deception were described from the IoT device point of view in Chapter 3 and network perspective in Chapter 4. In this chapter, we look at the schemes in the information processing layer that can provide insights into developing schemes aimed at wasting attacker resources and increasing the information asymmetry. Deception can help to reduce the likelihood of an attacker's success and cost of defense from deception-less situations. As the total space of attacker's options grow large, if far more deceptions than actual attack paths are presented, the workload of the attacker for detecting real targets and differentiating between real and fake systems increases. In practice, the defender can observe the attacker's activity, gain knowledge of the presence of an attacker and their characteristics, and direct deception tactics toward the attacker. Attackers can, in turn, seek to present different characteristics closely associated with legitimate users to make it harder for the deception system to detect them, differentiate between attackers and legitimate users, and increase the defender's workload. To model such strategic interactions between a system defender and an attacker, this chapter presents a game-theoretic model to analyze attack–defense scenarios that uses fake nodes (computing devices) for deception under consideration of the system deploying defense resources to protect individual nodes in a cost-effective manner. The developed model has important applications in the Internet of Battlefield Things (IoBT). Considering the nodes on which the defense resources are present to be public knowledge, the dilemma for the system becomes determining which node to use for performing a certain computational task (or, in other words, choosing which nodes should be used as fake nodes), and the dilemma for the attacker becomes choosing whether to attack nodes having defense resources (expecting the defender to use a node having defense resources) or whether to attack a node left unguarded without any defense resources installed. This chapter presents a game-theoretic model to analyze such a situation and develops Nash equilibrium-based strategic deception tactics.

Chapter 7: Deception for Cyber Adversaries: Status, Challenges, and Perspectives, *Abdullah Alshammari, Danda B. Rawat, Moses Garuba, Charles A. Kamhoua, and Laurent L. Njilla*

We conclude Part 1 with a chapter that provides an insight into the research challenges, current state-of-the-art solutions, and perspectives moving forward. Cybersecurity continues to be a major concern for many organizations. Losing private and critical information to attackers is often unthinkable for both individuals and organizations. Over the decades, different organizations have developed a wide range of techniques to secure their private information and infrastructure. Cyber deception is one technique used to learn and combat the malicious cyberattacks. The purpose of this chapter is to present the current status, challenges, and perspectives of cyber deception for cyber adversaries. This chapter also investigates the game-theoretic deception approach for cyber adversaries. We present unique advantages for using deception techniques for cybersecurity over traditional security defenses, where the system could learn and adapt the security parameters on the fly to combat malicious cyber operations.

Part 2: IoT Security Modeling and Analysis

The focus of Part 2 is to provide in-depth coverage of modeling formalisms and formal analytic techniques to understand threats to Internet of Things (IoT) and secure IoT solutions. The content is organized into nine chapters containing IoT security modeling techniques, such as multi-model-based approaches to understand the vulnerabilities in IoT devices used in smart homes and smart cities, moving target defense-based techniques to realize cyber deception in a networked context, and IoT security analysis techniques, such as cross-layer profiling to analyze attack surfaces across multiple layers; dynamic vulnerability graphs to understand the impact of stepping stone attacks; chain of attack vectors for specific threats; and anomalous behavior of IoT protocols, secure state estimation, and governmental applications of risk in IoT. Specifically, Part 2 of this volume is divided into the following chapters:

Chapter 8: Cyber-Physical Vulnerability Analysis of IoT Applications Using Multi-Modeling, *Ted Bapty, Abhishek Dubey, and Janos Sztipanovits*
We begin Part 2 by taking the smart home as a use case to systematically understand the vulnerabilities sensors, communication links, and devices used for video monitoring, storage, and HVAC control. The chapter discusses security challenges in Internet of Things (IoT)-based integrated systems that are frequently complex and incorporate a wide range of heterogeneous devices with many potential vulnerabilities. The starting point of the discussion is that potential effects of vulnerabilities can propagate across system layers causing impacts logically far away from the place of the exploit. This observation supports the argument that vulnerability mitigation and hardening methods restricted to networks and software are not sufficient – the considerations need to be extended to the full system stack from the physical layer to the human layer. With this justification, the chapter focuses on physical layer vulnerabilities as a potential source of system-wide impacts. The selected attack category is energy injection to electronic hardware that is widely used in IoT devices. Using a home automation system example, authors discuss a methodology for discovering vulnerabilities in hardware that can be exploited by electromagnetic (EM) energy injection. The methodology is based on a multi-modeling approach that represents the geometric structure of the hardware and correlates it with behavioral models. The analysis shows low-energy EM pathways for corrupting system behavior and argues the importance of EM injection analysis at behavior-critical inputs and outputs. Authors also discuss a methodology for system-level impact analysis. The chapter conclusion is that susceptibility to physical layer attacks increases the attack surface due to a large number of heterogeneous IoT devices. This brings in a physical dimension to vulnerability analysis and risk mitigation.

Chapter 9: Securing Smart Cities: Implications and Challenges, *Ioannis Agadakos, Prashant Anantharaman, Gabriela F. Ciocarlie, Bogdan Copos, Michael Emmi, Tancrède Lepoint, Ulf Lindqvist, Michael Locasto, and Liwei Song*
Extending beyond the smart home showcase example in Chapter 8, this chapter focuses on vulnerabilities in devices used in smart cities. Challenges and solutions are discussed for achieving reliability, safety, security, and privacy in the large and complex system-of-systems known as a smart city. As Internet of Things (IoT) systems form the core of smart-city technologies, this chapter focuses on approaches to monitoring the behavior and properties of IoT devices and protocols. Modeling allows us to reason and draw conclusions about the functioning and security of IoT devices as part of complex system-of-systems. Composition emerges as the key issue related to designing, implementing, and integrating the systems that

comprise a smart city. Security problems often occur in the seams between integrated system components, and they are especially important in complex, heterogeneous smart-city systems that affect so many aspects essential to ensuring a safe and functional community. A number of issues are related to anticipatory design in which system designers lack knowledge of how their systems will continue to be used and integrated in the future. How can components be designed to achieve interoperability while solving the much more difficult issues of reliability, safety, security, and privacy? The questions, trade-offs, and knowledge regarding the potential impact of decisions are extremely important. A loose process in which every system is developed in isolation and is not strong enough to robustly interact with other systems will not yield the smart-city future we want to reside in. New processes guided by sound principles and development integrity must take on complex unaddressed (and unanticipated) challenges. This chapter provides some knowledge about currently identified principles, issues, and possible solutions. Most of us will someday be living in technology-dependent smart cities, which will need to safely and efficiently support and enhance our individual and collective wellbeing as well as that of future generations.

Chapter 10: Modeling and Analysis of Integrated Proactive Defense Mechanisms for Internet of Things, *Mengmeng Ge, Jin-Hee Cho, Bilal Ishfaq, and Dong Seong Kim*

Chapters 8 and 9 focus on modeling vulnerabilities from a device point of view. This chapter explores a proactive defense mechanism for the Internet of Things (IoT) that exploits the interconnections between adversary and defender. Although many intrusion detection systems (IDSs) have been developed to detect intrusions, it is a reactive mechanism in nature by detecting the intrusions after they already penetrated into the system. This common situation shows a defender taking an action way beyond compared to the actions taken by agile and smart attackers. To complement inherent limitation as reactiveness of the IDS, intrusion prevention systems (IPSs) have been developed for mitigating the impact of the intrusions before they break into the system. In this chapter, we introduce an integrated defense mechanism for a software-defined IoT network based on cyber-deception technology (i.e. a decoy system) and moving target defense (MTD). In particular, for a given software-defined networking-based IoT environment with a decoy system, we propose a network topology shuffling-based MTD technique that is executed adaptively to be well aware with system security vulnerability and defense cost due to the execution of MTD operations. To validate the effectiveness and efficiency of the proposed MTD, we developed a graphical security model-based evaluation framework using a Hierarchical Attack Representation Model. To realize high adaptiveness and proactiveness of the proposed MTD technique, we employed a fitness function based on genetic algorithm for identifying an optimal network topology where a network topology can be shuffled based on the detected level of system vulnerability. The proposed defense mechanism provides an affordable defense service that meets the system goals of maximizing system security while minimizing the defense cost. From our simulation study, we proved that our proposed integrated defense mechanism outperforms among all other comparable counterparts considered in this work in terms of the system lifetime (i.e. mean time to security failure), the number of decoy nodes used in attack paths, and the defense cost introduced by the proposed MTD technique.

Chapter 11: Addressing Polymorphic Advanced Threats in Internet of Things Networks by Cross-Layer Profiling, *Hisham Alasmary, Afsah Anwar, Laurent L. Njilla, Charles A. Kamhoua, and Aziz Mohaisen*

Starting with this chapter and the rest of the chapters in Part 2, the focus is on exposing the readers to several analytical techniques to understand Internet of Things (IoT) attack surfaces and defense mechanisms. In this chapter, the motivation is to highlight the benefits from a fine understanding of the system and use models of IoT applications with the diversity of devices to address polymorphic threats through network-based cross-layer profiling. With IoT devices persistently connected to the Internet, security and privacy challenges become of paramount importance, and methods for understanding their usage toward mitigating security risks due to unauthorized communications that violate their confidentiality, integrity, or availability would be needed. Prior studies have suggested various security issues and concerns in the hardware and software of today's IoT devices and ways to mitigate those concerns, although a comprehensive approach is yet to materialize. One of the major threats is the stealthy advanced persistent threat (APT): a threat that would result in a volume of behavioral artifacts in IoT environments, such as smart home networks. With this work, we aim to assist the intrusion detection of APTs in the home network by proposing Security Layer for Smart Home (SLaSH), a cross-layer security mechanism that incorporates capabilities and features of the various layers to help secure the entire smart home network. To formulate a baseline for the intrusion detection system, the attack surface of each of the three layers – device, network, and service – is analyzed. To facilitate the cross-layer security, a security layer is designed to detect and prevent various threats emanating from the different system layers. A machine learning-based cross-layer intrusion detection system that uses both rule- and behavior-based features is suggested. The multilayered architecture of the system accommodates the heterogeneity of devices in a smart home environment and allows for modular implementation and continuous evolution.

Chapter 12: Analysis of Stepping Stone Attacks in Internet of Things using Dynamic Vulnerability Graphs, *Marco Gamarra, Sachin Shetty, Oscar Gonzalez, David M. Nicol, Charles A. Kamhoua, and Laurent L. Njilla*

In this chapter, the reader will learn about stepping stones attack surface, which unlike polymorphic threats described in Chapter 11, is exploited at the network level. Vulnerability graphs have been used as an effective tool for analyzing exploitability and the impact of chains of exploits in networked environments. A chain of "stepping stones" from the attacker origin to the desired target creates the attack graphs. The stepping stones not only provide the intermediate steps to reach the target but they also make it difficult to identify the attacker's true location. In this chapter, stepping stones are modeled and analyzed in Internet of Things using dynamic vulnerability graphs. Most analysis based on attack graphs assumes that the graph edges and weights remain constant during the attacker's attempt to propagate through the network. A biased min-consensus technique is proposed for dynamic graphs with switching topology as a distributed technique to determine the attack paths with more probable stepping stones in dynamic vulnerability graphs. A min-plus algebra to determine the necessary and sufficient convergence conditions is also discussed. The necessary condition for convergence to the shortest path in the switching topology is then presented.

Chapter 13: Anomaly Behavior Analysis of IoT protocols, *Pratik Satam, Shalaka Satam, Salim Hariri, and Amany Alshawi*

This chapter reviews the security of Internet of Things (IoT) communications protocols and presents the methodology to develop anomaly behavior analysis of these protocols. Cisco VNI forecasts that there will be 4.6 billion Internet users by the end of 2021, and 27.1 billion devices connected to the Internet by the end of 2021 [1]. This increase in the number of devices on the Internet can be attributed to the increasing number of IoT devices on the Internet. Our home/work environment is increasingly becoming "smart." This IoT revolution is brought about by introducing smart devices into all aspects of our life and economy. These devices allow the users to control and manage their environment remotely, thus increasing user comfort and acceptance. IoT devices communicate with each other or over cloud-based services via the Internet or wireless local networks like Wi-Fi, Bluetooth, and Zigbee. IoT devices have arrays of sensors and communication networks, thus giving them the capacity to track, monitor, and observe, which raises the questions of privacy/confidentiality, integrity, and availability of the IoT services. As IoT devices are becoming ubiquitous, there is a need to ensure security of these devices and the privacy of their smart services from attacks over the communication network. For instance, there was a large-scale attack in 2014 that targeted over 100 000 IoT devices. Researchers have been able to target and exploit several vulnerable IoT devices like baby monitors, light bulbs, power switches, and smoke alarms, as well as target sensor suits deployed in smart cars, thus highlighting that there is a critical need to secure IoT infrastructure and their services against cyber threats.

In this chapter, we present a systematic approach to model the behavior of IoT devices, perform threat modeling on IoT devices and their applications, and present solutions to detect and mitigate the impacts of attacks against their protocols such as Wi-Fi and Bluetooth. The presented IoT architecture divides IoT/cyber physical systems into four functional layers: End devices, Communications, Services, and End Users/Applications. Using this four-layer functional architecture, we use the IoT threat-modeling framework to identify existing IoT attack surfaces and develop mitigations and protection methods against potential exploitations. In this chapter, we show how to apply the IoT threat modeling framework to the Wi-Fi and the Bluetooth protocols, and how our Anomaly Behavior Analysis-based Intrusion Detection Systems (ABA-IDS) can detect attacks that exploit their vulnerabilities. The ABA approach uses machine learning techniques to model the normal behavior of IoT protocols. As all attacks trigger abnormal behaviors, the ABA-based IDS can detect not just known attacks but also new and modified attacks.

Chapter 14: Dynamic Cyber Deception Using Partially Observable Monte-Carlo Planning Framework, *Md Ali Reza Al Amin, Sachin Shetty, Laurent L. Njilla, Deepak Tosh, and Charles A. Kamhoua*

The adoption rate of Internet of Things (IoT) technologies in information technology (IT) and operational technology (OT) sectors has resulted in an abundance of various IoT solutions. Due to the static nature of organizations, IT/OT networks lead adversaries to perform reconnaissance activity and identify potential threats. The aim of the reconnaissance phase is to collect as much critical information about the network as possible, including network topology, open ports and services, and unpatched vulnerabilities. Having that critical information maximizes the intruder's chance of penetrating a system and gaining a foothold successfully, where a network administrator can minimize the likelihood of penetration by patching a vulnerability. Unfortunately, sometimes the vulnerability exposure window (vulnerability discovery and developing a patch for that weakness) lasts five to six months. This extended period puts the cyber network at higher risk. To address this issue, one needs an active defense system to thwart cyberattacks while considering information in real time and providing appropriate defense actions. One such defense system is cyber-deception technology, where a cyber defender can create and fortify the view of the network by revealing or concealing the information to the intruder. Fake networks are used to prompt the adversary to expend their resources and time, so the defender can gather critically important information about the adversaries' strategies, tactics, capabilities, and intent. The Partially Observable Monte-Carlo Planning (POMCP) algorithm balances the trade-off between security and availability costs while deploying network decoys. Deploying both network decoys and the real network nodes leads to a scalability issue, which we solved using the POMCP algorithm. In this chapter, we describe the POMCP algorithm based on an efficient and scalable deception approach that influences an attacker to take pursue the fake network while also capturing attacker progression using the vulnerability dependency graph.

Chapter 15: A Coding Theoretic View of Secure State Reconstruction,
Suhas Diggavi and Paulo Tabuada
Chapters 12–14 focus on modeling the attack surfaces from the adversary's perspective. This chapter shifts the viewpoint to system monitoring and observers to characterize their effectiveness in the presence of attacks on sensors and actuators in IoT. This chapter discusses secure state estimation and control mechanisms extending classical observers and Kalman estimation methods to the case when there are adversaries. It has been shown that although the tight coupling between physical and cyber components in cyber-physical systems (CPSs) brings new challenges, it also offers new solutions to security. On the one hand, physical dynamics offer a new attack surface against which existing cybersecurity mechanisms have no defense – for example, by spoofing physical signals before they reach the cyber domain. On the other hand, the physical dynamics offer additional degrees of redundancy that can be exploited to improve security. Therefore, one can turn the perceived vulnerabilities in physical systems into an advantage, as an adversary can attack the sensing, communication, or computation, but cannot alter physics.

In this chapter, we illustrate these ideas by reviewing some of our results in secure state estimation and control mechanisms in the presence of sensing and actuating attacks, extending classical observers and Kalman estimation methods to the case when there are adversaries. Central to any control algorithm for CPS is the estimation of the state of the physical system being controlled. In this context, it is crucial to understand the fundamental limits for state estimation in the presence of malicious attacks. We focus on securely estimating the state of a linear dynamical system from a set of noisy and maliciously corrupted sensor measurements and actuator commands. We restrict the attacks to be sparse in nature, i.e. an adversary can arbitrarily corrupt an unknown subset of sensors and actuators in the system but is restricted by an upper bound on the number of attacked sensors and actuators.

Chapter 16: Governance for the Internet of Things: Striving Toward Resilience, *S. E. Galaitsi, Benjamin D. Trump, and Igor Linkov*
Modeling security risks in the Internet of Things (IoT) from a technological perspective only provides half the story, as described in Chapters 8–15. It is equally crucial to understand risks to IoT from an organizational perspective. This chapter presents modeling secure IoT from a governance perspective that exploits organizational and regulatory policies to mitigate risks to IoT. IoT is quickly becoming ingrained within everyday life as digital interconnectivity enables centralized control of multiple systems through smartphones or computers. These technological developments also expose many systems to novel disruption through cyberattacks or hacking attempts, both from state-based and non-state-based actors. Such threats are constantly evolving to exploit new and unique vulnerabilities within complex information systems and digitally connected devices. Meeting the growing challenge of safe and efficient IoT development and use will require informed and adaptive governance strategies. Governance strategies differ across industries and between companies, but the critical question being asked of IoT developers and users centers upon how they intend to prevent, avoid, mitigate, manage, and recover from disruptions to IoT devices and interconnected information systems. This chapter explores the concept of resilience and its relationship to emerging scholarly and policy discussions of the digital economy and IoT governance. Despite a range of governance frameworks, the philosophical and methodological applications of resilience can be applied to support efficient IoT device recovery and adaptation from an evolving cyber threat landscape.

Part 3: IoT Security Design

The focus of Part 3 is to provide in-depth coverage of design techniques to secure IoT. We have organized the content into nine chapters, which includes design mechanisms focused on protecting 3D printers from advanced persistent threats, moving target defense (MTD)-based design to protect devices from hardware threats such as side-channel attacks, characterize intent in zero-day attacks on devices, and design of attestation schemes to protect IoT devices. Next, we shift gears and present design mechanisms at the network infrastructure layer, such as software-defined networking-based techniques to realize resilient cyber deception, scale MTD and improve cyber resilience, outlier detection, and distributed ledger design to create a decentralized trust layer. We conclude with chapters focusing on the design of anomaly-based intrusion detection and prevention systems.

Chapter 17: Secure and Resilient Control of IoT-based 3D Printers,
Zhiheng Xu and Quanyan Zhu
This chapter focuses on a specific advanced persistent threat (APT) plaguing 3D printers. Internet of Things (IoT)-based networks bring many revolutionary opportunities to traditional manufacturing systems, especially for the 3D printing system. In Chapter 1, we discuss the security issues of an IoT-based 3D printer. Due to the high cost of the 3D-printing process, people focus on outsourcing the production to a third party who specializes in 3D-printing technology. The IoT networks make this idea possible. However, integrated with cyber systems, IoT-based 3D printers introduce new challenges. Nowadays, APT can launch a zero-day exploration to find the vulnerability of the system. An APT-type attack may intrude the IoT-based 3D printer, modify the critical files, and damage the physical components. In this chapter, we aim to use a game-theoretic framework to capture the properties of the APT attack. We also use other games to describe the behaviors of the cyber and physical layers. We use the equilibrium of the game to develop optimal defense strategies and present numerical experiments to evaluate defensive performance.

Chapter 18: Proactive Defense Against Security Threats on IoT Hardware,
Qiaoyan Yu, Zhiming Zhang, and Jaya Dofe
This chapter shifts the focus on software security (explored in Chapter 18) to focus on hardware security for the Internet of Things (IoT). Hardware is the root of trust. Assurance of hardware integrity and security is increasingly important due to the globalized semiconductor supply chain. This chapter presents three typical hardware security threats: side-channel analysis attacks, hardware tampering, and hardware Trojan insertion from untrusted computer-aided design tools. A dynamic masking and error deflection method is introduced to address the fault attack. To thwart correlation power analysis attacks, the noise originated from the power distribution network is exploited to obscure the correlation between real power consumption and the input/key-dependent power estimation. The on-chip network router is hardened by dynamic permutation and error control coding to reduce the success of hardware Trojan insertion in application-specific integrated circuits. The principle of moving target defense is employed to the Trojan placement in the field-programmable gate array platform. Despite the emergence of various security threats, dynamic design parameters, system inherent variables, and configurable architecture have great potential to be applied to address hardware security issues.

Chapter 19: IoT Device Attestation: From a Cross-Layer Perspective, *Orlando Arias, Fahim Rahman, Mark Tehranipoor, and Yier Jin*

This chapter shifts the perspective from designing secure Internet of Things (IoT) schemes to protect against targeted attacks to a viewpoint of ensuring the authenticity of software running on IoT devices. In recent years, we have seen a rise in popularity of networked devices. From traffic signals in a city's busiest intersection and energy metering appliances to Internet-connected security cameras, these embedded devices have become entrenched in everyday life. As a consequence, a need to ensure secure and reliable operation of these devices has also risen. Device attestation is a promising solution to the operational demands of embedded devices, especially those widely used in IoT and cyber-physical systems.

In this chapter, we discuss the basics of device attestation. We then summarize attestation approaches by classifying them based on the functionality and reliability guarantees they provide to networked devices. Last, we discuss the limitations and potential issues current mechanisms exhibit and propose new research directions.

Chapter 20: Software-Defined Networking for Cyber Resilience in Industrial Internet of Things (IIoT), *Kamrul Hasan, Sachin Shetty, Amin Hassanzadeh, Malek Ben Salem, and Jay Chen*

We shift gears from secure Internet of Things (IoT) design at a device level to design mechanisms at the network level. Software-defined networking (SDN) is a networking paradigm to provide automated network management at run time through network orchestration and virtualization. SDN is primarily used for quality of service (QoS) and automated response to network failures. In the context of Energy Delivery System (EDS), SDN can enable Industrial Internet of Things (IIoT) that also enhances system resilience through recovery from failures and maintaining critical operations during cyberattacks. Researchers have proposed SDN-based architectures for autonomous attack containment, which dynamically modifies access control rules based on configurable trust levels. One of the challenges with such architectures is the lack of a cost model to select the countermeasure, which balances the trade-off between security risk and network QoS. Prior to choosing a particular countermeasure, which either quarantines the attack or mitigates the impact, it is also critical to assess the impact on the ability of the operator to conduct normal operations. In this chapter, an approach is presented to aid in the selection of security countermeasures dynamically in an SDN platform based on IIoT technology-enabled EDS and achieving trade-off between providing security and QoS. The modeling of security cost based on end-to-end packet delay and throughput is discussed. A non-dominated sorting-based multi-objective optimization framework is proposed, which can be implemented within an SDN controller to address the joint problem of optimizing between security and QoS parameters by alleviating time complexity.

Chapter 21: Leverage SDN for Cyber Security Deception in Internet of Things, *Yaoqing Liu, Garegin Grigoryan, Laurent L. Njilla, and Charles A. Kamhoua*
This chapter continues to explore additional advantages of software-defined networking (SDN) beyond achieving cyber resilience in Internet of Things (IoT). IoT is becoming an increasingly attractive target for cybercriminals. We observe that many attacks to IoTs are launched in a collusive way, such as brute-force hacking usernames and passwords, to target a particular victim. However, most of the time our defense mechanisms to such attacks are carried out individually and independently, which leads to ineffective and weak defense. To this end, we propose to leverage SDNs to enable cybersecurity deception and cooperative security for legacy IP-based IoT devices. In this chapter, we discuss the IoT security problems and challenges, and propose to leverage cybersecurity deception techniques and an SDN-based architecture to enable IoT security in a cooperative manner. Furthermore, we implemented a platform that can simulate cybersecurity deception techniques, quickly share the attacking information with peer controllers, and block the attacks. We carried out our experiments in both virtual and physical SDN environments with OpenFlow switches. Our evaluation results show that both environments can scale well to handle attacks, but hardware implementation is much more efficient than a virtual one. We also argue that it is very challenging to deploy SDN-based cyber-deception techniques in mobile and distributed networks with the traditional TCP/IP architecture. We propose to combine SDN and Named Data Networks to implement cyber deception in a mobile IoT environment. We also discuss the benefits and the challenges for this new research direction.

Chapter 22: Decentralized Access Control for Internet of Things based on Blockchain and Smart Contract, *Ronghua Xu, Yu Chen, and Erik Blasch*
This chapter shifts the focus from infrastructure-based security design in moving target defense and software-defined networking to distributed ledger design using blockchain. The Internet of Things (IoT) contributes significantly to improve our daily life by ubiquitously providing revolutionary applications and services covering transportation, healthcare, industrial automation, emergency response, and more. However, highly connected smart IoT devices built on highly heterogeneous platforms with insufficient security enforcement incur more concerns on security and privacy. This chapter provides readers a decentralized framework based on blockchain and smart contract methods to address security issues in IoT networks. By reviewing the state-of-the-art developments in access control (AC) solutions for IoT systems from both model and architecture aspects, readers will be aware of the

limitations of conventional AC models as well as many problems incurred by centralized architecture. Because of multiple attractive features like decentralization and anonymity, the blockchain technology is promising to enable a manageable and efficient security mechanism that meets the requirements raised by IoT networks. Current research on blockchain-based security solutions to IoT is introduced, and the main challenges are highlighted. To demonstrate the feasibility of a blockchain-enabled security mechanism, a case study of a decentralized AC mechanism for IoT, called BlendCAC, is illustrated. Through encapsulating AC logic as a smart contract that is deployed on a blockchain network, the BlendCAC scheme allows users to control their devices and resources instead of relying on a centralized third authority to maintain the trust relationship in a trustless network environment. Thanks to the blockchain and smart contract, the risks resulting from centralized architecture are eliminated. Furthermore, edge computing-driven intelligence is performed on smart things through transferring power and intelligence from the cloud server to the edge of network, which demonstrates the feasibility of decentralized, scalable, heterogeneous, and dynamic IoT-based applications.

Chapter 23: Intent as a Security Design Primitive, *Prashant Anantharaman, J. Peter Brady, Ira Ray Jenkins, Vijay H. Kothari, Michael C. Millian, Kartik Palani, Kirti V. Rathore, Jason Reeves, Rebecca Shapiro, Syed H. Tanveer, Sergey Bratus, and Sean W. Smith*

Chapters 18–22 focus on designing secure Internet of Things (IoT) mechanisms to combat against known attacks. There is a need to look beyond known attacks to zero-day attacks that plague a wide class of IoT devices and is not agnostic to any particular manifestation of IoT. This chapter introduces the concept of intent as a secure design primitive. The intents of the designer, developer, and user – and whether these intents are realized – lie at the heart of system security. Many security failures stem from a disconnect between what an actor intends for a technology to do and what it actually does. The field of language-theoretic security (LangSec) and the intent enforcement mechanism Executable and Linkable Format (ELF)-based access control (ELFbac) aim to preserve designer and developer intents. This chapter discusses the importance of preserving intents, provides primers on both LangSec and ELFbac, demonstrates how these two paradigms can be used to construct a secure implementation of the Advanced Message Queuing Protocol (AMQP), and examines evaluation techniques.

LangSec aims to help protocol designers, protocol implementers, and programmers to protect against exploits, i.e. crafted inputs that invoke and piece together unintended computation capabilities offered by the target. LangSec advocates separating the parser from the program internals, ensuring parsing is done in full before the remainder of the computation is performed; choosing protocols and parsers that are minimally expressive; and requiring parser equivalence for secure composition. ELFbac focuses on developer intent enforcement via intra-process memory access controls. Using the existing compiler infrastructure, i.e. the linker and the loader, security policies can be written to isolate portions of memory space within a single process. The enforcement of these policies then relies on kernel memory management, i.e. page table mechanisms. Illegal or unintended accesses then become paging or memory faults. In this way, the natural boundaries inherent in a programmer's mental model of code execution can be encoded at compile time and enforced at run time.

We provide a case study for LangSec and ELFbac by applying them to the AMQP. The AMQP is being used extensively in the IoT due to its ease of use and lightweight nature. We describe the functioning of the AMQP and implement a parser for its message formats. We use ELFbac to isolate the parser code and data from other system modules. While the importance of LangSec and ELFbac may be clear to advocates, an important area of study is on evaluation techniques. The chapter closes with discussion of established techniques as well as those in development.

Chapter 24: A Review of Moving Target Defense Mechanisms for Internet of Things Applications, *Nico Saputro, Samet Tonyali, Abdullah Aydeger, Kemal Akkaya, Muhammad Rahman, and Selcuk Uluagac*

This chapter explores Internet of Things (IoT) security from the perspective of effective defense and advocates a case for the utilization of moving target defense (MTD) paradigm that has been successfully applied to various domains in enterprise networks. The chapter starts by providing a background on MTD, specifying the categories of MTD, and surveys existing studies that explored the performance of the approaches under these MTD categories. Among such categories, one of the prevalent ones is based on the emerging software-defined networking (SDN) technology. SDN's deployment serves the purpose of more efficient centralized control of the switches for more agile MTD capabilities. The chapter then turns its focus to MTD techniques that can be applied to various IoT environments. The authors start with a discussion of the feasibility of applying MTD to resource-constrained IoT devices to make a compelling argument. They then review the existing studies that already applied MTD in IoT environments. A categorization of these approaches is provided in terms of the type of MTD studied. In this section, the authors emphasize SDN-based MTD approaches as the newly emerging area that would fit certain IoT applications. The authors argue that Internet of Battlefield Things (IoBT) and Industrial IoT (IIoT) are the two critical domains, which can benefit from SDN-based MDT. As these applications are subject to more attack and need continuous support, SDN could be a useful means to manage the devices in these domains that will facilitate more effective defense approaches through

MDT. As there is almost no work that specifically focuses on SDN-based MDT for IoT, the authors also discuss potential research challenges that could be studied on this topic before the chapter is concluded.

Chapter 25: Toward Robust Outlier Detector for Internet of Things Applications,
Raj Mani Shukla and Shamik Sengupta
In the Internet of Things (IoT) domain, sensors measure the data that are used for making optimal decisions by different applications. Poor-quality data, such as data being tampered by nefarious adversaries, affect the performance of the applications. Attacks where an adversary alters the data value and induces outliers in data are called data-falsification attacks. Outliers in time-series IoT data are frequent and may affect the performance of applications significantly. Therefore, their detection and mitigation are an important concern to make applications perform efficiently.

This chapter discusses various state-of-the-art anomaly detection techniques for IoT data. We cover different distributed and centralized approaches that have been used to detect outliers. We also discuss their limitations and emphasize the need to investigate a robust method to detect anomalies in time-series data for IoT applications. The robust method should be able to handle diverse time-series coming from different sources. It should be able to differentiate between whether the outliers are due to environmental factors or due to an attack. Also, the method should be scalable and able to handle huge amounts of data.

To solve the problem, we propose a technique that fuses different modules. The correlation module determines how diverse time-series are related to each other. It can be used to verify whether an anomaly is due to external conditions or an attack. The technique uses a combination of neural network, statistical analysis, and gradient analysis to determine the outliers in time-series data. Neural network can estimate the expected data and assist in rough estimation of the anomaly. Statistical and gradient analyses can provide more accurate information about the outliers.

This chapter further describes some of the open research issues that the current technique may pose. Specifically, we describe the threat due to open-source software in IoT and requirement of computing method for efficient implementation of outlier detector.

1.4 Summary and Future Work

This book provides several important insights to the area of IoT security. We use a defense-in-depth architecture to describe IoT security. Each layer in the defense-in-depth architecture represents cyber deception, modeling, and design as a proactive defense of IoT. Within each layer we first present techniques to secure the Things (devices, sensors, and actuators) from adversarial attacks. Next, we describe the methodologies to realize a secure interconnection between the Things. Finally, we provide the necessary distributed information processing techniques required to achieve scalable and robust protection of the Things. The book comprises foundational techniques and applied methods in both military (IoBT) and commercial domains (smart homes) to offer a flexible, low-cost, and secure means to provide resilient critical infrastructure, reduce risks of exploitable attack surfaces, and improve the survivability of physical processes.

References

1 Cisco Visual Networking Index: Global Mobile Data Traffic Forecast Update, 2017–2022 White Paper. https://www.cisco.com/c/en/us/solutions/collateral/service-provider/visual-networking-index-vni/white-paper-c11-738429.pdf

2 S. Shetty, C.A. Kamhoua, L. Njilla, editors. *Blockchain for Distributed System Security*. Washington, DC: Wiley-IEEE Computer Society Press; April 2019. ISBN:978-1-119-51960-7.

3 R.H. Campbell, C.A. Kamhoua, K.A. Kwiat, editors. *Assured Cloud Computing*. Washington, DC: Wiley-IEEE Computer Society Press; October 2, 2018. ISBN-10: 1119428637.

4 Proofpoint. Proofpoint Uncovers Internet of Things (IoT) Cyberattack; January 16, 2014 [accessed April 10, 2019]. https://www.proofpoint.com/us/proofpoint-uncovers-internet-things-iot-cyberattack

5 E. Blumenthal, E. Weise. Hacked home devices cause massive internet outage. *USA Today*; October 21, 2016 [accessed April 10, 2019]. https://eu.usatoday.com/story/tech/2016/10/21/cyber-attack-takes-down-east-coast-netflix-spotify-twitter/92507806

6 J. Finkle. J&J warns diabetic patients: Insulin pump vulnerable to hacking. *Reuters*; October 4, 2016 [accessed April 10, 2019]. https://www.reuters.com/article/us-johnson-johnson-cyber-insulin-pumps-e/jj-warns-diabetic-patients-insulin-pump-vulnerable-to-hacking-idUSKCN12411L

7 Jeep Cherokee hack. https://www.kaspersky.com/blog/blackhat-jeep-cherokee-hack-explained/9493

8 A. Bogle. A cyber attack may have caused a Turkish oil pipeline to catch fire in 2008. *Slate*; December 11, 2014 [accessed April 11, 2019]. http://www.slate.com/blogs/future_tense/2014/12/11/bloomberg_reports_a_cyber_attack_may_have_made_a_turkish_oil_pipeline_catch.html?via=gdpr-consent

9 R. Zhang, N. Zhang, C. Du, W. Lou, Y.T. Hou, Y. Kawamoto. From electromyogram to password: exploring the privacy impact of wearables in

augmented reality. *ACM Transactions on Intelligent Systems and Technology*. October 2017;**9**(1) 13:1–13:20.

10 Y. Zheng, A. Moini, W. Lou, Y.T. Hou, Y. Kawamoto. Cognitive security: securing the burgeoning landscape of mobile network. *IEEE Network*. July-August 2016;**30**(4).

11 A. Kott, A. Swami, B.J. West. The internet of battle things. *IEEE Computer*. 2016;**49** (12): 70–75.

Part I

Game Theory for Cyber Deception

2

Game-Theoretic Analysis of Cyber Deception

Evidence-Based Strategies and Dynamic Risk Mitigation

Tao Zhang, Linan Huang, Jeffrey Pawlick, and Quanyan Zhu

Department of Electrical and Computer Engineering, Tandon School of Engineering, New York University, Brooklyn, NY, USA

2.1 Introduction

Deception is a technique used to cause animals [1], humans [2, 3], or computer systems [4] to have false beliefs. The purpose of deception is to mislead the deceivees to behave against their interests but favorably to the deceiver. It is a fundamental type of interactions that can be found in applications ranging from biology [1] to criminology [3] and from economics [2] to the Internet of Things (IoT) [5]. Cyberspace creates particular opportunities for deception, since information lacks permanence, imputing responsibility is difficult [6], and some agents lack repeated interactions [7]. For instance, online interactions are vulnerable to identify theft [8] and spear phishing [9], and authentication in the IoT suffers from a lack of infrastructure and local computational resources [10].

Deception can be used as an approach for attacks. For example, phishing is a typical deception-based attack that is one of the top threat vectors for cyberattacks [11]. Phishing can be email-based in which a phisher manipulates the email to appear as a legitimate request for sensitive information [12, 13]. It can also be website-based in which the deceiver uses genuine-looking content to camouflage a legitimate website to attract target deceivees to reveal their personal data such as credit card information and social security number. Defenders can also implement deception. Defenders in the security and privacy domains have proposed, e.g. honeynets [14], moving target defense [15], obfuscation [16], and mix networks [17].

Modeling and Design of Secure Internet of Things, First Edition. Edited by Charles A. Kamhoua, Laurent L. Njilla, Alexander Kott and Sachin Shetty.

Published 2020 by John Wiley & Sons, Inc.

Using these techniques, defenders can obscure valuable data such as personally identifiable information or the configuration of a network. Using these approaches, defenders can send false information to attackers to waste their resources or distract them from critical assets. They can also obscure valuable data such as sensitive information or the configuration of a network to avoid direct accesses from the attackers. Both malicious and defensive deceptions have innumerable implications for cybersecurity. Successful deception fundamentally depends on the information asymmetry between the deceiver and the deceivee. Deceivees make indirect observations of the true state and then make decisions. Deceivers can take advantage of this by pretending to be a trustworthy information provider. It is possible to fool, mislead, or confuse the deceivees. But the deceivers need to plan their strategies and take actions that may be costly. Therefore, successful deception also requires the deceivers to have the ability to acquire information, accurately understand the goals of the deceivees, and make the induced actions predictable.

The deceivers strategically manipulate the private information to suit their own self-interests. The manipulated information is then revealed to the deceivees, who, on the other hand, make decisions based on the information received. It is important for the deceivee to form correct beliefs based on past observations, take into account the potential damage caused by deception, and strategically use the observed information for decision-making. If deception is necessary to achieve the deceivers' goal that would cause damages to the deceivees, the deceivees can then be prepared to invest resources in detecting and denying the deceptions as well as recovering the damage.

Modeling deceptive interactions online and in the IoT would allow government policymakers, technological entrepreneurs, and vendors of cyber-insurance to predict changes in these interactions for the purpose of legislation, development of new technology, or risk mitigation. Game-theoretic models are natural frameworks to capture the adversarial and defensive interactions between players [18–29]. It can provide a quantitative measure of the quality of protection with the concept of Nash equilibrium where both defender and an attacker seek optimal strategies, and no one has an incentive to deviate unilaterally from their equilibrium strategies despite their conflict for security objectives. The equilibrium concept also provides a quantitative prediction of the security outcomes of the scenario the game model captures. With the quantitative measures of security, game theory makes security manageable beyond the strong qualitative assurances of cryptographic protections. Recently, we have seen game-theoretic methods applied to deal with problems in cross-layer cyber-physical security [5, 21, 28, 30–33], cyber deception [19, 22, 29, 34–36], moving target defense [15, 37, 38], critical infrastructure protection [26, 27, 39–46], adversarial machine learning [16, 47–51], insider threats [52, 53], and cyber risk management [54–57].

This chapter presents a class of modeling of deception based on signaling games to provide a generic, quantitative, and systematic understanding of deceptions in the cyber domain. We show three variants of the model to illustrate the applications in different situations. We consider the cases when the deceivee is allowed to acquire knowledge through investigations or by deploying detectors. The baseline signaling game model is extended to include evidence through side-channel knowledge acquisition. We also show a multistage Bayesian game with two-sided incomplete information and present a dynamic belief update and an iterative decision process that are used to develop long-term optimal defensive policies to deter the deceivers and mitigate the loss.

2.1.1 Related Work

Deception game is related to a class of security games of incomplete information. For example, Powell in [58] has considered a game between an attacker and a defender, where the defender has private information about the vulnerability of their targets under protection. Powell models the information asymmetric interactions between players by a signaling game, and finds a pooling equilibrium where the defender chooses to pool, i.e. allocate resources in the same way for all targets of different vulnerabilities, and the attacker cannot know the true level of vulnerability of all targets. Brown et al. [59] have studied a zero-sum game between an attacker and a defender in the scenario of ballistic missile positioning. They have introduced the incomplete information to investigate the value of secrecy by restricting the players' access to information.

Previous literature has also considered deception in a variety of scenarios including proactive defense against advanced persistent threats (APTs) [18, 19, 22, 39, 60], moving target defense [15, 61, 62], and social engineering [23, 35, 36]. Horák et al. [19] have considered a class of cyber-deception techniques in the field of network security and studied the impact of the deception on attacker's beliefs using the quantitative framework of the game theory by taking into account the sequential nature of the attack and investigating how attacker's belief evolves and influences the actions of the players. Zhang and Zhu [22] have proposed an equilibrium approach to analyze the GPS spoofing in a model of signaling game with continuous type space. They have found a perfect Bayesian Nash equilibrium (PBNE) with pooling in low types and separating in high types, and provided an equilibrium analysis of spoofing. The hypothesis testing game framework in [63] has studied the influence of deceptive information on the decision-making and analyzed the worst-case scenario by constructing equilibrium strategies. The model proposed by Ettinger et al. [64] has used an equilibrium approach to belief deception in bargaining problems when the agents only have coarse information about their opponent's strategy.

2.1.2 Organization of the Chapter

The chapter is organized as follows. In Section 2.2, we briefly describe common game-theoretic approaches for security modeling and introduce the basic signaling game model and its associated solution concept PBNE. In Section 2.3, we formulate the deception using a signaling game model with a detector over a binary information space, and describe the equilibrium result. In Section 2.4, we present a signaling-game-based framework of a deception game to model the strategic behaviors over a continuous one-dimensional information space. We also consider the knowledge acquisition for the receiver through investigations. In Section 2.5, we show a multistage Bayesian game framework to model the deception in APTs. Section 2.6 discusses the results and provides concluding remarks.

2.2 Game Theory in Security

Game theory is the systems science that studies interactions between rational and strategic players (or agents) that are coupled in their decision-makings. Players are rational in the sense that they choose actions to optimize their own objectives (e.g. utility or cost), which capture varying interaction contexts. Being strategic in game theory refers to that players choose their own actions by anticipating the actions of the other agents. Their decision-makings are coupled because their objective functions depend both on their own actions and on the actions of the other players. Among the game-theoretic cybersecurity models, Stackelberg game, Nash game, and signaling game account for the most commonly used approaches [35].

Stackelberg games consist of a *leader* (L) and a *follower* (F). L has actions $a_L \in A_L$ and receives utility U_L, and F has actions $a_F \in A_F$ and receives utility U_F. Once both players have taken actions, L receives utility $U_L(a_L, a_F)$ and $U_F(a_L, a_F)$. In Stackelberg games, F acts after knowing L's action. Therefore, defensive cybersecurity models often take the defender as the leader and the attacker as the follower by considering the worst-case scenario that the attacker will observe and react to defensive strategies. Let $\mathcal{P}(A)$ denote the power set of the set A. Let $BR_F : A_L \to \mathcal{P}(A_F)$ denote the best response of F to L's action such that $BR_F(a_L)$ gives the optimal a_F to respond to a_L. Best response can be one single action or a set of equally optimal actions. The function BR_F is defined as

$$BR_F(a_L) = \underset{a_F \in A_F}{\text{argmax}}\, U_F(a_L, a_F).$$

By anticipating $BR_F(a_L)$, L chooses optimal action a_L^* which satisfies

$$a_L^* \in \underset{a_F \in A_F}{\text{argmax}}\, U_L(a_L, BR_F(a_L)).$$

The action profile (a_L^*, a_F^*) characterizes the equilibrium of the Stackelberg game, where $a_F^* \in BR_F(a_L^*)$.

In Nash games, on the other hand, players commit to his or her own strategy and move simultaneously or before knowing the other player's action [35]. Let H and T denote two players in a Nash game with actions $a_H \in A_H$ and $a_T \in A_T$, respectively. Define $BR_H(a_T)$ as the best response for H that optimally respond to T's action a_T. Similarly, let $BR_T(a_H)$ be the best response for T. *Nash equilibrium* is the solution concept of such games, which is defined by a strategy profile (a_H^*, a_T^*), where

$$a_H^* \in BR_H(a_T),$$
$$a_T^* \in BR_T(a_H).$$

In other words, Nash equilibrium requires each player to simultaneously choose a strategy that is optimal, given the other player's optimal strategy. In a *pure-strategy* Nash equilibrium, each player chooses one specific strategy (i.e. one pure strategy), while in a mixed-strategy equilibrium, at least one player randomizes over some or all pure strategies.

A signaling game is a two-player dynamic game of incomplete information. Signaling game typically names two players as sender (S, she) and receiver (R, he) [65]. Signaling game is information asymmetric because the sender privately possesses some information that is unknown to the receiver. The private information is usually referred to as state (or type) of the world. The sender communicates the receiver by sending a message, and the receiver only learns about the state through the message. Generally, the degree of conflict of interest between S and R may range from perfectly aligned (e.g. Lewis signaling game [66]) to completely opposite (e.g. zero-sum game [67]). The timing of the game is described as follows:

1) Nature randomly draws a state with a prior common to both players, and the sender privately observes the state.
2) The sender sends a message to the receiver.
3) The receiver takes an action upon observing the message.

One key feature of cyber deception is the multistage execution of the attack. A deceiver has to engage the deceivee in multiple rounds of interactions to gain the trust. The dynamic interactions have been observed in APT threats and the operation of honey devices [39]. Hence, signaling games provide a suitable description of essential features of the deception, and the basic signaling game models will be elaborated in the following section.

2.2.1 Signaling Game Model

Let $\theta \in \Theta$ denote the state privately possessed by S that is unknown to R. The state space Θ can be discrete (e.g. $\Theta_D \equiv \{\theta_0, \theta_1\}$) or continuous (e.g. $\Theta_C \equiv [\underline{\theta}, \overline{\theta}]$). For example, the state could represent, whether the sender is a malicious or benign actor, whether she has one set of preferences over another, samples of data stream, and location coordinate. For simplicity but without loss of generality, we focus on discrete state space in this introductory description of the signaling game. The state θ is drawn according to a prior distribution common to both players. Harsanyi conceptualized the state selection as a randomized move by a nonstrategic player called *nature*. Let p denote the probability mass function of the state, where $\sum_{\theta \in \Theta_D} p(\theta) = 1$. All aspects of the game except the value of the true state θ are common knowledge.

After privately observing the state θ, S chooses a message $m \in M$. Let $\sigma^S \in \Gamma^S$ denote the behavioral strategy of S to choose m based on θ. She could use pure strategy $\sigma^S(\theta)$ as well as mixed strategy $\sigma^S(m \mid \theta)$, such that $\sigma^S(\theta)$ chooses message m, given θ and $\sigma^S(m \mid \theta)$ gives probability with which S sends message m, given the state θ. We assume the pure strategy $\sigma^S(\theta)$ induces a conditional probability $q^S(m \mid \theta) \in [0, 1]$.

After receiving m, R forms a posterior belief μ^R of the true state such that $\mu^R(\theta \mid m) : \Theta \to [0, 1]$ gives the likelihood with which R believes that the true state is θ, given the message m. Based on the belief μ^R, R then chooses an action $a \in A$ according to a strategy $\sigma^R \in \Gamma^R$. Similarly, R may employ pure strategy $\sigma^R(m)$ or mixed strategy $\sigma^R(a \mid m)$, where $\sigma^R(m)$ yields the action R acts upon the message m, and $\sigma^R(a \mid m)$ produces the probability with which R takes action a, given message m. The action a is the final decision of R that represents the inference about the true state. Let $U^S : \Theta \times M \times A \to \mathbb{R}$ denote a utility function for S such that $U^S(\theta, m, a)$ yields the utility of the player when her type is θ, she sends message m, and R plays action a. Similarly, let $U^R : \Theta \times M \times A$ denote R's utility function so that $U^R(\theta, m, a)$ gives his payoff under the same scenario.

2.2.2 Perfect Bayesian Nash Equilibrium

In two player games, Nash equilibrium defines a strategy profile in which each player best responds to the optimal strategies of the other player [68]. Signaling games motivate the extension of Nash equilibrium in two ways. First, information asymmetry requires R to maximize his expected utility over the possible types of S. An equilibrium in which S and R best respond to each other's strategies, given some belief μ^R, is called a *Bayesian Nash equilibrium* [69]. Furthermore, R is required to update μ^R rationally. PBNE captures this constraint and is described in Definition 2.1.

Definition 2.1 (Perfect Bayesian Nash Equilibrium) A PBNE of a signaling profile $(\sigma^{S*}, \sigma^{R*})$ and a posterior belief system μ^R such that:

1) (C_1): $\forall \theta, \sigma^{S*} \in \arg\max_{\sigma^S} U^S(\theta, \sigma^S, \sigma^{R*})$,
2) (C_2): $\forall m \in M, \sigma^{R*} \in \operatorname{argmax}_{\sigma^R} \sum_{\theta} \mu^R(\theta|m) U^R(m, \sigma^R, \theta)$,
   and

3) (C_3): $\mu^R(\theta|m) = \frac{p(\theta)\sigma^{S*}(m|\theta)}{\sum_{\theta'} p(\theta')\sigma^{S*}(m|\theta')}$, if $\sum_{\theta'} p(\theta')\sigma^{S*}(m|\theta') > 0$; and $\mu^R(\theta \mid m)$ is any probability distribution on Θ if $\sum_{\theta'} p(\theta')\sigma^{S*}(m|\theta') = 0$.

In Definition 2.1, C_1 and C_2 are the perfection conditions for S and R, respectively, that characterizes the *sequential rationality* of both players. Specifically, C_1 captures that S optimally determines σ^{S*} by taking into account the effect of σ^S on σ^R. C_2 says that R responds rationally to S's strategy, given his posterior belief about the state θ. C_3 states that the posterior belief is updated based on Bayes' rule. If the observation is a probability-0 event, then Bayes' rule is not applicable. In this case, any posterior beliefs over the state space is admissible. C_3 also implies that the consistency between the posterior beliefs and strategies: belief is updated depending on the strategy that is optimal, given the belief. There are three categories of strategies: *separating*, *pooling*, and *partially pooling* equilibria, which are defined based on the strategy of S. In separating PBNE (S-PBNE), S chooses different strategies for different states. In this case, R is able to identify each state with certainty. In pooling PBNE (P-PBNE), S chooses the same strategy for different states. This strategy makes the corresponding message m uninformative to R. In partially pooling PBNE (PP-PBNE), however, S chooses messages with different, but not completely distinguishable, strategies for different states. This makes the belief of R remain uncertain.

2.3 Binary State Space: Leaky Deception Using Signaling Game with Evidence

In [34], Pawlick et al. have modeled the strategic interactions between the deceiver S and the deceivee R over a binary information space by extending signaling games by including a detector that gives off probabilistic warnings called *evidence* when S acts deceptively.

2.3.1 Game-Theoretic Model

With reference to Figure 2.1, the detector emits evidences based on whether the message m is equal to the state θ. The detector emits $e \in E \equiv \{0, 1\}$ by the probability $\lambda(e \mid \theta, m)$. Let $e = 1$ denote an alarm and $e = 0$ no alarm. The evidence e is assumed to be emitted with an exogeneous probability that neither R nor S can control. In this respect, the detector can be seen as a second move by nature. Let $\beta \in [0, 1]$ be the true-positive rate of the detector. For simplicity, both true-positive rates are set to be equal: $\beta = \lambda(1 \mid 0, 1) = \lambda(1 \mid 1, 0)$. Similarly, let $\alpha \in [0, 1]$ denote the false-positive rate of the detector with $\alpha = \lambda(1 \mid 0, 0) = \lambda(1 \mid 1, 1)$. A valid detector has

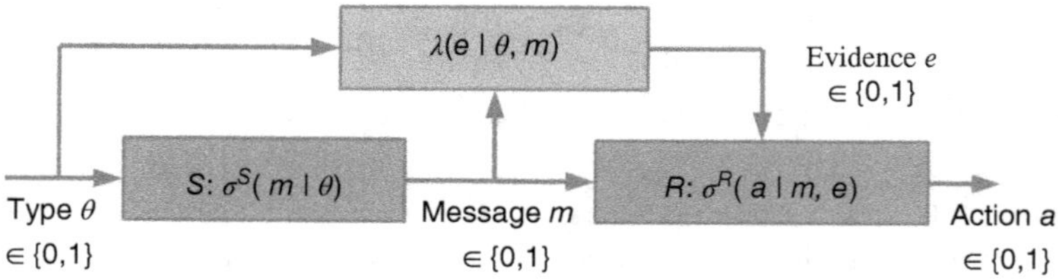

Figure 2.1 Signaling games with evidence add a detector block (shown on the top) to the S and R bloacks. The probability $\lambda(e|\theta, m)$ of emitting evidence e depends on S's type θ and the message m that she transmits [34].

$\beta \geq \alpha$. This is without loss of generality, because otherwise α and β can be relabeled. The timing of the game becomes:

1) Nature randomly draws state $\theta \in \{0, 1\}$ according to $p(\theta)$.
2) S privately observes θ and then chooses a message m based on strategy $\sigma^S(m \mid \theta)$.
3) The detector emits evidence $e \in \{0, 1\}$ with $\lambda(e \mid \theta, m)$.
4) After receiving both m and e, R forms a belief system $\mu^R(\theta \mid m, e)$ and then chooses an action $a \in \{0, 1\}$ according to strategy $\sigma^R(a \mid m, e)$.

The following assumptions characterize a *cheap-talk signaling game with evidence*. The message m is payoff-irrelevant in a cheap-talk signaling game.

Assumption 2.1 The utilities U^S and U^R satisfy the following assumptions:

1) U^S and U^R do not depend exogenously on m.
2) $\forall m, \widetilde{m} \in M,\ U^R(0, m, 0) > U^R(0, \widetilde{m}, 1)$.
3) $\forall m, \widetilde{m} \in M,\ U^R(1, m, 0) < U^R(1, \widetilde{m}, 1)$.
4) $\forall m, \widetilde{m} \in M,\ U^S(0, m, 0) < U^S(0, \widetilde{m}, 1)$.
5) $\forall m, \widetilde{m} \in M,\ u^S(1, m, 0) > u^S(1, \widetilde{m}, 1)$.

Assumption 2.1(1) implies that the interaction is a cheap-talk signaling game. Assumption 2.1(2) and 2.1(3) state that R receives higher utility if he correctly chooses $a = \theta$ than if he chooses $a \neq \theta$. Finally, Assumption 2.1(4) and 2.1(5) say that S receives higher utility if R chooses $a \neq \theta$ than if he chooses $a = \theta$.

2.3.1.1 Utilities

Define an expected utility function $\overline{U}^S : \Gamma \times \Gamma^R \to \mathbb{R}$ such that $\bar{U}(\sigma^S, \sigma^R \mid \theta)$ gives the expected utility to S when she plays strategy σ^S, given that the state is θ. This expected utility is given by

$$\bar{U}^S(\sigma^S, \sigma^R \mid \theta) = \sum_{a \in A} \sum_{e \in E} \sum_{m \in M} \sigma^R(a \mid m, e)\lambda(e \mid \theta, m)\sigma^S(m \mid \theta)U^S(\theta, m, a). \tag{2.1}$$

The involvement of evidence e in Eq. (2.1) is due to the dependence of R's strategy $\sigma^R(a \mid m, e)$. S must anticipate the probability of leaking evidence e by using $\lambda(e \mid \theta, m)$. Similarly, define $\bar{U}^R : \Gamma \to \mathbb{R}$ such that $\bar{U}^R(\sigma^R \mid \theta, m, e)$ gives the expected utility to R when he plays strategy σ^R, given message m, evidence e, and state θ. The expected utility function is given by

$$\bar{U}^R(\sigma^R \mid \theta, m, e) = \sum_{a \in A} \sigma^R(a \mid m, e)U^R(\theta, m, a).$$

2.3.2 Equilibrium Concept

The involvement of the evidence extends the PBNE in Definition 2.2 as follows.

Definition 2.2 (PBNE with Evidence) A PBNE of a cheap-talk signaling game with evidence is a strategy profile $(\sigma^{S*}, \sigma^{R*})$ and posterior beliefs $\mu^R(\theta \mid m, e)$ such that

$$\forall \theta \in \Theta, \sigma^{S*} \in \underset{\sigma^S \in \Gamma^S}{\operatorname{argmax}}\ \bar{U}^S(\sigma^S, \sigma^{R*} \mid \theta), \tag{2.2}$$

$$\forall m \in M, \forall e \in E,$$

$$\sigma^{R*} \in \underset{\sigma^R \in \Gamma^R}{\operatorname{argmax}} \sum_{\theta \in \Theta} \mu^R(\theta \mid m, e)\bar{U}^R(\sigma^R \mid \theta, m, e), \tag{2.3}$$

and if $\sum_{\tilde{\theta} \in \Theta} \lambda\left(e \mid \tilde{\theta}, m\right)\sigma^S\left(m \mid \tilde{\theta}\right)p\left(\tilde{\theta}\right) > 0$, then

$$\mu^R(\theta \mid m, e) = \frac{\lambda(e \mid \theta, m)\mu^R(\theta \mid m)}{\sum_{\tilde{\theta} \in \Theta} \lambda\left(e \mid \tilde{\theta}, m\right)\mu^R\left(\tilde{\theta} \mid m\right)}, \tag{2.4}$$

where

$$\mu^R(\theta \mid m) = \frac{\sigma^S(m \mid \theta)p(\theta)}{\sum_{\hat{\theta} \in \Theta} \sigma^S(m \mid \hat{\theta})p(\hat{\theta})}. \tag{2.5}$$

If $\sum_{\tilde{\theta} \in \Theta} \lambda\left(e \mid \tilde{\theta}, m\right)\sigma^S\left(m \mid \tilde{\theta}\right)p\left(\tilde{\theta}\right) = 0$, then $\mu^R(\theta \mid m, e)$ may be set to any probability distribution over Θ.

Equations (2.4)–(2.5) extend C_3 in Definition 2.2. First, R updates her belief according to m using Eq. (2.5); then R updates her belief according to e using Eq. (2.4). When S plays pooling strategy, i.e. $\forall m \in M$, $\sigma^S(m \mid 0) = \sigma^S(m \mid 1)$. In this case, the message m is uninformative, and R updates his belief only depending on the evidence e, i.e.

$$\mu^R(\theta \mid m, e) = \frac{\lambda(e \mid \theta, m)p(\theta)}{\sum_{\tilde{\theta} \in \Theta} \lambda\left(e \mid \tilde{\theta}, m\right) p\left(\tilde{\theta}\right)}. \tag{2.6}$$

2.3.3 Equilibrium Results

Under Assumption 2.1(1) to 2.1(5), the cheap-talk signaling game with evidence admits no S-PBNE. This results from the opposing utility functions of S and R. S wants to deceive R, and R wants to correctly guess the type. It is not incentive-compatible for S to fully reveal the type by choosing a separating strategy. For brevity, define the following notations:

$$\Delta_0^R \triangleq u^R(\theta = 0, m, a = 0) - u^R(\theta = 0, m, a = 1),$$
$$\Delta_1^R \triangleq u^R(\theta = 1, m, a = 1) - u^R(\theta = 1, m, a = 0).$$

Δ_0^R gives the benefit to R for correctly guessing the type when $\theta = 0$, and Δ_1^R gives the benefit to R for correctly guessing the type when $\theta = 1$. Lemmas 2.1 and 2.2 solve for σ^{R*} within five regimes of the prior probability $p(\theta)$ of each type $\theta \in \{0, 1\}$.

Lemma 2.1 For P-PBNE, R's optimal actions σ^{R*} for evidence e and messages m on the equilibrium path[1] vary within five regimes of $p(\theta)$. The top half of Figure 2.2 lists the boundaries of these regimes for detectors in which $\beta < 1 - \alpha$, and the bottom half of Figure 2.2 lists the boundaries of these regimes for detectors in which $\beta > 1 - \alpha$.

The regimes in Figure 2.2 shift toward the right as Δ_0^R increases. Intuitively, a higher $p(1)$ is necessary to balance out the benefit to R for correctly identifying a type $\theta = 0$ as Δ_0^R increases. The regimes shift toward the left as Δ_1^R increases for the opposite reason.

Lemma 2.2 gives the optimal strategies of R in response to pooling behavior within each of the five parameter regimes.

---

1 In P-PBNE, the message "on the equilibrium path" is the one that is sent by both types of S. Messages "off the equilibrium path" are never sent in equilibrium, although determining the actions that R *would play* if S were to transmit a message off the path is necessary in order to determine the existence of equilibria.

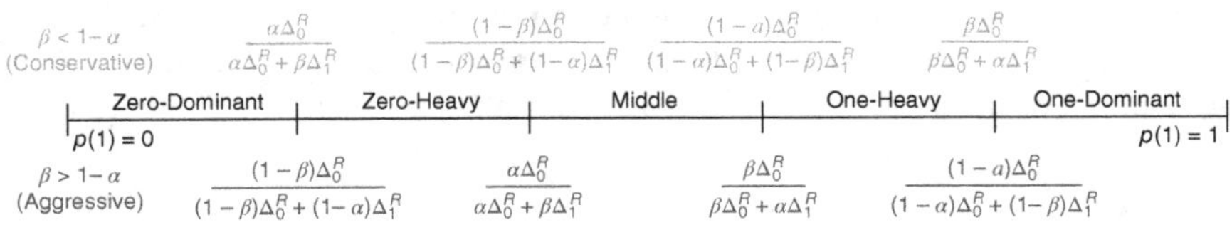

Figure 2.2 PBNE differs within five prior probability regimes. In the Zero-Dominant regime, $p(\theta = 1) \approx 0$, i.e. type 0 dominates. In the Zero-Heavy regime, $p(\theta = 1)$ is slightly higher, but still low. In the Middle regime, the types are mixed almost evenly. The One-Heavy regime has a higher $p(\theta = 1)$, and the One-Dominant regime has $p(\theta = 1) \approx 1$. The definitions of the regime boundaries depend on whether the detector is conservative or aggressive.

Lemma 2.2 For each regime, σ^{R*} on the equilibrium path is listed in Table 2.1 if $\beta < 1 - \alpha$ and in Table 2.2 if $\beta > 1 - \alpha$. The row labels correspond to the Zero-Dominant (0-D), Zero-Heavy (0-H), Middle, One-Heavy (1-H), and One-Dominant (1-D) regimes.

Lemmas 2.3 and 2.4 give conditions under which the beliefs μ^R exist such that each pooling strategy is optimal for both types of S.

Table 2.1 $\sigma^{R*}(1 \mid m, e)$ in pooling PBNE with $\beta > 1 - \alpha$.

	$\sigma^{R*}(1 \mid 0, 0)$	$\sigma^{R*}(1 \mid 0, 1)$	$\sigma^{R*}(1 \mid 1, 0)$	$\sigma^{R*}(1 \mid 1, 1)$
0-D	0	0	0	0
0-H	0	1	0	0
Middle	0	1	1	0
1-H	1	1	1	0
1-D	1	1	1	1

Table 2.2 $\sigma^{R*}(1 \mid m, e)$ in pooling PBNE with $\beta > 1 - \alpha$.

	$\sigma^{R*}(1 \mid 0, 0)$	$\sigma^{R*}(1 \mid 0, 1)$	$\sigma^{R*}(1 \mid 1, 0)$	$\sigma^{R*}(1 \mid 1, 1)$
0-D	0	0	0	0
0-H	0	0	1	0
Middle	0	1	1	0
1-H	0	1	1	1
1-D	1	1	1	1

Lemma 2.3 Let m be the message on the equilibrium path. If $\sigma^{R*}(1 \mid m, 0) = \sigma^{R*}(1 \mid m, 1)$, then there exists a μ^R such that pooling on message m is optimal for both types of S. For brevity, let $a^* \triangleq \sigma^{R*}(1 \mid m, 0) = \sigma^{R*}(1 \mid m, 1)$. Then, μ^R is given by

$$\forall e \in E, \mu^R(\theta = a^* \mid 1 - m, e) \geq \frac{\Delta^R_{1-a^*}}{\Delta^R_{1-a^*} + \Delta^R_{a^*}}.$$

Lemma 2.4 If $\sigma^{R*}(1 \mid m, 0) = 1 - \sigma^{R*}(1 \mid m, 1)$ and $\beta \neq 1 - \alpha$, then there does not exist a μ^R such that pooling on message m is optimal for both types of S.

The P-PBNE of the cheap-talk signaling game with evidence is summarized by Figure 2.3. For $\beta \neq 1 - \alpha$, the Middle regime does not admit any P-PBNE. This result is not found in conventional signaling games for deception, in which all regimes support P-PBNE. It occurs because R's responses to message m depends on e, i.e. $\sigma^{R*}(1 \mid 0, 0) = 1 - \sigma^{R*}(1 \mid 0, 1)$ and $\sigma^{R*}(1 \mid 1, 0) = 1 - \sigma^{R*}(1 \mid 1, 1)$. One of the types of S prefers to deviate to the message off the equilibrium path. Intuitively, for a conservative detector, S with type $\theta = m$ prefers to deviate to message $1 - m$, because his deception is unlikely to be detected. On the other hand, for an aggressive detector, S with type $\theta = 1 - m$ prefers to deviate to message $1 - m$, because his honesty is likely to produce a false-positive alarm, which will lead R to guess $a = m$.

For $\beta \neq 1 - \alpha$, since the Middle regime does not support P-PBNE, we search for partially separating PBNE. In this PBNE, S and R play mixed strategies. In mixed-strategy equilibria in general, each player chooses a mixed strategy that makes the other players indifferent between the actions that they play with positive probability. Theorems 2.1 and 2.2 give the results.

Theorem 2.1 (Partially Separating PBNE for Conservative Detectors) For $\beta < 1 - \alpha$, within the Middle regime, there exists an equilibrium in which the sender strategies are

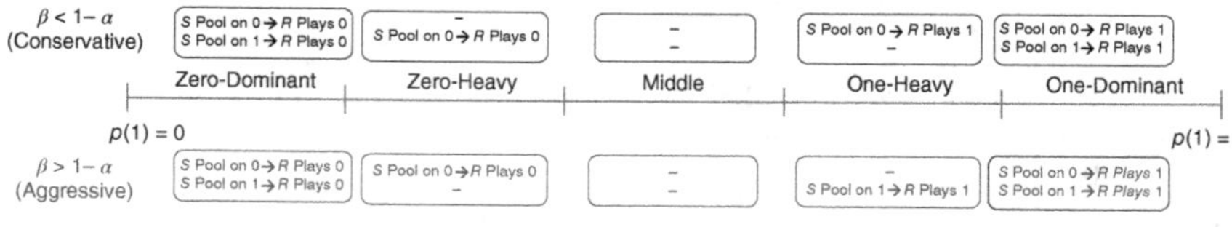

Figure 2.3 PBNE in each of the parameter regions defined in Figure 2.2. For $m \in \{0, 1\}$, "*S Pool on m*" signifies $\sigma^{S*}(m \mid 0) = \sigma^{S*}(m \mid 1) = 1$. For $a \in \{0, 1\}$, "*R Plays a*" signifies $\sigma^{R*}(a \mid 0, 0) = \sigma^{R*}(a \mid 0, 1) = \sigma^{R*}(a \mid 1, 0) = \sigma^{R*}(a \mid 1, 1) = 1$. Lemma 2.3 gives μ^{R*}. The Dominant regimes support pooling PBNE on both messages. The Heavy regimes support pooling PBNE on only one message. The Middle regime does not support any pooling PBNE.

$$\sigma^{S*}(m=1\,|\,\theta=0) = \frac{\beta^2}{\beta^2-\alpha^2} - \frac{\alpha\beta\Delta_1^R}{(\beta^2-\alpha^2)\Delta_0^R}\left(\frac{p(1)}{1-p(1)}\right),$$

$$\sigma^{S*}(m=1\,|\,\theta=1) = \frac{\alpha\beta\Delta_0^R}{(\beta^2-\alpha^2)\Delta_1^R}\left(\frac{1-p(1)}{p(1)}\right) - \frac{\alpha^2}{\beta^2-\alpha^2},$$

the receiver strategies are

$$\sigma^{R*}(a=1\,|\,m=0,e=0) = \frac{1-\alpha-\beta}{2-\alpha-\beta},$$

$$\sigma^{R*}(a=1\,|\,m=0,e=1) = 1,$$

$$\sigma^{R*}(a=1\,|\,m=1,e=0) = \frac{1}{2-a-b},$$

$$\sigma^{R*}(a=1\,|\,m=1,e=1) = 0,$$

and the beliefs are computed by Bayes' law in all cases.

Theorem 2.2 (Partially Separating PBNE for Aggressive Detectors) For any $g \in [0, 1]$, let $\bar{g} \triangleq 1-g$. For $\beta > 1-\alpha$, within the Middle regime, there exists an equilibrium in which the sender strategies are

$$\sigma^{S*}(m=1\,|\,\theta=0) = \frac{\bar{\alpha}\bar{\beta}\Delta_1^R}{\left(\bar{\alpha}^2-\bar{\beta}^2\right)\Delta_0^R}\left(\frac{p(1)}{1-p(1)}\right) - \frac{\bar{\beta}^2}{\bar{\alpha}^2-\bar{\beta}^2},$$

$$\sigma^{S*}(m=1\mid\theta=1) = \frac{\bar{\alpha}^2}{\bar{\alpha}^2-\bar{\beta}^2} - \frac{\bar{\alpha}\bar{\beta}\Delta_0^R}{\left(\bar{\alpha}^2-\bar{\beta}^2\right)\Delta_1^R}\left(\frac{1-p(1)}{p(1)}\right),$$

the receiver strategies are

$$\sigma^{R*}(a=1\,|\,m=0,e=0) = 0,$$

$$\sigma^{R*}(a=1\,|\,m=0,e=1) = \frac{1}{\alpha+\beta},$$

$$\sigma^{R*}(a=1\,|\,m=1,e=0) = 1,$$

$$\sigma^{R*}(a=1\,|\,m=1,e=1) = \frac{\alpha+\beta-1}{\alpha+\beta},$$

and the beliefs are computed by Bayes' law in all cases.

In Theorem 2.1, S chooses the σ^{S*} that makes R indifferent between $a=0$ and $a=1$ when he observes the pairs $(m=0, e=0)$ and $(m=1, e=0)$. This allows R to choose mixed strategies for $\sigma^{R*}(1\,|\,0, 0)$ and $\sigma^{R*}(1\,|\,1, 0)$. Similarly, R chooses $\sigma^{R*}(1\,|\,0, 0)$ and $\sigma^{R*}(1\,|\,1, 0)$ that make both types of S indifferent between sending $m=0$ and $m=1$. This allows S to choose mixed strategies. A similar pattern follows in Theorem 2.2 for σ^{S*}, $\sigma^{R*}(1\,|\,0, 1)$, and $\sigma^{R*}(1\,|\,1, 1)$.

2.4 Continuous State Space: Knowledge Acquisition and Fundamental Limits of Deception

In [70], Zhang and Zhu proposes a game-theoretic framework of a deception game to model the strategic behaviors of the deceiver S and the deceivee R and construct strategies for both attacks and defenses over a continuous one-dimensional state space. R is allowed to acquire probabilistic evidence about the deception through investigations, and misrepresenting the state is costly for S. The deceivability of the deception game is analyzed by characterizing the PBNE.

2.4.1 Game-Theoretic Model

We assume that the state θ is continuously distributed over $\Theta \equiv [\underline{\theta}, \bar{\theta}]$ according to a differentiable probability distribution $F(\theta)$, with strictly positive density $f(\theta)$, for all $\theta \in \Theta$. Again, all aspects of the game except the value of the true state θ are common knowledge.

2.4.1.1 Message and Report

In this game model, we use the message to describe the format of information about the state S communicates to R. We introduce a notion report to represent the value of state carried by the message. After privately observing the state θ, S first determines a report $r \in \Theta$ for the true state θ, and then sends R a message $m \in M$, where M is a Borel space of messages. Let $\Omega : M \to \Theta$ denote the report interpretation function such that $\Omega(m)$ gives the report r carried in m. Given the true state θ, we say m tells the truth if $\Omega(m) = \theta$. We assume that for each state $\theta \in \Theta$, there is a sufficiently large number of messages that yields the same report, and each $m \in M$ has a unique value of report $\Omega(m)$. In other words, the message space can be partitioned as $M = \cup_r M_r$, with $|M_r| \to \infty$ for all r and $M_r \cap M_{r'} = \emptyset$ if $r \neq r'$, and $\forall m \in M_r$, $\Omega(m) = r$. This assumption can capture the feature of rich language in practical deceptions. We further assume that message m is formed by "common language" that can be understood precisely by both S and R. In other words, function Ω is commonly known by both players.

2.4.1.2 Strategies and Actions

Let $\sigma^S : \Theta \to \Theta$ be the strategy of S such that $r = \sigma^S(\theta)$ determines the report r of the true state θ. Let $\eta^S : \Theta \times \Theta \to M$ be the message strategy of S associated with σ^S such that $m = \eta^S(r)$ selects the message m from M_r when the strategy $\sigma^S(\theta)$ determines the report r and the true state is θ. Given θ, the strategy $\sigma^S(\theta)$ determines the set of messages $M_{\sigma^S(\theta)}$ for η^S to choose from, and η^S determines which specific message $m \in M_{\sigma^S(\theta)}$ to send. We assume $\sigma^S(\theta)$ associated with η^S induces a conditional probability $q^S(m \mid \theta)$. After receiving m, R chooses an action $a \in A \equiv \Theta$ according to a strategy $\sigma^R : \Theta \times M \to A$ using $r = \Omega(m)$. $\sigma^R(r, m)$ gives the action R acts upon the message m (and thus $r = \Omega(m)$). The action a is the final decision of R that represents the inference about the true state.

2.4.1.3 Utilities

The utility functions of S is given by $U^S(a, \theta, r) \equiv U^A(a, \theta) - kU^D(r, \theta)$ where $U^A(a, \theta) \equiv -(a - (\theta + b))^2$ is the utility depending on the induced action a in R, $U^D \equiv -(r - \theta)^2$ is the utility related to the misrepresentation of the true state, and $k \geq 0$ quantifies the intensity of U^D. On the deceivee's side, his utility is given by $U^R \equiv -(a - \theta)^2$, which takes into account the risk induced by R's misinference of the true state θ via his action a. Define, for all $\theta \in \Theta$, $\alpha^S(\theta) \equiv \arg\min_a C^S(a, \theta, r)$, and $\alpha^R(\theta) \equiv \arg\min_a C^R(a, \theta)$; i.e. $\alpha^R(\theta)$ and $\alpha^S(\theta)$ are two actions taken by R as functions of θ that are the most preferred by R and S, respectively.

2.4.1.4 Beliefs

Based on m (and thus $r = \Omega(m)$) and his prior belief $f(\theta)$, R forms a posterior belief $\mu^R : \Theta \to [0, 1]$ of the true state $\theta \in \Theta$. The posterior belief $\mu^R(\theta \mid m)$ gives the likelihood with which R believes that the true state is θ based on m. R then determines which action to choose based on his belief μ^R.

2.4.2 Deceivability

We restrict attention to a class of monotone inflated deception, in which the strategy profile (σ^S, σ^R) satisfies conditions in the following definition.

Definition 2.3 A deception with S's strategy σ^S and R's belief μ^R is monotone if

1) $\sigma^S(\theta)$ is a nondecreasing function of θ.
2) $\sigma^R(r, m)$ is a nondecreasing function of r.

The deceivability can be quantified as follows.

Definition 2.4 Given the state $\theta \in [\theta'', \theta']$, S's strategy $\sigma^S(\theta) = r$, and message strategy $\eta^S(r) = m$,

- R is *undeceivable* over $[\theta'', \theta']$ if $\sigma^R(r, m) = \alpha^R(\theta) \equiv \theta$, for all $\theta \in [\theta'', \theta']$. Here, $\sigma^S(\theta) \neq \sigma^S(\theta')$ for all $\theta \neq \theta' \in [\theta'', \theta']$, and $\eta^S(r) \in M_r$. The corresponding μ^R is *informative*. The interval $[\theta'', \theta']$ is called *undeceivable region* (UR).
- R is *deceivable* over $[\theta'', \theta']$ if the only knowledge she has is that θ lies in $[\theta'', \theta']$. R chooses $\sigma^R(r, m) = \hat{a}^R(\theta'', \theta')$, by maximizing the expected utility over $[\theta'', \theta']$, i.e.

$$\hat{a}^R(\theta'', \theta') \in \arg\max_{\sigma^R \in A} \int_{\theta'}^{\theta''} U^R(\sigma^R, \theta) f(\theta) d\theta.$$

Here, $\sigma^S(\theta)$ and $\eta^S(r)$, respectively, choose the same report r and the same message m, for all $\theta \in [\theta'', \theta']$. Thus, given m, $q^S(m \mid \theta)$ is the same for all $\theta \in [\theta'', \theta']$, where $q \in (0, 1)$. The corresponding μ^R is *uninformative*. The interval $[\theta'', \theta']$ is called *deceivable region* (DR).

2.4.3 Knowledge Acquisition: Evidence

We allow R to acquire additional knowledge through investigations when the state is in a DR, $[\theta'', \theta']$, by partitioning it into multiple intervals, denoted by a strictly increasing sequence, $<\theta_0 = \theta'', \theta_1, \ldots, \theta_J = \theta'>$. Then, R conducts investigations for each interval. Here, we consider the case when there are two investigation intervals to simplify the analysis. Let $\sigma^c \in (\theta'', \theta')$ be the *investigation partition state* such that $[\theta'', \theta']$ is partitioned into two nonoverlapping investigation regions $\Theta^0 = [\theta'', \theta^c]$ and $\Theta^1 = [\theta^c, \theta']$. Let $\Psi \in \Gamma = \{\Psi^0, \Psi^1\}$, where Ψ^i denote the event $\{\theta \in \Theta^i\}$, for $i = 0, 1$, with the probability $\frac{P(\Psi^i) = \int_{\Theta^i} f(\tilde{\theta}) d\tilde{\theta}}{\int_{[\theta'', \theta']} f(\tilde{\theta}) d\tilde{\theta}}$. The investigation for Θ^0 and Θ^1 generates noisy evidence $e \in E = \{0, 1\}$, where $e = i$ represents Ψ^i, for $i = 0$, 1. Suppose that the investigation emits evidence by the probability $\gamma(e \mid \Psi, m)$. Let $x = \gamma(e = 0 \mid \Psi^0, m)$ and $y = \gamma(e = 1 \mid \Psi^1, m)$ be the two true positive rates, which are private information of R.

With a slight abuse of notation, let $\sigma^R(\Psi, m, e) : \Gamma \times M \times E \to A$ be the strategy of R with evidence e. Figure 2.4 depicts the signaling game model for the deception with knowledge acquisition through investigation.

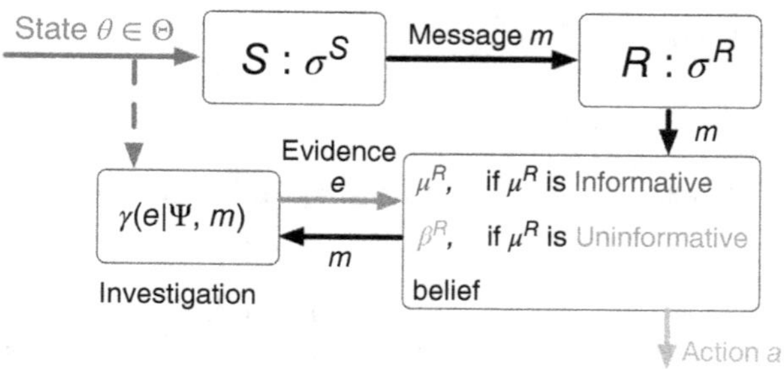

Figure 2.4 Signaling games with evidence acquisition by investigation. The probability $\gamma(e \mid \Psi, m)$ of emitting evidence e depends on the event Ψ and the message m sent by S. If the belief μ^R is informative, μ^R is used; if μ^R is uninformative, β^R is used as the posterior.

2.4.4 Equilibrium Concept

The knowledge acquisition of the signaling game over continuous state space extends the PBNE in Definition 2.1 to the following.

Definition 2.5 (Perfect Bayesian Nash Equilibrium) A PBNE of the game is a strategy profile (σ^S, σ^R) and a posterior belief system (μ^R, β^R) that satisfy the following conditions:

- *(Deceiver's Sequential Rationality)* S maximizes her expected utility, given the deceivee's strategy σ^R and the distribution of the evidence e: for each $\theta \in \Theta$,

$$\sigma^{S*}(\theta) \in \arg\max_{\sigma^S} U^S\left(\sigma^{R*}, \theta, \sigma^S\right).$$

- *(Deceivee's Sequential Rationality)* R maximizes his expected utility, given S's strategy σ^{S*} and his posterior belief $\mu^R(\theta \mid m)$: for any $m \in M$,
  - if $\mu^R(\theta \mid m)$ is informative, $\sigma^{R*}(r, m) \in \arg\max_{\sigma^R \in A} \int_{\theta\in\Theta} U^R(\sigma^R, \theta)\mu^R(\theta \mid m)d\theta$;
  - if $\mu^R(\theta \mid m)$ is uninformative over $\Theta_U \equiv [\theta'', \theta'] \subseteq \Theta$, $\sigma^{R*}(\Psi, m, e) \in \arg\max^i_{\hat{a}} \sum_{i=0}^{1} \int_{\Theta^i_U} \beta(\Psi^i \mid m, e) U^R\left(\hat{a}^i, \tilde{\theta}\right) f\left(\tilde{\theta}\right) d\tilde{\theta}$,
  where $\theta^0_U \equiv [\theta'', \theta^c]$, $\theta^1_U \equiv [\theta'', \theta^c]$, and $\hat{a}^i \equiv \arg\max \int_{\theta^i_U} U^R\left(a, \tilde{\theta}\right) d\tilde{\theta}$.

- *(Consistent Belief)* The posterior belief of R is updated according to Bayes' rule (i.e. $\mu^R(\theta \mid m)$ is *informative*), as

$$\mu^R(\theta \mid m) = \frac{f(\theta) q^S(m \mid \theta)}{\int_\Theta f\left(\tilde{\theta}\right) q^S\left(m \mid \tilde{\theta}\right) d\tilde{\theta}}.$$

- If $\int_\Theta f\left(\tilde{\theta}\right) q^S\left(m \mid \tilde{\theta}\right) d\tilde{\theta} = 0$, $\mu^R(\theta \mid m)$ may be set to any probability distribution over Θ. If μ^R is *uninformative*, i.e.

$$\mu^R = \frac{f(\theta)}{\int_{[\theta'', \theta']} f\left(\tilde{\theta}\right) d\tilde{\theta}}.$$

- R acquires evidence through investigation, and updates belief using evidence as

$$\beta^R(\Psi \mid e, m) = \frac{\gamma(e \mid \Psi, m) P(\Psi)}{\sum_{j=0}^{1} \gamma(e \mid \Psi^j, m) P(\Psi^j)},$$

- and if $\sum_{j=0}^{1} \gamma(e \mid \Psi^j, m) P(\Psi^j) = 0$, $\beta^R(\Psi \mid e, m)$ may be set to any probability distribution over Θ.

In separating equilibrium, the deceiver sends message m with different values of report $\Omega(m)$ for different states. Separating equilibria are also called *revealing* equilibria because the strategic deceivee can infer the true state even if $\Omega(m)$ does not tell the truth. In pooling equilibrium, the deceiver sends message $m \in M_r$ with the same value of report $\Omega(m) = r$ for all states. In partial-pooling equilibrium, however, the deceiver sends the message with the same report for some states and different reports for other states. Clearly, the PBNE strategy σ^{S*} associated with a DR (resp. UR) is pooling (resp. separating) strategy.

2.4.5 Equilibrium Results

From the definition of UR, the equilibrium strategy of R gives the most preferred action, $\alpha^R(\theta) \equiv \theta$, for all θ in the UR. Therefore, in any differentiable S-PBNE, the utility U^S and the strategy σ^S have to satisfy the following first-order optimality condition, given $\sigma^{R*}(\theta) = \alpha^R(\theta)$ according to the sequential rationality: $U_1^S(\alpha^R(\theta), \theta, \sigma^S(\theta)) \frac{d\alpha^R(\theta))}{d\theta} + U_3^S(\theta, \theta, \sigma^S(\theta)) \frac{d\sigma^S(\theta)}{d\theta} = 0.$

Lemma 2.5 summarizes the property of the strategy σ^{S*} in any UR.

Lemma 2.5 If $[\theta_s, \theta_l]$ is an UR, then for each $\theta \in [\theta_s, \theta_l]$, the equilibrium strategy $\sigma^{S*}(\theta) > \theta$ and it is a unique solution of the differential equation

$$\frac{d\sigma^S(\theta)}{d\theta} = \frac{b}{k(\sigma^S(\theta) - \theta)},$$

with initial condition $\sigma^{S*}(\theta_s) = \theta_s$.

Lemma 2.5 underlies the following proposition.

Proposition 2.1 With initial condition $\sigma^{S*}(\underline{\theta}) = \underline{\theta}$, there exists a cutoff state $\hat{\theta} < \bar{\theta}$ such that a unique solution σ^{S*} to the differential equation in Lemma 2.5 is well defined on $[\underline{\theta}, \hat{\theta}]$ with $\sigma^{S*}(\hat{\theta}) = \bar{\theta}$, and there is no solution on $(\hat{\theta}, \bar{\theta}]$.

Proposition 2.1 notes that in S-PBNE, the optimal strategy σ^{S*} of S has to choose a report r that is strictly larger than the true state θ, but eventually σ^{S*} runs out of such report for $\theta > \hat{\theta}$. Proposition 2.1 implies that there is no S-PBNE strategy of S for all $\theta > \hat{\theta}$, because there are not enough states to support the monotone S-PBNE strategy of S for the state in $(\hat{\theta}, \bar{\theta}]$. This suggests a class of PP-PBNE for the state space Θ, which is separating in low states and pooling in higher states. For convention, let $\sigma^{S,\,P}: \Theta \rightarrow \Theta$ and $\eta^{S,\,P}$, respectively, denote the P-PBNE strategy and the associated message strategy of S. We define this class of PP-PBNE by introducing a boundary state in the following precise sense.

Definition 2.6 We say that the strategy σ^S is a Separating in Low states And Pooling in High states (SLAPH) strategy if there exists a boundary state $\theta_B \in [\underline{\theta}, \hat{\theta}]$ such that

1) *(S-PBNE)* $\sigma^{S*}(\theta) = r$ with $\eta^{S*}(r) \in M_r$, for all $\theta \in [\underline{\theta}, \theta_B)$, and $\sigma^{S*}(\theta) \neq \sigma^{S*}(\theta')$, for all $\theta \neq \theta' \in [\underline{\theta}, \theta_B)$;
2) *(P-PBNE)* $\sigma^{S*,p}(\theta) = \bar{\theta}$ with $\eta^{S*,p}(\theta) \in M_{\bar{\theta}}$, for all $\theta \in [\theta_B, \bar{\theta}]$.

In any SLAPH equilibrium, both players have no incentive to deviate from the equilibrium strategies. This requires the boundary state θ_B to be consistent in the sense that the equilibrium at θ_B is well defined. Specifically, the utility of S has to satisfy the following *boundary consistency* (BC) condition at θ_B:

$$U^S\left(\sigma^{R*}\left(\sigma^{S,p}(\theta_B), m_p\right), \theta_B, m_p\right) = U^S\left(\alpha^R(\theta_B), \theta_B, m_s\right),$$

where $m_p \in M_{\bar{\theta}}$ and $m_s \in M_{\sigma^{S*}(\theta_B)}$. The BC condition implies that S is indifferent between sending $m_p \in M_{\bar{\theta}}$ with $a^* = \sigma^{R*}(\sigma^{S*}(\theta_B), m_p)$ and sending $m_s \in M_{\sigma^{S*}(\theta_B)}$ with action $a^* = \theta_B$.

The conflict of interest, b, is a utility-relevant parameter for S that can induce incentives for S to reveal partial information about any state $\theta \in [\theta_B, \bar{\theta}]$ to R while her utility-maximizing P-PBNE strategy σ^{S*} is maintained. This can be achieved based on the assumption $|M_{\bar{\theta}}| \to \infty$ and the fact that C^D is equally expensive for all the messages chosen for all state $\theta \in [\theta_B, \bar{\theta}]$. Specifically, the P-PBNE region $[\theta_B, \bar{\theta}]$ can be further partitioned into multiple pools. First, some notations for describing the multiple pools are needed. Let $\Theta^P \equiv (\theta_0, \theta_1, ..., \theta_{K-1}, \theta_K)$ be a partition of $[\theta_B, \bar{\theta}]$, with $\theta_0 = \theta_B < \theta_1 < ... < \theta_K = \bar{\theta}$. We call each interval $\Theta_{j,j+1} = [\theta_j, \theta_{j+1}]$ is a pool. With an abuse of notation, let $\eta^{S*,p}(\sigma^{S*}(\theta), \theta)$ denote the message strategy that chooses a message $m \in M_{\sigma^{S*}(\theta)}$ for a state θ. In each pool $\Theta_{j,j+1}$, $\eta^{S*,p}(\bar{\theta}, \theta)$ chooses the same message $m \in M_{\bar{\theta}}$, for all $\theta \in \Theta_{j,j+1}$, $j = 0, ..., K-1$. After R determines a pool $\Theta_{j,j+1}$, she acquires evidence $e \in \{e_0, e_1\}$ through investigations by dividing $\Theta_{j,j+1}$ into two subintervals $\Theta^0_{j,j+1} \equiv \left[\theta_j, \theta^I_{j,j+1}\right]$ and $\Theta^1_{j,j+1} \equiv \left[\theta^I_{j,j+1}, \theta_{j+1}\right]$. Let $\Psi_{j,j+1} \in \Gamma_{j,j+1} \equiv \left\{\Psi^0_{j,j+1}, \Psi^1_{j,j+1}\right\}$ such that $\Psi^i_{j,j+1}$ represents the event $\left\{\theta \in \Theta^i_{j,j+1}\right\}$, with probability $P\left(\Psi^i_{j,j+1}\right) = \frac{\int_{\Theta^i_{j,j+1}} f(\theta)d\theta}{\int_{\Theta_{j,j+1}} f(\theta)d\theta}$, for $i = 0, 1$. On the equilibrium path, R must play $\sigma^{R*}(\Psi_{j,j+1}, m_j, e)$ as defined in Definition 2.5 for any m_j such that $\eta^{S*,p}(\bar{\theta}, \theta) = m_j$, for all $\theta \in \Theta_{j,j+1}$. Define

$$\hat{a}^i(\theta_j, \theta_{j+1}) \equiv \arg\max_a \int_{\Theta^i_{j,j+1}} U^R(a, \theta) f(\theta) d\theta,$$

for $i = 0, 1$.

The necessary and sufficient conditions for the existence of SLAPH equilibrium are summarized in the following theorem.

Theorem 2.3 (Necessary Condition) In any SLAPH equilibrium, there exists a boundary state θ_B such that the pooling interval $[\theta_B, \bar{\theta}]$ can be partitioned into multiple pools denoted by a strictly increasing sequence (θ_0, θ_1, ..., θ_{K-1}, θ_K) with $\theta_0 = \theta_B$ and $\theta_K = \bar{\theta}$, such that, for all $j = 0, ..., K-1$,

$$U^A\left(\bar{a}\left(\theta_j, \theta_{j+1}\right), \theta_{j+1}\right) = U^A\left(\bar{a}\left(\theta_{j+1}, \theta_{j+2}\right), \theta_{j+1}\right), \tag{2.7}$$

$$U^S(\bar{a}(\theta_0, \theta_1), \theta_B, \bar{\theta}) = U^S\left(\theta_B, \theta_B, \sigma^{S*}(\theta_B)\right), \quad \text{if } \theta_B > \bar{\theta}, \tag{2.8}$$

where $\bar{a}\left(\theta_j, \theta_{j+1}\right) = \sum_{i=0}^{1} P(\Psi^i)\hat{a}^i\left(\theta_j, \theta_{j+1}\right)$, for all $j = 0, ..., K-1$. (Sufficient Condition) Given the multiple-pool PBNE characterized by Eq. (2.7) and (2.8), and if $\theta_B = \underline{\theta}$ and

$$U^S(\bar{a}(\theta_0, \theta_1), \underline{\theta}, \bar{\theta}) \leq U^S(\alpha^R(\underline{\theta}), \underline{\theta}, \sigma^{S*}(\underline{\theta})), \tag{2.9}$$

there exists an SLAPH equilibrium.

Note that U^D is equal for all $\theta \in [\theta_B, \bar{\theta}]$. Eq. (2.7) says that at each link state θ_{j+1} connecting $\Theta_{j,\ j+1}$ and $\Theta_{j+1,\ j+2}$, the utilities induced by $\bar{a}\left(\theta_j, \theta_{j+1}\right)$ and $\bar{a}\left(\theta_{j+1}, \theta_{j+2}\right)$, respectively, should keep the same. Otherwise, S has incentives to deviate from the current partition to combine these two consecutive pools by sending the message that induces more profitable U^A but the same U^D. This is not ideal for R because larger pools make the posterior less informative that could decrease the utilities for R. Similarly, Eq. (2.8) says that at the boundary state θ_B, S should be indifferent between playing S-PBNE strategy and inducing action $\sigma^{R*}(\theta_B)$ versus playing P-PBNE and introducing action $\bar{a}(\theta_0, \theta_1)$. Inequality (2.9) notes that if the boundary state $\theta_B = \underline{\theta}$, then S is indifferent between pooling with $[\underline{\theta}, \theta_1]$ and reporting $\bar{\theta}$ for $\underline{\theta}$ versus separating at $\underline{\theta}$. The existence of SLAPH requires (2.7)–(2.9) to be jointly satisfied.

2.5 Adaptive Strategic Cyber Defense for APT in Critical Infrastructure Network

2.5.1 Multistage Dynamic Bayesian Game Model

In [39, 71], Huang et al. have proposed a multistage Bayesian game framework to capture the stealthy, dynamic, and adaptive natures of the APTs as shown in Figure 2.5.

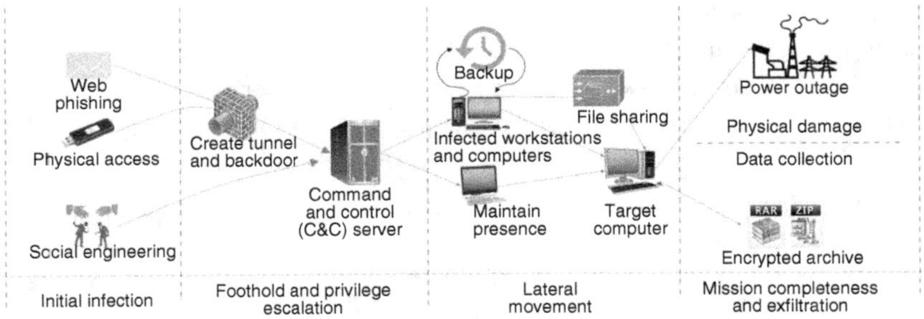

Figure 2.5 The APTs' life cycle includes a sequence of phases and stages such as the initial entry, privilege escalations, and lateral movements. APTs use each stage as a stepping stone for the next and aim to cause physical damages or collect confidential data.

2.5.1.1 Continuous Type

Due to the cyber deception, the system defender R cannot directly determine whether a user is legitimate or not even he can observe the user's apparent behaviors. Thus, user S has a type $\tilde{\theta}$ which is a random variable and the realization $\theta \in \Theta := [0, 1]$ is private information of S. The value of the type indicates the strength of the user in terms of damages that she can inflict on the system. A user with a larger type value indicates a higher threat level. At each stage $k \in \{0, 1, ..., K\}$, S chooses an action $m_k \in M_k$ and R chooses an action $a_k \in A_k$. The user's actions represent the apparent behaviors and observable activities from log files such as a privilege escalation request and sensor access. A defender cannot identify the user's type from observing his actions. The defender's action includes prevention and proactive behaviors such as restricting the escalation request or monitoring the sensor access. The action pair (m_k, a_k) is known to both players after stage k and forms a *history* $h_k := \{m_0, ..., m_{k-1}, a_0, ..., a_{k-1}\}$. The state $x_k \in \mathcal{X}_k$ shows the system status at each stage k such as the location of the APTs. Since the initial state x_0 and history h_k uniquely determine the state, x_k contains information of history up to k and has the transition kernel described by $x_{k+1} = g_k(x_k, m_k, a_k)$. The behavior mixed strategies $\sigma_k^S \in \Gamma_k^S : \mathcal{X}_k \times \Theta \to \triangle M_k$ and $\sigma_k^R \in \Gamma_k^R : \mathcal{X}_k \to \triangle A_k$.

2.5.1.2 Believe Update

To strategically gauge the user's type, the defender specifies a belief $\mu_k^R : \mathcal{X}_k \mapsto \triangle \Theta$ as a distribution over the type space according to the information available at stage k. The prior distribution of the user's type is known to be f and the belief of the type updates according to the Bayesian rule.

$$\mu_{k+1}^R(\theta | x_{k+1}) = \frac{\mu_k^R(\theta | x_k)\sigma_k^R(a_k | x_k, \theta)}{\int_0^1 \mu_k^R(\hat{\theta} | x_k)\sigma_k^R(a_k | x_k, \hat{\theta})d\hat{\theta}}.$$

2.5.1.3 Utility Function

The user's type influences P_i's immediate payoff received at each stage k, i.e. $U_k^S : \mathcal{X}_k \times M_k \times A_k \times \Theta \mapsto \mathbb{R}$. For example, a legitimate user's access to the sensor benefits the system while a pernicious user's access can incur a considerable loss. Define $\sigma_{k':K}^S := \{\sigma_k^S \in \Gamma_k^S\}_{k=k',\ldots,K} \in \Gamma_{k':K}^S$ as a sequence of policies from k' to K.

The defender has the objective to maximize the cumulative expected utility:

$$\bar{U}_{k':K}^R(\sigma_{k':K}^S, \sigma_{k':K}^R, x_{k'}) := \sum_{k=k'}^{K} \mathrm{E}_{\theta\sim\mu_k^R, m_k\sim\Gamma_k^S, a_k\sim\Gamma_k^R} U_k^R(x_k, m_k, a_k, \theta),$$

and the user's objective function is

$$\bar{U}_{k':K}^S(\sigma_{k':K}^S, \sigma_{k':K}^R, x_{k'}, \theta) = \sum_{k=k'}^{K} \sum_{a_k \in A_k} \sigma_k^R(a_k | x_k) \sum_{m_k \in M_k} \sigma_k^S(m_k | x_k, \theta) U_k^S.$$

2.5.2 Equilibrium Concept

The insider threats of APTs lead to the following definition of PBNE where the defender chooses the most rewarding policy to confront the attacker's best-response policies.

Definition 2.7 In the two-person multistage game with a sequence of beliefs $\mu_k^R, k \in \{k', \ldots, K\}$ satisfying the Bayesian update shown in Section 2.5.1.2 and the cumulative utility function $\bar{U}_{k':K}^S$, the set $R_2^{\theta, x_{k'}}(\sigma_{k':K}^R) := \{\gamma \in \Gamma_{k':K}^S : \bar{U}_{k':K}^S(\sigma_{k':K}^R, \gamma, x_{k'}, \theta) \geq \bar{U}_{k':K}^S(\sigma_{k':K}^R, \sigma_{k':K}^S, x_{k'}, \theta), \forall \Gamma_{k':K}^R \in \Gamma_{k':K}^R, \forall x_{k'} \in \mathcal{X}_{k'}, \theta \in \Theta\}$ is S's **best-response set** to R's policy $\sigma_{k':K}^R \in \Gamma_{k':K}^R$ under state $x_{k'}$ and type θ.

Definition 2.8 In the two-person multistage Bayesian game with R as the principal, the cumulative utility function $\bar{U}_{k':K}^S, \bar{U}_{k':K}^R$, the initial state $x_{k'} \in \mathcal{X}_{k'}$, the type $\theta \in \Theta$, and a sequence of beliefs defined in Definition 2.7, a sequence of strategies $\sigma_{k':K}^{R*} \in \Gamma_{k':K}^R$ is called a **PBNE** for the principal, if

$$\bar{U}_{k':K}^{R*}(x_{k'}) := \inf_{\sigma_{k':K}^S \in R_2^{\theta, x_{k'}}(\sigma_{k':K}^{R*})} \bar{U}_{k':K}^{R*}(\sigma_{k':K}^{R*}, \sigma_{k':K}^S, x_{k'}).$$

A strategy $\sigma_{k':K}^{S*} \in \arg\max_{\sigma_{k':K}^S \in \Gamma_{k':K}^S} \bar{U}_{k':K}^S(\sigma_{k':K}^{R*}, \sigma_{k':K}^S, x_{k'}, \theta)$ is a PBNE for the agent S.

A conjugate-prior method allows online computation of the belief and reduces Bayesian update into an iterative parameter update. The forwardly updated parameters are assimilated into the backward dynamic programming computation to characterize a computationally tractable and time-consistent equilibrium solution based on the expanded state space.

2.5.3 Case Study

We consider a four-stage transition with the first three stages related to cyber transition of the APTs and the last stage related to the benchmark Tennessee Eastman (TE) process as the targeted physical plant. The TE process involves two irreversible reactions to produce two liquid (liq) products. The process shuts down when the safety constraints are violated such as a high reactor pressure, a high/low separator/stripper liquid level. The attacker can revise the sensor reading, trigger an undesired feedback-control, and cause a loss. The state x_k has five possible values representing the working status of the system where state 1 is the most desirable and state 5 is the least desirable. We can obtain the normal operation reward and the reward of attacks from the simulation. The belief state $\{\alpha_k, \beta_k\}$ uniquely determines the belief distribution which is assumed to take the form of the beta distribution. The larger α_k means that the user is more likely to have a smaller type value and have lower threats to the system.

As shown in Figure 2.6, a high-value $\bar{U}^{R*}_{k':K}(x_{k'})$ for the defender is the result of a healthy system state x_k as well as a belief of a low-threat user. At the most desirable system state $x_k = 1$, attackers will not attack because the reward incurred is insufficient. Then, the defender does not need to defend and obtains the maximum utility.

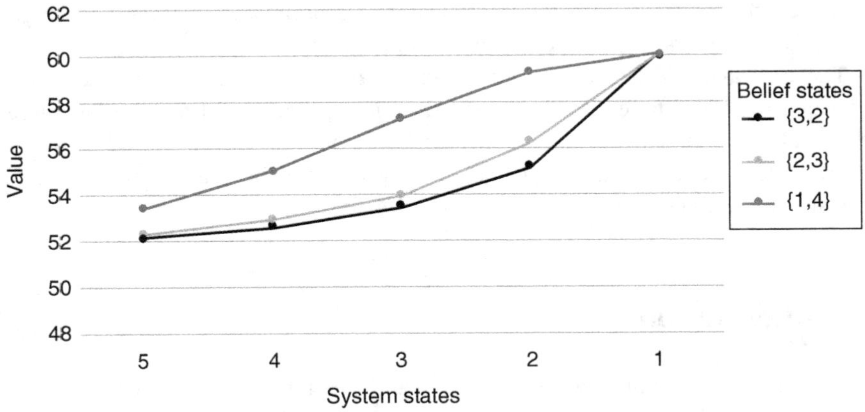

Figure 2.6 Value function $\bar{U}^{R*}_{k':K}(x_{k'})$ under different expanded states $\{x_k, \alpha_k, \beta_k\}$.

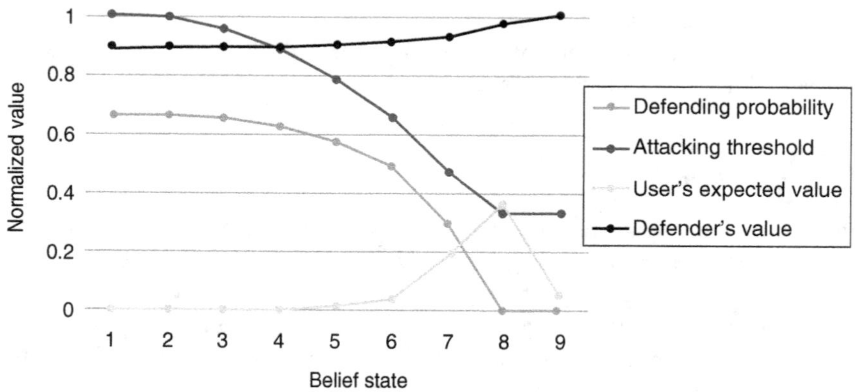

Figure 2.7 The effect of the defender's belief.

To investigate the effect of the defender's belief, we change the belief state (α_k, β_k) from (9, 1) to (1, 9), which means that the defender grows optimistically that the user is of a low threat level with a high probability. Since players' value functions are of different scales in terms of the attack threshold and the probability, we normalize the value functions with respect to their maximum values to illustrate their trends and make them comparable to the threshold and the probability as shown in Figure 2.7. When α_k is large, the defender chooses to protect the system with a high probability, which completely deters attackers with any type values because the probability to attack is 0.

As the defender trusts more about the user's legitimacy, the defending probability decreases to 0 when $\alpha_k = 1$. Since the defender is less likely to defend, the attacker bears a smaller threshold to launch the attack. The resulting defending policy captures a trade-off between security and economy and guarantees a high value for defenders at most of the belief states.

The central insight from the multistage analysis is the adversary's trade-off between the instantaneous reward and the hiding to arrive at a more favorable state in the future stages. The higher the belief of the defender in S as a legitimate user, the less probability he will act defensively and thus the attacker has a smaller threshold to launch the attack.

2.6 Conclusion

Deception is a technique that can be viewed as an advanced approach to secure the devices or attacks. Understanding deception quantitatively is pivotal to provide rigor, predictability, and design principles. In this chapter, we have formulated signaling-game-theoretic models of deceptions over discrete and continuous information spaces. We have studied leaky deception models and extended the baseline PBNE to versions involving knowledge acquisitions characterized by evidences. We have analyzed the impacts of evidence on the belief updates and strategy constructions on the equilibrium path.

In the binary state space, the leaky deception game with evidence admits an equilibrium that includes a regime in which the deceivee should choose whether to trust the deceiver based on the evidence, and regimes in which the deceivee should ignore the message and evidence and merely guess the private information based only on the prior probabilities. For the deceiver, the equilibrium results imply that it is optimal to partially reveal the private information in the former regime.

We have also studied leaky deception games over a continuous one-dimensional information space. We have studied the PBNE as the solution concept to analyze the outcome of the deception game and characterize the deceivability of the game. The proposed deception game admits a class of PBNE called SLAPH. The necessary and sufficient conditions for the existence of such PBNE are given. However, a full UR does not exist and there exists a DR. We have also shown that the DR can be partitioned into multiple sub-DRs without decreasing total utilities for the deceiver when the conflict of interest is insignificant.

Furthermore, we have explored a multistage incomplete-information Bayesian game model for defensive deception frameworks for critical infrastructure networks with the presence of APTs. With the multistage and multiphase structure of APTs, the belief of the defender is formed dynamically using observable footprints. The conjugate priors are used to reduce Bayesian updates into parameter updates, which leads to a computationally tractable extended-state dynamic programming that admits an equilibrium solution consistent with the forward belief update and backward induction. The TE process has been used as a case study to demonstrate the multistage deception game. The numerical simulations have shown that the game-theoretic defense strategies have significantly improved the security of the critical infrastructures.

References

1 H. Cott, *Adaptive Coloration in Animals*, London: Methuen, 1940.

2 U. Gneezy, "Deception: the role of consequences," *The American Economic Review*, vol. **95**, no. 1, pp. 384–394, 2005.

3 A. Vrij, S.A. Mann, R.P. Fisher, et al., "Increasing cognitive load to facilitate lie detection: The benefit of recalling an event in reverse order," *Law and Human Behavior*, vol. **32**, no. 3, pp. 253–265, 2008.

4 S. Bodmer, D.M. Kilger, G. Carpenter, et al., *Reverse Deception: Organized Cyber Threat Counter-Exploitation*, New York, NY: McGraw-Hill, 2012.

5 J. Pawlick, Q. Zhu, "Strategic trust in cloud-enabled cyber-physical systems with an application to glucose control," *IEEE Transactions on Information Forensics and Security*, vol. **12**, no. 12, pp. 2906–2919, 2017.

6 L.J. Janczewski, A.M. Colarik, *Cyber Warfare and Cyber Terrorism*, New York, NY: Information Science Reference, 2008.

7 U.F. Minhas, J. Zhang, T. Tran, et al., "A multifaceted approach to modeling agent trust for effective communication in the application of mobile ad hoc vehicular networks," *IEEE Transactions on Systems, Man, and Cybernetics, Part C: Applications and Reviews*, vol. **41**, no. 3, pp. 407–420, 2011.

8 E. Harrell, "Victims of identity theft, 2014," US Department of Justice, Office of Justice Programs, Bureau of Justice Statistics, 2015.

9 Symantec Corporation, "Internet security threat report. Tech. rep.," *Symantec Corporation*, 2016. [Online]. Available: www.symantec.com. [Accessed Nov. 10, 2018].

10 L. Atzori, A. Iera, G. Morabito, "The Internet of Things: a survey," *Computer Networks*, vol. **54**, pp. 2787–2805, 2010.

11 Phishlabs, "2018 Phishing trends intelligence report," *Phishlabs*, 2018. [Online]. Available: https://info.phishlabs.com/. [Accessed Nov. 10, 2018].

12 I. Bose, A.C.M. Leung, "Unveiling the mask of phishing: threats, preventive measures, and responsibilities," *Communications of the Association for Information Systems*, vol. **19**, no. 1, pp. 24, 2007.

13 I. Bose, A.C.M. Leung, "Assessing anti-phishing preparedness: a study of online banks in Hong Kong," *Decision Support Systems*, vol. **45**, no. 4, pp. 897–912, 2008.

14 T.E. Carroll, D. Grosu, "A game theoretic investigation of deception in network security," *Security and Communication Networks*, vol. **4**, no. 10, pp. 1162–1172, 2011.

15 Q. Zhu, T. Başar, "Game-theoretic approach to feedback-driven multi-stage moving target defense," in Proc. International Conference on Decision and Game Theory for Security, Oct. 2013, pp. 246–263.

16 J. Pawlick, Q. Zhu, "A Stackelberg game perspective on the conflict between machine learning and data obfuscation," in Proc. IEEE International Workshop on Information Forensics and Security, Dec. 2016, pp. 1–6.

17 N. Zhang, W. Yu, X. Fu, et al., "gPath: A game-theoretic path selection algorithm to protect tor's anonymity," in Proc. International Conference on Decision and Game Theory for Security, Dec. 2010, pp. 58–71.

18 S. Farhang, M.H. Manshaei, M.N. Esfahani, et al., "A dynamic Bayesian security game framework for strategic defense mechanism design," in Proc. Decision and Game Theory for Security, Oct. 2014, pp. 319–328.

19 K. Horák, Q. Zhu, B. Bošanskỳ, "Manipulating adversarial belief: A dynamic game approach to deception by design for proactive network security," in Proc. International Conference on Decision and Game Theory for Security, Oct. 2017, pp. 273–294.

20 M.H. Manshaei, Q. Zhu, T. Alpcan, et al., "Game theory meets network security and privacy," *ACM Computing Surveys (CSUR)*, vol. **45**, no. 3, no. 25, 2013.

21 F. Miao, Q. Zhu, M. Pajic, et al., "A hybrid stochastic game for secure control of cyber-physical systems," *Automatica*, vol. **93**, pp. 55–63, 2018.

22 T. Zhang, Q. Zhu, "Strategic defense against deceptive civilian GPS spoofing of unmanned aerial vehicles," in Proc. International Conference on Decision and Game Theory for Security, Dec. 2017, pp. 213–233.

23 Q. Zhu, A. Clark, R. Poovendran, et al., "Deployment and exploitation of deceptive honeybots in social networks," in Proc. IEEE 52nd Annual Conference on Decision and Control (CDC), Dec. 2013, pp. 212–219.

24 S. Roy, C. Ellis, S. Shiva, et al., "A survey of game theory as applied to network security," in Proc. 43rd Hawaii International Conference on System Sciences, Jan. 2010, pp. 1–10.

25 T. Alpcan, T. Basar, "A game theoretic approach to decision and analysis in network intrusion detection," in Proc. 42nd IEEE International Conference on Decision and Control, Dec. 2003, pp. 2595–2600.

26 S. Farhang, M.H. Manshaei, M.N. Esfahani, et al, "A dynamic Bayesian security game framework for strategic defense mechanism design," in Proc. International Conference on Decision and Game Theory for Security, Nov. 2014, pp. 319–328.

27 S. Rass, A. Alshawish, M.A. Abid, et al, "Physical intrusion games – optimizing surveillance by simulation and game theory," *IEEE Access*, vol. **5**, pp. 8394–8407, 2017.

28 Q. Zhu, S. Rass, "On multi-phase and multi-stage game-theoretic modeling of advanced persistent threats," *IEEE Access*, vol. **6**, pp. 13958–13971, 2018.

29 J. Zhuang, V.M. Bier, O. Alagoz, "Modeling secrecy and deception in a multiple-period attacker–defender signaling game," *European Journal of Operational Research*, vol. **203**, no. 2, pp. 409–418, 2010.

30 A. Monga, Q. Zhu, "On solving large-scale low-rank zero-sum security games of incomplete information," in Proc. IEEE International Workshop on Information Forensics and Security (WIFS), Dec. 2016, pp. 1–6.

31 Q. Zhu, T. Basar, "Game-theoretic methods for robustness, security, and resilience of cyberphysical control systems: games-in-games principle for optimal cross-layer resilient control systems," *IEEE Control Systems*, vol. **35**, no. 1, pp. 46–65, 2015.

32 J. Chen, Q. Zhu, "Security as a service for cloud-enabled internet of controlled things under advanced persistent threats: A contract design approach," *IEEE Transactions on Information Forensics and Security*, vol. **12**, no. 11, pp. 2736–2750, 2017.

33 Z. Xu, Q. Zhu, "Secure and practical output feedback control for cloud-enabled cyber-physical systems," in Proc. IEEE Conference on Communications and Network Security (CNS), Oct. 2017, pp. 416–420.
34 J. Pawlick, E. Colbert, Q. Zhu, "Modeling and analysis of leaky deception using signaling games with evidence," *IEEE Transactions on Information Forensics and Security*, vol. **14**, no. 7, pp. 1871–1886, 2019.
35 J. Pawlick, E. Colbert, Q. Zhu, "A game-theoretic taxonomy and survey of defensive deception for cybersecurity and privacy," arXiv:1712.05441[cs], 2017.
36 J. Pawlick, Q. Zhu, "Deception by design: Evidence-based signaling games for network defense," arXiv:1503.054 58[cs], 2015.
37 S. Jajodia, A.K. Ghosh, V. Swarup, et al., *Moving Target Defense: Creating Asymmetric Uncertainty for Cyber Threats*, New York, NY: Springer, 2011.
38 H. Maleki, S. Valizadeh, W. Koch, et al, "Markov modeling of moving target defense games," in Proc. of the ACM Workshop on Moving Target Defense, Oct. 2016, pp. 81–92.
39 L. Huang, Q. Zhu, "Adaptive strategic cyber defense for advanced persistent threats in critical infrastructure networks," *ACM SIGMETRICS Performance Evaluation Review*, vol. **46**, pp. 52–56, 2019.
40 T. Basar, "The Gaussian test channel with an intelligent jammer," *IEEE Transactions on Information Theory*, vol. **29**, no. 1, pp. 152–157, 1983.
41 J. Pita, M. Jain, J. Marecki, et al. "Deployed ARMOR protection: the application of a game theoretic model for security at the Los Angeles International Airport," in Proc. of the 7th International Joint Conference on Autonomous Agents and Multiagent Systems: Industrial Track, May 2008, pp. 125–132.
42 E. Shieh, B. An, R. Yang, et al., "Protect: A deployed game theoretic system to protect the ports of the United States," in Proc. of the 11th International Conference on Autonomous Agents and Multiagent Systems, June 2012, pp. 13–20.
43 J. Chen, C. Touati, Q. Zhu, "A dynamic game analysis and design of infrastructure network protection and recovery," *ACM SIGMETRICS Performance Evaluation Review*, vol. **45**, no. 2, pp. 128, 2017.
44 J. Chen, Q. Zhu, "Interdependent network formation games with an application to critical infrastructures," in Proc. American Control Conference (ACC), July 2016, pp. 2870–2875.
45 Y. Hayel, Q. Zhu, "Resilient and secure network design for cyber attack-induced cascading link failures in critical infrastructures," in Proc. 49th IEEE Annual Conference on Information Sciences and Systems (CISS), Mar. 2015, pp. 1–3.
46 J. Pawlick, Q. Zhu, "Proactive defense against physical denial of service attacks using Poisson signaling games," in Proc. International Conference on Decision and Game Theory for Security, Oct. 2017, pp. 336–356.
47 R. Zhang, Q. Zhu, "Secure and resilient distributed machine learning under adversarial environments," in Proc. 18th International Conference on Information Fusion (Fusion), Jul. 2015, pp. 644–651.

48 J. Pawlick, Q. Zhu, "A mean-field stackelberg game approach for obfuscation adoption in empirical risk minimization," in Proc. IEEE Global Conference on Signal and Information Processing (GlobalSIP), Nov. 2017, pp. 518–522.

49 W. Wang, Q. Zhu, "On the detection of adversarial attacks against deep neural networks," in Proc. ACM Workshop on Automated Decision Making for Active Cyber Defense, Nov. 2017, pp. 27–30.

50 R. Zhang, Q. Zhu, "A game-theoretic defense against data poisoning attacks in distributed support vector machines," in Proc. IEEE 56th Annual Conference on Decision and Control (CDC), December 12–15, 2017, pp. 4582–4587.

51 R. Zhang, Q. Zhu, "A game-theoretic approach to design secure and resilient distributed support vector machines," *IEEE Transactions on Neural Networks and Learning Systems*, vol. **99**, pp. 1–16, 2018.

52 W. Casey, J.A. Morales, E. Wright, et al., "Compliance signaling games: Toward modeling the deterrence of insider threats," *Computational and Mathematical Organization Theory*, vol. **22**, no. 3, pp. 318–349, 2016.

53 W.A. Casey, Q. Zhu, J.A. Morales, et al., "Compliance control: Managed vulnerability surface in social-technological systems via signaling games," in Proc. 7th ACM CCS International Workshop on Managing Insider Security Threats, Oct. 2015, pp. 53–62.

54 J. Chen, Q. Zhu, "Security investment under cognitive constraints: A Gestalt Nash equilibrium approach," in Proc. IEEE Annual Conference on Information Sciences and Systems (CISS), March 2018, pp. 1–6.

55 C.J. Fung, Q. Zhu, "FACID: A trust-based collaborative decision framework for intrusion detection networks," *Ad Hoc Networks*, vol. **53**, pp. 17–31, 2016.

56 Y. Hayel, Q. Zhu, "Epidemic protection over heterogeneous networks using evolutionary Poisson games," *IEEE Transactions on Information Forensics and Security*, vol. **12**, no. 8, pp. 1786–1800, 2017.

57 R. Zhang, Q. Zhu, Y. Hayel, "A bi-level game approach to attack-aware cyber insurance of computer networks," *IEEE Journal on Selected Areas in Communications*, vol. **35**, no. 3, pp. 779–794, 2017.

58 R. Powell, "Allocating defensive resources with private information about vulnerability," *American Political Science Review*, vol. **101**, no. 4, pp. 799–809, 2007.

59 G. Brown, M. Carlyle, D. Diehl, et al., "A two-sided optimization for theater ballistic missile defense," *Operations Research*, vol **53**, no. 5, pp.745–763, 2005.

60 L. Huang, Q. Zhu, "Analysis and computation of adaptive defense strategies against advanced persistent threats for cyber-physical systems," in Proc. International Conference on Decision and Game Theory for Security, Oct. 2018, pp. 205–226.

61 A. Clark, Q. Zhu, R. Poovendran, et al., "Deceptive routing in relay networks," in Proc. International Conference on Decision and Game Theory for Security, Oct. 2012, pp. 171–185.

62 Q. Zhu, A. Clark, R. Poovendran, et al., “Deceptive routing games,” in Proc. IEEE 51st Annual Conference on Decision and Control (CDC), Dec. 2012, pp. 2704–2711.

63 T. Zhang, Q. Zhu, “Hypothesis testing game for cyber deception,” in Proc. International Conference on Decision and Game Theory for Security, Oct. 2018, pp. 540–555.

64 D. Ettinger, P. Jehiel, “A theory of deception,” *American Economic Journal: Microeconomics*, vol. **2**, no. 1, pp. 1–20, 2010.

65 D. Fudenberg, J. Tirole, *Game Theory*, Cambridge, MA: The MIT Press, 1991.

66 D. Lewis, *Convention*, Cambridge, MA: Harvard University Press, 1969.

67 J.P. Ponssard, S. Zamir, “Zero-sum sequential games with incomplete information,” *International Journal of Game Theory*, vol. **2**, no. 1, pp. 99–107, 1973.

68 J.F. Nash, “Equilibrium points in n-person games,” in *Proceedings of the National Academy of Sciences of the United States of America*, vol. **36**, no. 1, pp. 48–49, 1950.

69 J.C. Harsanyi, “Games with incomplete information played by “Bayesian” players,” *Management Science*, vol. **50**, no. 12, pp.1804–1817, 1967.

70 T. Zhang, Q. Zhu, “A game-theoretic foundation of deception: Knowledge acquisition and fundamental limits,” arXiv:1810.00752[cs], 2018.

71 L. Huang, J. Chen, Q. Zhu, “Factored Markov game theory for secure interdependent infrastructure networks,” in S. Rass, S. Schauer, ed. *Game Theory for Security and Risk Management*, Cham, Switzerland: Springer, 2018, pp. 99–126.

3

A Hypergame-Based Defense Strategy Toward Cyber Deception in Internet of Battlefield Things (IoBT)

Bowei Xi[1] *and Charles A. Kamhoua*[2]

[1] *Department of Statistics, Purdue University, West Lafayette, IN, USA*
[2] *US Army Research Laboratory, Adelphi, MD, USA*

3.1 Introduction

Internet of Things (IoT) is a network of a very large number of physical objects interconnected through communication protocols such as Wi-Fi, Bluetooth, and IEEE 802.15.4. IoT sees widespread applications including smart home, smart city, smart grids, smart buildings, and smart healthcare among others [1]. Many IoT devices often operate with low power supply and have low computing capabilities, whereas data on the network are transmitted over lossy links. Although communication among IoT devices is secured by a variety of network layer and transport layer security protocols, many IoT devices are not secured themselves. Security and privacy issues are a big challenge for IoT. Malwares such as Mirai and its variants are able to infect a huge number of IoT devices [2] and launch massive distributed denial of service (DDoS) attacks with very low cost.

In the 1980s, Defense Advanced Research Projects Agency (DARPA) funded the Distributed Sensor Network project, which became one of the forerunners of today's IoT technologies [3]. Today's Internet of Battlefield Things (IoBT) is an IoT of connected devices on a military network, such as sensors, wearable devices on soldiers, devices in weapon systems, drones, vehicles, equipment, field command and control systems, and the central command system. IoBT devices have the ability to collect, compute, and upload operational field data to the network. Compared with low-capacity IoT devices, it is desirable for IoBT devices to carry substantial edge-computing capabilities. Meanwhile, as observed from the Internet and IoT, adversaries would exploit vulnerabilities in IoBT devices and attempt

Modeling and Design of Secure Internet of Things, First Edition. Edited by Charles A. Kamhoua, Laurent L. Njilla, Alexander Kott and Sachin Shetty.

Published 2020 by John Wiley & Sons, Inc.

to compromise the critical nodes on an IoBT network [4, 5, 6, 7, 8]. Hence, it is an important task to secure the IoBT network communication and devices.

We develop a cyber-deception strategy, based on a hypergame framework, to protect IoBT from cyberattacks using one unique property of IoBT – a military network is established and stays alive for only a short time period, whereas civilian networks, such as the Internet, are in service for an extended period of time. Hence, for IoBT, time is a crucial factor to be included in a defense strategy. The defender successfully defeats a cyberattack as long as the defender prevents the adversary from reaching the target nodes while a military network is in service.

We use a hypergame to model the interaction between two players, the defender and the adversary. In this chapter, we assume the defender employs one deceptive technique – installing honeypots – to slow down the attack and gather information about the adversary.[1] Although we restrict our analysis to honeypots, we recognize that cyber-deception schemes extend beyond honeypots. The defender decides where to install honeypots, and the adversary decides which node to compromise next based on the hypergame solution. In the hypergame, each player follows their own perceived game, which may or may not be the same one as the actual game. Each player chooses the next action based on the mixed equilibrium strategy of their own perceived game. At least one player, the adversary, is fooled by the deceptive technique. Since the hypergame is only played for a short time (i.e. when an ad hoc military network is in service), this is a realistic assumption. During the short time period, the adversary is not able to discover which type of deceptive technique is used by the defender or where the deceptive technique is deployed. If both players are fully rational, have complete information, and both know the actual game, the defender is not able to use any type of cyber-deception technique. Hence, hypergame provides a suitable model to create cyber deception and understand the effects.

We then conduct experiments that provide important quantitative measures for moving target defense, including the time needed for the adversary to explore the entire network and discover the target nodes. Chapter 2 in this book uses Bayesian games to model deceptions. Chapter 6 considers a game where an attacker chooses whether to attack a node with defense resource, and a defender decides which node should be used as a fake node. Compared with existing work and other chapters, the major contributions of this chapter are the following:

- We develop a hypergame framework to address IoBT security concerns.
- In a dynamic fashion, the defender chooses where to install honeypots based on the hypergame framework.

---

1 Honeypot is a useful tool to detect IoT botnet. A sophisticated new IoT botnet, Torii, which can run on almost all computers, smart phones, and tablets, was first detected by a security researcher's honeypots. (https://www.zdnet.com/article/meet-torii-a-new-iot-botnet-far-more-sophisticated-than-mirai).

- We model how an attack spreads on an IoBT using a time-varying attack graph.
- We estimate the amount of time needed for an attacker to reach the target nodes through experiments, which provides a quantitative measure to determine when it is necessary to disable all the connections to the target nodes.
- We are able to estimate the amount of time needed for an attacker to compromise an IoBT through experiments, which provides important guidance for the defender to determine when it is necessary to dissolve a military network or apply a moving target defense.
- Based on the hypergame, the defender can decide how often to periodically clean the IoBT devices in order to contain a cyberattack (i.e. to ensure the attacker fails to reach any critical node while a military network is in service).

3.1.1 Hypergames

In this chapter, we follow the definition of hypergames as in [9, 10]. Hypergame was developed to study the effect of players' misperceptions on the decision-making process in conflicts – in a hypergame, one or more players have misperceptions of the conflict. A hypergame consists of the following components:

- N players
- A set of strategies A_i for each player, $i = 1, ..., N$
- Each player's payoff functions F_i, $i = 1, ..., N$, over the outcome space $A_1 \times A_2 \times \cdots \times A_N$
- Each player's preference vector $PREF_i$, which is an ordered vector of the outcomes in $A_1 \times A_2 \times \cdots \times A_N$ from the least preferred to the most preferred, determined by the player's payoff functions F_i

Player a's preference vector perceived by another player b $PREF_{a,\,b}$ for all the possible (a, b) pairs

In an ordinary game, all the players see a set of common preference vectors, $PREF_{a,\,b} = PREF_a$. In a hypergame, at least one player's perceived preference vector is different from the actual preference vector. A player may not be aware of the existence of another player. For a two-player game, the first-level hypergame is $G^1 = \{G_a^0, G_b^0\}$, where G_a^0 and G_b^0 are the two zero-level games played by players a and b. We can write $G_a^0 = \{PREF_a, PREF_{b,a}\}$ and $G_b^0 = \{PREF_{a,b}, PREF_b\}$. For the first-level hypergame, $G_a^0 \neq G_b^0$. Players are not aware that they play different games. If $G_a^0 = G_b^0$, it becomes an ordinary game where two players have complete information and both correctly perceive the conflict.

In a second-level hypergame, $G^2 = \{G_a^1, G_b^1\}$. Both players now realize they play different games and try to estimate the game played by the other player. However, they still play different first-level games, $G_a^1 \neq G_b^1$. Essentially, players play different games in a hypergame based on how each player perceives the conflict. Each player makes a decision based on their own perceived game. In this chapter, we let the adversary and the defender play a first-level hypergame, assuming during the short time period when an IoBT is alive, the adversary is not aware of the deceptive technique.

3.1.2 Related Work

Game-theoretic models provide important tools for security applications in adversarial environment. In [11], a two-player zero-sum Markov game model was used to solve for the optimal strategy by disconnecting vulnerable devices to enhance power grid security. Gutierrez et al. [12] modeled the Insider Threat as a hypergame and conducted a stability analysis. Cho et al. [13] used hypergame to model Advanced Persistent Threat attacks with attack behavior derived from cyber kill chain. House and Cybenko [14] used hidden Markov model and maximum entropy to learn the probability of each player's subgame under the envelope of an overall hypergame and proposed a concept – a learning versus obfuscation equilibrium for the hypergame under such a structure. Vane [15] applied hypergame to evaluate and perform preplanning for simulated terrorist-caused emergencies. Takahashi et al. [16] used hierarchical hypergames for a case study of Allied invasion of Normandy. Kiekintveld and Wellman [17] developed meta-game models incorporating observation noises to make outcome predictions in multi-agent domains.

Attack graphs are a common tool used to model and measure network security. Kamdem et al. [11] used a vulnerability multi-graph where nodes are hosts and edges represent the vulnerabilities between a pair of hosts. Wang et al. [18] used attack graphs to model the causal relationship of different vulnerabilities and proposed a probabilistic metric for network security. Frigault and Wang [19] used Bayesian attack graphs to measure network security. Based on Bayesian attack graphs, Poolsappasit et al. [20] proposed an approach to assess and mitigate the security risks of a network.

This chapter is organized as follows. Section 3.2 describes the time-varying attack graph and the hypergame for modeling the spread of an attack on an IoBT network. Section 3.3 presents the experimental results. Section 3.4 concludes this chapter.

3.2 Modeling the Spread of an Attack

We present a first-level hypergame in this section to model the cyber-deception technique employed by the defender to deter the spread of attack on an IoBT network. One important component of our hypergame model is the time factor needed for the adversary to exploit the vulnerabilities of the IoBT devices and compromise these devices. An effective deceptive technique, such as employing a honeypot, misleads the adversary to spend an excessive amount of time on stepping stone devices, and slows down the spread of attack. At the same time, the defender is able to gather information about the adversary.

3.2.1 Time-Varying Attack Graph

In this chapter, we consider a targeted attack scenario, where the adversary compromises an edge device (i.e. an entry device) and through exploitation of vulnerabilities of connected IoBT devices, launches a targeted attack aiming to compromise several critical devices on the network. We use a time-varying attack graph to demonstrate the potential attack paths and the associated times for the attacks to spread from a compromised edge device to a target critical device. Figure 3.1 shows a sample network topology. We use a directed acyclic graph to model a network. We assume the defender knows the actual network topology, and the adversary always moves from a compromised node to a clean node.

In Figure 3.1, node n_0^A is the compromised edge device. Nodes $n_{2,2}^T$ and $n_{2,4}^T$ are the adversary's targets. Compromising a target node carries a significant payoff for the attacker. Payoffs may be different for different target nodes. A directed edge between two nodes $n_{i',\,j'}$ and $n_{i,j}$ indicates that adversary can exploit the vulnerabilities of node $n_{i,j}$, and compromise node $n_{i,\,j}$ from node $n_{i',\,j'}$. There are four potential attack paths shown in Figure 3.1: (n_0^A, $n_{1,1}$, $n_{2,1}$), (n_0^A, $n_{1,1}$, $n_{2,2}^T$), (n_0^A, $n_{1,2}$, $n_{2,3}$), and (n_0^A, $n_{1,2}$, $n_{2,4}^T$). Two attack paths lead to a target node: (n_0^A, $n_{1,1}$, $n_{2,2}^T$) and (n_0^A, $n_{1,}$, $n_{2,4}^T$).

We let the time T required to compromise a clean node $n_{i,\,j}$ follow an exponential distribution, $\exp\{1/\lambda\}$. We have $E(T) = \lambda$. The average time, λ, is determined by a

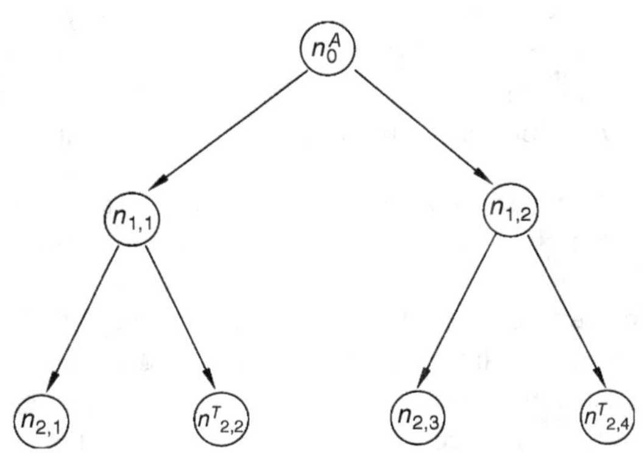

Figure 3.1 A sample network topology.

Table 3.1 October 2018 CVSS score distribution published by NVD.

Severity	Number	Percentage
Critical	5 175	15.70
High	14 887	45.17
Medium	12 342	37.45
Low	553	1.68

score learned from the Common Vulnerability Scoring System (CVSS) [21]. CVSS v3.0 scores range from 0 to 10. A score of 0 means the severity level of a vulnerability is none; a score between 0.1 and 3.9 means the severity level is low; a score between 4.0 and 6.9 means the severity level is medium; a score between 7.0 and 8.9 means the severity level is high; and a score between 9.0 and 10.0 means the severity level is critical [22]. Let the CVSS score for a node $n_{i,j}$ be $cvss_{i,j}$. We generate $cvss_{i,j}$ for the nodes on the network following the distribution of CVSS scores published by National Vulnerability Database (NVD) [23] in October 2018, shown in Table 3.1. For node $n_{i,j}$, we set $\lambda_{i,j} = k \times cvss_{i,j}$, where k is a predetermined constant.

3.2.2 Cyber Deception Modeled as a First-Level Hypergame

In this chapter, we use a hypergame to model the cyber-deception technique employed by the defender against the adversary. Unlike other game-theoretic models, where both players share a common perception of the game, a hypergame allows the players in the game to have their own perceived game. Here, we define a first-level hypergame as $H = (A, D, S_A \otimes S_D, P_A \otimes P_D)$, where

- A is the adversary. D is the defender. A and D are the two players in the game
- S_A is the set of strategies available to player A, the adversary
- S_D is the set of strategies available to player D, the defender
- P_A is player A's preference vector of the strategies in S_A
- P_D is player D's preference vector of the strategies in S_D

A preference vector can be replaced by a player's utility matrix, as shown later in this section. In a first-level hypergame, each player is playing their own perceived game. Let $G(A)$ be the game perceived by the adversary A, and $G(D)$ be the game perceived by the defender D. We may have $G(A) \neq G(D)$, but one or both players are not aware of the fact. Hence, in the first-level hypergame, the defender can choose to employ a cyber deceptive technique, and the adversary is not aware of the existence of such a deceptive technique. This is a reasonable assumption since the adversary may not discover the deceptive technique in a short time period when an IoBT is in service.

The first-level hypergame is played multiple rounds. At a newly compromised node, another round of the hypergame is played to determine the equilibrium strategies for the two players. Two players take the next actions according to their own perceived equilibrium strategies. This equilibrium of the hypergame at each node is a snapshot equilibrium [9, 10].

3.2.2.1 Defender's Strategies S_D

Let C_A be the set of nodes currently compromised by the adversary A, and R_A be the set of the nodes that has been compromised by the adversary A at least once. C_A maps the current spread of attack on the network, whereas R_A is the portion of the network topology learned by the adversary. We assume the defender has perfect information about the adversary's attack through frequently running an intrusion detection service. Hence, the defender maintains the complete and accurate lists of nodes in both C_A and R_A. Let E be the set of target nodes and N be the set of all the nodes on the IoBT network. We let the defender to choose from three strategies.

- D_1: No action
- D_2: Periodically cleaning up all the nodes every K time units, removing viruses and malware
- $D_{i,j}$: Install a honeypot on a clean node $n_{i,\,j}$, which is not a target node and not in R_A. $n_{i,\,j} \in N - E \cup R_A$

After a periodic cleaning, $C_A = \{n_0^A\}$ contains only the entry device. The adversary restarts from the entry device to enter the network. Meanwhile, after a periodic cleaning, the portion of the network known to the adversary R_A remains unchanged. We assume the defender has a finite budget of O honeypots. Honeypots installed on the nodes remain there after periodic cleaning, whereas the nodes in R_A without a honeypot installed cannot be considered for a honeypot in the future hypergames.

3.2.2.2 Adversary's Strategies S_A

The adversary also maintains the same two lists, C_A and R_A. The adversary chooses to move to one of the current clean nodes that has a direct edge with one of the current compromised nodes in C_A. The adversary stops only when it has explored all the nodes on the network, $R_A = N$. Figures 3.2 and 3.3 illustrate the adversary's strategies in each round of the game.

In the first round of the hypergame, $C_A = \{n_0\}$ and $R_A = \{n_0\}$. The adversary has two strategies available, $S_A = \{a_{1,\,1}, a_{1,\,2}\}$, where $a_{1,\,1}$ is to compromise $n_{1,\,1}$; and $a_{1,\,2}$ is to compromise $n_{1,\,2}$. Depending on which node is chosen by the adversary, the

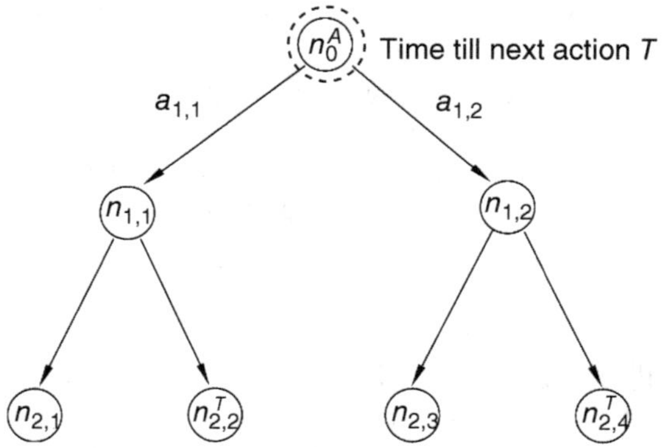

Figure 3.2 The adversary's strategies in Round 1.

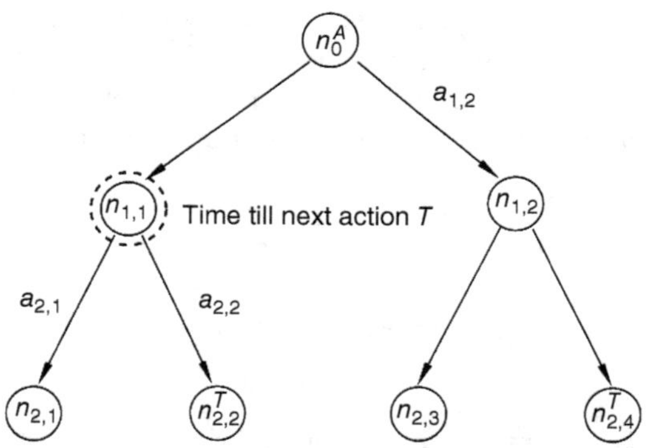

Figure 3.3 The adversary's strategies in Round 2.

time it takes to compromise that node follows an exponential distribution with $\lambda_{i,j}$ linked to the node's CVSS score, as discussed in Section 3.2.1.

Assume the adversary chooses to compromise $n_{1,\,1}$. In the second round of the hypergame, $C_A = \{n_0, n_{1,\,1}\}$ and $R_A = \{n_0, n_{1,\,1}\}$. The adversary has three strategies available, $S_A = \{a_{2,\,1}, a_{2,\,2}, a_{1,\,2}\}$, where $a_{2,\,1}$ is to compromise $n_{2,\,1}$; $a_{2,\,2}$ is to compromise $n_{2,\,2}$; and $a_{1,\,2}$ is to compromise $n_{1,\,2}$. The amount of time the adversary spends on the next node now depends on whether that node has a honeypot installed at the end of the first round. A honeypot causes the adversary to spend an excessive amount of time. Without a honeypot, the amount of time equals the time it takes the adversary to compromise the next node, determined by its CVSS score. If the adversary hits a honeypot, the adversary is deceived but does not stop until it explores all the nodes on the network.

After a periodic cleaning, the adversary restarts from the entry device, $C_A = \{n_0\}$. The portion of the network learned by the adversary remains unchanged, $R_A = \{n_0, n_{1,1}\}$. If a node chosen to be compromised next is in R_A (i.e. it has been compromised before), the time it takes to re-compromise that node follows exp{1/λ} with a very small λ, which is no longer related to the node's CVSS score.

Remark 3.2.2 Building a honeypot on the nodes chosen through the hypergame may deter the adversary long enough, so that it is never able to reach the true target nodes before the network dissolves or is randomized. Hence, the defender contains the attack. This is one outcome of playing the hypergame. On the other hand, the adversary stops only after it has compromised all the nodes on the entire network at least once. The adversary wants to explore the entire network to ensure it can discover all the target nodes. This is the other potential outcome. By playing the following hypergames multiple rounds, one round at a newly compromised node, we have a quantitative model of how an attack spread over a network. Through experiments we can obtain the estimated probabilities and the time needed for the adversary to hit the target node(s).

3.2.3 Equilibrium of the First-Level Hypergames

To solve for an equilibrium of a first-level hypergame, each player plays their own perceived game. We first solve for an equilibrium of the adversary A's perceived game $G(A)$. Let $(a^e(A), d^e(A))$ be the pair of equilibrium strategies for the two players in the adversary's perceived game $G(A)$. We then solve for an equilibrium of the defender D's perceived game $G(D)$. Let $(a^e(D), d^e(D))$ be the equilibrium strategy for the two players in the defender's perceived game $G(D)$. The adversary is not aware that $G(A) \neq G(D)$. The hypergame has a snapshot equilibrium $(a^e(A), d^e(D))$.

3.2.3.1 Adversary's Perceived Game $G(A)$

The adversary plays a zero-sum matrix game in each round. Stepping stone devices are nodes on the graph located between the entry node and the target nodes. Let M_S be the set of stepping stone devices without honeypots that are directly linked to one of the current compromised nodes (C_A). Let M_H be the set of stepping stone devices with honeypots installed that are directly linked to one of the current compromised nodes. Let M_T be the set of clean target nodes that are directly linked to one of the current compromised nodes. The number of strategies available to the adversary equals to the number of nodes in $M_S \cup M_H \cup M_T$. The matrix in Table 3.2 shows the payoffs for the two players' strategy pairs. The adversary plays the column strategies, whereas the defender plays the row strategies. The payoff is from the row player to the column player. The adversary wants to maximize its payoff (utility gain), whereas the defender wants to minimize its payoff (utility loss). The adversary solves for a mixed equilibrium strategy for its perceived matrix game according to Table 3.2.

Table 3.2 The adversary's perceived matrix game.

	Node $n_{i,\,j} \in M_S$	Node $n_{l,\,k} \in M_H$	Node $n_{u,\,v} \in M_T$
D_1	$U_S \times \gamma(n_{i,\,j})$	$U_T \times \gamma(n_{l,\,k})$	$U_T \times \gamma(n_{u,\,v})$
D_2	$-U_{restart}$	$-U_{restart}$	$-U_{restart}$

Table 3.3 The reduced game for the adversary.

	Node $n_{i,\,j} \in M_S$	Node $n_{l,\,k} \in M_H$	Node $n_{u,\,v} \in M_T$
D_1	$U_S \times \gamma(n_{i,\,j})$	$U_T \times \gamma(n_{l,\,k})$	$U_T \times \gamma(n_{u,\,v})$

U_S is a stepping stone device's payoff, a small positive value. U_T is a target node's payoff, a large positive value. $-U_{restart}$ is a large negative payoff for the adversary since it has to restart compromising the network from the entry device. $\gamma(n_{i,\,j})$, $\gamma(n_{l,\,k})$, and $\gamma(n_{u,\,v})$ are time discount factors. The adversary's payoff of a node is penalized by the time spent to compromise it. We set $\gamma(n_{i,\,j})$ equal to 1 for a node $n_{i,\,j}$ in R_A (a node compromised before). The adversary wants to spread over as many nodes as possible within K time units before a periodic cleaning happens. We can choose the time discount factors to favor certain type of nodes in an experiment. For example, the time discount factors can be chosen to favor nodes not in R_A or vice versa. In this hypergame the adversary is deceived by the honeypots. Hence, its perceived payoff of a node with a honeypot is the same as a true target node.

Notice the defender plays D_2 only at fixed time points. Hence, when we solve for a mixed equilibrium strategy, we can eliminate D_2 and let $G(A)$ be reduced to the matrix game in Table 3.3.

3.2.3.2 Defender's Perceived Game *G*(*D*)

Let $M_D = N - T \cup R_A \cup M_H$ be the set of nodes that are not target nodes, do not have a honeypot installed, and have never been compromised before. The defender plays its perceived matrix zero-sum game, as shown in Table 3.4. Again, D_2 is played only at fixed time points and can be eliminated. Hence, we solve for a defender's mixed equilibrium strategy for the matrix game according to Table 3.5. The payoffs $L(n_{i',j'}, n_{i,j})$, $L(n_{i',j'}, n_{l,k})$, and $L(n_{i',j'}, n_{u,v})$ depend on the pair of the nodes from the column strategy and the row strategy. Since the defender knows where honeypots are installed, the adversary receives a large negative payoff $-U_H$ if it hits a honeypot.

Table 3.4 The defender's perceived matrix game.

	Node $n_{i,j} \in M_S$	Node $n_{l,k} \in M_H$	Node $n_{u,v} \in M_T$
D_1	$U_S \times \gamma(n_{i,j})$	$-U_H$	$U_T \times \gamma(n_{u,v})$
D_2	$-U_{restart}$	$-U_{restart}$	$-U_{restart}$
Node $n_{i',j'} \in M_D$	$L(n_{i',j'}, n_{i,j})$	$L(n_{i',j'}, n_{l,k})$	$L(n_{i',j'}, n_{u,v})$

Table 3.5 The reduced game for the defender.

	Node $n_{i,j} \in M_S$	Node $n_{l,k} \in M_H$	Node $n_{u,v} \in M_T$
D_1	$U_S \times \gamma(n_{i,j})$	$-U_H$	$U_T \times \gamma(n_{u,v})$
Node $n_{i',j'} \in M_D$	$L(n_{i',j'}, n_{i,j})$	$L(n_{i',j'}, n_{l,k})$	$L(n_{i',j'}, n_{u,v})$

3.3 Experiments

We conduct the first experiment on a two-layer binary network as shown in Figure 3.4, where there is one target node (same network topology as Figure 3.1 with only one target node). The defender has a budget of one honeypot. We set 0 equal to the number of layers minus 1.

For the players' payoffs, we set $U_S = 1$ and $U_H = U_T = 10$. We let $\gamma_{i,j} = 1$ for all the nodes. Hence, there is no discount to a node's payoff even if it takes a long time to compromise the node. Here, the time for compromising a node itself becomes a penalty since the attacker may not reach the target before the defender periodically cleans up all the nodes. Because the defender wants to minimize the payoff in the matrix, we set $L(n_{i',j'}, n_{i,j})$ to be -10 if $n_{i',j'} = n_{i,j}$ and $+1$ if two nodes are different. We set $L(n_{i',j'}, n_{l,k})$ to be -10 for all $n_{l,k} \in M_H$ and $L(n_{i',j'}, n_{u,v})$ to be $+10$ for all $n_{u,v} \in M_T$. We set $\lambda = 30$ for moving to a target node or a node with a honeypot installed, both for the first time and for hitting the node again after periodic cleaning. Hence, we assume the attacker always spends a long time on its perceived target node. We generate λ for moving to a stepping stone node the first time following the CVSS score distribution in Table 3.1, from 0.1 to 10. We set $\lambda = 0.1$ for moving to a stepping stone node in R_A after periodic cleaning – it is much faster to compromise a node again. Tables 3.6 and 3.7 show the games played by the adversary and the defender in this experiment. In the defender's perceived game shown in Table 3.7, the columns for $n_{l,k} \in M_H$ have a payoff of -10. They are strictly dominated strategies. The defender's perceived game is first reduced by eliminating the strategy $n_{l,k} \in M_H$. Then, each player repeatedly solves for its

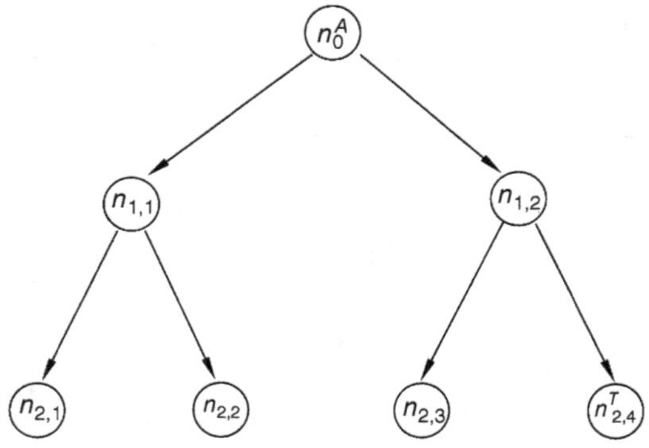

Figure 3.4 Two-layer binary network with one target node.

Table 3.6 Adversary's perceived game in the experiment.

	Node $n_{i,j} \in M_S$	**Node $n_{l,k} \in M_H$**	**Node $n_{u,v} \in M_T$**
D_1	1	10	10

Table 3.7 Defender's perceived game in the experiment.

	Node $n_{i,j} \in M_S$	**Node $n_{l,k} \in M_H$**	**Node $n_{u,v} \in M_T$**
D_1	1	−10	10
Node $n_{i',j'} \in M_D$	−10 or 1	−10	10

mixed equilibrium strategy. Players choose their actions based on their perceived mixed equilibrium strategies, until the adversary has explored all the nodes on the network and the game stops.

We set the periodic cleaning to take place every 35 time units, $K = 35$, and perform 100 runs of the experiment. Figure 3.5 shows the boxplots of (a) the time for the attacker to explore the entire network; (b) the time it takes to first hit the target node; and (c) the proportion of the total time used to hit the target node the first time. Table 3.8 shows the corresponding percentiles computed from the 100 runs. In less than 50% of the runs, the attacker hit the target node before the first periodic cleaning (every 35 time units). In over 10% of the runs, it took up to 158.8 time units or longer to reach the target node the first time. In more than 75% of the runs, it took the attacker over 381.9 time units to explore the entire network.

Next, we set $K = 20$ and perform another 100 runs. Figure 3.6 shows the corresponding boxplots, and Table 3.9 shows the percentiles from the 100 runs under

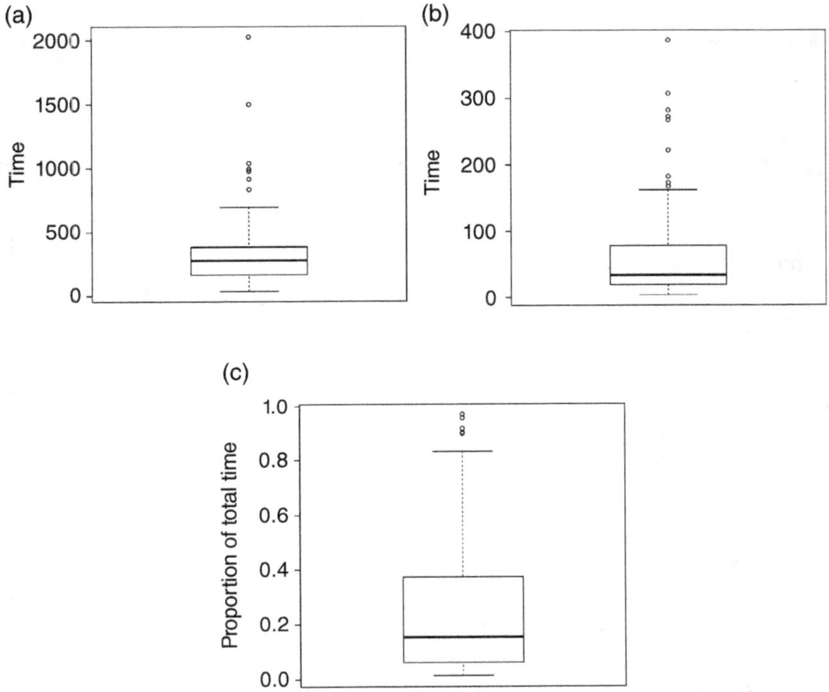

Figure 3.5 Boxplots with $K = 35$. From left to right: (a) total time to explore the entire network; (b) first time to hit the target node; and (c) the proportion of total time.

Table 3.8 Percentiles of time and proportion with $K = 35$.

	25th	50th	75th	90th
Total time	164.1	275.8	381.9	564.0
First time on target	19.1	33.7	72.9	158.8
Proportion	0.06	0.15	0.36	0.72

more frequent network cleaning. In less than 25% of the runs, the attacker hit the target node before the first periodic cleaning (every 20 time units). In over 10% of the runs, it took up to 188.7 time units or longer to reach the target node the first time. In more than 75% of the runs, it took the attacker over 986.2 time units to explore the entire network. The attacker needs significantly more time to explore the entire network, and longer time to discover the target node under more frequent network cleaning.

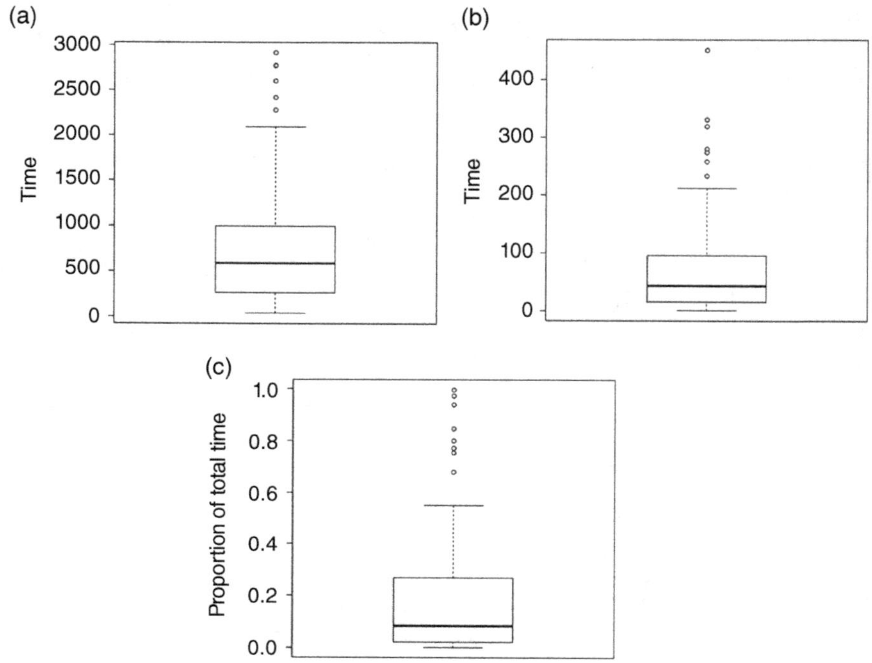

Figure 3.6 Boxplots with $K = 20$. From left to right: (a) total time to explore the entire network; (b) first time to hit the target node; and (c) the proportion of total time.

Table 3.9 Percentiles of time and proportion with $K = 20$.

	25th	50th	75th	90th
Total time	266.7	588.2	986.2	1504.1
First time on target	16.1	43.5	94.2	188.7
Proportion	0.02	0.08	0.27	0.52

We conduct a second experiment on a six-layer binary network as shown in Figure 3.7 with three target nodes, $n_{6,\,7}$, $n_{6,\,20}$, and $n_{6,\,48}$. The defender has a budget of five honeypots. We again set 0 equal to the number of layers minus 1. We set the utilities and λs to be the same as in the first two experiments. We conduct 100 runs on the six-layer network with $K = 420$, and another 100 runs with $K = 200$. We examine the total time it takes the attacker to compromise the entire network,

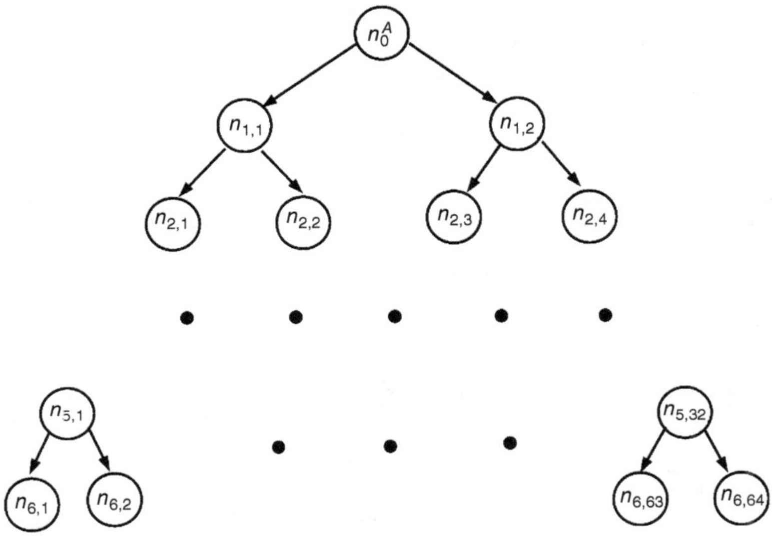

Figure 3.7 Six-layer binary network.

Table 3.10 Percentiles of time and proportion with $K = 420$.

	25th	50th	75th	90th
Total time	1598.9	1874.2	2033.3	2385.9
All targets	1036.1	1206.9	1552.5	1671.1
Proportion	0.53	0.69	0.82	0.92

the time needed to discover all three target nodes, and the proportion of total time needed to discover all three target nodes. Tables 3.10 and 3.11 show the 0.25, 0.50, 0.75, and 0.90 percentiles. Figures 3.8 and 3.9 show the corresponding boxplots.

The experimental results demonstrate that K, the time interval to conduct periodic cleaning, is also an important component of the defense strategy. When a network is scanned and cleaned of viruses and malware more often, it takes longer time for the attacker to discover the target nodes and compromise the entire network. This is more evident as the size of the network increases. Provided how long a military network will stay in service is known, through experiments, our hypergame model is able to provide a recommended time interval K to periodically clean the network.

Table 3.11 Percentiles of time and proportion with $K = 200$.

	25th	50th	75th	90th
Total time	3982.8	4754.7	5394.3	5998.9
All targets	1891.3	2499.7	2986.1	3527.9
Proportion	0.39	0.52	0.65	0.78

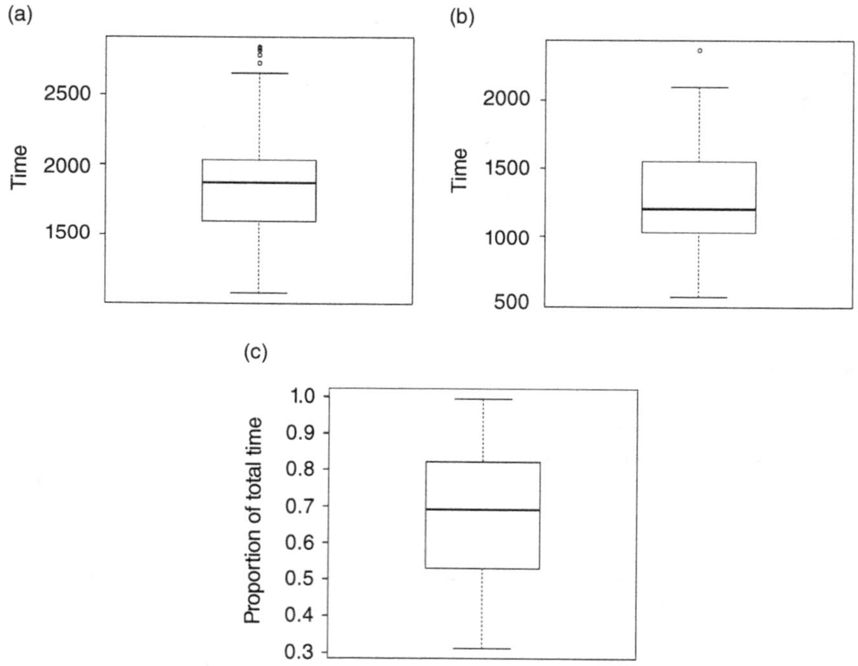

Figure 3.8 Boxplots with $K = 420$. From left to right: (a) total time to explore the entire network; (b) first time to hit the target node; and (c) the proportion of total time.

3.4 Conclusion

The hypergame framework in this chapter provides an important theoretical model for us to examine how an attack spreads on a network. Here, we allow the defender to take one simple action to deter the attack – installing honeypots on stepping stone nodes. The defender can decide on where to install the honeypots based on the hypergames. The utilities in players' perceived matrix games are the key factors to determine how the players choose their next move. We could set the utilities so the attacker prefers to cover as many nodes as possible before the

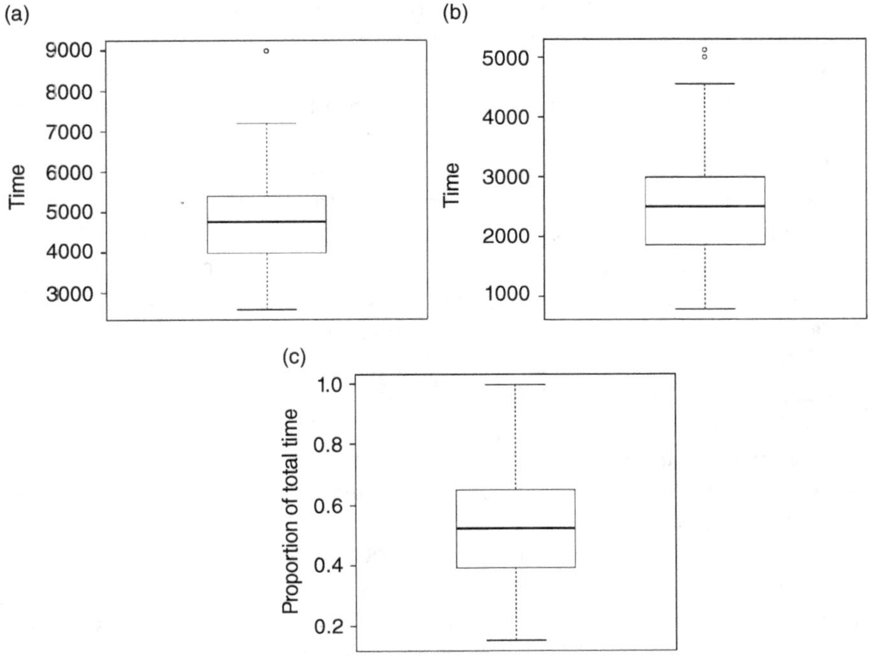

Figure 3.9 Boxplots with $K = 200$. From left to right: (a) total time to explore the entire network; (b) first time to hit the target node; and (c) the proportion of total time.

next periodic cleaning. On the other hand, we could choose the utilities so the attacker prefers to explore the nodes that it has not compromised before. Similarly, we can set the utilities so the defender prefers certain nodes to install the honeypots. Based on the hypergame, we obtain quantitative measures for moving target defense. For future work, we plan to incorporate more potential deceptive techniques beyond honeypot in the hypergame so the game-theoretic model would closely mimic the real-world applications. Moreover, we will compare our result to the Bayesian game with incomplete information. Furthermore, we plan to extend the hypergame framework to mobile ad hoc networks, to more closely match the properties of military networks.

References

1 Al-Fuqaha, A., Guizani, M., Mohammadi, M., Aledhari, M. and Ayyash, M., 2015. "Internet of Things: A Survey on Enabling Technologies, Protocols, and Applications." *IEEE Communications Surveys & Tutorials*, 17(4), pp. 2347–2376.

2 Kolias, C., Kambourakis, G., Stavrou, A. and Voas, J., 2017. "DDoS in the IoT: Mirai and Other Botnets." *Computer*, 50(7), pp. 80–84.

3 MIT Lincoln Laboratory, 1986, September. "Distributed Sensor Networks, Final Report to DARPA." https://apps.dtic.mil/dtic/tr/fulltext/u2/a204719.pdf.

4 Abuzainab, N. and Saad, W., 2018. "Dynamic Connectivity Game for Adversarial Internet of Battlefield Things Systems." *IEEE Internet of Things Journal*, 5(1), pp. 378–390.

5 Azmoodeh, A., Dehghantanha, A. and Choo, K.K.R., 2018. "Robust Malware Detection for Internet of (Battlefield) Things Devices Using Deep Eigenspace Learning." *IEEE Transactions on Sustainable Computing*, 4(1), 88–95, 2019.

6 Farooq, M. J. and Zhu, Q., 2018. "On the Secure and Reconfigurable Multi-Layer Network Design for Critical Information Dissemination in the Internet of Battlefield Things (IoBT)." *IEEE Transactions on Wireless Communications*, 17(4), pp. 2618–2632.

7 Wang, L., Islam, T., Long, T., Singhal, A. and Jajodia, S., 2008, July. "An Attack Graph-Based Probabilistic Security Metric." In *IFIP Annual Conference on Data and Applications Security and Privacy,* Berlin, Heidelberg, pp. 283–296.

8 Kott, A., Swami, A. and West, B.J., 2016. "The Internet of Battle Things." *Computer*, 49(12), pp. 70–75.

9 Wang, M., Hipel, K.W. and Fraser, N.M., 1988. "Modeling Misperceptions in Games." *Behavioral Science*, 33(3), pp. 207–223.

10 Wang, M., Hipel, K.W. and Fraser, N.M., 1989. "Solution Concepts in Hypergames." *Applied Mathematics and Computation*, 34(3), pp. 147–171.

11 Kamdem, G., Kamhoua, C., Lu, Y., Shetty, S. and Njilla, L., 2017, June. "A Markov Game Theoritic Approach for Power Grid Security." In *2017 IEEE 37th International Conference on Distributed Computing Systems Workshops (ICDCSW)*, pp. 139–144.

12 Gutierrez, C.N., Bagchi, S., Mohammed, H. and Avery, J., 2015, March. "Modeling Deception in Information Security as A Hypergame – A Primer." In *Proceedings of the 16th Annual Information Security Symposium,* p. 41. CERIAS-Purdue University.

13 Cho, J.H., Zhu, M. and Singh, M., 2019. "Modeling and Analysis of Deception Games Based on Hypergame Theory", In *Autonomous Cyber Deception*, pp. 49–74. Springer.

14 House, J.T. and Cybenko, G., 2010, May. "Hypergame Theory Applied to Cyber Attack and Defense." In *Sensors, and Command, Control, Communications, and Intelligence (C3I) Technologies for Homeland Security and Homeland Defense IX*, Vol. 7666, p. 766604. International Society for Optics and Photonics.

15 Vane III, R.R., 2005, December. "Planning for Terrorist-Caused Emergencies." In *Proceedings of the 37th conference on Winter simulation*, pp. 972–978. Winter Simulation Conference.

16 Takahashi, M.A., Fraser, N.M. and Hipel, K.W., 1984. "A Procedure for Analyzing Hypergames." *European Journal of Operational Research*, 18(1), pp. 111–122.

17 Kiekintveld, C. and Wellman, M.P., 2008, May. "Selecting Strategies Using Empirical Game Models: An Experimental Analysis of Meta-Strategies." In *Proceedings of the 7th international joint conference on Autonomous agents and multiagent systems*, Vol. 2, pp. 1095–1101.

18 Wang, L., Islam, T., Long, T., Singhal, A. and Jajodia, S., 2008, July. "An Attack Graph-Based Probabilistic Security Metric." In *IFIP Annual Conference on Data and Applications Security and Privacy*, pp. 283–296. Springer, Berlin, Heidelberg.

19 Frigault, M. and Wang, L., 2008, July. "Measuring Network Security Using Bayesian Network-Based Attack Graphs." In *Annual IEEE International Computer Software and Applications Conference*, pp. 698–703. IEEE.

20 Poolsappasit, N., Dewri, R. and Ray, I., 2012. "Dynamic Security Risk Management Using Bayesian Attack Graphs." *IEEE Transactions on Dependable and Secure Computing*, 9(1), pp. 61–64.

21 Common Vulnerability Scoring System, https://www.first.org/cvss.

22 Vulnerability Metrics, https://nvd.nist.gov/vuln-metrics/cvss.

23 National Vulnerability Database, https://nvd.nist.gov/general/nvd-dashboard.

4

Cooperative Spectrum Sharing and Trust Management in IoT Networks

Fatemeh Afghah[1], Alireza Shamsoshoara[1], Laurent L. Njilla[2], and Charles A. Kamhoua[3]

[1] School of Informatics, Computing and Cyber Systems, Northern Arizona University, Flagstaff, AZ, USA
[2] Cyber Assurance Branch, US Air Force Research Laboratory, Rome, NY, USA
[3] US Army Research Laboratory, Adelphi, MD, USA

4.1 Introduction

One of the key challenges toward wide-scale implementation of new wireless communication technologies, in particular growing Internet of Things (IoT) with potentially billions of wireless devices, is spectrum scarcity. A considerable portion of this required radio spectrum is provided through dedicated IoT spectrum or unlicensed spectrum bands. However, in order to facilitate the required throughput, latency, and reliability of several emerging IoT applications, a new generation of spectrum management techniques is required to enhance the utilization efficiency of radio spectrum through reliable and practical dynamic spectrum sharing methods [1]. The majority of current spectrum management efforts focus on common spectrum sharing mechanisms where the licensed users are unaware of unlicensed users' presence (e.g. database control and spectrum sensing mechanisms) [2, 3]; however, these methods are not the best choice for spectrum sharing with IoT devices. Energy constraint and limited spectrum mobility of tiny IoT devices restrict their performance in spectrum sensing mechanisms. These methods involve a considerable energy exhaustion for the device to monitor different frequency bands and identify the spectrum holes. Moreover, the devices are required to have the capability of working in different frequencies (spectrum mobility) [4–7]. In database control spectrum sharing mechanisms, the IoT devices need

Modeling and Design of Secure Internet of Things, First Edition. Edited by Charles A. Kamhoua, Laurent L. Njilla, Alexander Kott and Sachin Shetty.

Published 2020 by John Wiley & Sons, Inc.

to have an access to geolocation databases, or they require to have access to a central controller or an enabler in the network. This enabler can keep a record of licensed users' spectrum usage and allowable interference range as required in database control technologies (e.g. TV white spaces). Hence, they may not be practical in all IoT networks.

Another drawback of the majority of common spectrum sharing models is their overly conservative approach to protect the incumbent users against harmful interference caused by unlicensed entities. Thus, such approaches that enforce a low level of transmission power for unlicensed users, or allocate a wide and static protection zone around the incumbents, are unable to address the increasing demand for radio spectrum from IoT networks [8, 9]. Furthermore, in common-model paradigm, the licensed users do not usually benefit from allowing the unlicensed users to share their spectrum, but they may even suffer due to interference from secondary users (SUs). That being said, flexible spectrum sharing models (i.e. right models) in which the primary users (PUs) can willingly lease a portion of their spectrum access to unlicensed users (e.g. IoT devices) in exchange for remuneration or some sort of compensation (e.g. energy harvesting or cooperative relaying service) can offer a win-win solution for both parties [10–16].

One main concern toward enabling shared access techniques to the radio spectrum, particularly the federal bands, is security. The security paradigm in IoT involves certain challenges related to the characteristics of IoT devices, including low operational power, low computation capability, and the difficulties of embedding security capabilities in a wide range of diverse devices. Some common security threats in wireless networks include: (i) concerns related to privacy of the users; (ii) authentication, physical attacks, and device cloning; and (iii) confidentiality of communications among the users in the presence of eavesdroppers. In addition to these conventional security threats, the networks with dynamic spectrum allocation mechanism are also vulnerable to unauthorized access to the spectrum as well as potential selfish behavior of users taking advantage of the ad hoc nature of such networks [17–20]. Therefore, the cognitive radio networks (CRNs) are prone to suffer from various exogenous, malicious, and selfish attacks. Some of these attacks are denial of service attacks to deny the unlicensed users from spectrum access, sensing data falsification attacks, and PU emulation attacks, only to name a few [21–24]. Different security mechanisms have been proposed to address these attacks, which among these information theoretic secrecy methods aim to secure the communication of the nodes from potential eavesdroppers by exploiting the physical characteristics of wireless channels [25]. This can be achieved by adding artificial noise, using beamforming techniques, or employing multiple-input multiple-out (MIMO) techniques [26, 27]. Conventional cryptographic methods such as well-known public key infrastructure rely on private keys that need to be produced by the server and safely distributed and stored in nonvolatile memory (NVM) of devices. Due to the large number of devices in IoT networks, generation and distribution of these private keys among the IoT devices is a challenging task. Moreover, the NVMs are highly vulnerable to physical attacks that can reveal the stored private keys to adversaries and jeopardize the security of IoT [28–30]. Therefore, one key advantage of information-theoretic secrecy-based techniques to enhance IoT security compared to the cryptographic-based methods is that they do not rely on encryption keys, and hence do not involve key distribution and key management complexities.

Cooperative jamming in CRNs has been implemented by employing the SUs to create artificial noise [31], or transmit structured codewords to reduce the eavesdropper's capability in decoding the PU's information. The potentially noncooperative SUs can be encouraged to provide such a service if they are granted with a chance of spectrum access for their own transmission, as introduced in [31]. Providing cooperative jamming for the PUs, in addition to other previously studied compensation techniques such as cooperative relaying and energy harvesting, can offer a practical solution for implementation of property-right CRNs without involving money exchange among the parties or other regulatory issues [31]. In [27], a Stackelberg game model is defined to model the interactions between a legitimate source node and a non-altruistic SU in the presence of an eavesdropper. The SU can be compensated with a fraction of the legitimate source node's access time to radio spectrum if it provides cooperative jamming to enhance the secrecy rate of the source node. This model is further extended to a scenario with multiple potential cooperative jammers where the competition among them is modeled as an inner Vickrey auction. The proposed model [27] favors the PU by defining the game in such a way that the time allocation to access the spectrum as well as the ratio of the power, which the SUs spend on cooperative jamming and the one they utilize for their own transmission, are all determined by the PU.

4.2 Problem Statement

Clearly in property-right models for spectrum sharing, including the aforementioned cooperative spectrum leasing mechanisms, the PUs deserve more benefits as the spectrum owner. Therefore, the majority of the previously reported works are designed based on the fact that the PUs make all the decisions regarding the spectrum assignment, time allocation of spectrum access, and the allowable power of the SUs. However, in reality there is no guarantee that the cognitive SUs will follow rules determined by the PUs as they may act selfishly to maximize their own benefits. For instance, in current cooperative spectrum leasing scenarios, the PUs determine the time allocation of access to spectrum as well as the power of the SU during the cooperative relaying. However, the cognitive SUs may deviate from the requested power allocation ratio requested by the PU after being granted with spectrum access to improve their own throughput. The authors in [32] study a similar model with the objective of enhancing the secrecy rate of both the primary and secondary nodes, where the secondary nodes are assumed to be fully trustable.

While cooperative jamming provided by the SUs can potentially enhance the secrecy rate of the licensed users, it involves the assumption of having unconditionally cooperative and trustable SUs, as considered in the majority of previously reported studies [27, 31–33]. However, this assumption is far too optimistic in communication networks, noting the non-altruistic nature of cognitive SUs (e.g. IoT devices). This frequent assumption has been revisited in some work considering different aspects of potential selfish or malicious behavior by SUs.

In this chapter, we consider the problem of spectrum sharing among a group of cognitive primary and cognitive IoT (C-IoT) users as the secondary unlicensed users. These C-IoT users have the capabilities of understanding, reasoning, and learning to ideally mimic the human decision-making [34–36]. Thus, these users are very likely expected to act selfishly. Noting the cognitive capability of these IoT users, we aim to develop a spectrum leasing mechanism that encourages the spectrum owners to share their spectrum motivated by the fact that they can improve their secrecy rate and quality of communication, while they can ensure that potential selfish behavior from the unlicensed users is penalized. This model also discourages the C-IoT users from any potential misbehavior knowing that their reputation will be highly impacted, hence it would be less likely for them to have the chance of spectrum access.

Current literature usually considers the misbehavior of the C-IoT users as acting as eavesdroppers and studies whether cooperation can still improve the secrecy [37, 38]. In this chapter, we consider a more general aspect of potential misbehavior of the SUs, where they can be selfish but not malicious. The selfish users are self-interested and their preference toward maximizing their own benefits can result in some sort of misconduct such as not participating in cooperation; however, they do not intend to impede the network performance as do the malicious users [39, 40]. To focus on the selfish behavior of the unlicensed users, it is assumed that the SUs are pre-authenticated, and therefore the intruding and malicious SUs are already filtered out using the authentication mechanism. We propose a game-theoretic reputation-based mechanism to identify such untrustable users and improve the performance of spectrum sharing mechanisms by enabling the licensed users to filter out potential selfish users.

The majority of previously proposed reputation-based methods in communication systems assumes that the users self-report their reputation or they consider an audit unit in the network that keeps a record of all the users' reputation [41, 42].

Authors in [43] proposed a weighted cooperative spectrum sensing framework for infrastructure-based networks, which requires a base station to receive reports and updates from SUs. However, the presence of an audit unit may not be practical in distributed system. The self-reporting reputation methods where each user reports its own reputation can be implemented in distributed systems, but they can impose a heavy signaling load to the network [44]. Therefore, such methods are prone to the security and reliability of communication channels to report these reputations. Another vulnerability of such methods is the possibility of reporting false reputations by selfish/malicious nodes [45, 46]. Therefore, the reputation-based method we proposed here in which the licensed users record the reputation of the local SUs can offer a practical solution for reputation management in distributed systems and can be implemented in any infrastructure-less or ad hoc networks. The fact that in our proposed model, the reputation of the secondary nodes is directly observed by the PU can prevent the impact of false reports. If such a first-hand reputation of an SU was not available to a PU, it can inquire this reputation from other PUs in the neighborhood (second-hand reputation).

In summary, we consider the problem of spectrum leasing to secondary IoT devices in the presence of a passive eavesdropper, in which these IoT devices provide cooperative relaying and cooperative jamming services to enhance the secrecy rate and quality of service of the PU in exchange for spectrum access. Since the IoT devices often have a limited energy, the natural tendency of such non-altruistic users can lead to selfish attacks where they violate their commitments to the PUs after they are granted the spectrum access. In this chapter, we propose a reputation-based method to monitor the cooperative behavior of these IoT SUs in terms of the power they dedicate to requested services from the PU, including cooperative jamming and cooperative relaying to prevent potential selfish behavior.

The structure of this chapter is organized as follows. Section 4.3 provides a brief introduction to physical layer security. Section 4.4 reviews the Stackelberg game theory. The proposed reputation-based mechanism is described in Section 4.5, and its performance in cooperative spectrum leasing is evaluated in this section. Finally, Section 4.6 includes the conclusion for this chapter.

4.3 Overview to Physical Layer Secrecy

Current communication systems commonly rely on cryptographic methods to enhance their security. These methods depend on secure storage of cryptographic keys in the device memories, involve several initial stages of key generation and key distributions, and also require a presence of a certification authority in the network. One important challenge toward implementation of current key-based cryptographic methods in IoT networks is the vulnerability of key storage mechanism to physical attacks. Another main concern related to cryptographic systems is that they rely on computational hardness of a mathematical problem that can be broken in future by quantum computing. These items motivated the use of physical layer secrecy methods in order to protect the communications systems from malicious users and eavesdroppers [47]. The idea behind the physical layer secrecy techniques is to exploit the randomness in noise and communication channels to hide the message from eavesdroppers. The physical layer secrecy techniques can be applied in addition to cryptography methods to increase the secrecy performance for all layers in the network. Figure 4.1 compares the advantages and disadvantages of cryptographic and information-secrecy-based security methods.

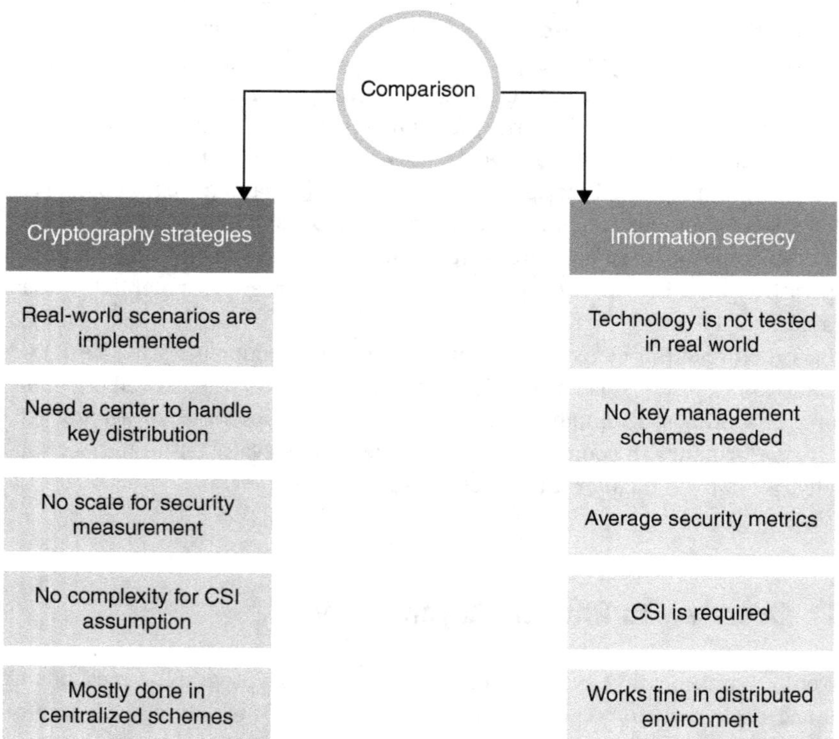

Figure 4.1 Comparison of information secrecy and cryptography methods.

4.3.1 Information-Theoretic Secrecy

Information-theoretic secrecy can offer the strongest notion of security by exploiting the randomness of communication channels to ensure that the message cannot be decoded by an eavesdropper. Wyner introduced the simplest wireless model for secrecy in communication systems as depicted in Figure 4.2 [48]. A more general model of wiretap channels has been analyzed by Csiszar and Korner in [49]. In this model, the legitimate source and destination (Alice and Bob) want to send and receive messages to and from each other. Eve or eavesdropper is an illegal node which tends to gather information from this communication. The secrecy rate is defined as a highest rate that the messages can be sent in a secure and reliable channel between Alice and Bob [50]. Several coding techniques, communication techniques such as MIMO, and cooperative relaying methods have been proposed to investigate the achievement of information-theoretic secrecy [25, 51–55].

Let us assume that a source message W is chosen to be sent to the legitimate receiver through a discrete memoryless channel, and $y^n$and z^n denote the two output sequences received at the receiver and the eavesdropper, respectively.

The secrecy rate can be defined as the equivocation rate $R_e^{(n)} = \frac{1}{n}H(W \mid Z^n)$. This equivocation rate is the level of confusion of the eavesdropper given its observations, Z^n. Now based on this rate, the rate-equivocation pair (R, R_e) is defined as the secrecy rate R attained under the secrecy level R_e. If $R_e < R$, it means that the system has some leakage to the eavesdropper, and $R_e = R$ indicates achieving a perfect secrecy rate. If the secrecy capacity is defined as $C_s = \max_{(R, R \in \ell)} R$, where ℓ is the capacity-equivocation region, the secrecy capacity for the wiretap channel can be calculated as

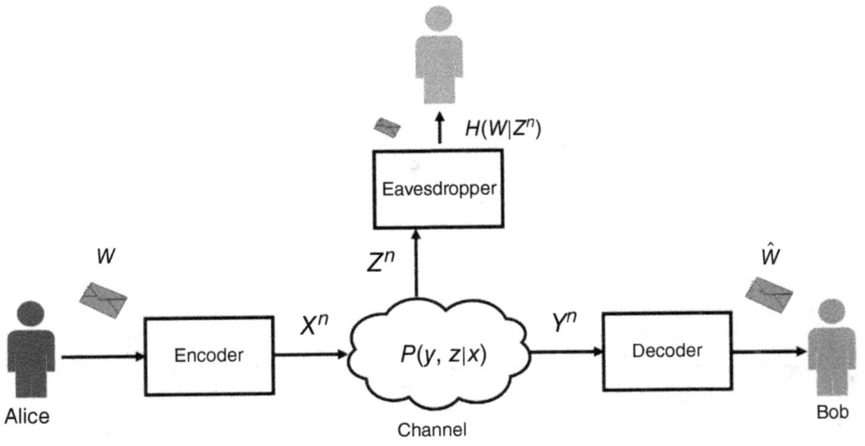

Figure 4.2 Wiretap model.

$$C_s = \max_{P_U X P_{Y Z \mid X}} [I(U;Y) - I(U;Z)]^+ . \tag{4.1}$$

In this equation, U is an auxiliary random variable satisfying Markov chain $U - X - (Y, Z)$, and $I(X, Y)$ is the mutual information of two discrete random variables X and Y. This equation implies that if the communication rate R is less than C_s, a reliable transmission between the source and the destination can be achieved.

In a wireless wiretap channel with additive Gaussian noise, depicted in Figure 4.3, the received signals for Bob and Eve are $y_i = h_{sd}x_i + n_{d,\ i}$and $z_i = h_{se}x_i + n_{e,i}$, respectively. If the noise sources are complex and symmetric such that $n_{d,i}\tilde{C\mathcal{N}}(0, \sigma_d^2)$ and $n_{e,i}\tilde{C\mathcal{N}}(0, \sigma_d^2)$, and h_{sd}, $h_{se} \in \mathbb{C}$, the secrecy capacity for the Gaussian wiretap model is obtained as

$$\begin{aligned} C_s &= \left(\log\left(1 + \frac{|h_{sd}|^2 P_s}{\sigma_d^2}\right) - \log\left(1 + \frac{|h_{se}|^2 P_s}{\sigma_e^2}\right)\right)^+ \\ &= (C_d - C_e)^+ = \text{legitimate rate} - \text{leakage rate}, \end{aligned} \tag{4.2}$$

where C_d denotes the capacity of the receiver channel and C_e is the capacity of the eavesdropper's channel. As a result, the secrecy rate is equal to the difference of the legitimate capacity rate and the leakage rate.

Cooperative relaying is a commonly used tool to increase the transmission rate in communication channels [56–62]. Figures 4.4 and 4.5 illustrate the simplest form of cooperative communication networks, which is a relay channel [56]. Different relaying modes of decode-and-forward (DF) relaying, compress-and-forward (CF), and amplify-and-forward (AF) can be utilized at the relay node depending on the application of interest and its desired performance metrics such

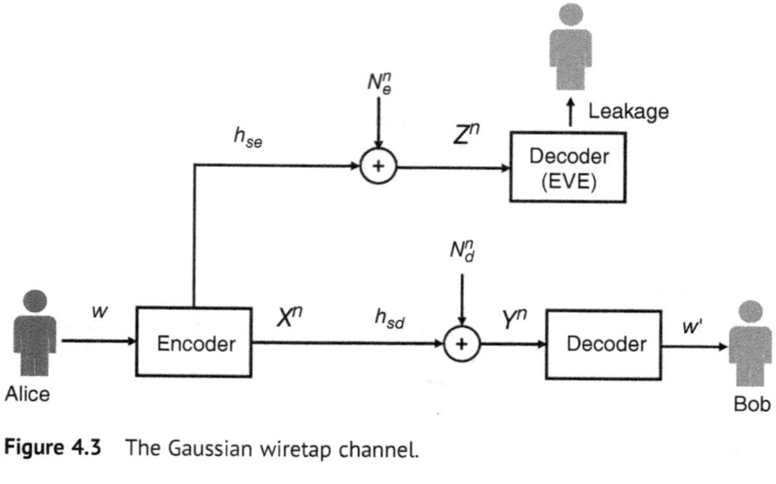

Figure 4.3 The Gaussian wiretap channel.

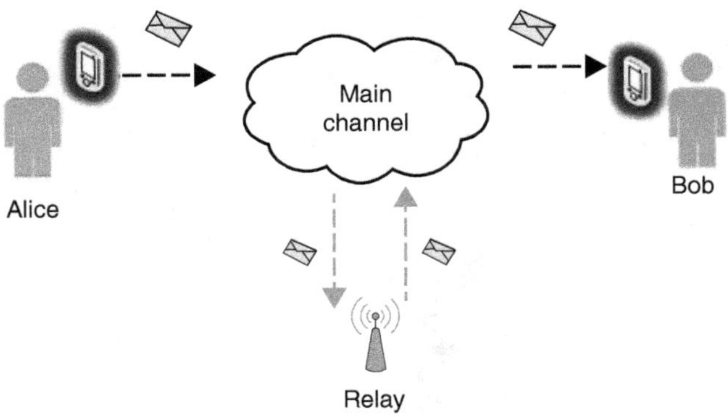

Figure 4.4 The relay channel.

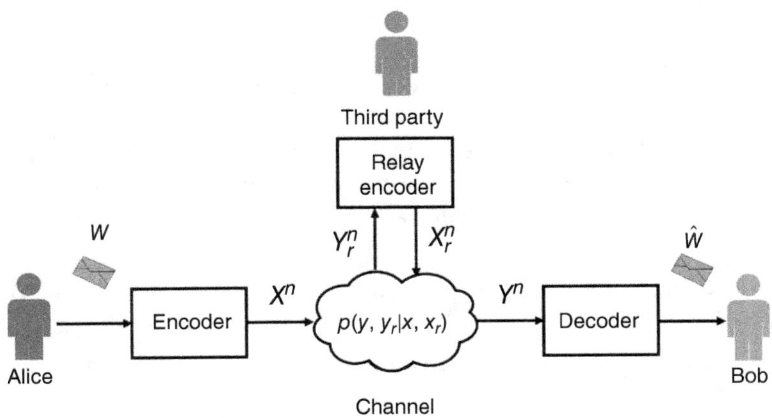

Figure 4.5 The relay channel using encoder and decoder.

as quality of service, bit error rate, and the complexity level of relaying operation. In the following, we describe the rate for the DF and AF schemes.

In the DF scheme, the transmitted message is decoded by the relay and it will be encoded again before retransmitting to the receiver. The rate of this scheme can be written as

$$R_{(DF)} = \min\left\{C\left(h_{sr}^2 P_r\right), C\left(h_{sd}^2 P_s\right) + C\left(h_{rd}^2 P_r\right)\right\}, \tag{4.3}$$

where h_{sr}, h_{sd}, and h_{rd} denote the channel coefficient between the source–relay, source–destination, and relay–destination, respectively. P_r and P_s denote the transmit power for relay and the source, respectively. This rate is obtained based on the

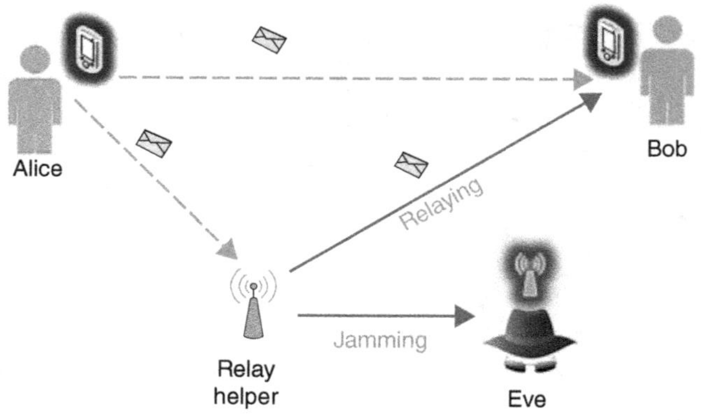

Figure 4.6 Relaying for secrecy.

cutsets on the receiver. In this rate, the weaker link determines the rate for the DF [63].

However, in the AF scheme, the relay node only amplifies and forwards the received signal. If the relay node scales the signal in a way to keep the relay power constraint P_r, the rate for AF can be found as

$$R_{(AF)} = C\left(h_{sd}P_s + \frac{h_{sr}^2 P_s h_{rd}^2 P_r}{1 + h_{sr}^2 P_s + h_{rd}^2 P_s}\right). \tag{4.4}$$

Here, the notations of h_{ij} and P_i where $i \in \{s, r\}$ and $j \in \{r, d\}$ are the same as defined earlier for the DF scheme.

Cooperative communication techniques can be also utilized to integrate secrecy in wireless network with relaying and jamming as demonstrated in Figure 4.6. Several strategies and methods have been proposed in the literature to enhance the secrecy of communication system. The taxonomy of these tactics is categorized into two branches. In the first approach, the relay enhances the level of secrecy of the legitimate nodes by interfering the eavesdropper signal. The relay sends artificial noise to the eavesdropper to confuse it. This strategy is also called as noise-forwarding (NF) or cooperative jamming. In this scheme, the codeword of the noise is independent of the transmitter codeword. Many researchers worked based on this phenomena and they denominated their work as NF [64], cooperative jamming [55, 65], or interface-assisted secret communication [66]. On the other hand, in the second branch, the idea is improving the quality of the main signal and link by employing one of the cooperation schemes such as DF, AF, and CF [25, 65–69]. Figures 4.7 and 4.8 illustrate the concepts of these two strategies, respectively.

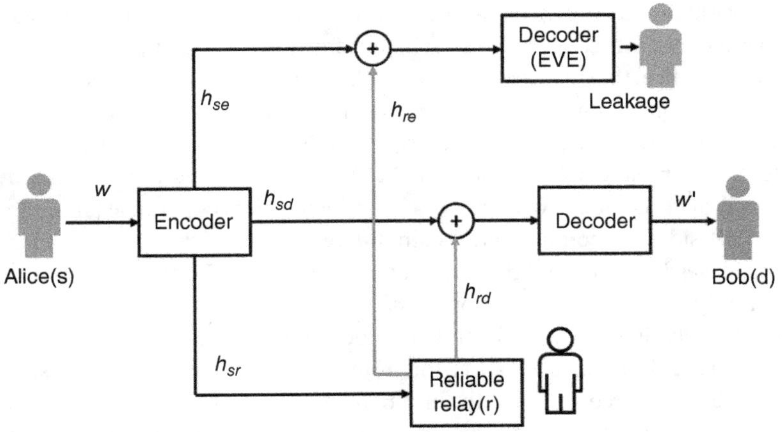

Figure 4.7 Cooperative jamming for enhancing the secrecy.

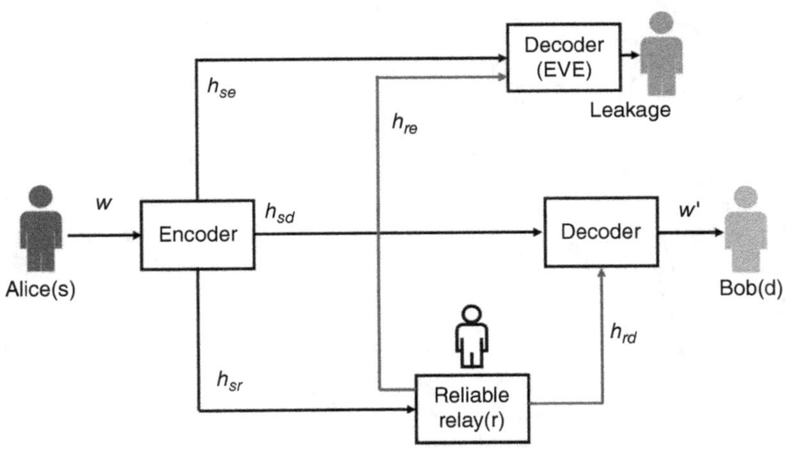

Figure 4.8 Active cooperation (DF or CF).

4.3.2 Cooperation for Enhancing Secrecy in CRNs

Due to the fast growth of CRNs in recent years, security challenges in this area have attracted increased attention. In the majority of previously reported models for spectrum sharing in CRNs, the licensed users are oblivious to the presence of unlicensed SUs. Moreover, they usually do not benefit from sharing their spectrum with the SU. In some cases they may even suffer from harmful interference caused by the SU's communications. In property-right models for spectrum sharing, the

PUs can decide about leasing their spectrum with the unlicensed users in exchange for some benefits such as cooperative relaying, energy harvesting, and cooperative jamming to enhance their secrecy rate with the help of the unlicensed users. Therefore, the property-right spectrum leasing mechanisms can motivate both parties to participate in dynamic spectrum sharing. Figure 4.9 demonstrates a system model for a simple CRN in which the SUs pay off for access to the licensed spectrum by providing the PU with cooperative relaying and cooperative jamming services [70]. Besides the common attacks in different layers of the network, such as jamming, eavesdropping, and routing attacks, these networks are highly prone to the possibility of selfish and malicious behavior by the SUs because there is no backbone in CRNs. For instance, in the described cooperative spectrum leasing scenario, a selfish SU can lease the spectrum from the PU but does not compensate the PU with the requested services. In this chapter, we investigate such scenario and develop a distributed reputation-based game-theoretic mechanism to prevent

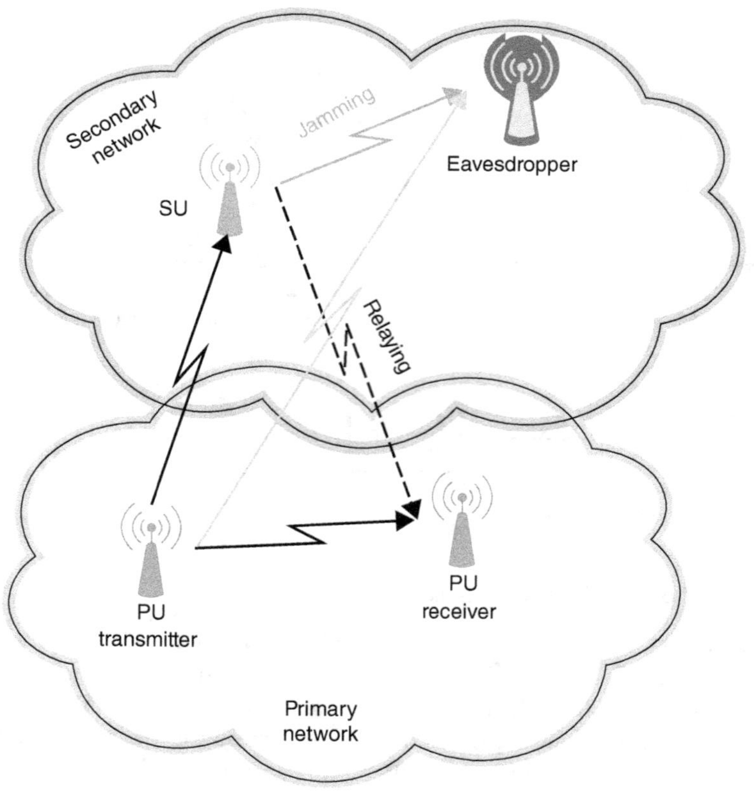

Figure 4.9 Enhancing secrecy with cooperation and game theory.

the unlicensed IoT users from acting selfishly. To protect the rights of the PUs as the spectrum owners, the spectrum leasing process in property-right models is commonly modeled by a Stackelberg game in which the game leader can have a higher priority by enforcing its action first.

4.4 Introduction to Stackelberg Games

Game theory is a powerful mathematical tool to model the interactions among the cognitive agents when they compete with one another or cooperate with each other to achieve the same goal. Game theoretical modeling has been widely used in communication networks to address a diverse range of optimization problems, as shown in Table 4.1 [71–73]. The capability of several game theoretical models to implement excessive punishment mechanisms can effectively prevent cognitive users from contingent malicious and selfish behavior.

Stackelberg games belong to a class of noncooperative strategic games in which the players take their actions sequentially rather than simultaneously. In these games, one or some of the players called the "game leaders" have a higher priority in decision-making and can declare their strategy first. The rest of the players called the "game followers" have a lower priority and need to take their actions after observing the leaders' strategy. Because of this hierarchical structure of Stackelberg games, they are an appropriate game model to analyze the interactions between the licensed and unlicensed users during radio spectrum sharing.

Table 4.1 Applications of game theory in communication networks.

Game theory application in communication networks [71]	Other networks	Internet networks
		Wireless local area networks (WLANs)
		Wireless access networks
		Multi hop networks
	Cooperative networks	Power control
		Resource allocation
		Relay selection
	Cognitive radio networks	Medium access control
		Security mechanisms
		Power allocation [72]
		Spectrum sharing [73]

The shaded section focuses on the path of the proposed game-theoretic spectrum leasing model.

To describe the Stackelberg games in more detail, let us assume a simple game model with two rational players, including one leader and one follower. We denote the action set of the leader and the follower by A_l and A_f, respectively. In this game, the follower selects an action from its action set after observing the leader's action. The solution of Stackelberg game called as "Stackelberg equilibrium solution" can be obtained by finding the best strategy of the leader considering that the lower priority players, follower, is rational and will respond to the leader's action with its best response. The Stackelberg equilibrium solution can be obtained by backward induction described as follows [74, 75]:

$$\begin{aligned} &\max{}_{(a_l,a_f)\in(A_l\times A_f)} U_l(a_l,a_f),\\ &s.t. a_f \in \arg\max{}_{a'_f\in A_f} U_f\left(a_l,a'_f\right), \end{aligned} \tag{4.5}$$

where U_l and U_f denote the utility functions of leader and follower, respectively. Compared to games in which the players simultaneously select their actions, the leader in Stackelberg games can always obtain a better payoff because the leader knows that the follower players are rational, and therefore always respond to its action by their best strategy. In this chapter, we utilize a Stackelberg game model to develop a spectrum leasing mechanism that enables the licensed users to take advantage of cooperative jamming and cooperative relaying services by the SUs in exchange for spectrum access.

4.5 Proposed Game-Theoretic Spectrum Leasing Model to Enhance Secrecy Rate

In this section, we describe the system model of the c-IoT system studied in this work. Moreover, we propose a reputation-based game-theoretic model to enhance the secrecy rate for the primary network and improve its transmission rate through cooperative spectrum leasing to unlicensed IoT users. The cooperative performance of the secondary IoT users is monitored by the licensed users to identify the most reliable SUs. Finally, the performance of this method in enhancing the secrecy rate and transmission rate of the primary network is evaluated. We also compare its performance in terms of determining the selfish SUs with conventional relay selection technique.

4.5.1 System Model

A cognitive radio system with a single PU and N secondary IoT users (SUs) is considered. The SUs seek spectrum access from the PU. The PU's transmitter, PT, aims to send a private message to its legitimate receiver, which is denoted as PR.

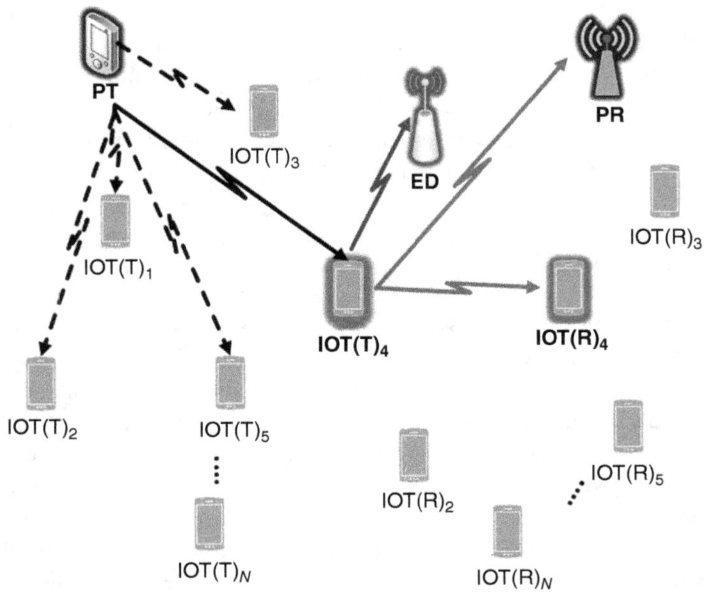

Figure 4.10 An example of spectrum leasing to secondary IoT devices in exchange for cooperative jamming and relaying. In this example, ST_4 is the selected trustable secondary IoT.

However, the PU's message can be eavesdropped by a passive malicious eavesdropper denoted as ED. Let us indicate the SUs' transmitter and receiver as ST_i and SR_i, respectively. Also, we assume that all the transceivers use a single antenna. This sample system model is depicted in Figure 4.10.

Slow Rayleigh fading with constant coefficients in each time slot is considered for all communication channels in the system. The channel coefficients are defined as follows: (i) h_{PS_i} denotes the channel parameters between the PU's transmitter and ith SU's transmitter; (ii) h_{S_iP}, h_{S_i}, and h_{S_iE} indicate the channel parameters between ith SU transmitter and the PU's receiver, the ith SU receiver, and the eavesdropper, respectively. In addition, we assume that there is no direct transmission between the PT and PR, as a realistic case where there is a strong shadowing or blockage in the primary's communication link. Moreover, we assume that the instantaneous channel state information (CSI) of the channels is available at the transmitters using common channel estimation methods [27, 31, 32]. The noise at the receivers is modeled by a complex and circularly symmetric Gaussian distribution (i.e. $n \sim CN(0, \sigma^2)$). It is assumed that the PU's transmitter sends data with constant power of P_P, and the maximum available energy at each SU is limited E_i^{max}.

To describe the cooperative spectrum leasing mechanism between the PUs and SUs, each time slot of the PU's access to the radio spectrum, denoted by T, is divided to three phases. Phase I, $(1-\alpha)T$, is the time portion in which the PU broadcasts its message to the SU. We assume that the PU and eavesdropper are not in the same range for direct transmitting and receiving. If the transmitted signal at PT is denoted by S, the received signal at SU ith transmitter during the first phase is

$$X_{ST_i} = \sqrt{P_P}\, h_{PS_i}\, S + n_{ST}\,. \tag{4.6}$$

In Eq. (4.6), n_{ST} is the noise at the SU's i transmitter. In phase II, a trustable SU is selected by the PU to relay its message and provide cooperative jamming. If the SU i is selected by the PU, this secondary allocates a cooperation power of P_{i_C} to forward the PU's message and also adds artificial noise to this message to confuse the eavesdropper using a jamming power of P_{i_J}. The proposed cooperative relaying and jamming can be deployed in any relaying schemes; here we assumed that a DF relaying mode is utilized at the SUs, where the selected SU can fully decode the received message from the PU. In addition, based on a common assumptions in works with a cooperative jamming mechanism for physical layer secrecy [32, 76, 77], it is assumed that the PU receiver has a priori knowledge about artificial noise added by the relay SU to extract the relayed message. One practical method to obtain this a priori knowledge of artificial noise is using a pseudorandom generator with finite states at the relay and legitimate primary receiver, while the state of the pseudorandom generator is sent to the primary receiver via a secure control channel. As a result, it is possible to enhance the transmission without any significant overhead. The extracted signal at the PU's receiver can be written as

$$X_{PR} = \sqrt{P_{i_C}}\, h_{S_iP}\hat{S} + n_{PR}\,, \tag{4.7}$$

where $\hat{S}$indicates the re-encoded signal after DF and n_{PR} is the noise at the PR. The received signal at the eavesdropper is calculated in Eq. (4.8):

$$X_{ED} = \sqrt{P_{i_C}}\, h_{S_iE}\hat{S} + \sqrt{P_{i_J}}\, h_{S_iE}Z + n_{ED}, \tag{4.8}$$

where Z~$CN(0, 1)$ is the artificial noise added by ST_i.

During phase II, the legitimate and selected SU finds time to borrow a portion of spectrum and sends its own message to the intended receiver with the power of P_{i_S}. Figure 4.11 illustrates the timing allocation for these three phases.

The received signal at the ith SU's receiver, during the third phase can be obtained as follows:

$$X_{S_iR} = \sqrt{P_{i_S}}\, h_{S_i}\hat{S}_i + n_{S_iR}\,, \tag{4.9}$$

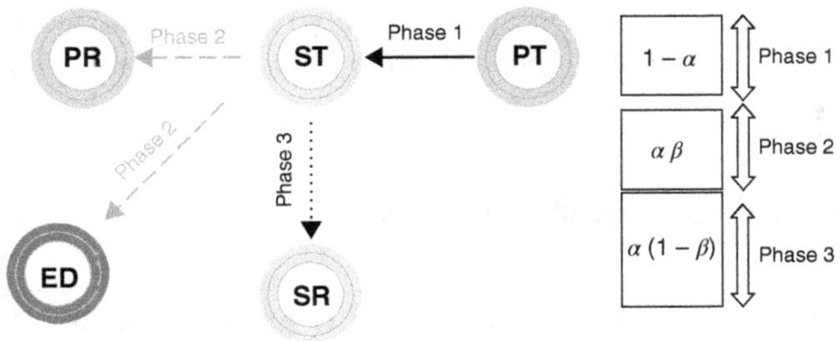

Figure 4.11 System model of proposed spectrum leasing scheme between the PU and the selected trustable SU.

where $n_{S_iR}\tilde{CN}(0,\sigma^2)$ models the noise at the SU's receiver, ST_i, and $\hat{S}_i$ denotes the transmitted signal by the ith SU's transmitter.

Given this spectrum leasing scheme and noting the definition of secrecy rate for a DF relay scheme as mentioned in Section 4.3.1, the secrecy rate of the PU, $R_{P_{SEC}}$, which is the communication rate at which no information is revealed to the eavesdropper, can be obtained as

$$R_{P_{SEC}} = [R_P - R_E]^+. \tag{4.10}$$

This secrecy rate is equal to the difference between the accumulated rate of the primary receiver (R_P) and the accumulated rate at the eavesdropper (R_E). $[x]^+$ denotes the max(x, 0). In a DF relaying mode, the rate of the primary receiver is calculated as the minimum of the two rates of relaying process. The first one is the rate between the primary transmitter and the selected SU's transmitter, R_{PS_i}, and the second one is the rate between the selected SU's transmitter and the primary receiver, R_{S_iP}; hence, $R_P = \min(R_{PS_i}, R_{S_iP})$ [60]. These two rates are defined in Eqs. (4.11) and (4.12).

$$R_{PS} = (1-\alpha)T\log_2\left(1 + \frac{P_P|h_{PS_i}|^2}{\sigma^2}\right). \tag{4.11}$$

$$R_{S_iP} = \alpha\beta T\log_2\left(1 + \frac{P_{i_C}|h_{S_iP}|^2}{\sigma^2}\right). \tag{4.12}$$

The information leakage at the eavesdropper can be formulated as

$$R_E = \alpha\beta T\log_2\left(1 + \frac{P_{i_C}|h_{S_iE}|^2}{\sigma^2 + P_{i_J}|h_{S_{iE}}|^2}\right) \tag{4.13}$$

Therefore, the secrecy rate can be obtained as

$$R_{P_{SEC}} = [\min(R_{PS}, R_{S_iP}) - R_E]^+. \tag{4.14}$$

To ensure achieving a nonzero secrecy rate at the PU's receiver, the conditions of $|h_{PS_i}|^2 > |h_{S_iE}|^2$ and $|h_{S_iP}|^2 > |h_{S_iE}|^2$ should be satisfied for the CSIs. There is no direct link between the PUs and the eavesdropper [78].

4.5.2 Proposed Reputation-based Game-Theoretic Model to Enhance the PU's Secrecy

Noting the higher priority of the PUs as the spectrum owner, we used a Stackelberg game model to model the interactions between the PUs and SUs. As defined in Section 4.4, the Stackelberg game is a type of sequential game, in which the game leader takes its strategy first, and then the game followers can observe the leader's action and respond by their best strategies. Therefore, the PU is considered as the game leader and the secondary IoT users are considered as the game followers.

In the proposed cooperative spectrum leasing scenario, the PU's objective is to select the most reliable SUs to relay its message to the destination since there is no direct link between the PU's transmitter and the receiver and also to enhance its secrecy rate using the cooperative jamming service provided by the secondary IoTs. One of the threats that the PU faces in such spectrum leasing scenario is that the SUs may refuse to forward the PU's message or refuse to add artificial noise to the message to hide it from the eavesdropper after being granted the spectrum. This is because the non-altruistic IoT devices would prefer to save up their limited energy to prolong their lifetime. The majority of the existing spectrum leasing models assume that the secondary IoT will assign an equal power to their own transmission as well as the cooperative task, including the cooperative jamming and cooperative relaying [27, 31–33]. However, this is an unrealistic assumption when dealing with self-interested users and can put the PUs at the risk of offering their spectrum while not receiving any service in return if there is no trust mechanism in place to enforce the secondary IoT to keep their commitments. Therefore, we develop a reputation-based mechanism to prevent the SUs from selfish misbehavior during the interactions with the PU and recognize the ones who have shown a selfish act in the past to abandon them from further interaction with the PU. This reputation mechanism has been designed based on the fact that the secondary IoT users will be selected to obtain the chance of spectrum access from the PU noting their cooperative behavior history. Otherwise, the secondary IoT users need to continuously maintain a good reputation to enhance the likelihood of being trusted by the PU. Therefore, although the message forwarding and cooperative jamming services come with the cost of consuming additional power, they will allocate a fair portion of their power to these services.

Utilizing the proposed reputation-based mechanism, the SUs face a trade-off between allocating their power to their own transmission with the objective of achieving a higher transmission rate, and hence having less remained power for relaying and jamming, $(P_{i_C} + P_{i_J})$ versus allocating enough power to these services so that they can sustain a competitive reputation for the next time slots when the PU selects the trustable SUs.

We formulate the proposed reputation-based Stackelberg game model by providing the strategy set and utility function of the game's leader and followers.

The PU's strategy set is as follows: (i) selecting the trustable SU and (ii) assigning an optimal time allocation for three transmission phases by determining the values of parameters α and β. Moreover, the PU can choose the ratio of the power assigned for cooperative jamming and cooperative relaying by the SU, denoted by $= \frac{P_{i_J}}{P_{i_C}}$. This statement does not counter with the assumption of having fully unrestricted SUs and does not enforce any constraints to the SU's preferences. Both jamming and relaying services requested by the PU occur during the second-phase $\alpha\beta T$; hence, the selected SU does not benefit from changing the ratio of the power spent on relaying or jamming (ρ) simply because its overall cost of service to the PU remains as $(P_{i_J} + P_{i_C})\alpha\beta T$.

4.5.2.1 Definition of Reputation

The reputation of each SU is defined as an accumulative factor that captures the difference between the power that each SU assigns to its transmission and the ones it allocates to the services requested by the PU. The reputation value increases when the SU consumes enough power for the cooperation and jamming. Vector $R^k(n) = (r_1(n), r_2(n), \ldots, r_N(n))$ shows the reputation of all SUs at time n that is monitored and updated by the PU k. The value of each reputation is in the range of (0, 1]. We consider two general scenarios to calculate the reputation of the SUs, named first-hand reputation and second-hand reputation as described in Sections 4.5.2.2 and 4.5.2.3, respectively.

4.5.2.2 First-hand Reputation

If PU k has a record of previous interaction with the ith SU, then the reputation of this SU at time (n) is updated as

$$r_i^k(n) = \min\left(r_i^k(n-1) + \eta_3 \ln\left(\epsilon_i^n\right), 1\right), \tag{4.15}$$

where η_3 is a predefined parameter. For each SU i, parameter ϵ_i^n is defined as a function of power values assigned for individual transmission and cooperative services as well as the channel coefficients of this user, as follows:

$$\epsilon_i^n = \frac{P_{i_J}|h_{S_iE}|^2 + P_{i_C}|h_{S_iP}|^2}{P_{I_S}|h_{S_i}|^2}. \tag{4.16}$$

The reason behind considering the CSIs in this definition is to assign a change of reputation proportional to actual channel condition that an SU is experiencing.

With this definition, the reputation of misbehaving SUs declines rapidly due to the characteristic of the ln function while it would take a fairly long time to restore the reputation in case of misbehaving. This mechanism can potentially prevent the SUs from oscillating between good behavior and misbehavior due to the long period of time required to restore their reputation.

4.5.2.3 Second-hand Reputation

In order to address the highly dynamic nature of IoT communication systems, where the users are usually mobile, we define a mechanism to compute second-hand reputation in cases where a PU and SU do not have a history of direct interactions with one another. If an SU recently moves to a proximity of a PU or there does not exist a prior record of interactions between a primary and secondary pair, then the PU can inquire the SU's reputation from its neighbor. In these cases, the second-hand reputation is defined as

$$r_i^k(n) = \frac{\sum\limits_{j \in P_i^k} r_i^j(n)}{\mid P_i^k \mid}, \tag{4.17}$$

where P_i^k denotes the set of PUs in proximity of the PU k.In case a secondary IoT user is totally new in a network, its reputation will be set to the minimum value.

4.5.2.4 Utility of the Game Players

The utility of the PU is defined as its secrecy rate as described in Eq. (4.18).

$$U_P(\alpha,\beta) = R_{P_{SEC}} = \left[\min\left(R_{PS}, R_{S_iP}\right) - R_E\right]^+. \tag{4.18}$$

The utility function for SU i is defined as

$$\begin{aligned} U_{S_i}(P_{i_S}, P_{i_C}, P_{i_J}) &= \alpha(1-\beta)T\log_2(1+SNR_i) \\ &- \eta_1\alpha(1-\beta)TP_{i_S} - \eta_2\alpha\beta T(P_{i_C}+P_{i_J}) \\ &+ \ln\left(P_{i_J}|h_{S_iE}|^2 + P_{i_C}|h_{S_iP}|^2 - P_{i_S}|h_{S_i}|^2 + 1\right), \end{aligned} \tag{4.19}$$

where SNR_i represents the signal-to-noise ratio at the received point for the ith SU. The logarithmic term in Eq. (4.19) is designed to capture the change in the SUs' reputation to motivate them to enhance their reputation. Hence, there is a trade-off in secondary utility between increasing its own transmission power to increase its transmission rate versus increasing the cooperation and jamming power to keep a good reputation for the next time slots. The limited energy available at the SUs can be formulated as

$$\alpha\beta TP_{i_C} + \alpha\beta TP_{i_J} + \alpha(1-\beta)TP_{i_S} \le \alpha TP_{i_{max}}, \tag{4.20}$$

where $P_{i_{max}}$ represents the maximum available power at each IoT, which is constant for all users. The self-interested IoT devices will indeed exhaust their maximum available power in each time slot; therefore, the above budget limitation inequality can be converted to $\alpha\beta TP_{i_C} + \alpha\beta TP_{i_J} + \alpha(1-\beta)TP_{i_S} = \alpha TP_{i_{max}}$.

Using this equality and the ratio of the jamming and cooperation power, we can simplify the utility of SU i as a function of a single strategy metric (e.g. P_{i_S}), as follows:

$$\begin{aligned} U_{S_i}(P_{i_S}) &= \alpha(1-\beta)T\log_2\left(1+\frac{P_{iS}|h_{S_i}|^2}{\sigma^2}\right) \\ &-\eta_1\alpha(1-\beta)TP_{i_S} - \eta_2\alpha\beta T\left(\frac{P_{i_{max}} - P_{i_S}(1-\beta)}{\beta(1+\rho)}\right) \\ &-\eta_2\alpha\beta T\rho\left(\frac{P_{i_{max}} - P_{i_S}(1-\beta)}{\beta(1+\rho)}\right) \\ &+\eta_4 \ln\left(P_{i_J}|h_{S_iE}|^2 + P_{i_C}|h_{S_iP}|^2 - P_{i_S}|h_{S_i}|^2 + 1\right). \end{aligned} \tag{4.21}$$

In this equation, η_1, η_2, and η_3 are three normalization factors to adjust the energy-related terms in this equation in the same range as the transmission rate.

4.5.2.5 Solution of the Proposed Game

Figure 4.12 demonstrates the interactions between players (PUs and SUs) for the proposed Stackelberg game. Finding the timing parameters (α, β), and selecting the reliable SU are the strategies of the PU, while the strategy of the SU is to set its transmission power P_{i_S}. A backward induction method can be utilized to find the solution of this game denoted by $\left(\alpha^*, \beta^*, P_{i_S}^*\right)$. In each iteration of the game, the selected SU observes the strategy of the PU (i.e. α, β), and responds to this by selecting its optimum power allocation to maximize its utility.

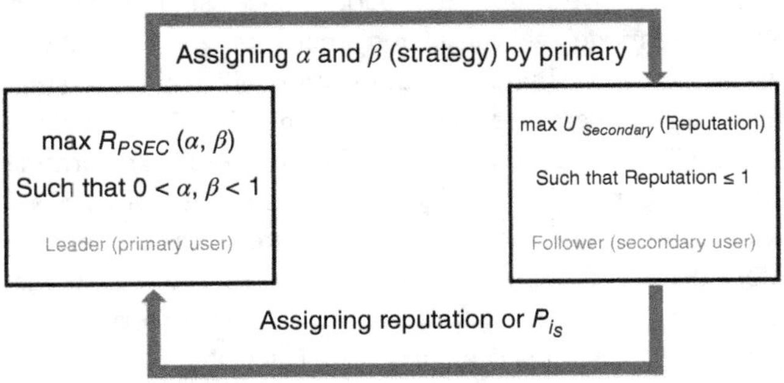

Figure 4.12 The proposed Stackelberg game model.

Theorem 4.1 The Stackelberg solution of the proposed game is the unique Nash equilibrium. ■

Proof: Based on the backward induction process, the PU first selects its strategy by maximizing its utility and derives the optimum value for α and β such that α^*, $\beta^* = \arg\max U_P(\alpha, \beta)$. Then, the selected SU observes this strategy and in response selects its best power allocation, which is the optimum value of P_{i_S}. Because the SU's utility is strictly concave in terms of P_{i_S} for any given value of the PU's strategy, the SU's response will be unique. Since the proposed game is a single-leader, single-follower one, the solution converges to the obtained unique solution. In addition, because the strategy of each user is determined based on a best response to the other opponent, the solution for this game is the Nash equilibrium.

Lemma 4.1 The utility function for each SU (4.21) is concave in terms of P_{i_S}. ■

Proof: To prove the concavity of secondary utility, the second derivative of Eq. (4.21) with respect to P_{i_s} is driven as

$$\frac{\partial^2 U_s(P_{i_S})}{\partial P_{i_S}^2} = \frac{-\alpha(1-\beta)T}{\ln 2} \frac{\left(\frac{|h_{S_i}|^2}{\sigma^2}\right)^2}{\left(1+\frac{P_{i_S}|h_{S_i}|^2}{\sigma^2}\right)^2} + \frac{-A^2}{(P_{i_C}|h_{S_iP}| + P_{i_J}|h_{S_iE}| - P_{i_S}|h_{S_i}| + 1)^2}, \tag{4.22}$$

where A^2 is equal to $\left(-\left(\frac{1-\beta}{\beta(1+\rho)}\right)|h_{S_iP}| - \left(\rho\frac{(1-\beta)}{\beta(1+\rho)}\right)|h_{S_iE}| - |h_{S_i}|\right)^2$. As seen in Eq. (4.22), U_{S_i} is strictly concave in terms of P_{i_S} for given values of α and β.

Since, the SU's utility is concave, it has a maximum value. Moreover, because of the Nash equilibrium solution for the primary, the final solution of this game will be the result to the best solution for each scenario [79, 80].

To demonstrate the effectiveness of the proposed reputation method to protect the PUs from selfish attacks by the SUs, we evaluated its performance using simulations in MATLAB. In this simulation model, we considered a network consisting of one PU and multiple SUs, where we assume that 70% of the SUs are selfish, meaning that they tend to consume most of their energy for their own transmission. The rest of the SUs are assumed as reliable IoT devices. A uniform distribution is used to determine the initial position of the SUs. We intentionally placed the selfish nodes much closer to the PU to evaluate the performance of our model in an extreme case. While the majority of relay selection methods prefer to choose the nodes close to the source, the proposed method considers the reliability factor of the users to protect the primary from the selfish attacks. The channel coefficients are calculated based on the distance between node i and node j, which can be defined as $h_{ij} \sim CN\left(0, d_{ij}^{-2}\right)$. For the sake of simplicity, we assume that each time slot (T) is equal to 1. The other values for the simulation are mentioned in Table 4.2.

Table 4.2 Simulation parameters.

Parameter	Value	Parameter	Value
η_1	0.004	Noise variance (σ^2)	1
η_2	0.0005	ρ	0.7
T	1	Primary transmit power	3 mW

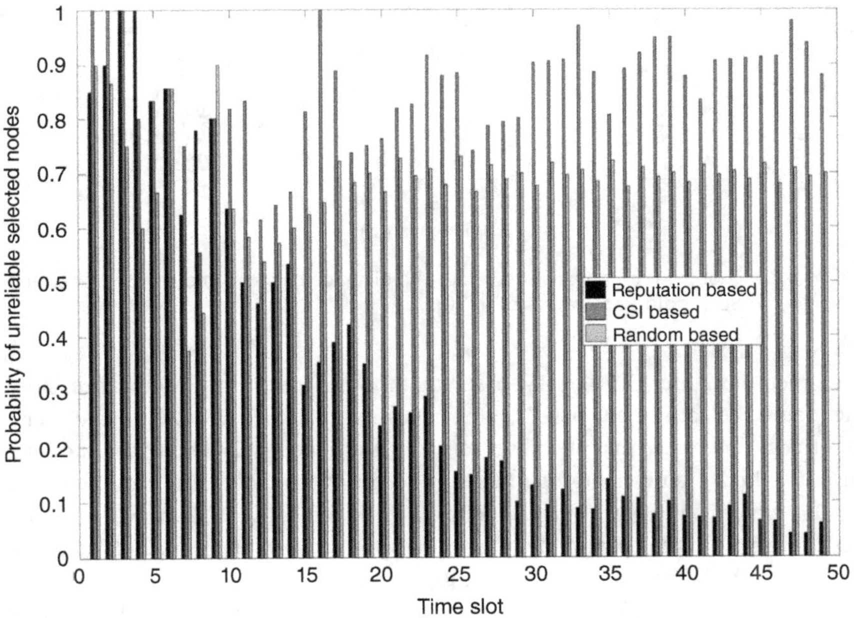

Figure 4.13 Probability of selecting unreliable nodes over time [70].

The value of η_1 is chosen to be larger than η_2 to enforce more strict punishment for the selfish behavior.

The performance of this reputation-based model is compared with two common relay selection methods of CSI-based (distance-based) and random selection. In all three methods, a Stackelberg game is utilized to extract the best solution for each run time. Moreover, we account for a real scenario where the cognitive selfish IoT users may show a normal behavior over several time slots and then act selfishly at some point. Therefore, we assumed that the selfish SUs show a misbehavior in 20% of the time slots. For comparison, we demonstrate the probability of selecting the unreliable nodes for three different selection methods in Figure 4.13.

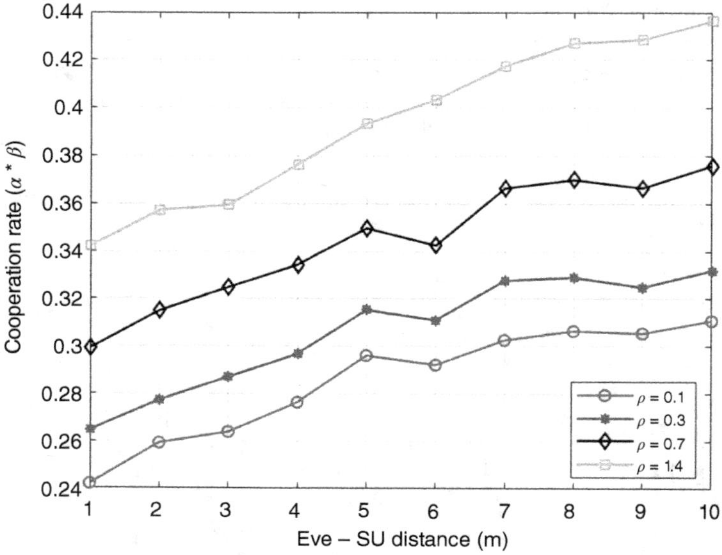

Figure 4.14 Cooperation phase ($\alpha\ \beta$) versus the distance between the eavesdropper and the selected SU for different values of ρ [70].

As seen in Figure 4.13, the CSI-based method has the worst performance, since the method selects the SU based on their location (distance to the PU), and therefore the selfish nodes that are closer to the PUs are often selected. In the random-based method, after converging time, the average of the probability is getting close to 70%, which is the percentage of selfish SUs in the network. However, the proposed method has the capability of identifying the unreliable users and categorizing them based on their channel condition and their reputation.

In order to show that the cooperation time for the PU is reinforced by this reputation-based method, we studied the time assigned to phase II, $\times\beta$. The allocated time is depicted versus the distance for different values of ρ. The increasing cooperation time in Figure 4.14 demonstrates that in scenarios where the link between the SU and eavesdropper is weaker, the PU is inclined to dedicate more time to cooperation.

4.6 Conclusion

In this chapter, we reviewed some of the security threats and countermeasures in IoT networks, including the information-secrecy-based solutions and cryptographic techniques. We discussed the potential impacts of these techniques, noting the unique characteristics of IoT networks such as the huge number of users, heterogeneity of the users in terms of their level of energy, computation, and cognition. Here, we focus on c-IoT networks where the users have the capability of reasoning and decision-making and develop a cooperative spectrum leasing method that can benefit both licensed and unlicensed users to establish reliable long-term interactions between them. The idea behind this model is to ensure the licensed users that they have the advantage of making several profits from leasing their spectrum while the misbehaviors from the unlicensed users are controlled and penalized by the system. The main contribution of this model is to come up with a reputation-based technique that allows the spectrum owner (PU) to observe the behavior of potentially selfish c-IoT devices in terms of their record of cooperative service. In the majority of previous spectrum sharing solutions, the SUs were assumed to be fully reliable. In emerging IoT networks, the SUs are cognitive and therefore self-interested; therefore, the spectrum owner can often suffer from the selfish attacks performed with these self-interested unlicensed users. Our proposed method using the Stackelberg game provides the spectrum owner the chance to only interact with the reliable C-IoT devices.

The future direction of this research is to establish a model for a long-term interaction among the licensed and unlicensed users considering the limited capabilities of IoT users and limiting the level of signaling and negotiation among the users.

A. Appendix

An initial version of this work was presented in IEEE Conference on Computer Communications Workshops (INFOCOM WKSHPS 2018) [70].

References

1 European Commission, "A spectrum roadmap for IoT opinion on the spectrum aspects of the internet-of-things (IoT) including M2M," Radio Spectrum Policy Group, European Commission, Directorate-General for Communications Networks, Content and Technology, 2013.

2 I. F. Akyildiz, W.-Y. Lee, M. C. Vuran, and S. Mohanty, "NeXt generation/dynamic spectrum access/cognitive radio wireless networks: A survey," *Computer Networks*, vol. 50, no. 13, pp. 2127–2159, 2006.

3 E. Z. Tragos, S. Zeadally, A. Fragkiadakis, and V. A. Siris, "Spectrum assignment in cognitive radio networks: A comprehensive survey," *IEEE Communications Surveys and Tutorials*, vol. 15, no. 3, pp. 1108–1135, 2013.

4 T. Yucek and H. Arslan, "A survey of spectrum sensing algorithms for cognitive radio applications," *IEEE Communications Surveys and Tutorials*, vol. 11, no. 1, pp. 116–130, 2009.

5 V. Patil and S. Patil, "A survey on spectrum sensing algorithms for cognitive radio," In *2016 International Conference on Advances in Human Machine Interaction (HMI)*, Doddaballapur, India, 3–5 March, 2016.

6 I. Christian, S. Moh, I. Chung, and J. Lee, "Spectrum mobility in cognitive radio networks," *IEEE Communications Magazine*, vol. 50, no. 6, pp. 114–121, 2012.

7 X. Huang, D. Lu, P. Li, and Y. Fang, "Coolest path: Spectrum mobility aware routing metrics in cognitive ad hoc networks," In *31st International Conference on Distributed Computing Systems (ICDCS)*, Minneapolis, MN, 20–24 June, 2011, IEEE, pp. 182–191.

8 I. F. Akyildiz, W.-Y. Lee, M. C. Vuran, and S. Mohanty, "A survey on spectrum management in cognitive radio networks," *IEEE Communications Magazine*, vol. 46, no. 4, pp. 40–48, 2008.

9 M. Zaeri Amirani, F. Afghah, and J. Ashdown, "Optimal relaying beamforming in MABC bidirectional cognitive radio networks in presence of interferers," *Springer International Journal of Wireless Information Networks*, vol. 25, no. 4, pp. 383–398, 2018.

10 B. Zhang and C. Kai, "Selective spectrum leasing in Internet of Things via nash bargaining solutions," In *Cloud Computing and Intelligent Systems (CCIS)*, Hangzhou, China, 30 October to 1 November, 2012.

11 O. Simeone, I. Stanojev, S. Savazzi, Y. Bar-Ness, U. Spagnolini, and R. Pickholtz, "Spectrum leasing to cooperating secondary ad hoc networks," *IEEE Journal on Selected Areas in Communications*, vol. 26, no. 1, pp. 203–213, 2008.

12 A. R. Korenda, M. Zaeri-Amirani, and F. Afghah, "A hierarchical stackelberg-coalition formation game theoretic framework for cooperative spectrum leasing," In *Information Sciences and Systems (CISS)*, Baltimore, MD, 22–24 March, 2017.

13 N. Namvar and F. Afghah, "Spectrum sharing in cooperative cognitive radio networks: A matching game framework," In *49th Annual Conference on Information Sciences and Systems (CISS)*, Baltimore, MD, 18–20 March, 2015.

14 F. Afghah, M. Costa, A. Razi, A. Abedi, and A. Ephremides, "A reputation-based Stackelberg game approach for spectrum sharing with cognitive cooperation," In *IEEE 52nd Annual Conference on Decision and Control (CDC)*, Florence, Italy, 10–13 December, 2013.

15 F. Afghah and A. Razi, "Cooperative spectrum leasing in cognitive radio networks," In *Wireless Research Collaboration Symposium (NWRCS)*, Idaho Falls, ID, 15–16 May, 2014.

16 M. Baidas, M. Afghah, and F. Afghah, "Distributed simultaneous wireless information and power transfer in multi-user amplify-and-forward ad-hoc wireless

networks," *International Journal of Communication Systems*, vol. 31, no. 1, p. e3411, 2017.

17 J.-M. Park, J. H. Reed, A. A. Beex, T. C. Clancy, V. Kumar, and B. Bahrak, "Security and enforcement in spectrum sharing," *Proceedings of the IEEE*, vol. 102, no. 3, pp. 270–281, 2014.

18 A. Attar, H. Tang, A. V. Vasilakos, F. Yu, and V. C. Leung, "A survey of security challenges in cognitive radio networks: Solutions and future research directions," *Proceedings of the IEEE*, vol. 100, no. 12, 2012.

19 A. Valehi and A. Razi, "Maximizing energy efficiency of cognitive wireless sensor networks with onstrained age of information," *IEEE Transactions on Cognitive Communications and Networking*, vol. 3, no. 4, pp. 643–654, 2017.

20 F. Afghah, B. Cambou, M. Abedini, and S. Zeadali, "A resistive random access memory physically unclonable functions (ReRAM PUF)-based approach to enhance authentication security in software defined wireless networks," *Springer International Journal of Wireless Information Networks*, vol. 52, no. 2, pp. 117–129, 2018.

21 A. G. Fragkiadakis, E. Z. Tragos, and I. G. Askoxylakis, "A survey on security threats and detection techniques in cognitive radio networks," *IEEE Communications Surveys and Tutorials*, vol. 15, no. 1, pp. 428–445, 2013.

22 F. R. Yu, H. Tang, M. Huang, Z. Li, and P. C. Mason, "Defense against spectrum sensing data falsification attacks in mobile ad hoc networks with cognitive radios," In *Military Communications Conference*, Boston, MA, 18–21 October, 2009.

23 V. Ramani and S. Sharma, "Cognitive radios: A survey on spectrum sensing, security and spectrum handoff," *China Communications*, vol. 14, no. 11, pp. 185–208, 2017.

24 J. Marinho, J. Granjal, and E. Monteiro, "A survey on security attacks and countermeasures with primary user detection in cognitive radio networks," *EURASIP Journal on Information Security*, vol. 2015, p. 4, 2015.

25 Y. Liang, H. V. Poor, and S. Shamai, "Information theoretic security," *Foundations and Trends in Communications and Information Theory* vol. 5, no. 4–5, pp. 355–580, 2009.

26 V.-D. Nguyen, T. Q. Duong, O. A. Dobre, and O.-S. Shin, "Joint information and jamming beamforming for secrecy rate maximization in cognitive radio networks," *IEEE Transactions on Information Forensics and Security*, vol. 11, pp. 2609–2623, 2016.

27 I. Stanojev and Y. Aylin, "Improving secrecy rate via spectrum leasing for friendly jamming," *IEEE Transactions on Wireless Communications* vol. 12, no. 1, pp. 134–145, 2013.

28 A. Korenda, F. Afghah, and B. Cambou, "A secret key generation scheme for internet of things using ternary-states ReRAM-based physical unclonable

functions," In *14th International Wireless Communications and Mobile Computing Conference (IWCMC)*, Limassol, Cyprus, 25–29 June, 2018.

29 A. Shamsoshoara, "Overview of Blakley's secret sharing scheme," arXiv preprint arXiv:1901.02802, Flagstaff, 2019.

30 A. Shamsoshoara, "Ring oscillator and its application as Physical Unclonable Function (PUF) for password management," arXiv preprint arXiv:1901.06733, 2019.

31 Y. Wu and K. R. Liu, "An information secrecy game in cognitive radio networks," *IEEE Transactions on Information Forensics and Security*, vol. 6, pp. 831–842, 2011.

32 A. Al-Talabani, Y. Deng, A. Nallanathan, and H. X. Nguyen, "Enhancing secrecy rate in cognitive radio networks via Stackelberg game," *IEEE Transactions on Communications*, vol. 64, no. 11, pp. 4764–4775, 2016.

33 C. A. Kamhoua, K. A. Kwiat, and J. S. Park, "*Surviving in cyberspace: A game theoretic approach*," Air Force Research Lab, New York, 2012.

34 A. Sheth, "Internet of things to smart IoT through semantic, cognitive, and perceptual computing," *IEEE Intelligent Systems*, vol. 2, pp. 108–112, 2016.

35 P. Vlacheas, R. Giaffreda, V. Stavroulaki, D. Kelaidonis, V. Foteinos, G. Poulios, P. Demestichas, A. Somov, A. R. Biswas, and K. Moessner, "Enabling smart cities through a cognitive management framework for the internet of things," *IEEE Communications Magazine*, vol. 51, no. 6, pp. 102–111, 2013.

36 A. Aijaz and A. H. Aghvami, "Cognitive machine-to-machine communications for internet-of-things: A protocol stack perspective," *IEEE Internet of Things Journal*, vol. 2, no. 2, pp. 103–112, 2015.

37 X. He and A. Yener, "Cooperation with an untrusted relay: A secrecy perspective," *IEEE Transactions on Information Theory*, vol. 56, no. 8, pp. 3807–3827, 2010.

38 M. Yuksel, X. Liu, and E. Erkip, "A secure communication game with a relay helping the eavesdropper," *IEEE Transactions on Information Forensics and Security*, vol. 6, no. 3, pp. 818–830, 2011.

39 R. Etkin, A. Parekh, and D. Tse, "Spectrum sharing for unlicensed bands," *IEEE Journal on Selected Areas in Communications*, vol. 25, no. 3, pp. 517–528, 2007.

40 S. Buchegger and J. Le Boudec, "Self-policing mobile ad hoc networks by reputation systems," *IEEE Communications Magazine*, vol. 43, no. 7, pp. 101–107, 2005.

41 Q. He, D. Wu, and P. Khosla, "SORI: A secure and objective reputation-based incentive scheme for ad-hoc networks," In *Wireless Communications and Networking Conference (WCNC)*, Atlanta, GA, 21–25 March, 2004.

42 S. Ganeriwal, L. K. Balzano, and M. B. Srivastava, "Reputation-based framework for high integrity sensor networks," *ACM Transactions on Sensor Networks (TOSN)*, vol. 4, no. 3, p. 15, 2008.

43 Y. Zhao, M. Song, and C. Xin, "A weighted cooperative spectrum sensing framework for infrastructure-based cognitive radio networks," *Computer Communications*, vol. 34, no. 12, pp. 1510–1517, 2011.

44 T. Zhang, R. Safavi-Naini, and Z. Li, "ReDiSen: Reputation-based secure cooperative sensing in distributed cognitive radio networks," In *International Conference on Communications ICC*, Budapest, Hungary, 9–13 June, 2013.

45 S. Soltanali, S. Pirahesh, S. Niksefat, and M. Sabaei, "An efficient scheme to motivate cooperation in mobile ad hoc networks," In *Third International Conference on Networking and Services (ICNS)*, Athens, Greece, 19–25 June, 2007.

46 Y. Zhang, L. Lazos, and W. Kozma, "AMD: Audit-based misbehavior detection in wireless ad hoc networks," *IEEE Transactions on Mobile Computing*, vol. 15, pp. 1893–1907, 2016.

47 J. Katz, A. J. Menezes, P. C. Van Oorschot, and S. A. Vanstone, *Handbook of applied cryptography*, CRC Press, 1996.

48 A. D. Wyner, "The wire-tap channel," *Bell System Technical Journal*, vol. 54, no. 8, pp. 1355–1387, 1975.

49 I. Csiszar and J. Korner, "Broadcast channels with confidential messages," *IEEE Transactions on Information Theory*, vol. 24, no. 3, pp. 339–348, 1978.

50 M. Bloch and J. Barros, *Physical-layer security: From information theory to security engineering*, Cambridge University Press, 2011.

51 M. Andersson, "*Coding for the wiretap channel*," KTH Royal Institute of Technology, 2011.

52 F. Oggier and B. Hassibi, "The secrecy capacity of the MIMO wiretap channel," In *International Symposium on Information Theory*, Toronto, Canada, 6–11 July, 2008, IEEE, pp. 524–528.

53 S. Shafiee, N. Liu, and S. Ulukus, "Towards the secrecy capacity of the Gaussian MIMO wire-tap channel: The 2-2-1 channel," arXiv preprint arXiv:0709.3541, 2007.

54 T. Liu and S. Shamai, "A note on the secrecy capacity of the multiple-antenna wiretap channel," *IEEE Transactions on Information Theory*, vol. 55, no. 6, pp. 2547–2553, 2009.

55 R. Bassily, E. Ekrem, X. He, E. Tekin, J. Xie, M. R. Bloch, S. Ulukus, and A. Yener, "Cooperative security at the physical layer: A summary of recent advances," *IEEE Signal Processing Magazine*, vol. 30, no. 5, pp. 16–28, 2013.

56 E. C. Van Der Meulen, "Three-terminal communication channels," *Advances in Applied Probability*, vol. 3, no. 1, pp. 120–154, 1971.

57 A. El Gamal and Y.-H. Kim, "*Network information theory*," Cambridge University Press, 2011.

58 G. Kramer, I. Marić, and R. D. Yates, "Cooperative communications," *Foundations and Trends in Networking*, vol. 1, no. 3–4, pp. 271–425, 2007.

59 A. Shamsoshoara and Y. Darmani, "Enhanced multi-route ad hoc on-demand distance vector routing," In *23rd Iranian Conference on Electrical Engineering (ICEE)*, Tehran, Iran, 10–14 May, 2015.

60 J. N. Laneman, D. N. Tse, and G. W. Wornell, "Cooperative diversity in wireless networks: Efficient protocols and outage behavior," *IEEE Transactions on Information Theory*, vol. 50, no. 12, pp. 3062–3080, 2004.

61 G. Kramer, M. Gastpar, and P. Gupta, "Cooperative strategies and capacity theorems for relay networks," *IEEE Transactions on Information Theory*, vol. 51, no. 9, pp. 3037–3063, 2005.

62 A. Shamsoshoara, M. Khaledi, F. Afghah, A. Razi, and J. Ashdown, "Distributed cooperative spectrum sharing in UAV networks using multi-agent reinforcement learning," In *IEEE Consumer Communications and Networking Conference (CCNC2019)*, Las Vegas, NV, 11–14 January, 2019.

63 A. El Gamal, M. Mohseni, and S. Zahedi, "Bounds on capacity and minimum energy-per-bit for AWGN relay channels," *IEEE Transactions on Information Theory*, vol. 52, no. 4, pp. 1545–1561, 2006.

64 L. Lai and H. El Gamal, "The relay eavesdropper channel: Cooperation for secrecy," *IEEE Transactions on Information Theory*, vol. 54, no. 9, pp. 4005–4019, 2008.

65 E. Tekin and A. Yener, "The general Gaussian multiple-access and two-way wiretap channels: Achievable rates and cooperative jamming," *IEEE Transactions on Information Theory*, vol. 54, no. 6, pp. 2735–2751, 2008.

66 X. Tang, R. Liu, P. Spasojevic, and H. V. Poor, "Interference-assisted secret communication," In *Information Theory Workshop, 2008 (ITW'08)*, Porto, Portugal, 5–8 May, 2008.

67 L. Dong, Z. Han, A. P. Petropulu, and H. V. Poor, "Improving wireless physical layer security via cooperating relays," *IEEE Transactions on Signal Processing*, vol. 58, no. 3, pp. 1875–1888, 2010.

68 T. T. Kim and H. V. Poor, "On the secure degrees of freedom of relaying with half-duplex feedback," *IEEE Transactions on Information Theory*, vol. 57, no. 1, pp. 291–302, 2011.

69 E. Ekrem and S. Ulukus, "Secrecy in cooperative relay broadcast channels," *IEEE Transactions on Information Theory*, vol. 57, no. 1, pp. 137–155, 2011.

70 F. Afghah, A. Shamsoshoara, L. Njilla, and C. Kamhoua, "A reputation-based stackelberg game model to enhance secrecy rate in spectrum leasing to selfish IoT devices," In *IEEE INFOCOM 2018 – IEEE Conference on Computer Communications Workshops (INFOCOM WKSHPS)*, Honolulu, HI, 15–19 April, 2018.

71 Z. Han, D. Niyato, W. Saad, T. Bacsar, and A. Hjorungnes, *Game theory in wireless and communication networks: Theory, models, and applications*, Cambridge University Press, 2012.

72 W. Saad, Z. Han, M. Debbah, A. Hjorungnes, and T. Basar, "Coalitional game theory for communication networks: A tutorial," arXiv preprint arXiv:0905.4057, 2009.

73 Z. Han and K. R. Liu, *Resource allocation for wireless networks: Basics, techniques, and applications*, Cambridge University Press, 2008.

74 S. P. Anderson and M. Engers, "Stackelberg versus Cournot oligopoly equilibrium," *International Journal of Industrial Organization*, vol. 10, no. 1, pp. 127–135, 1992.

75 R. Amir and I. Grilo, "Stackelberg versus Cournot equilibrium," *Games and Economic Behavior*, vols. 1–21, no. 1, p. 26, 1999.

76 H. Koorapaty, A. A. Hassan, and S. Chennakeshu, "Secure information transmission for mobile radio," *IEEE Communications Letters*, vol. 4, no. 2, pp. 52–55, 2000.

77 L. Dong, H. Yousefi'zadeh, and H. Jafarkhani, "Cooperative jamming and power allocation for wireless relay networks in presence of eavesdropper," In *International Conference on Communications (ICC)*, Kyoto, Japan, 5–9 June, 2011.

78 P. Z. Yang, J. Yuan, J. Chen, J. Wang, and J. Yang, "Analyzing amplify-and-forward and decode-and-forward cooperative strategies in Wyner's channel model," In *2009 IEEE Wireless Communications and Networking Conference*, Budapest, Hungary, 5–8 April, 2009.

79 R. Amir and I. Grilo, "Stackelberg versus Cournot equilibrium," *Games and Economic Behavior*, vol. 26, no. 1, pp. 1–21, 1999.

80 J. Zhang and Q. Zhang, "Stackelberg game for utility-based cooperative cognitiveradio networks," In *Proceedings of The enth ACM International Symposium on Mobile Ad Hoc Networking and Computing*, New Orleans, LA, May, 2009, ACM, pp. 23–32.

5

Adaptation and Deception in Adversarial Cyber Operations

George Cybenko

Dorothy and Walter Gramm Professor of Engineering, Dartmouth College, Hanover, NH, USA

5.1 Introduction

Computational game theory (CGT) has accomplished remarkable successes in recent years by learning to playing nontrivial games at superhuman levels. Those games include Chess, Go, Shogi, Poker, some forms of Poker, and multiple video games [1–6]. These successes are thoroughly documented in both the popular press and technical literature so there is no need to review them in any detail here.

Given these truly remarkable breakthroughs, one can easily be seduced into thinking that CGT can immediately deliver similar breakthroughs in other domains where adversarial exchanges and behaviors are encountered. Because game theory has been closely tied with economics, it is not surprising that new game models, analyses, and computations are being proposed for novel financial situations such as arise in high-frequency trading, for example [7].

Moreover, these kinds of game models and analyses are increasingly being proposed for military and security domains such as:

- cognitive radio and spectrum allocation [8]
- unmanned combat air vehicle (UCAV) engagements [9, 10]
- security patrols [11]
- cyber security [12–19]
- deception [20]

Modeling and Design of Secure Internet of Things, First Edition. Edited by Charles A. Kamhoua, Laurent L. Njilla, Alexander Kott and Sachin Shetty.

Published 2020 by John Wiley & Sons, Inc.

to list just a few examples among many others. The main points we wish to make here are:

- Real-world adversarial engagements will not typically lend themselves to traditional learning algorithms as have been recently used successfully in playing simple, well-defined games such as Chess, Go, Shogi, and Poker.
- Learning and rational play (in the sense of myopically optimizing play given the current model of an adversarial engagement) are mutually exclusive in the sense that to learn, one has to sacrifice optimal play, and to operate "rationally," one has to sacrifice learning.
- Whether an agent is fully rational or not, it is possible for an opponent to exploit the adaptive play of the agent to the opponent's advantage.
- There is a growing community in both economics and computer science that questions whether classical concepts of game "solution" such as based on Nash equilibria (NE), for example, are the right solution concepts in dynamically evolving situations.

As a result of these observations, extreme caution is advised when considering the use of currently popular and success machine learning algorithms that are based on well-defined board games. Going further, we believe that a whole new generation of machine learning and game playing logics are required to succeed in dynamic, real-world adversarial engagements.

In Section 5.2, we identify some critical characteristics and assumptions behind the recent successes of CGT and explore the extent to which those assumptions can hold in military and security domains as identified above. After discussion of those assumptions, in Section 5.3, we briefly review some aspects of real-world adversarial situations as opposed to stylized artificial games that question the validity of the assumptions required for current CGT to succeed. Section 5.4 compares several recent research results which at first blush might seem contradictory but are in fact not as we will see. However, they raise questions and concerns about what kinds of solutions are effective for real-world adversarial engagements. Finally, Section 5.5 is a summary with some suggestions for ways forward.

5.2 Key Aspects of Recent Game Theory Successes

In this section, we review recent reinforcement learning and related approaches used to produce high-performance game playing systems such as have been reported for Chess, Go, and other games mentioned above. However, note that a major recent Poker success does not evidently use reinforcement learning algorithms [1]. Nonetheless, some of the issues raised here will be applicable to those techniques as well.

At a high level of abstraction, reinforcement learning is an iterative algorithm for computing solutions to discrete stochastic control problems called Markov decision processes (MDP). As such they can be viewed as iterative approaches for computing solutions to the Bellman optimality equation that optimal solutions are well known to satisfy.

Numerous excellent tutorials and surveys on reinforcement learning exist [21, 22] and the reader unfamiliar with the concepts is encouraged to review those. An extension within the reinforcement learning framework is the so-called Q-Learning family of algorithms [23–26]. Q-Learning algorithms learn a different function through iteration than does classical reinforcement learning but both approaches compute approximate solutions to MDPs and both require the same basic assumptions to converge those solutions to optimal ones.

A key challenge in MDPs is the "curse of dimensionality" – a term coined by Richard Bellman to capture the fact that the size of memory required to effectively compute solutions to MDPs typically grows as the product of the process' state space sizes times the action space size. In problems where that memory is infeasibly large, function approximation methods based on neural network and deep neural network representations have been introduced resulting in approaches like "neurodynamic programming" and more recently "deep reinforcement learning" and "deep Q-Learning" [27]. These embellishments however are not relevant to the basic application of reinforcement learning to compute effective strategies for playing games. They are essentially implementation enhancements.

The baseline theory behind reinforcement learning requires the following nontrivial conditions for the learning algorithms to converge to optimal solutions:

1) a stationary (that is, statistically static) environment
2) an effective encoding of the states and actions
3) a model or at least some way to realize or simulate the result of performing an action in a particular state
4) a method for computing a reward or cost for applying a given action in a given state
5) infinite repeated plays of every state-action possibility

Under such conditions, the convergence to an optimal policy for controlling the MDP can be shown in the limit as the number of state-action updates goes to infinity.

The recent successes of CGT to games such as Go, Chess, and Shogi use reinforcement learning to learn effective play against a fixed opponent. That is, an existing program is used as an opponent and that opponent is "stationary" in the sense that the opponent is not learning or adapting during a reinforcement learning stage. Playing a game against a nonadaptive, stationary opponent (or opponents for that matter if there are multiple players) is a Markov decision process control problem and so reinforcement learning algorithms can be and are used.

In this fixed opponent setup, the opponent program is effectively a model of the adversary and large numbers of state-action updates can be performed because they are computer simulated. As a result, millions of games and resulting state-action pairs can be evaluated in hours or days, to the point where the strategy is effectively converged and/or the opponent is reliably beaten in full game simulations.

Once the MDP approximate solution is considered converged or at least adequate enough, the learned strategy then becomes the opponent and reinforcement learning is restarted with the learning program playing against the just-learned strategy which stops learning and adapting, thus resulting in a new MDP control problem to solve. This leap-frog approach to game play learning is typically called "self-play" and while some formal convergence analyses to accepted game-theoretic solutions such as NE are known [28], in practice, heuristics are used to evaluate effectiveness and decide when to stop the self-play iterations [6, 29].

5.3 Real-World Adversarial Situations

By "real-world" adversarial situations, we mean engagements between agents or players that occur in actual human activity, not artificial games created for entertainment, pedagogy, or research experimentation. Certainly, games such as Go, Chess, and Poker are adversarial but, as we argue below, they are completely defined in terms of rules and notions of winning, which is typically not the case in real-world situations.

For example, consider possible future adversarial encounters between or among autonomous vehicles (which could be air, space, cyber, marine, and/or land based) operating on behalf of different nations or groups. To define a game in order to use current CGT for modeling and computing effective strategies for such encounters, one would need to approximately know the maneuverability, range, weaponry, sensing capabilities, and objectives/utilities of all vehicles as well as models of their control logics. Such adversarial encounters are being contemplated already, so this kind of scenario is not entirely academic [30, 31].

Surely, it is possible to make educated guesses about various adversaries' autonomous vehicle capabilities and hypothesize a variety of utilities the adversaries might have for operating them. A simulation can then be created to model adversary autonomous vehicles. Such a simulation can then be used to power reinforcement learning as described above so that an effective, near-optimal control strategy is computed for the blue force vehicles, individually and even in collaborative swarms.

Self-play is no longer possible because of the asymmetry between blue and red force vehicles but that is not a problem. We can use reinforcement learning on the red vehicle control logic to improve their performance against blue forces which are fixed during that iteration. Then we fix the improved red force logic and learn an improved blue force logic against it using reinforcement learning.

We note that there have been relatively simple experiments done on optimizing predator strategies in predator–prey games [32]. In those models and associated experiments, when the prey control logic was held fixed, the predators dominated and were able to always capture the prey. However, when both predators and prey were allowed to improve their strategies through adaptation, the prey were able to evade the predators.

This is an important consideration for adversarial engagements of autonomous systems. If one side assumes their adversary cannot adapt through learning, they could be in for a major surprise! Alternatively, if they deploy a nonadaptive autonomous system for prolonged adversarial engagements without a learning or adaptive capability themselves, but their adversaries' have, another possible big surprise.

The point being that modeling adversaries and deploying nonadaptive autonomous systems based on optimal control strategies derived from adversary modeling can only be the first generation of autonomous systems. This seems to be where we are now.

We propose that the second generation of autonomous systems will have the capability to learn and adapt autonomously during a mission but assuming that adversary systems are not adapting, learning, or otherwise changing their control logics quickly enough to matter during a mission. One would think that such learning and adaptation could be done using existing reinforcement learning algorithms, but this is not the case.

This is because the current baseline reinforcement learning approaches require visiting all states and performing all actions infinitely often to achieve convergence to optimal solutions. Now the state of the system is not a simulation in a real engagement but rather the actual state that the blue and red force's vehicles or systems put themselves into. That is, the system state is governed by the actual system evolution and cannot in general be put into an arbitrary state for learning purposes.

By analogy, imagine playing Chess against an opponent and hoping to learn while playing. In general, you cannot force your opponent into a state that you desire simply to learn his playing style – you have to accept whatever state the opponent puts the game into after your move and proceed from there.

To make matters worse, not only can you not put the state of the game into any desired state, you are required to take suboptimal moves just to explore your own action space. That is, in classical economics terminology, you are not playing rationally with respect to the specific engagement utilities. In rational play, you would make the possibly myopic move specified by the learned value or Q-function being used.

It is, of course, possible to define a different game in which, in addition to the standard moves in the underlying game, you add "learning" moves that allow you to do suboptimal, irrational moves with respect to the underlying game but which have value in the more strategic game of learning about your opponent and thereby improving your own play against them. But this begs the question of how to value such "learning" moves, which if you have to learn and then optimize, you are back to the same problem, leading to a hierarchy of learning problems in theory in principle.

Perhaps in the future, the players in real-world adversarial engagements who can create and effectively use more levels of such a hierarchy will prevail in adaptive, autonomous systems arms races.

Returning to the assumptions required to make classical, vanilla reinforcement applicable to real-world engagements, we offer the following scorecard:

1) a stationary environment – these cannot be guaranteed if the adversary is adaptive as well
2) an effective encoding of the states and actions – these require exquisite knowledge of your own and your adversaries' systems
3) a model or at least some way to realize or simulate the result of performing an action in a particular state – this is accomplished in the course of the engagement
4) a method for computing a reward or cost for applying a given action in a given state – this is a basic advantage of reinforcement learning in that the rewards and costs are recursively computed by the iteration itself; however, it requires being able to compute or estimate values of some states concretely (such as a huge penalty for a system being destroyed and a huge reward for destroying or degrading an adversary system)
5) infinite repeated plays of every state-action possibility – as discussed above, this is difficult given the dynamics of the engagement environment and the look ahead capabilities of the system

In spite of the above challenges, it is important to note that there are many interesting and powerful algorithmic approaches for multi-agent reinforcement learning [15, 33–37]. However, all of these results specifically dictate what algorithms all agents are using which are patently not feasible in real-world adversarial engagements. I cannot require my adversary or competitor to use a specific algorithm for learning.

The same can be said about modeling "rational" best-response learning. In best-response and related adversarial learning procedures such as minimal-regret, all agents are being fully rational myopically in the sense that they select an action which is best in some way according to their current model and computed beliefs [28, 38, 39]. We will say more about this subject in the next section.

5.4 Paradoxes and Paradoxes Resolved

A series of results over the past 20 years have striking and sobering results and implications for learning in adversarial engagements.

First off, we quote a paper with an early and powerful multi-agent reinforcement result:

> "We design a multiagent Q-learning method under this framework, and prove that it converges to a Nash equilibrium under specified conditions." [35]

This and subsequent results in reinforcement learning for games appear to be quite powerful. But in the past two decades, contradicting results have appeared. One such result from a paper titled "On the impossibility of predicting the behavior of rational agents" by Foster and Young [40] is summarized by the authors as follows:

> "We demonstrate that there is an inherent tension between rationality and prediction when players are uncertain about their opponent's payoff functions. Specifically, there are games in which it is impossible for perfectly rational players to learn to predict the future behavior of their opponents (even approximately) no matter what learning rule they use." [40]

The point being, as we have earlier mentioned, that reinforcement learning and its variants are technically not rational in the economic meaning of the term, namely that an agent will optimize their moves to maximize their utility function. Reinforcement learning requires performing suboptimal moves infinitely in fact so suboptimal play is not transient but must persist for the duration of the interaction.

We can surely define a new game in which there is a family of new moves associated with precisely performing suboptimal moves with the goal of learning and thereby improving play in a nonstationary adversarial environment. But this approach begs the question of determining the utility of learning. That could be done through reinforcement learning but then this leads obviously to an infinite regress of learning to learn, to learn, to learn, and so on. Early work in this direction is presently ongoing [41].

There have been several recent results showing that simple games in which players adapt to each other or change the game to improve their performance, non-convergence, cyclic behavior, and even chaos are all possible [18, 38, 42, 43].

In addition, going back almost 30 years, economists have been questioning the concept of NE as being the right solution formalism [44]. Computer scientists and engineers like equilibria as solution concepts because they are static, well-defined, and either computable or approximatable. However, in recent years even computer scientists have started to explore solution concepts and game semantics in terms of play dynamics, not convergence to some abstract set of strategies for playing [45].

Finally, we point out other very recent results concerning the ability to exploit a player who is using an adaptive game playing or control strategy. In particular, an agent can formulate an MDP, solvable by dynamic programming or reinforcement learning, based on the adaptation logic their opponent is using whose solution offers them the ability to exploit the opponent. That is, one player, knowing how another is learning, can operate in an optimal way to elicit exploitable behaviors [39]. While at an intuitive level this is not surprising, such results provide formal, algorithmic foundations for how such exploitation can be performed.

5.5 Summary

This chapter has outlined a variety of challenges to designing systems to operate adaptively in adversarial environments where the adversaries themselves are adapting. This should be considered the norm when operating against capable adversaries in rapid tempo engagements such as might arise in autonomous cyber, IoT, unmanned vehicles, and other national security operations.

As a result of recent research, operations where all agents are learning and adapting can lead to exploitation, chaotic operations, and, ultimately, defeat. This suggests that new directions in machine learning for dynamic, adversarial operations should be explored in the near future.

Acknowledgments

This work was supported in part by US Army Research Office (ARO) MURI grant W911NF-13-1-0421 and the Air Force Research Laboratory Autonomous Capabilities Team (ACT3) effort. The views and opinions expressed in this article are solely those of the author and do not necessarily reflect the official policy or position of any agency of the U.S. government.

References

1 N. Brown and T. Sandholm, "Superhuman AI for heads-up no-limit poker: Libratus beats top professionals," *Science*, vol. **359**, no. 6374, pp. 418–424, 2018.

2 E. Gibney, "Deepmind algorithm beats people at classic video games," *Nature*, vol. **518**, no. 7540, pp. 465–466, 2015.

3 V. Mnih, K. Kavukcuoglu, D. Silver, A. Graves, I. Antonoglou, D. Wierstra, and M. Riedmiller, "Playing Atari with deep reinforcement learning," arXiv preprint arXiv:1312.5602, 2013.

4 V. Mnih, K. Kavukcuoglu, D. Silver, A.A. Rusu, J. Veness, M.G. Bellemare, A. Graves, M. Riedmiller, A. K. Fidjeland, G. Ostrovski, et al., "Human-level control through deep reinforcement learning," *Nature*, vol. **518**, no. 7540, p. 529, 2015.

5 D. Silver, A. Huang, C. J. Maddison, A. Guez, L. Sifre, G. Van Den Driessche, J. Schrit-Twieser, I. Antonoglou, V. Panneershelvam, M. Lanctot, et al., "Mastering the game of go with deep neural networks and tree search," *Nature*, vol. **529**, no. 7587, p. 484, 2016.

6 D. Silver, T. Hubert, J. Schrittwieser, I. Antonoglou, M. Lai, A. Guez, M. Lanctot, L. Sifre, D. Kumaran, T. Graepel, et al., "A general reinforcement learning algorithm that masters chess, shogi, and go through self-play," *Science*, vol. **362**, no. 6419, pp. 1140–1144, 2018.

7 A. Lachapelle, J.-M. Lasry, C.-A. Lehalle, and P.-L. Lions, "Efficiency of the price formation process in presence of high frequency participants: A mean field game analysis," *Mathematics and Financial Economics*, vol. **10**, no. 3, pp. 223–262, 2016.

8 J. Lunden, V. Koivunen, and H. V. Poor, "Spectrum exploration and exploitation for cognitive radio: Recent advances," *IEEE Signal Processing Magazine*, vol. **32**, no. 3, pp. 123–140, 2015.

9 A. M. Campbell, "Enabling tactical autonomy for unmanned surface vehicles in defensive swarm engagements," Ph.D. dissertation, Massachusetts Institute of Technology, 2018.

10 H. Duan, P. Li, and Y. Yu, "A predator-prey particle swarm optimization approach to multiple UCAV air combat modeled by dynamic game theory," *IEEE/CAA Journal of Automatica Sinica*, vol. **2**, no. 1, pp. 11–18, 2015.

11 Y. Qian, C. Zhang, B. Krishnamachari, and M. Tambe, "Restless poachers: Handling exploration-exploitation tradeoffs in security domains," In *Proceedings of the 2016 International Conference on Autonomous Agents and Multiagent Systems*. International Foundation for Autonomous Agents and Multiagent Systems, pp. 123–131, 2016.

12 J. T. House and G. Cybenko, "Hypergame theory applied to cyber attack and defense," In *Sensors, and Command, Control, Communications, and Intelligence (C3I) Technologies for Homeland Security and Homeland Defense IX*, vol. **7666**. International Society for Optics and Photonics, p. 766604, 2010.

13 J. T. House, "Exploiting exploration strategies in repeated normal form security games," In *Sensors, and Command, Control, Communications, and Intelligence (C3I) Technologies for Homeland Security and Homeland Defense XI*, vol. **8359**. International Society for Optics and Photonics, p. 835907, 2012.

14 Z. Hu, M. Zhu, P. Chen, and P. Liu, "On convergence rates of game theoretic reinforcement learning algorithms," arXiv preprint arXiv:1612.04724, 2016.

15 M. Panfili, A. Giuseppi, A. Fiaschetti, H. B. Al-Jibreen, A. Pietrabissa, and F. D. Priscoli, "A game-theoretical approach to cyber-security of critical infrastructures based on multiagent reinforcement learning," In *26th Mediterranean Conference on Control and Automation (MED)*, IEEE, pp. 460–465, 2018.

16 T. H. Nguyen, M. Wright, M. P. Wellman, and S. Singh, "Multistage attack graph security games: Heuristic strategies, with empirical game-theoretic analysis," *Security and Communication Networks*, vol. **2018**, 2018.

17 M. Rasouli, E. Miehling, and D. Teneketzis, "A scalable decomposition method for the dynamic defense of cyber networks," In *Game Theory for Security and Risk Management*. Springer, pp. 75–98, 2018.

18 P. Sweeney and G. Cybenko, "An analytic approach to cyber adversarial dynamics," In *Sensors, and Command, Control, Communications, and Intelligence (C3I) Technologies for Homeland Security and Homeland Defense XI*, vol. **8359**. International Society for Optics and Photonics, p. 835906, 2012.

19 T. Nguyen, M. P. Wellman, and S. Singh, "A Stackelberg game model for botnet data exfiltration," In *International Conference on Decision and Game Theory for Security*. Springer, pp. 151–170, 2017.

20 E. Al-Shaer, J. Wei, K. W. Hamlen, and C. Wang, "Modeling and analysis of deception games based on hypergame theory," In *Autonomous Cyber Deception*. Springer, pp. 49–74, 2019.

21 L. P. Kaelbling, M. L. Littman, and A. W. Moore, "Reinforcement learning: A survey," *Journal of Artificial Intelligence Research*, vol. **4**, pp. 237–285, 1996.

22 R. S. Sutton and A. G. Barto, *Reinforcement Learning: An Introduction*. MIT Press, 2018.

23 G. Cybenko, R. Gray, and K. Moizumi, "Q-learning: A tutorial and extensions," In *Mathematics of Neural Networks*. Springer, pp. 24–33, 1997.

24 J. Hu and M. P. Wellman, "Nash Q-learning for general-sum stochastic games," *Journal of Machine Learning Research*, vol. **4**, pp. 1039–1069, 2003.

25 J. N. Tsitsiklis, "Asynchronous stochastic approximation and Q-learning," *Machine Learning*, vol. **16**, no. 3, pp. 185–202, 1994.

26 C. J. Watkins and P. Dayan, "Q-learning," *Machine Learning*, vol. **8**, no. 3–4, pp. 279–292, 1992.

27 D. P. Bertsekas and J. N. Tsitsiklis, "Neuro-dynamic programming: an overview," In *Proceedings of the 34th IEEE Conference on Decision and Control*, vol. **1**. IEEE Publications, Piscataway, NJ, pp. 560–564, 1995.

28 V. Conitzer and T. Sandholm, "Awesome: A general multiagent learning algorithm that converges in self-play and learns a best response against stationary opponents," *Machine Learning*, vol. **67**, no. 1–2, pp. 23–43, 2007.

29 J. Heinrich and D. Silver, "Deep reinforcement learning from self-play in imperfect-information games," arXiv preprint arXiv:1603.01121, 2016.

30 "Defense Science Board Task Force on Counter Autonomy," https://www.acq.osd.mil/dsb/TORs/2018_TOR_CounterAutonomy18June2018.pdf, 2018.

31 Secretary of the Air Force Public Affairs, "AF releases vision for development of autonomous systems," https://www.af.mil/News/Article-Display/Article/601443/af-releases-vision-for-development-of-autonomous-systems/, June 23, 2015.

32 T. Haynes and S. Sen, "Evolving behavioral strategies in predators and prey," In *International Joint Conference on Artificial Intelligence*. Springer, pp. 113–126, 1995.

33 L. Busoniu, R. Babuska, and B. De Schutter, "A comprehensive survey of multiagent reinforcement learning," *IEEE Transactions on Systems, Man, and Cybernetics-Part C: Applications and Reviews*, vol. **38**, no. 2, pp. 156–172, 2008.

34 K. Tuyls and A. Nowé, "Evolutionary game theory and multiagent reinforcement learning," *The Knowledge Engineering Review*, vol. **20**, no. 1, pp. 63–90, 2005.

35 J. Hu, M. P. Wellman, "Multiagent reinforcement learning: Theoretical framework and an algorithm." In *ICML*, vol. **98**. Citeseer, pp. 242–250, 1998.

36 Y. Ishiwaka, T. Sato, and Y. Kakazu, "An approach to the pursuit problem on a heterogeneous multiagent system using reinforcement learning," *Robotics and Autonomous Systems*, vol. **43**, no. 4, pp. 245–256, 2003.

37 P. Stone and M. Veloso, "Multiagent systems: A survey from a machine learning perspective," *Autonomous Robots*, vol. **8**, no. 3, pp. 345–383, 2000.

38 Y. Sato, E. Akiyama, and J. D. Farmer, "Chaos in learning a simple two-person game," *Proceedings of the National Academy of Sciences*, vol. **99**, no. 7, pp. 4748–4751, 2002.

39 B. C. Schipper, "Dynamic exploitation of myopic best response," *Dynamic Games and Applications*, pp. 1–25, 2018.

40 D. P. Foster and H. P. Young, "On the impossibility of predicting the behavior of rational agents," *Proceedings of the National Academy of Sciences*, vol. **98**, no. 22, pp. 12848–12853, 2001.

41 O. Besbes, Y. Gur, and A. Zeevi, "Optimal exploration-exploitation in a multi-armed-bandit problem with non-stationary rewards," Available at *SSRN* 2436629, 2018.

42 T. Börgers and R. Sarin, "Learning through reinforcement and replicator dynamics," *Journal of Economic Theory*, vol. **77**, no. 1, pp. 1–14, 1997.

43 P. Mertikopoulos, C. Papadimitriou, and G. Piliouras, "Cycles in adversarial regularized learning," In *Proceedings of the Twenty-Ninth Annual ACM-SIAM Symposium on Discrete Algorithms*. SIAM, pp. 2703–2717, 2018.

44 D. M. Kreps, *Game Theory and Economic Modelling*. Oxford University Press, 1990.

45 C. Papadimitriou and G. Piliouras, "Game dynamics as the meaning of a game," *ACM SIGecom Exchanges*, vol. **16**, no. 2, pp. 53–63, June 2018.

6

On Development of a Game-Theoretic Model for Deception-Based Security

Satyaki Nan[1], Swastik Brahma[1], Charles A. Kamhoua[2], and Laurent L. Njilla[3]

[1] *Department of Computer Science, Tennessee State University, Nashville, TN, USA*
[2] *US Army Research Laboratory, Adelphi, MD, USA*
[3] *Cyber Assurance Branch, US Air Force Research Laboratory, Rome, NY, USA*

6.1 Introduction

Deception has found use in warfare and politics in the past. Recently, researchers and practitioners have started to investigate the use of deception for designing secure networks. Intuitively, it can become easier to launch a successful attack if the targeted system has no deception-based defense tactics, granting the attacker clear path and access to gain sensitive information as they explore the space of real systems. Without deception, all paths are real and the attacker can relatively easily gain information as they explore the space of real systems. Deception strategies can help to reduce the likelihood of an attacker's success and cost of defense from deception-less situations.

Attackers, in practice, have a certain level of knowledge of systems as they attempt to attack; deception exploits the knowledge, trust of attackers, and attempts to mislead the attacker toward expending their resources for attacking fake nodes and pursuing unsuccessful attack paths. As the total space of attacker's options grow large, if far more deceptions than actual attack paths are presented, the workload of the attacker for detecting real targets and differentiating between real and fake systems increases. At the same time, the defender who is able to observe the attacker's activity gains knowledge of the presence of an attacker and their characteristics and can direct further deception toward the attacker. Attackers can, in turn, seek to present different characteristics closely associated

Modeling and Design of Secure Internet of Things, First Edition. Edited by Charles A. Kamhoua, Laurent L. Njilla, Alexander Kott and Sachin Shetty.

Published 2020 by John Wiley & Sons, Inc.

with legitimate users in order to make it harder for the deception system to detect them, differentiate between attacker's and legitimate users, and increase the difficulty of designing defense tactics.

To model such strategic interactions between a system defender and an attacker, this chapter presents a game-theoretic model to analyze attack–defense scenarios that use fake nodes (computing devices) for deception under consideration of the system deploying defense resources to protect individual nodes in a cost-effective manner. The developed model has important applications in the Internet of Battlefield Things (IoBT). Considering that the nodes on which the defense resources are installed to be known to the attacker (thereby considering the attacker to be at an advantage), the dilemma for the system becomes determining which node to use for performing a certain computation task (with the nodes not chosen to perform the task used as fake nodes). Again, the dilemma for the attacker becomes choosing whether to attack nodes having defense resources (expecting the defender to use a node having defense resources) or whether to attack a node left unguarded without any defense resources installed. Specifically, the main contributions of the chapter are as follows:

- A game-theoretic model is presented to analyze attack–defense scenarios that use fake nodes (computing devices) for deception under consideration of the system deploying defense resources to protect individual nodes in a cost-effective manner.
- The Nash equilibrium (NE) of the game-theoretic model is characterized in terms of the attack–defense strategies that utilize deception under the consideration that the attacker is aware of system-level information (which is in sharp contrast to previous work [1]), leading to worst-case consideration of attack dynamics with the attacker being at an advantage.
- This chapter also presents a technique for strategic deception when a computing resource can be accessed via a network exhibiting a tree structure.
- Extensive simulation results are provided to gain insights into the deception model presented in this chapter.

The rest of the chapter is organized as follows: Section 6.2 describes the related research in the area of deception. Section 6.3 illustrates the game-theoretic model considering availability of two computing devices. Section 6.3.1 illustrates the game-theoretic model considering availability of three computing devices. Section 6.4 generalizes our result to the scenario where an arbitrary number of nodes are available for performing a computing task. Section 6.5 proposes a strategy for deception in a tree network. Section 6.6 provides simulation results to gain insights into the game-theoretic deception strategies presented in this chapter. Finally, Section 6.7 concludes the chapter.

6.2 Related Research

The use of game theory for security has been studied extensively [2–5]. Deception [6, 7] plays a critical role in many interactions in communication and network security. Deception has also been widely studied as a means to improve the protection of enterprise networks from potential hackers and intruders [8, 9]. For example, a game-theoretic model, called cheap-talk signaling game, captures the dynamic and information asymmetric nature of deceptive interaction. Another example includes [10] where a signaling game describes a game between two players (agents) with incomplete information: a sender S and a receiver R. The sender is aware of its own type T (assigned by nature: cooperative or deceptive) and sends messages to a receiver R. The receiver R, unaware of the sender's type, uses the received message to select an action leading to different payoffs depending on the outcomes of type, signal, and action. Schlenker et al. [1] presents the study of a signaling game where the defender signals to an adversary if a system is either real or fake when the adversary performs a scan.[1]

One technique for deception is honeypot [12–17], which aims to detect, deflect, or, in some manner, counteract attempts at unauthorized use of information systems. Schlenker et al. [1] uses honeypot to gather as much information as possible about the attack, which can be used to protect the real systems. The purpose of a honeypot is to determine tools and techniques used by hackers in order to prevent these hackers from gaining access to the real systems. A honeypot is involved in deception activities if its responses can deceive an attacker into thinking that the responses are from the real system. A honeypot can have a low, medium, or high interaction level, depending on the functionality and availability of system services [18, 19]. A low-interaction honeypot is one where an interaction with the system service is kept to a minimum [19]. However, attackers can relatively easily discover the presence of such a honeypot. A honeypot medium-interaction level implements a reduced set of services. The services at this level do, however, have some intelligence built into them so that they are not immediately recognizable as honeypots. High-interaction honeypots implement a full operating system with the entire range of system functionality available to the attacker [18–20]. This means that such a honeypot can potentially reveal information about the real system, which can be used by an attacker to launch attacks. However, honeypots implementing this level of interaction can hide their true nature from their attackers

---

1 It should be noted that, in this chapter, we consider the attacker to be aware of which nodes have the defense resource installed, unlike some previous work where the attacker has uncertainty regarding system-level information. Thus, in this chapter, we consider the attacker to be at an advantage, leading to consideration of worst-case attack dynamics, which is in contrast to scenarios considered by past literature (such as [1, 10, 11]).

better than low- and medium-interaction-level implementations. In fact, when the attacker discovers weaknesses in the honeypot design, the honeypot can be made to reconfigure itself and adapt to the situation.

A dynamic honeypot refers to a honeypot that learns from the type of attack in order to keep itself relevant. One of the most important features of such a honeypot is its ability to capture and record new attacks on the system [21]. In reconnaissance mode, the honeypot is used to determine the tools and techniques used by the attacker. This information is used to implement heuristics-based rules that can be applied in intrusion detection and prevention systems. The study of honeypot selection, where a defender chooses the properties of a network when the attacker can use probe actions to test the network, is presented in [5].

In the next section, we introduce our game-theoretic model for deception considering two computing devices.

6.3 Game-Theoretic Deception Model

This section presents our game-theoretic model for enabling strategic deception. Consider a system (say S), an attacker (say A), and two computing devices (nodes), say Node 1 and Node 2, with Node 1 (without loss of generality) having a defense resource installed in it (with the attacker being aware of which node has the defense resource installed). The cost for deploying the defense resource,[2] as well as the cost of attacking a node having a defense resource installed is, say C, with the cost of deploying a defense resource and the cost of attacking a node with a defense resource considered to be the same for simplicity of exposition. Further, suppose that the probability of successfully compromising a node having a defense resource installed is P_C. Also, suppose that the benefit of the system from performing the computation task successfully is B^S and the benefit of the attacker from a successful attack (which requires the attacker to successfully compromise the node being used by S) to be B^A. Table 6.1 shows the notations used. The payoff matrix of the game is shown in Table 6.2, where P_S is the probability of the defender[3] choosing to execute the task on a node having the defense resource (DR) installed (i.e. Node 1), and P_A is the probability of the attacker choosing to attack the node having the DR installed (i.e. Node 1).

Clearly, from the payoff matrix shown in Table 6.2, the game does not have a pure strategy NE unless $P_C \leq \frac{C}{B^A}$, which yields a trivial NE. In the next lemma, we present the mixed strategy NE of the game.

2 The defense resource, for example, can correspond to an anti-malware software.

3 We use the terms "defender" and "system" interchangeably in this chapter, unless otherwise mentioned explicitly.

Table 6.1 Notations used in analysis.

Notation	Description
S	System
A	Attacker
P_S	Probability of the system using a node having defense resource
P_A	Probability of the attacker attacking a node having defense resource
$1-P_S$	Probability of the system using a node without defense resource
$1-P_A$	Probability of the attacker using a node without defense resources
B^A	Benefit of the attacker from a successful attack
B^S	Benefit of the system from performing the computation task successfully
C	Cost of deploying/attacking the defense resource
P_C	Probability of successfully compromising a node having a defense resource

Table 6.2 Payoff matrix for system(S) and attacker (A).

System/Attacker	DR installed (P_A)	DR not installed ($1 - P_A$)
DR installed (P_S)	$((1-P_C)*B^S-C), P_C*B^A-C$	$B^S-C, 0$
DR not installed ($1-P_S$)	$B^S-C, -C$	C, B^A

P_s is the probability of the system choosing to execute a computation task on a node having the defense resource (DR) installed, and P_A is the probability of the attacker choosing to attack the node having the DR installed.

Lemma 6.1 *When two computing devices are available, with one device having a defense resource installed, the system (S) should use the device with the defense resource for performing a task with probability* $P_S = \left(\frac{1}{P_C+1}\right)\left(1+\frac{C}{B^A}\right)$, *while the attacker (A) attacks the device having the defense resources with probability* $P_A = \left(\frac{1}{P_C+1}\right)$.

Proof: The expected utility of S from choosing Node 1 for the computation task is

$$E_S^1 = ((1-P_c)*B^S-C)P_A + (B^S-C)(1-P_A) \tag{6.1}$$

Similarly, the expected utility of S from using Node 2 for the computation task is

$$E_S^2 = (B^S-C)P_A + (-C)(1-P_A) \tag{6.2}$$

Equating (6.1) and (6.2), we obtain the mixed strategy NE (for the attacker) as follows:

$$P_A^* = \frac{1}{1 + P_C} \tag{6.3}$$

Now, the expected utility of A from attacking Node 1 is

$$E_A^1 = \left(P_C * B^A - C\right)P_S + (-C)(1 - P_S) \tag{6.4}$$

Similarly, the expected utility of A from attacking Node 2 is

$$E_A^2 = (0)P_S + B^A\,(1 - P_S) \tag{6.5}$$

Equating (6.4) and (6.5), we obtain the mixed strategy NE (for the system) as follows:

$$P_S^* = \left(\frac{1}{P_C + 1}\right)\left(1 + \frac{C}{B^A}\right) \tag{6.6}$$

This proves the lemma. □

Next, to obtain further insights, we analyze the game considering the availability of three computing devices.

6.3.1 Deception Using Three Computing Devices

Consider a system (say S), an attacker (say A), and three computing devices (nodes), say Node 1, Node 2, and Node 3, with Node 1 (without loss of generality) having a defense resource installed in it, and with the attacker being aware of which node has the defense resource installed (see Figure 6.1). As before, consider C to be the cost of deploying the defense resource and P_C to be the probability with which the defense resource can be compromised. Suppose that P_S is the probability of the defender using Node 1 for the computation task, and that $1 - P_S$ is the probability of the defender using Node 2 or Node 3 (which do not have defense resources) for running the computation task. For simplicity of exposition, consider the defender to uniformly choose between Node 2 and Node 3 for running the task in the case where the defender refrains from using Node 1. Further, suppose that P_A is the probability of the attacker attacking Node 1 and that $1 - P_A$ is the probability of the attacker attacking either Node 2 or Node 3. Again, consider the attacker to uniformly choose between Node 2 and Node 3 in the case where the attacker does not attack Node 1. In the next lemma, we present the mixed strategy NE of the game when three computation devices are available.

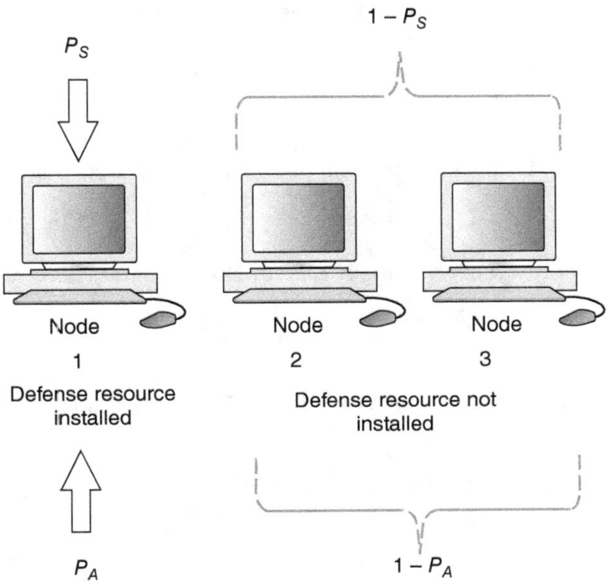

Figure 6.1 Three-node structure depicting the attack–defense strategies.

Lemma 6.2 *When three computing devices are available, with one device having a defense resource installed, the system (S) should use the devices with the defense resource for performing a task with probability* $P_S^* = \left(\frac{1}{2P_C + 1}\right)\left(1 + 2 * \frac{C}{B^A}\right)$, *while the attacker (A) attacks the device having the defense resource with probability* $P_A^* = \frac{1}{2P_C + 1}$.

Proof: The expected utility of *S* (system) from choosing Node 1 for performing the task is

$$E_S^1 = ((1 - P_C)B^S - C) * (P_A) + (B^S - C) * (1 - P_A) \tag{6.7}$$

Again, the expected utility of *S* (system) from not choosing Node 1 (i.e. running the task on Node 2 or Node 3) is

$$E_S^{-1} = (B^S - C)P_A + \binom{2}{1}(-C)\left(\frac{1}{4}\right)(1 - P_A) + \binom{2}{1}(B^S - C)\left(\frac{1}{4}\right)(1 - P_A) \tag{6.8}$$

Equating (6.7) and (6.8), we obtain the mixed strategy NE (for the attacker) as follows:

$$P_A^* = \frac{1}{2P_C + 1} \tag{6.9}$$

Now, the expected utility of A (attacker) from choosing Node 1 for attacking is

$$E_A^1 = \left(P_C * B^A - C\right) * P_S + (-C) * (1 - P_S) \tag{6.10}$$

Again, the expected utility of A from choosing Node 2 or Node 3 for attacking is

$$E_A^{-1} = (0)P_S + \left(B^A\right)(1 - P_S)\binom{2}{1}\left(\frac{1}{4}\right) + \left(B^A\right)(1 - P_S)\left(\frac{1}{4}\right)\binom{2}{1} \tag{6.11}$$

Equating (6.10) and (6.11), we get the mixed strategy NE (for the system) as follows:

$$P_S^* = \left(\frac{1}{2P_c + 1}\right)\left(1 + 2 * \frac{C}{B^A}\right) \tag{6.12}$$

This proves the lemma. □

In the next section, we generalize our result for the scenario where N computation devices are available for performing the task with some devices used as fake nodes.

6.4 Deception Using *N* Computing Devices

Consider a system (say S), an attacker (say A), and N computing devices (nodes), say Node 1 to Node N, with Node 1 (without loss of generality) having a defense resource installed in it (with the attacker being aware of which node has the defense resource installed), and Node 2 to Node N having no defense resources installed (see Figure 6.2). Consider the attacker to be aware of which node has the defense resource installed. Consider C to be the cost of deploying the defense resource and P_C to be the probability with which the defense resource can be compromised. Consider that P_S is the probability of the defender using Node 1 for the computation task and that $1 - P_S$ is the probability with which the defender executes the task on one of the remaining $N - 1$ nodes. Again, suppose that P_A is the probability of the attacker attacking Node 1 and that $1 - P_A$ is the probability with which the attacker attacks one of the remaining $N - 1$ nodes. For simplicity of exposition, consider the system (S) to uniformly choose between the $N - 1$ nodes in the case where Node 1 is not chosen for running the task. Similarly, also

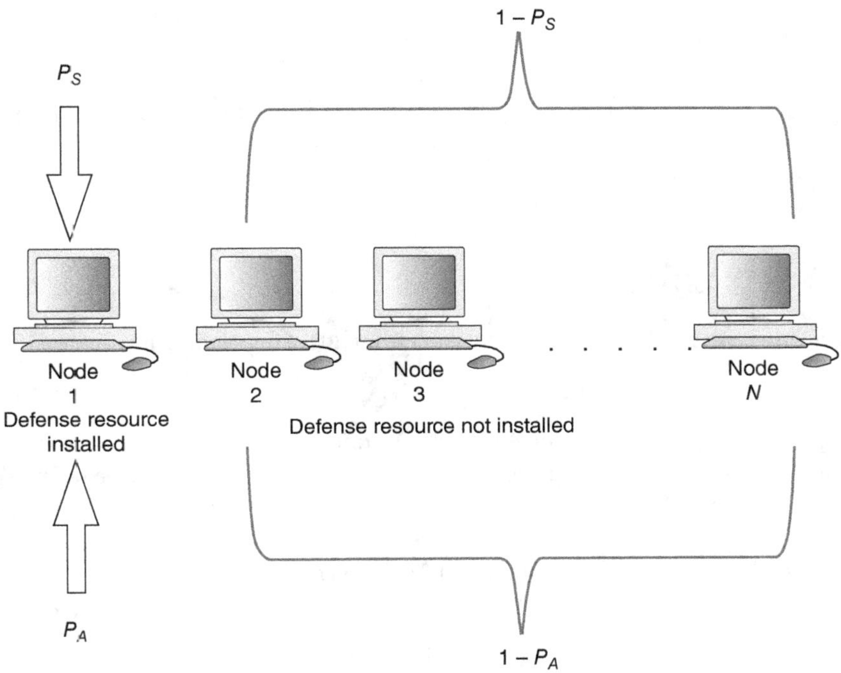

Figure 6.2 *N* node structure depicting the attack–defense strategies.

consider the attacker to uniformly choose between the $N-1$ nodes in the case where the attacker (A) does not attack Node 1.

In the next theorem, we present the NE of the aforementioned game.

Theorem 6.1 When N computing devices are available, with one device having a defense resource installed, the system should use the device with the defense resource for performing a task with probability $P_S = \left(\frac{1}{(N-1)P_C+1}\right)\left(1+(N-1)\frac{C}{B^A}\right)$, while the attacker attacks the device having the defense resource with probability $P_A = \frac{1}{1+(N-1)P_C}$.

Proof: The expected utility of S (system) from choosing Node 1 to perform the computation task is

$$E_s^1 = ((1-P_C)*B^S - C)P_A + (B^S - C)*(1-P_A) \tag{6.13}$$

Again, the excepted utility of S (system) from choosing a node from among Node 2 to Node N is

$$E_S^{-1} = (B^S - C)P_A + \left\{ \binom{N-1}{1} \left(\frac{1}{N-1}\right) \left(\frac{1}{N-1}\right)(-C) + \binom{N-1}{1} (B^S - C) \left(\frac{1}{N-1}\right) \left(\frac{N-2}{N-1}\right) \right\} * (1 - P_A) \tag{6.14}$$

Equating (6.13) and (6.14), we obtain the mixed strategy NE (for the attacker) as follows:

$$P_A^* = \frac{1}{1 + (N-1)P_C} \tag{6.15}$$

Now, the expected utility of A (attacker) from attacking Node 1 is

$$E_A^1 = \left(P_C B^A - C\right) * P_S + (-C)(1 - P_S) \tag{6.16}$$

Again, the expected utility of A (attacker) from attacking a node from among Node 2 to Node N, i.e. from among the $N-1$ nodes (which do not have defense resources installed) is

$$E_A^{-1} = (0)P_S + \binom{N-1}{1} \left(\frac{1}{N-1}\right) \left(\frac{1}{N-1}\right) (B^A)(1 - P_S) + \binom{N-1}{1} \left(\frac{1}{N-1}\right) \left(\frac{N-2}{N-1}\right) (B^A)(1 - P_S) \tag{6.17}$$

Equating (6.16) and (6.17), we obtain the mixed strategy NE (for the system) as follows:

$$P_S^* = \left(\frac{1}{(N-1)P_C + 1}\right) \left(1 + (N-1)\frac{C}{B^A}\right) \tag{6.18}$$

This proves the theorem. □

In the next section, we present a game-theoretic model for deception considering the availability of a computing device which is accessible via a network exhibiting a tree structure.

6.5 Deception in a Tree-Based Network Structure

This section presents our game-theoretic model for enabling strategic deception when a computation device is accessible via a network. Specifically, consider a computation device, referred to as the Target (T), which is accessible via a two-level tree network. This is shown in Figure 6.3. Suppose that every level of the tree

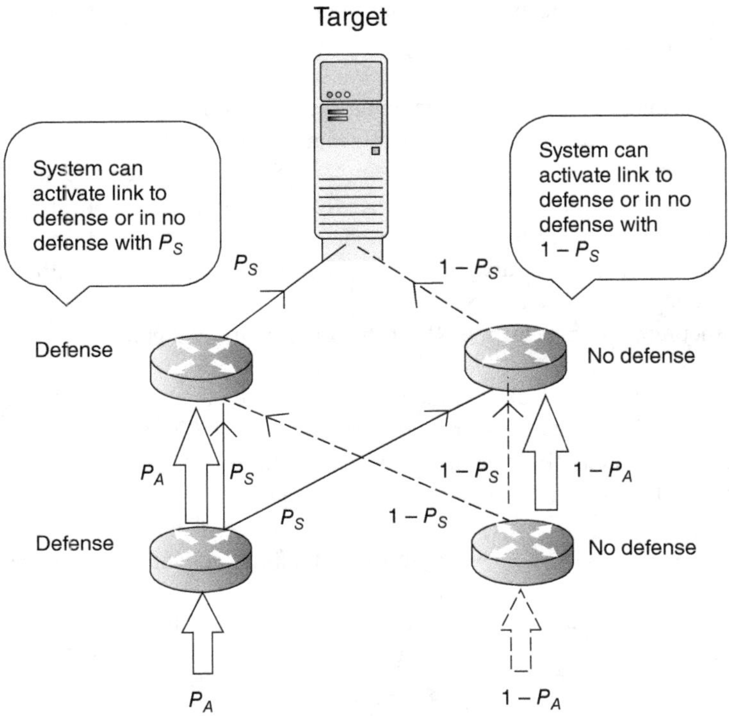

Figure 6.3 System model for deception in a tree-based network structure.

has a node on which a defense resource is installed, with the attacker, at every level, facing the dilemma of deciding whether to attack a node with a defense resource installed or whether to attack a node that does not have a defense resource. Again, the defender, at every level, faces the dilemma of deciding whether to activate links to the next level from a node that has a defense resource installed or whether to activate links to the next level from a node that does not have defense resources installed.

To analyze the problem, suppose that, at every level, the defender with probability P_S activates links arising from a node having a defense resource installed to enable going toward the target (and with probability $1 - P_S$ activates links originating from the defenseless node). Again, suppose that, at every level, the attacker attacks nodes with a defense resource installed with probability P_A (and attacks the node that does not have any defense resource with probability $1 - P_A$). Consider C to be the cost of deploying the defense resource at a node and P_C to be the probability with which the defense resource can be compromised. Suppose also that the benefit of the system from performing the computation task successfully at the target is B^S and the benefit of the attacker from a successful attack (which requires the attacker to successfully compromise the target) to be B^A.

In the aforementioned problem, it can be noted that activation of links from the defender's side acts as a decoy to deceive the attacker to use a route that does not lead to the target. In a strategic context, the defender would want to activate links in an optimal manner that prevents the attacker from reaching (and disrupting) the computation task executed in the target node. The strategy of the defender is, of course, dependent on the strategy of the attacker, requiring characterization of a game-theoretic deception strategy.

In the next theorem, we present the NE of the aforementioned game.

Theorem 6.2 When a computation device, referred to as the target T, is accessible via a two-level tree network, with every level of the tree having a defense resource installed, the system, at every level, should activate links from the node having the defense resource with probability $P_S = \sqrt{\frac{(B^A)^2 - C^2}{(B^A)^2 * (P_C)^2 + (B^A)^2 - 2P_c * B^A * C}}$ to go toward T (and activates links from the node without defense resource with probability $1 - P_S$). Again, the attacker, at every level, should attack the node having the defense resource installed with probability $P_A = \sqrt{\frac{2B^S * C - (B^S)^2}{\left((B^S)^2 * (1 - P_c)\right)^2 + 2B^S * C - 2P_c * B^s * C - 2(B^s)^2}}$ (and attacks the node without defense resource with probability $1 - P_A$).

Proof: The expected utility of the system (S) from activating links at every level from the node having the defense resource installed is

$$E_S^C = ((1 - P_C) * B^S - C)P_A\ ((1 - P_C) * B^S - C)P_A + (B^S - C) * (1 - P_A)\ (B^S - C) * (1 - P_A) \tag{6.19}$$

Again, the expected utility of the system from activating links at every level from the node that does not have the defense resource installed is

$$E_S^{-C} = (B^S - C)P_A * (B^S - C)P_A + (-C)(1 - P_A)(-C)(1 - P_A) \tag{6.20}$$

Equating (6.19) and (6.20), we obtain the mixed strategy NE (for the attacker) as follows:

$$P_A^* = \sqrt{\frac{2B^S * C - (B^S)^2}{\left((B^S)^2 * (1 - P_c)\right)^2 + 2B^S * C - 2P_c * B^s * C - 2(B^s)^2}} \tag{6.21}$$

The expected utility of the attacker (A) from attacking the node having the defense resource installed at every level is

$$E_A^C = (P_C * B^A - C)P_S \, (P_C * B^A - C)P_S + (C^2)(1 - P_S)(1 - P_S) \tag{6.22}$$

Similarly, the expected utility of the attacker (A) from attacking the node that does not have the defense resource installed at every level is

$$E_A^{-C} = 0 + B^A(1 - P_S) * B^A \, (1 - P_S) \tag{6.23}$$

Equating (6.22) and (6.23), we obtain the mixed strategy NE (for the system) as follows:

$$P_S^* = \sqrt{\frac{(B^A)^2 - C^2}{(B^A)^2 * (P_C)^2 + (B^A)^2 - 2 * P_c * B^A * C}} \tag{6.24}$$

This proves the theorem. □

In the next section, we provide simulation results.

6.6 Simulation Results

In this section, we provide simulation results to gain insights into the game-theoretic analysis presented in this chapter. Figure 6.4 shows how the system utility varies with the number of computing devices. For the figure, we consider $B^S = 5$, $B^A = 0.5$, $C = 2$, $P_C = 0.5$, and $P_A = 0.4$. In the figure, for a given number of computing devices on the X-axis, we show the utility of the system from adopting different strategies, with the white bar showing the system's utility from always running the computation task on the computing device having the defense resource installed, the grey bar showing the system's utility from always running the task on a device that does not have a defense resource installed (with a device chosen uniformly from among the devices that do not have a defense resource), and the black bar showing the system's utility from choosing a device to run the task based on the NE strategy proposed in Theorem 6.1. For each of the three strategies of the system, the attacker plays its best response. As can be seen from the figure, the system's utility is best when the system chooses a device based on our proposed NE strategy which shows the effectiveness of our approach. Also, note that as we increase the number of computing devices, the utility of the system increases for the case when the system chooses the device for performing the task based on the NE derived in this chapter. This is due to the fact that as we increase

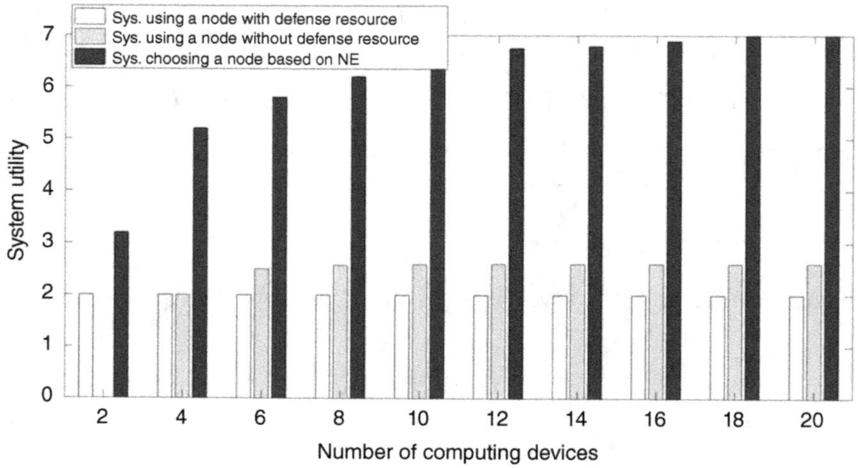

Figure 6.4 System utility versus number of computing devices. From the figure, we observe the advantage of using our proposed NE-based deception strategy.

the number of computing devices, it becomes harder for the attacker to know which node is being used by the system to execute the program and which are the fake nodes.[4]

Figure 6.5 shows the expected utility of the system and the attacker with varying number of computing devices when both choose their strategies based on NE. From the figure, we see that as the number of computing devices increases, expected utility of the attacker decreases while the expected utility of the system increases. This is because as the number of computing devices increases, it becomes harder for the attacker to decide where to attack. However, for the system, an increase in the number of computing devices is beneficial since more nodes enable the use a greater number of fake devices to deceive the attacker, thereby resulting in the increase of the system's expected utility.

To gain insights into the attack–defense dynamics based on the game-theoretic deception strategies derived in this chapter, we show how the NE-based strategy of the system (in Figure 6.6) and the attacker (in Figure 6.7) vary with increasing probability of compromising the defense resource (P_C). For the figures, $P_S = 0.5$, $C = 5$, and $B^A = 50$. As can be seen from Figure 6.6, the probability (P_S) with which the system uses the device with the defense resource installed for performing a task decreases as the probability with which the defense resource can be compromised (P_C) increases. This is because with increasing P_C, the defense resource

4 Note that as the number of computing devices increases, system's utility increases (but starts to saturate eventually).

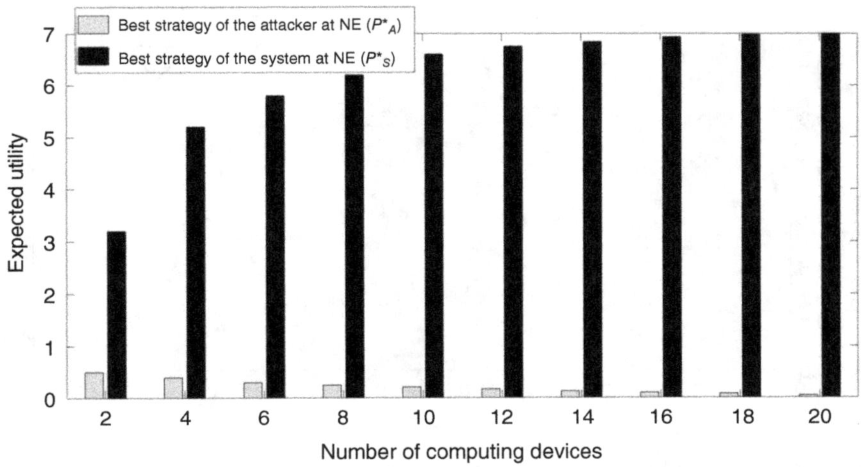

Figure 6.5 Expected utility of the system and the attacker with varying number of computing devices at Nash equilibrium (NE).

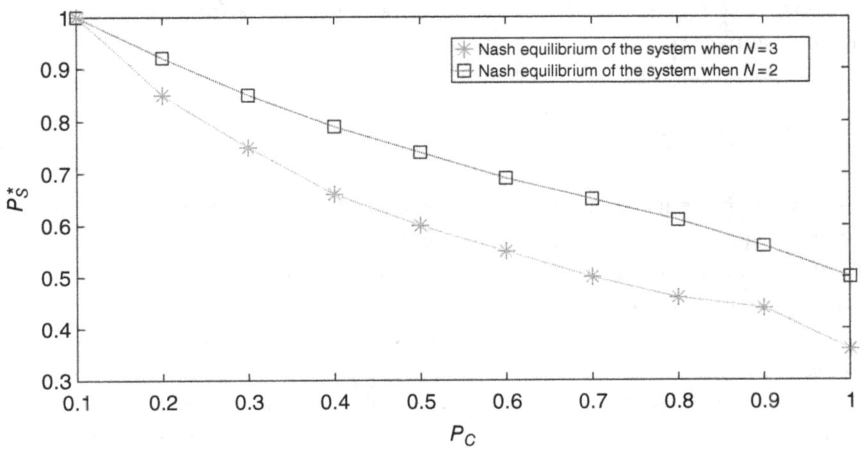

Figure 6.6 Optimal NE-based strategy (P_S^*) of the system with varying probability of compromising the defense resource (P_C).

can be more easily compromised by the attacker. Accordingly, as a best response, with increasing P_C, the attacker also decreases its probability P_A of attacking the device with the defense resource installed, which can be seen from Figure 6.7. Moreover, as the number of available computing devices increases, the system takes advantage of the increase in devices to try deceiving the attacker into

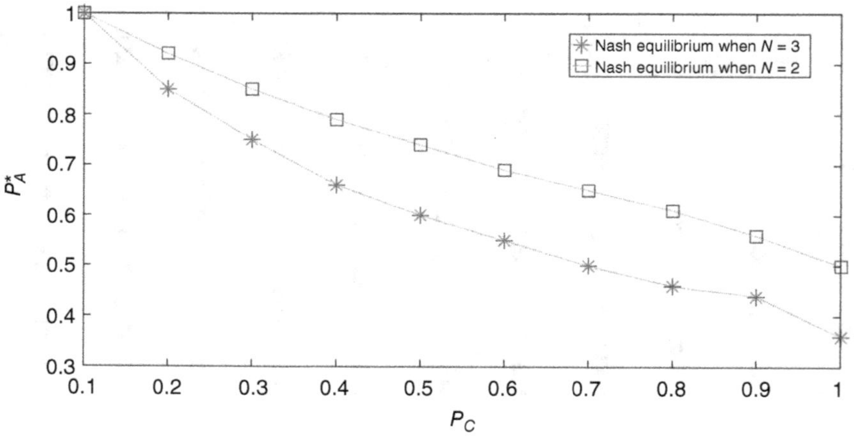

Figure 6.7 Optimal NE-based strategy (P_A^*) of the attacker with varying probability of compromising the defense resource (P_C).

attacking fake nodes by decreasing P_S, as can be seen from Figure 6.6. Accordingly, the attacker, at NE, also decreases its probability (P_A) of attacking the device with the defense resource installed.

6.7 Conclusion

This chapter presented a game-theoretic model for strategic deception, which has important applications in IoBT. The chapter developed a game-theoretic model for deception considering the availability of two as well as three computing devices. Our model was also generalized to a scenario where the defender can use any arbitrary number of fake nodes for deception. Our model considered the fact that defense resources may become compromised under an attack and showed that the defender, in a probabilistic manner, may utilize unprotected nodes for performing a computation while the attacker is deceived into attacking a node with defense resources installed. The chapter also presented a game-theoretic model for strategic deception when a computing device is available via a network exhibiting a tree structure. Numerical results were presented to provide insights into our strategic deception technique. In the future, we will consider scenarios where the system and the defender face different costs for deploying and attacking a defense resource, respectively, as well as consider optimization of the number of computing devices used by the system. Moreover, we will also seek to develop a Nudge theory-based approach for luring (nudging) attackers into attacking fake nodes to further enhance network security using deception.

References

1 Aaron Schlenker, Omkar Thakoor, Haifeng Xu, Fei Fang, Milind Tambe, Long Tran-Thanh, Phebe Vayanos, and Yevgeniy Vorobeychik. Deceiving Cyber Adversaries: A Game Theoretic Approach, In *Proceedings of the 17th International Conference on Autonomous Agents and MultiAgent Systems* (AAMAS '18). International Foundation for Autonomous Agents and Multiagent Systems, Richland, SC, pp. 892–900., July 2018.

2 Swastik Brahma, Kevin Kwiat, Pramod Kumar Varshney, and Charles Kamhoua. Diversity and System Security: A Game Theoretic Perspective, In IEEE Military Communications Conference, 2014.

3 Tansu Alpcan and Tamer Baar. *Network Security: A Decision and Game Theoretic Approach*, Cambridge University Press, 2010.

4 Aron Laszka, Yevgeniy Vorobeychik, and Xenofon D. Koutsoukos. Optimal Personalized Filtering Against Spear-Phishing Attacks. In AAAI Conference on Artificial Intelligence, pp. 958–964.

5 Aaron Schlenker, Haifeng Xu, Mina Guirguis, Chris Kiekintveld, Arunesh Sinha, Milind Tambe, Solomon Sonya, Darryl Balderas, and Noah Dunstatter. Don't Bury your Head in Warnings: A Game-Theoretic Approach for Intelligent Allocation of Cyber-security Alerts. In IJCAI 2017, pp. 381–387, 2017.

6 David Ettinger and Philippe Jehiel. A Theory of Deception. *American Economic Journal: Microeconomics*, vol. 2, no. 1, pp. 1–20, February 2010.

7 Mohammed H. Almeshekah and Eugene H. Spafford. Planning and integrating deception into computer security defenses. In Proceedings of the 2014 Workshop on New Security Paradigms Workshop. ACM, pp. 127–138, 2014.

8 Massimiliano Albanese, Ermanno Battista, and Sushil Jajodia. A deception based approach for defeating OS and service fingerprinting. In IEEE Conference on Communications and Network Security (CNS), Florence, Italy, pp. 317–325, 2015.

9 Massimiliano Albanese, Ermanno Battista, and Sushil Jajodia. Deceiving Attackers by Creating a Virtual Attack Surface. In S. Jajodia, V.S. Subrahmanian, V. Swarup, C. Wang, eds., *Cyber Deception*, Springer, pp. 169–201, 2016.

10 William Casey, Parisa Memarmoshref, Ansgar Kellner, Jose Andre Morales, and Bud Mishra. Identity Deception and Game Deterrence via Signaling Games. In 9th International Conference on Bio-inspired Information and Communications Technologies (BICT14), New York, December, 2015.

11 Christopher Kiekintveld, Viliam Lisy, and Radek Pibil. Game-Theoretic Foundations for the Strategic Use of Honeypots in Network Security. In S. Jajodia, P. Shakarian, V.S. Subrahmanian, V. Swarup, and C. Wang, eds., *Cyber Warfare*. Springer, p. 81101, 2015.

12 Brain Scottberg, William Yurcik, and David Doss, Internet honeypots: Protection or entrapment? Technology and Society International Symposium (ISTAS02), Raleigh, NC, pp. 387–391, 2002.

13 Cristine Hoepers, Klaus Steding-Jessen, and Antonio Montes. Honeynets Applied to the CSIRT Scenario. In Proceedings of the 15th Annual Computer Security Incident Handling Conference, Ontario, Canada, 2003.

14 Jeremy Briffaut, Jean-Francois Lalande, and Christian Toinard, Security and results of a largescale high-interaction honeypot. *Journal of Computers* vol. 4, no. 5, pp. 395–404, 2009.

15 Shiva Azadegan and Vanessa McKenna. Use of honeynets in computer security education. In Fourth Annual ACIS International Conference on Computer and Information Science, pp. 320–325, 2005.

16 Lance Spitzner. *Honeypots: Tracking Hackers*, Addison Wesley, 2002.

17 Honeypot (Computing). Available at: https://en.wikipedia.org/wiki/Honeypot_(computing).

18 Ronald M. Campbell, Keshnee Pad, and Themba Masombuka. A Survey of Honeypot Research: Trends and Opportunities, In The 10th International Conference for Internet Technology and Secured Transactions (ICITST), London, UK, 2015.

19 Radek Pibil, Viliam Lisy, Christopher Kiekintveld, Branislav Boansk, and Michal Pechoucek. Game theoretic model of strategic honeypot selection in computer networks. *Decision and Game Theory for Security*, vol. 7638, pp. 201–220, 2012.

20 Radek Pibil, Viliam Lisy, Christopher Kiekintveld, Branislav Bosansky, and Michal Pechoucek. Game Theoretic Model of Strategic Honeypot Allocation in Computer Networks. In Third International Conference on Decision and Game Theory for Security (GameSec), Budapest, Hungary, 2012.

21 Max H. Bazerman and Don A. Moore. *Judgment in Managerial Decision Making*, 6th edition, Wiley, 2005.

7

Deception for Cyber Adversaries

Status, Challenges, and Perspectives

Abdullah Alshammari[1], Danda B. Rawat[1], Moses Garuba[1], Charles A. Kamhoua[2], and Laurent L. Njilla[3]

[1] *Data Science and Cybersecurity Center (DSC2), Department of Electrical Engineering and Computer Science, Howard University, Washington, DC, USA*
[2] *US Army Research Laboratory, Adelphi, MD, USA*
[3] *Cyber Assurance Branch, US Air Force Research Laboratory, Rome, NY, USA*

7.1 Introduction

Most organizations store data in their servers, making them vulnerable to external attacks. The advent of corporate espionage, persistent threats, and other forms of attacks has caused many organizations to develop security systems that integrate deception. Because of the introduction of such systems, organizations have reported unsuccessful attacks on their network systems [1]. The introduction of deception security mechanisms is necessary to ensure that protection and controls are deployed to safeguard information in an organization. Deceit and negative information are important in enhancing the defensive capabilities of security systems in smart cities. Organizations could use the kill chain model to introduce more effective security controls at each stage to examine the effects of individual controls. Since security mechanisms cannot be achieved using a single strategy, most organizations have developed a mechanism that allows them to prevent potential incidents of attacks [1–3]. Effectively, this has been used to prevent unauthorized access of information stored in the system and hide the existence of data within security systems. The use of deception security systems is also considered important for slowing down attackers and significantly reduces the probability of adversaries attaining sensitive data from the organization. In effect, this ensures that valuable information is protected, and if it is acquired by hackers, that its

Modeling and Design of Secure Internet of Things, First Edition. Edited by Charles A. Kamhoua, Laurent L. Njilla, Alexander Kott and Sachin Shetty.

Published 2020 by John Wiley & Sons, Inc.

utility is reduced. As such, using negative information and deception enhances an organization's intrusion detection capabilities by deploying detection methods [2, 3]. These security controls protect the system from attacks by introducing doubt and risk into the data that can be obtained by an adversary. Furthermore, such deception techniques ensure that attackers are frustrated and unsuccessful in attaining their objectives. In some cases, the offensive strategies introduced enable the organization to establish a reliable network and system defense mechanisms.

This chapter analyzes the use of deception techniques in the establishment of security systems in an organization. To understand the wide use of deception techniques in the formulation of security systems, the chapter presents the background underpinning the use of deception techniques in security systems. In the digital world, deception techniques are considered pivotal in protecting systems and ensuring that the mechanisms set to protect systems are effective. Some deceptive techniques that have been adopted include honeypots and honey words. The intention of honeypots is to protect critical systems from external attacks [3]. Furthermore, the kill chain strategy can be introduced as a deception technique to make sure that system defenders can create a link comprising a number of steps, which have to be followed, thus identifying attacks being carried out on a system. This link enables an organization to break the chain of attack at any moment and protect critical assets within the system.

The motivation of introducing deception systems in organizations is to ensure that all potential attacks are monitored and prevented from occurring, as shown in Figure 7.1. Deception security mechanisms ensure that digitized data stored are not vulnerable to external attacks. This is mainly because such systems aim to prevent persistent attacks, corporate espionage, and other forms of attacks [1–3]. The motivation of system defenders is to ensure that deception techniques act as a preventative mechanism and provide a response. To ensure that infiltration is unsuccessful, planned actions are taken to mislead attackers. Deception security systems establish an approach using multiple security applications, such as detecting and stopping spam by analyzing existing malware. This is done in a manner to ensure that honeypots are put in place to secure databases, the motivation being to create a server honeypot and client honeypots that can be used as decoys to protect an organization from potential attacks [1, 3]. These honeypots create a detection mechanism, which provides an advantage to an organization, mainly because using them, an organization can assess hacking attempts trying to infiltrate the system and recommend practical solutions to ensure the system is not vulnerable. Furthermore, using deceptive security strategies is considered to be effective for detecting outbreaks of computer attacks in the industry. In this manner, attacks are prevented by deterring attackers. One advantage of using honeypots is that Wi-Fi providers, which use independent systems, can be disconnected and analyzed after successful attacks are staged without hindering the functionality of

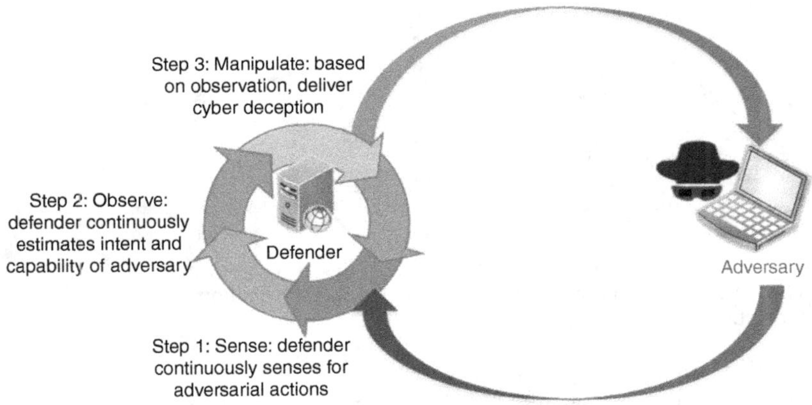

Figure 7.1 Typical lifecycle of cyber deception for cyber adversaries.

production systems [2, 3]. This ensures that hackers' information is analyzed within an organization and the tools can be used to prevent the latest attacks.

The rest of the chapter is organized as follows: Section 7.2 presents the background of the cyber-deception security. Section 7.3 reviews the game-theoretic models used to study cybersecurity. Section 7.4 shows game-theoretic approaches to defensive deception for cybersecurity. Section 7.5 presents some research directions followed by the summary in Section 7.6.

7.2 Background

Deceptive techniques have been used to solve human conflicts throughout history. Digital conflict employs deceptive techniques to ensure that the computing issues are resolved. Computer defenses that use deception are created to ensure that the deceptive element is introduced when creating a security structure [3]. There are unique advantages to using deception-based mechanisms to improve traditional computer security defenses. Figure 7.2 illustrates how deception can be used in the creation of security systems to ensure that multilayered protection is adopted in a given organization [1].

Traditionally, deception has been used to ensure that security systems are created in a manner that is more effective so system developers can exploit the biases of their adversaries. Secure digital systems incorporate deception to create effective security structures. Honeypots, deployed around 2000, were considered to be the first techniques [1] to bolster information security. Now, cyber defenses can be created using a broad range of deception techniques.

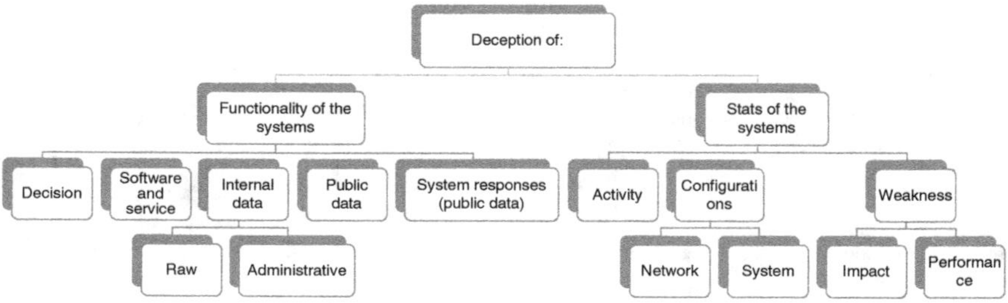

Figure 7.2 Typical cyber infrastructure components where deception can be integrated with.

Many current and common attacks use deceptive techniques to ensure their success. Traditionally, phishing techniques have often used deceptive techniques where users are deceived into clicking links that appear to be legitimate. This technique has been used to lure users into thinking they are dealing with legitimate websites while encouraging them to provide their credentials using a decoy to mimic legitimate links [3]. In 2003, various papers illustrated how honeypots could be used to enhance computer defenses. Following the idea of honeypots, components were introduced to ensure that values are attained by organizations to prevent them from being attacked by adversaries [4]. As a consequence, Honeywell words have been used to confuse attackers preventing them from cracking a stolen password file. Effectively, honeypots confuse attackers and enhance the security of critical systems by presenting honeypot links as real links. This approach of deception security originates from the development of anti-honeypot techniques, which use advanced methods to detect honeypots [5]. Several deception techniques have been proposed to enhance the overall security of computer systems. The main idea behind security systems is that deception intends to gain critical information on the organization. Figure 7.3 illustrates how organizations should develop a secure and integrated security system using deception tactics and shows the four major categories of protection mechanisms [5].

As discussed, as part of a successful deception security plan, multiple security applications have used honeypots to detect and stop spam. This result is achieved by analyzing malware and ensuring the security of the databases [4]. Specifically, honeypots have been used to entice attackers and potential hackers. The information gathered through honeypots can be used to prevent future attacks on servers as well as ensure that servers are not compromised. When security attacks occur, client accounts report to the server with an intention of preventing any potential attacks. Honeypots present a distinct advantage over traditional detection mechanisms since they generate less logging data that are not used in normal operations. This protection mechanism is preferred mainly because the rate of

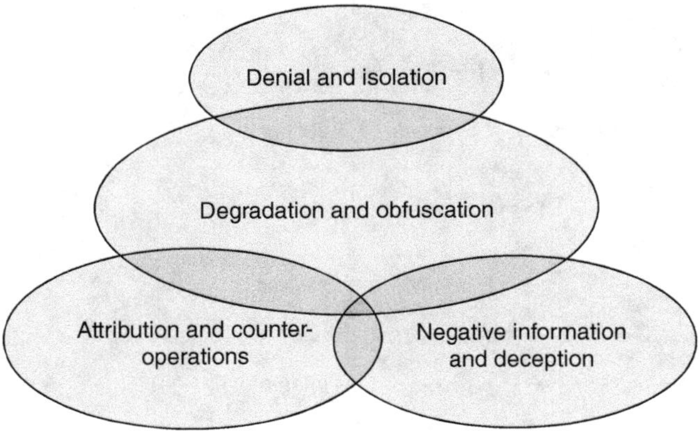

Figure 7.3 Taxonomy of information protection mechanisms using deception techniques.

false positives is low, thereby creating an effective security breach-prevention mechanism [2, 5]. The scheme of using deception as part of security systems is considered to be effective since it is also used to shadow systems for further investigations. Furthermore, honeypots are considered to be helpful in detecting outbreaks in industry-wide attacks [2, 5]. Traditional honeypots have been used to slow down attackers and deter them from infiltrating security systems. One form, stick honeypots, is specifically used to slow down attacks. The positive response to using deception systems is mainly due to them being totally independent systems, which can be analyzed after a successful attack. In the analysis process, operational and functional systems simplify the duty of forensic analysts who ensure that the security system is preserved [6]. Current research demonstrates that these tools have been previously used to defend computer systems against malware.

Deception can also be used in the cyber kill chain, which uses an intelligence-driven security model. This important model has been used to prevent attackers from succeeding by creating a sequence that makes it difficult for them to crack the system. Being able to break the chain at any step is an important strategy to prevent attackers from infiltrating the system. The cyber kill chain model is considered a good framework since it demonstrates the effectiveness of implementing deception at different levels within the chain. This is made possible through early detection of adversaries with an intention of creating deceptive strategies and learning about effective techniques and methods [6]. To a large extent, deception techniques can and have been used as an intelligence gathering method. The kill chain is also effective at detecting potential attacks aimed at reducing the ability of individual systems. There has been a general consensus that deception techniques can be used to prevent hackers from eliminating vulnerabilities in the system. The

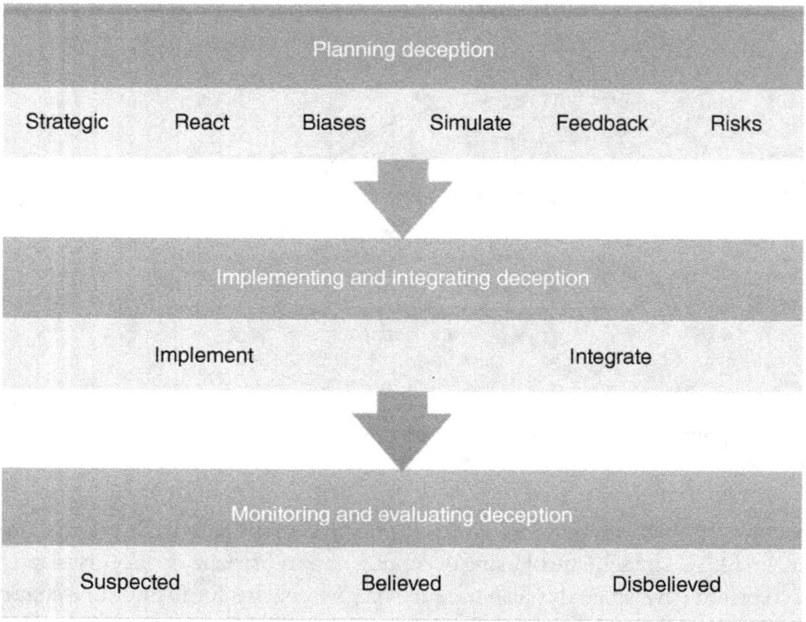

Figure 7.4 Framework to incorporate deception for cyber defense.

design of these systems is used to augment systems with enhanced deception-based techniques [5]. Organizational biases and cognitive biases affect the design of security systems that integrate deception. Such biases can be easily exploited, leading to the failure of security systems. Research on deception security measures is designed to ensure that cognitive biases are reduced, thus reducing the risk of prediction. The implementation of deception in the creation of security systems has been done to ensure that the organization can conduct a risk assessment before they recommend and adopt the most appropriate and effective strategies. In this way, the organization clearly understands and attains a degree of security focus [5, 7]. Technical preparation is done by creating a network map to identify the most appropriate security strategies. Traditional cyber-deception solutions were decentralized, even as primary components were separated. However, the cyber-deception solutions that have been adopted have been highly centralized to ensure that security tools are used together [7]. Critical assets must be identified in an organization to ensure that they are protected. Using bread crumbs to lure attackers by navigating them away from these assets ensures that believable decoys have been established. Figure 7.4 shows a typical framework to incorporate cyber deception for cyber defense.

Deception security systems have been deployed to create a strategic mechanism ensuring that the decoys established were relevant to the architecture of the organization. This deployment mechanism is a critical stage in protecting critical assets [8]. Periodic health checks are performed on the deception security mechanisms put in place as a system check. Penetration testing projects are also performed with an intention of applying prior knowledge of the system.

7.2.1 Design and Implementation of Cyber Deception

The design and the implementation of deception mechanisms in security systems of an organization vary according to the security challenges and needs of that organization. The applications also vary based on the way the organization intends to deploy security techniques. Threat assessment and security strategies are critical in ensuring that the security system is properly designed [9]. This is accomplished by using deception strategies that allow an organization to adopt effective mechanisms to address specific organizational concerns. Technical preparations are important to properly design the system and create modern cyber-deception solutions to develop existing security tools [8]. The implementation of these mechanisms involves the protection of critical assets through the placement of breadcrumbs. The deception security systems create a scenario where information is gathered in previous stages and a strategic creation protects critical systems within an organization to ensure that weaknesses in the system are properly understood.

7.2.2 Methods of Decision Process for Cyber Deception

It is necessary to be secretive when implementing any kind of deception to prevent the target from suspecting anything and failing to behave in the expected manner. Therefore, it is very important to decide on the people who should know about the use of deception as a cybersecurity measure within an organization. As the number of people with that knowledge increases, the likelihood of leakage to the target also increases. However, it is also risky informing only very few people as this could lead to mistakes. For example, during World War II in Operation Overload, the Allied forces used deception as a technique to prevent information from leaking to the enemy. However, they were so secretive that they fooled their commanders into making mistakes. Security is costly and generates substantial challenges, especially in the implementation of technology [10]. For instance, if a device can only be effective in a secretive environment, this prevents it from being applied widely. A device created with a set of operational modes that are kept secret can be relatively easy to distribute and use. The most critical issue in deception is identifying the elements that should be kept secret and the items that should be revealed. Revealing too many details undermines the effectiveness of the

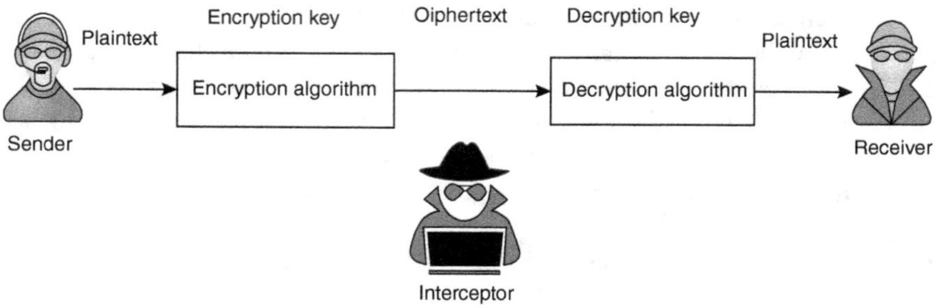

Figure 7.5 A typical cryptographic system.

deception, but withholding too much information makes the deception less effective as only a limited number of people can apply it. Even when an organization is using deception as a cybersecurity measure, it is paramount to be aware that its device designs and implementations will probably leak out. This recognition triggers the soundness for a cryptographic system to be based on the hypothesis that only the keys are kept secret [10]. The same principle is applied in various deception technologies. To avoid interception, a cryptographic system is used, as shown in Figure 7.5.

As illustrated, the sender creates a plain text to be sent to the receiver. The plaintext message is encrypted by an encryption algorithm to a ciphertext to make it impossible to intercept. When the text reaches the receiver, it is decrypted using a decryption algorithm to make it legible to the receiver. It is vital to take into account the deterrent impact of extensive published use of deception in cybersecurity. Potential attackers are deterred by high quality and widespread use of deceptions because they do not know how to differentiate deceptions from non-deceptions. The need for differentiation increases the workload for potential attackers, which most often keeps them away. It was this reasoning that informed the introduction of the Deception Toolkit (DTK) [10]. It was argued that use of DTK by the large number of people would deter attacks as attackers would be distracted by so many deception systems. Norbert Weiner's system theory describes many systems with feedback, where feedback information (possibly received from deception module) will help stabilize the system and provide resiliency, as shown in Figure 7.6.

The usual objective is to identify the difference between desired inputs and actual inputs, and adjust the outputs to restore stability. Expectations substantially influence the vulnerability of the target to deception. If the deception generates outcomes that are materially beyond the usual range of expectation, it becomes difficult for the target to ignore it. Deception with a common pattern lures the target to follow the expectations of that pattern in the absence of a reason not to do so.

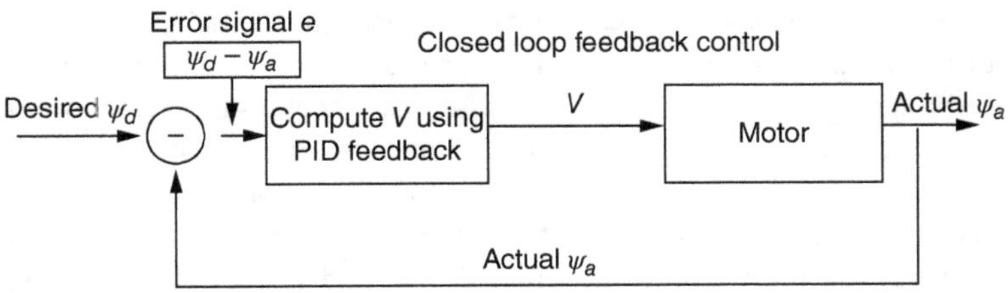

Figure 7.6 Control theory model where feedback helps stabilize the system.

The chances of drawing attention to the deception can be increased by changing the patterns from time to time. Of course, the target will examine a new pattern of deception more keenly and with higher skepticism [10]. Hence, the most effective strategy is to develop something that does not deviate much from expectations, but does not exactly resemble a previous pattern. It is not easy to predict how a specific target will react to a pattern of deception. Each target reacts differently to a variety of sorts and degrees of variations from expectations. Different factors influence the reactions of targets such as sensor limitations, focus of attention, cognitive structure, skills, training, intelligence, and predisposition. It is imperative to gather as much data as possible about the target to determine the nature and complexity of deception to use. Unfortunately, many systems do not have capacity to carry out a comprehensive systematic analysis of all situations as they arise. They only match common patterns quickly and then apply the important deep processing resources when the pattern matching does not reconcile the variations between expectations and interpretation. Consequently, the system can be deceived by avoiding its logical reasoning and preferring pattern matching.

Reduced thoughtfulness and increased automatic responses in people is caused by rushing, stress, vagueness, apathy, distraction, and fatigue. Similarly, human reasoning can be enhanced through steadiness, sobriety, certainty, attention, and alertness. The experiment of a valet parking person can be used to elaborate this explanation. A person who looked like a valet parking person was standing outside a pizza shop. It was found that wealthy customers were often giving them their car keys. The customers did not use reason to question the probably of the presence of a valet parking person at a common pizza place [10]. They did not question because their minds were on food and conversation, or possibly they just missed it. The results of this experiment can also be applied in computers. It is possible to restrain high-level cognitive functions by triggering a driver-level response to inbound information or force high-level attention to devastate reasoning by triggering conditions that produce increased processing regimens.

The nature of the interaction with targets in a deception environment can be described as recursive. The interaction involves presentation of observables from and to the targets. The impact of the deception can only be judged using the observables from the targets, and predetermined expectations affect how the observables are interpreted. In fact, the target can also use deception by presenting observables that are expected from them [10]. Therefore, the deception process is continuous and can be properly explained by a story about American and Russian ambassadors. The ambassadors met at a dinner party and started to discuss different issues as was the norm. A listening device had just been discovered and their conversation also touched on that subject. The Russian diplomat explained to their American counterpart that they had known about the device for some time. On the other hand, the American clarified that they knew that Russia had known about it for a quite a while. The Russian responded that they knew that the Americans knew they knew. The American confirmed that they knew the Russians knew that the Americans knew they knew. The same pattern of conversation went on for a while until the American exclaimed that they did not know that.

Dealing with a recursive situation can be difficult as it requires characterizing what happens at a single level and incorporating the links to the recursion. It is impossible to directly make desired and significant modifications without being discovered. The most effective strategy is to apply an indirect approach that facilitates changes that generate the desired effects. The approach should not only be executed in unexpected manner but also indirectly. For example, when deception is used in a complex system with multiple users, it is not mandatory for all participants to be affected such that the entire system behaves differently. The members can be categorized into four groups: zealots in favor, zealots opposed, neutral parties, and willing participants [10]. A small group of zealots will oppose an idea trying to prevent it, fearing its controversy. Thus, small deceptions should be used to build larger deceptions. This is the same method commonly used in a "big con" plan. The perpetrator identifies a victim, strives to gain their confidence, and then shows the victim the money. They tell the victim a pleasant story that is supported by a sample return on investment. They calculate the benefits, and the victim sends for more money, which is all is taken away. The perpetrator pays off the victim to keep them quiet. Out of all these acts, only the first one does not involve deception. It is especially interesting that this common deception sequence is very complicated, but it is very reliable in producing results. There are people who have perfected its use: they know how to maneuver every stage and they have techniques for limiting damage if something does not go as planned [10]. They are very flexible to accommodate variations to keep the target engaged in the activity.

It is difficult to understand the level of intelligence required to successfully implement deception in cybersecurity. First, the target can also use deception to mislead the attacker's intelligence efforts. Second, there are seemingly minor items that can have a massive impact on the ability to understand and foretell the behavior of the target. Nonetheless, intelligence is very important in the success of deception.

This intelligence is not only applied to the target but also in anticipating and constraining their behavioral patterns. The computer and software can theoretically be used to predict exact behavior through comprehensive design knowledge. Complexity is driven up by the use of large and complicated mechanisms and it is not easy to obtain data on specific mechanisms. Generic deceptions are effective in detecting a wide range of attacks, but there is always an attack that goes unnoticed, either by design or by accident. The aim of deceptions in an imperfect knowledge environment is to augment the odds [10]. By answering the question about which techniques increase or decrease the odds in a specific situation, it is possible to identify deceptions that drive up the computational complexity of differentiating between deceptions and non-deceptions for large categories of situations.

7.2.3 Game-Theoretic Models

In this section, we discuss some of the most common game-theoretic models used in defensive deception for cybersecurity. Game theory is the study of the interactions among rational participants in a complete analysis.

7.2.3.1 Signaling Games

The signaling game is a game theory model characterized by an interaction between two players with incomplete information. In a signaling game, the players are categorized either as the sender or the receiver, depending on their role [11]. Signaling game theory operates under the assumption that the sender has unlimited access to information that is unknown to the target/receiver. The payoff function of the target/receiver is determined by the attacker/sender and is known to both players [12]. In the signaling game, the outcomes account for the costs and benefits associated with the participants and the outcomes of the analysis of the interaction. It also serves as a lifesaver in situations when one actor is unaware of the kind of individual they are tackling, particularly in the presence of imperfect information. Game theory indicates the finest course of action that should be considered based on the notions of the target/receiver. The sender releases the signal to the receiver for analysis and makes decisions on what to do. A good example of a signaling game between two players is one in which one player is the employer and the other is the job seeker. The employer is the sender of signal while the job seeker is the receiver. There are two extremes of jobs seekers: the good one and the bad one. However, the employer does have an idea the kind of a job seeker they are dealing with. They only know that 40% of job seekers in the market are good and 60% are bad. Fundamentally, the employer is restricted to two actions: hire and do not hire [13]. Lack of knowledge about the kind of job applicant that the employer is dealing with compels them to develop techniques to separate the candidates into good and bad ones. Signaling games have two types of equilibrium: pooling and separating. Pooling equilibrium is attained when every kind of sender sends the same message regardless of their type. Separating equilibrium is attained when every sender sends a message that corresponds with that type. It is difficult to differentiate the two types when the signal is the same. The threshold level of probability based on prior belief is determined using the payoff to all situations. The determined probability with respect to the threshold is denied for the optimal action for the employer [13].

In the game from the defender's side, the decision will be made by the fake avatar to either raise the alert or not. The reason for the employment of the fake avatar is to deal with the external user. Internal users are aware of the presence of a fake avatar and all the necessary details. The sender acts as the defender by utilizing the services of the fake avatar and the receiver is an outsider, producing two types of parties: a normal user and an attacker. At the same time, the avatar does not know the kind of external user they are dealing with and this creates a gap that can be utilized by the external user to send a suspicious or non-suspicious signal [13]. The defender will decide on whether to raise the alert or not.

Traditional game theory can also be used to explain the use of deception in cybersecurity. It represents a two-player game whose objective is to locate and destroy the server of the opponent that stores valuable information. The players equip their computers with attack packages (if the players are attackers) and defense packages (if the players are defenders) [14]. The differences in the defense packages are influenced by the kinds of attacks they block. There are defense packages for blocking worms and spoof; others are designed to prevent attacks from denial of services and modification messages. Each turn in the game is characterized by the player making a single move. If the player is an attacker, a move can involve initiating an attack or moving an attack one unit. Similarly, if the player is a defender, the move can be learning the attack using cyber deception and applying the defense solution to real systems.

7.2.3.2 Stackelberg Game

The Stackelberg security game is a game theory model characterized by players, actions, and utilities. The Stackelberg game is mainly employed to formulate interactions between the attacker and the defender [15], which rely on the states of the attackers and the defenders. In the Stackelberg game, the defender is perpetually defending several targets using limited resources and the attacker has the capacity to survey and learn the strategies employed by the defender and can attack the targets after undertaking careful planning [16]. In the Stackelberg game, the pure strategy of the attacker is to attack a target, while the pure strategy of the defender is to optimize the available resources to protect vulnerable targets. The Stackelberg game further holds that the payoff or reward of an outcome is dependent on the value of the attacked target and the level of protection provided by the defender [17]. However, it is worth noting that the payoff of an outcome does not depend on the strategies employed by the defender to protect other targets. The targets in the Stackelberg security game are accompanied with a set of values, which define the utilities of the attacker and the defender in case of a failed or successful attack. The Stackelberg game holds that it is essential for the defender to constantly provide protection to the targets whereas it is important for the attacker to aim at unprotected targets to maximize payoff. The solution of the Stackelberg game is obtained using the Stackelberg equilibrium [18].

The Stackelberg equilibrium utilizes probabilistic and optimizing strategies to obtain the utility of the defender. In essence, the solution of the Stackelberg security game problem is a mixed strategy that maximizes the utility of the defender. As noted previously, the mixed strategy for the defender is defined by a probability distribution of all pure strategies [18]. In other words, the mixed strategy of the defender is equal to the probability that each target is protected by the defender. The pure strategy of the defender refers to the allocation of the security resources to targets. The strong Stackelberg equilibrium is the most commonly used solution for Stackelberg security game problems. The defender's utility for an action profile (C, t) is obtained using Eq. (7.1):

$$U_d\,(t, C) = ct\,U_{dc}(t) + (1 - ct)\,U_{du}\,(t) \tag{7.1}$$

The use of the Stackelberg Security Game (SSG) in addressing real-world security issues is hampered by several challenges, which include the inability of dealing with the bounded rationality of human adversaries, handling a high number of security uncertainties, and scaling up to security problems.

7.3 The Taxonomy for Cyber Deception

This section presents important classification of cyber deception used to provide cybersecurity that includes moving target defense (MTD), obfuscation method, perturbation method, mixing method, honey-x method, and game-theoretic models, as shown in Figure 7.7.

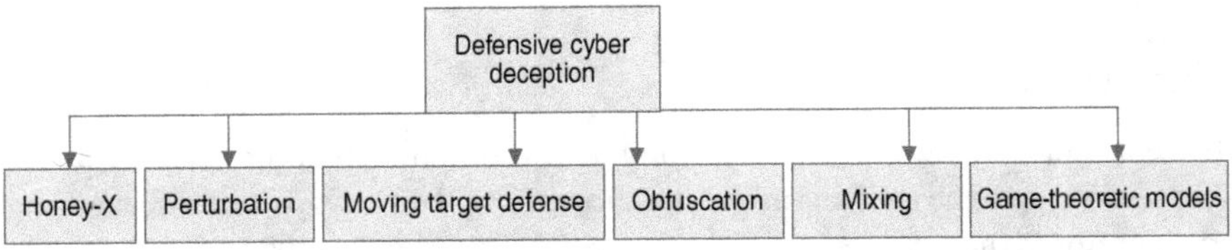

Figure 7.7 Taxonomy of cyber-deception techniques.

7.3.1 Moving Target Defense

MTD is one way of deploying cyber deception for security defense. MTD uses different strategies such as diversification, randomization, and dynamism to reduce the predictability of defenses [19]. MTD helps defenders by increasing the resiliency of the cyber systems and the workload of the attackers. To reduce the static and deterministic nature of computer systems [20], MTD uses different dynamic techniques such as dynamic applications techniques, dynamic software applications, dynamic runtime environments, and dynamic platforms. A typical design of MTD uses a Markov decision process and mixed strategies [19, 20]. The effectiveness of the mixed strategies in MTD is achieved by the inability of attackers inferring the costs when moving from one strategy to another.

7.3.2 Obfuscation

In obfuscation-based cybersecurity, the system should mask data/application by employing IP hopping capabilities and encryption techniques. For instance, the legitimate sender sends data through complex pathways by scrambling the IP address, location, the identity of the users, and other data [21]. The obfuscation method results are semantically and functionally equal to the original code but limit reverse engineering of the code by making it unintelligible to the attackers. The obfuscation of the code makes the attack process hard and costly to attack the protected code. Similarly, the program flow can obfuscate the bogus insertions that use dummy codes, dummy blocks, redundant operands, dummy classes, and new segments, among others, which are never executed but confuse the tools used to analyze the flow of the code [22].

7.3.3 Perturbation

The data perturbation helps to enhance the data/system security by adding random noise to databases. The perturbation process masks the content of sensitive data by applying a random modification of the data using additive noise [23]. Perturbation in data can be done using a probability distribution approach to enhance privacy by replacing the data within the same distribution sample or a value distortion approach by encapsulating confidential data with additive or multiplicative noise [24].

7.3.4 Mixing

The concept of mixing involves sending data through a chain of network nodes that allow data to be sent anonymously throughout the network without exposing the identity of the senders and receivers [24]. Furthermore, mixing can be implemented by encrypting and sending encrypted messages along with dummy messages. Another variety of mixing is threshold mixing, which utilizes an algorithm that sends messages to their destination in a random manner after the input to the mix surpasses the lower threshold but remains below the upper limit [25].

The concept of mixing involves sending information through a chain of network nodes to prevent linkability. Mall mixing allows data to be sent anonymously throughout the network without exposing the identity of the people who are communicating. Mixing networks is achieved by padding, creating messages of equal size and sending all messages at the same rate [24]. Mixing utilizes batching strategies to determine the steps to undertake to deliver messages to their destination anonymously. The batching strategies utilized by mixing networks to send data anonymously over the network include the threshold mix, the threshold pool mix, the timed pool mix, and the timed dynamic pool mix.

The threshold mix uses an algorithm that sends messages to their destination in random order after the input of the mix surpasses a set limit. The timed mix employs an algorithm to send information in a random manner after a certain period of time has dissipated. The threshold pool mix uses an algorithm to forward messages randomly to their destination after the input to the mix surpasses the lower threshold but remains below the upper limit [25]. The timed pool mix utilizes a unique algorithm to send all messages present in a pool to their destination in a random manner after a certain amount of time is surpassed. The timed dynamic pool mix uses a special algorithm to randomly send a fraction of the messages present in a pool to their destination.

7.3.5 Honey-X

The honey-x deception method disguises honeypots as real systems to learn about the attacks and protect the real systems [26]. The misrepresentation of the real systems as the honeypots makes attackers spend their resources on honeypots while leaving the real systems intact. Furthermore, logs and traces of attacks in honeypots are used to learn about the types and strengths of the attacks, which are then used to prepare defense solutions to protect real systems. Typically, honeypots are of two types: client honeypot and server honeypot. The client honeypots detect and report any attacks to servers so that the system can prepare to combat the attacks. The server honeypot is designed with some vulnerabilities to attract attackers so that the real system can use the learned lessons.

7.4 Game-Theoretic Model for Cyber Deception

The defender aims at securing a set S of Internet of Things (IoT) devices by using a set of original configurations C and has an associated utility U_C, which captures the loss caused by attacks from adversaries. $N_c = |S|$ represents the cardinality of the IoT devices with configurations C. The deceiving servers have the configurations D. The adversary attacks the servers to learn about or damage the servers. From the perspective of the adversary, original and deceiving systems look like the same. However, there is a cost associated with deceiving, denoted as P (C, D), which is needed to deploy deception servers and configurations by the defender. Naturally, C, D, and N_c are known to the defender. Thus, possible defending strategies for the defender can be represented as a non-negative $|C| \times |D|$ matrix, called P_{CD} with an element in $P_{CD} = 1$ when deceiving server and its original server have indistinguishable configuration, and $P_{CD} = 0$ otherwise. Based on the defender's strategy and D deceiving servers with indistinguishable configurations to the adversary, the adversary must be indifferent among all servers when deciding which server to attack. The utility observed by the adversary is the sum of all nonzero or non-negative values of the matrix P_{CD}.

When the given defender chooses a strategy P_D, upon attacking the deceiving server, the adversary's expected utility is given by

$$U_d = E[U_c \mid P_{CD}, c], \quad c \in C \text{ and } d \in D$$

We can consider the game as a zero-sum, since the loss of the defender is the gain of the attacker. Thus, the defender's expected utility is $-U_d$ when d is attacked.

Figure 7.8 shows the main concept of cybersecurity deception.

The network reconnaissance domain in which an adversary scans the defender's IoT devices and the defender deceptively alters the system's responses is shown in Figure 7.9.

We simulated the IoT scenario by considering the original configurations with and without a cyber-deception system and plotted the loss (and gain) of IoT system (and of attacker), as shown in Figure 7.10, where we can see that the loss of information of the IoT system could be as high as 100%, which is the gain of the attacker

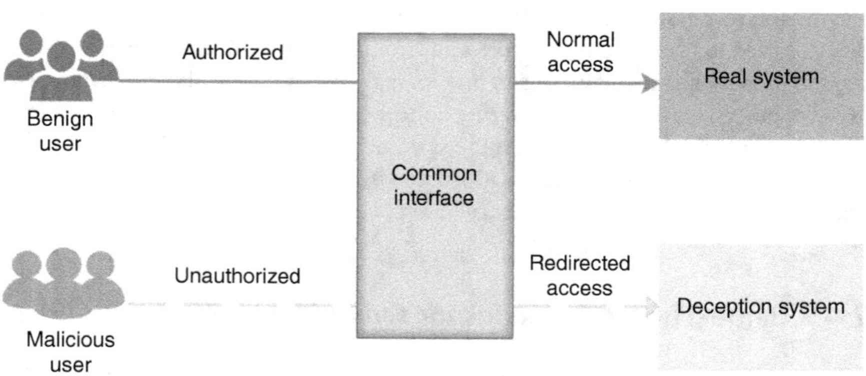

Figure 7.8 Cybersecurity deception concept.

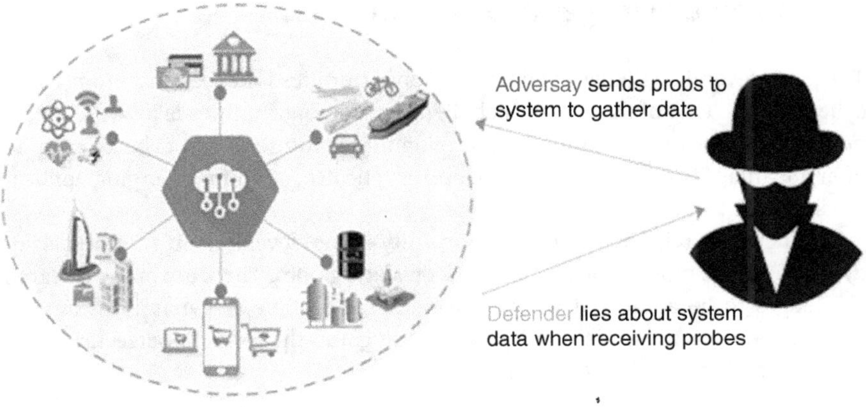

Figure 7.9 An adversary scans the defender's IoT devices and the defender deceptively alters the system's responses.

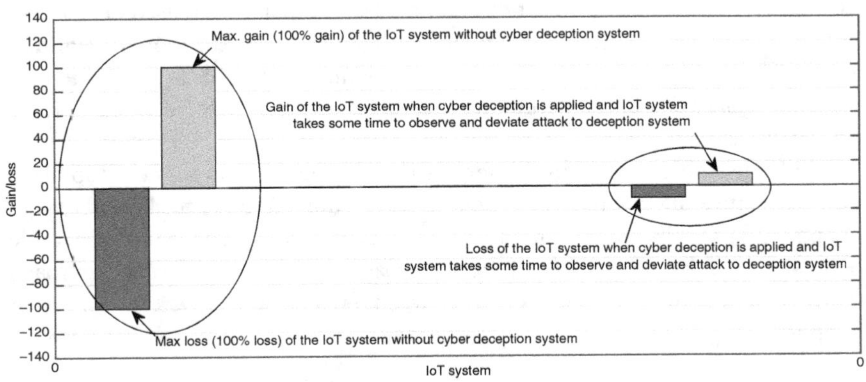

Figure 7.10 Loss of the IoT system and gain of the attacker when there is no cyber-deception system implemented as well as when a cyber-deception system is implemented. Loss of information is high for the IoT system when cyber deception is not implemented.

when there is no cyber-deception system implemented to hide the actual system being attacked. Whereas when a cyber-deception system is implemented, the IoT system could lose a fraction of information during sensing/observing and manipulating of the system by the IoT system for the attacker, as shown in Figure 7.10. Note that we did not consider the cost of the equipment (hardware and software) to implement cyber deception for Figure 7.10. We can say that the equipment investment of the attacker to attack the system and the equipment investment of the legitimate IoT systems to prepare cyber-deception systems could be more or less the same and can be relaxed while analyzing the impact of cyberattacks.

7.5 Open Challenges and Opportunities

There are several challenges and possible opportunities for cyber deception-based cybersecurity. One of the research challenges and opportunities is *mimetic deception*, which is worth exploring and developing for future use in cybersecurity to offer advantages over static deception and cryptic deception for emerging applications [27, 28].

Another research challenge or opportunity is the development of *evolution or dynamic games*; most of the studies focus on static games. There are many research challenges to be addressed when considering evolution or dynamic games for cyber deception, thus, the application of the game theory in cybersecurity and cyber deception is in its early stage.

Some of the techniques such as game theory are well studied in the research but there is an urgent need to *realize them in practical scenarios*. Furthermore, a lack of collaboration between cybersecurity analysts and game theorists hinders the implementation of game theory for cyber defense.

Deception is regarded as a multidisciplinary topic and has been used in a diverse set of applications to achieve domain-specific goals such as deception in criminology, economics, and trade business. *Interdisciplinary teamwork* is needed to successfully implement cyber deception to combat attacks targeted at vulnerable cyber systems.

The ultimate goal of any cyber defense solutions including cyber deception is to protect sensitive information in an organization from habitual attacks and ensure that potential attacks are prevented [9] and that the cyber system is not vulnerable to outside attacks [8]. Among other approaches, a game-theoretic approach can help to model the interactions between victims and attackers for cyber deception.

7.6 Summary

This chapter presented the current status, challenges, and perspectives of cyber deception for cybersecurity. The chapter has also presented the unique advantages of using deception techniques for cybersecurity over traditional security defense solutions, allowing the system to learn and adapt the security parameters on the fly to combat malicious cyber operations. This chapter also highlights some open challenges and research directions. Note that cybersecurity cannot be achieved using a single strategy. Thus, a good cybersecurity solution needs to be developed using security by design, and involving different stakeholders and disciplines.

References

1 Almeshekah, M. H. and E. H. Spafford, "Cyber security deception." In *Cyber Deception* (pp. 23–50). Springer, 2016.

2 Wang, C. and Z. Lu, "Cyber deception: Overview and the road ahead," *IEEE Security & Privacy*, vol. **16**, no. 2, pp. 80–85, 2018.

3 Cardenas, A., S. Amin, B. Sinopoli, A. Giani, A. Perri, and S. Sastry, Challenges for securing cyber physical systems. In *Workshop on Future Directions in Cyber-Physical Systems Security* Vol. **5**, DHS, 23 July, 2009.

4 Cymmetria Inc., *Introduction to Cyber Deception*. Palo Alto. Cymmetria Solutions, 2016.

5 Almeshekah, M. H. and E. H. Spafford, Planning and integrating deception into computer security defenses. In *Proceedings of the 2014 New Security Paradigms Workshop* pp. 127–138, Canada, 15–18 September, 2014.

6 Kwon, C., W. Liu, and I. Hwang, Security analysis for cyber-physical systems against stealthy deception attacks. In *American Control Conference (ACC), 2013* (pp. 3344–3349), Washington, DC, 17–19 June, 2013.

7 Almeshekah, M. H., M. Atallah, and E. H. Spafford, The center for education and research information assurance and security, Purdue University, *CERIAS Tech Report, 13*, 2013.

8 Jafarian, J. H., *Cyber agility for attack deterrence and deception* (Doctoral dissertation, The University of North Carolina at Charlotte), 2017.

9 Almeshekah, M. H., "*Using deception to enhance security: A taxonomy, model, and novel uses* (Doctoral dissertation, Purdue University)," *Science*, vol. **1**, p. 219, 2015.

10 Cohen, F., D. Lambert, C. Preston N. Berry, C. Steward, and E. Thoman, A framework for deception. pp. 1–67, 2001. https://pdfs.semanticscholar.org/3dc3/22bb0c04e8c177accd66d3fb547073d2c06b.pdf

11 Dassiou, X. and X. Glycopantis, "A tree formulation for signaling games," *Game Theory*, vol. **1**, no. 1, pp. 1–12, 2013.

12 Bangerter, A., N. Roulin, and C. König, "Personnel selection as a signaling game," *Journal of Applied Psychology*, vol. **97**, no. 4, pp. 719–738, 2012.

13 Lucideus Research, *Using game theory in deception strategy for cyber security*, 2018. https://lucideustech.blogspot.com/2018/02/using-game-theory-in-deception-strategy.html.

14 Zahir, S., J. Pak, J. Singh, J. Pawlick, and Q. Zhu, Protection and deception: Discovering game theory and cyber literacy through a novel board game experience. pp. 1–8, 2015.

15 Chua, F., N. Vasnani, L. Pacio, and L. Ocampo, "A Stackelberg game in multi Period planning of make-to-order production system across the supply chain," *Journal of Manufacturing Systems*, vol. **46**, pp. 231–246, 2018.

16 Sinha, A., T. Nguyen, D. Kar, M. Brown, M. Tambe, and A. Jiang, "From physical security to cybersecurity," *Journal of Cybersecurity*, vol. **1**, no. 1, pp. 19–38, 2015.

17 Liang, X. and Y. Xiao, "Game theory for network security," *IEEE Communications Surveys & Tutorials*, vol. **15**, no. 1, pp. 472–486, 2013.

18 Chaudhary, G. and Y. Narahari, "A Stackelberg game approach for secure and efficient surveillance," *Procedia Computer Science*, vol. **24**, no. 1, pp. 205–216, 2013.

19 Lei, C., H. Zhang, J. Tan, Y. Zhang and X. Liu, "Moving target defense techniques: A survey," *Security and Communication Networks*, vol. **1**, no. 1, pp. 1–25, 2018.

20 Xu, P. Guo, M. Zhao, R. Erbacher, M. Zhu, and P. Liu, "Comparing different moving target defense techniques," *Proceedings of the First ACM Workshop on Moving Target Defense - MTD '14*, vol. **1**, no. 1, pp. 97–107, 2014.

21 Behera, K. and D. Bhaskari, "Different obfuscation techniques for code protection," *Procedia Computer Science*, vol. **70**, no. 1, pp. 757–763, 2015.

22 Hashemzade, B. and A. Maroosi, "Hybrid obfuscation using signals and encryption," *Journal of Computer Networks and Communications*, vol. **1**, no. 1, pp. 1–6, 2018.

23 Chasparis, J. Shamma, and A. Rantzer, "Perturbed learning automata in potential games," *IEEE Conference on Decision and Control and European Control Conference*, vol. **1**, no. 1, pp. 2453–2458, 2011.

24 Jackson, M., T. Rodriguez-Barraquer, and X. Tan, "Epsilon-equilibria of perturbed games," *Games and Economic Behavior*, vol. **75**, no. 1, pp. 198–216, 2012.

25 Keeling, M. J. and K. Eames, "Networks and epidemic models," *Journal of The Royal Society Interface*, vol. **2**, no. 4, pp. 295–307, 2005.

26 Mphago, B. "Deception in web application honeypots: Case of Glastopf," *International Journal of Cyber-Security and Digital Forensics*, vol. **6**, no. 4, pp. 179–185, 2017.

27 Rawat, D. B., R. Doku, and M. Garuba, "Cybersecurity in big data era: From securing big data to data-driven security," *IEEE Transactions on Services Computing*, Early Access Available 2019. DoI: https://doi.org/10.1109/TSC.2019.2907247

28 Rawat, D. B. and K. Z. Ghafoor, *Smart Cities Cybersecurity and Privacy*, Elsevier Press, ISBN: 9780128150320, November 2018.

Part II

IoT Security Modeling and Analysis

8

Cyber-Physical Vulnerability Analysis of IoT Applications Using Multi-Modeling

Ted Bapty, Abhishek Dubey, and Janos Sztipanovits

Institute for Software Integrated Systems, Vanderbilt University, Nashville, TN, USA

8.1 Introduction

Cyber-physical systems (CPS) are engineered systems where functionality is emerging from the networked interaction of physical and computational processes. The appearance and wide-scale introduction of CPS during the past decade has stimulated a pervasive technology advancement, which is impacting all industrial sectors and almost all aspects of society. The progress was further accelerated during the past five years with emerging industrial platforms such as the Internet of Things (IoT) [1], Industrial Internet of Things (IIoT) [2], and Fog Computing (FC) [3]. These platforms have factored out critical and complex methods, services, and universal hardware/software components from the application space and opened up the opportunity for developing CPS applications faster and without specialized expertise.

CPS examples span a wide range of systems from small devices, such as surgical micro-robots [4], to large geographically distributed systems such as smart grid and traffic control networks. IoT platforms are primarily used for implementing distributed CPS, formed by networks of physical and computation processes connected by sensors and actuators. This system category has several distinguishing characteristics that deeply influence their design and operation.

- *Open Architecture*: IoT-based systems take advantage of ubiquitous connectivity to collect information about physical systems, utilize this information in closed

Modeling and Design of Secure Internet of Things, First Edition. Edited by Charles A. Kamhoua, Laurent L. Njilla, Alexander Kott and Sachin Shetty.

Published 2020 by John Wiley & Sons, Inc.

loop real-time control, data analytics, higher level decision processes, and integration with enterprise systems and human participants. System boundaries, functionalities, and use cases though are evolving. Key system characteristics, such as security, safety, reliability, and resilience, emerge from component interactions and potentially change as components are added or removed.

- *Physicality*: Physical properties are essential in all CPS. This makes most IoT-based systems safety critical, because they need to be operated without risking physical injury of people or damage to the environment. Traditional design approaches in safety-critical physical systems do not assume that safety hazards may emerge via connectedness, especially software. At the same time, traditional safety-critical software approaches by and large neglect the physical impacts that may influence correct operation of software.
- *Scale*: Ubiquitous and inexpensive connectivity enables the emergence of societal-scale systems exemplified by networks of connected vehicles. Even in their relatively simple form, such as Google Map's dynamic, traffic-aware routing, the scale has unexpected consequences. The real-time sensing of traffic flow, congestion weighted routing service, and drivers' decisions on accepting or rejecting the routing recommendation form a complex, closed-loop networked control system, affecting the overall traffic flow. Dynamics emerge from the networked interactions among physical systems (cars), physical processes (traffic flow), routing algorithms, human decisions (drivers), and network delays. While in normal operation the service increases the resilience of the traffic networks against physical disturbances (e.g. road blocks caused by accidents), it also opens up a new attack surface for creating large disturbances via cyberattacks on the routing service.[1] Furthermore, unsafe use of these services can have dramatic consequences.[2]

In this chapter, we examine security challenges that are tightly linked to these distinguishing characteristics of IoT-based CPS systems. Our focus will be on *Physicality*, because identifying and understanding physical vulnerabilities and their potential propagation paths across system layers is frequently overlooked. Vulnerability analysis for physical attacks is also challenging because it requires models, tools, and analysis methods that are strongly different from those used in software or network security.

In Section 8.2, we provide an overview of sources of vulnerabilities and their potential impact on notional IoT-based system architecture. We use this overview for highlighting the place of our example in the overall architecture that we

1 https://www.popsci.com/article/gadgets/israeli-students-spoof-waze-app-fake-traffic-jam
2 https://www.washingtonpost.com/news/worldviews/wp/2018/02/12/israeli-soldiers-using-waze-attacked-by-palestinians-after-taking-wrong-route/

elaborate on. In Section 8.3, we discuss low-level physical vulnerabilities and present methods for vulnerability analysis.

8.2 Vulnerabilities in IoT

CPS and Human-CPS (H-CPS) systems are typically deeply layered, with each layer providing distinct suites of services that are used for implementing the functions pictured above. Figure 8.1 shows the adaptation of the CPS/H-CPS architectural layer representation in [5] to IoT-based applications.

The five fundamentally different layers are: The Physical Layer, the Network Layer, Service Platform Layer, Application Layer, and the Human Layer (see Figure 8.1). The Physical Layer embodies physical components and interactions and includes the physical environment and the physical aspect of computing and networking. The physical layer interactions are modeled in continuous (physical) time. The mathematics of composition is based on linear algebra, topology, differential equations, and discrete event systems. The design requirements and key properties are described using well-understood abstractions of physical systems such as continuous (usually multi-physics), lumped, or distributed parameter dynamic models and geometry. The Network, Service Platforms, and Application

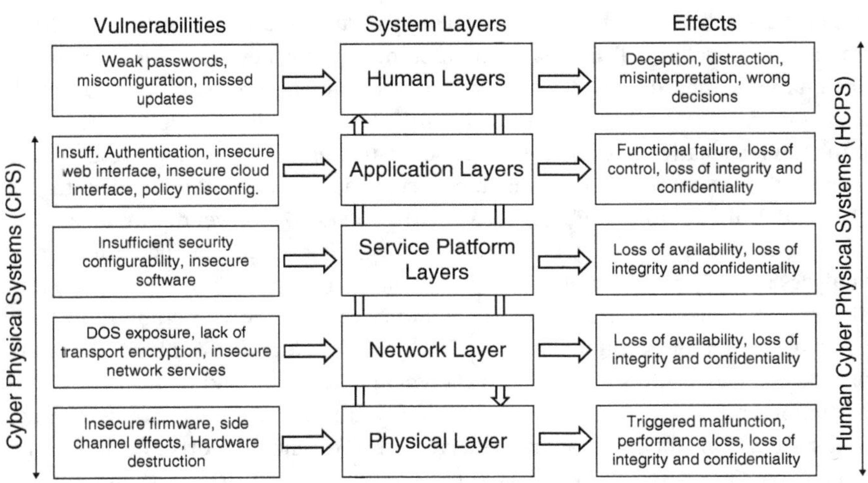

Figure 8.1 Vulnerability propagation across the different layers of a CPS.

Layers comprise the cyber side of CPS and include software for networking, middleware, and applications. Execution of software applications on physical processors and transfer of data on physical communication links "translate" their abstract behavior into physical, real-time behavior. The Human Layer incorporates humans involved in an H-CPS as decision-making entities that are directly involved in the overall system dynamics. In many large-scale H-CPS such as human drivers in traffic flows or human consumers on the power grid, the overall operation concept may require the modulation of human behavior via incentives using the theoretical framework of mechanism design [6].

Security risks in an IoT-based system arise from a security threat (an agent attempting to do harm), a vulnerability in the targeted layer that the threat will exploit, and the insufficiency of countermeasures that attempt to reduce the impact of attacks. Vulnerabilities are specific to the technology used in the different system layers. In Figure 8.1, we demonstrate this by placing some of the IoT vulnerabilities identified by the OWASP project [7] in the relevant system layers. The mapping between vulnerabilities and effects are many-to-many due to cross-layer propagation of interactions. For example, deception of system users can be achieved by intentional misconfiguration of device connections on the Human Layer or by spoofing sensors on the Physical Layer. The potential implications of the exploits of vulnerabilities are significantly amplified by salient characteristics of IoT systems described before.

8.2.1 Open Architecture and IoT Security

A significant challenge in IoT is that the open architecture allows dynamic addition and removal of IoT devices from different vendors, devices made with different quality standards, and security robustness, which continually interact with each other. This makes mitigating identified vulnerabilities complicated, and impossible to statically evaluate. As an example, assume that a vulnerability is identified in a common building block such as `libcurl` and added to the National Vulnerability Database (CVE-2018-10000007). A vendor A of an IoT device determines that they use `libcurl` in their car multimedia systems and analyzes the effect of the vulnerability on their device, which is used in home theater systems. The vendor then declares and advertises the vulnerability to their customers. Vendor B, a home automation systems integrator, analyzes the effect of the vendor A's vulnerability on their home theater setup and then notifies their customers, offering a mitigation process to avoid problems in the home automation system. This model is labor-intensive and slow when it works. Furthermore, not all suppliers can/will expend labor resources to perform this constant monitoring of prior-generation systems, and, in addition, few end-users are able or willing to enact on the mitigation on otherwise "working" systems.

If Vendor B had an appropriate CPS engineering model of their home automation system, they could evaluate the effect of the vulnerability in an automated way, as soon as it is posted to the National Vulnerability Database. The result of the analysis could be distributed to the end-users and with the help of automated update or a deployable checking tool, thus mitigating the problem. This automated analysis would not only decrease the time required to receive and evaluate problems but also generate a more comprehensive list of the effects of the vulnerability. The option of identifying potential mitigation strategies is even more powerful. In this case, Vendor B would not only be notified of the potential effects on their home automation systems due to the vulnerability but they would be offered range of mitigation strategies. At one extreme, `libcurl` could be updated and the updates posted to all of the deployed home automation systems. The time required to implement this solution would be the time required to develop a `libcurl` patch plus the time required to distribute that patch through Vendor B's network. An alternative would be to disable the web interface functionality that uses `libcurl`. The impact of disabling the web interface would depend on the configuration of the product. A comprehensive list of mitigation strategies and their trade-offs would allow for informed decisions.

8.2.2 Effects of Inter-Layer Interactions on IoT Security

As Figure 8.1 shows, while attack vectors are layer-specific, their impacts are not restricted to a single layer. Cross-layer interactions can propagate anomalies and failures up and down in the hierarchy. For example, a network attack on a wireless communication link may prevent a network service to send temperature updates to a temperature control application, which in turn may fail to cool down a server room resulting in physical failure due to overheating. It means that a Network Layer attack may propagate to cause a physical layer fault not through direct physical interactions but through crossing cyber layer boundaries. It is easy to extrapolate that in open, ad-hoc IoT system configurations with intractable cross-layer interactions and frequently partially known components, security risks are significant and their impact may extend to each layer.

A typical IoT system incorporates many system elements coming from different sources. The overall trustworthiness of the system depends on trust in each of these elements, the way they are integrated, and the manner they interact with each other. Propagation of trust is a hierarchical flow of trust within a system from its usage to all its components. In carefully constructed systems, each layer of the trust model depends on the one below it: each element builds a trust relationship with the elements below. Trust is achieved in the operational system when assurance that the operational requirements of the system have been met. In

IoT systems with many heterogeneous components in ad-hoc configurations, integration platforms play fundamental role in maintaining trustworthiness.

8.2.3 Effects of Physicality on IoT Security

Since dominant use cases of IoT platforms are cyber-physical, identifying physical layer vulnerabilities and impacts have large significance in IoT security. A large category of risks relates to the infiltration and exfiltration of information via physical side-channels. One simple example is the decoding of wired network traffic from the flashing pattern of light from the LED status light of older routers. Recently, the idea has been modified to allow malicious code running on the router to intentionally modulate the LEDs to exfiltrate data from an air-gapped network [3]. More seriously, energy injection using, for example, radio frequency (RF) interference, can stimulate internal communication signals to change nominal behavior from outside the system, totally bypassing the strongest cyber security measures (e.g. firewalls, encryption, etc.).

The need to understand the system in which a component operates is well understood in the vulnerability community. It is an essential part of the Common Vulnerability Scoring System (CVSS); however, the definitions in CVSS do not always apply directly to cyber-physical vulnerabilities [6]. The problem is that existing methods require expert knowledge across many physical and cyber design domains to determine potential paths through a system, and the ability to identify these non-designed, side-channels where energy or information can "leak" through nominally disconnected paths. It is not just a theoretical possibility that experts will miss these complex paths, it has been shown repeatedly in everything from the Target data breach [7], which combines social and software vulnerabilities, to the recently announced Meltdown vulnerability that builds upon a cache timing vulnerability that was first described in 1995 [8], or hacking sensors using a COTS software-defined radio device http://console-cowboys.blogspot.com/2017/10/hacking-everything-with-rf-and-software.html.

In the rest of this chapter, we use examples of smart home systems created using IoT to describe the various vulnerabilities, focusing on the vulnerabilities introduced due to the physical channels specially. Our approach in this paper is to describe the methodology to model these vulnerabilities and their analysis.

We believe that the ability to focus on modeling and discussion of physical vulnerabilities is important as most vulnerability analysis studies focus only on the cyber-elements often ignoring the threats caused due to physical layers, which subsequently lead to anomalies in system operation. The other aspect we focus on in the chapter is the difficulty in studying and analyzing the vulnerability of the systems' both cyber and physical layers. To mitigate the complexity, we use an approach called as the multi-modeling approach.

Multi-modeling, in its simplest terms, is an ensemble of models (and corresponding modeling languages) representing physical and computational artifacts, and energy and information flows. In a typical model-based systems engineering (MBSE) approach, multiple, loosely coupled models are created manually, per domain and/or mathematical abstraction. For example, a thermal model is created separately from the electrical circuit model. Further, multiple thermal models are developed based on a circuit model vs. a finite element/geometrical model. With our multi-modeling approach, a single integrated model is used to tightly couple the design domain and abstractions. This permits more cross-domain analysis to occur in an automated manner. The approach has been used in CPS design and optimization for multiple applications, including an amphibious military vehicle, a missile system, and multiple electronics designs [8].

In the home-automation system example, essential modeling views from the threat mitigation point of view are the basic modalities in system operation, services, energy and information flows together with their configurable attributes, the physical architecture and equipment, and the relationship among them.

8.3 Multi-modeling Approach for CPS Vulnerability Analysis

Consider a smart home system consisting of several IoT devices interacting with a smart home hub. Let us consider that the IoT devices include a thermostat, a smart door lock, an in-home monitoring camera, and a security alarm system. The ideal operation of this system will lock the doors when the system is alarmed, turn on the in-home monitoring camera, and lower the thermostat temperature. The system can also provide temperature statistics to a community smart system, which provides socio-comparative benchmarks for encouraging lowering the temperature setting in winters and increasing the temperature settings in summers to reduce the energy consumption. When designed properly, such a system can improve the quality of life of the residents and encourage lower energy usage throughout the community, making it a smart community.

IoT sensors are prone to a number of attacks due to their lower security protections. For example, the software in many IoT sensors is not updated frequently. Such an attack can often allow a malicious adversary to find a way from one IoT sensor; for example, the thermostat which was communicating to the outside world to upload the aggregated temperature statistics to a critical equipment such as door lock or cameras. If an information flow exists between the thermostat and the camera, the camera can be turned on maliciously, impacting privacy and in worse cases the locks can be opened while the residents are way. In such a system, ensuring that we have strict information flows is very important.

The other not so obvious problem in these systems is that of side-channels. While most researchers, including us, focus on the information flow and security as a primarily cyber-problem, we ignore the inherent risks due to the physical elements and interactions in the system. Most of these systems are electrical in nature and information can be injected into the connected architecture even if it firewalled in the cyber-domain through the RF channels. For example, if a sensor calibration input can be manipulated externally, it can be coerced into producing an invalid set of readings, or disabled altogether.

Therefore, a complete analysis requires modeling the cyber elements, modeling the physical elements, and modeling the integration and then analyzing them. In the next section, we describe our modeling language that allows us to express the system in this manner and show how we can do the cyber-physical analysis for identifying security vulnerabilities in these systems.

8.3.1 The OpenMETA Multi-Modeling Environment for CPS Security

OpenMETA is a component-based, multi-domain modeling and analysis framework developed for the purpose of rapid design and analysis of cyber-physical systems [8]. The tool functions around the following representation precepts:

- Components are the basis for capturing the system. Components, at the atomic level, have:
  - Well-specified interfaces that describe potential interactions with other parts (components) of the system. These interfaces represent information flow, power flow, and geometric interactions with adjacent/connected components.
  - Interfaces are grouped together in Connectors, which organize and consolidate the set of interfaces/interactions that are implied by a connection between a set of components (typically a pair). Interfaces can be causal (e.g. a signal with a defined source and destination).
  - A set of domain artifacts that capture the behavior, structure, or of the implementation of the component. Domain artifacts are electronically readable, in a form that are readily usable by specific engineering tools.
- Parameters and properties allow a component to be adjusted by external variables, to adjust behaviors and structure, e.g. the diameter of a cylinder or the gain of an amplifier.
- Component Assemblies capture instances of components and the graphs of component interactions. Component Assemblies represent subsystems or entire systems. Special forms of the Component Assembly, Design Containers, allow the designer to capture optional implementations of systems, to represent discrete variability in a design space.

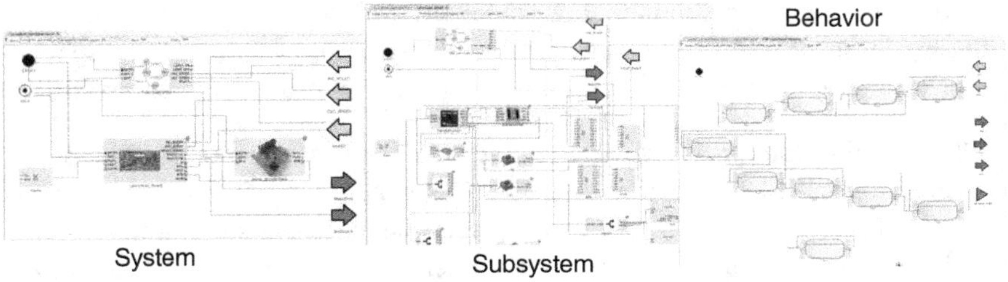

Figure 8.2 Example multi-domain system model hierarchy. A system captures the various components and subsystems as block diagrams. Each component's behavior is modeled in another aspect. The behaviors and block diagrams can capture the interactions across different domains, e.g. electrical, mechanical.

- Component Assemblies can be nested to any level, to capture design hierarchy, and manage complexity. Connectors in a Component Assembly allow the internal component connections to be exposed and accessed at higher levels of the design hierarchy.

The architectural specification aspects of the OpenMETA design language are similar to SysML blocks, block diagrams, and Internal Block Diagrams (IDBs). The block diagrams, as illustrated in Figure 8.2, allow specification of overall system/subsystem content breakdown along with the interconnectivity among the blocks. Other views, e.g. Figure 8.2 right, capture the behavioral view in states and events, as in SysML State Machine diagrams. These models can capture a behavior associated with a set of physical or cyber components, or behavior at any level of the system. They can also be used to specify expected environmental or human participation.

Models exist for the purpose of design specification, analysis and evaluation, and/or otherwise reasoning about the system. The OpenMETA tools provide a framework and automation system for these purposes. These processes are also specified as a Model, termed the testbench. Testbench models can be automatically evaluated, elaborated, or synthesized for user-specified specific purposes within the framework, exploiting a range of external tools in the workflow.

A testbench typically contains:

- The system under test, a reference to a design model.
- A set of parameters that can be used to control environments or system parameters.
- A set of metrics that are to be calculated via execution of the testbench (typically critical attributes of the target system, associated with requirements of the system).

- A workflow describing the steps to be executed to elaborate, compose, transform the model to an executable form, execute the transformed model in a specific tool, and procedures to extract the metrics from the result of model execution.

When a multi-domain analysis is required, an analytics model of the Parametric Exploration Tool (PET) prescribes the set of testbenches to execute, the information flow of computed data among connected testbenches. It also includes a Driver function that can execute a design of experiments (DOE) process or an optimization of the target system, which is then used to identify the set of optimal component configurations or parameters.

In addition to the function of design automation, the multi-domain nature of OpenMETA provides a strong base for analyzing security-relevant aspects of the system. This is accomplished using a range of testbench capabilities, specifically:

- RF Analysis: producing accurate 3D models of the target, for analysis by ANSYS SiWave[3] and HFSS.[4]
- Electrical Analysis: producing fully interconnected models of the underlying electrical components and networks interconnecting the electronic components. This analysis can be done with the commonly used SPICE simulator such as NGSPICE [9], or with Modelica simulator tools.
- MultiDomain Simulation: producing power-flow models that represent coupling of physical power between components/subsystems of the system. Modelica simulators, such as OpenModelica [10], are well suited for combined electrical, thermal, fluid, and mechanical power simulation.
- Behavioral Analysis: representing the behavior of the software, at a high level, to determine the impact of an input to the state and functional trajectory of the control/business logic of the software. State machine abstractions, when probability is included, are well suited for simulation and reachability analysis. We have used tools such as PRISM [11].

RF analysis involves modeling geometry of a target (with proper materials properties assigned to components of the system), and control of the implied electrical connectivity among components. OpenMETA satisfies requirement using a workflow capable of composing material-attributed complex geometries, consisting of:

- Creating material-accurate geometries of the system, using the PTC Creo CAD [12] tool to instantiate component solid models, accurately constraining the

---

3 ANSYS SIwave is a specialized design platform for power integrity, signal integrity, and EMI analysis of IC packages and PCBs: https://www.ansys.com/products/electronics/ansys-siwave

4 ANSYS HFSS is a 3D simulation software for designing and simulating high-frequency electronic products such as antennas, antenna arrays, RF or microwave components, high-speed interconnects, filters, connectors, IC packages, and printed circuit board: https://www.ansys.com/products/electronics/ansys-hfss

component models into their proper geometric relationships, and producing a range of target files that are compatible with 3D RF analysis tools, such as those from ANSYS (mentioned earlier).

- Incorporating accurate, detailed PCB geometry models into the physical structures (mounts, cases, external wiring, antennas).
- Assigning analysis ports for signal injection or measurement, and network termination impedances for conductive structures, and establishing a ground for the system. Analysis ports are efficiently assigned so as to minimize unnecessary analysis, selecting only behaviorally relevant inputs.
- Executing the analysis for the generated system and extracting pertinent results, including frequencies, polarizations, and angles of incidence where energy can be transferred onto a trace/wire at significant powers and at inputs to the circuit where desired behaviors can be externally influenced.

Figure 8.3 shows an example analysis of a printed circuit board (PCB). The 2D analysis, left side, shows an analysis on a PCB in isolation. Induced voltages are computed across a set of input ports (each line) over a range of frequencies. This establishes a set of frequencies that should be considered for further analysis (i.e. those above the logic threshold, indicated by the red line). The 3D analysis, right side, performs the analysis in the context of the physical enclosure, which will further subset the available attack surface based on the shielding effects achieved by the surrounding conductive surfaces.

In addition, the tool supports electrical analysis using SPICE for the electrical components of the IoT system. Electrical and RF analyses done using the tools mentioned earlier allow the electrical behavior of a target system to be assessed for impact of electrical disturbances throughout the system. The modeling tools automatically compose SPICE netlist for the system based on the block diagram, importing component SPICE models, creating nets, and unrolling connections within the block diagram topology. For normal design processes, typical inputs are applied and outputs can be observed, transfer functions computed, etc. For cyber-physical security assessment, any network that has been observed to be sensitive to RF (or other energy) insertion can be stimulated by inserting signal sources with the frequency content and amplitudes derived from the RF analysis.

Computing the impact of an effective RF stimulation is assessed in both the continuous electrical system energy transfer/lumped parameter modeling space and in a discrete/cyber behavioral space. Generalized, multi-domain behavior can be assessed with the modeling tool by composing a full-system Modelica model: (i) unrolling system hierarchy, (ii) tracing model connections, and (iii) synthesizing a composite model of the system. Just as with SPICE, the electrical/physical disturbances can be inserted with signal sources matching the potential inserted energy. This allows computation of the signal that can be created at an active electrical input, such as a logic gate, or a sampled electrical signal.

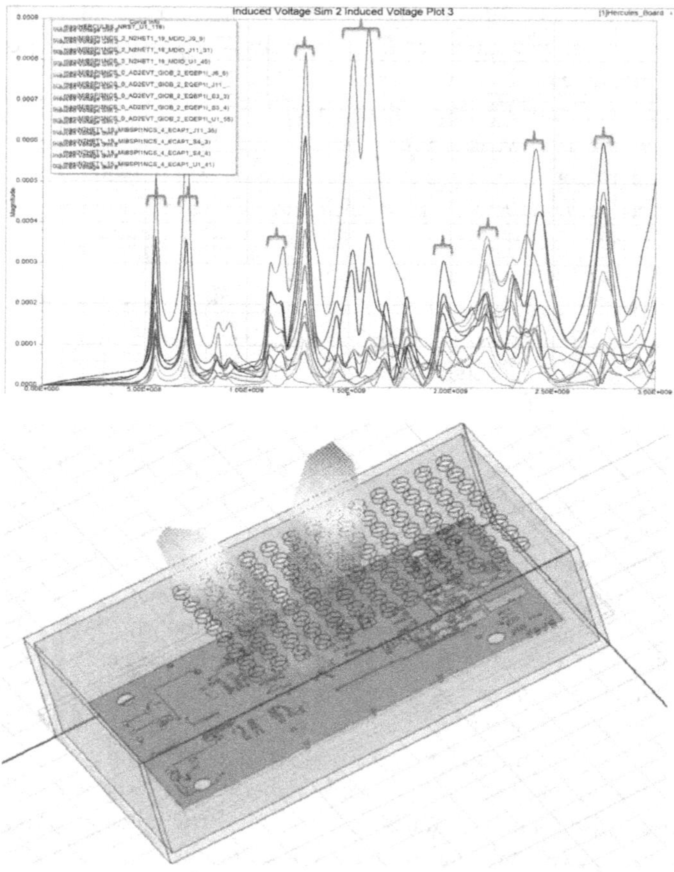

Figure 8.3 Example 2D and 3D analysis of a circuit board in a system context.

For a logic gate, a discrete event can be created, provided that: (i) the logic threshold (positive or negative) is exceeded; (ii) the frequency is not filtered out by the circuitry; and (iii) the event occurs with a "sensitive" timing (e.g. the noise is not removed by a debouncing circuit). If all conditions are satisfied, then a logical event can be created. Note that for sampled analog signals, the impact is more of a continuous nature, and should be computed by propagating the energy into the digital (numerical) domain. An example of this is shown in Figure 8.4, where an RF signal modifies the logic state on a communications bus.

Given a valid event (typically with an estimated probability of success) the impact on discrete behaviors can be analyzed. With an accurate state machine

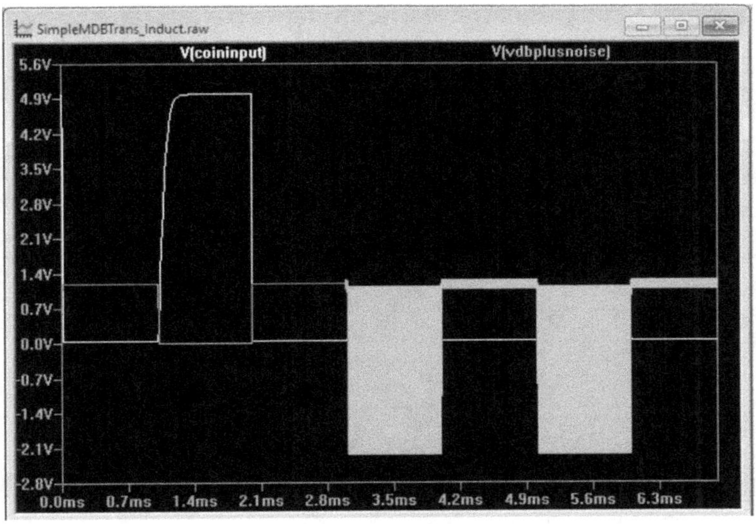

Figure 8.4 Example signal with RF energy insertion.

behavior, a reachability analysis can be performed. Probability can be assessed with PRISM, using an assessment of injected signal voltages w.r.t. the logic thresholds. Resultant impacts correspond to the attacker's desired system states.

The reachable are logged and can be evaluated against allowable deviations, as determined by system specifications. Any serious behavior alterations can be flagged for further evaluation.

8.3.2 Example: Smart Home Sensor – Physical Vulnerability Analysis

To understand how these modeling tools help in identifying vulnerabilities, let us discuss their use in identifying the physical vulnerabilities in an IoT sensor, say thermostat. Let us assume that, to keep installation costs minimal, the unit is to be battery powered and report data wirelessly. A low-power wireless networking technology, Bluetooth Low Energy (BLE), has been chosen, with the assumption that the sensor setup security (pairing) issues are managed securely.

The IoT sensor design is modeled in OpenMETA as shown in Figure 8.5. The IoT edge device includes the following subsystems:

1) a Bluetooth microprocessor (A Cypress PSOC 4, Subckt_Surrogate_BLE_Platform)
2) a temperature sensor (Temp Sensor), based on an integrated MEMs device
3) a light sensor (Light Sensor), based on an integrated optoelectronic device

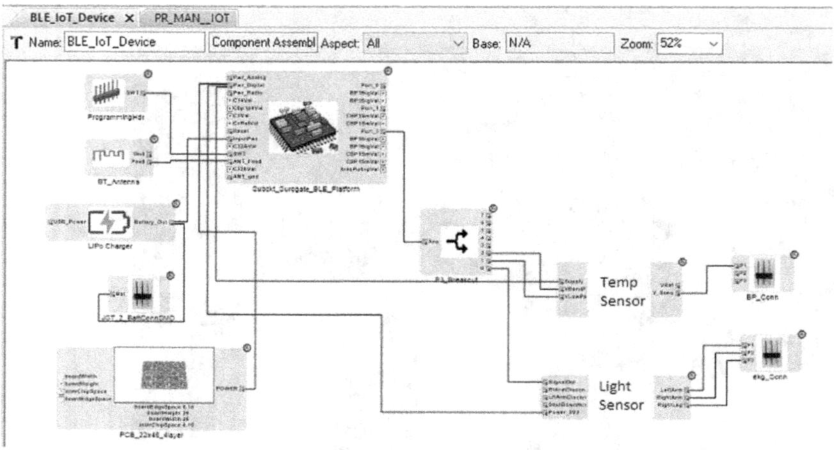

Figure 8.5 Functional blocks of the IoT sensor node.

4) a LIPO Battery subsystem(LIPO Charger), for managing an external battery, both charging and energy usage
5) a PCB (PCB_32x46_4layer) with integrated Bluetooth antenna
6) additionally, the physical PCB is modeled, along with all connectors

The model contains subsystems that implement the major functional blocks of the IoT Sensor node. Within each block, the necessary circuits are modeled to implement the functionality, down to the atomic component (integrated circuit, resistor/capacitor/inductor, connector, etc.). The detailed implementation of the PSOC Core is shown in Figure 8.6, consisting of the Cypress PSOC4 chip, surrounding capacitors, resistors, and crystals necessary to provide power conditioning, clocks, reset, etc., along with creating a somewhat abstracted view of the low-level pins of the chips.

These models are used directly in the design flow for the system, producing the manufacturable device in a set of automated steps where the design for the IoT sensor is transformed into a physical implementation by what we term a Model Composer.

1) Flatten out the circuit graph, creating part instances and signal nets.
2) Synthesize a Schematic and PCB design file, compatible with EagleCAD. (The EagleCAD PCB is shown in Figure 8.7.)
3) Add subsystem pre-routes, produced either by hand, from a reference circuit design, or with an autorouter.
4) Complete routing of signals between subcircuits circuits, typically done with the assistance of the autorouter.
5) Manufacture the PCB and add the electronics devices, connectors, etc.

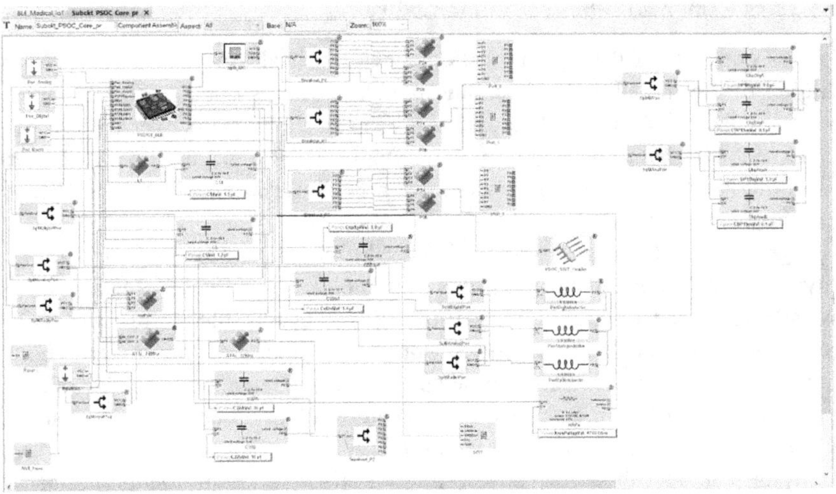

Figure 8.6 Internal model of the core BLE platform.

PCB

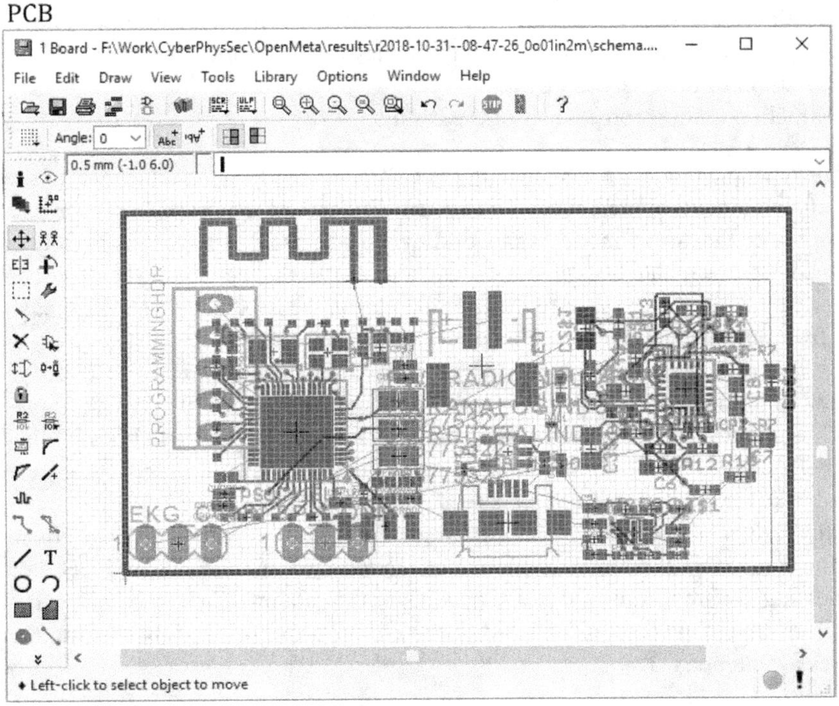

Figure 8.7 PCB design.

8.3.3 Assessment of Physical Vulnerability

To assess potential of RF injection, the behavior-critical inputs and outputs are identified in the model as Events (Blue Arrows). This is done by the designer, who is cognizant of the hardware/software interactions. Effectively, these are the pins that are used for any system functionality. Typically, a multipurpose chip such as the PSOC microcontroller will have many unused pins in any particular design implementation.

Figure 8.8 shows the model of the behavior, linking hardware events to specific behavioral transitions and states. In this example, we capture events which have active signals. In the absence of this identification of events, the entire board must be analyzed for physical vulnerabilities. Complex boards can contain many thousands of metallic traces, vias, and connectors. Focusing the target analysis to a limited set of events used in a particular application can reduce required analysis compute load by several orders of magnitude.

Analysis is achieved via a RF susceptibility testbench. The testbench performs the following steps:

- Creation of an Ansys SI Wave project.
- Importation of the target system to be analyzed.
- Creation/activation of the ports of interest, based on the models shown above.
- Setup the analysis job to sweep the frequency bands of interest.
- Execution of the analysis job.
- Extraction and postprocessing of the results.

The example testbench is shown in Figure 8.9, which allows specification of the RF analysis in a graphical form. The System Under Test (PSOC_IoT_Sensor) is a reference to the design model. The workflow object specifies that any assembly of 3D geometry associated with the system be executed and model prepared. The other parameter objects specify the location of preprocessed SI Wave PCB model, the frequencies to sweep, and the target nets to analyze.

Execution of the testbench produces an SI Wave project, shown in the image below. Automatically creating the SI Wave model within the OpenMETA framework is still a work in progress, due to tool file format conversions of the synthesized EagleCAD files into an Ansys Model. Currently, several manual steps are involved, using other external tools (e.g. Altium).

The results of the analysis model setup are shown in the SI Wave tool in Figure 8.10. Execution of the Induced Voltages analysis in SI Wave produces data indicating the amplitude and phase of voltages that are predicted to be induced if an electromagnetic wave is directed at the circuit board, effectively treating the traces as antennas [13].

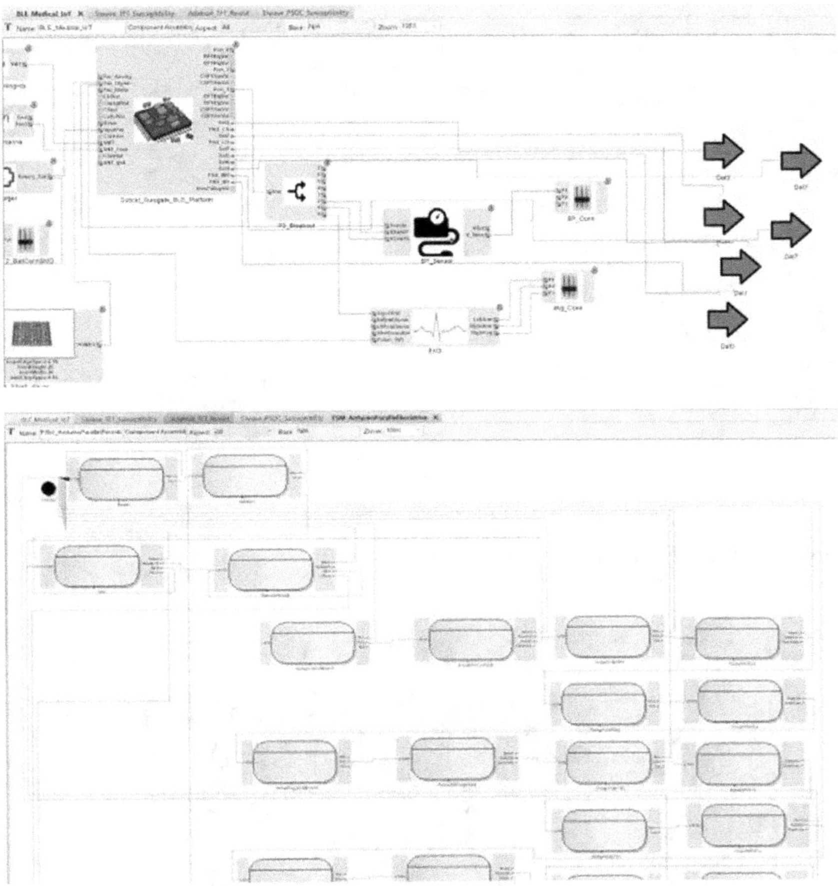

Figure 8.8 Links between hardware events and behavior models.

We can compute a gross/first-order estimate of probability of attack success on a particular trace via a weighted summation of induced voltages across the anticipated spectrum, evaluation of the peak induced voltage levels, or some combination thereof. Figure 8.11 shows an example of the raw computed induced voltages. The analysis spanned a nominal range (100 MHz to 5 GHz) and shows peak induction around 1.3 and 2.6 GHz. Experimental data for RF model validation was not able to be collected for the design study due to the cost of instrumenting the board.

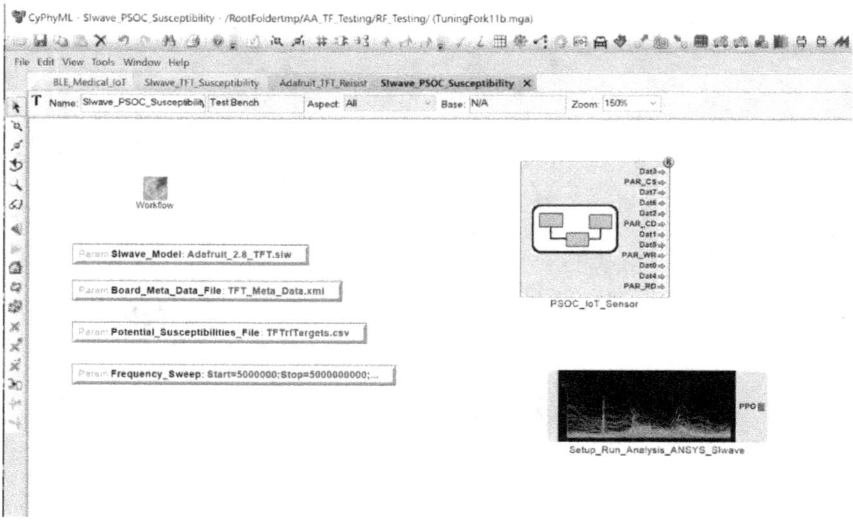

Figure 8.9 Example testbench for RF susceptibility.

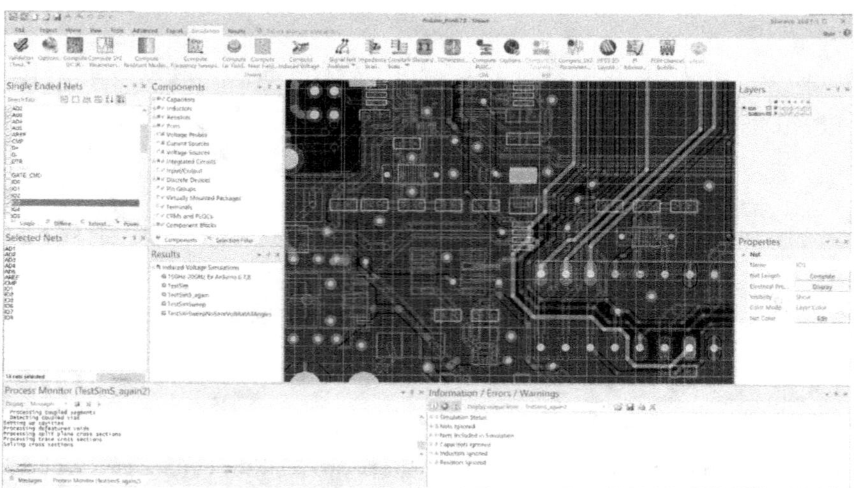

Figure 8.10 Model for SI wave analysis.

To properly measure received RF energy, specialized probes are required that must be carefully shielded, electronically isolated, and attached to the PCB trace. Typical scope probes, and/or lead wires bonded to the board, will receive RF energy and artificially inject the voltage/current into the probe point, drastically changing

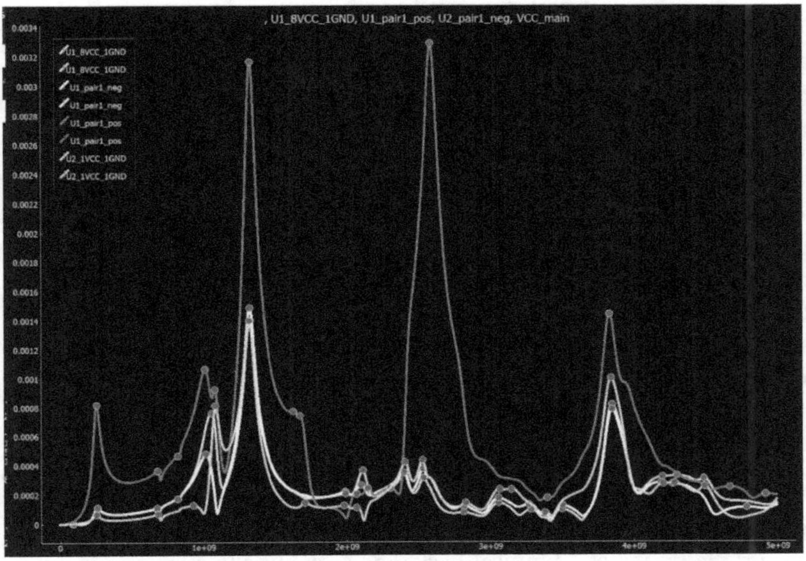

Figure 8.11 Evaluation results.

the RF behavior. Given sufficient resources, high-bandwidth RF probes can be designed into the PCB and routed to external, isolated instrumentation.

This evaluation example has been performed using 2D, PCB-only analysis tools. If the physical housing, cabling, and external mounting are available, they can be modeled and evaluated using full 3D analysis tools (Ansys HFSS) for induced voltages in a similar manner.

Once the physical RF analysis determines a set of traces that can be energized at specific frequencies, the impact of the energy on the circuit must be determined. For this, we can use circuit analysis tools, e.g. Spice, to determine how the energy is transformed and can appear at a device input. Passive and active devices can amplify, filter, and act on voltages/currents in nonlinear ways to modulate and transform between frequencies. Unfortunately, estimating the exact equivalent circuit relies on chip I/O circuits, which are often proprietary. Deriving their properties from measurements is inexact and expensive. Given this, we propose using an estimate of probability of impact based on the magnitude of the induced voltages and the type of input. Experimental data are needed to determine how to weigh induced voltages, how to determine the impact of multiple frequencies, and the impact of various waveforms on the target. Creation and evaluation of these probability functions are a work-in-progress.

8.3.4 System-Level Impact Analysis

Assessing the consequence of a potential physical impact uses the behavior state machine abstractions captured in the system models. The thread is as follows:

1) The models contain a mapping of physical pins to events perceivable in the discrete hardware and software.
2) The computed probabilistic events are used to annotate all the transitions of the behavior state machines.
3) A set of anomalous states are annotated by a system designer, corresponding to potential undesired hacking targets.
4) A set of nominal states are identified as typical behavioral points occurring in normal operation.
5) The system models are transformed to the PRISM tool,[5] with full states/transitions and probabilities.

Cyber analysis can then be performed by assessing the probability of reaching each of the anomalous states from each of the nominal states. Note that this does not consider the timing of a physical hack, and sequences thereof, rather it performs a pessimistic approximation of a properly sequenced attack on a device or subsystem.

8.3.5 Conclusions

Once single device is compromised, the other devices connected to it can be compromised. Typical cyber-information flow-based security analysis can be used to explore system impacts once an entry point is achieved.

Cyber-physical attacks can bridge these boundaries in ways that are not constrained in the software design. Real systems have real physical implementations. Physical components react to energy in predictable ways, but often are not considered in the security design of the system, given the myriad interactions possible over the available power conduction mechanisms.

For this reason, automated searching and reasoning tools are needed, much as system performance testing tools are needed in software development. Computational searching of the physical/behavioral device can be done before the system is implemented, at a lower cost and to a greater variety than is possible in a physical laboratory. In the prior section, we presented a method for assisting in the assessment of physical entry points and the associated behavioral impacts. While the tools are still in their infancy, the results are promising.

5 https://www.prismmodelchecker.org/

8.4 Open Questions and Future Research Directions

Open architecture, physicality, and scale make security of IoT systems an open challenge. The sheer volume of information required for managing devices makes centralized solutions all but impossible in many current and future IoT applications. As the example described in this chapter showed, susceptibility to physical layer attacks increases the attack surface to IoT devices, and brings in new dimensions to vulnerability analysis and risk mitigation. This brings increasing importance to research targeting the development and deployment of IoT composition platforms that enforce the use of security frameworks in all potentially safety-critical IoT applications.

As noted in prior sections, accurate vulnerability analysis is dependent on detailed information of designs, which is often proprietary. Constructing surrogate models, based on circuit technologies, will assist in the electrical/physical analysis. The approaches proposed here involve large search problems, across multiple domains. Creating efficient solutions which minimize computational time is highly desirable. Furthermore, reasoning about the temporal and sequencing impacts will make vulnerability assessment more accurate.

References

1 James Manyika, Michael Chui, Peter Bisson, Jonathan Woetzel, Richard Dobbs, and Jacques Bughin, Dan Aharon: *The Internet of Things: Mapping the Value Beyond the Hype*, McKinsey Global Institute, June 2015.

2 Peter Evans and Marco Annunziata: *Industrial Internet: Pushing the Boundaries of Minds and Machines*, GE, November 26, 2012.

3 https://www.openfogconsortium.org/.

4 Stephen Ornes: Inner workings: Medical microrobots have potential in surgery, therapy, imaging, and diagnostics. *Proceedings of the National Academy of Sciences of the United States of America* 2017; **114**(47):12356–12358.

5 Gabor Karsai, Xenofon Koutsoukos, Himanshu Neema, Peter Volgyesi, and Janos Sztipanovits: Transportation networks, in Alexander Kott and Igor Linkov (Eds) *Cyber Resilience of Systems and Networks*, pp. 425–446, Springer, 2018.

6 Steven Williams: *Communication in Mechanism Design*, Cambridge University Press, 2008.

7 Open Web Application Security Project (OWASP): The free and open software security community, Internet of Things Project, retrieved 2016-09-26 http://www.owasp.org/index.php/OWASP_Internet_of_Things_Project.

8 Janos Sztipanovits, Ted Bapty, Ethan Jackson, Xenofon Koutsoukos, Zsolt Lattman, and Sandeep Neema: Model and tool integration platform for cyber-physical system design, *Proceedings of the IEEE* 2018; **106**(9):1501–1526.

9 http://ngSPICE.sourceforge.net/.

10 https://openmodelica.org/.

11 https://www.prismmodelchecker.org/.

12 https://www.ptc.com/en/products/cad/creo.

13 G. Kevin Zhu, Werner Thiel, and J. Eric Bracken: A method to assess the radiated susceptibility of printed circuit boards. Proceedings of the 2016 IEEE 25th Conference on Electrical Performance of Electronic Packaging And Systems (EPEPS), October 2016, San Diego, USA 10.1109/EPEPS.2016.7835431

9

Securing Smart Cities

Implications and Challenges

Ioannis Agadakos[1], Prashant Anantharaman[2], Gabriela F. Ciocarlie[1], Bogdan Copos[3], Michael Emmi[4], Tancrède Lepoint[5], Ulf Lindqvist[6], Michael Locasto[1], and Liwei Song[7]

[1] *SRI International, New York, NY, USA*
[2] *Department of Computer Science, Dartmouth College, Hanover, NH, USA*
[3] *Google Inc., Mountain View, CA, USA*
[4] *Amazon Inc., New York, NY, USA*
[5] *Google Inc., New York, NY, USA*
[6] *SRI International, San Luis Obispo, CA, USA*
[7] *Princeton University, Princeton, NJ, USA*

9.1 Foreword

Smart cities are designed and envisioned as cities augmented with technological capabilities facilitating various types of actuation, sensing, and control of their infrastructures. Countless possible applications include traffic monitoring and control, energy consumption, parking, lighting, building health and automation, waste management and recycling, air quality, noise monitoring, hazard detection, and auditing. The level of detail at which a smart city may operate and control its infrastructure varies, but the common vision is that it will be able to do so in high resolution, having embedded network-capable devices in as many relevant components as possible.

Although the quality of collected information depends on the safe and effective deployment and maintenance of various essential devices and software capabilities, the benefits a smart city brings to its residents and visitors are unquestionable. The aggregated live and accurate image of a city's infrastructure and the ability to automatically respond to changing conditions will allow city governments to serve and protect citizens more effectively than ever before.

Modeling and Design of Secure Internet of Things, First Edition. Edited by Charles A. Kamhoua, Laurent L. Njilla, Alexander Kott and Sachin Shetty.

Published 2020 by John Wiley & Sons, Inc.

However, existing smart technologies are not yet fully designed to cope with disruptions such as power or Internet outages, intentional cyberattacks, natural disasters, or warfare (megacities are envisioned as future battlefields). Overcoming risk and skepticism regarding the guaranteed safe implementation of pervasive and broadly public smart-city technology requires engineering tools and approaches, algorithms, models, and analytics that reflect the intricate and ever-changing details of smart cities and so that the resulting technology can fully sense Internet of Things (IoT) environments, connect and federate devices to create self-awareness, respond to threats and error conditions, and in all ways guarantee the safety and rights of the population.

9.1.1 Smart City Origins

While the term "smart city" has been used with increasing frequency over the past decade, it is natural to wonder where it came from and what it really means. This section briefly examines the history and evolution of the current vision of a smart city. The roots of a smart-city concept first emerged in the realm of science fiction, appeared as Urban Cybernetics [1] in the 1970s, included the use of information and communication technologies (ICTs) in cities of the mid-1990s, and evolved to the current view of the term.

One of the first instances of an advanced smart city appeared in the iconic 1927 film *Metropolis* by Fritz Lang. In this unconventional film, sophisticated machinery turns an urban environment into a futuristic habitat where the central authority of the city pervasively monitors and controls both the infrastructure and the people. The possibly most well-known version of a smart city is found in *1984*, George Orwell's uniquely influential novel. This groundbreaking work, authored in 1948, features a dystopian futuristic city where the population is under constant scrutiny and control, and emphasizes the deprivations and dangers for the inhabitants of such environments. Both of these works, although dysfunctional science fiction, presciently included a common denominator that exists in today's scientific view of smart cities: the surveillance and control of both infrastructure and the inhabitants.

The realm of urban planning in the 1970s introduced the term "Urban Cybernetics" [1] in an extensive work that summarized the research at that time, established the notion of systems and their dynamics that involved modeling cybernetic cities, and specified a framework for the scientific study of this novel idea. In the 1980s, as discussed by Hollands [2] and Söderström et al.[3], the research community adopted the term "smart" from the Smart Growth and New Urbanism movements that were prevalent in the United States. In the 1990s, with the proliferation of the Internet to everyday life, more cities earned the smart designation by implementing information technology (IT) services such as e-governance and augmented ICT infrastructure [3]. The increasing demand for digitization and e-governance has provided significant incentive for business and industry to advance a more positive vision and viable technology for a smart city, and to establish themselves as authorities in the field and garner the associated profits.

More recently, a rapid proliferation of publications has revolved around the notion of smart cities. As Komninos and Mora [4] indicate, the count of published documents grew from 20 documents (1992–2001) to 290 (2002–2009) and to 916 (2010–2012). Komninos and Mora also found an unfortunate trend that papers are not cited, leading to a disconnected and fragmented research landscape. Söderström et al [3] noted an interesting connection between rapid publication growth and industry interest in IBM's trademark of the term "Smarter Cities" in 2011, following a campaign to establish its view of the smart city notion and its expertise in the field.

While our historical and literature review is far from comprehensive, it highlights the evolution and complexity of the smart-city concept, which is a deeply interdisciplinary problem involving scientists from fields such as urban planning, electrical engineering, computer science, environmental engineering, and others. This chapter focuses on the security and privacy perspectives of smart cities and outlines the crucial steps for a successful secure smart-city implementation.

9.1.2 Chapter Layout

The remainder of this chapter introduces the concepts of detection, fingerprinting, identification, and various proposed techniques that can be used to aid in the difficult but crucial task of mapping the IoT scape in a smart-city environment. Section 9.2 discusses the crucial first steps in smart-city surveillance: the detection, identification, fingerprinting, and behavioral monitoring of system components. Section 9.3 discusses how this collected information can be used to reason about and model the smart-city environment. Section 9.4 outlines the principles governing the tasks related to connecting and federating IoT technologies in smart cities.

9.2 Detection, Identification, Fingerprinting, and Behavioral Monitoring in Smart Cities

The first step in evaluating a smart environment's status and identifying potential problems is the ability to sense the devices in a given location. From a security perspective, sensing can also detect the presence of malicious devices. While discovery and characterization of fixed wired assets has received significant attention over the last decades, the algorithms and limits of the large-scale discoverability and characterization of cyber-physical, mobile, and wireless assets remain less well-studied. In contrast with conventional systems, the diversity and heterogeneity of smart devices, which can include intermittently connected sensors or actuators that do not appear at consistent topological locations, prohibit instrumentation. Because each device may run custom and/or proprietary software that cannot be modified, it is impossible to instrument the software of each smart device.

However, continuous noninvasive monitoring methods are becoming more scalable and practical in a smart-city scenario. Sensing devices in smart cities need to use hybrid techniques that combine cyber-discovery with physical discovery by using methods to exploit observations of physical signals such as radio frequency (RF) sensing, the over-the-air capture and processing of RF emanations from devices. Passively capturing RF communications is noninvasive and can provide significant details about the identity and behavior of a device. The advent of the IoT generated a plethora of protocols that can transmit on a variety of frequencies (e.g. BLE, Wi-Fi, Zigbee, and Z-Wave). While the most proliferated protocols (BLE, Wi-Fi, Zigbee) broadcast in 2.4 GHz, different modulation schemes require the capture and processing of specialized radios; for example, BLE uses RF hopping, while Wi-Fi uses quadrature amplitude modulation (QAM). To provide an accurate view of the devices in a location, a sensing system must capture the full spectrum of potential signals. A dedicated RF sniffer with specialized detection algorithms for each known communication channel (Zigbee, Wi-Fi, BLE) combined with additional analysis capabilities would provide insight into the origin of each transmission.

Similar to a living organism with various organs that work in tandem to sustain and power its functions, a smart city is composed of devices forming a loosely coupled distributed network. Devices comprising the smart-city network must be monitored to ensure that they behave as expected and that only authorized devices are part of the smart city's network. New devices introduced to the network must be promptly detected and identified. If a new device is introduced by an unauthorized source, the system must isolate it and ensure that it does not contaminate the rest of the system by providing false sensor data or interfering with other devices (e.g. MAC spoofing, jamming, impersonating).

Protecting a smart city from potentially malicious introduced devices requires a continuous monitoring system that passively (to avoid performance degradation of functional smart-city tasks) searches for and detects new devices. The detection process has several granularities; each level reveals more information and depends on the previous level.

The research literature contains some ambiguity in the definitions of identification and fingerprinting. This chapter uses the terms *identification* and *fingerprinting* as defined in the following categories outlining the different levels of sensing devices' capabilities.

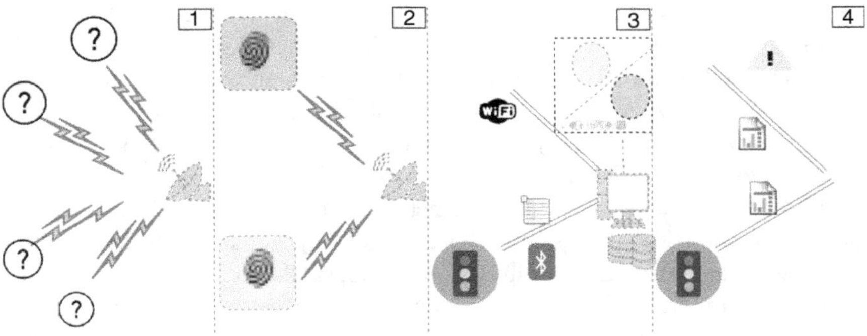

Figure 9.1 Different granularities of analyzing smart-city devices. (1) Detection starts with perceiving the existence of a device based on transmitted signals. (2) Fingerprinting results in a link between a signal and a specific device, enabling the assessment of the number of devices. (3) Fingerprinting uses higher level analysis based on exposed protocol characteristics or a MAC address and other collected information to identify and link a device to a particular type. (4) Behavioral monitoring determines the type, specific instance, and capabilities of a device and reasons about whether the device has been compromised, discriminating between valid states of that particular device and its dangerous or invalid states based on its perceived behavior.

1) Detect that a device is present: sense the presence of an RF signal emanated from a device.
2) Fingerprint a device: reason about the characteristics of the RF signal originating from a device and attribute the received signals to corresponding physical devices and/or transmitters.
3) Identify a device: exploit the collected information to reason about a detected device's physical and cyber capabilities (operating system (OS), kernel version, etc.), including the type and number of output/input channels (BLE, Wi-Fi, sound, etc.). More sophisticated identification methods could match collected information against an offline analytical database to discover a superset of capabilities augmenting the perceived information (see Figure 9.1 step 4). A sound and complete identification step is crucial for more advanced analysis.
4) Monitor the behavior of an identified device: reason about changes in its operational profile that could signal that the device was compromised or verify that it continues to operate as intended.

9.2.1 State of the Art

The research literature contains a plethora of methodologies and techniques that have been used to achieve device fingerprinting, identification, and, more recently, behavioral monitoring based on RF signals. While an exhaustive literature review is beyond the scope of this chapter, we can classify various **active** and **passive identification methods** by the way they interact with the device, and **physical**- and **network-layer approaches** based on the type of information they use. While we list a small number of systems proposed in the literature for each category; a comprehensive study of recent fingerprinting methodologies can be found in [5].

Active identification methods interact with a device over a series of steps to elicit information. The downside of this methodology is that it induces unnecessary network load that, in environments with thousands of devices, could prohibitively interfere with network operation and performance. On the other hand, these methods can actively search for and interact with devices to discover more information about them. The work of Bratus et al. [6] performs identification by polling the target device with a sequence of specially crafted or malformed 802.11 frames and observing its responses.

Passive identification methods listen to network traffic and RF signals without actively initiating any communication. These methods have minimal or zero impact on the monitored network, but they are limited in observing traffic generated from other parties.

Physical-layer-based approaches look for protocol-agnostic features such as the estimated clock skew of a device or the transient on/off behavior of a transmitter; they do not rely on exchanged information. These approaches have been studied extensively and have high accuracy, but they fail to identify cyber features such as the type of software running on a device. The earliest methods in the literature (during World War II) were used to determine allied versus enemy radar signatures; matching was done by visually (and manually) comparing the oscilloscope-generated RF signatures [7]. More recently, due to the importance of device sensing, various physical traits are used for device fingerprinting based on unique transient signatures [8], amplitude and phase-generated fingerprints [9], each device's unique clock skew when compared to a certain point of reference [10], and modulation characteristics [11].

Network-layer or **data-dependent** [5] **approaches** exploit *a priori* knowledge of the observed networking protocol (UDP, TCP, HTTP, etc.) and attempt to retrieve parts (e.g. MAC address, packet identifiers, source-target IP) of the exchanged messages to elicit information about devices [12]. They can either require access to the router [13] and operate on all traffic, or, without requiring access to the router, they can use unencrypted visible packet information (e.g. Franklin et al. [14]) and beacon requests to identify drivers, and use MAC addresses (e.g. source and target address, packet size) to fingerprint devices. **Hybrid** methodologies can take advantage of both the signal-agnostic benefits and the detail and qualitative information that can be elicited through the data-layer.

9.2.2 Our Approach

Our view of a smart city includes a variety of devices with different interfaces (e.g. Zigbee, BLE, and Wi-Fi) and various protocols. While methods have been proposed for device identification and fingerprinting, these were deployed in individual settings (e.g. only Wi-Fi and BLE) or addressed only parts of the entire identification process such as identification, but not behavioral monitoring or fingerprinting [13]. In our in-house prototype system, our RF sniffer tackles the problem as a whole by monitoring networks composed of devices that communicate over any and all of the previously described interfaces, while simultaneously performing fingerprinting, device capability identification, and behavioral monitoring.

9.2.2.1 Cyber Characteristics

The first step to identifying devices is capturing their transmitted signals. This is handled by a sniffer device with receiver antennae and the appropriate demodulation hardware to capture given types of signal. A captured signal can be used unmodulated as a sequence of in-phase and quadrature (IQ) pairs or demodulated with the digital sequence of bits extracted in some data format. Physical-layer approaches rely on the raw signal characteristics or on attributes extracted by some calculation such as the estimated transmitter frequency or clock. At least one sniffer is required for each RF communication type being monitored. RF communication type denotes the complex technological construct that makes up a given communication method. For example, Wi-Fi uses 2.4 GHz as a center frequency and QAM modulation, while Bluetooth uses frequency hopping around 2.4 GHz. For some protocols, such as Wi-Fi, a single sniffer could monitor multiple devices, but other types of RF communications could require more than one sniffer. For example, the complex communication scheme of the Bluetooth protocol requires a dedicated sniffer per connection. Besides dealing with the complexity of a tracked RF scheme, using multiple sniffers for the same protocol serves other purposes, such as estimating the physical location of the transmitter and eliciting higher level information such as coalescing different interfaces to one device. The SRI International Sniffer strategy uses multiple sniffers to extract location-sensitive features.

9.2.2.2 Physical Characteristics

Our scheme uses the relative received signal strength information (RSSI) from receivers with multiple antennae to extract direction and distance metrics, and uses those metrics to separate devices of the same class. Extracting these metrics does not require specialized equipment, which would be prohibitive for a large-scale deployment; common-off-the-shelf (COTS) Wi-Fi routers with more than one antenna can provide the required information with adequate accuracy for small distances. Specifically, this information is extracted from the RFTAP 802.11 header; each of the router's antennae reports the received signal strength on every collected packet. (This information is also available without associating the router with a network by operating in monitor mode and observing beacon messages.)

As cyber-physical systems, IoT devices have the potential to exhibit implicit interactions through their actuation and sensing capabilities. The probability for such interactions increases in areas with a high density of IoT devices. Often subtle and unanticipated, these interactions can have unwanted consequences and should be considered equivalent to explicit interactions via standard protocols and radio-based technologies. However, unlike traditional communications, these interactions are difficult to observe and monitor.

Our prior work [57] introduces a cyber-physical IoT monitor that uses various radios to monitor communications between IoT devices. The monitor is equipped with sensors designed to capture changes in the physical environment, such as ambient light, temperature, and sound. Providing this level of monitoring in IoT environments allows the identification of implicit interactions between devices through sensing and actuation actions. This capability introduces transparency into the complexity of the interactions among IoT devices and enables administrators or users to identify the root causes of activities in the IoT. Additionally, such monitoring enables the composition of more complete behavioral models for IoT devices. Behavioral models can be used to identify anomalies and to better understand IoT deployments and their activities.

While monitoring the physical environment is valuable in IoT, several challenges are associated with this approach. Scalability and coverage are challenging in smart cities where devices are geographically dispersed; capturing implicit interactions through sensing and actuation requires the distribution of sensors. Similar to scalability and coverage challenges, it is difficult to reason about the completeness of such monitoring. It is trivial to identify the intended actuation and sensing capabilities for a given set of IoT devices; however, electronic devices can experience involuntary electromagnetic emanations and/or sound. It is reasonable to imagine a scenario where such emanations are sensed by another neighboring device equipped with the appropriate hardware. The added complexity and cost for the infrastructure required to monitor IoT devices and their impact on the physical environment and other neighboring devices is a formidable challenge.

9.2.2.3 IoTHound

The infrastructure of smart cities is tightly coupled to the technology that drives the infrastructure. A challenge arises when comparing the longevity of the cities to that of device manufacturers. The devices of a smart city will include new devices constructed by established manufacturers, proof-of-concept prototypes from young ventures or enthusiasts, and unsupported devices from defunct manufacturers. Such dynamism and heterogeneity introduce unique challenges for security monitoring with respect to identification and fingerprinting of devices. Specifically, security-monitoring mechanisms must be able to identify and characterize devices of all types and maturity levels. An incomplete understanding of assets and their activities elevates the risks for cyber-security incidents.

Previous research efforts [16] have considered defining distributed blockchain-based frameworks for the identification and management of devices. Although such frameworks have many benefits for asset enumeration, security, and management, it is important to consider complementary approaches that support the identification of new and rogue devices.

IoTHound is our novel technique for characterizing and tracking IoT devices based on the characteristics of network traffic. Previous approaches [13] rely on the organizationally unique identifier (OUI) of MAC addresses or IP-level details (destination IPs, protocol, etc.) and apply supervised machine-learning techniques to generate fingerprints for various device types. In a future smart city, where new devices can be introduced seamlessly without coordination, supervised machine-learning approaches will fail to accurately identify devices. Additionally, some features, such as MAC address, may not be reliable in the presence of moving target defense mechanisms such as MAC address randomization. IoTHound is designed to combat these limitations. It requires no previous knowledge regarding devices and device types; instead, the proposed approach applies the hierarchical density-based spatial clustering of applications with a noise algorithm (HDBSCAN) to cluster devices into classes (e.g. sensor, actuator, and router) and roles (e.g. hub, end-device) based on the representative timing and volume attributes of their communications, such as inter-packet time, packet length, and packet payload length. Observations are sliced into time windows and, for each time window, the mean, standard deviation, minimum, maximum, and quartiles (Q1, median, Q3) of these features are computed. The measurements are agnostic to the communication technology (e.g. Wi-Fi, Bluetooth, and Zigbee). As depicted by Figures 9.2 and 9.3, evaluation of the approach on commercial IoT devices produced promising results of 95% classification accuracy.

The benefits of black-box characterization approaches as presented by IoTHound lie in their dual-use for security. Specifically, performing the clustering analysis over time can provide useful information regarding behavior deviations of the devices. Such deviations can be benign, caused by firmware updates, configuration changes, or the introduction of new devices that causes role changes. However, deviations from normal behavior can also indicate cyberattacks (e.g. the Mirai distributed denial of service (DDoS) attack significantly changed the

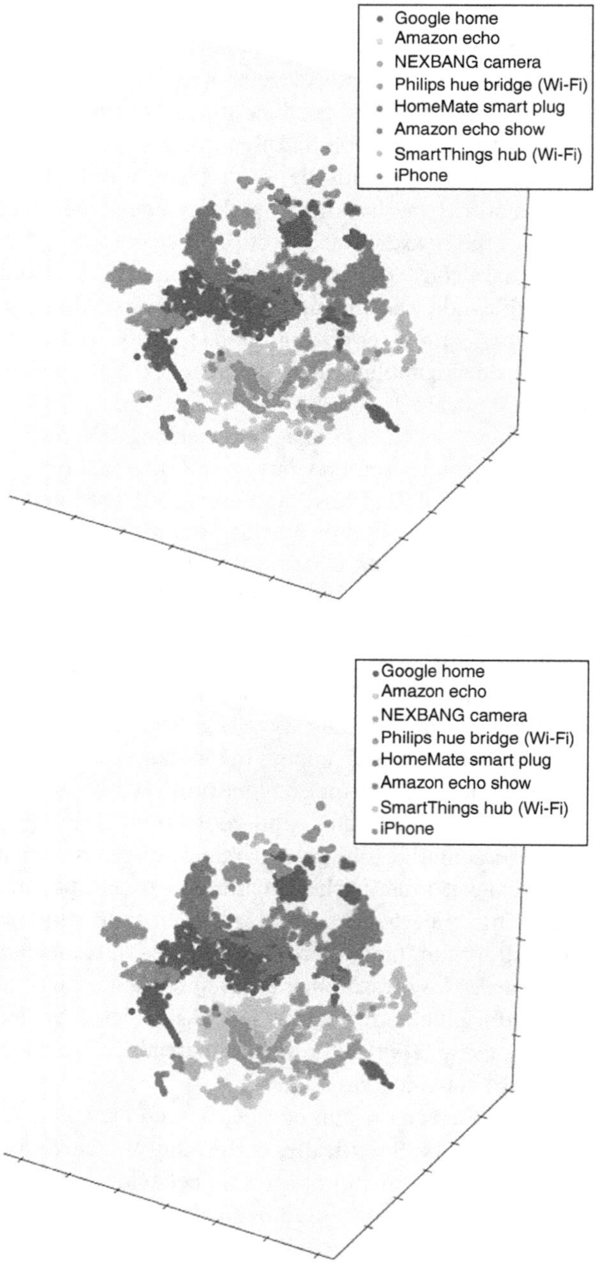

Figure 9.2 Ground truth (top) versus clustering results (bottom). Despite many similarities between the clusters, the approach is able to classify devices with high accuracy.

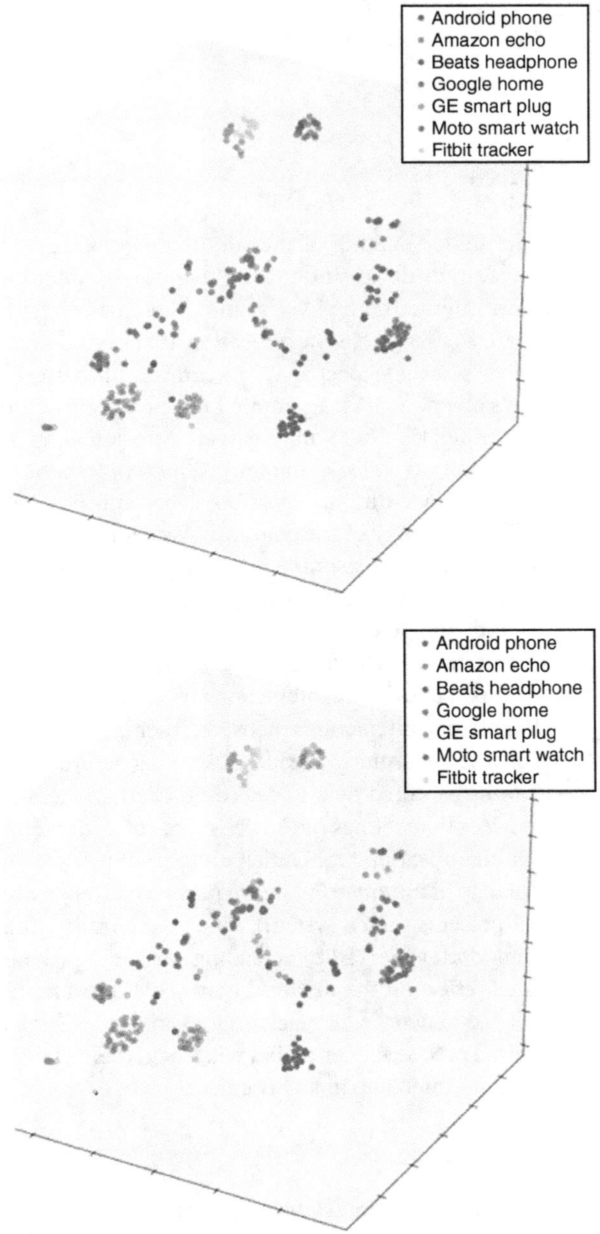

Figure 9.3 Ground truth (top) versus clustering results (bottom). The clustering approach generally works very well although there are some misclassification results (e.g. top right cluster).

network behavior of IoT devices). Regardless of the intent, the insight produced by such analysis can be beneficial and introduce transparency in the otherwise complex and obscured nature of the IoT.

9.3 Modeling Smart Cities

Going beyond sensing toward the understanding of the smart-city environment requires the creation of a real-time, inter-device interaction map. The numerous interfaces with which smart devices interact, both physical and cyber, make it hard to depict such an interaction map. For example, in a location that contains an Amazon Echo and a smart TV, besides the expected Wi-Fi connection between them (if they are connected to the same network), an implicit physical connection exists over the physical acoustic channel; i.e. the TV may activate Amazon Echo by playing a video containing the appropriate voice command. When a plethora of sensors and actuators exist in the same vicinity, the possible cyber and physical interactions can affect the behavior and safety of the composite system. To tackle this problem, we leverage automated formal reasoning engines.

The inherent rigor of formal reasoning engines implies that, given a realistic model, our results will be sound and complete within specified bound limits (e.g. number of devices or the type of modifications an attacker may perform). Our prototype framework is able to capture possible device interactions based on their provided interfaces and depict all possible interactions between them. To model interactions, our system is based on a set of formal logic rules describing natural laws (e.g. sound is broadcasted and can be captured by any device in a certain vicinity based on range) and a model of devices that defines the key elements of device entities (e.g. interfaces, size, location, authentication capabilities, and movement potential). Creating an accurate physical map of the environment requires comprehensive multichannel sensing (Section 9.2 presents such a system). Finally, the framework can also reason with linear temporal logic (LTL), enabling reasoning about sequences of discrete steps and their effect on the modeled network. For example, our model can determine whether the network will reach an invalid state within a given number of steps and identify the precise sequence that will lead it there, which will be particularly useful for causality inference in smart cities.

9.3.1 State of the Art

Modeling systems and their behavior have been extensively studied. Researchers have demonstrated the use of languages [17] to describe systems and how such descriptions can be leveraged to formally verify (security) properties of the systems.

Significant work has also been done in threat modeling. Threat modeling aims to analyze the security of a system and identify potential threats by approaching the problem from the perspective of a hypothetical attacker. A popular method for threat modeling is by leveraging attack trees [18]. Attack trees represent attacks and countermeasures as a tree structure, where the root node represents the goal of the attack and the rest of the nodes denote attacker moves. Other approaches leverage attack graphs to evaluate the security of various systems [19–21], while others study methods for automatically generating attack graphs [22, 23].

More recently, researchers have begun to apply some of the previously mentioned methods to cyber-physical systems and specifically to IoT systems. Kovatsch et al. [24] proposed a technique to build a web-like mesh of IoT devices by composing the representational-state-transfer (REST)-based application programming interfaces (APIs) of various IoT devices using a rule-based reasoning engine. In their work, the authors found that their model led to a state-space explosion for large networks. Unlike previous work [24] that focused on the application layer, we use a satisfiability-based approach and look at the protocols from both the data link and application layer. Mohsin et al. [25] proposed IoTSAT, a formal framework for the security analysis of IoT networks. They also follow a satisfiability-based approach to model and analyze security. These papers, however, report different assumptions in their respective models. First, IoTSAT assumes that an attack on actuators can affect sensors only through the sensors' observations of the environment [25] (Figure 9.3). We do not make such assumptions (i.e. interactions of only one type and direction: Sensors → Controllers → Actuators), thus accounting for arbitrary ways in which actuators can affect sensors (or other types of devices); hence, our model can detect attacks that are undetectable in IoTSAT. Our model also considers controllers and actuators as potential starting points of attacks and does not assume them to be adequately hardened.

Guilly et al. [26] applied an event-condition-action (ECA), language-based approach to extend ECA using timed automata to identify faults and unsafe conditions in a home automation system. Corno and Sanaullah [27] proposed a design-time modeling and formal verification methodology for smart environments using ontologies and state charts. Augusto and Hornos [28] presented a methodological guide on the use of LTL to model, simulate, and verify intelligent environments. Coronato and Pietro [29] proposed guidelines for requirement specifications and verification through extensions to ambient calculus and ambient logic. All this research focused on verifying the inherent and non-malicious behavior of specific systems and lacks generality and security analysis against different attack vectors.

Nonformal techniques have also been applied to evaluate and enforce the security and privacy of IoT systems. IoT Sentinel [13] uses software-defined networks

(SDNs) to restrict the communications of devices according to a set of predefined policies. Both IoT Sentinel and our work perform network traffic analysis to automatically identify devices on a given IoT network. Our work, however, aims to determine potential attack paths, while IoT Sentinel focuses on mitigation and isolation techniques. Mavropoulos et al. [30] developed a tool named ASTo that uses a domain-specific modeling language to generate a visualization of IoT systems, this approach facilitates security analysis during the design and implementation phases of an IoT system. Rullo et al. [31] applied a game-theoretic approach to IoT security modeling to allow a defender to efficiently allocate resources to protect an IoT system, but this approach abstracts the heterogeneity of IoT systems and lacks system-level evaluation capabilities. Tekeoglu and Tosun [46] described a testbed for evaluating the security and privacy of IoT systems by analyzing layer-2 and -3 packets. Although our work also analyzes network traffic, we go beyond the observed network traffic to draw conclusions about the security and privacy of an IoT system. Geneiatakis et al. [33] studied the interactions between devices in a smart-home testbed and used the information to make inferences about general IoT security and privacy flaws, while considering both internal and external attackers with varying capabilities. While their output is similar, our approach automatically verifies the security properties of an IoT system through model checking.

9.3.2 Our Approach

To ensure the predictable operation of smart-city environments, automated formal reasoning engines facilitate the creation of mathematical models that abstract the properties and interactions of devices participating in smart-city networks. To be effective, the design of mathematical modeling choices must satisfy several properties. First, the constructed models must faithfully capture the crucial properties of the real-world devices and the interactions they represent, and the properties that are subject to prediction. For example, the models would capture cyber and physical channels among networked devices and predict whether a cyber channel could be established among a given pair of devices in the network. Unsound modeling choices undermine the ability for prediction. For instance, ignoring the physical audio channels on a smartphone could prevent the prediction of attacks that penetrate smart assistants.

Additionally, these models must be expressible in the input language of effective automated reasoning engines. Popular choices include logical satisfiability solvers and temporal-logic model checkers. This is an important consideration which can also undermine predictions, since performance requirements may preclude the use of certain classes of formal reasoning engines that address intractable decision problems.

The remainder of this section outlines the general design of models that capture cyber-physical interactions among devices, the general form of queries that capture the system properties for prediction, and the possible formal reasoning engines that can answer these queries.

9.3.2.1 Modeling Devices and Interactions

The basic abstractions of smart-city environments are presented as labeled graphs.[1] Nodes capture physical entities like devices and locations, while edges capture cyber and physical relations like Bluetooth connections, proximity, and earshot. For instance, in the smart-city scenario in Figure 9.4, the properties of devices and relations are captured by labels. Node labels capture device characteristics and state, for example, if a device includes Bluetooth capabilities, and if the Bluetooth receiver is switched on. Edge labels capture channel characteristics and state, for example, if a channel is Bluetooth or Wi-Fi and if encryption is enabled. This simple characterization allows for great flexibility in modeling choices, since graphs with labeled nodes and edges can capture arbitrary binary relations among the constituent nodes; further flexibility could be achieved by extending the characterization to hypergraphs, whose edges relate arbitrary tuples of nodes, not only pairs.

Besides characterizing momentary snapshots of smart-city environments, the ability to predict motivates an interest in capturing temporal properties [15], i.e.

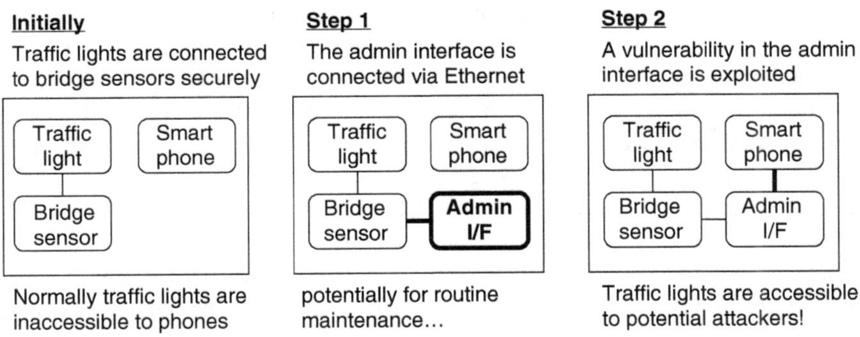

Figure 9.4 Graph-based temporal modeling of a scenario where traffic lights are controlled by bridge sensors. During periodic maintenance, a vulnerable administrator interface is connected to the sensors, at which point potential attackers may transitively gain control over traffic lights.

1 A labeled graph $G = \langle N, E, L \rangle$ is a set N of nodes, along with a set E of edges connecting pairs of nodes, and a function L mapping nodes and edges to labels.

the manner in which a given network may evolve over time. Modeling these temporal aspects considers the representation of transitions between graphs. Each transition consists of preconditions and postconditions on the labels of graph nodes and edges. A transition relates a source and a target graph as long as the source satisfies the preconditions and the target satisfies the postconditions. Transitions can be represented in many ways, although logical constraints and rewriting rules are typical. An execution is a sequence of graphs in which each pair of successive graphs are related by some transition. This characterization allows for great modeling flexibility because transitions can be designed to represent arbitrary interactions among any number of devices and channels.

9.3.2.2 Modeling Temporal Properties

While there are many kinds of properties one may care to predict in smart-city environments, the typical classification of temporal properties [15] includes safety and liveness properties. Safety properties capture the phenomena that should hold along every step of an execution or should never be violated on any step of an execution. Liveness properties capture the phenomena that should hold at some future time from given points in an execution. Canonical examples of safety and liveness properties are avoiding error states and responsiveness, respectively.

This material focuses on two general classes of safety properties: connectivity and device-state reachability. The intent of connectivity is to characterize potential cyberattack paths from malicious source devices to vulnerable target devices. Formally, two devices are connected in a given graph if there is a sequence of nodes from the source to the target so that each pair of successive nodes is connected by an edge characterizing a cyber channel. This definition allows the characterization of vulnerable situations, such as supposedly air-gapped devices being connected to the Internet. A graph is reachable if there is an execution from a given initial state to that graph. Accordingly, the connectivity problem is to determine whether a given pair of devices is connected in some reachable graph. Besides connectivity, device-state reachability, whether a given label appears in some reachable graph, captures many other relevant prediction problems.

The next section discusses limited variations of the connectivity and device-state reachability problems, such as bounded reachability problems, which consider connectivity and device-state only in graphs reachable from a given initial configuration in executions that are bounded by a given length. The bounded variations of safety properties are unsound for prediction since executions that exceed a given bound may violate the property, even if all executions with a given bound do not.

9.3.2.3 Automated Reasoning Engines

While the general graph-based approach to modeling temporal properties of smart-city environments is not tied to any particular formal reasoning engine, the choice of engines is restricted by the complexity of the constructed models and queries. A very high-level discussion considers three types of reasoning engines. First, deductive reasoning engines [34] can prove unbounded safety properties in finite time, but they require an *a priori* inductive invariant that satisfies the given safety property. In other words, they require an *a priori* characterization of the set of graphs that includes the initial graph, so that the targets of all transitions from this set are included in this set. Since providing inductive invariants is generally a manual process, and we seek automatic methods, we exclude deductive reasoning engines.

Logical satisfiability solvers [35] and finite- and infinite-state model checkers [36] provide automated solutions, but the choice among these engines involves a trade-off. Logical satisfiability solvers can handle models that include predicates from any decidable theory, while model checkers are restricted to either finite-state descriptions or infinite-state descriptions that satisfy certain structural conditions to guarantee termination of the reachable-states computation [37]. However, logical satisfiability solvers handle only bounded safety properties, as the size of logical encodings grows with the bound on executions, while model checkers naturally consider arbitrarily long executions.

9.4 The Future: Connecting and Federating IoT Technologies in Smart Cities

Smart cities are systems of systems: federated collections of interrelated complex IoT systems that securely and efficiently interact to automate a variety of community infrastructure services. This section presents the challenges and requirements for federated smart cities and suggests potential solutions.

The current instantiations of smart cities raise a number of questions. Who chooses these devices and what criteria are applied? How are maintenance and lifecycle management handled? How do devices interact in a given environment? What potentially unwanted interactions could occur between devices? Even if each device is certified, is the combination of devices safe? Given devices that are connected to third parties (e.g. Google, Amazon, Samsung), what sort of data do they share and upload to these third parties? What weaknesses could lead to violations of personal privacy? Will a compromised third-party server wreak havoc over the entire infrastructure? What dangers arise if devices are abused? Under what circumstances could a device turn into a powerful weapon in the hands of an adversary? What capabilities would an adversary gain? These questions highlight that introducing un-sanitized and insufficiently protected smart devices to augment a city without careful planning and meticulous security analysis may open a Pandora's box, leading to a myriad of novel problems. The unique value of IoT in smart cities is in the unpredictable emergent interactions among otherwise unassociated devices and infrastructures, and the main challenge for devices and owners is trustworthiness assurance at scale. Realizing the potential benefits of these kinds of interactions demands a clear, detailed understanding of the safety and security properties required for all devices and a commonly adopted discipline and defined set of requirements governing future relationships between complex systems and devices that arise from unintended or serendipitous composition.

Based on the questions presented in [38], the remainder of the section outlines a set of recommendations for how to design and deploy security mechanisms specialized to enforce security and provide resiliency among and across network components with non-coordinated designs. The recommendations are organized across the five phases of the system development life cycle (SDLC), which encompasses initiation, acquisition/development, implementation/assessment, operations/maintenance, and sunset/disposal [39]. Each phase defines the security requirements, discusses existing research, and recommends potential future directions.

9.4.1 Initiation

The federation of IoT technologies needed to build a smart city begins with defining the business requirements in terms of the three traditional cyber security pillars: confidentiality, integrity, and availability.

9.4.1.1 Security Requirements

During the planning process, it is important to explicitly identify and define the security requirements. These security requirements are driven by current laws, regulations, and standards. Despite numerous efforts to standardize the aspects of IoT, no standards have currently emerged as dominant. On the other hand, data privacy laws, such as the EU General Data Protection Regulation (GDPR) [40] must be considered. During the initiation phase, it is important to determine the data flow process of the IoT systems, particularly focusing on sensitive data that require special handling. Smart cities encompass a wide variety of systems whose data are considered sensitive. IoT systems that collect information about users (e.g. location, physical characteristics, and sound) should be carefully assessed. Private information must be properly secured. Additionally, it is important to consider side-channel attacks that may reveal sensitive information to a passive eavesdropper [41–43]. Previous research efforts have demonstrated how encrypted network traffic originating from IoT devices can be used to infer the state of the devices [43–47] and consequently the state of the users [41].

9.4.1.2 Unanticipated Compositions

The main challenge for devices and owners is trustworthiness estimation at scale to ensure the safety of compositions. The heterogeneity of IoT systems raises challenges in terms of managing and controlling their security. In a smart city setting, it is possible that sensors, actuators, and even various hubs are not owned by the same organization. However, their autonomy may enable interactions between these devices. Moreover, places with a high density of IoT devices present a significant potential for implicit interactions between devices over physical channels.

9.4.1.3 Security Responsibility

Despite best efforts and implementation of best practices, cyber security incidents are likely to occur. The initiation phase should be used to establish a locus of security responsibility and control during initiation. Smart cities' systems should be assessed to determine how (and to what extent) control can be shared with another entity. This will enable smart-city officials and administrator to compose an emergency response plan, comprised of fail-safe and isolation-imposing configurations. Moreover, we propose that this plan is assessed such that its effectiveness and consequences are well understood. Formal models, such as the ones presented in Section 9.3, can be applied to determine, for example, unreachability to critical assets caused by a configuration change.

9.4.2 Acquisition/Development

Once the requirements are well understood and established, the acquisition or development stage provides an opportunity to conduct a risk assessment and use its results to supplement the baseline security controls.

9.4.2.1 Risk Assessment and Scenario Prediction

An initial risk assessment should include a formal analysis of the consequences of the composition of devices. The model presented in Section 9.3 can be used to perform differential analysis. Differential analysis provides a mechanism for reasoning about the potential consequences of modifications to the IoT components of a smart city. This approach compares the model of potential interactions of the current configuration of devices with a future possible configuration of devices that may include new devices as well as removal of old devices. Additionally, for every interaction, the model identifies characteristics that make the interaction possible, such as physical proximity of devices or configuration settings. This information may be analyzed and leveraged to redesign or reconfigure IoT components of a smart city in a way that minimizes risks.

The formal model of potential interactions and the differential analysis enable the prediction of changes and impacts to security and privacy state of the smart cities. Complementarily, an assessment of device characteristics is critical toward ensuring safe autonomy. A result of such an assessment entails an understanding of how autonomy of the various components can be controlled and under what conditions. These configuration options should be carefully considered during the designing of the security architecture. When systems have the potential to significantly impact the physical world, it is important to have the ability to limit and monitor the autonomy of systems. In congruence with the Saltzer and Schroeder's security design principle of continuous complete mediation [48], autonomy should be modeled in a way that enables a calculation about the potential physical side effects. While this introduces the possibility of undecidable computation, the challenge is to define the security decision's limits and how this limit attests to the controls provided.

Devices with actuation capabilities should be assessed for feedback-loop mechanisms that provide additional guarantees, or measures, about the safety of the system. We believe that actuation, especially when it pertains to safety, should be tightly controlled by feedback. Additionally, it is important to consider how data from various components of the smart city is redundant and can be used to introduce checks and balances against malfunctioning or compromised sensors.

9.4.3 Implementation/Assessment

Security and privacy play a significant role during the implementation and assessment of IoT components of a smart city. Similar to previous stages, it is important to consider the autonomy and interactions between devices and their impact on the security of the IoT deployment. The implementation and assessment phases focus on providing mechanisms for attestation, monitoring, and modeling the smart city's components.

9.4.3.1 Safe and Secure Communications

Enabling autonomy and communication between systems requires an agreement among all parties on syntax and semantics. For example, fields pertaining to time or dates must have the same meaning (semantics) and follow the same format (syntax). Differences in the assumed meaning and interpretation of such data can lead to faults and are often exploited by attackers. One solution to this problem is for all systems to agree in advance on a global namespace. However, devices are built by different manufacturers and achieving consensus is nontrivial. An alternative solution is to construct a high-assurance translator function (or system, e.g. hub) that serves as the authoritative entity. This function would be responsible for understanding the syntax and semantics of all parties involved in the communications and providing translation as a service. Additionally, this entity could also be responsible for leveraging language-theoretic security

(LangSec)[2] parsers to enhance the security of the communications. LangSec parsers aim to eliminate the ad-hoc programming of input handling at all layers of the network stacks and replace it with a validation of inputs using a formal language that serves as the recognizer for that language. As a result, LangSec parsers take untrusted inputs and provide only parsed and validated inputs that can be trusted according to the specifications of the protocol. LangSec parsers protect against exploitation of unsafe parsing of network input. Our previous work [49] presents a methodology for building LangSec parsers for two IoT protocols: XMPP and MQTT, and shows that the performance cost and the human effort in building such parsers are very reasonable. It is important to note that LangSec parsers make no guarantees about bugs or faults present in the logic of the program operating on the validated input.

Attention must be paid to side-channel attacks targeting device communications. Previous research efforts [44, 45] demonstrate how patterns in encrypted traffic originating from IoT devices can be used to infer their states and other potentially sensitive information. Secure communications should not leak information about devices and their states.

9.4.3.2 Provenance

During implementation, it is important to consider data and action provenance. Given the potential catastrophic impact IoT components may have on the physical world, data and actions should carry provenance data showing their origin and history. To achieve interoperability, devices need to exchange data. As data flow seamlessly across many devices, they are passed through a variety of communication technologies and network stacks. Data often undergo transformations and, as they cross boundaries, they are often stripped of meta information or other provenance data. However, in case of an event, lack of provenance information makes it nontrivial to (i) determine what led to the event and (ii) determine the origin of data and actions.

A rich body of literature addresses data provenance in various settings. Specifically focused on IoT, Wang et al. [50] present a platform-centric approach to centralized IoT auditing. ProvThings instruments IoT applications and device APIs to generate data provenance information. This information can then be used to provide a holistic explanation of system activities. ProvThings' limitation is in the granularity level and completeness of the instrumentation. API instrumentations and IoT applications may only capture partial information, i.e. other interactions between devices may occur without triggering APIs or changes to IoT applications. An alternative solution would be the construction of authoritative hubs that record information as they receive it. In current IoT networks, devices rarely interact directly, but rather through a hub, creating a star network topology. In case of

2 http://langsec.org/

an event, such hubs could provide a ledger or append-only record of all the information exchanges. The hub would provide a well-defined format or data dictionary service for storing provenance data. This data may be made available to third parties as a service. It is important to limit the access to such a service due to the potential sensitivity of the provenance data. Again, the challenge and limitation of platform-based approaches is in the heavy reliance on instrumentation, a luxury that is not often attainable in third-party IoT components.

9.4.3.3 Minimizing Attack Surface

Many IoT devices are bloated with features and functionalities that are not critical to their core purpose. Such unnecessary functionalities introduce complexity and increase the attack surface. This problem is further emphasized by the attempts of other parties to interoperate with existing, partially compliant, partially observable, descriptions of protocols, services, or interfaces to achieve some goal. This encourages poorly built and unprincipled combinations of data flows.

During the implementation phase, it is important to understand how these extra functionalities can be controlled through configurations. Mohsin et al. [51] introduce IoTChecker, a data-driven framework for semantically modeling IoT configurations and leveraging the model to identify configuration anomalies and analyze threats. The configuration analytics rely on information describing the context of IoT interactions and dependencies through rule-supported reasoning and queries. These analytics leverage ontology-based security classification methods.

An alternative approach is to construct a configuration analysis engine that can (i) extract configurations from devices via available interfaces; and (ii) understand the impact of those configurations on device functionality. Given a mapping of configuration parameters to functionality and a goal, the analysis would be able to determine an ideal configuration setting that minimizes the attack surface and disables extraneous code.

9.4.3.4 Security Assessment

Systems in development or revision must be tested and evaluated before implementation. Assessing the security of an IoT system presents several challenges. First, IoT systems are often distributed systems composed of many agents (e.g. cloud software, mobile applications, physical IoT devices, and their corresponding software). The second challenge involves testing IoT systems in combination with other systems. While the security assessment of an independent IoT system is necessary and advised, its devices may serve as bridges to other IoT systems and enable unanticipated interactions and causal relationships. These two characteristics require thorough testing that involves many different possible physical deployments and organizations of devices. For example, many implicit interactions between IoT devices over physical channels (such as heat or audio) are only possible when devices are in close proximity. Although such interactions are often conspicuous, they can serve as data channels and have unanticipated consequences. One such publicized example is the interaction between two virtual personal assistant (VPA) systems over the voice channel. The response of a VPA can trigger an action in another VPA. While this interaction could be mitigated by physically separating devices beyond the range of the physical channel, that is not always a viable approach.

Alrawi et al. [52] proposed a methodology for analyzing the security properties of IoT devices that includes all the possible components of IoT systems (i.e. cloud, mobile applications, other devices and instruments). The methodology uses the Nessus Scanner [53] to scan and assess devices for service discovery, profiling, and vulnerability. Active network scanners can be used to find problems regarding device setup, software updates, and other configurations. Other dynamic and static analysis techniques can be applied on the mobile applications to identify problems pertaining to over-permissions, data leakage, or incorrect use of cryptography.

IoTFuzzer [54] presents an automatic white-box fuzzing framework for IoT that focuses on identifying memory-corruption vulnerabilities in IoT devices. The framework leverages IoT mobile applications to learn how to interact with devices and leverages program-specific logic found in applications to generate probes. Other fuzzing techniques can be used to test available interfaces of IoT devices.

9.4.4 Operations/Maintenance

Operation and maintenance focus on managing and monitoring the system of systems and instituting processes and procedures for assured operations. Monitoring of IoT systems of systems presents unique challenges in contrast to traditional IT monitoring. First, IoT devices rely on a variety of communication technologies, among which the most popular are Wi-Fi, Bluetooth, Bluetooth Low Energy, and Zigbee. Information can flow across these communication technologies seamlessly. Effective monitoring must be able to identify all communication channels used and track information and activity across them. This presents a key challenge: How can an observer determine that the data coming into a device through one channel (e.g. Wi-Fi) is related to the output generated by another channel (e.g. Bluetooth)? While the heterogeneity of IoT manufacturers indicates that not all devices can explicitly communicate, even those devices have the potential to implicitly interact through various causal relationships triggered by sensing and actuating actions.

9.4.4.1 Detecting Causality

Unfortunately, IoT automation often occludes the understanding of such causal relationships in an IoT network. As the effects of a particular device's actions on its environment are observed (e.g. the light turns on), it is important to capture the full causality path to the originating device. This information can be useful for security monitoring, as well as debugging complex system of systems.

Our previous work [57] presented a promising approach that enables the inference of a potential sequence of actions (or any deviation) that led to a particular observation. Compared to alternative methods, the proposed approach is nonintrusive, i.e. it does not require any instrumentation or modification of the IoT devices. The approach passively monitors network traffic, side-channel modalities, and the physical environment to create representations of behaviors for each device. The patterns observed are captured in a probabilistic tree-based model representing the observed behaviors. The underlying model relies on Probabilistic Suffix Tree (PST) [55] as the predictive model, similar to the work proposed by Yoon and Ciocarlie [56]. The nodes of the tree represent events, while the leaves indicate potential causes. The edges of the model connect events and depict the temporal order of the events. Each edge is associated with a probability representing the likelihood of the transition between the source and destination events.

The analysis begins at the root of the tree with the device that triggered the event and its observed behavior. Starting at the root, the approach performs three recursive steps: (i) compare observed behavior against the model; (ii) identify which nodes may have caused the behavior (using the potential interaction model); and (iii) analyze suspected causal sources for behavior that matches the model. The potential interaction model discussed in Section 3 serves two essential purposes. First, it is used to determine, for any given node, which neighboring nodes could have caused the change in its behavior. Second, by identifying a subset of devices for further analysis, it reduces the dimensions of the search space, improving the efficiency of the approach. The observed behavior for the suspected causal devices is compared against the expected behavior. This process is repeated until it no longer reaches unvisited neighboring nodes or it reaches a leaf node in the expected behavior model. The last node(s) traversed represents the cause(s) that triggered the event. When no predefined behavior is known, a generic root cause "change" is associated with the input.

9.4.4.2 Validation of Physical Properties

As cyber-physical systems, IoT devices marry the cyber and physical worlds. Computations and data transfers executed by devices drive physical processes through actuation. Given this potential impact on the physical environment and the critical need for safety, it is advised to employ methods that enable checks and balances between the semantics of the information shared between devices, their digital states, as well as the state of the physical environment.

Sensors are often an attractive target for malicious actors, primarily because of the central role the data produced plays. Protecting against sensor attacks is non-trivial. Previous research efforts have explored error-correcting methods and byzantine agreement frameworks.

We believe it is important to have an external validation and attestation of the physical properties. This external information about the physical environment can be verified against devices' behavior, which encompasses their states and data transmissions. Discrepancies may be indicative of malfunctions or compromises. The challenge of this approach is twofold. First, the attestation must be performed in a trustworthy and secure manner that imposes minimal overhead on the system. Second, appropriate mechanisms must be established to safely intervene when anomalies are identified.

Every autonomous action must be referred to an independent competent authority, which must receive and process an action credential that attests to the propriety of the action to be performed and its binding to the entity proposing to perform it.

9.4.4.3 Isolation

An essential component of incident response is minimizing the impact of an attack. One method for minimizing the impact of an incident (e.g. worms, botnet) is to impose flexible isolation boundaries.

The challenge with IoT is determining the optimal boundaries that isolate the intended devices. As previously mentioned, IoT devices rely on a variety of communication standards and protocols. To be effective, isolation boundaries must be able to account for all potential interaction paths. The model presented in Section 1.3 enables the reasoning about potential interaction paths as well as the attributes of the paths that make the interaction possible. The proposed model can be leveraged to identify characteristics of the isolation boundaries (i.e. what channels need to be isolated?), to determine isolation techniques (e.g. physically moving devices null the Bluetooth channel), and to compute the optimal placement of boundaries to accomplish the necessary isolation configuration.

SDNs have been studied for the purpose of security, specifically for imposing dynamic isolation boundaries. Current approaches only support IP networks and often require device instrumentation and/or additional components (e.g. hardware or software SDN controllers).

Complete isolation must also consider the implicit interactions between devices through the physical environment. The natural mitigation approach involves physical removal or deactivation of devices; however, this may not always be achievable in a timely manner, if at all (e.g. devices embedded into walls, underground, or under water).

9.4.5 Sunset/Disposal

Since individual designs do not anticipate collective weaknesses, different subsystems have opportunities to compensate for others in case of disruptions. Finally, smart-city systems can have long planned lifetimes, which present challenges related to managing obsolescence and sunsetting; however, infrastructure systems tend to remain in use much longer than originally planned. Plans and considerations for deploying a new system should include plans for safe sunsetting and possible replacement when the system has reached its anticipated useful lifetime.

The recommendations for sunsetting include:

- Build and execute a disposal/transition plan
- Archive critical information
- Sanitize media
- Dispose of hardware and software

It is important that, during sunsetting, critical information is archived for historical purposes. However, this information should be protected and sanitized not to reveal any sensitive information in the event of an incident. Moreover, the removal of hardware and software prevents the pollution with unsupported and vulnerable devices.

9.5 Epilogue

Just like you can no longer purchase a new car that does not rely on dozens of built-in computers and external connectivity, it will soon be impossible to find a city without some degree of smart technology. Our daily lives, which already depend heavily on electric power, data communication, transportation, and other critical infrastructures, will also come to depend on the information and services a smart city will provide. Therefore, sound principles are required to guide the design, implementation, and operation of smart-city technologies to achieve the expected levels of reliability, safety, security, and privacy.

The federation of systems into larger smart-city systems could create many problems without the strict adherence to guiding principles. With the implementation of new digital infrastructure that will be in place for decades and affect many millions of people, it is crucial to do everything possible to lay a solid technological foundation.

To provide information and guidance to those who are engaged in designing and implementing smart-city technology, this chapter has described many technology challenges for smart cities and suggested some solutions to those challenges at all stages of the process lifecycle.

Acknowledgment

Research was sponsored by the Army Research Laboratory and was accomplished under Cooperative Agreement Number W911NF-17-2-0196. The views and conclusions contained in this document are those of the authors and should not be interpreted as representing the official policies, either expressed or implied, of the Army Research Laboratory or the U.S. Government. The U.S. Government is authorized to reproduce and distribute reprints for Government purposes notwithstanding any copyright notation herein.

References

1 J. Brian McLoughlin and J. N. Webster, "Cybernetic and general-system approaches to urban and regional research: A review of the literature," *Environment and Planning*, vol. 2, pp. 369–408, 1970. http://citeseerx.ist.psu.edu/viewdoc/download?doi=10.1.1.458.8320&rep=rep1&type=pdf

2 R. G. Hollands, "Will the real smart city please stand up?," *City*, vol. 12, no. 3, pp. 303–320, 2008.

3 O. Söderström, T. Paasche, and F. Klauser, "Smart cities as corporate storytelling," *City*, vol. 18, no. 3, pp. 307–320, 2014. https://www.tandfonline.com/doi/abs/10.1080/13604810802479126

4 N. Komninos and L. Mora, "Exploring the big picture of smart city research," *Scienze Regionali: Italian Journal of Regional Science*, vol. 1, pp. 15–38, 2018.

5 B. Danev, D. Zanetti, and S. Capkun, "On physical-layer identification of wireless devices," *ACM Computing Surveys*, vol. 45, no. 1, Article 6, 2012. https://dl.acm.org/citation.cfm?id=2379782

6 S. Bratus, C. Cornelius, D. Kotz, and D. Peebles, "Active behavioral fingerprinting of wireless devices," in *Proceedings of the First ACM Conference on Wireless Network Security*, 2008. https://dl.acm.org/citation.cfm?id=1352543

7 D. Margerum, "Pinpointing location of hostile radars," *Microwaves*, vol 8, no. 11, pp. 60–73. November, 1969.

8 J. Hall, M. Barbeau, and E. Kranakis, "Enhancing intrusion detection in wireless networks using radio frequency fingerprinting," *Communications, Internet, and Information Technology*, pp. 201–206, 2004.

9 K. Ellis and N. Serinken, "Characteristics of radio transmitter fingerprints," *Radio Science*, vol. 36, pp. 585–597, 2001.

10 F. Lanze, A. Panchenko, B. Braatz, and A. Zinnen, "Clock skew based remote device fingerprinting demystified," in *Global Communications Conference*

(GLOBECOM), IEEE, 2012. https://ieeexplore.ieee.org/abstract/document/6503213

11 V. Brik, S. Banerjee, M. Gruteser, and S. Oh, "Wireless device identification with radiometric signatures," in *14th ACM International Conference on Mobile Computing and Networking* (MobiCom'08), 2008. https://dl.acm.org/citation.cfm?id=1409959

12 J. Pang, B. Greenstein, R. Gummadi, S. Seshan, and D. Wetherall, "802.11 user fingerprinting," in *Proceedings of the 13th Annual ACM International Conference on Mobile Computing and Networking*, 2007. https://dl.acm.org/citation.cfm?doid=1287853.1287866

13 M. Miettinen, S. Marchal, I. Hafeez, N. Asokan, A.-R. Sadeghi, and S. Tarkoma, "IoT Sentinel: Automated device-type identification for security enforcement in IoT," in *Distributed Computing Systems (ICDCS), 2017 IEEE 37th International Conference*, 2017. http://www.icri-sc.org/fileadmin/user_upload/Group_TRUST/PubsPDF/iot-sentinel-miettinen.pdf

14 J. Franklin, D. McCoy, P. Tabriz, V. Neagoe, J. V. Randwyk, and D. Sicker, "Passive data link layer 802.11 wireless device driver fingerprinting," in *USENIX Security Symposium*, 2006. http://damonmccoy.com/papers/wireless-fingerprinting.pdf

15 A. Pnueli, "The temporal logic of programs," in *18th Annual Symposium on Foundations of Computer Science*, pp. 46–57. IEEE Computer Society, 1977. https://ieeexplore.ieee.org/document/4567924

16 Z. Gao, L. Xu, G. Turner, B. Patel, N. Diallo, L. Chen, and W. Shi, "Blockchain-based identity management with mobile device," in *Proceedings of the 1st Workshop on Cryptocurrencies and Blockchains for Distributed Systems*, pp. 66–70, 2018. https://doi.org/10.1145/3211933.3211945.

17 A. Tiwari, "HybridSAL relational abstracter." In: Madhusudan P., and Seshia S. A. (eds) *Computer Aided Verification*. CAV 2012. *Lecture Notes in Computer Science*, vol. 7358, pp. 725–731, 2012, https://doi.org/10.1007/978-3-642-31424-7_56.

18 B. Schneier, "Attack trees," *Dr. Dobb's Journal*, vol. 24, no. 12, pp. 21–29, 1999.

19 R. P. Lippmann, K. W. Ingols, C. Scott, K. Piwowarski, K. Kratkiewicz, M. Artz, and R. Cunningham, "Evaluating and strengthening enterprise network security using attack graphs," Lincoln Laboratory, Massachusetts Institute of Technology Lexington, Massachusetts, Project Report ESC-TR-2005-064, 2005.

20 S. Noel and S. Jajodia, "Optimal IDS sensor placement and alert prioritization using attack graphs," *Journal of Network and Systems Management*, vol. 16, no. 3, pp. 259–275, 2008. https://doi.org/10.1007/s10922-008-9109-x

21 L. Wang, A. Liu, and S. Jajodia, "Using attack graphs for correlating, hypothesizing, and predicting intrusion alerts," *Computer Communications*, vol. 29, no. 15, pp. 2917–2933, 2006. https://doi.org/10.1016/j.comcom.2006.04.001

22 O. Sheyner, J. Haines, S. Jha, R. Lippmann, and J. M. Wing, "Automated generation and analysis of attack graphs," in *IEEE Symposium on Security and Privacy*, pp. 273–284, 2002. https://doi.org/10.1109/SECPRI.2002.1004377

23 X. Ou, W. F. Boyer, and M. A. McQueen, "A scalable approach to attack graph generation," in *Proceedings of the 13th ACM Conference on Computer and Communications Security*, pp. 336–345, 2006. https://doi.org/10.1145/1180405.1180446.

24 M. Kovatsch, Y. N. Hassan, and S. Mayer, "Practical semantics for the Internet of Things: Physical states, device mashups, and open questions," in *5th International Conference on the Internet of Things (IOT)*, pp. 54–61, 2015. https://doi.org/10.1109/IOT.2015.7356548

25 M. Mohsin, Z. Anwar, G. Husari, E. Al-Shaer, and M. A. Rahman, "IoTSAT: A formal framework for security analysis of the internet of things (IoT)," in *IEEE Conference on Communications and Network Security (CNS)*, pp. 180–188, 2016. https://doi.org/10.1109/CNS.2016.7860484

26 T. L. Guilly, J. H. Smedegård, T. Pedersen, and A. Skou, "To do and not to do: Constrained scenarios for safe smart house," in *International Conference on Intelligent Environments*, pp. 17–24, 2015. https://doi.org/10.1109/IE.2015.11

27 F. Corno and M. Sanaullah, "Modeling and formal verification of smart environments," *Security and Communication Networks*, vol. 7, no. 10, pp. 1582–1598, 2014. https://doi.org/10.1002/sec.794

28 J. C. Augusto and M. J. Hornos, "Software simulation and verification to increase the reliability of Intelligent Environments," *Advances in Engineering Software*, vol. 58, no. 18–34, 2013. https://doi.org/10.1016/j.advengsoft.2012.12.004

29 A. Coronato and G. D. Pietro, "Formal design of ambient intelligence applications," *Computer*, vol. 43, no. 12, pp. 60–68, 2010. https://doi.org/10.1109/MC.2010.335

30 O. Mavropoulos, H. Mouratidis, A. Fish, and E. Panaousis, "ASTo: A tool for security analysis of IoT systems," in *IEEE 15th International Conference on Software Engineering Research, Management and Applications (SERA)*, pp. 395–400, 2017. https://doi.org/10.1109/SERA.2017.7965757

31 A. Rullo, D. Midi, E. Serra, and E. Bertino, "A game of things: Strategic allocation of security resources for IoT," in *Proceedings of the Second International Conference on Internet-of-Things Design and Implementation*, 2017. https://ieeexplore.ieee.org/document/7946874

32 A. Tekeoglu and A. Tosun, "A testbed for security and privacy analysis of IoT devices," in *IEEE 13th International Conference on Mobile Ad Hoc and Sensor Systems (MASS)*, pp. 343–348, 2016. https://doi.org/10.1109/MASS.2016.051

33 D. Geneiatakis, I. Kounelis, R. Neisse, I. Nai-Fovino, G. Steri, and G. Baldini, "Security and privacy issues for an IoT based smart home," in *40th International Convention on Information and Communication Technology, Electronics and*

Microelectronics (MIPRO), pp. 1292–1297, 2017. https://doi.org/10.23919/MIPRO.2017.7973622

34 C. A. R. Hoare, "An axiomatic basis for computer programming," *Communications of the ACM*, vol. 12, no. 10, pp. 576–580, 1969.

35 M. Davis, G. Logemann, and D. W. Loveland, "A machine program for theorem-proving," *Communications of the ACM*, vol. 5, no. 7, pp. 394–397, 1962.

36 E. M. Clarke, O. Grumberg, and D. A. Peled, *Model Checking*. MIT Press, 2001.

37 A. Finkel and P. Schnoebelen, "Well-structured transition systems everywhere!," *Theoretical Computer Science*, vol. 256, no. 1–2, pp. 63–92, 2001.

38 U. Lindqvist and M. Locasto, "Building code for the internet of things," IEEE, 2017. https://ieeecs-media.computer.org/media/technical-activities/CYBSI/docs/Building_Code_IoT_online.pdf

39 R. L. Kissel, K. M. Stine, M. A. Scholl, H. Rossman, J. Fahlsing, and J. Gulick, "Security considerations in the system development life cycle," in *NIST Special Publication 800-64 Revision 2*, 2008. https://nvlpubs.nist.gov/nistpubs/Legacy/SP/nistspecialpublication800-64r2.pdf

40 General Data Protection Regulation (GDPR). 2006. https://eur-lex.europa.eu/legal-content/EN/TXT/PDF/?uri=CELEX:32016R0679

41 G. Ateniese, B. Hitaj, L. V. Mancini, N. V. Verde, A. Villani, "No place to hide that bytes won't reveal: Sniffing location-based encrypted traffic to track a user's position." In: Qiu M., Xu S., Yung M., and Zhang H. (eds) *Network and System Security. Lecture Notes in Computer Science*, vol. 9408. Springer, Cham, 2015.

42 P. Kocher, J. Jaffe, and B. Jun, "Differential power analysis." In: Wiener M. (ed) *Advances in Cryptology — CRYPTO' 99. Lecture Notes in Computer Science*, vol. 1666. Springer, Berlin, Heidelberg, 1999.

43 M. Vuagnoux and S. Pasini, "Compromising electromagnetic emanations of wired and wireless keyboards," in *USENIX Security Symposium*, 2009. https://www.usenix.org/legacy/event/sec09/tech/full_papers/vuagnoux.pdf

44 B. Copos, K. Levitt, M. Bishop, and J. Rowe, "Is anybody home? Inferring activity from smart home network traffic," in *Security and Privacy Workshops (SPW), 2016 IEEE*, pp. 245–251, IEEE, 2016. https://ieeexplore.ieee.org/document/7527776

45 A. Acar, H. Fereidooni, T. Abera, A. K. Sikder, M. Miettinen, H. Aksu, M. Conti, A.-R. Sadeghi, and A. S. Uluagac, "Peek-a-boo: I see your smart home activities, even encrypted!," *arXiv preprint arXiv*:1808.02741, 2018. https://arxiv.org/pdf/1808.02741.pdf

46 N. Sehatbakhsh, A. Nazari, A. Zajic, and M. Prvulovic, "Spectral profiling: Observer-effect-free profiling by monitoring EM emanations," in *49th Annual IEEE/ACM International Symposium on Microarchitecture*, IEEE Press, 2016. https://ieeexplore.ieee.org/document/7783762

47 A. Nazari, N. Sehatbakhsh, M. Alam, A. Zajic, and M. Prvulovic, "Eddie: EM-based detection of deviations in program execution," in *ACM/IEEE 44th Annual*

International Symposium on Computer Architecture (ISCA), 2017. https://ieeexplore.ieee.org/document/8192483

48 J. H. Saltzer and M. D. Schroeder, "The protection of information in computer systems," *Proceedings of the IEEE*, vol. 63, no. 9, pp. 1278–1308, 1975.

49 P. Anantharaman, M. Locasto, G. F. Ciocarlie, and U. Lindqvist, "Building hardened internet-of-things clients with language-theoretic security," in *IEEE S&P Workshop on Language-Theoretic Security (LangSec)*, 2017. https://ieeexplore.ieee.org/document/8227297

50 Q. Wang, W. U. Hassan, A. Bates, and C. Gunter, "Fear and logging in the Internet of things," in *Network and Distributed Systems Security (NDSS) Symposium*, 2018. http://dx.doi.org/10.14722/ndss.2018.23282

51 M. Mohsin, Z. Anwar, F. Zaman, and E. Al-Shaer, "IoTChecker: A data-driven framework for security analytics of Internet of things configurations," *Computers and Security*, vol. 70, pp. 199–223, 2017.

52 O. Alrawi, C. Lever, M. Antonakakis, and F. Monrose, "Sok: Security evaluation of home-based IoT deployments," in *2019 IEEE Symposium on Security and Privacy (SP)*, pp. 208–226, 2009. https://csdl.computer.org/csdl/proceedings/sp/2019/6660/00/666000a208.pdf

53 Nessus Scanner. https://www.tenable.com/products/nessus/nessus-professional

54 J. Chen, W. Diao, Q. Zhao, C. Zuo, Z. Lin, X. Wang, W. C. Lau, M. Sun, R. Yang, and K. Zhang, "IoTFuzzer: Discovering memory corruptions in IoT through app-based fuzzing," in *Network and Distributed Systems Security (NDSS) Symposium*, 2018. http://dx.doi.org/10.14722/ndss.2018.23159

55 D. Ron, Y. Singer, and N. Tishby, "Learning probabilistic automata with variable memory length," in *Proceedings of the Seventh Annual Conference on Computational Learning Theory*, pp. 35–46, ACM, 1994. http://citeseerx.ist.psu.edu/viewdoc/download?doi=10.1.1.52.598&rep=rep1&type=pdf

56 M.-K. Yoon and G. Ciocarlie, "Communication pattern monitoring: Improving the utility of anomaly detection for industrial control systems," in *Proceedings of the 2014 NDSS Workshop on Security of Emerging Networking Technologies (SENT)*, 2014. https://www.ndss-symposium.org/wp-content/uploads/2017/09/02_4-paper_0.pdf

57 I. Agadakos, G. Ciocarlie, B. Copos, T. Lepoint, M. Locasto, and U. Lindqvist, "Butterfly effect: Causality from chaos in the IoT," in *1st International Workshop on Security and Privacy for the Internet-of-Things [IoTSec]*, 2018. https://www.researchgate.net/profile/Bogdan_Copos/publication/327515023_Butterfly_Effect_Causality_from_Chaos_in_the_IoT/links/5b92edf9a6fdccfd5423b208/Butterfly-Effect-Causality-from-Chaos-in-the-IoT.pdf

10

Modeling and Analysis of Integrated Proactive Defense Mechanisms for Internet of Things

Mengmeng Ge[1], *Jin-Hee Cho*[2], *Bilal Ishfaq*[3], *and Dong Seong Kim*[4]

[1] *School of Information Technology, Deakin University, Geelong, Victoria, Australia*
[2] *Department of Computer science, Virginia Tech, Falls Church, VA, USA*
[3] *Department of Computer Science and Software Engineering, University of Canterbury, Christchurch, New Zealand*
[4] *School of Information Technology and Electrical Engineering, University of Queensland, Brisbane, Queensland, Australia*

10.1 Introduction

Internet of Things (IoT) has received significant attention due to its enormous advantages. Advances in IoT technologies can be easily leveraged to maximize effective service provisions to users. However, due to the high heterogeneity and the constrained resources of composed entities, and its large-scale networks, we face the following challenges [1]: (i) distributed technologies for communications, data filtering, processing, and dissemination with an enormous amount of various forms of data (e.g. text, voice, haptics, image, video) in large-scale networks with heterogeneous entities (i.e. devices, humans); (ii) severely restricted resources in battery, computation, communication (e.g. bandwidth), and storage, causing significant challenges in resource allocation and data processing capabilities; (iii) highly adversarial environments, introducing compromised, deceptive entities and data, which may result in detrimental impacts on the capabilities of critical mission-related decision-making; and (iv) highly dynamic interactions between individual entities, data, and environmental factors (e.g. network topology or resource availability), where each factor is highly dynamic in time/space. Due to these characteristics of IoT environments, highly secure, lightweight defense mechanisms are in need to protect and defend a system (or network) against potential attacks.

Modeling and Design of Secure Internet of Things, First Edition. Edited by Charles A. Kamhoua, Laurent L. Njilla, Alexander Kott and Sachin Shetty.

Published 2020 by John Wiley & Sons, Inc.

As a solution to protect and defend a system against inside attacks, many intrusion detection systems (IDSs) have been developed to identify and react to the attacks. However, the core idea of an IDS is reactive in nature and even though it detects intrusions which have already been in the system. Hence, this reactive mechanism would be way behind and not effective for the actions by agile and smart attackers. Due to the inherent limitation of an IDS with these reactive nature, intrusion prevention systems (IPSs) have been developed to thwart potential attackers and/or mitigate the impact of the intrusions before they penetrate into the system [2]. In this work, we are interested in developing an integrated intrusion prevention mechanism based on the technologies of cyber deception (i.e. a decoy system) and moving target defense, namely MTD (i.e. network topology shuffling), and evaluating their effectiveness and efficiency by a graphical security model (GSM)-based evaluation framework.

10.1.1 Research Goal and Contributions

This work aims to propose an integrated proactive defense based on intrusion preventive mechanisms, such as cyber deception and MTD techniques, to minimize the impact of potential attackers trying to penetrate into IoT systems via multiple entries. We make the following ***key contributions*** in this book chapter:

- We developed an *integrated proactive defense system* by proposing an adaptive MTD technique by shuffling a network topology where a network consists of both decoy nodes and real nodes. As decoy nodes are the part of a decoy system, which is a common cyber-deception technology, this work integrates an MTD technique with cyber deception to propose an adaptive, proactive intrusion defense mechanism that maximizes hurdles and/or complexity for attackers to launch their attacks while minimizing the defense cost to execute MTD operations. The key goal of the proposed network topology shuffling-based MTD (NTS-MTD) with decoy nodes is to generate a network topology that can maximize disadvantages against the attackers. Little work has integrated both cyber deception and MTD techniques particularly in terms of network topology shuffling in software-defined networking (SDN)-based IoT environments.
- We took a metaheuristic approach based on a genetic algorithm (GA) by devising fitness functions that can achieve the objective of our proposed proactive defense mechanism in terms of minimizing defense cost as well as security vulnerability in attack paths. For example, given a set of network topologies that are uniformly selected at random, we use the GA and devise fitness functions that identify an optimal network topology to minimize security vulnerability and defense cost (e.g. network shuffling cost). In particular, the devised fitness functions estimate the utility of triggering NTS-MTD operation based on a generated network topology in an adaptive way, instead of triggering the MTD operation

based on a fixed time interval. This allows our proposed integrated defense mechanism to provide a *secure, affordable defense service.*

- We considered a *SDN-based IoT* as our network environment. The merits of SDN technology in terms of programmability and controllability are highly leveraged to design the proposed proactive defense mechanism equipped with MTD and cyber deception techniques and examine its performance in terms of security and performance metrics, including the number of attack paths toward decoy targets, mean time to security failure (i.e. MTTSF or system lifetime), and defense cost.

We adopted a GSM [3, 4] to evaluate the proposed deception and MTD techniques. The GSM offers design solutions to consider attack graphs (AGs) and/or attack trees (ATs) which can provide efficient methods to calculate the potential security (or vulnerability) levels of attack paths. In order to deal with large-scale networks, we will develop a hierarchical attack representation model (HARM) [3, 4] as a GSM model. The HARM allows us to evaluate the proposed proactive mechanisms including both the cyber deception and MTD techniques.

10.1.2 Structure of this Chapter

The rest of this chapter is organized as follows. Section 10.2 provides the brief overview of the related work in terms of MTD and cyber deception techniques for IoT environments, security models, and metrics, and SDN technology and its use for IoT environments. Section 10.3 gives the overview of the system model, including models of describing a targeted network environment, node characteristics, attack behaviors, and defense mechanisms in place and security failure conditions. Section 10.4 describes the design features of the proposed integrated proactive defense mechanism in terms of NTS-MTD, optimization techniques for NTS-MTD based on GA, and GSM for NTS-MTD in IoT environments. Section 10.5 shows the experimental results and discusses the overall trends of the results observed. Section 10.6 concludes this chapter and suggests future work directions.

10.2 Related Work

In this section, we provide the overview of the state-of-the-art approaches with the topics as the basis of the proposed work, including: (i) MTD and cyber deception techniques for IoT; (ii) security models and metrics; and (iii) SDN technology for IoT.

10.2.1 MTD and Cyber Deception Techniques for IoT

MTD is an emerging technique aiming at constantly changing the attack surface of the networks to increase the attack complexity [3]. Based on [3], MTD approaches are classified into three categories: shuffling, diversity, and redundancy. Shuffling changes the network configurations (e.g. IP address randomization, device migration, or topology reconfiguration). Diversity employs a variety of different implementations for the same functionalities (e.g. choosing various operating systems for the web server). Redundancy provides replications of the devices in the network to increase network reliability in the presence of attacks.

MTD approaches have been proposed to protect resource-constrained IoT devices. In our prior work [4], we examined the performance of address space layout randomization (ASLR) for wireless sensor nodes as an MTD technique and evaluated its effectiveness using the proposed HARM in one of the use cases. Casola et al. [5] proposed an MTD framework to reconfigure the IoT devices by switching among different cryptosystems and firmwares, and then evaluated the framework via a case study of wireless sensor networks (WSNs). Two metrics, attack probability and attack time, are used to assess the effectiveness of the reconfiguration; but there is still a lack of system-level metrics to quantify the proposed approach. Sherburne et al. [6] proposed a dynamically changing IPv6 address assignment approach over the IoT devices using low-powered wireless personal area networks (LPWPANs) protocol to defend against various network attacks. Zeitz et al. [7] extended [6] and presented a design of fully implementing and testing the MTD approach, which uses the address rotation to obscure the communications among IoT devices. However, both works [6, 7] do not have any experimental validation of the design. Mahmood and Shila [8] developed an MTD security framework based on context-aware code partitioning and code diversification for IoT devices to obfuscate the attackers; but it also does not perform any analysis to validate the framework.

Cyber deception techniques are proactive approaches in cyber defense which add an extra layer of defense onto the traditional security solutions (e.g. IDS, firewalls, or endpoint antivirus software). Once attackers are inside a network, they start probing to acquire information to determine the potential valuable assets and then move laterally in the network to launch attacks based on the information they gather during the probes. Cyber deception aims at luring the attackers into the decoy systems within the network and interacting with the attackers to monitor and analyze attackers' behaviors. Honeypot is one of the commonly used technologies in cyber deception. It is created as a fake asset and deployed around the valuable assets to divert attackers. However, the management complexity and scalability issues of the honeypots hinder the wide usage by the enterprises. Modern deception technology uses basic honeypot technology along with visualization and automation techniques [9]. It allows distributed deployment and update of decoy systems to achieve adequate coverage but still cost-effective.

The state-of-the-art approaches in this domain have focused on developing and deploying honeypot technology for IoT. La et al. [10] introduced a game-theoretic method to model the interaction between an attacker who can deceive the defender with suspicious or seemingly normal traffic and a defender in honeypot-enabled IoT networks. Anirudh et al. [11] used honeypots for online servers to mitigate distributed denial of service (DDoS) attacks launched from IoT devices. Dowling et al. [12] created a ZigBee honeypot to capture attacks and used it to identify the DDoS attacks and bot malware. However, none of the work analyzed the impact of the deception techniques on the system-level security. Besides, little study analyzed the modern deception technology for an IoT system which allows distributed deployment and the update of decoys to achieve adequate coverage and provide cost-effective defense service [13].

In our prior work [2], we looked into an integrated defense system to identify what components of each defense mechanism can provide a best solution for achieving "defense in breadth" considering both enhanced security and defense cost. However, it is purely a model-based analysis based on an abstract model in which a granularity of each defense mechanism is omitted to capture a system-level performance to some extent. In addition, there is no current work on developing an integrated defense system equipped with two proposed defense techniques in an IoT, including deception and MTD techniques. Therefore, this work proposes integrated and proactive defense mechanisms that can effectively and efficiently mitigate the adverse effect of attackers before the attackers penetrate into a target IoT system.

10.2.2 Security Models and Metrics

GSMs (AGs [14], ATs [15]) have been widely employed for security analysis in various types of networks and combined with security metrics to evaluate different defense mechanisms. In the graph-based attack models, an AG shows all possible sequences of the attackers' actions that eventually reach the target. With the increasing size of the network, calculation of a complete AG has exponential complexity, thus causing a scalability issue. In the tree-based attack models, an AT is a tree with nodes representing the attacks and the root representing the goal of attacks. It systematically presents potential attacks in the network. However, an AT also has the scalability issue when the size of the network increases. In order to address the above issues, the two-layered HARM [3] was introduced, which combines various GSMs onto different layers. In the two-layered HARM, the upper layer captures the network reachability information and the lower layer represents the vulnerability information of each node in the network. The layers of the HARM can be constructed independently of each other. This decreases the computational complexity of calculating and evaluating the HARM compared with the calculation and evaluation of the existing single-layered GSMs.

In our prior works [3, 4], we investigated the effectiveness of defense mechanisms based on the GSM, particularly using HARM. In [4], we developed a framework to automate the security analysis of the IoT system in which HARM along with various security metrics (e.g. attack cost, attack impact) is used to assess the effectiveness of both device-level and network-level defense mechanisms in three scenarios. Similarly, in [3], we evaluated the proposed MTD techniques in a virtualized system based on HARM with a risk metric; but we analyzed the performance of three different MTD techniques, including shuffling, diversity, and redundancy, separately while leaving the investigation of an integrated defense system as the future work, which is studied in this work.

Some existing approaches have adopted a risk-based or vulnerability-based security model to assess the effectiveness of defense mechanisms [16–18]. Abie and Balasingham [16] proposed a risk-based security framework for IoT environments in the eHealth domain to measure expected risk and/or potential benefits by taking a game-theoretic approach and context-aware techniques. Savola et al. [18] proposed an adaptive security management scheme considering security metrics (e.g. metrics representing authentication effectiveness, authorization metrics) to deal with the challenges in eHealth IoT environments. However, both works [16, 18] only proposed high-level ideas about the metrics without any formulation and did not consider the key characteristics of IoT environments where lightweight defense mechanisms are vital to securing a large-scale, resource-constrained IoT system. Rullo et al. [17] proposed an approach to compute the optimal security resource allocation plan for an IoT network consisting of mobile nodes and introduced a risk metric inspired by economics to evaluate the allocation plans. However, it only considered the device-level mechanisms and did not show the system-level evaluation.

Unlike the existing approaches above [16–18], we proposed a lightweight, affordable method to evaluate the deployment of an integrated defense mechanism for an IoT environment by meeting both security and performance goals of a system.

10.2.3 SDN Technology for IoT

SDN is a promising technology to flexibly manage complex networks. In the SDN-based architecture, the control logic is decoupled from the switches and routers and implemented in a logically centralized controller; the controller communicates with the data-forwarding devices via the southbound application programming interface (API) and provides the programmability of network applications using the northbound API. OpenFlow (OF) is the most widely used southbound API which provides the specifications for the implementation of OF switches (including the OF ports, tables, channel, and protocol) [19].

Some SDN solutions are applied to IoT networks to control data flows among IoT devices [20] while others focused on applying an SDN to WSNs for managing sensing devices [21]. Trevizan de Oliveira et al. [20] designed and implemented an SDN architecture consisting of SDN-enabled sensor nodes and one or multiple controllers in a TinyOS environment. Galluccio et al. [21] proposed a stateful SDN solution for WSNs to reduce the amount of exchanged data between the sensor nodes and the SDN controller and to enable a variety of operations that are not supported by stateless solutions. Some SDN approaches were proposed in heterogeneous wireless networks, such as wireless access networks [22] or mobile networks [23], to manage end-to-end flows. On the other hand, other approaches provided a software-defined IoT architecture for managing IoT devices in different application domains, e.g. SD-IoT architecture for smart urban sensing in [24]. Current SDN solutions show the feasibility to control traffics within IoT networks by dynamically updating forwarding rules in either switches or SD end devices [25]. This feature could be utilized by network-level defense mechanisms to improve the response time, which can better deal with potential security issues arising from IoT networks. In our prior work [26], we designed a topology reconfiguration method for the SD WSNs with non-patchable vulnerabilities to change the attack surface of the network in order to increase the attack complexity. In this work, we consider a general IoT network with the support of SDN functionality for the network topology shuffling where an IoT network consists of both decoy and real nodes.

10.3 System Model

We discuss our system model in terms of (i) the network model based on SDN technology for IoT environments; (ii) the attack model describing attack behaviors considered in this work; and (iii) the defense model addressing defense mechanisms placed in the given network.

10.3.1 Network Model

In this work, we concern an IoT environment consisting of servers and IoT nodes. In a given IoT environment, nodes gather data and periodically deliver them to the servers via single or multiple hop(s) for further processing to provide a queried service. IoT nodes of different types/functionalities are placed in different virtual local area networks (VLANs) in the given network. Servers are also deployed in a separate VLAN within the network. We assume that each VLAN has a certain number of IoT nodes with similar types regarding their functionalities.

In this work, we leverage the SDN technology [20–22, 27] in order to effectively and efficiently manage and control nodes in an IoT network. There exists an SDN controller placed in a remote server. The SDN controller communicates with the SDN switches. The servers and IoT nodes in the IoT network are connected to the SDN switches. The SDN controller manages the flows between IoT nodes and servers.

10.3.2 Node Model

In a given network, we characterize a node's attributes in terms of four aspects: (i) whether a node is compromised or not (i.e. $n_i . c = 1$ for compromised; $n_i . c = 0$ otherwise); (ii) whether a node is a real node or a decoy node (i.e. $n_i . d = 1$ for a decoy; $n_i . d = 0$ otherwise); (iii) whether a node is a critical node that has confidential information where the information should not be leaked out to unauthorized parties (i.e. $n_i . r = 1$ for a critical node; $n_i . r = 0$ otherwise); and (iv) a list of vulnerabilities that a node contains (i.e. $n_i . v = \{v_1, \ldots, v_m\}$ where m is the total number of vulnerabilities). Hence, node i's attributes are represented by:

$$A_{n_i} = [n_i.c, n_i.d, n_i.r, n_i.v] \tag{10.1}$$

10.3.3 Attack Model

In this work, we consider the following attacks that may lead to breaching system security goals:

- *Reconnaissance Attacks*: Outside attackers can perform scanning attacks in order to identify vulnerable targets (e.g. a server) to break into a system (or a network). If this attack is successful, the outside attacker successfully identifies the vulnerable target and compromises it. This attack success leads to the *loss of system integrity* because the system has its system component compromised as well as the attacker from the outside. The success of this attack type is related to triggering the system failure based on the security failure condition 1 (SFC1), explained in Section 10.3.5.
- *Data Exfiltration Attacks*: After an outside attacker becomes an inside, legitimate attacker by using the credentials (e.g. login credentials or a legitimate key to access the system/network resources) obtained from the compromised, target node, it can leak confidential information out to unauthorized, outside parties. If this attack is successful, this results in the *loss of confidentiality* because confidential information to be only shared by legitimate users is leaked out to the unauthorized party. When this attack is successful, it will lead to the security failure based on SFC2 in Section 10.3.5.

In this work, the following attack behaviors are assumed to characterize the considered attackers:

- An attacker is assumed to have limited knowledge on whether a given node is decoy (i.e. a fake node mimicking a real, normal node) or not. The attacker's capability to detect the deception depends on the knowledge gap between the attacker and the real system state or how effectively the developed decoy system mimics the real system in a sophisticated manner. For simplicity, we use the probability of an attacker interacting with a decoy to represent the level of the attacker's intelligence in detecting a decoy node, as described in Section 10.3.4.
- After the attacker interacts with a decoy, the attacker's behavior is monitored. If the attacker realizes that the node with which it interacted is a decoy, it terminates the interactions with the decoy node immediately and attempts to find a new target to get into the system.
- An attacker's ultimate goal is to compromise the servers to leak the confidential information to unauthorized entities outside the IoT network.
- An attacker is capable of identifying exploitable, unpatched vulnerabilities or unknown vulnerabilities and compromising the vulnerable IoT nodes in a given IoT network.
- It is hard for the attacker to compromise the servers directly as each server has strong protection mechanisms. Thus, the attacker is able to exploit vulnerable IoT nodes as entry points. Once the attacker breaks into the network by using IoT nodes, it can move laterally within the network and eventually compromise the servers by identifying and exploiting unpatched or unknown vulnerabilities.
- The SDN controller is well protected where the communications between the SDN controller and the SDN switches are assumed to be secure [27].

10.3.4 Defense Model

We assume that traditional defense mechanisms are in place on the IoT network, including a network-based IDS, firewalls, and antivirus software on the servers. The IDS is capable of monitoring the whole IoT network and creates alerts once an intrusion is being detected for incident response. In addition, we have two types of intrusion prevention mechanisms in place, cyber deception and MTD, to dynamically change the attack surface of the IoT network so that they can make attackers hard to launch their attacks, resulting in high attack complexity.

10.3.4.1 Decoy System as Defensive Cyber Deception

This defense mechanism allows defenders to capture and analyze malicious behaviors by luring attackers into a decoy system within a given network and interacting with the attackers. The decoy system is deployed independently from the real system. Accordingly, we assume that normal, legitimate users are not aware of the existence of the decoy system while the defenders will only get alerts caused by the malicious intrusions if an attacker breaks into the decoy system.

We consider two types of decoys utilized throughout an IoT network considered in this work:

- *Emulation-Based Decoys*: This type of decoys allows defenders to create a variety of fake assets and to provide a large-scale coverage across the network.
- *Full OS-Based Decoys*: This type of decoys enables the replication of actual production devices to increase the possibility that the attacker engages the decoys and exposes its intention and/or strategy.

Both emulation-based and full OS-based decoys can be autonomously created to fit within the environment with no changes to the existing infrastructure. They can provide various types of interactive capabilities. To increase the overall chances of accessing decoys by attackers, the decoys can be created by combining multiple, diverse forms of decoys that look like real, legitimate nodes. In addition, the decoys can be deployed in every VLAN of the network. Besides, there exists an intelligence center performing the following tasks: (i) create, deploy, and update a distributed decoy system; (ii) provide automated attack analysis, vulnerability assessment, and forensic reporting; and (iii) integrate the decoy system with other prevention systems (e.g. security incident and event management platform, firewalls) to block attacks. The module for the decoy node deployment can be implemented and placed in a remote server.

For these two types of decoy nodes to be considered in this work, we create a design parameter, P_d, indicating the probability that an attacker interacts with a decoy node. Since full-OS-based decoys are considered as having more sophisticated, real-node like quality with more cost, we consider P_d^{em} as the probability that the attacker interacts with an emulation-based decoy while using P_d^{OS} as the probability based on the assumption that the attacker will more likely to interact with an OS-based decoy with $P_d^{em} \leq P_d^{OS}$.

10.3.4.2 Network Topology Shuffling-based MTD

In this work, we consider NTS-MTD, to change the topology of the given IoT network where the network consists of both real, legitimate nodes and decoy nodes. NTS-MTD is to be triggered following the concept of event-based MTD in that the network topology changes upon a certain event indicating that the network is at risk due to too many nodes being compromised. For example, following the concept of Byzantine Failure [27], if the system is close to a security failure state, such as more than one-third of nodes being compromised, the network topology is being shuffled in order to stop or mitigate the spread of nodes being compromised.

In this work, we assume that the SDN controller can control the traffic of the decoy nodes in an SDN-based decoy system and manage to change the network topology upon a certain event. We combine the cyber deception and NTS-MTD and propose a network topology shuffling with decoy nodes to change the attack surface of the IoT network. The details of the proposed decoy system and the NTS-MTD are described in Section 10.4.

10.3.5 Security Failure Conditions

A system fails when either of the below two conditions is met:

- *Security Failure Condition 1 (SFC1)*: This system failure is closely related to the attacker's successful reconnaissance attacks and accordingly their successful compromise of system components (or capture of IoT nodes). We define this system failure based on the concept of Byzantine Failure [27]. That is, when more than one third of legitimate, member nodes are compromised, we define it as the system failure due to the *loss of system integrity*.
- *Security Failure Condition 2 (SFC2)*: This system failure occurs when confidential information is leaked out to unauthorized parties by inside attackers (or compromised nodes), which perform the so-called *data exfiltration attack*. This type of system failure occurs due to the *loss of data confidentiality*.

10.4 Proposed Proactive Defense Mechanisms

In this section, we provide the overview of our proposed NTS-MTD in terms of the decoy-based network topology shuffling, optimization method, and GSM for security analysis.

10.4.1 Network Topology Shuffling with Decoy Nodes

In this section, we describe the details of the proposed NTS-MTD in terms of how to deploy the initial set of nodes, how to select a network topology to be shuffled, and when to shuffle a network topology.

10.4.1.1 Initial Deployment of Nodes

We consider both servers and IoT node decoys to be deployed in the IoT network. As the network is divided into different VLANs, a certain number of decoy nodes can be placed into each VLAN based on the real nodes placed to the corresponding VLAN. At least one decoy server needs to be deployed to interact with the attacker and reveal the attacker's intent. We distribute the IoT decoy nodes into each VLAN based on the deployment of real nodes in the VLAN. Note that we can deploy more decoys if the VLAN has a large number of real nodes with different types. When adding the decoy nodes, we link real IoT nodes with the decoy nodes to lure attackers into the decoy system. The flows from the real IoT nodes to the decoy nodes or from the decoy nodes to the decoy nodes are controlled by the SDN controller. There will be no flows from decoy nodes to real nodes as the decoy nodes are used to divert the attackers from the real system; once the attacker is lured into the decoy system, it will be diverted to other decoys within the decoy system and the behavior will be monitored; if the attacker finds out a node it interacted with is a decoy node, it will terminate the interaction with the decoy node and look for a new target to break in. The directional flows between real and decoy nodes may reveal some information to the attackers in the long term. We will consider changing flows from real nodes to real nodes in the future work to increase the complexity of the connection changes.

We assume that newly added flows will not affect normal flows from the IoT nodes to servers for the service delivery. In practice, the IoT nodes will consume more energy to deliver more flows and may delay the time to send normal packets toward the server. Thus, the service availability provided by the server may be affected. This will be examined in our future work.

10.4.1.2 Selection of a Network Topology to Shuffle

Given the IoT network with real nodes, we first decide the number of decoys to be potentially deployed in each VLAN and then randomly generate a set of deployments of the decoy nodes with added connections to some randomly chosen real nodes. The set of these randomly generated network topologies will be used as the initial set of topologies and they will be used in the shuffling optimization algorithm to identify an optimal network topology.

10.4.1.3 Adaptive Shuffling of a Network Topology as an MTD

As a baseline MTD, we can simply choose a fixed time interval to be used to execute a given MTD mechanism. Alternatively, we can design an intelligent MTD to be executed in an *adaptive* manner based on the system security level detected by the defender. For example, a network topology needs to be changed when the defender perceives the system security vulnerability level larger than a given threshold ρ. The system security vulnerability level at time t, denoted by $SSV(t)$, is measured by two dimensions: (i) how many legitimate, inside nodes are compromised until time t, associated with SFC1; and (ii) how many neighboring nodes of a critical node (i.e. $n_i.\ r = 1$) within k hops from the critical node i are compromised until time t, which is related to detecting SFC2. Note that when the system meets either SFC1 or SFC2, then the system fails, leading to $SSV(t) = 1$. Otherwise, $SSV(t)$ is computed by:

$$SSV(t) = w_1 \frac{CN(t)}{N} + w_2 \frac{CN_{ck}(t)}{N_{ck}(t)} \tag{10.2}$$

Here, w_1 and w_2 are weights to consider SFC1 and SFC2, respectively, where $w_1 + w_2 = 1$. N is the total number of real nodes initially deployed and $CN(t)$ is the number of compromised, real nodes at time t. $N_{ck}(t)$ is the total number of real nodes within k hops from given critical nodes at time t while $CN_{ck}(t)$ is the total number of compromised, real nodes among the real nodes within the k hops from the given critical nodes. Since there may be multiple critical nodes which have confidential information that should not be leaked out to outside, non-authorized nodes, we estimate $CN_{ck}(t)$ by:

$$CN_{ck}(t) = \sum\nolimits_{i \in L_k(t)} n_i \, . \, c(t) \tag{10.3}$$

where $L_k(t)$ means the number of real nodes that belong to neighbors of any critical nodes within k hops from them at time t. Recall that $n_i.\ c(t)$ refers to whether node i is compromised ($n_i.\ c(t) = 1$) or not ($n_i.\ c(t) = 0$) at time t. Thus, the cardinality of $L_k(t)$ (i.e. $|\,L_k(t)|$) is the same as $N_{ck}(t)$. Note that as a network topology keeps changing due to executing the MTD to change the network topology, both $N_{ck}(t)$ and $CN_{ck}(t)$ are the functions of time to reflect their dynamic changes. Note that the set $L_k(t)$ may include any critical nodes being compromised. If this happens, it means the system meets SFC2 and the system failed. That is, $SSV(t) = 1$ and no further detection of system security level is needed.

For each $SSV(t)$, we calculate the mean time to compromise (MTTC) associated with it. MTTC refers to the total amount of time that the attacker takes to compromise a series of nodes within the network until the system reaches a certain security vulnerability level. The computation of MTTC is detailed in Section 10.5.3.

10.4.2 GA-based Network Shuffling Optimization

In this section, we discuss a GA-based network shuffling optimization technique in a given SDN-based IoT network.

To breach system security goals, an attacker may be able to find multiple attack paths to reach a target node via one or multiple entry points. An attack path specifies a sequence of nodes that the attacker can compromise to reach the target node. We consider a set of attack paths AP for the attacker to reach all the targets from all possible entry points. Each attack path ap is a sequence of nodes over the attack path. We use AP_r to represent a set of attack paths with real nodes as targets and AP_d to indicate a set of attack paths with decoy nodes as targets. AP_r only contains the real nodes while AP_d contains both real nodes and decoy nodes. To be specific, in the attack model, we assume the attacker could exploit any IoT nodes as entry points. Once the attacker successfully compromises a node, it could use this node as the stepping stone to compromise other nodes and further compromise servers as their targets. The attacker may find a real node as the entry point and then is diverted to a decoy node. Once the attacker is lured into the decoy system, it will be diverted to other decoy nodes within the decoy system. If the attacker reaches the decoy server, this is accounted as an attack path in AP_d; but if the attacker figures out the decoy node and terminates its interaction with the decoy node, this is not counted as an attack path because the attacker does not reach the decoy server (i.e. target). Besides, the decoy nodes could be easily updated or cleared once being compromised via the central management portal thus the attacker will not recognize the same decoy node during the subsequent attacks.

We design three metrics to be optimized: (i) the number of attack paths toward the decoy targets (N_{DT}^{AP}); (ii) MTTSF; and (iii) defense cost (C_D). The computations of these metrics are described in Section 10.5.2.

10.4.3 GSM for the IoT MTD

We apply the GSM to assess the security of an SDN-based IoT network with the proposed proactive defense mechanism. Figure 10.2 describes the workflow consisting of network generation, topology generation, security model generation, shuffling mechanism evaluation, and shuffling optimization.

The workflow of this security analysis consists of the following five phases:

- *Phase 1*: The security decision maker provides the IoT Generator with the system information (i.e. an initial network topology and node vulnerability) to construct an IoT network.
- *Phase 2*: Given the network and initial deployment of the decoys, the Topology Generator randomly generates a set of different topologies (i.e. add connections from real nodes to decoys) based on the shuffling algorithm. Each shuffling is presented in an integer format and passed onto the Optimization Module.
- *Phase 3*: The Security Model Generator takes the shuffled network as input and automatically generates the HARM which captures all possible attack paths. We use the three-layered HARM [3, 28] as our GSM. In the three-layered HARM, the upper layer captures the subnet reachability information, the middle layer represents the node connectivity information (i.e. nodes connected in the topological structure), and the lower layer denotes the vulnerability information of each node.

- *Phase 4*: The Shuffling Evaluator takes the HARM as input along with the evaluation metrics and computes the results which are then fed into the Optimization Module.
- *Phase 5*: Based on the initial set of shuffled topologies and the associated evaluation results, the Optimization Module applies the multi-objective GA to compute the optimal topology for the IoT network based on the termination conditions (e.g. the maximum number of generations defined by the security decision maker).

10.5 Numerical Results and Analysis

In this section, we describe our experimental setup and simulation results, along with the analysis of the observed trends from the obtained results.

10.5.1 Experimental Setup

We use the example SD-IoT network shown in Figure 10.1. Specifically, the network consists of four VLANs. There are two Internet of Medical Things (i.e. MRI and CT scan) in $VLAN_1$; a smart thermostat, a smart meter, and a smart camera in $VLAN_2$; a smart TV and a laptop in $VLAN_3$; and a server located in $VLAN_4$. $VLAN_4$ is connected with other three VLANs as IoT devices need to deliver information to the server for further processing. $VLAN_2$ is also connected to $VLAN_3$ for the applications on smart TV and laptop to control the smart sensors as well as receiving videos from the smart camera.

For the software vulnerability analysis, we will be using Common Vulnerabilities and Exposures (CVE)/National Vulnerability Database (NVD) [29] for IoT networks. We assume each real node has one vulnerability that could be exploited by the attacker to gain a root privilege. More vulnerabilities could be chosen for the nodes in the future work as the focus of the research is to propose and evaluate the proactive defense mechanism, rather than demonstrating the capabilities of the GSM to analyze the security posture of the IoT network with multiple vulnerabilities. The vulnerability information of real nodes (i.e. CVE ID) is assumed following Table 10.1.

We also assume the compromise rate of each vulnerability. The compromise rate represents the frequency that an attacker could successfully exploit the vulnerability to gain root privilege per time unit (i.e. hour). We estimate the value according to the base score from the Common Vulnerability Scoring System (CVSS). Specifically, we estimate the compromise rate as once per day (i.e. 0.042) if the base score is 10.0, twice per week (i.e. 0.012) with the base score of around 8.0, once per week (0.006) when the base score is around 7.0, and once per 10 days (i.e. 0.004) under

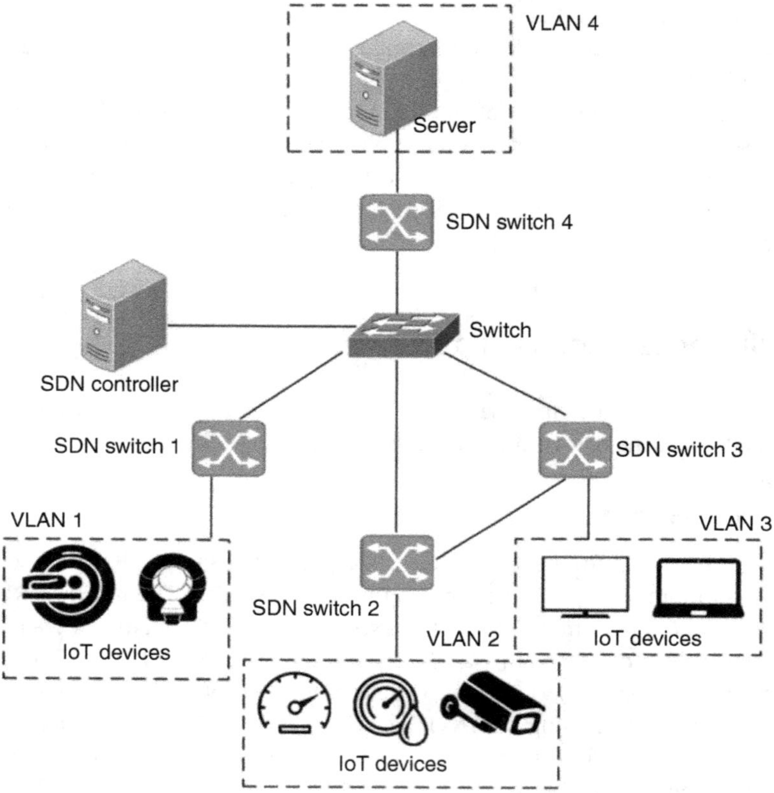

Figure 10.1 Example of a software-defined IoT network.

Table 10.1 Real node and vulnerability information.

Real node	VLAN	CVE ID	Compromise rate
MRI	VLAN1	CVE-2018-8308	0.006
CT scan	VLAN1	CVE-2018-8308	0.006
Smart Thermostat	VLAN2	CVE-2018-11315	0.006
Smart Meter	VLAN2	CVE-2017-9944	0.042
Smart Camera	VLAN2	CVE-2018-10660	0.042
Smart TV	VLAN3	CVE-2018-4094	0.012
Laptop	VLAN3	CVE-2018-8345	0.004
Server	VLAN4	CVE-2018-8273	0.006

Table 10.2 Decoy node and vulnerability information.

Decoy node	VLAN	CVE ID	Compromise rate
CT scan	VLAN1	CVE-2018-8308	0.006
		CVE-2018-8136	0.012
Smart Camera	VLAN2	CVE-2018-6294	0.042
		CVE-2018-6295	0.042
		CVE-2018-6297	0.042
Smart TV	VLAN3	CVE-2018-4094	0.012
		CVE-2018-4095	0.012
Server	VLAN4	CVE-2016-1930	0.042
		CVE-2016-1935	0.012
		CVE-2016-1962	0.042

the score of around 5.0. This value will be used to calculate MTTSF and MTTC in the simulations based on the HARM.

We consider using one type of decoys in each VLAN in the initial deployment of the decoy system. In order to lure the attackers, each decoy is assumed to be configured to have multiple vulnerabilities. The attacker could exploit any vulnerability to gain the root permission of the node. We assume to use the vulnerabilities of the decoys based on Table 10.2.

10.5.2 Metrics

We use the following metrics to measure security and performance of the proposed proactive defense mechanism:

- *Number of Attack Paths Toward Decoy Targets* (N_{DN}^{AP}): This metric indicates the level of deception that diverts the attacker from the real system. N_{DT}^{AP} is calculated by $|AP_d|$ to sum the attack paths toward the decoy targets.
- *MTTSF*: This metric measures the system lifetime indicating how long the system prolongs until the system reaches either SFC1 or SFC2 (described in Section 10.3.5). That is, this measures the system uptime without any security failure occurring. MTTSF is measured by:

$$MTTSF = \sum_{i \in S} (1 - SF_i) \int_{t=0}^{\infty} P_i(t) dt \tag{10.4}$$

where S is a set of all system states and SF_i returns 1 when system state i reaches either SFC1 or SFC2; 0 otherwise. $P_i(t)$ indicates the probability that the system is at system state i.

- *MTTC*: This metric refers to the total amount of time that the attacker takes to compromise a series of nodes within the network until the system reaches a certain security vulnerability level, *SSV*. This metric is used to detect the system security vulnerability level upon the number of nodes being compromised by the attacker and employed to determine when to trigger an MTD operation. Similar to the computation of MTTSF as above, MTTC is estimated by:

$$MTTC = \sum_{i \in S} S_i \int_{t=0}^{\infty} P_i(t)dt \tag{10.5}$$

where S refers to a set of all system states that do not reach the given *SSV* and S_i returns 1 when system state i didn't reach the *SSV*; 0 otherwise. $P_i(t)$ is the probability of the system being at system state i.
- *Defense Cost* (C_D): This metric captures the cost associated with the shuffling operations. That is, we count the number of edges shuffled (i.e. from connected to disconnected or from disconnected to connected) by:

$$C_D = \int_{t=0}^{MTTSF} C_S(t) \tag{10.6}$$

where $C_S(t)$ refers to the number of shuffled edges at time t. Note that a same edge can be shuffled multiple times over time and each shuffling is counted as a separate MTD operation during the system uptime.

10.5.3 Comparing Schemes

We have two aspects of MTD to investigate: (i) when to shuffle a network topology; and (ii) how to select the network topology. As described in Section 10.4.2, (i) is to investigate an interval or when to execute the proposed MTD operation in an adaptive manner while (ii) is to investigate the effect of a selected network topology generated by a GA-based network topology (GANT) or a random network topology (RNT).

The two types of ***"when to shuffle a network topology" strategies*** are:

- *Fixed Shuffling (FS)*: This shuffling represents a baseline scheme using a fixed time interval to shuffle a given network topology.
- *Adaptive Shuffling (AS)*: This shuffling is designed to execute the MTD in an *adaptive* manner based on the system security vulnerability level ($SSV(t)$), detected by the defender with two given thresholds: (i) β to check the decrease of the *SSV* during the time Δ; and (ii) ρ to check the current system security vulnerability, $SSV(t)$, as described in Section 10.4.1. The NTS-MTD is executed if the condition, $(SSV(t) - SSV(t - \Delta) > \beta) \wedge (SSV(t) < \rho)$, is met where Δ is a checking interval.

The two types of ***"how to select a network topology" strategies*** are:

- *RNT*: This scheme is used as a baseline model that can simply select a network topology based on a simple random selection of a node's edges to other nodes based on a rewiring probability P_r which represents the probability that a node is connected with another node in a given network. Here P_r is critical in determining the overall network density in a given network.
- *GANT*: This scheme selects a network topology based on the method proposed in Section 10.4.1. That is, a network topology to be used for a next round of shuffling is selected based on the network topology that maximizes the objective functions used in the GA, as discussed in Section 10.4.1.

Since we have two strategies under each category of the proposed MTD mechanisms, we investigate the following four schemes as follows:

- *FS-RNT*: This scheme executes the NTS-MTD based on a certain fixed time interval, γ, with a selection of a network topology based on RNT.
- *AS-RNT*: This scheme adaptively executes the NTS-MTD based on the level of system security vulnerability, *SSV*(t), detected by the defender selecting a network topology based on RNT.
- *FS-GANT*: This scheme changes a current network topology to a selected network topology based on the decision of the fitness functions using GA and executes the NTS-MTD based on a certain fixed time interval.
- *AS-GANT*: This scheme adaptively changes a network topology based on the system security vulnerability level, *SSV*(t), detected by the defender and selects a network topology to change based on the decision of the fitness functions using GA.

10.5.4 Simulation Steps and Parameter Details

We implement the simulations based on the workflow shown in Figure 10.2. The `Shuffling Evaluator` could be either a GA-based shuffling algorithm or a random shuffling algorithm.

We consider MTTSF as an expected mean MTTSF, $E\left(\overline{MTTSF}\right)$, in computing the shuffled network topology based on the assumptions of the potential attacker (see Section 10.3.3). For each shuffled network, we construct HARM and calculate potential attack paths. We assume that the attacker randomly selects one entry point and compromises the nodes on the attack path at each time until either security failure condition (see Section 10.3.5) is met. We assume the defender will clear the decoy nodes once it detects the attacker's interaction with the decoy target. Therefore, the attacker will not recognize the same decoy node during its following action. The attacker's intelligence, estimated by the probability to interact with the

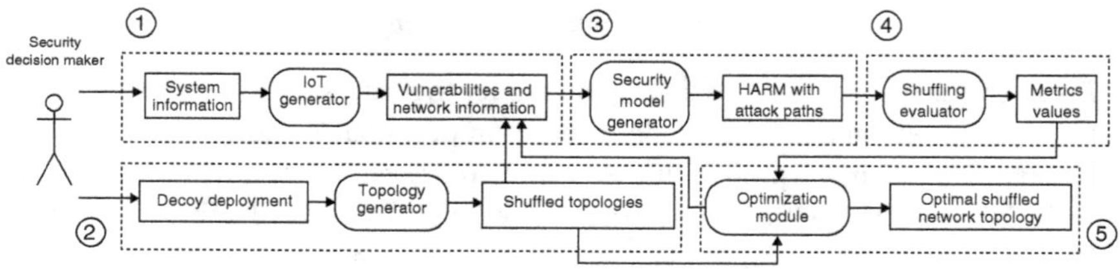

Figure 10.2 Workflow for security analysis.

decoy, P_d^{em} and P_d^{OS} (see Section 10.3.4) is incorporated into the calculation of the expected mean MTTSF, $E(\overline{MTTSF})$, as well as actual MTTC in AS. We consider the defender's error probabilities in estimating the attacker's interaction probability with a decoy with the ranges of [−0.05, 0.05]. This affects the defender's ability to estimate $E(\overline{MTTSF})$ because the defender needs to compute the actual MTTC based on its detection probability, $\alpha = 0.95$ (i.e. a defender's confidence about the attacker's intelligence). The detailed calculation steps for these two metrics are found in [4]. As the entry points are randomly chosen, we run the attacker model for 100 times and calculate $E(\overline{MTTSF})$ from the HARM for the given shuffled network.

In GANT, we encode each shuffling solution for the whole network as a binary valued vector where 1 represents the existence of an edge between two nodes (i.e. two nodes are connected) while 0 represents no edges (i.e. two nodes are not connected). We limit the potential connections to be the edges from real IoT nodes to decoy nodes. Hence, to optimize the defense cost, we aim to maximize $C_T(t) - C_D(t)$, where $C_T(t)$ refers to the total defense cost that counts the total number of potential changes of the edges at time t while $C_D(t)$ is the number of edges changed by the used NTS-MTD (see Section 10.5.2) at time t. Here, we aim to solve a multi-objective optimization (MOO) problem with three objectives to maximize N_{DN}^{AP} and $E(\overline{MTTSF})$ while minimizing C_D (or maximizing $C_T(t) - C_D(t)$). The optimization problem is to compute a set of Pareto optimal solutions (or Pareto frontier) [30]. In order to choose one optimal solution among the Pareto frontier, we first normalize the three metrics, denoted by $\widetilde{N_{DN}^{AP}}$, $\widetilde{E(\overline{MTTSF})}$, and $\widetilde{C_D}$, and assign a weight for each metric based on scalarization-based MOO technique to make the MOO problem to a single-objective optimization (SOO) problem [30], respectively. The metric is normalized by:

$$\widetilde{X} = \frac{X}{X_{max}} \tag{10.7}$$

where $\widetilde{X}$ is the normalized metric value, X is the original metric value, and X_{max} is the maximum metric value of the corresponding fitness function among the final population in the GA-based algorithm.

The objective function we aim to maximize is represented by:

$$\max\ w_N \widetilde{N_{DN}^{AP}} + w_M E\left(\widetilde{\overline{MTTSF}}\right) + w_C \widetilde{C_D} \tag{10.8}$$

where w_N, w_M, and w_C are weights for each metric with $w_N + w_M + w_C = 1$. The optimal solution is the network topology with the maximum objective value.

In this study, we consider an equal weight for w_N, w_M, and w_C, respectively, with 1.0/3.0.

We assume the following algorithm parameters for the simulations with GANT: population size $(N) = 100$, maximum number of generations $(N_g) = 100$, crossover rate $(r_c) = 0.8$, and mutation rate $(r_m) = 0.2$. In RNT, we randomly change the edges between the real IoT nodes and decoy nodes. We also considered the probability that an edge will be shuffled (i.e. add/remove an edge) with $P_r = 0.5$.

When calculating $SSV(t)$ for AS, we use the following default values for the weights (w_1, w_2): two SSV-related thresholds to execute the NTS-MTD (β and ρ) and the k number of hops to determine the neighbor nodes to a critical node: $w_1 = w_2 = 0.5$, $\beta = 0.01$, $\rho = 0.1$ and $k = 1$. The checking interval Δ is determined upon detecting a compromised node by the defender.

Whenever the defender detects a compromised real node, it will check whether the system reaches the given $SSV(t)$, reflecting the nature of an event-driven adaptive MTD.

We assume there is an attacker during each simulation run. The attacker randomly chooses entry points and compromises nodes along the attack paths with the behaviors defined in Section 10.3.3. By using the FS schemes, the network may be shuffled while a node is under attacks. We assume the attacker is forced to quit the network due to lost connections and needs to find other ways to break into the network. During the subsequent attack after shuffling, the attacker could continue its previous attack action once it encounters the same real node next time (i.e. MTTC for the real node is accumulated throughout the MTTSF). By using the AS schemes, the network is shuffled due to the system security vulnerability level detected by the defender. The attacker is also forced to quit the network after each shuffling due to lost connections and needs to find ways to re-enter the network. For both schemes, the decoy node is cleared at each shuffling.

For each simulation, we calculate actual MTTSF under the existence of real attackers, average N_{DT}^{AP}, denoted as $\overline{N_{DT}^{AP}}$, by summing N_{DT}^{AP} from each shuffled network and dividing it by the total number of shuffling times within MTTSF, and defense cost per time unit, $\widehat{C_D}$, by summing C_D from each shuffling network and dividing it by actual MTTSF.

We run the entire simulation for 100 times and calculate the average metric value by summing the metric value from each simulation and dividing the

Table 10.3 Design parameters, their meanings, and the default values.

Parameter	Meaning	Default value
N	The total number of network topologies with initial decoy deployment and randomly generated connections between real and decoy nodes	100
N_{DN}^{AP}	The number of attack paths toward the decoy targets	Metric
MTTSF	Mean time to security failure, representing the system lifetime	Metric
MTTC	Mean time to compromise a fraction of nodes in a network until the system reaches a certain security vulnerability level	Metric
C_D	The number of edges changed from a previous network topology to a current network topology due to the network shuffling-based MTD	Metric
w_N	A weight to consider N_{DN}^{AP}	1/3
w_M	A weight to consider MTTSF	1/3
w_C	A weight to consider C_D	1/3
w_1	A weight to consider the security vulnerability associated with SFC1	0.5
w_2	A weight to consider the security vulnerability associated with SFC2	0.5
N_g	The maximum number of the generation used in GANT	100
r_c	Crossover rate used in GANT	0.8
r_m	Mutation rate used in GANT	0.2
P_r	The probability of an edge being shuffled	0.5
β	The threshold used to estimate the decrease of the system security vulnerability level during the time Δ used in GANT	0.01
ρ	The threshold of tolerating system security vulnerability used in GANT	0.1
k	The number of hops to determine a node's ego network	1
γ	The fixed shuffling time interval used in RNT (hour)	24

summed value by the total number of simulations. We use $\overline{MTTSF}$, $\overline{\overline{N_{DT}^{AP}}}$, and $\widehat{C_D}$ to represent the mean metric value in our experimental results shown in the following section. Table 10.3 summarizes the design parameters, their meanings, and corresponding default values used in our simulation experiments.

10.5.5 Simulation Results

In our simulation experiments, we consider three scenarios by varying the number of decoys in each VLAN and the level of attackers' intelligence in detecting decoy nodes (i.e. P_d^{em} and P_d^{OS}): (i) one decoy node assigned for each VLAN in the presence of low-intelligent attackers (i.e. $P_d^{em} = 0.9$ and $P_d^{OS} = 1.0$); (ii) one decoy node assigned for each VLAN in the presence of medium-intelligent attackers (i.e. $P_d^{em} = 0.5$ and $P_d^{OS} = 0.9$); and (iii) two decoy nodes assigned for each VLAN in the presence of low-intelligent attackers (i.e. $P_d^{em} = 0.9$ and $P_d^{OS} = 1.0$). For other key design parameters, we will follow default values summarized in Table 10.3.

10.5.5.1 One Decoy Node for Each VLAN with Low-Intelligent Attackers

Figure 10.3 shows how the considered four MTD schemes perform in terms of mean values of the three metrics described in Section 10.5.2 over 100 simulations. In Figure 10.3a, with the random shuffling, both FS-RNT and GA-RNT have a similar number of attack paths toward decoy targets, 29; with the GA-based shuffling, $\overline{\overline{N_{DT}^{AP}}}$ is much higher with 54 for both schemes. $\overline{\overline{N_{DT}^{AP}}}$ is determined by the network topology; thus, either random shuffling or GA-based schemes have a similar performance in $\overline{\overline{N_{DT}^{AP}}}$.

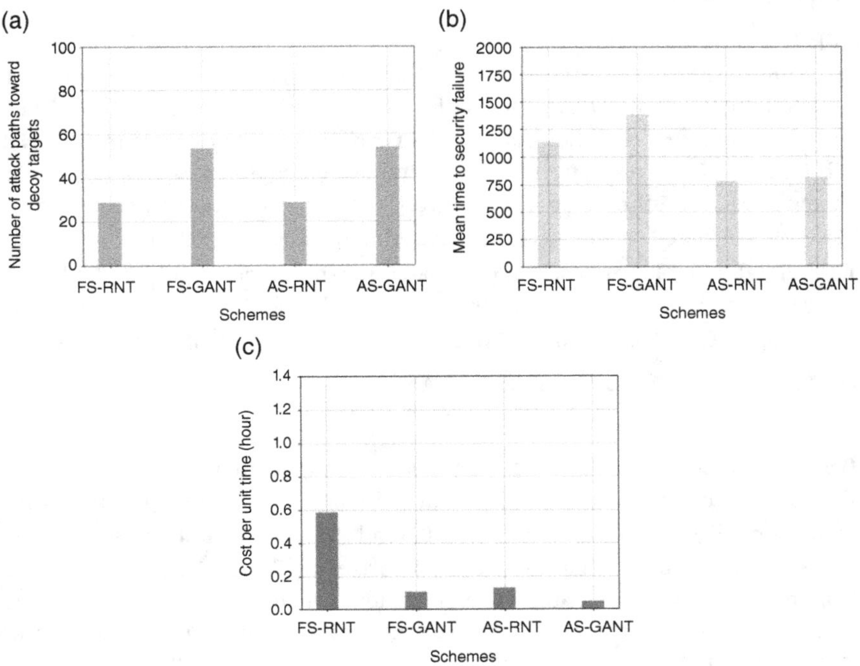

Figure 10.3 Performance comparison of the four MTD schemes with low-intelligent attackers where one decoy node is deployed for each VLAN. (a) Average number of attack paths toward decoy targets ($\overline{\overline{N_{DT}^{AP}}}$). (b) Average MTTSF ($\overline{MTTSF}$). (c) Average defense cost per time unit ($\widehat{\overline{C_D}}$).

Figure 10.3b shows the values of the average MTTSF. With FS, both schemes have higher $\overline{MTTSF}$ compared with AS schemes based on the following reason. A node may still be under attacks while the network is shuffled because the fixed interval is much smaller than MTTC for the node. However, the attacker is forced to quit the network due to lost connections and needs to re-enter the network by randomly choosing entry points to compromise.

During the subsequent attack after each shuffling, the attacker could continue its previous attack action once it encounters the same real node next time (i.e. MTTC for the real node is accumulated throughout MTTSF). After then, the attacker needs to launch a new attack for decoys as they are cleared at each shuffling. However, although MTTC is accumulated for the real node, the time to meet either security failure condition (i.e. compromise a certain number of nodes or the critical node) increases over time. In addition, FS-GANT has higher $\overline{MTTSF}$ (i.e. 1381 hours) than that of FS-RNT (i.e. 1131 hours) while AS-GANT also has slightly higher $\overline{MTTSF}$ than that of AS-RNT, showing 782 and 817 hours, respectively.

Now we investigate the effect of the four schemes in terms of the average defense cost per time unit which shows the trade-off for defense cost per time unit and MTTSF. Figure 10.3c shows the average defense cost per time unit $\widehat{\overline{C_D}}$ under the four schemes. As expected, FS-RNT scheme incurs the highest cost (i.e. 0.59) due to frequent executions of network shuffling while AS-GANT has the lowest cost (i.e. 0.05). Surprisingly, FS-GANT incurs less cost than AS-RNT with 0.11 and 0.13, respectively. This is due to less edge changes of GA-based scheme in each shuffling even if the network topology changes at a fixed interval.

Overall, GA-based schemes can preserve a higher security level in terms of N_{DN}^{AP} and maintaining a lower cost while FS-based schemes incur higher $\overline{MTTSF}$. We can see there is a balance between MTTSF and defense cost per time unit. $\overline{MTTSF}$ of FS-GANT is roughly 1.7 times higher than $\overline{MTTSF}$ of AS-GANT while $\widehat{\overline{C_D}}$ of FS-GANT is twice of that of AS-GANT.

10.5.5.2 One Decoy Node for Each VLAN with Medium-Intelligent Attacker

We use the same initial decoy deployment in Section 10.5.5.1, except considering medium-intelligent attackers with $P_d^{em} = 0.5$ and $P_d^{OS} = 0.9$. For other design parameters, we follow their default values summarized in Table 10.3.

Figure 10.4 shows the performance of the four schemes in terms of the mean value of each metric in Section 10.5.2 based on 100 simulation runs when the attackers have the medium-intelligent levels. Figure 10.4a shows the similar trend observed in Figure 10.3a. With the random shuffling, both FS-RNT and GA-RNT have a similar number of attack paths toward decoy targets which is about 29; for the GA-based shuffling, much higher N_{DT}^{AP} is observed, showing 54 for both schemes.

Figure 10.4b shows a similar trend as Figure 10.3b. With FS, both schemes have higher $\overline{MTTSF}$ compared with AS schemes. Besides, FS-GANT has higher $\overline{MTTSF}$ (i.e. 1414 hours) than FS-RNT (i.e. 1139 hours) while AS-GANT also has slightly higher $\overline{MTTSF}$ than AS-RNT, showing 720 and 766 hours, respectively.

Figure 10.4c shows the similar trend in Figure 10.3c. As expected, FS-RNT incurs the highest cost (i.e. 0.59) among all due to frequent executions of network shuffling while AS-GANT has the lowest cost (i.e. 0.05). Surprisingly, FS-GANT incurs less cost than AS-RNT, showing 0.1 and 0.14, respectively, due to less edge changes of GA-based scheme in each shuffling.

Overall, we observe GA-based schemes can maintain a higher security level in terms of N_{DN}^{AP} and generate lower cost while FS schemes incur higher $\overline{MTTSF}$. We can also see there is a balance between MTTSF and defense cost per time unit. $\overline{MTTSF}$ of FS-GANT is roughly 1.8 times higher than that of AS-GANT while $\overline{\widehat{C_D}}$ of FS-GANT is twice of that of AS-GANT. Additionally, FS-based schemes could keep $\overline{MTTSF}$ similar under low and medium attack intelligence. This is because the fixed interval is much smaller than MTTC for a node. Therefore, a node may still be under attacks while the network is shuffled. However, the attacker is forced to quit the network and could continue its previous attack when encountering the same real node thus causing MTTC for the real node being accumulated throughout the MTTSF. Although the attacker's intelligence increases causing shorter MTTC for decoy nodes, the time to meet either security failure condition (i.e. compromise a certain number of nodes or the critical node) remains similar. For AS-based schemes, $\overline{MTTSF}$ with low-intelligent attackers is slightly higher than that with medium-intelligent attackers. This implies that potential attackers with higher intelligence in detecting the decoys hurts the system lifetime as measured based on MTTSF under AS. However, we prove that AS-GANT is resilient under high-intelligent attacks without much reduction of MTTSF when compared with the case with low-intelligent attacks as shown in Section 10.5.5.1.

10.5.5.3 Two Decoy Nodes for Each VLAN with Non-Intelligent Attackers

We use two decoy nodes for each VLAN as the initial deployment. We use the low-intelligent attackers and follow the default values of other design parameters in Table 10.3.

Figure 10.5 shows the performance of the four schemes under this scenario (i.e. two decoy nodes for each VLAN with low-intelligent attackers) in terms of the mean values of three metrics based on the results collected from 100 simulation runs. As expected, Figure 10.5a shows the results similar to what we observed in Figures 10.3a and 10.4a. The random shuffling-based schemes, both FS-RNT and GA-RNT, have a similar number of attack paths toward decoy targets which is roughly 52; with the GA-based shuffling, $\overline{\overline{N_{DT}^{AP}}}$ is much higher, showing 85 with

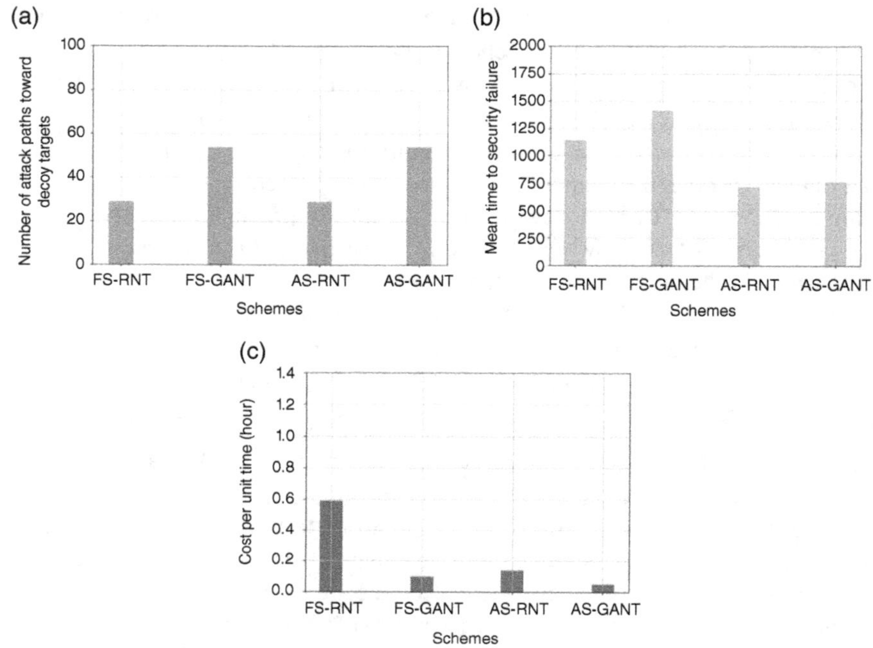

Figure 10.4 Performance comparison of the four MTD schemes with medium-intelligent attackers where one decoy is deployed for each VLAN. (a) Average number of attack paths toward decoy targets ($\overline{\overline{N_{DT}^{AP}}}$). (b) Average MTTSF ($\overline{MTTSF}$). (c) Average defense cost per time unit ($\widehat{\overline{C_D}}$).

FS-GANT and 88 with AS-GANT. As more decoys are deployed in the network, all four schemes have higher $\overline{\overline{N_{DT}^{AP}}}$ compared with the schemes when only one decoy is deployed in each VLAN.

Figure 10.5b also follows the similar trends to the results shown in Figures 10.3b and 10.4b. With FS, both schemes have higher $\overline{MTTSF}$ compared with AS schemes. Besides, the FS-GANT scheme has higher $\overline{MTTSF}$ (i.e. 1757 hours) than the FS-RNT scheme (i.e. 1423 hours) while the AS-GANT scheme also has slightly higher $\overline{MTTSF}$ than the AS-RNT scheme, demonstrating 809 and 831 hours, respectively. Due to more decoys deployed in the network, all four schemes have higher $\overline{MTTSF}$ than a single decoy being deployed in each VLAN.

Figure 10.5c shows a slightly different trend compared to the results in Figures 10.3c and 10.4c. As expected, FS-RNT incurs the highest cost (i.e. 1.03) among all while FS-GANT incurs the second highest cost (i.e. 0.44) due to more frequent executions of network shuffling than AS. AS-GANT has the lowest cost

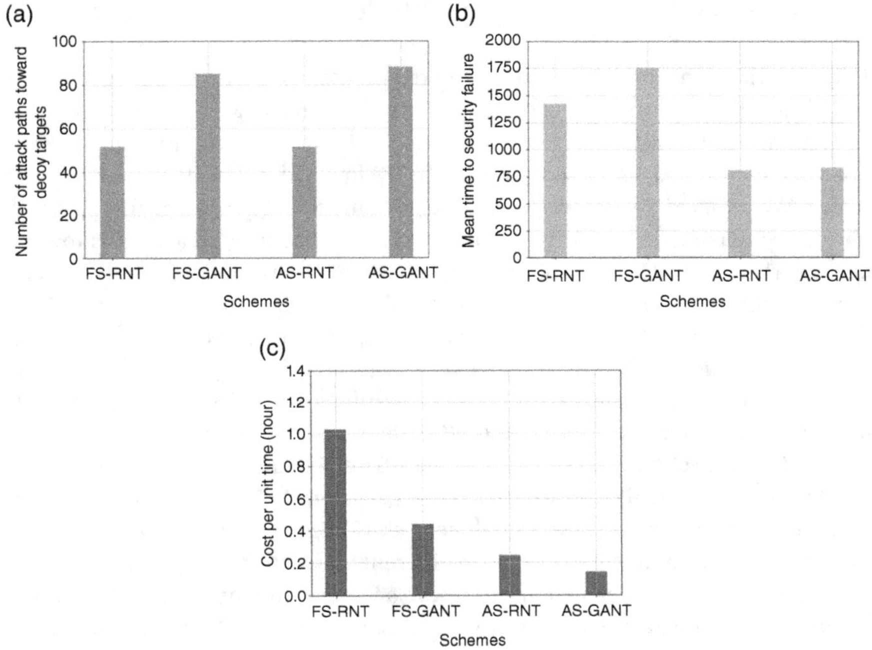

Figure 10.5 Performance comparison of the four MTD schemes with low-intelligent attackers where two decoy nodes are deployed for each VLAN. (a) Average number of attack paths toward decoy targets ($\overline{\overline{N_{DT}^{AP}}}$). (b) Average MTTSF ($\overline{\overline{MTTSF}}$). (c) Average defense cost per time unit ($\overline{\widehat{C_D}}$).

(i.e. 0.14) among all. As more edges are available to be changed due to the increased number of decoy nodes, both FS-RNT and FS-GANT schemes incur higher cost than AS counterparts. Due to more connections that could be shuffled between decoys and real nodes, all four schemes under two decoys in each VLAN incur higher cost than the corresponding schemes under one decoy deployed in each VLAN.

Overall, GA-based schemes can provide a higher security level in terms of N_{DN}^{AP}. AS-based schemes incur lower cost while FS-based schemes incur higher $\overline{MTTSF}$. We can see there is a balance between MTTSF and defense cost per time unit. $\overline{MTTSF}$ of FS-GANT is roughly 2.1 times higher than $\overline{MTTSF}$ of AS-GANT while $\overline{\widehat{C_D}}$ of FS-GANT is 3.1 times higher than that one of AS-GANT. Compared with the results in Section 10.5.5.1, the network with an increasing number of decoys has higher security level but requires more cost due to a greater number of potential connections for shuffling.

10.6 Conclusions and Future Work

In this work, we have proposed an integrated proactive defense mechanism for an SDN-based IoT environment based on cyber deception and MTD to improve security and mitigate the impact of potential attacks. More specifically, we have developed a network topology-based shuffling (NTS) in a network deployed with decoy nodes. We adopted a GA and devised fitness functions to consider three system objectives: maximizing security by maximizing a number of attack paths toward decoy targets and MTTSF while minimizing defense cost derived from network topology shuffling operations. We also considered AS by introducing a metric to detect the system security vulnerability level based on the number of detected, compromised nodes and the number of critical nodes being compromised. For the assessment of our proposed NTS-MTD, we employed a GSM, namely HARM, based on the combination of AGs and ATs. Via our extensive simulation study, we devised the four schemes considering either AS or FS to determine when to trigger an MTD operation while answering how to trigger an MTD operation by using a random shuffling or GA-based intelligent shuffling. We compared the performance of the four schemes under three different scenarios by varying the number of decoys and the attacker's intelligence levels in detecting the decoys. Finally, we observed the outperformance of GA-based shuffling regarding the number of attack paths toward decoy targets and the balance between MTTSF and defense cost for fixed and adaptive GA-based shuffling schemes under the extensive performance analysis [31].

In our future work, we plan to explore: (i) how to apply our proposed scheme in large-scale IoT networks with a variety of decoy nodes by showing high scalability; (ii) carrying out sensitivity analysis by varying the values of other key design parameters (e.g. weights to consider each system objective, a more number of decoy nodes deployed in the network, a more number of attackers, and/or system security vulnerability thresholds); (iii) investigating the effect of deception and/or MTD on service availability, such as delay introduced by the deployment of decoy nodes and/or network topology shuffling; and (iv) identifying an optimal setting of adaptive network topology shuffling algorithms in terms of thresholds in detecting system vulnerabilities.

References

1 R. Roman, J. Zhou, and J. Lopez, "On the features and challenges of security and privacy in distributed internet of things," *Computer Networks*, vol. **57**, no. 10, pp. 2266–2279, 2013.

2 J.-H. Cho and N. Ben-Asher, “Cyber defense in breadth: Modeling and analysis of integrated defense systems,” *The Journal of Defense Modeling and Simulation*, vol. **15**, no. 2, pp. 147–160, 2018.

3 J. B. Hong and D. S. Kim, “Assessing the effectiveness of moving target defenses using security models,” *IEEE Transactions on Dependable and Secure Computing*, vol. **13**, no. 2, pp. 163–177, 2016.

4 M. Ge, J. B. Hong, W. Guttmann, and D. S. Kim, “A framework for automating security analysis of the internet of things,” *Journal of Network and Computer Applications*, vol. **83**, pp. 12–27, 2017.

5 V. Casola, A. De Benedictis, and M. Albanese, “A moving target defense approach for protecting resource-constrained distributed devices,” in *2013 IEEE 14th International Conference on Information Reuse and Integration (IRI)*, 2013, pp. 22–29: IEEE.

6 M. Sherburne, R. Marchany, and J. Tront, “Implementing moving target ipv6 defense to secure 6lowpan in the internet of things and smart grid,” in *Proceedings of the 9th Annual Cyber and Information Security Research Conference*, 2014, pp. 37–40: ACM.

7 K. Zeitz, M. Cantrell, R. Marchany, and J. Tront, “Designing a micro-moving target IPv6 defense for the Internet of Things,” in *2017 IEEE/ACM Second International Conference on Internet-of-Things Design and Implementation (IoTDI)*, 2017, pp. 179–184: IEEE.

8 K. Mahmood and D. M. Shila, “Moving target defense for Internet of Things using context aware code partitioning and code diversification,” in *2016 IEEE 3rd World Forum on Internet of Things (WF-IoT)*, 2016, pp. 329–330: IEEE.

9 T. Miyazaki *et al.*, “A software defined wireless sensor network,” in *2014 International Conference on Computing, Networking and Communications (ICNC)*, 2014, pp. 847–852: IEEE.

10 Q. D. La, T. Q. Quek, J. Lee, S. Jin, and H. Zhu, “Deceptive attack and defense game in honeypot-enabled networks for the internet of things,” *IEEE Internet of Things Journal*, vol. **3**, no. 6, pp. 1025–1035, 2016.

11 M. Anirudh, S. A. Thileeban, and D. J. Nallathambi, “Use of honeypots for mitigating DoS attacks targeted on IoT networks,” in *2017 International Conference on Computer, Communication and Signal Processing (ICCCSP)*, 2017, pp. 1–4: IEEE.

12 S. Dowling, M. Schukat, and H. Melvin, “A ZigBee honeypot to assess IoT cyberattack behaviour,” in *Signals and Systems Conference (ISSC), 2017 28th Irish*, 2017, pp. 1–6: IEEE.

13 L. Pingree, “*Emerging Technology Analysis: Deception Techniques and Technologies Create Security Technology Business Opportunities*,” Gartner, Inc., 2015.

14 O. Sheyner, J. Haines, S. Jha, R. Lippmann, and J. M. Wing, “Automated generation and analysis of attack graphs,” in *null*, 2002, p. 273: IEEE.

15 V. Saini, Q. Duan, and V. J. J. o. C. S. i. C. Paruchuri, "Threat modeling using attack trees," in *Journal of Computing Sciences in Colleges*, vol. **23**, no. 4, pp. 124–131, 2008.

16 H. Abie and I. Balasingham, "Risk-based adaptive security for smart IoT in eHealth," in *Proceedings of the 7th International Conference on Body Area Networks*, 2012, pp. 269–275: ICST (Institute for Computer Sciences, Social-Informatics and Telecommunications Engineering).

17 A. Rullo, E. Serra, E. Bertino, and J. Lobo, "Shortfall-based optimal security provisioning for Internet of things," in *2017 IEEE 37th International Conference on Distributed Computing Systems (ICDCS)*, 2017, pp. 2585–2586: IEEE.

18 R. M. Savola, H. Abie, and M. Sihvonen, "Towards metrics-driven adaptive security management in e-health IoT applications," in *Proceedings of the 7th International Conference on Body Area Networks*, 2012, pp. 276–281: ICST (Institute for Computer Sciences, Social-Informatics and Telecommunications Engineering).

19 OpenFlow Switch Specification (Version 1.3.0). Technical Report, ONF, 2012.

20 B. T. De Oliveira, L. B. Gabriel, and C. B. Margi, "TinySDN: Enabling multiple controllers for software-defined wireless sensor networks," *IEEE Latin America Transactions*, vol. **13**, no. 11, pp. 3690–3696, 2015.

21 L. Galluccio, S. Milardo, G. Morabito, and S. Palazzo, "SDN-WISE: Design, prototyping and experimentation of a stateful SDN solution for WIreless SEnsor networks," in *2015 IEEE Conference on Computer Communications (INFOCOM)*, 2015, pp. 513–521: IEEE.

22 T. Lei, Z. Lu, X. Wen, X. Zhao, and L. Wang, "SWAN: An SDN based campus WLAN framework," in 2014 *4th International Conference on Wireless Communications, Vehicular Technology, Information Theory and Aerospace & Electronic Systems (VITAE)*, 2014, pp. 1–5: IEEE.

23 C. J. Bernardos *et al.*, "An architecture for software defined wireless networking," *IEEE Wireless Communications*, vol. **21**, no. 3, pp. 52–61, 2014.

24 J. Liu, Y. Li, M. Chen, W. Dong, and D. Jin, "Software-defined internet of things for smart urban sensing," *IEEE Communications Magazine*, vol. **53**, no. 9, pp. 55–63, 2015.

25 I. Farris, T. Taleb, Y. Khettab, and J. S. Song, "A survey on emerging SDN and NFV security mechanisms for IoT systems," *IEEE Communications Surveys & Tutorials*, vol. **21**, no. 1, pp. 812–837, 2019.

26 M. Ge, J. B. Hong, S. E. Yusuf, and D. S. Kim, "Proactive defense mechanisms for the software-defined Internet of Things with non-patchable vulnerabilities," *Future Generation Computer Systems*, vol. **78**, pp. 568–582, 2018.

27 F. C. Gärtner, "Byzantine failures and security: Arbitrary is not (always) random," Swiss Federal Institute of Technology (EPFL), 2003.

28 F. Jia, J. B. Hong, and D. S. Kim, "Towards automated generation and visualization of hierarchical attack representation models," in 2015 *IEEE International Conference on Computer and Information Technology; Ubiquitous Computing and*

Communications; Dependable, Autonomic and Secure Computing; Pervasive Intelligence and Computing (CIT/IUCC/DASC/PICOM), 2015, pp. 1689–1696: IEEE.

29 NIST, "National Vulnerability Database (NVD)." 2005, Available: https://nvd.nist.gov/.

30 J. Cho, Y. Wang, I. Chen, K. S. Chan, and A. Swami, "A survey on modeling and optimizing multi-objective systems," *IEEE Communications Surveys Tutorials*, vol. **19**, no. 3, pp. 1867–1901, 2017.

31 K. S. Trivedi, *Software packages: SHARP & SPNP*, 2011. Available: http://trivedi.pratt.duke.edu/software-packages.

11

Addressing Polymorphic Advanced Threats in Internet of Things Networks by Cross-Layer Profiling

Hisham Alasmary[1], Afsah Anwar[1], Laurent L. Njilla[2], Charles A. Kamhoua[3], and Aziz Mohaisen[1]

[1] *Department of Computer Science, University of Central Florida, Orlando, FL, USA*
[2] *Cyber Assurance Branch, US Air Force Research Laboratory, Rome, NY, USA*
[3] *US Army Research Laboratory, Adelphi, MD, USA*

11.1 Introduction

Internet of Things (IoT) devices and their use are on the rise, and they are expected to revolutionize our way of living [1], especially in applications such as home networks [2]. However, with this promise and potential come various security challenges, among the most important, Advanced Persistent Threats (APT), which include malware [3, 4]. With APT, an adversary can not only breach the security and privacy of users and home network devices but affect their safety as well. Unlike mass-market malicious software, which can be easily identified using the signatures of their behavior [5], APTs are stealthy, requiring new capabilities that can better meet the specific needs of the home network.

To address the distressing need of security in smart homes, we propose an effective mechanism for detection and prevention of intrusion due to APT, cross-layer security. Our approach relies on four building blocks, including our major contribution, Security Layer for Smart Home (SLaSH), a security layer for the smart home. First, we identify a plausible and generic system model that consists of three layers: device, network, and service, with actual device settings to explore security issues. Second, we formally analyze the attack surface, based on both confirmed and suspected vulnerabilities that constitute an actual threat and can be exploited by the adversary to launch attacks using the vulnerable device or orchestrating attacks amplified by other devices in the smart home, and study their implications.

Modeling and Design of Secure Internet of Things, First Edition. Edited by Charles A. Kamhoua, Laurent L. Njilla, Alexander Kott and Sachin Shetty.

Published 2020 by John Wiley & Sons, Inc.

Such analysis is a first step toward realistically establishing a baseline of what is normal and what is abnormal in a networked smart home environment.

Third, we design a security layer for the smart home that addresses various security threats at the various layers of the system model. At the core of our security layer is a system's cross-layer security function, which features an intrusion detection and prevention system. For our system, we utilize the behavioral features of devices and their communication patterns. With the heterogeneity of devices, not all security functions can be implemented on the device. Thus, we construct a multilayered architecture for intrusion detection, with on-host profiling and in-network detection and feedback capabilities. We enrich our detection engine with both rules- and behavior-based detection capabilities. Our rules-based detection is based on the (nonconclusive) characterization of the adversary through the attack surface analysis, and our behavior-based detection is based on both the baselines created for benign execution and the behavior of smart home devices, as well as the abnormalities detected through logging and statistical analysis. Our intrusion detection system (IDS) utilizes a machine-learning component that uses both supervised and unsupervised algorithms to detect unexpected behavior or identify intrusions. These unexpected or abnormal behaviors of devices in a network do not behave as intended and are referred to as anomalies.

The organization of the rest of this chapter is as follows. In Section 11.2, we identify the system model and objectives of this work. In Section 11.3, we identify the attack surface and analyze it across multiple layers of the home network. In Section 11.4, we introduce the design of our security layer building components across the multiple layers of the system model, including a cross-layer security design. Concluding remarks are provided in Section 11.5.

11.2 System Model

Our system model consists of three layers, namely, the device, the network, and the service layer. In the following sections, we elaborate on each of those lawyers.

11.2.1 Device Layer

At the device layer reside various smart home network devices with various capabilities and functionalities. Understanding the capabilities of the devices guides the type of security primitives that can be implemented on them and the security functions they can operate in order to secure the smart home network as a whole. Smart home devices range in capabilities and include RFID tags, sensors, home appliances, wearable technologies, mobile devices and tablets, and personal computers. A preliminary listing of such devices and their computer, memory, and network capabilities is provided in Table 11.1 [6].

Table 11.1 An enumeration of the various components (computational capabilities, storage, and power control of the security functions that can be implemented on the device) in the device layer of a typical home network system.

Device type	Chipset	Core frequency	RAM	Flash memory	Power
HID glass tag ultra (RFID)	EM 4305	134.2 kHz	512 bit RW	NA	NA
HID Piccolino tag (RFID)	I-Code SLIx, SLIx-S	13.56 MHz	2048 bit RW	NA	NA
Sensor devices	Microcontroller	4–32 MHz	4–16 kB	16–128 kB	Battery
Google Chromecast 2nd generation	ARM Cortex-A7	1.2 GHz	512 MB	256 MB	NA
NETGEAR router	Broadcom BCM4709A	1.0 GHz	256 MB	128 kB	AC power
Gateway WISE-3310	ARM Cortex-A9	1.0 GHz	NA	4 GB	AC power
Elster REX2 smart meter	Teridian 71M6531F SoC	10 MHz	4 kB	256 kB	Battery
Philips hue light bulb	TI CC2530 SoC	32 MHz	8 kB	256 kB	Battery
Nest smoke detector	ARM Cortex-M0	48 MHz	16 kB RAM	128 kB	Battery
Nest learning thermostat	ARM Cortex-A8	800 MHz	512 MB RAM	2 GB	Battery
Samsung smart Cam	GM812x SoC	Up to 540 MHz	N/A	Up to 64 GB	AC power
Samsung smart TV	ARM-based Exynos SoC	1.3 GHz	1 GB	N/A	AC power
OORT Bluetooth smart controller	ARM Cortex-M0	50 MHz	16 kB/32 kB	Up to 256 kB	Battery
Dacor Android Oven	PowerVR SGX 540 graphics	1 GHz	512 MB	NA	AC power
Fitbit smart wrist band flex	ARM Cortex-M3	32 MHz	16 kB	128 kB	Battery
LG Watch Urbane 2nd Edition	Snapdragon 400 chipset	1.2 GHz	768 MB	4 GB	Battery
Samsung watch gear S2	MSM8x26	1.2 GHz	512 MB RAM	4 GB	Battery
Apple watch	S1	520 MHz	512 MB RAM	8 GB	Battery
iPhone 6s Plus	A9/64-bit/M9 coprocessor	1.85 GHz	2 GB	Up to 128 GB	Battery
12.9-inch iPad Pro	A9X/64-bit/M9 coprocessor	1.85 GHz	4 GB	Up to 256 GB	Battery

11.2.2 Network Layer

At the network layer are various networking protocols and capabilities for internetworking between the various devices, among themselves and other management infrastructure and devices of the smart home, or even among devices in the smart home network and other services that fall out of the smart home network premise. Examples of network layer protocols and capabilities include Wi-Fi (IEEE 802.11), Bluetooth, ZigBee, NFC, etc., each of which has its own security features and issues [7]. Some of the network layer devices are listed in Table 11.1.

11.2.3 Service Layer

The service layer lays on top of the network layer, where data collected from various devices and transmitted over the network are being stored or processed for further use in the service layer. Examples of service layer components include cloud (for storage), web browsers, mobile apps, multimedia services, etc.

We note that a good understanding of the system model mandates the kind of security risks and functions that can be implemented in the home network. For example, when a specific software is utilized by various devices, existing vulnerabilities in that software affect the attack surface of the system. Second, given the various hardware capabilities of the devices, one can envision the type of profiling capabilities that can be implemented on those devices, and perhaps even an optimal strategy to tackle profiling off-device. Third, with energy being of paramount importance in home networks, particularly with energy-limited devices, one can envision a scenario of analyzing profiling and usage of the various devices as a potential feature for understanding actual and anomalous use. Energy is a major contributor to the continuous operational cost of those devices. In this work, we capitalize on those questions arising from the understanding of the system model.

11.3 Attack Surface Analysis

While creating a conclusive set of capabilities that an adversary may have with regard to a complex and interconnected system, like a smart home, is a tedious task and may not always be possible or result in a perfect characterization, understanding the estimated attack surface of today's smart home devices, networks, and services is the first step toward building effective defenses, including intrusion detection and prevention capabilities and systems.

The attack surface analysis is concerned with enumerating the devices, protocols, and services typically used in smart homes; the potential and confirmed vulnerabilities (based on actual or hypothesized attack scenarios); and the attacks on those vulnerabilities that can be used to launch other attacks. The implications of attacks can be multifaceted, and a successful attack surface analysis foresees such implications and consequences in order to build a baseline for what is considered a normal and abnormal behavior or characteristic.

An attack on a smart home device may have privacy and security implications at the same device being exploited with the vulnerability or through instrumentation with other devices. An example of the single device attack is sending out information associated with the device, e.g. what is in a refrigerator, or using that device to attack other devices in the system, like sending spam from the refrigerator [8, 9]. Another example of multiple devices is an oven attacking [10, 11] or stealing information from other devices in the smart home network and sending it out on the wire, or using the other devices as zombies to attack other networks and resources on the Internet. In the following, we review a preliminary attack surface of the three layers in our system model.

11.3.1 Attacks Due to Devices

As we have seen earlier, devices in the home network are diverse in terms of their software and hardware capabilities. However, the various devices share common aspects, which make it easier to analyze them. Ideally, devices in the device layer should provide several guarantees: (i) have secure software, (ii) allow for authorized access only, and (iii) securely store and transmit data to other devices in the home network. As a result, the attack surface in the smart home network is a result of the violation of one or more of those requirements.

The susceptibility of the devices at risk can be misused to abuse the attack surface, putting the network in jeopardy. Factors include software being vulnerable due to poor software design and language use, like buffer overflow, injection vulnerabilities, failure to handle errors correctly allowing for cross-site scripting; improper use of security packages like TLS and SSL at the device layer; information leakage; and poor usability, among others. On the other hand, authorized access might be violated by a lack of authentication at the device layer, use of weak passwords, or unauthorized key exchange. Last, the secure storage and transmission of data might be violated by failing to implement the proper encryption algorithm or falling for one or more of the vulnerabilities above that would make encryption insecure (e.g. by exposing keys).

For example, voice assistants such as Amazon Alexa and Google Home can be exploited by adversaries to launch attacks using malicious voice commands. Zhang et al. [12] demonstrated attacks on voice assistants using voice signals using action phrases that are inaudible to human ears and then forcing the assistants to obey their commands. Similarly, Carlini et al. [13] used hidden voice samples to control targeted IoT devices. Additionally, design and implementation flaws in the Software Guard eXtensions (SGX) in modern microprocessors were exploited to launch attacks such as Spectre, Meltdown, and Foreshadow [14–16]. Xiao et al. [17] analyzed the side-channel traces to detect SSL/TLS vulnerabilities in the SGX.

We emphasize that the device layer contributes greatly to the shape and size of the attack surface of the entire smart home network, which we tackle in this work. Table 11.1 shows examples of devices and their contribution to the attack surface of a smart home. Although some of the attacks listed in this table hint at issues with the network layer, we are concerned with vulnerabilities originating at the device level. For example, the unsecured channel between the network and a coffee maker [18–20] is the result of not implementing proper encryption at the device.

At the device layer, there could be various vulnerabilities. Those vulnerabilities can result in undesirable events such as accessing a sensor (via network) or destroying it (remotely) by unauthorized persons. Smart devices can be lost or stolen, especially if they are mobile and not bound to the smart home environment. Furthermore, constrained devices, such as sensors or other appliances, which have limited functionality, are widely used in the smart home network. However, applying the proper encryption and vaccination (from unwanted software) is out of reach for many of those devices, calling for unconventional protection techniques.

The number of those devices in home networks is also large, making it difficult for an untrained administrator to patch them for bugs and monitor them for misbehavior. In summary, Table 11.2 shows some examples of the vulnerabilities, attack methods, and impacts for the smart home devices [23–25]. These devices do not use a secure password, channel, or encryption function, which are supposed to be part of their requirements, as highlighted previously. Thus, they can be easily hacked (as done in the past, as the vulnerabilities of many of those devices were not yet fixed).

11.3.2 Attacks on Network

The network layer comes second in importance after the device layer of the home network. Here, we distinguish between the network device, which is subject to all issues of the device layer, and the network protocol, which is used for transmitting data among devices in the home network. Vulnerabilities of the network protocols may include not supporting the security options, allowing for distributed denial of service attacks (DDoS), and having a poor implementation of key exchange and password transmission.

For example, due to constraints on various devices in the home network, some options of the network layer security might not be used, i.e. passwords are exchanged among smart home devices in an unencrypted format. Using such

Table 11.2 Attack surface: enumeration of vulnerabilities, attacks, and impacts at the device layer.

Device	Vulnerability	Attack	Impact
Smart light bulb	Static password	MitM, password stealing, unauthorized execution	Bulb controlled by remote
Wall pad	Buffer overflow	Value manipulation, shellcode exe.	Housebreaking, unauthorized monitoring
Network camera	Firmware integrity	Firmware modulation, force an update of the device firmware [21, 22]	CCTV as a zombie, damage peripheral devices
Google Chromecast	Rickrolling	Disconnect due to attacker-controlled device flooding; reconnects to attacker	Content is streamed from a source owned by an attacker
Coffee machine	Non-protected channel	Device plugged in, creates a non-encrypted hotspot; listens to Universal Plug and Play (UPnP)	Hijack password of local Wi-Fi
Fridge	Accepting generic authentication	Sniff packets, malicious code infection	Email proxies, send malicious mail
Oven	Unsecured Wi-Fi	MitM attack	Access to a temperature controller, a garage door, car, etc.

vulnerability, an attacker can eavesdrop on the unencrypted traffic and recover user passwords. Also, with the likelihood that many users use the same password for multiple devices and systems, an adversary might be able to utilize eavesdropped passwords to access other devices in the home network.

Not enabling the security functions at the network level, including a lack of strong identification and authentication, might also enable additional vulnerabilities, like packets injection, DDoS, and data forgery. Similarly, choosing a transport protocol over another for the resource constraints (e.g. UDP over TCP) may result in similar security vulnerabilities at the network layer.

For example, IoT devices have been a target of malware authors recently. These botnets employ brute force attacks using a dictionary of user credentials generated by collecting default credentials of different device manufacturers. They then use the access to the devices to infect, propagate, and inflict damage to the devices under attack. For example, the massive DDoS attack generating 1.1 Tbps, launched by a Mirai botnet targeted 400 000 devices over TCP ports 23 and 2323 [26, 27]. Pa et al. [28] created a honeypot, IoTPOT, and analyzed the propagation behavior of DDoS attacks occurring over the Telnet protocol for a period of 39 days.

11.3.3 Attacks on Service

To manage and control smart home devices, an administrator may use various cloud services, including network storage, web browsers, mobile applications, and others. For example, various online services have been under attack by massive DDoS attacks; GitHub, for instance, was targeted for about 10 minutes by a DDoS attack with generated traffic of 1.35 Tbps [29]. Those technologies are no different than today's web and cloud technologies, and all security applicable today can also be applied to them. For example, a web service in the service layer might be vulnerable to broken authentication sessions, cross-site scripting, insecure direct object referencing, security misconfiguration, sensitive data exposure, function-level access control, cross-site request forgery, components with known vulnerabilities, and invalidated redirects and forwards. All of those vulnerabilities could be used directly or indirectly for launching attacks on the service layer, and impacting the smart home by stealing and manipulating smart home data (which could have both security and privacy implications).

Besides the layer level of attack analysis, another type of surface layer classification could shed light on the capabilities to be implemented in the home network, which is as follows.

11.3.3.1 Horizontal Attack Surface

The horizontal exploration of the attack surface is concerned to understand the capabilities of different devices, networks, and services in the home network. These devices include not only new devices (hardware), but also the devices already employed, with different versions of firmware, operating system, or applications (as with smartphones and tablets).

11.3.3.2 Vertical Attack Surface

The vertical exploration of the attack surface is related to exploring potential vulnerabilities present in the devices connected to the home network. Such vulnerabilities may not necessarily be the result of the software on the device, but also due to the way that the devices in the home network interact among themselves and with the external networks. Benefiting from our understanding of contemporary network profiling and capabilities, we consider attacks not commonly known or well understood on the network layer. One approach that we advocate in the rest of this work, given that the attack surface might be hard to quantify, is limiting the entities with which a home network device can interact, which is justifiable in the model highlighted previously. With that in mind, we use these design principles to understand the various issues raised in the attack surface.

11.3.3.3 In-the-Wild Analysis

This category deals with idealized attacks, which are applicable in the wild on the various devices and software versions. One challenge associated with these attacks is creating the proper metric that is operationally relevant for quantifying the attack surface. Using the analysis of the attack surface, one may explore metrics for the "coverage" of the attack surface, which captures its size. Using automation tools, we envision methods for automated vulnerability testing to understand and potentially to minimize attacks.

11.4 A Security Layer

In this research work, we benefit from a fine understanding of the system and use models with a diversity of devices, as well as attack surface analysis, to propose a security layer for the smart home (SLaSH). In this section, we highlight the design of this layer with the various functions and capabilities implemented in it.

11.4.1 Design

Our security layer addresses security needs at the various layers of the system, comprehensively, and partly incorporating a finer understanding of the attack surface of those layers. SLaSH also has a cross-layer security mechanism that incorporates capabilities and features of the various layers to ultimately secure the entire smart home network. The main design of our security layer is shown in Figure 11.1.

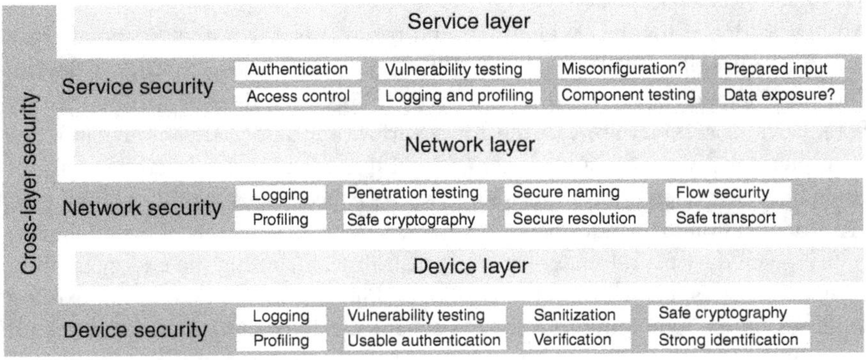

Figure 11.1 Security layer for smart home (SLaSH).

In the subsequent sections, we outline at length the various functions and capabilities in each of the components of our security layer. In particular, our security layer relies on implementing various monitoring, profiling, logging, and sanitization capabilities in all layers, as well as safe use, based on the finer understanding of the attack surface, in the upper layers. Borrowing strengths from the different layers, our system further implements the cross-layer function, which aids the capabilities of the individual layers.

11.4.2 Functions and Capabilities

Again, our security layer is a system cross-layer that implements various functions at the various layers of the system. In the following, we elaborate on those functions.

11.4.2.1 Device Layer Level Security

At the device layer level, we envision various security functions and whether they are implemented in a given device would depend on the hardware and software capabilities of that device (see Tables 11.1 and 11.2 for more details). In principle, a device in the home network would ideally support logging and profiling of events (like access, usage, etc.), strong identification, usable authentication, and safe cryptographic operations. Furthermore, the device's security should be facilitated by vulnerability testing, sanitization of input (to address unexpected states in the software due to, for example, injection attacks), and verification. Some of the capabilities implemented in the device, such as logging and profiling, are an integral part of the security of the device and used mainly for collecting behavioral patterns that could be used for identifying normal use and later for identifying abnormal use that could be used for detection.

11.4.2.1.1 Authentication

Authentication using device identifier and passwords, the authentication certificate, the subscriber identity module (SIM), etc., and access control to devices and their data at the device level are needed. Those requirements are necessary to reduce misuse, and minimize the attack surface and associated anomalies. However, given the diversity of devices in a typical home network, there are various challenges that need to be addressed. First, we expect those various mechanisms of authentication to be required. Second, finding the security mechanism that best suits a given device, and under various constraints, including the energy requirements of devices, is an additional challenge. Finally, understanding the existing mechanisms and providing additional mechanisms that would allow for improved security, with reasonable assumptions about their operation and cost, as well as the attack model they are deployed in, is a third challenge.

In this layer, we address the aforementioned security functions, requirements, and challenges in the various home network devices at the device layer. On the one hand, while many of those capabilities are expected to be well established and implemented in various advanced and powerful devices like smartphones and tablets, other constrained devices will not provide some of them. For example, safe cryptography, strong identification, logging, and profiling, among others, are not well supported by sensors with both hardware and software limitations. To this end, the following tasks are envisioned to address the device layer security requirements. To facilitate the equivalent of such function, we propose various innovative ideas building on constrained access and split architectures. In the following, we elaborate on those innovations.

11.4.2.1.2 Constrained Access

For constrained access and to reduce the amount of misuse, we create a baseline for the various devices for resources that they are supposed to communicate with or access for their normal operation. We then use such constrained access to create a baseline profile in the logging and profiling of the device, which we can then use to perform anomaly detection. This approach, as opposed to other approaches that take into account open access to devices, would have various benefits. First, it addresses the issue of zero-day vulnerabilities, with any previously unapproved access triggering a log alarm in the anomaly detection system. Second, such an approach would reduce the number of rules to be checked: instead of checking all rules for resources that are not possible to access a device, a network gear, or a service, the new approach minimizes that number into the set that is possible to use, which is a small subset. However, such an approach will not work for all devices; while it is possible to employ for a sensor network, a smart TV, a smart refrigerator, or an oven, it will certainly not work for an advanced device like a smartphone, a tablet, or a personal computer, which has a rich set of entities that it will communicate with and often those entities and resources are not predictable in advance.

11.4.2.1.3 Split Architecture

We envision a split architecture that facilitates security services at the device layer through added security primitives. This approach is inspired by the well-established and widely used offloading technique, in which a third party performs certain functions of behalf of devices. We extend such idea to use in the context of home networks. In principle, for devices that are too constrained to perform some of the security functions mentioned previously at the device layer, we use a security gateway to assist with the security operations, where possible. For example, we envision that profiling and logging are addressed in the various devices based on their constraints: for devices that are more powerful and can handle logging and profiling of information (extraction of features that could be later used for understanding usage, creating baseline profiles, etc.), we perform such functions at the device level. For those devices that cannot perform such functions at the device level, assisting devices, such as a security gateway, are used to help with such primitives. We note that such a split architecture would be facilitated only if strong identification is in place, which we assume and would explore the extent to which it exists in a variety of home network devices.

11.4.2.1.4 Improved Operational Realities

As a secondary capability in this work, an improvement over the status quo and ways with respect to practices associated with various security primitives at the device layer is suggested, including authentication. For example, constraints for home network devices are not limited to processing power, but also to display capabilities. A lot of the devices in the home network will lack a large screen that can view a fully-fledged keyboard for freely authenticating a user. Thus, we explore how authentication in home networks, in particular, and IoT applications, in general, could be facilitated by new, innovative, and operationally relevant authentication mechanisms. Such mechanisms build on our prior work in this field designing usable authentication mechanisms even under security constraints (i.e. when the device on which authentication is being performed is compromised).

11.4.2.2 Network Layer Level

While the network layer shares a lot of commonalities with respect to the device layer, as shown in Figure 11.1, and would benefit from the various research tasks proposed in the previous section, two functions at the network layer could greatly improve its security: strong authentication with confidentiality and intrusion detection.

The first function is needed because the network layer level requires integration of the security primitives and functions for heterogeneous protocol operation and security technique development for both the heterogeneous and low specification network environment. Furthermore, encryptions utilizing the likes of standard security mode (SSM) and high-security mode (HSM) for ZigBee, and Wi-Fi Protected Access (WPA) and WPA2 for Wi-Fi are important at the network layer level, but whether they are enabled in various devices in actual use or not is unclear [30–32].

How much those primitives actually cost and how they could be used for understanding intrusion (e.g. by failed encryption and initialization) are unclear, and could potentially be used to facilitate a better IDS. In addition to those primitives, we envision that the network layer would include functions for analyzing flow security, ensuring secure naming, secure name resolution, sage transportation, and applied penetration testing. This layer addresses several of those primitives and functions as follows.

11.4.2.2.1 Authentication

Most of the today's network-layer protocols (with respect to the layering explained earlier) support one form or other of authentication. However, how authentication is being implemented or enabled in the various network devices, and whether such devices are using best practices of authentication, has not been studied. In this task, we explore the status quo of authentication and provide best practices for authentication that should help improve the overall security posture of the home network by authenticating the network user.

11.4.2.2.2 Profiling and Logging

In this layer, profiling and logging in the network are used for improving the security of the home network. In particular, our approach to profiling the network includes logging and collecting information (i) on behalf of the devices, in the split architecture advocated for the devices layer security above; and (ii) on the use and access of the various network layer devices and resources, utilizing various protocols. One candidate set of artifacts includes the classical five-tuples utilized for the TCP and TCP-like connection profiling: source IP and port, destination IP and port, and the protocol in use. We further explore the use of such artifacts in the cross-layer security for anomaly detection. Many other pieces of information can also be derived from this simple information, including geospatial feature and information (associated with the location of access at the AS, city, or organization-level), temporal features and information (associated with the time when such networks are accessed), among others.

11.4.2.2.3 Flow Security

Flow security in the context of the home network is facilitated by having a secure transport mechanism in place, which we explore in this work. In particular, we explore the operational and security issues associated with enabling flow security by requiring a secure and reliable transport.

11.4.2.2.4 Secure Naming and Resolution

Naming in IoT, in general, and home networks, in particular, is a challenging issue. IoT devices are expected to be large in number, overwhelming current naming infrastructures. Furthermore, having a consistent, easily accessible, and strong naming (that could potentially result in strong forms of identification) is required. To this end, in this task, we explore new forms of secure naming and resolution. A starting point of enabling this function is a technique inspired by the DNS-based

Authentication of Named Entities (DANE), which allows one to securely authenticate entities by delivering authentication credentials in the public DNS.

11.4.2.2.5 Network-Aware Notification

With network authentication being of paramount importance to the overall security of home networks and default passwords being the status quo for many networks, we envision that some of the insight learned about network authentication and access control would contribute toward better security by keeping the human (operator) in the loop. This, for example, includes alerting users when passwords are expired, when passwords have been attempted for authentication multiple times, or where (location, at various granularities) authentication has been attempted.

11.4.2.3 Service Layer Level

The extensive listing of the various security functions and primitives we envision in this layer are outlined in Figure 11.1. Similar to the network layer, the service layer ideally requires authentication and authorization services, as well as profiling and logging, although the former two functions are out of the scope of this work. As in the two other layers, logging and profiling of access and use patterns at the service layer would be an enabler for the secure operation of the home network by providing us with the wealth of information for anomaly detection. However, more related to this work are configuration and component testing, and data exposure, where the latter can be particularly useful in identifying the various legitimate uses as a baseline. To this end, we highlight the various capabilities in this layer.

11.4.2.3.1 Logging and Profiling

Profiling and logging in this layer is similar to that in the network layer. Capabilities in this layer are developed for generating access log information that would be useful in the service layer. In particular, the service layer's logging and profiling functions would extract patterns from actual behavior, including service access, data usage, and data transfer, and generate related features in both the time and space domains (temporal and spatial).

11.4.2.3.2 Fine Granularity Versus Minimization

With the service layer being accessed by multiple devices, utilizing multiple stacks of protocols – that might not be the same across all devices – we see several challenges. Atop those challenges is the granularity level of the logged data and extract features. We follow two fundamental approaches to address this challenge: data minimization and objective-aware granularity adjustment. In the first approach, we aim to reduce the amount of the information being logged and collected to the specific problem being addressed at the time, such as flow security analysis.

Only the minimal amount of information without which such a goal could not be achieved would be collected. Defining and realizing such a minimal set would be a challenge that we address through experimentation and by inducing various benign and malicious or anomalous behaviors. On the other hand, the objective-aware granularity adjustment technique takes into account a broader set of applications and use patterns and wants to achieve a reasonable accuracy of detection through the same set of features. The challenge would to be to enumerate such features and strike a cost-effectiveness balance between the size of such features and the utility of the service.

11.4.3 Cross-Layer Security

The cross-layer security embodies several of the primitives highlighted in the three layers discussed. It aims to allow intrusion and anomalies detection through comprehensive profiling at the device, in the network, and at the service. Comprehensive profiling and logging will provide a wealth of information for the anomaly detector that then can be used to raise alarms about a potential breach and anomalies with high confidence. The key idea of why cross-layer security would work better than a single layer-based security mechanism is that the multiple avenues of observation and profiling would more comprehensive, and thus, inclusive of anomalies and security weaknesses in an end-to-end manner. In the following, we discuss the motivations for such a cross-layer security service and provide its details.

11.4.3.1 Motivation and Benefits

To highlight the benefits of a cross-layer security function, we provide an example. An attacker can access smart home devices through various points on the Internet. For example, an attacker can eavesdrop the payload by network monitoring of user communication, obtain the user's unencrypted password, and then exert control over the user's smart home network. An adversary can also eavesdrop on data being sent from a home network device or a network gadget to the service, and vice versa. Furthermore, an attacker can breach the service layer by either compromising an account that is guarded by a weak password or utilizing an advanced attack, such as injection. In all of those cases, the security primitives advocated in this proposed approach, which we aim to explore further in the various tasks, will ensure security against those attacks.

Alternatively, we simplify the operation of our system in the home network by exploiting a cross-layer security function that takes into account events, behaviors, and artifacts that occurred at the various layers at the same or different times. We realize such a vision as a comprehensive cross-layer IDS that utilizes the

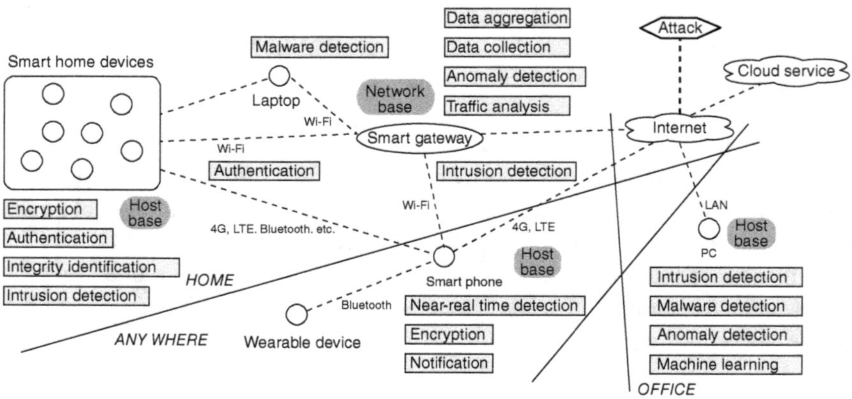

Figure 11.2 Cross-layer security to thwart advanced persistent threats in the home network.

capabilities of various layers, feature extractors, detection combiners, and an alarm notification service. In the following, we highlight some of the technical aspects of this architecture.

Figure 11.2 shows our IDS, which can detect any intrusion attempts at any point in the three-layered home network system (similar although primitive system utilizing similar insight is used in [33–35]). The main role of the cross-layer security is to collect the data, detect the attacks, and trigger appropriate alerts in all locations where users may intervene to interrupt and act upon those alerts. For the build of our IDS, we envision two different approaches: in-network profiling and decision, and device-assisted IDS. In the following, we elaborate on those techniques.

11.4.3.2 In-Network Profiling and Decision

In this approach and at the smart gateway, in order to analyze the abnormal behavior of smart home devices, our IDS gathers the raw data from smart home devices and performs a statistical pattern analysis according to the user's lifestyle, a baseline, and an unexpected usage pattern based on that baseline according to the following steps:

1) *Data Collection:* Traffic and status information sent periodically by the smart home devices are often logged in the smart gateway, which operates as the central network point for enabling in-network profiling and decision/alert notification. The smart gateway collects such information to potentially understanding anomalies.
2) *Features Extraction:* From the various pieces of data collected by the smart gateway, we extract various features. The features are extracted using a similar insight as in the network logging by utilizing an objective-aware method that

maximizes detection rate while reducing the amount of overhead. The features extracted concern logging sessions, data flow, access patterns, access attempts, etc. The actual values of the features include frequency and statistics concerning the data volume, time information, device identifier, IP addresses and associated attributes (geospatial), traffic volume of the smart home devices, etc.

3) *Baseline Establishment:* Using the actual norms of users, we establish a vector space of normal and peace-time behavior of the user to establish a statistical baseline and profile. The behavioral, user, and other device artifacts, such as header information of traffic, destination IP address, the flow of data, firmware details, interaction with the applications, etc., are used to create a profile, e.g. benign versus malicious. To do so, a set of features are created depending on the execution of the device. The extracted features are then labeled as benign feature or anomalous feature. These baselines are then used to train a model based on which the intrusion detector identifies the anomalous behavior.
4) *Features Storage:* As features of behavior over time for various entities are learned (e.g. logging session and data flow), both summaries of the raw data and features get stored for later analysis, use, and model/baseline retraining. Also, such information is stored for consumption in the notification service. The features are stored on a cloud service, as in Figure 11.2.
5) *Real-Time Detection:* As feature vectors for the various entities highlighted previously are represented in a compact and normalized form, they get compared to the baseline. Based on the outcomes of the comparison and how far it is from the baseline, an intrusion alert is assigned to them, an intrusion is raised, and the owner (user) is notified. This information is also placed in storage.

11.4.3.3 Assumptions of Operation

The basic assumption in our real-time intrusion, in general, is that the behavior of the user can be characterized and anything out of the norm can call for alerts. In particular, upon receiving such information, our IDS recognizes abnormal events by comparing such event to a knowledge-base that is driven by the normal usage pattern of the homeowner (user): e.g. if the devices are operating at unusual times, accessing previously unknown resources, or generating abnormal artifacts and events. Such unusual events include all possibilities addressed in the profiling and logging activities above: our intrusion system monitors connection attempts and identifies unusual traffic. For example, an attacker can try to connect to the smart home network using some default passwords or simple passwords repeatedly, and our IDS gathers information such as the IP addresses related to the repeated login attempts to connect to the smart home network as potential features of entities that are anomalous. Our detection system will succeed if, and only if, such behavior is not exhibited by actual and normal user behavior.

Table 11.3 A complete description of the set of features used in our system.

Feature	Class	Subclass	Description
Fc	File system	File operation	Count of files created
Fm	File system	File operation	Count of files modified
Fd	File system	File operation	Count of files deleted
sz1	File system	File size	Count of normalized file sizes created that are less than 25% of the files created
sz2	File system	File size	Count of normalized file sizes created that are less than 50% and more than 25%
sz3	File system	File size	Count of normalized file sizes created that are less than 75% and more than 50%
sz4	File system	File size	Count of normalized file sizes created that are more than 75% of the files created
Nue	File system	File extension	Count of unique extensions
Au	File system	File path	Count of files created under ALLUSERPROFILE path
Ad	File system	File path	Count of files created under APPDATA path
Cp	File system	File path	Count of files created under COMMONPROGRAMFILES path
Pf	File system	File path	Count of files created under PROGRAMFILES path
Wd	File system	File path	Count of files created under WINDIR path
Up	File system	File path	Count of files created under USERPROFILE path
Tm	File system	File path	Count of files created under TEMP path
Rc	Registry	Key operation	Count of created registry keys
Rm	Registry	Key operation	Count of modified registry keys
Rd	Registry	Key operation	Count of deleted registry keys
Rs	Registry	Key type	Count of registry keys with REG_SZ type
Rb	Registry	Key type	Count of registry keys with REG_BINARY type
Rw	Registry	Key type	Count of registry keys with REG_DWORD type
Ipn	Network	IP address	Count of unique destination IPs
p0	Network	Port number	Count of connections to port 20

p1	Network	Port number	Count of connections to port 21
p2	Network	Port number	Count of connections to port 22
p3	Network	Port number	Count of connections to port 25
p4	Network	Port number	Count of connections to port 53
p5	Network	Port number	Count of connections to port 80
p6	Network	Port number	Count of connections to port 102
p7	Network	Port number	Count of connections to port 110
p8	Network	Port number	Count of connections to port 143
p9	Network	Port number	Count of connections to port 389
p10	Network	Port number	Count of connections to port 443
p11	Network	Port number	Count of connections to port 465
p12	Network	Port number	Count of connections to port 587
p13	Network	Port number	Count of connections to port 636
p14	Network	Port number	Count of connections to port 993
p15	Network	Port number	Count of connections to port 995
p16	Network	Port number	Count of connections to port 6347 to 6665 or 6679, 6697 (IRC)
p17	Network	Port number	Count of connections to port 8080
p18	Network	Port number	Count of connections to other unaccounted for ports
Tcp	Network	Connection type	Count of TCP connections
Udp	Network	Connection type	Count of UDP connections
Raw	Network	Connection type	Count or RAW connections
rz1	Network	Request size	Count of the normalized network request size less than 25%
rz2	Network	Request size	Count of the normalized network request size less than 50% greater than 25%

(*Continued*)

Table 11.3 (Continued)

Feature	Class	Subclass	Description
rz3	Network	Request size	Count of the normalized network request size less than 75% greater than 50%
rz4	Network	Request size	Count of the normalized network request size greater than 75%
Pst	Network	Request type	Count of POST requests
Get	Network	Request type	Count of GET requests
Hed	Network	Request type	Count of HEAD requests
Th	Network	Response type	Count of response code 200s
Thh	Network	Response type	Count of response code 300s
Fh	Network	Response type	Count of response code 400s
fvh	Network	Response type	Count of response code 500s
hz1	Network	Response size	Count of the normalized reply size less than 25%
hz2	Network	Response size	Count of the normalized reply size less than 50% greater than 25%
hz3	Network	Response size	Count of the normalized reply size less than 75% greater than 50%
hz4	Network	Response size	Count of the normalized reply size greater than 75%
mx	Network	DNS type	Count of DNS MX
ns	Network	DNS type	Count of DNS NS
A	Network	DNS type	Count of DNS A
ptr	Network	DNS type	Count of DNS PTR
soa	Network	DNS type	Count of DNS SOA
Cn	Network	DNS type	Count of DNS CNAME

A sample of the features used for characterizing behavior is shown in Table 11.3. The file system and registry are those features used on the device, particularly those where such features could be extracted, while the final set (network) is in the network used in the network layer while profiling.

11.4.3.4 Feature Vectors Comparison

In our system, we use a simple statistical distance comparison metric to tell whether the behavior characterized by the feature vector of an event or entity is an anomaly or not. In particular, given normalized feature vectors (with respect to a fixed vector index) $a = (a_1, a_2, ..., a_n)$ and $b = (b_1, b_2, ..., b_n)$, we compute the distance between the two vectors as the statistical difference defined as

$$SD(a, b) = \frac{1}{2}\sum_{i=1}^{n}|a_{0i} - b_i|$$

We note that many other alternatives could be used for the same purpose. We use this technique for its simplicity.

11.4.3.5 Offline Learning and Mining

In step 4 above, we offload the features collected by the gateway (corresponds to storing a similar step in the device-assisted profiling technique). Such information provides a wealth of opportunities for offline learning and mining for malicious activities and intrusion patterns. Our IDS utilizes such data for two aspects: (i) revising the baseline model used in our simple statistical IDS and (ii) uncovering new and unseen anomalies, like zero-days.

11.5 Conclusion

In this chapter, we looked at the problem of APT in home networks. Unlike existing systems that perform security, profiling, detection, and attribution in one entity, our system implements such security services across the system, benefiting from capabilities in the system model and addressing issues in the attack surface.

So far, our work has mainly focused on the design aspect of our system. In the future, we will look into evaluating the system across multiple evaluation criteria, including detection rate, performance metrics (such as responsiveness and delay), among others. We will also look into improving the system design and evaluating it by incorporating a device-assisted profiling and decision function and multi-detector composition.

References

1 L. Atzori, A. Iera, and G. Morabito, "The Internet of Things: A survey," *Computer Networks*, vol. 54, no. 15, pp. 2787–2805, 2010.

2 N. Xu, "A survey of sensor network applications," *IEEE Communications Magazine*, vol. 40, pp. 102–114, 2002.

3 A. Sajid, H. Abbas, and K. Saleem, "Cloud-assisted IoT-based SCADA systems security: A review of the state of the art and future challenges," *IEEE Access*, vol. 4, pp. 1375–1384, 2016.

4 H. Alasmary, A. Anwar, J. Park, J. Choi, D. Nyang, and A. Mohaisen, "Graph-based comparison of IoT and android malware," in *Proceedings of the 7th International Conference on Computational Data and Social Networks, CSoNet*, 2018, pp. 259–272.

5 A. Mohaisen and O. Alrawi, "AV-Meter: An evaluation of antivirus scans and labels," in *Proceedings of the Detection of Intrusions and Malware, and Vulnerability Assessment*, DIMVA, 2014, pp. 112–131.

6 C. Lee, L. Zappaterra, K. Choi, and H. Choi, "Securing smart home: Technologies, security challenges, and security requirements," in *Proceedings of the IEEE Conference on Communications and Network Security, CNS*, 2014, pp. 67–72.

7 Q. Gou, L. Yan, Y. Liu, and Y. Li, "Construction and strategies in IoT security system," in *Proceedings of the IEEE International Conference on Green Computing and Communications (GreenCom) and IEEE Internet of Things (iThings) and IEEE Cyber, Physical and Social Computing (CPSCom)*, 2013, pp. 1129–1132.

8 Chosun Biz, "Smart TV and fridge can send SPAMs," Available at [Online]: http://bit.ly/28xJY3J, 2014.

9 Proof Point, "Your fridge is full of SPAM: Proof of an IoT-driven attack," Available at [Online]: http://bit.ly/1XSyAMJ, 2014.

10 CIO from IDG, "IoT hacking scenario," Available at [Online]: http://www.ciokorea.com/news/26406, 2015.

11 C. Roberts, "From the oven to the power station 'security hopscotch'," in *RSA Conference*, 2015.

12 G. Zhang, C. Yan, X. Ji, T. Zhang, T. Zhang, and W. Xu, "Dolphinattack: Inaudible voice commands," in *Proceedings of the Conference on Computer and Communications Security, CCS*, 2017, pp. 103–117.

13 N. Carlini, P. Mishra, T. Vaidya, Y. Zhang, M. Sherr, C. Shields, D. A. Wagner, and W. Zhou, "Hidden voice commands," in *Proceedings of the 25th USENIX Security Symposium*, 2016, pp. 513–530.

14 P. Kocher, J. Horn, A. Fogh, D. Genkin, D. Gruss, W. Haas, M. Hamburg, M. Lipp, S. Mangard, T. Prescher, M. Schwarz, and Y. Yarom, "Spectre attacks: Exploiting speculative execution," in Proceeding of the IEEE Symposium on Security and Privacy, SP, 2019. pp. 1–19.

15 M. Lipp, M. Schwarz, D. Gruss, T. Prescher, W. Haas, A. Fogh, J. Horn, S. Mangard, P. Kocher, D. Genkin, Y. Yarom, and M. Hamburg, "Meltdown: Reading kernel memory from user space," in *Proceedings of the 27th USENIX Security Symposium*, 2018, pp. 973–990.

16 J. V. Bulck, M. Minkin, O. Weisse, D. Genkin, B. Kasikci, F. Piessens, M. Silberstein, T. F. Wenisch, Y. Yarom, and R. Strackx, "Foreshadow: Extracting the keys to the intel SGX kingdom with transient out-of-order execution," in *Proceedings of the 27th USENIX Security Symposium*, 2018, pp. 991–1008.

17 Y. Xiao, M. Li, S. Chen, and Y. Zhang, "STACCO: Differentially analyzing side-channel traces for detecting SSL/TLS vulnerabilities in secure enclaves," in *Proceedings of the Conference on Computer and Communications Security, CCS*, 2017, pp. 859–874.

18 Boannews, "Coffee maker can hack. IoT generation," Available at [Online]: http://bit.ly/1U9oypa, 2015.

19 Developer, "Vulnerable coffee machine demonstrates brewing security challenges of IoT," Available at [Online]: http://ubm.io/20y7l9n, 2015.

20 Secure List, "Surviving in an IoT-enabled world," Available at [Online]: https://tinyurl.com/yay5l9uv, 2015.

21 D. D. Chen, M. Woo, D. Brumley, and M. Egele, "Towards automated dynamic analysis for Linux-based embedded firmware," in *Proceedings of the 23rd Annual Network and Distributed System Security Symposium, NDSS*, 2016.

22 H. Ham, H. Kim, M. Kim, and M. Choi, "Linear SVM-based android malware detection for reliable IoT services," *Journal of Applied Mathematics*, vol. 2014, pp. 594 501:1–594 501:10, 2014.

23 LG CNS, "IoT security threats," Available at [Online]: http://blog.lgcns.com/896, 2017.

24 Hankook Kyungjae, "House remote control and IoT devices," Available at [Online]: http://bit.ly/1S2IBxP, 2016.

25 Joongang Sunday, "Security system for IoT," Available at [Online]: http://sunday.joins.com/archives/124654, 2016.

26 M. Antonakakis, T. April, M. Bailey, M. Bernhard, E. Bursztein, J. Cochran, Z. Durumeric, J. A. Halderman, L. Invernizzi, M. Kallitsis, D. Kumar, C. Lever, Z. Ma, J. Mason, D. Menscher, C. Seaman, N. Sullivan, K. Thomas, and Y. Zhou, "Understanding the mirai botnet," in *Proceedings of the 26th USENIX Security Symposium*, 2017, pp. 1093–1110.

27 C. Kolias, G. Kambourakis, A. Stavrou, and J. M. Voas, "DDoS in the IoT: Mirai and other botnets," *IEEE Computer*, vol. 50, no. 7, pp. 80–84, 2017.

28 Y. M. P. Pa, S. Suzuki, K. Yoshioka, T. Matsumoto, T. Kasama, and C. Rossow, "IoTPOT: A novel honeypot for revealing current IoT threats," *Journal of Information Processing, JIS*, vol. 24, pp. 522–533, 2016.

29 S. Kottler, "February 28th DDoS incident report," Available at [Online]: https://tinyurl.com/y93rpucd, 2018.

30 A. Riahi, Y. Challal, E. Natalizio, Z. Chtourou, and A. Bouabdallah, "A systemic approach for IoT security," in *Proceedings of the IEEE International Conference on Distributed Computing in Sensor Systems, DCOSS*, 2013, pp. 351–355.

31 R. Roman, J. Zhou, and J. Lopez, "On the features and challenges of security and privacy in distributed Internet of Things," *Computer Networks*, vol. 57, no. 10, pp. 2266–2279, 2013.

32 S. Raza, L. Wallgren, and T. Voigt, "SVELTE: Real-time intrusion detection in the Internet of Things," *Ad Hoc Networks*, vol. 11, pp. 2661–2674, 2013.

33 P. Kasinathan, G. Costamagna, H. Khaleel, C. Pastrone, and M. A. Spirito, "An IDS framework for Internet of Things empowered by 6LoWPAN," in *Proceedings of the Conference on Computer and Communications Security*, CCS, 2013, pp. 1337–1340.

34 D. Chen, G. Chang, L. Jin, X. Ren, J. Li, and F. Li, "A novel secure architecture for the Internet of Things," in *Proceedings of the 5th International Conference on Genetic and Evolutionary Computing, ICGEC*, 2011, pp. 311–314.

35 X. M. Wei, X. S. Jiang, and X. G. Wang, "Research of IOT intrusion detection system based on hidden Markov model," *Applied Mechanics and Materials*, vol. 263, pp. 2949–2952, 2013.

12

Analysis of Stepping-Stone Attacks in Internet of Things Using Dynamic Vulnerability Graphs

Marco Gamarra[1], Sachin Shetty[2], Oscar Gonzalez[1], David M. Nicol[3], Charles A. Kamhoua[4], and Laurent L. Njilla[5]

[1] *College of Engineering, Old Dominion University, Norfolk, VA, USA*
[2] *Virginia Modeling Analysis and Simulation Center, Old Dominion University, Norfolk, VA, USA*
[3] *Department of Electrical and Computer Engineering, University of Illinois at Urbana-Champaign, Champaign, IL, USA*
[4] *US Army Research Laboratory, Adelphi, MD, USA*
[5] *Cyber Assurance Branch, US Air Force Research Laboratory, Rome, NY, USA*

12.1 Introduction

The ubiquitous adoption of the Internet of Things (IoT) has resulted in an exponential growth in the number of IoT devices. Though IoT has made great impacts to systems in commercial and military domains, there is a growing concern about the security risks introduced by IoT devices. The pervasive networked computing capabilities provided by IoT increases its security risks as attack enablers rather than attack targets. IoT devices can be unwitting participants in a botnet that could lead to secondary attacks. In a 2008 attack on a Turkish oil refinery, security analysts found out that vulnerabilities in the surveillance camera's communication software enabled attackers to gain entry and penetrate deeper into the internal network. Thus, cameras were used as stepping stones to gain access to the network that housed the critical assets.

The growth in the number of IoT devices increases the potential of using these devices as stepping stones to assist attacks by providing additional attack platforms, more opportunities for attack indirection, and eventually increasing the attribution complexity. Security risk assessment in IoT hinges on the ability to quantify the probability of lateral propagation by attackers through a network

Modeling and Design of Secure Internet of Things, First Edition. Edited by Charles A. Kamhoua, Laurent L. Njilla, Alexander Kott and Sachin Shetty.

Published 2020 by John Wiley & Sons, Inc.

of stepping stones. The choice of stepping stones is influenced by the attacker's goal, increased difficulty of attribution, and exertion of minimum effort. The identification of attacker stepping stones is important for security administrators to aid the mitigation process. Vulnerability graphs have been used to model and analyze stepping stones in networked systems [1]. The construction of vulnerability graphs hinges on the availability of network connectivity; identification of software, operating systems, and network protocols; network and system access control rules; and vulnerability information. National Vulnerability Database (NVD) and Common Vulnerability Scoring System (CVSS) scores are typically used to quantify the exploitability of vulnerabilities. In Nicol and Mallapura [1], the authors propose an approach to determine the most vulnerable paths corresponding to stepping-stone attacks by calculating the shortest path in a vulnerability graph with fixed topology. An heuristic analysis using Monte Carlo simulations has been developed in the case of switching topology, assuming that when the attacker is at any host $\boldsymbol{h}$ and the graph topology changes, the path from the host $\boldsymbol{h}$ to the target is invariant. However, there is no guarantee that the edge weights will always remain the same during the attacker's lateral propagation due to defensive mechanisms that can result in modification to the firewall rules, patching of vulnerabilities, and application of security controls. In this chapter, we develop a mathematical model to explore whether it is possible for attackers and/or network administrators to identify stepping stones when the edge weights in the vulnerability graph change.

The shortest path problem in a directed graph was formulated as a linear equation in a min-plus algebra, which can be solved using the Bellman–Ford algorithm [2, 3]. A variant of the Bellman–Ford algorithm for single-source shortest paths in graphs that optimize the algorithm, compared with the previously best variant by Yen [4], has been reported in Bannister and Eppstein [5]. In all of these works, the graph topology is fixed. Average consensus in networks with switching topologies was investigated in Olfati-Saber and Murray [6]. In Nejad et al. [7], max-consensus in graphs with switching topology was investigated using max-plus algebra.

This chapter is complementary to Nicol and Mallapura [1] and attempts to calculate the minimum cost path between source and target nodes when the graph topology changes. Assuming that there is a path from every source to any target in an attack graph at every time instant, the challenge is to calculate the minimum cost path from a source node to a target node when the vulnerability graph changes according to a switching signal that is triggered by the network system's defenses and then analyze scenarios when the problem is not NP-hard (nondeterministic polynomial-time hardness). The main contribution of this chapter is twofold:

1) A mathematical model that describes the stepping-stone attack cost as a dynamical system in a vulnerability graph with switching topology is provided,

where an interplay between min-consensus and min-plus algebra is used for modeling and analysis.

2) The necessary and sufficient conditions are provided for finite-time convergence in the fixed topology case, and a necessary condition for the time interval between two consecutive switching signals that ensure the convergence of the minimum path in finite time is also provided.

The rest of the chapter is organized as follows: Section 12.2 discusses the background, Section 12.3 develops the model of the stepping-stone cost attack in a vulnerability dynamic graph with fixed and switching topology, and Section 12.4 introduces the min-plus algebra and its interplay with the biased min-consensus to analyze the shortest path cost convergence. Finally, Section 12.5 discusses the chapter's results, and Section 12.6 provides the chapter's conclusions and future research.

12.2 Background

12.2.1 Graphs

Given a finite set $V = \{v_1, \ldots, v_n\}$.

Definition 12.1 A **directed Graph** over V is an ordered pair $G = (V, E)$, where E is a subset of the Cartesian product $V \times V$. In this context, V is called Vertex set and E is called Edge set. Every $v_i \in V$ is called **vertex** or **node** and every ordered pair (v_i, v_j) of E is called **directed edge**, where v_i is called the tail and v_j is called the head of the edge.

Definition 12.2 A **directed weighted Graph** over V is an ordered triple $G = (V, E, w)$, where (V, E) is a directed graph and $w : E \rightarrow \mathbb{R}$ is a function that associates a value to each edge.

Definition 12.3 Given a directed weighted graph $G = (V, E, w)$.

1) A vertex v_j is said **adjacent** to v_i if, and only if, $(v_i, v_j) \in E$.
2) For every vertex, v_i is defined as the set of all its **neighbors** as $N_i = \{v_j \in V/(v_i, v_j) \in E\}$.
3) A **path** C of length m in G is a sequence of $m + 1$ vertices $v_{i_1}, v_{i_2}, \cdots, v_{i_{m+1}}$, such that $(v_{i_k}, v_{i_{k+1}}) \in E$ for all $k = 1, \ldots, m$. If $v_{i_1} = v_{i_{m+1}}$, then C is called a **cycle** of length m. A **cycle** of length 1 is called a self-loop.
4) The weighted **adjacency** matrix associated to the weighted graph $G = (V, E, w)$ is defined as the square $n \times n$ matrix A, such that $[A]_{ij} = w_{ij} > 0$ if $(v_i, v_j) \in E$ and $[A]_{ij} = 0$ in all other cases.

Definition 12.4 An undirected Graph over V is a directed graph $G = (V, E)$, such that for every directed edge (v_i, v_j) in V, there is a directed edge (v_j, v_i) in E. These two edges are denoted in a compact way as the unordered pair $\{v_i, v_j\}$ and are called an undirected edge.

Definition 12.5 A directed graph is said to be strongly connected if there is a path connecting every two nodes, and it is called weakly connected if the graph obtained by adding an edge (v_j, v_i) for every existing edge (v_i, v_j) in the original graph is strongly connected. An undirected graph is connected if, and only if, it is strongly connected.

Definition 12.6 A directed multigraph is a directed graph that is permitted to have multiple directed edges between any two vertices, that is, there are multiple edges that have the same head and tail. If the edges are undirected, the graph is called undirected multigraph. If the edges are weighted, the graph is called weighted multigraph.

12.2.2 Vulnerability Graphs and Stepping Stones

Stepping-stone attacks are modeled and analyzed in Nicol and Mallapura [1], which describes a stepping-stone attack as a path through a multigraph where nodes represent hosts, and each edge represents one way in which an attacker, who is resident on the source host, can gain a foothold through an exploit on the destination host. According to the CVSS, a scoring function for every edge in the stepping-stone path has been constructed, called the *exploit complexity score*, which ranges between 0 and 10 defined as $\epsilon = 10 - \varepsilon / A_v$, where ε is the *exploitability score* and A_v is the *access vector*. The *exploit complexity score* quantifies the difficulty of compromise for each node. The smaller the score, the easier it is to exploit the vulnerability. See Nicol and Mallapura [1] for more details in this construction.

A **vulnerability multigraph** has been constructed such that given a set of n node hosts $\{h_1, \cdots, h_n\}$, a weighted edge exists from h_i to h_j if, and only if, there is a vulnerability that allows an attacker on h_i to compromise h_j. This weight ε_{ij} is the *exploit complexity score*, so $0 < \varepsilon_{ij} < 10$. A *stepping-stone path* is a path through a vulnerability multigraph, where the edges denote the vulnerabilities exploited by the attacker to reach the last host in the path from the first.

The **cost** of a stepping-stone path is defined as "the sum of the costs of the edges of the path." The problem of finding a min-cost stepping-stone path between any two hosts comes when the vulnerability graph topology change is NP-hard [1]. As a feasible solution to this problem, Monte Carlo simulations have been used in which stochastic sample paths are generated and the low-cost ones are saved. The challenge posed by varying the edge weights is **solved** under the assumption that the as-yet unseen edge costs are invariant; the standard shortest path algorithm can be used to **estimate** the minimum remaining cost. For more information on this approach, see Nicol and Mallapura [1].

Example 12.1 To illustrate the idea of vulnerability graphs and stepping-stone attacks, we are considering a network that is presented in [8]. Including two virtualization servers and one physical server, the network topology is presented in Figure 12.1. The configuration and function information of the servers is shown in Table 12.1. In this example, it is assumed that an attacker takes the root permissions in the target database PM1 as the final goal to obtain business data. To achieve this goal, an attacker can follow many ways and means. As a first path, the attackers can find the SQL injection vulnerability CVE-2011-2688 on the web server VM21. Through this vulnerability, the attacker gets the user rights of the VM21 and establishes a connection with the database server PM1 on the VM21 with a legitimate identity. Then, through the CVE-2012-2122 and CVE-2010-2693 vulnerabilities on the server PM1, the access mechanism is bypassed and the local authority is carried out. Finally, the attacker gets the root permissions

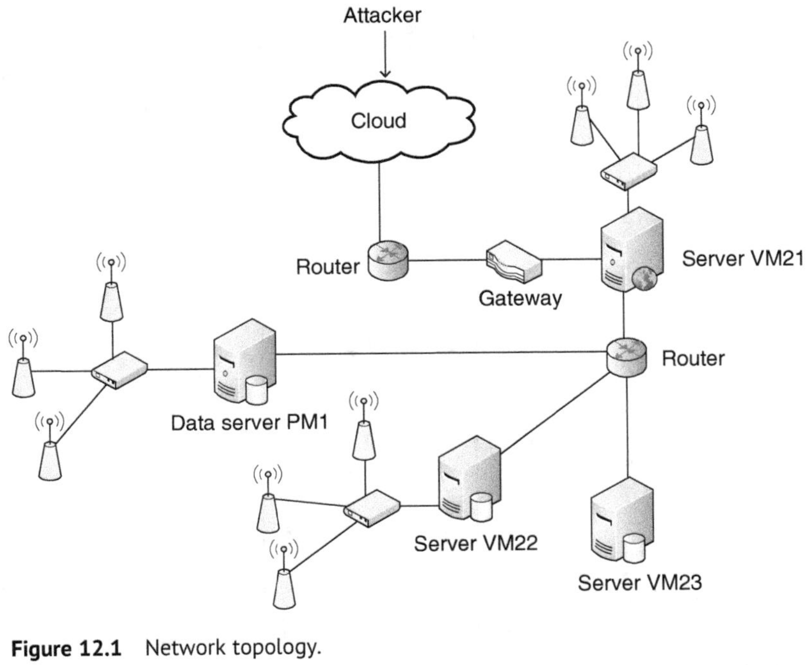

Figure 12.1 Network topology.

Table 12.1 Host configuration, function information, and vulnerability scores information.

Host	OS	Function	Server	Vulnerability CVE-ID	Exploitability subscore ε	A_v	Exploit complexity score $10 - \varepsilon/A_v$	NVD last modified
VM21	Redhat 5.4	Web server	HTTP, SSH	CVE-2011-2688	10.0	1	0	28 August 2017
PM1	Redhat 5.4	Database server	SSH	CVE-2012-2122,	4.9	1	5.1	20 February2014
				CVE-2010-2693	3.9	0.395	0.1	14 July 2010
VM22	Redhat 5.4	File server	FTP, SSH	CVE-2011-4130,	8.0	1	2	12 August 2011
				CVE-2012-6067	10.0	1	0	12 May 2012
VM23	Redhat 5.4	Host	SSH	CVE-2008-3234	8.0	1	2	28 September 2017

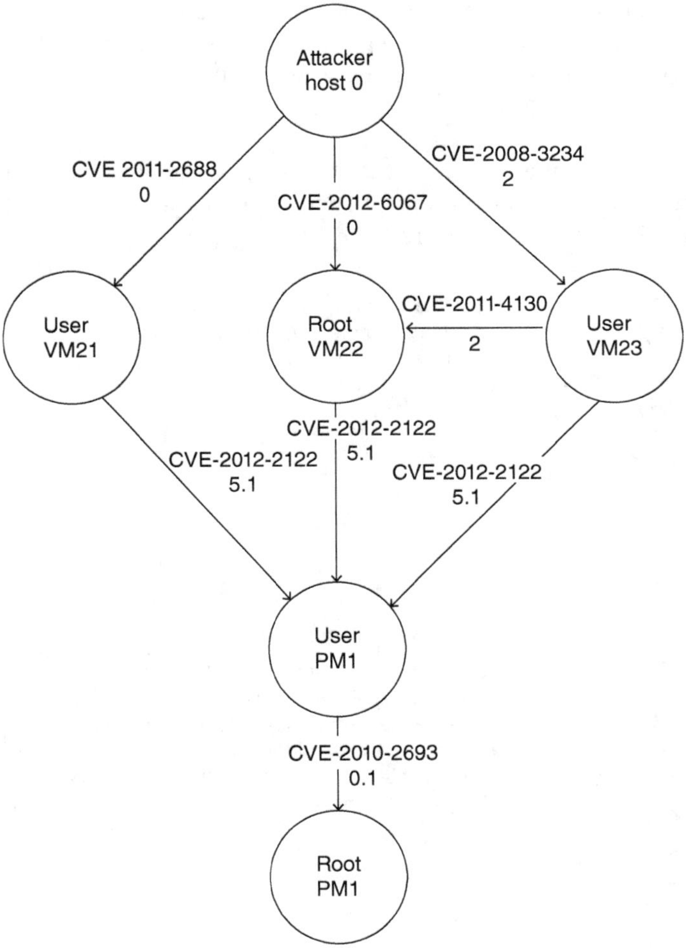

Figure 12.2 A vulnerability graph derived from the network topology presented in Figure 12.1. The weight in the edges is the exploit complexity score associated with its vulnerability. For example, there is a vulnerability in VM23 (CVE-2008-3234) that an attacker in Host 0 can exploit and gain access to VM23, then there is an edge from Host 0 to VM23 with weight 2.

of the database server PM1. On the second path, an attacker can also improve VM22 permissions through a vulnerability, CVE-2012-6067 and CVE-2011-4130 on the file server VM22, then infiltrate to the database server PM1, and finally get the root permissions. In a third path, attackers reach the final goal by directly attacking the database server PM1, see [8] for more details. The associated vulnerability graph is presented in Figure 12.2.

In this context, the most probable path that an attacker will follow is the one with minimum cost. For example, the stepping-stone attack Host 0, VM21, PM1 (User), PM1(Root) that has a cost $0 + 5.1 + 0.1 = 5.2$, or Host 0, VM22, PM1(User), PM1(Root) that has a cost $0 + 5.1 + 0.1 = 7.2$.

Example 12.2 Figure 12.3 presents an abstraction of a physical network as an undirected graph G, and two vulnerability graphs derived from G for two different times, t_1 and t_2, are presented in Figures 12.4 and 12.5, respectively.

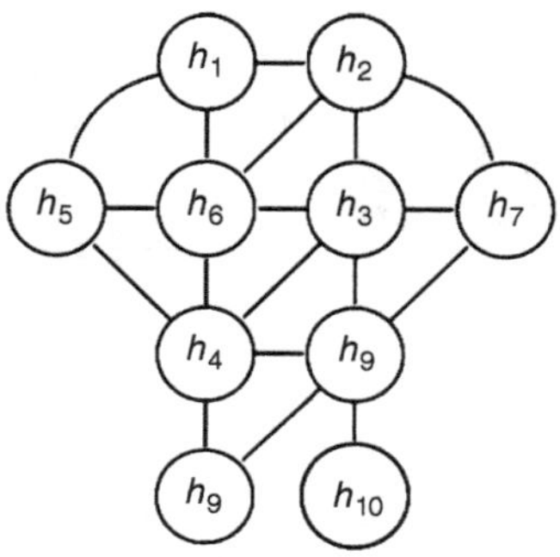

Figure 12.3 Physical network represented with a 10-node undirected graph G. The undirected edge between two nodes means that the communication flow is bidirectional.

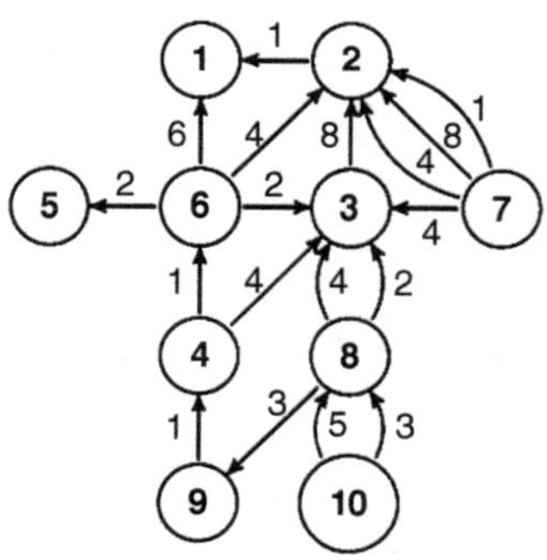

Figure 12.4 A vulnerability multigraph derived from G at any time t_1. For example, there are three directed weighted edges from node 7 to node 2, which means that there are three vulnerabilities with respective exploit complexity scores that allow an attacker on host h_7 to compromise h_2.

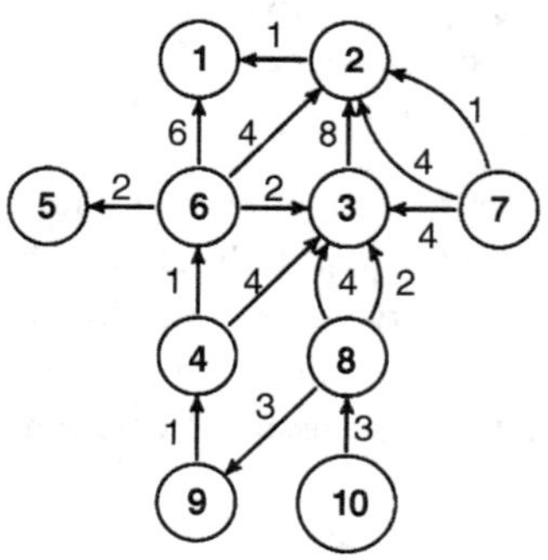

Figure 12.5 Vulnerability multigraph derived from G at any time t_2. In this case, there are two directed weighted edges from node 7 to 2.

12.2.3 Consensus Protocol in a Dynamic Graph

Let $G = (V, E)$ be a directed graph with a vertex set $V = \{v_1, \ldots, v_n\}$ and edge set E. Let $x \in \mathbb{R}$ denote the value of vertex v_i. Definitions 12.7 and 12.8 are introduced according to Olfati-Saber and Murray [6].

Definition 12.7 The pair (G, x) with $x = [x_1, \ldots, x_n]^T$ is called an **algebraic graph** with value $x \in \mathbb{R}^n$ and topology (or information flow) G. The value x_i of a vertex v_i represents any physical quantities or any other attribute of the network.

Assume that every vertex-value x_i in the algebraic graph (G, x) is a dynamic agent with dynamics $\frac{dx_i}{dt} = \dot{x}_i(t) = f_i\,(x(t), u_i),\quad i \in I = \{1, \ldots, n\}$.

Definition 12.8 A **dynamic graph** is a dynamical system with a state (G, x) in which the value $x = [x_1, \ldots, x_n]^T$ evolves according to the network dynamics

$$\dot{x}_i = f_i(x, u_i), \quad \forall i = 1, 2, \ldots, n, \tag{12.1}$$

where

$$u_i = g_i(x) \tag{12.2}$$

is called a **distributed protocol** with topology G. In matrix form

$$\dot{x} = f(x, u), \tag{12.3}$$

where

$$\dot{x} = [\dot{x}_1, \ldots, \dot{x}_n]^T, \quad f(x, u) = [f_1(x, u_1), \ldots, f_n(x, u_n)]^T,$$

and

$$u = [u_1, \ldots, u_n]^T = [g_1(x), \ldots, g_n(x)]^T.$$

In the literature, an algebraic graph is usually called an algebraic network and a dynamical graph is called a dynamic network. In the rest of the chapter, we illustrate a simple case of the dynamical network (12.1) when $f_i(x, u_i) = u_i$ is considered, that is,

$$\dot{x}_i(t) = u_i(t), \quad \forall i = 1, 2, \ldots, n. \tag{12.4}$$

The next two examples are an illustration of a distributed calculation with the linear and nonlinear functions:

$\chi_1(x) = \text{Ave}\{x_1(t), \ldots, x_n(t)\} = \text{Ave}\{x(t)\}$, $\chi_2(x) = \min\{x_1(t), \ldots, x_n(t)\} = \min\{x(t)\}$, respectively, using a dynamic graph (G, x), where G is assumed strongly connected and $N_i = \{x_j/(x_i, x_j)$ is an edge in $G\}$ is the set of all the neighbors of the vertex x_i.

Example 12.3 (Average consensus). The following distributed linear protocol,

$$u_i = \sum_{j \in N_i} (x_j - x_i), \tag{12.5}$$

asymptotically solves the average consensus problem and is called the average consensus protocol, that is, replacing (12.5) in (12.4) yields

$$\dot{x}(t) = \sum_{j \in N_i} (x_j - x_i), \tag{12.6}$$

as is proved in [6],

$$\begin{aligned} \lim_{t \to \infty} x_i(t) &= \frac{1}{n} \sum_{i=1}^{n} x_i(0), \quad i = 1, \dots, n \\ &= \text{Ave}\{x(0)\}. \end{aligned}$$

Example 12.4 (Min-consensus). The following distributed nonlinear protocol,

$$u_i = \min_{j \in N_i} (x_j - x_i), \tag{12.7}$$

asymptotically solves the min-consensus problem. Replacing (12.7) in (12.4) yields

$$\dot{x}_i = \min_{j \in N_i} (x_j - x_i) = -x_i + \min_{j \in N_i} (x_j) \tag{12.8}$$

as is proved in [6],

$$\begin{aligned} \lim_{t \to \infty} x_i(t) &= \min \{x_1(0), \dots, x_n(0)\}, \quad i = 1, \dots, n \\ &= \min \{x(0)\}. \end{aligned}$$

12.2.4 Biased Min-Consensus

The min-consensus protocol is perturbed [9], yielding the following distributed nonlinear protocol,

$$u_i = -x_i + \min_{j \in N_i} (x_j + w_{ij}), \tag{12.9}$$

where w_{ij} is the weight of edge (x_i, x_j), which is called *biased min-consensus protocol.* Closing the loop, we have

$$\dot{x}_i = -x_i + \min_{j \in N_i} (x_j + w_{ij}), \tag{12.10}$$

that asymptotically converges to the equilibrium point x_i^* [9], that is,

$$\lim_{t \to \infty} x_i(t) = x_i^*,$$

which satisfies the following equation:

$$x_i^* = \min_{j \in N_i} \left(x_j^* + w_{ij} \right), \quad i = 1, \ldots, n.$$

12.2.5 Leader–Follower Strategy

In many applications of multi-agent systems, a leader–follower strategy is applied. In this approach, a subset of agents is called *leader set* N_l, and the remainder agents are called *follower set* N_f. In this context, the average consensus, min-consensus, and biased min-consensus are

$$\begin{cases} \dot{x}_i = v_i & \text{if } i \in N_l \\ \dot{x}_i = \sum\limits_{j \in N_i} (x_j - x_i) & \text{if } i \in N_f \end{cases}, \tag{12.11}$$

$$\begin{cases} \dot{x}_i = v_i & \text{if } i \in N_l \\ \dot{x}_i = -x_i + \min\limits_{j \in N_i} (x_j) & \text{if } i \in N_f \end{cases}, \tag{12.12}$$

$$\begin{cases} \dot{x}_i = v_i & \text{if } i \in N_l \\ \dot{x}_i = -x_i + \min\limits_{j \in N_i} (x_j + w_{ij}) & \text{if } i \in N_f \end{cases}, \tag{12.13}$$

respectively, where v_i is an exogenous input. If $v_i = 0$, the systems (Eqs. 12.11–12.13) are called *static leaders system*. The following theorem has been stated and proven in Zhang and Li [9].

Theorem 12.1 Let G be an undirected connected graph, and suppose that the dynamical network (G, x) evolves according to the protocol (Eq. 12.13) with static leaders. Then the system asymptotically converges to the equilibrium point x^* of (Eq. 12.13), which satisfies the following equation [9]:

$$\begin{cases} x_i^* = x_i(0) & \text{if } i \in N_l \\ x_i^* = \min\limits_{j \in N_i} \left(x_j^* + w_{ij} \right) & \text{if } i \in N_f \end{cases}. \tag{12.14}$$

12.2.6 Biased Min-Consensus and Shortest Path

The relationship between the biased min-consensus protocol in a dynamical network (G, x) and the shortest path in G has been developed in Zhang and Li [9] as follows: The "leader" agents are static, that is, $\dot{x}_i(t) = 0$, $\forall \dot{x}_i \in N_l$, and are called *destination nodes*. The "follower" agents are called *source nodes*.

- If there is an edge between x_i and x_j, the weight w_{ij} of this edge is the *length* between these agents.

- The system evolves according to the protocol (Eq. 12.13) with static leaders. Note that according to the optimality principle of Bellman's dynamic programming [10], the solution of the considered shortest path problem satisfies the following nonlinear equations:

$$\begin{cases} x_i^* = 0 & \text{if } i \in N_l \\ x_i^* = \min\limits_{j \in N_i} \left(x_j^* + w_{ij} \right) & \text{if } i \in N_f \end{cases}, \tag{12.15}$$

which are the equilibrium points of (Eq. 12.13) with static leaders and $x_i(0) = 0,\ \forall x_i \in N_l$.

The following theorem has been stated and proven in Zhang and Li [9].

Theorem 12.2 If $x_i(0) = 0,\ \forall x_i \in N_l$, then the equilibrium of the system (Eq. 12.13) with statics leaders is given by (Eq. 12.15), which forms a solution to the corresponding shortest path problem [9].

Remark 12.1

1) When the states have reached the equilibrium point (Eq. 12.15), the shortest path can be found by recursively finding the parent nodes [9].
2) According to Theorems 12.1 and 12.2, all the state values of the system globally converge to the lengths of the corresponding shortest path independently of the initial state values [9].

12.3 Stepping-Stone Dynamics

12.3.1 Fixed Topology Case

Given a vulnerability graph G with n hosts $\{h_1, \ldots, h_n\}$, a dynamic graph (G, x) is assigned such that for every h_i, there is a state $x_i \in \mathbb{R}$ and the weight of the edge (x_i, x_j) is its exploit complexity score ϵ_{ij}. A leader–follower strategy is used where the state x_i evolves according to biased min-consensus dynamics (Eq. 12.13) with static leaders; in this model, the leaders are called ***valuable targets*** and the followers are called ***source nodes***. If x_l is a valuable target, then the state of the source x_i is defined as the stepping-stone cost from x_i to x_l, that is, the length of the path from x_i to x_l. Mathematically, the stepping-stone dynamics can be written as

$$\begin{cases} \dot{x}_i(t) = 0, \quad x_i(0) = 0 & \text{if } i \in N_l \\ \dot{x}_i(t) = -x_i(t) + \min\limits_{j \in N_i} \left(x_j(t) + \epsilon_{ij} \right) & \text{if } i \in N_f \end{cases}, \tag{12.16}$$

then, according to Theorem 12.2, in the equilibrium $x_i^* \in N_f$ is the minimum stepping-stone cost from x_i to any valuable target x_l, that is, the most probable stepping-stone attack from host h_i to any valuable target h_l.

As claimed in Nicol and Mallapura [1], according to experience, as an attacker penetrates more deeply into a system, the exploit complexity score should change and, as a consequence, the graph topology will change as well. In the following section, we develop a more realistic model in which the *exploit complexity score* changes as the attack penetrates more deeply in the system.

12.3.2 Switching Topology Case

Let $\{G_1, ..., G_m\}$ be a finite collection of vulnerability multigraphs with the same n hosts $\{h_1, ..., h_m\}$, and $s: \mathbb{R} \rightarrow \{1, ..., m\}$ a switching signal. For every $s(t) = l$, a vulnerability multigraph $G_l \in \{G_1, ..., G_m\}$ and its associated dynamic graph (G_l, x) are well defined, and then the stepping-stone dynamics with switching topology is equivalent to the hybrid system

$$\begin{cases} \dot{x}_i(t) = 0, \quad x_i(0) = 0 & \text{if } i \in N_l \\ s(t) = l & \\ \dot{x}_i(t) = -x_i(t) + \min\limits_{j \in N_i} \left(x_j(t) + \epsilon_{ij}(i)\right) & \text{if } i \in N_f \end{cases}, \tag{12.17}$$

where $\epsilon_{ij}(l)$ is the exploit vulnerability score of the edge $\{x_i, x_j\}$ in the dynamic graph (G_l, x). Hence, if the network has a vulnerability graph G_p and at any time $t > 0$:

1) An attack from h_i is detected in h_j, then the network's vulnerability graph switches to $G_q = s(t)$ such that $\epsilon_{ij}(q) > \epsilon_{lm}(p)$ for all $\epsilon_{lm}(p) = \epsilon_{ij}(p)$.
2) An attack from h_i compromises h_j, then the network's vulnerability graph switches to $G_q = s(t)$ such that $\epsilon_{ij}(q) < \epsilon_{lm}(p)$ for all $\epsilon_{lm}(p) = \epsilon_{ij}(p)$.

The progress of the stepping-stone dynamics is monitored at δ time intervals; hence, the stepping-stone dynamics at a discrete time for the fixed topology case is

$$\begin{cases} x_i[k+1] = x_i[k], \quad x_i[0] = 0 & \text{if } i \in N_l \\ x_i[k+1] = \min\limits_{j \in N_i} \left(x_j[k] + \epsilon_{ij}(l)\right) & \text{if } i \in N_f \end{cases}, \tag{12.18}$$

where

$$x_i[k] \stackrel{\text{def}}{=} x_i(\delta k), \delta > 0 \text{ and } k \in \mathbb{Z}_0{}^+, \tag{12.19}$$

and for the switching-topology case, is

$$\begin{cases} x_i[k+1] = x_i[k], \quad x_i[0] = 0 & \text{if } i \in N_l \\ l = s[k] & \\ x_i[k+1] = \min\limits_{j \in N_i} \left(x_j[k] + \epsilon_{ij}(l)\right) & \text{if } i \in N_f \end{cases}. \tag{12.20}$$

12.4 Min-Plus Algebra

Min-plus algebra consists of two binary operations, $\oplus$ and $\otimes$, on the set $\mathbb{R}_{\min} = \mathbb{R} \cup \{+\infty\}$, which is defined as follows:

$$a \oplus b = \min\{a, b\}, \tag{12.21}$$

$$a \otimes b = a + b. \tag{12.22}$$

The neutral element with respect to the min-plus addition $\oplus$ is $+\infty$, denoted as θ, and with respect to the min-plus multiplication $\otimes$ is 0, denoted as e. Both operations are associative and commutative, and the multiplication is distributive over the addition. Both operations are extended to matrices as follows:

Given $A, B \in \mathbb{R}_{\min}^{m \times n}$,

$$[A \oplus B]_{ij} = a_{ij} \oplus b_{ij}, \quad i = 1, \ldots, m \quad j = 1, \ldots, n.$$

Given $A \in \mathbb{R}_{\min}^{m \times q}, B \in \mathbb{R}_{\min}^{q \times n}$,

$$\begin{aligned}[A \otimes B]_{ij} &= \bigoplus_{k=1}^{q} \left(a_{ik} \otimes b_{kj}\right), \quad i = 1, \ldots, m \quad j = 1, \ldots, n \\ &= \min_{k=1}^{q} \left\{a_{ik} + b_{kj}\right\}.\end{aligned}$$

The identity matrix of size n is a square matrix denoted by I_n and given by

$$[I_n]_{ij} = \begin{cases} e & \text{for } i = j \\ \theta & \text{for } i \neq j \end{cases}.$$

If $A \in \mathbb{R}_{\min}^{n \times n}$ for any integer $k \geq 1$, $A^k = \underbrace{A \otimes A \otimes \cdots \otimes A}_{k-1 \text{ multiplications}}$ and $A^0 = I_n$. For more properties and application of min-plus algebra, see Cohen et al. [3] and the references therein. If G is a vulnerability graph with n nodes $\{1, \ldots, n\}$, then a modified weighted adjacency matrix, $A \in \mathbb{R}_{\min}^{n \times n}$, is associated to G and defined as

$$A = \begin{cases} [A]_{ii} = e, & \text{if } i \text{ is a target node} \\ [A]_{ij} = \epsilon_{ij}, & \text{if } (i, j) \text{ is an edge} \\ [A]_{ij} = \theta, & \text{in other cases} \end{cases}. \tag{12.23}$$

Notice that in this matrix, $[A]_{ij} \neq 0$ means that there is one path of length **1** from i to j in G. In $A^2 = A \otimes A$, if $[A \otimes A]_{ij} \neq \theta$, then there is a path of length **2** in G with a minimum cost from node i to node j, that is, there is a node l in G such that (i, l) and (l, j) are edges in G, such that $\min_{k=1}^{n}\{a_{ik} \otimes a_{kj}\} = a_{il} \otimes a_{lj} = a_{il} + a_{lj}$, but also $a_{il} \otimes$

$a_{lj} \neq \theta$ implies that $a_{il} \neq \theta$ and $a_{lj} \neq \theta$ simultaneously, and $a_{il} \otimes a_{lj} = \theta$ implies that $a_{il} = \theta$ or $a_{lj} = \theta$. Using the min-plus formalism, the stepping-stone dynamics with fixed topology (Eq. 12.18) can be rewritten as

$$\begin{cases} x_i[k+1] = x_i[k], \quad x_i[0] = 0 & \text{if } i \in N_l \\ x_i[k+1] = \underset{j \in N_i}{\oplus} \left(x_j[k] \otimes \epsilon_{ij}\right) & \text{if } i \in N_f \end{cases},$$

or in matrix form as

$$x[k+1] = A \otimes x[k] = A^{k+1} \otimes x[0], \tag{12.24}$$

and the stepping-stone dynamics with switched topology (Eq. 12.20) can be written as

$$\begin{cases} x_i[k+1] = x_i[k], \quad x_i[0] = 0 & \text{if } i \in N_l \\ l = s[k] & \\ x_i[k+1] = \underset{j \in N_i}{\oplus} \left(x_j[k] \otimes \epsilon_{ij}(l)\right) & \text{if } i \in N_f \end{cases},$$

or in matrix form as

$$x[k+1] = A_l \otimes x[k], \quad l = s[k], \tag{12.25}$$

where A_l is the matrix associate with the vulnerability graph, $G_l \in \{G_1, \ldots, G_m\}$.

Theorem 12.3 A necessary and sufficient condition for which the stepping-stone dynamics (Eq. 12.18) converge to equilibrium (Eq. 12.14) is that $A^{k+1} = A^k$ for any integer $k \geq 1$.

Proof: **Sufficiency**

$$x[k+1] = A \otimes x[k] = A^{k+1} \otimes x[0] = A^k \otimes x[0] = x[k]$$

hence, $x[k+1] - x[k] = 0$, which implies (Eq. 12.14).

Necessity: If the system is in equilibrium for any integer $k \geq 1$, then this implies that $x[k+1] - x[k] = 0$,

or equivalently

$$\begin{cases} x_i[k+1] = x_i[k] = x_i[0] = 0 & \text{if } i \in N_l \\ x_i[k+1] = x_i[k] = \min\limits_{j \in N_i} \left(x_j[k] + \epsilon_{ij}\right) & \text{if } i \in N_f \end{cases},$$

hence $[A^{k+1}]_{ii} = [A^k]_{ii} = 0$ if $i \in N_l$, because in the vulnerability graph there are no self-loops in the source nodes. If $i \in N_f$, there is a path of length k from node i to any target node j with minimum cost $x_i[k] = [A^k]_{ij} \otimes x_j[0] = [A^k]_{ij}$, and there is a path of length $k+1$ from node i to the same target node j with minimum cost

$x_i[k+1] = [A^{k+1}]_{ij} \otimes x_j[0] = [A^{k+1}]_{ij} = x_i[k] = [A^k]_{ij}$, because $x_j[0] = 0$ is the cost for a target node, and therefore, $A^{k+1} = A^k$. □

Corollary 12.1 If $A \in \mathbb{R}_{\min}^{n \times n}$ is the matrix associated with the vulnerability graph G and there is an integer $k \geq 1$ such that $A^{k+1} = A^k$, then $k \leq n - 1$.

Proof: $x_i[k+1] = [A^{k+1}]_{ij} = x_i[k] = [A^k]_{ij}$ implies that there is a path of length k from node i to node j with minimum cost, and equivalently there is a simple path from i to j with length k. Because the maximum length of a simple path in a graph with n nodes is $n - 1$, then $k \leq n - 1$. □

Remark 12.2 Corollary 12.1 implies that the stepping-stone dynamics (Eq. 12.18) converge to equilibrium in finite time $\tau = k\delta$ with $k \leq n - 1$ instant communications.

Theorem 12.4 The stepping-stone dynamics with switching topology (Eq. 12.20) converges to equilibrium if the time interval between two consecutive switching signals is large enough, as $k = n - 1$ instant communications.

Proof: By Corollary 12.1, $n - 1$ instant communications are the finite time period that guarantees the equilibrium for any fixed graph topology, and if the time interval between two switching signals is larger than $n - 1$, then the system reaches equilibrium and the shortest path is calculated, which proves the theorem. □

Remark 12.3 In Theorem 12.4, notice that a convergence to equilibrium is possible for any time interval between switching signals in less than $n - 1$ communication instants, as shown in Example 12.6. We will discuss this point in Section 12.5.

The model was developed for a directed graph with the condition that there is a path from every source node to any target. In the following examples, the vulnerability graph is considered to be strongly connected, assuming that an attacker has the advantage of compromising any node in any layer of the vulnerability graph.

Example 12.5 (Fixed topology). Consider the 10-node vulnerability graph G presented in Figure 12.6. The exploit complexity scores are encoded in its weighted adjacency matrix presented in Eq. (12.26), and the modified adjacency matrix, as defined in Eq. (12.23), can be derived immediately, which is presented in Eq. (12.27).

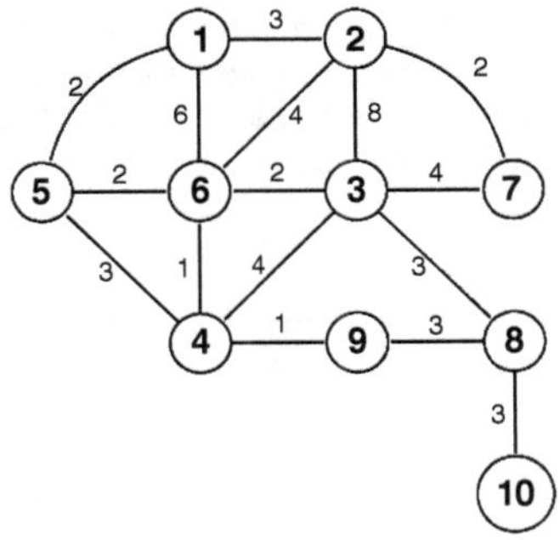

Figure 12.6 A 10-node vulnerability graph $G = G_p$ with node 1 as a target node.

$$\epsilon = \begin{bmatrix} 0 & 3 & 0 & 0 & 2 & 6 & 0 & 0 & 0 & 0 \\ 3 & 0 & 8 & 0 & 0 & 4 & 2 & 0 & 0 & 0 \\ 0 & 8 & 0 & 4 & 0 & 2 & 4 & 3 & 0 & 0 \\ 0 & 0 & 4 & 0 & 3 & 1 & 0 & 0 & 1 & 0 \\ 2 & 0 & 0 & 3 & 0 & 2 & 0 & 0 & 0 & 0 \\ 6 & 4 & 2 & 1 & 2 & 0 & 0 & 0 & 0 & 0 \\ 0 & 2 & 4 & 0 & 0 & 0 & 0 & 0 & 0 & 0 \\ 0 & 0 & 3 & 0 & 0 & 0 & 0 & 0 & 3 & 3 \\ 0 & 0 & 0 & 1 & 0 & 0 & 0 & 3 & 0 & 0 \\ 0 & 0 & 0 & 0 & 0 & 0 & 0 & 3 & 0 & 0 \end{bmatrix} \tag{12.26}$$

$$A = \begin{bmatrix} e & 3 & \theta & \theta & 2 & 6 & \theta & \theta & \theta & \theta \\ 3 & \theta & 8 & \theta & \theta & 4 & 2 & \theta & \theta & \theta \\ \theta & 8 & \theta & 4 & \theta & 2 & 4 & 3 & \theta & \theta \\ \theta & \theta & 4 & \theta & 3 & 1 & \theta & \theta & 1 & \theta \\ 2 & \theta & \theta & 3 & \theta & 2 & \theta & \theta & \theta & \theta \\ 6 & 4 & 2 & 1 & 2 & \theta & \theta & \theta & \theta & \theta \\ \theta & 2 & 4 & \theta & \theta & \theta & \theta & \theta & \theta & \theta \\ \theta & \theta & 3 & \theta & \theta & \theta & \theta & \theta & 3 & 3 \\ \theta & \theta & \theta & 1 & \theta & \theta & \theta & 3 & \theta & \theta \\ \theta & \theta & \theta & \theta & \theta & \theta & \theta & 3 & \theta & \theta \end{bmatrix} \tag{12.27}$$

Table 12.2 Stepping-stone cost evolution from every source node to the target, where every numerical column shows the stepping-stone cost from the respective source; for example, the row labeled with the agent x_8 shows that in the third iteration, its stepping-stone cost is 6.

		Stepping-stone cost/iteration						
Source	x2	3	3	3	3	3	3	3
	x3	3	4	5	6	6	6	6
	x4	2	3	4	5	5	5	5
	x5	2	2	2	2	2	2	2
	x6	2	3	4	4	4	4	4
	x7	3	5	5	5	5	5	5
	x8	4	5	6	7	8	9	9
	x9	2	3	4	5	6	6	6
	x10	4	7	8	9	10	11	12

If x_1 is a valuable target, then the stepping-stone dynamics are

$$\begin{cases} x_1[k+1] = x_1[k] = x_1[0] = 0 \\ x_i[k+1] = x_i[k] = \min\limits_{j \in N_i} \left(x_j[k] + \epsilon_{ij}\right) \qquad \text{if } i \in \{2, \ldots, 10\} \end{cases} \cdot \tag{12.28}$$

The simulation is presented in Table 12.2, where the initial states are $x_i(0) = 1, \quad \forall i = 2, \ldots, 10$, and the most vulnerable stopping-stone path for all the sources (shortest path) has been reached after seven iterations. For example, the most vulnerable stepping-stone path from node 10 to node target 1 have cost 12, and there are two: $10 \rightarrow 8 \rightarrow 3 \rightarrow 6 \rightarrow 5 \rightarrow 1$ and $10 \rightarrow 8 \rightarrow 9 \rightarrow 4 \rightarrow 6 \rightarrow 5 \rightarrow 1$. The most vulnerable stepping-stone path from node 6 to node target 1 has cost 4 and is unique: $6 \rightarrow 5 \rightarrow 1$.

Example 12.6 (Switching topology). Assume that at any time $t_1 > 0$, a 10-node vulnerability graph has topology $G_p = G$ (presented in Figure 12.6). The exploit complexity scores are encoded in its weighted incidence matrix $\epsilon(p) = \epsilon$ (presented in Eq. (12.26)). Assume that at any time $t_2 > t_1$, and an attack from host h_6 is detected in host h_5, then the network defense is activated with a switching signal $s(t_2) = q$ yielding a new network topology G_q (presented in Figure 12.7). The new exploit complexity scores are encoded in its weighted incidence matrix $\epsilon(q)$ as presented in Eq. (12.29), where $\epsilon_{ij}(q) = 9$ if $\epsilon_{ij}(p) = 2$ and $\epsilon_{ij}(q) = \epsilon_{ij}(p)$ if $\epsilon_{ij}(p) \neq 2$. The new modify incidence matrix A_q, as defined in Eq. (12.23), is presented in Eq. (12.30).

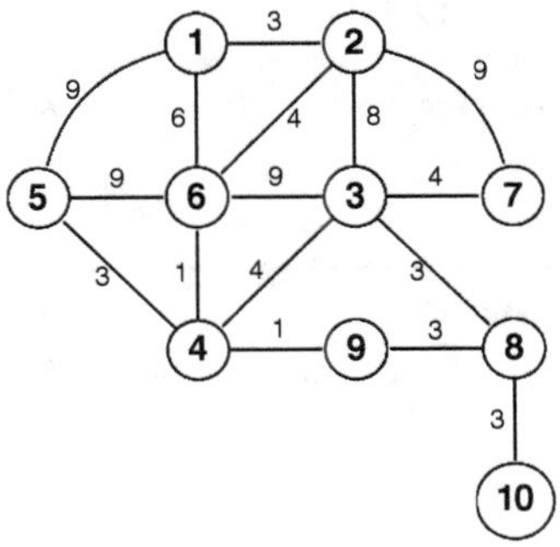

Figure 12.7 A 10-node vulnerability graph G_q derived from graph G_p where the edges with weight 2 have been switched with edges with weight 9.

$$\epsilon(q) = \begin{bmatrix} 0 & 3 & 0 & 0 & 9 & 6 & 0 & 0 & 0 & 0 \\ 3 & 0 & 8 & 0 & 0 & 4 & 9 & 0 & 0 & 0 \\ 0 & 8 & 0 & 4 & 0 & 9 & 4 & 3 & 0 & 0 \\ 0 & 0 & 4 & 0 & 3 & 1 & 0 & 0 & 1 & 0 \\ 9 & 0 & 0 & 3 & 0 & 9 & 0 & 0 & 0 & 0 \\ 6 & 4 & 9 & 1 & 9 & 0 & 0 & 0 & 0 & 0 \\ 0 & 9 & 4 & 0 & 0 & 0 & 0 & 0 & 0 & 0 \\ 0 & 0 & 3 & 0 & 0 & 0 & 0 & 0 & 3 & 3 \\ 0 & 0 & 0 & 1 & 0 & 0 & 0 & 3 & 0 & 0 \\ 0 & 0 & 0 & 0 & 0 & 0 & 0 & 3 & 0 & 0 \end{bmatrix} \tag{12.29}$$

$$A_q = \begin{bmatrix} e & 3 & \theta & \theta & 9 & 6 & \theta & \theta & \theta & \theta \\ 3 & \theta & 8 & \theta & \theta & 4 & 9 & \theta & \theta & \theta \\ \theta & 8 & \theta & 4 & \theta & 9 & 4 & 3 & \theta & \theta \\ \theta & \theta & 4 & \theta & 3 & 1 & \theta & \theta & 1 & \theta \\ 9 & \theta & \theta & 3 & \theta & 9 & \theta & \theta & \theta & \theta \\ 6 & 4 & 9 & 1 & 9 & \theta & \theta & \theta & \theta & \theta \\ \theta & 9 & 4 & \theta & \theta & \theta & \theta & \theta & \theta & \theta \\ \theta & \theta & 3 & \theta & \theta & \theta & \theta & \theta & 3 & 3 \\ \theta & \theta & \theta & 1 & \theta & \theta & \theta & 3 & \theta & \theta \\ \theta & \theta & \theta & \theta & \theta & \theta & \theta & 3 & \theta & \theta \end{bmatrix} \tag{12.30}$$

In the simulation presented in Table 12.3 where the initial states are $x(0) = [0\ 4\ 7\ 4\ 9\ 8\ 7\ 8\ 3\ 9]^T$, the graph topology has changed from G_p to G_q in the fifth iteration, that is, for $k = 5$, $t_2 = \delta 5$, then $s[5] = q$. The most vulnerable stepping-stone path for all the sources (shortest path) has been reached after nine iterations; for example, the most vulnerable stepping-stone path from node 10 to

Table 12.3 Stepping-stone cost evolution from every source; notice that after the double vertical line, when $k = 5$, the graph topology has changed, and the stepping-stone cost has been recalculated with the new information.

		Stepping-stone cost/iteration								
Source	$x2$	3	3	3	3	3	3	3	3	3
	$x3$	44	8	6	6	9	9	11	11	11
	$x4$	31	5	5	5	5	7	7	7	7
	$x5$	2	2	2	2	8	8	9	9	9
	$x6$	6	4	4	4	6	6	6	6	6
	$x7$	42	5	5	5	10	12	12	12	12
	$x8$	33	44	11	9	9	9	9	11	11
	$x9$	41	32	6	6	6	6	8	8	8
	$x10$	83	36	47	14	12	12	12	12	14

node target 1 has cost 14 and is unique: $10 \to 8 \to 9 \to 4 \to 6 \to 1$. The most vulnerable stepping-stone path from node 6 to node target 1 has cost 6 and is unique: $6 \to 1$.

12.5 Discussion

If r is the minimum cost from node i to node j, and s is the minimum cost from node i to node l with $r < s$, then equations (Eq. (12.18)) that are used in the calculation of the shortest path provide r and the paths themselves are calculated according to Remark 12.1, which means that the most probable stepping-stone attack from the source node i is toward the target node j.

In the model, a path between every source and any target has been assumed, but also, if there is a source without a path to any target, then the equations (Eq. (12.18)) converge to the minimum path for all the other sources.

Theorem 12.4 provides a time interval $\tau = n - 1$ of instant communication between two consecutive switching signals that ensures the convergence of the system to the minimum path. Notice that the convergence to the minimum path is by Corollary 12.1 $k \leq n - 1$ instant communications, so the convergence to the minimum path could be observed and detected with the equilibrium condition for $k \leq n - 1$. As previously reported in Watanabe and Watanabe [2] and Bannister and Eppstein [5], the Bellman–Ford algorithm can be optimized by reducing the iteration in more than $n/2$. However, this analysis is out of the scope of this chapter and will be studied in future research.

12.6 Conclusions

In this chapter, the stepping-stone cost is modeled as a multi-agent dynamical system in a vulnerability graph with fixed and switching topology. A biased min-consensus protocol is used for distributed calculation of the shortest path, and because the network is monitored in discrete time, the model is discretized using min-plus algebra for modeling and analysis. Theorem 12.3 and Corollary 12.1 prove that stepping-stone dynamics in a vulnerability graph with fixed topology converge to the shortest path in finite time, given $k = n - 1$ instant communications. Theorem 12.4 provides a metric for the time interval between switching signals that guarantees convergence to the minimum path for switching topology. In future work, we will develop a metric for computing the minimum time interval between two switching signals that ensures convergence to the shortest path for the vulnerability graphs with switching topology. The biased min-consensus protocol used as a model for the stepping-stone cost dynamics is deterministic. We will develop a stochastic model to characterize the dynamics in vulnerability graphs.

Acknowledgment

This material is based upon work supported by the Department of Energy under Award Number DE-OE0000780 and Office of the Assistant Secretary of Defense for Research and Engineering agreement FA8750-15-2-0120.

References

1 D. M. Nicol and V. Mallapura. "Modeling and analysis of stepping stone attacks." In: *Proceedings of the Winter Simulation Conference 2014*, pages 3036–3047, Dec 2014.

2 S. Watanabe and Y. Watanabe. "Min-plus algebra and networks." *Novel Development of Nonlinear Discrete Integrable Systems. RIMS Ko^kyu^roku Bessatsu B*, 47, 2014.

3 G. Cohen, F. Baccelli, G. J. Olsder, and J. P. Quadrat. *Synchronization and Linearity: An Algebra for Discrete Event Systems*. Wiley. http://www-rocq.inria.fr/metalau/cohen/DES/book-online.html, 2001.

4 J. Y. Yen. "An algorithm for finding shortest routes from all source nodes to a given destination in general networks." *Quarterly of Applied Mathematics*, **27**:526–530, 1970.

5 M. J. Bannister and D. Eppstein. "Randomized speedup of the Bellman-Ford algorithm." In: *Proceedings of the Meeting on Analytic Algorithmics and Combinatorics, ANALCO '12*, pages 41–47. Society for Industrial and Applied Mathematics, 2012.

6 R. Olfati-Saber and R. M. Murray. "Consensus problems in networks of agents with switching topology and time-delays." *IEEE Transactions on Automatic Control*, **49** (9):1520–1533, Sept 2004.

7 B. M. Nejad, S. A. Attia, and J. Raisch. "Max-consensus in a max-plus algebraic setting: the case of switching communication topologies." *IFAC Proceedings Volumes*, **43**(12):173–180, 2010. 10th IFAC Workshop on Discrete Event Systems.

8 H. Wang, Z. Chen, J. Zhao, X. Di, and D. Liu. "A Vulnerability Assessment Method in Industrial Internet of Things Based on Attack Graph and Maximum Flow," *IEEE Access*, vol. **6**, pp. 8599–8609, 2018.

9 Y. Zhang and S. Li. "Distributed biased min-consensus with applications to shortest path planning." *IEEE Transactions on Automatic Control*, **62**(10):5429–5436, Oct 2017.

10 R. Bellman. "On a routing problem." *Quarterly of Applied Mathematics*, **16**(1):87–90, 1958.

13

Anomaly Behavior Analysis of IoT Protocols

Pratik Satam[1], Shalaka Satam[1], Salim Hariri[1], and Amany Alshawi[2]

[1] *Department of Electrical and Computer Engineering, University of Arizona, Tucson, AZ, USA*
[2] *National Center for Cyber Security Technology, King Abdulaziz City for Science and Technology, Riyadh, Saudi Arabia*

13.1 Introduction

Over the last few decades, the Internet has grown from a network that connected two research universities to a juggernaut that encompasses the whole world with over 4.1 billion users as of July 2018 with a growth rate of 1052% from 2000 to 2018 [1]. Modern Internet supports a wide variety of features and services like cloud computing and storage, social networking, content delivery services, blogs and social interactions, online banking and shopping, just to name a few. Deployment of these technologies has made the Internet a household commodity. Alongside the development of the Internet, technologies relating to sensors [2–5], wireless communications [6, 7], and mobile computing have seen an unprecedented growth [8], which has contributed to the introduction of the new paradigm: Internet of Things (IoT).

IoT is a paradigm that will enable all devices and objects to have the capability of being all connected, can sense their surroundings, communicate over multiple communication networks, and connect with other Internet-enabled devices and services [2–5, 8]. The core concept of IoT involves interfacing all devices with sensor data and consequently making the device to be "smart" by providing a wide range of intelligent services that cover all aspects of our life and economy. Smart devices are proliferating our daily lives rapidly, wherein wearable

Modeling and Design of Secure Internet of Things, First Edition. Edited by Charles A. Kamhoua, Laurent L. Njilla, Alexander Kott and Sachin Shetty.

Published 2020 by John Wiley & Sons, Inc.

Figure 13.1 Rise of wearables and future wearable technology.

technology (see Figure 13.1) like smart watches, shoes, glasses, and devices like smart phones, smart refrigerators, smart cars, etc., are becoming an integral part of our lives [9].

The core technologies of IoT are not new. Sensing technologies have been used in the past on factory manufacturing floors, for monitoring and tracking livestock, etc. The idea of machine-to-machine communication is also not new as that is the core concept of the Internet. IoT is an evolution in the use of these technologies in terms of number of devices, the type of devices, and the services provided via these devices [3]. IoT promises will lead to the development of a new generation of devices that will provide personalized services; services that are tailored and modified to each user's needs and demands.

For instance, let us consider smart speakers like Google Home [10] or Amazon Alexa [11] that are traditional speakers, with a processor (attached to a microphone) connected to the Internet (wired or wireless). These smart speakers interface with the user with voice and provide services like playing music of the Internet that is tailored to the music favored by the user. Smart devices like the apple watch and fitness trackers will soon find their way into the healthcare market, allowing doctors to monitor the health of their patients, perform predictive data analytics, and thus help save lives and significantly improve the quality of healthcare services.

The number of IoT smart objects is expected to reach 212 billion entities by the end of the year 2020 (see Figure 13.2) [12]. This growth will constitute an

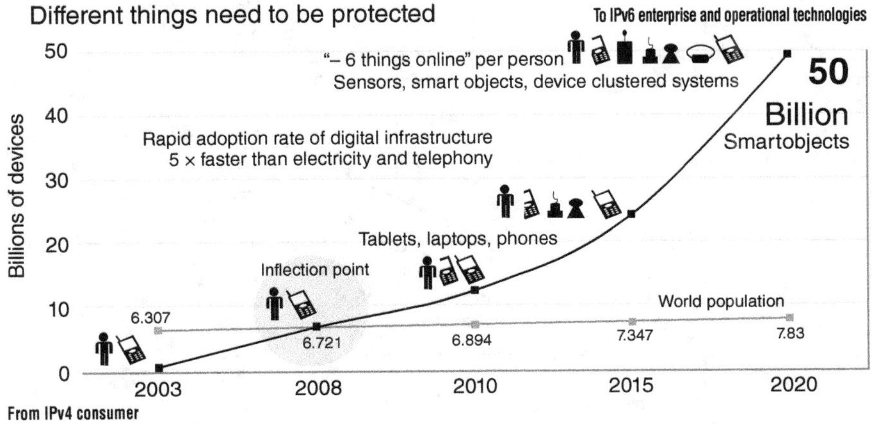

Figure 13.2 Growth of IoT by 2020.

unprecedented increase in Internet traffic and wireless network traffic. It is predicted that IoT will impact the economy in the range of $3.9 trillion to $11.1 trillion by 2025 [13]. All these statistical observations point to a fast-paced growth of the IoT industry till 2025, which will result in ubiquitous presence of smart devices that will be monitoring their users in different capacities, sharing data with each other and with services on the Internet. The downside to this fast-paced growth in the IoT industry is the fact that the vulnerability of IoT devices has grown exponentially and made their security a major research challenge.

Cyber attackers can compromise IoT devices to execute attacks to hinder the operations of large computer networks, factories, nuclear power plants, destroy global financial systems, power grids, water distribution systems, etc. To make the situation worth, we are observing an alarming trend of increasing attack sophistication while reducing the knowledge required to launch sophisticated attacks as shown in Figure 13.3.

Furthermore, the attack propagation time used to range in weeks in the 1980s now ranges to a fraction of a second to reach sizable Internet-enabled devices [14] as shown in Figure 13.4. For instance, in May 2017, WannaCry attack was able to target over 200 000 computers in 150 countries in four days, propagating at the speed of 3600 computers per hour [15]. This attack would have had targeted more systems at a faster rate if aimed at the deployed IoT devices and sensors.

To address the security challenges associated with IoT infrastructure and services, we need a paradigm shift in how we monitor, analyze, and protect their operations and services.

The aim of this chapter is to discuss a methodology to address the security challenges of IoT infrastructure. The organization of this chapter is as follows.

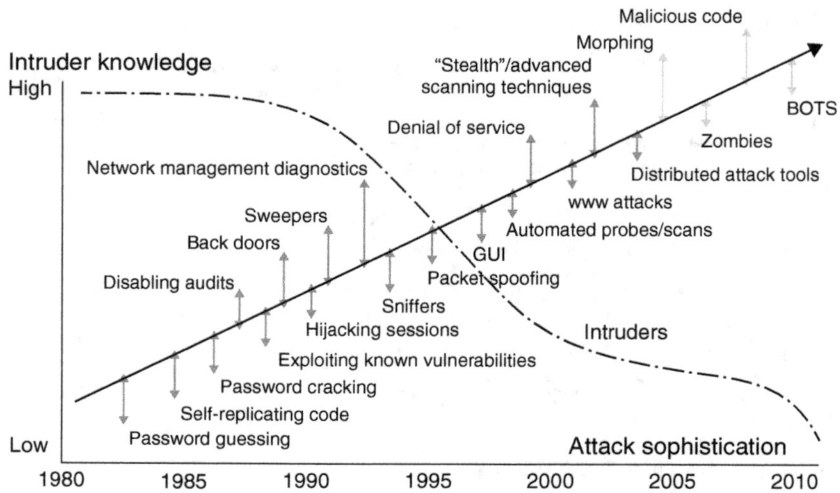

Figure 13.3 Attack sophistication versus intruder technical knowledge.

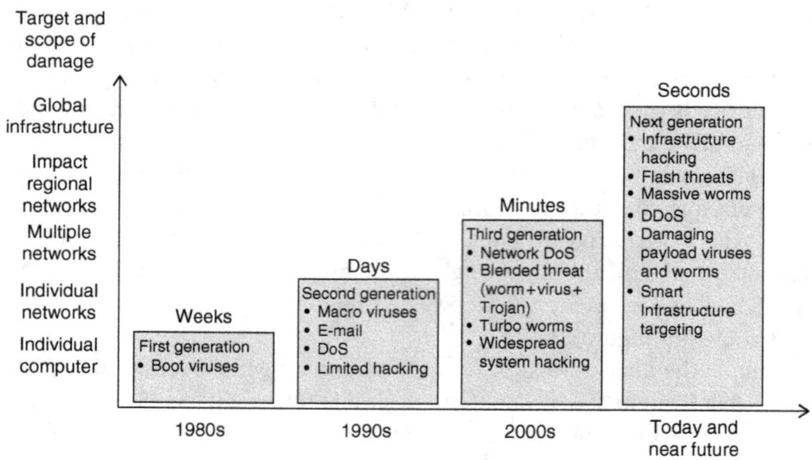

Figure 13.4 The timeline of cyber-threads scope of damage and impact time.

In Section 13.2, we introduce the IoT architecture. In Section 13.3, we introduce a general methodology for IoT threat modeling. In Section 13.4, we apply the IoT threat modeling concepts to different protocols. In Section 13.5, we present a conclusion and future work.

13.2 IoT Architecture

In this section, we present our IoT architecture that will be used to effectively model the behavior of any cyber-physical systems. We then introduce IoT threat modeling methodology to perform systematic threat modeling analysis of the cyber-physical system followed by some examples of the threat modeling analysis.

The IoT architecture [16] helps model the IoT/cyber-physical devices and helps understand the behavior of the IoT device, model its strength and weaknesses [17, 18]. As shown in Figure 13.5, the IoT modeling architecture consists of four layers: end devices, communications, services, and applications. Cyberattacks can be launched against the functions and services at each layer.

The first layer (bottom layer) is composed of end devices and controllers that control these end devices. Devices like RFID tags, sensors like temperature sensors and actuators interact with the IoT users, collect data, and pass them to the processing services or present the results of the data processed [19, 20]. The second layer (communications layer) is responsible for providing the required connectivity and communications among all the sensors and actuators associated with IoT services or applications [20, 21]. The communication technologies that can be used include Internet, satellite, mobile cellular networks, wireless sensor networks, and internal

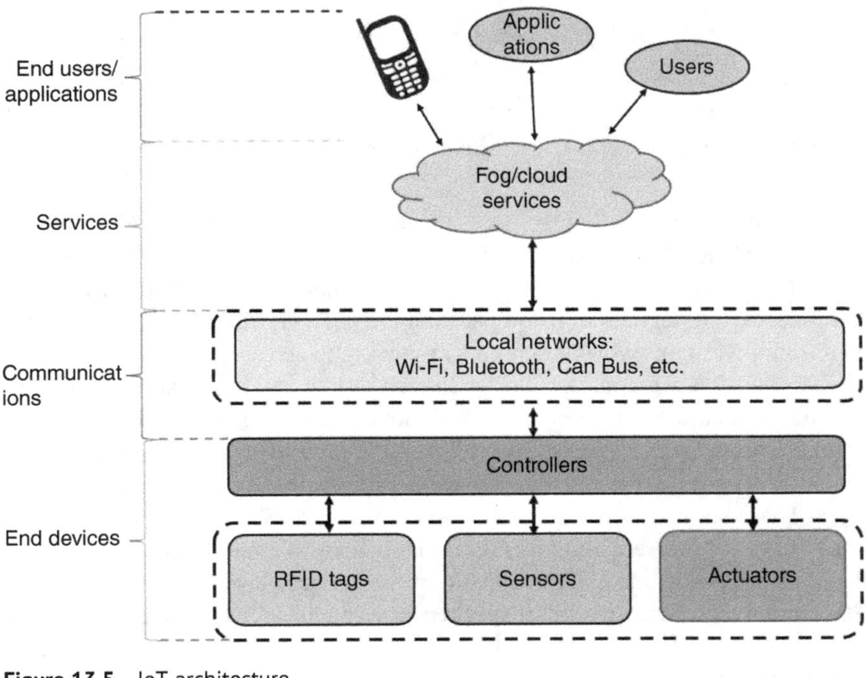

Figure 13.5 IoT architecture.

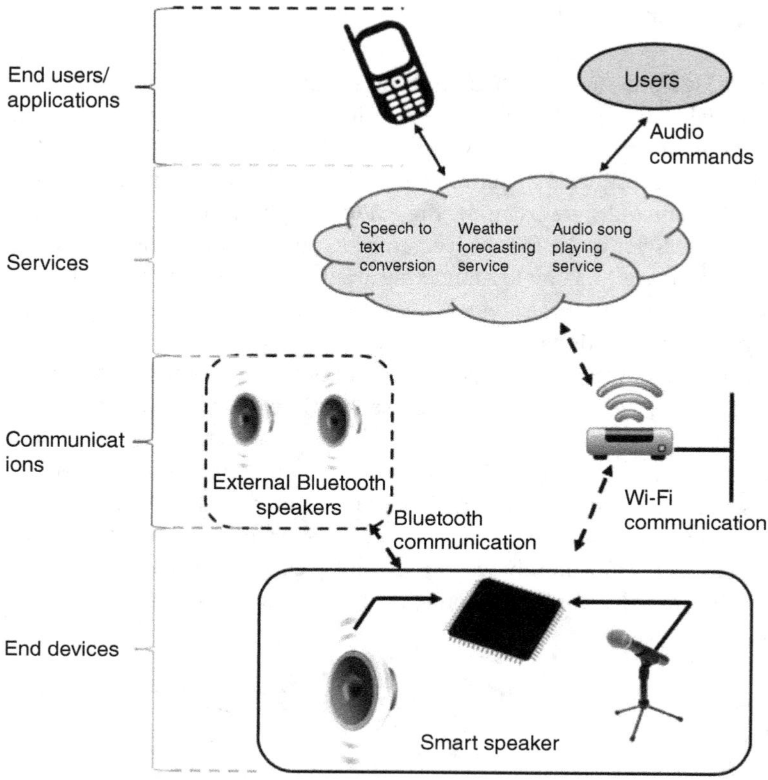

Figure 13.6 Application of IoT architecture to a Smart speaker.

car network infrastructures (e.g. CAN, Bluetooth, I2C, and MOST). Various functionalities of the IoT devices are provided through the services layer, i.e. the third layer. The services layer provides common middleware and functions to build sophisticated IoT services in the application layer. The applications layer provides end-user applications which are used by the users to access the IoT devices.

In Figure 13.6, we apply our IoT architecture to a smart speaker system. The speakers, onboard controllers, and microphones constitute the End-devices layer of the IoT framework. These speakers use Wi-Fi, Bluetooth, and Ethernet to communicate with other devices and web services, constituting the communications layer. These speakers use cloud-based speech to text (STT) conversion services [22–24], weather forecasting services, song-playing platforms like YouTube or Pandora or Spotify, etc., which constitute the services layer of the framework. All commercially available smart speakers have applications that allow the users to control these devices through their smart phones, and other smart devices which constitute the services layer.

13.3 IoT Threat Modeling Methodology

Figure 13.7 shows our IoT threat modeling framework, which helps analyze the vulnerabilities in cyber-physical systems like IoT devices and help develop methodologies to remove or mitigate the impact of these vulnerabilities if they were exploited by attackers in the cyber-physical systems. As shown in Figure 13.7, the IoT threat modeling consists of four layers and each layer has the following five tasks: modeling the target layer's function, investigating the attack surfaces (ASs) for the model, investigating the targets of the AS, investigating the impact of the attack if executed, and investigation of the attack mitigation strategies. For each layer, we first develop a model that captures the functions to be provided by that layer. The layer model will then be analyzed to identify the AS that characterizes the entry points that can be exploited by attackers to inject malicious events or behaviors into that layer functions. For each vulnerability or AS entry point, we then identify the potential target(s) that can be exploited by attackers, and the risk and impact if the attack was successful. The final step in each row of the IoT threat modeling framework is to develop methods to mitigate or eliminate these vulnerabilities in order to achieve the required secure operations of the functions provided by each layer.

13.3.1 AS Identification, Impact Analysis, and Mitigation Strategies

After modeling the operations of a given layer, we identify the ASs for each layer model, perform risk and impact analysis, and then develop mitigation methods to protect against each detected vulnerability. This analysis helps identify the imminent threats to the IoT system, and help securing the system in a systematic

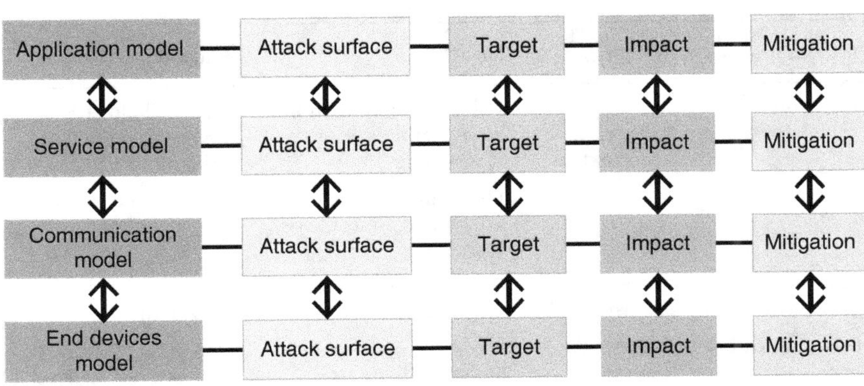

Figure 13.7 IoT threat modeling framework.

fashion. In the subsequent subsections, we perform AS identification, impact analysis, and mitigation methods for the IoT system shown in Figure 13.6.

13.3.1.1 End-Devices Layer

Table 13.1 shows the AS, impact, mitigation, and mitigation mechanisms associated with the end-devices layer.

13.3.1.2 Communication Layer

Table 13.2 shows the AS, target, impact, and mitigation mechanisms associated with the communication layer.

Table 13.1 End nodes layer.

Attack surface	Target	Impact	Mitigation mechanism
Controllers	Control, information	Lost control, human life safety, failures	Anomaly behavior analysis (ABA) to detect abnormal control information, encryption
Sensors	Information, access to the system	Lost control, human life safety, failures	Lightweight encryption, ABA, secure sensor identification and authentication
Actuators	Control	Lost control, human life safety, failures	Lightweight encryption, ABA, anti-jamming
Entertainment	Access to the system	Long time, waste energy, cyber crime	Encryption, moving target defense, ABA

Table 13.2 Communications plane.

Attack surface	Target	Impact	Mitigation mechanism
Protocols	Access control, information	Lost control, human life safety, failures, money, longer time	Authentication, ABA, moving target defense, anti-jamming
Firewalls	Access control to the system	Lost control, human life safety, long time, energy waste	IDS, behavior analysis, authentication
Routers	Access, information	Control, human life safety, time	ABA, anti-jamming, encryption
Communication bus	Information, control	Privacy, money, human life safety, long time, failures	Encryption, IDS, moving target defense, ABA

Table 13.3 Services layer.

Attack surface	Target	Impact	Mitigation mechanism
Cloud storage	Personal and confidential information	Data loss, money, long time, safety	Encryption, ABA, moving target defense, behavior analysis, selective disclosure, data distortion, big data analysis
Web services	Control, monitor	Control, human life safety, money, cyber crime	Authentication, secure identity management, ABA, encryption

Table 13.4 Applications layer.

Attack surface	Target	Impact	Mitigation mechanism
Mobile devices (cellphone, tablets)	Information, control	Human life safety, personal information, money	Authentication, access control, ABA, moving target defense
Programs/ applications	Access to the system, control, information	Time, money, safety, reputation	ABA, encryption, white listing, continuous authentication

13.3.1.3 Services Layer

Table 13.3 shows the AS, target, impact, and mitigation mechanisms associated with the services layer.

13.3.1.4 End Users/Applications Layer

Table 13.4 shows the AS, target, impact, and mitigation mechanisms associated with the end users/applications layer.

13.4 Application of the Threat Modeling Concepts to Different Protocols

In the following section, we apply the IoT threat modeling methodology to secure IoT protocols like Wi-Fi protocol and Bluetooth protocol.

13.4.1 Wi-Fi Protocol

Wi-Fi protocol also known as the IEEE 802.11 [25–30] is a wireless local area network protocol. It is the Ethernet's equivalent for wireless networks. The protocol was first formalized in the year 1997. This protocol has been upgraded and has

Table 13.5 IEEE 802.11 physical layer standards.

Release date	Standard	Frequency band	Bandwidth	Modulation	Data rate
1997	802.11	2.4 GHz	20 MHz	DSSS, FHSS	2 Mbps
1999	802.11b	2.4 GHz	20 MHz	DSSS	11 Mbps
1999	802.11a	5 GHz	20 MHz	OFDM	54 Mbps
2003	802.11g	2.4 GHz	20 MHz	DSSS, OFDM	542 Mbps
2009	802.11n	2.4 GHz,	20 MHz, 40 MHz	OFDM	600 Mbps
2013	802.11ac	5 GHz	40 MHz, 80 MHz, 160 MHz	OFDM	6.93 Gbps

faced many changes over the years of its existence. But most of these upgrades have been to enhance the data rate and the link quality of the network and little has been done to improve the security of the network.

The Wi-Fi protocol [25] operates in the 2.4 GHz UHF and the 5 GHz SHF bands; both of them fall under the category of ISM bands and hence have been sanctioned for unlicensed use. The original Wi-Fi protocol that was declared in the year 1997 specified bit rates of 1 or 2 Mbits/s while specifying three alternate physical layer configurations which were diffuse infrared at 1 Mbps, frequency-hopping spread spectrum (FHSS) at 1or 2 Mbps, and direct-sequence spread spectrum (DSSS) at 1or 2 Mbps.

- *IEEE 802.11b:*
  This specification of the protocol [27] supports a maximum data rate of 11 Mbps using the same physical specifications as the original Wi-Fi protocol. This protocol supported the use of the DSSS in the 2.4 GHz UHF ISM band. Carrier Sense Multiple Access/Collision Avoidance (CSMA/CA) is the protocol that is used to manage the media access.
- *IEEE 802.11a:*
  This standard was added to the original 802.11 protocol in the year 1999 [26]. This standard specified the operations of the Wi-Fi in the 5 GHz SHF supporting 52 subcarrier orthogonal frequency-division multiplexing (OFDM) supporting data rates up to 54 Mbps.
- *IEEE 802.11g:*
  This standard was ratified to the original 802.11 protocol in the year 2003 [29]. This standard increased the data rate in the 2.4 GHz UHF ISM band to 54 Mbps. This was done by the adoption of the use of OFDM.
- *IEEE 802.11n:*
  This enhancement to the 802.11 protocol enhanced the data rates up to 600 Mbps [29]. This increase in data rate is achieved by the use of multiple data

streams with channel widths of 40 MHz. It can operate in the 2.4 GHz UHF ISM band and the 5 GHz SHF band. Moreover, one of the enhancements that have been added includes the use of multiple antennas that allows seamless maintenance of simultaneous data streams.

- *IEEE 802.11ac:*
  This standard was approved in the year 2014 [30]. This protocol processes that each single link to have a throughput of 500 Mbps while overall throughput for a multi-station WLAN would be at least 1 Gbps. This increase in throughput is achieved by the use of wider channels, and up to 8 MIMO spatial streams and use of 256 QAM.

Figure 13.8 shows a Wi-Fi frame, which consists of a Preamble, Wi-Fi Header, and Data. The preamble is the first set of bit sequence that follows the actual header. The preamble marks the start of the frame. The preamble allows the receiver frequency synchronization and receiver time synchronization. This helps the receiver extract the header and the data from the modulated signals that are sent over the channel. The Wi-Fi Header is made up of a 2 byte frame control field, a 2 byte duration ID, 48 bit long address fields which include the source and the destination address fields. The structure of the Wi-Fi frame header is as shown in Figure 13.8. When security is enabled in the Wi-Fi protocol, the data in the Wi-Fi frame are encrypted and transmitted over the wireless channels. The preamble and

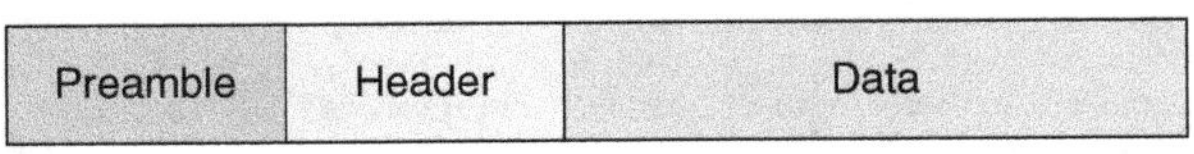

Figure 13.8 Wi-Fi frame structure.

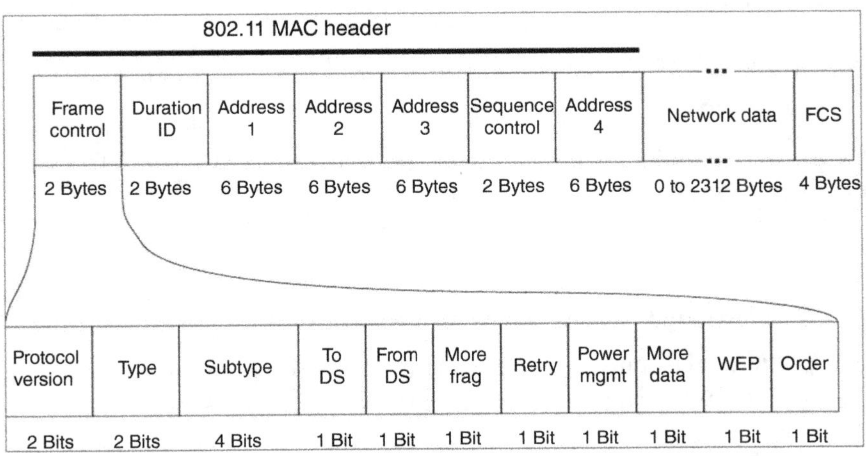

Figure 13.9 Wi-Fi header.

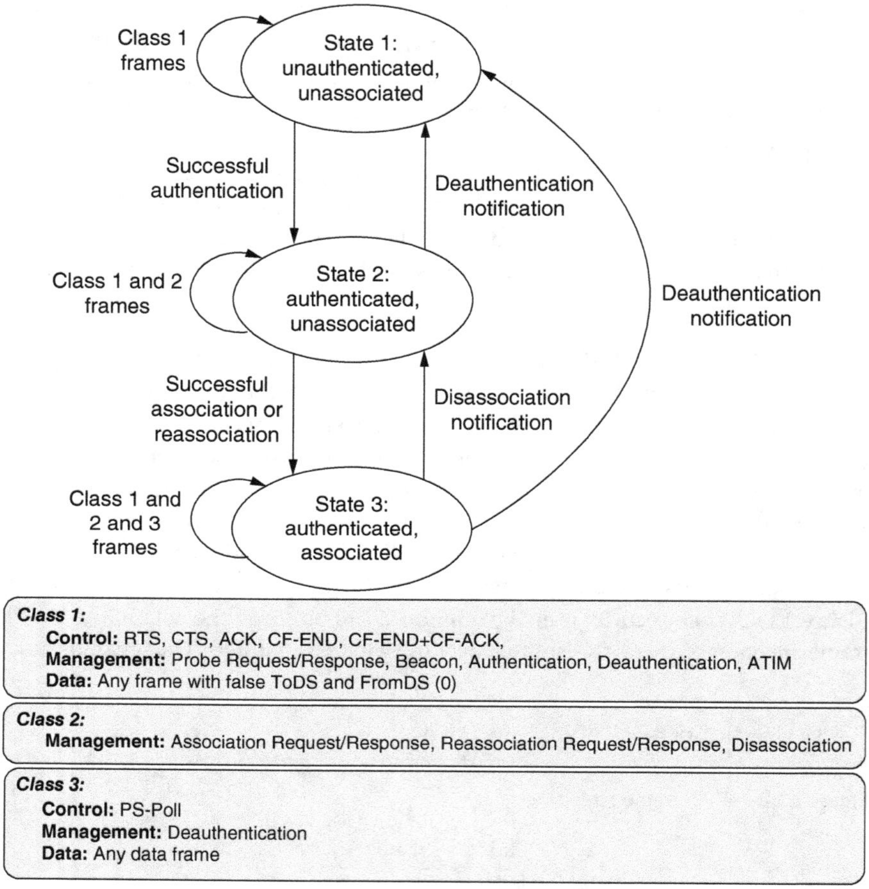

Figure 13.10 Wi-Fi protocol state machine.

the header of the Wi-Fi frame goes unencrypted and are exploited by attackers to execute attacks on the Wi-Fi protocol, leaving it vulnerable to data-link layer attacks (Figure 13.9).

13.4.2 Modeling the Data-Link Layer Behavior of the Wi-Fi Protocol

The data-link layer of the Wi-Fi protocol can be modeled by using the Wi-Fi state machine. The Wi-Fi protocol is a stateful protocol and follows the state machine shown in Figure 13.10.

As shown by the Wi-Fi state diagram or Wi-Fi protocol model in Figure 13.10, Wi-Fi protocol has three states. A device is in state 1 or unauthenticated and unassociated when it is not connected to the Wi-Fi network. The device transitions from state 1 to state 2 on successful authentication. In state 2, the device is authenticated to the network but is not associated to a particular access point. On successful association, the device transitions to state 3, i.e. the authenticated and associated state. In state 3, the device exchanges data frames with the Wi-Fi network. When a Wi-Fi device switches to another access point in the same Wi-Fi network, it transitions from state (authenticated and associated state) to state 2 (authenticated and unassociated state) for that access point and then transitions from state 2 to state 3 for the new access point. On disconnections from the Wi-Fi network, the device transitions to state1 and is no longer a part of the Wi-Fi network. Thus, the state machine in Figure 13.10 helps us effectively model the normal behavior of the Wi-Fi protocol.

13.4.2.1 Identification of ASs, Impact, and Mitigation Analysis for the Wi-Fi Protocol

Figure 13.10 highlights the state machine used by the Wi-Fi protocol. A careful analysis of this state machine also highlights the different ASs that can be used to target the Wi-Fi protocol. Wi-Fi protocol is vulnerable to jamming attacks. Such jamming attacks pose a threat to all wireless protocols and can be easily detected by systems monitoring signal strengths in frequencies. Attacks on the data-link layer of the Wi-Fi protocol performed through frame spoofing pose a bigger threat to the Wi-Fi protocol. These attacks could be used to perform de-authentication attacks, man in the middle attacks, denial of service (DoS) attacks, Fake authentication attacks, etc. The above-mentioned ASs exist on any device using Wi-Fi networks including IoT devices.

Notwithstanding the above-mentioned attack vectors, a particular IoT device might have possible attack vectors as a result of bad system design. For instance, Sivaraman et al. [31] in their work have identified IoT devices like the Withings Baby Monitor which allows the user to use an IP camera to monitor his/her baby, exchanges image data over Wi-Fi in plain text, allowing the researchers to execute a man in the middle attack on the baby monitor; in Withings Smart Body Analyzer, a weighing scale to measure fat, heart rate, and BMI, the researchers observed that the user's personal information was exchanged on Wi-Fi channels unencrypted. These attack vectors are a result of bad design or low computation power of the IoT device add on to the attack vectors already working on the Wi-Fi protocol and make it easier for the attacker to execute attacks like man in the middle attacks.

In the next section, we present an anomaly behavior analysis-based intrusion detection system (ABA-IDS) that is able to detect attacks targeting the different attack vectors for the Wi-Fi protocol.

13.4.3 Wi-Fi IDS to Detect Attacks on the Wi-Fi Protocol

Depending on the detection mechanism used in an IDS, they can be classified into two types: signature-based IDS and anomaly detection. Signature-based IDS rely on the security expert's knowledge on system attacks and attack signatures provided by these experts. As signature-based IDS' monitor the target system for these attack signatures, they are easy to design and develop but are unable to detect new or modified attacks. Signature-based systems depend on regular updates to the signature data set as some of the archaic rules might not be valid in changing network environment.

Anomaly-based IDS model the normal behavior of the system using normal models. This allows them to detect any transition from the normal behavior and thus detect attacks. Anomaly-based IDS are able to detect new and modified attacks and do not have to depend on regular updates from domain knowledge experts. But anomaly-based IDS are hard to design and generally have high false positive rates.

In this section, we present an anomaly IDS for the Wi-Fi networks. The IDS uses machine learning to model the normal behavior of the Wi-Fi network.

13.4.3.1 Anomaly Behavior Analysis

The normal operations of a system can be represented by an n-dimensional data structure as shown in Figure 13.11. As long as the system is performing normally, its operation point is confined to the normal operation regions. When an anomalous event occurs, the operation point of the system moves outside the cube and is detected by the ABA module and countermeasures are taken to bring the

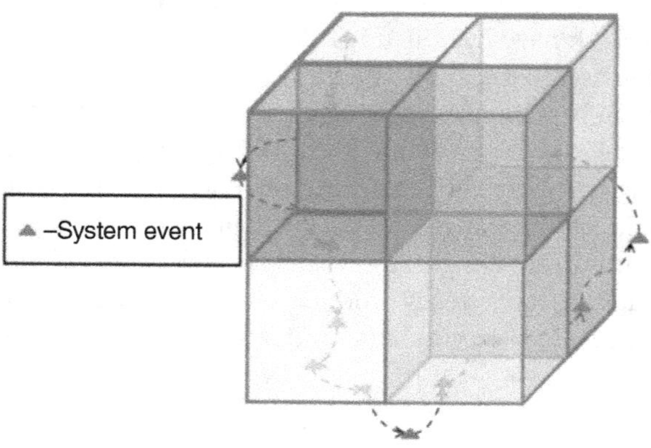

Figure 13.11 Anomaly behavior analysis.

operational point back inside the cube. We apply the concept of ABA analysis to detect intrusions on computer networks or attacks on different protocols [16, 32–37].

Our methodology can be summarized in the following steps:

- Step 1: Collecting Wi-Fi traffic when the network is behaving normally
  This step helps map the complete normal event space for the Wi-Fi protocol. Wi-Fi traffic is collected by monitoring the operational Wi-Fi networks for long periods of time to obtain complete normal event set. The Wi-Fi traffic is captured in the form of Wi-Fi header and data shown in Figure 13.9. Wi-Fi data are in plain text or encrypted depending on the network being monitored. The Wi-Fi header is always in plain text and is used for characterizing the normal event space.
- Step 2: Mapping the observed Wi-Fi traffic into a representation for analysis
  It is difficult to characterize the normal behavior of the Wi-Fi protocol from the raw Wi-Fi traffic collected in step 1. In step 2 are represented data structures like n-grams and observation-flow to help model the normal behavior of the Wi-Fi protocol.
- Step 2.1: Feature Extraction
  While designing and implementing an ABA analysis system, it is important to identify the operational point of the system. The process of identification of the operation point of the Wi-Fi network begins with collecting Wi-Fi frames in the network. Literature review and study of the Wi-Fi protocol characterized by Figure 13.10 allowed us to extract a feature set that helped us characterize the operational point of the Wi-Fi protocol. This feature set is listed in Figure 13.12.
- Step 2.2: Mapping of Wi-Fi traffic into observation-flow and n-grams
  A fingerprinting data structure is needed to capture the state machine transitions made by the protocol. Wireless n-grams (Wgram) and observation-flow are two data structures that effectively fingerprint the behavior of the W-Fi protocol.
  a) Observation-flow
     It is a continuous flow of frames or packets between a source–destination pair sampled at specific intervals of time "t." The observation-flow characterizes

Figure 13.12 Feature set.

Sr. no.	Features	Description
1.	frame_epoch_time	Epoch time
2.	frame_src_address	Frame source time
3.	frame_dst_address	Frame destination time
4.	frame_type	Frame type
5.	frame_sub_type	Frame sub-type

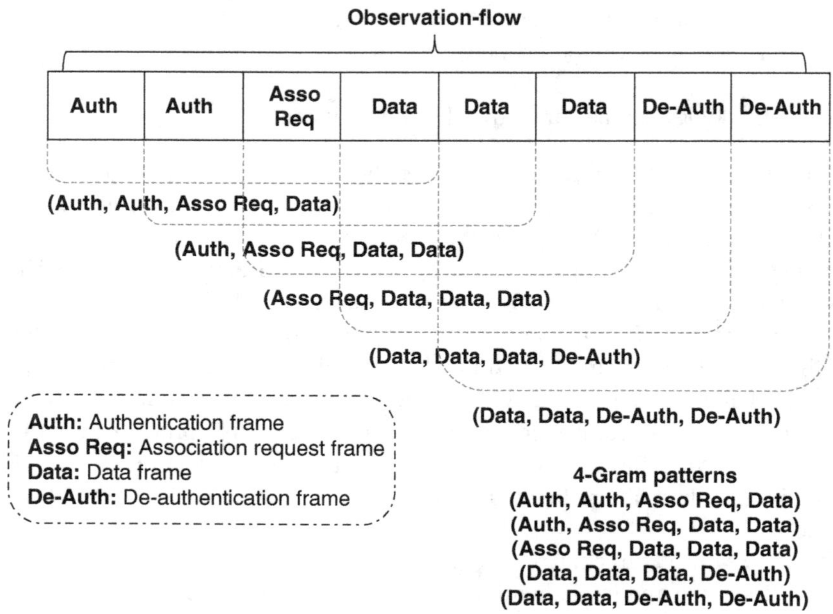

Figure 13.13 Wireless n-gram (W-gram).

the state transitions made by the protocol. Figure 13.12 shows a simplified example of an observation-flow for the Wi-Fi protocol. The flow begins with receiving Auth frame and ends with De-Auth frame.

b) Wireless n-gram (Wgram):
An n-gram is a sliding window of a pre-defined size, sampled from the observation-flow as shown in Figure 13.13. The n-grams are used to model the temporal behavior of the target protocol, in this case the Wi-Fi protocol as shown in the Figure 13.13. {Auth, Auth, Asso Req, Data} form an n-gram of size 4. An n-gram can be of any size and n-gram size analysis experiment has to be performed to determine the size of the n-gram to be adopted for use in an IDS for a protocol.

Figure 13.14 shows how the feature set is converted into a flow of n-grams. As shown in the diagram, the features frame_type and frame_sub_type are hashed using a hashing function. For a source–destination address pair in the same master–slave role, the hashed characteristics are used to form the n-grams, while the data observed into "t" second windows according to frame_epoch_time to form observation-flow.

- Step 3: Building models to characterize the normal behavior of Wi-Fi protocol
We train the IDS on the normal operations of the Distributed Wi-Fi network in

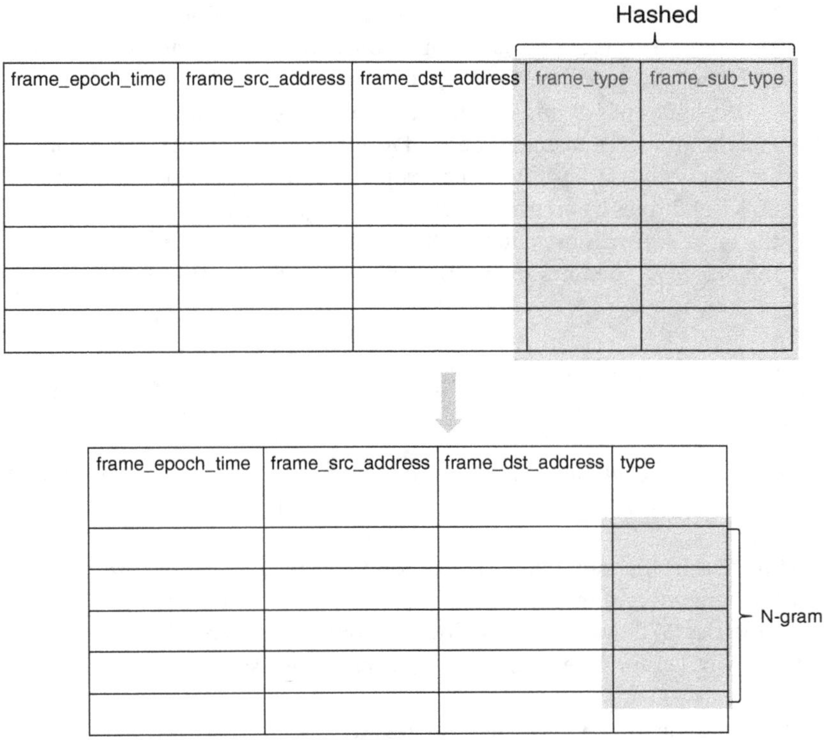

Figure 13.14 Feature set to fingerprinting data-structure mapping.

order to differentiate between the normal and the abnormal operations. The training takes place in two stages.

a) Stage 1

In this stage, we monitor the network to obtain all the possible ***n-grams*** in the Wi-Fi network. This stage is used to build two databases. The first database is for all the normal ***n-grams*** in the network. To ensure that attack traffic is not passed as normal traffic while building the database, the network is carefully monitored to filter out any known attack traffic.

The second database is for all then-grams obtained from known attack libraries. This approach of having partial attack traffic helps us better shape the event space to differentiate between normal and abnormal traffic.

In addition to storing the n-grams that characterize the normal and abnormal behavior of the Wi-Fi frames, we also store information about:

1) Occurrence Rate (OR): It measures the ratio of the number of occurrences of the ***n-gram*** during the training to the total number of ***n-grams*** obtained during the training.

2) Modulo of total n-grams (MG): It measures the modulo (Remainder of Division) of number of occurrences of the ***n-gram*** during the training with the total number of ***n-grams***.

   The values of OR and MG for each of the ***n-gram*** in both the normal and the attack database can be used to accurately characterize the normal and abnormal traffic of Wi-Fi network. Let us denote receiving an ***n-gram*** of Wi-Fi frames by an event $\boldsymbol{\varepsilon_i}$ where the ***n-gram*** is **the ith *n-gram***, and the set of ***n-grams*** used during the training stage is denoted as **E** wherein the *Size of (**E**)* denotes the number of ***n-grams*** present in the set **E.**

   Then, $\boldsymbol{\varepsilon_i \in \mathbf{E}}$

$$OR_i = \frac{\text{Number of times}\,\boldsymbol{\varepsilon_i}\,\text{occurs}}{\text{Size of}(\boldsymbol{E})} \tag{13.1}$$

$$MG_i = \text{Number of times}\,\boldsymbol{\varepsilon_i}\,\text{occurs}\ \%\ \text{Size of}\ (\boldsymbol{E}) \tag{13.2}$$

b) Stage 2

In this stage of the training, the network is monitored first with normal traffic in the network and then with abnormal traffic in the network to calculate four metrics (Wflow_1, Wflow_2, Wflow_3, Wflow_4) for each of the wireless flows (***Wflows***) that are observed. The four wireless flowscores are computed as follows:

If we assume that a particular wireless flow has n ***n-grams.***

Then:

$$\text{Wflow}_1 = \left(\sum_{i=1}^{i=n} OR_i\right)_{\text{normal}} \tag{13.3}$$

$$\text{Wflow}_2 = \left(\sum_{i=1}^{i=n} MG_i\right)_{\text{normal}} \tag{13.4}$$

$$\text{Wflow}_3 = \left(\sum_{i=1}^{i=n} OR_i\right)_{\text{attack}} \tag{13.5}$$

$$\text{Wflow}_4 = \left(\sum_{i=1}^{i=n} MG_i\right)_{\text{attack}} \tag{13.6}$$

where the values of the *OR* and MG for each ***n-gram*** are obtained from the normal and abnormal ***n-gram*** databases obtained in Stage 1.

This obtained set of four metrics for each of the ***Wflows*** is stored in a database called ***flow_score*** database with a binary entry added in front of each flow marking if the traffic that was observed was normal or abnormal. The abnormal traffic is generated as mentioned earlier by the use of known attack libraries.

Machine learning models are trained on the ***flow_score*** to obtain the conditions in which the four Wflow metrics represent normal and abnormal traffic. In the system that was developed as a proof of concept, a simple Conjunctive Rule algorithm was used to get the conditions of abnormal operations. This algorithm learns a single rule that predicts the nominal class value. In this algorithm, the uncovered test instances are assigned the default class, and the knowledge gain on occurrence of each of the antecedent is computed. The rule pruning is done by the use of reduced error pruning.

These machine learning models are used to detect abnormal wireless traffic during the operational phase of the Wi-Fi IDS. The IDS classify the different types of attacks on the Wi-Fi network, by counting the different types of frames that are present in the anomalous wireless flow and associating the type of frame with the type of attack.

Figure 13.15 shows the general architecture of the Wi-Fi IDS designed using the described methodology. As shown in Figure 13.15, the Wi-Fi IDS has two main modules: Sniffer Module and the Behavior Authentication Module. The

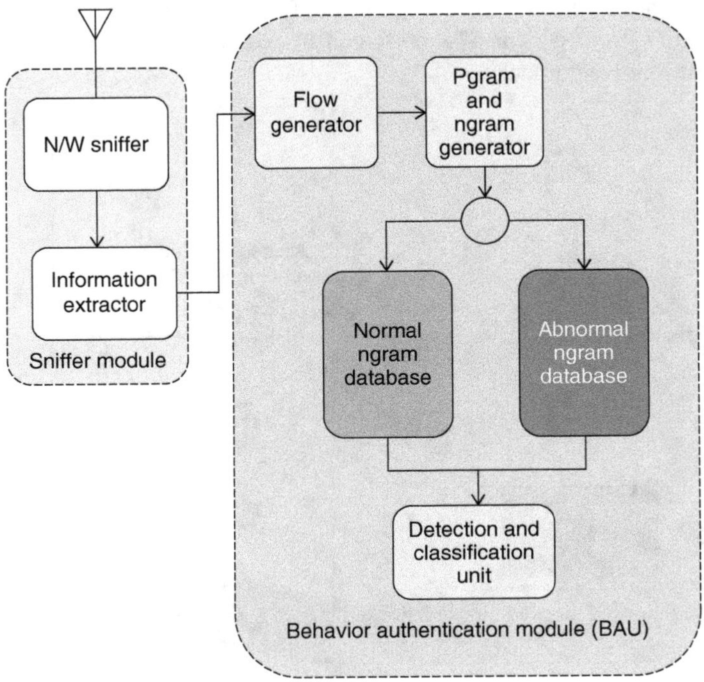

Figure 13.15 IEEE 802.11 behavior analysis module.

Sniffer Module monitors the Wi-Fi networks and collects the Wi-Fi frames. The Behavior Authentication Module generates the ***W-grams***, observation-flow, and performs behavior analysis to detect and classify Wi-Fi attacks.

13.4.4 Experimental Results and Evaluation

Figure 13.16 shows the Wi-Fi testbed used in the Autonomic Computing Lab (ACL) at the University of Arizona. The Wi-Fi access point is a Linux machine that is running the Wi-Fi IDS. Training was performed in the ACL at the University of Arizona by monitoring the supervised traffic on channel 1. The traffic is supervised to ensure that no attack traffic enters the training stream.

13.4.4.1 Wi-Fi Attacks

Wireless networks are prone to attacks as the attacker has an easy access to the medium of communications. Most wireless communications including Wi-Fi protocol are prone to jamming and interference attacks. Wi-Fi protocol is also susceptible to link-layer attacks, wherein an attacker can spoof specific Wi-Fi frames to disrupt the normal functioning of the Wi-Fi protocol, irrespective of employment of encryption schemes prescribed by the protocol. Table 13.6 lists some such attacks on the Wi-Fi protocol. The IDS presented in Section 13.4.3.1 has been designed to detect link-layer attacks.

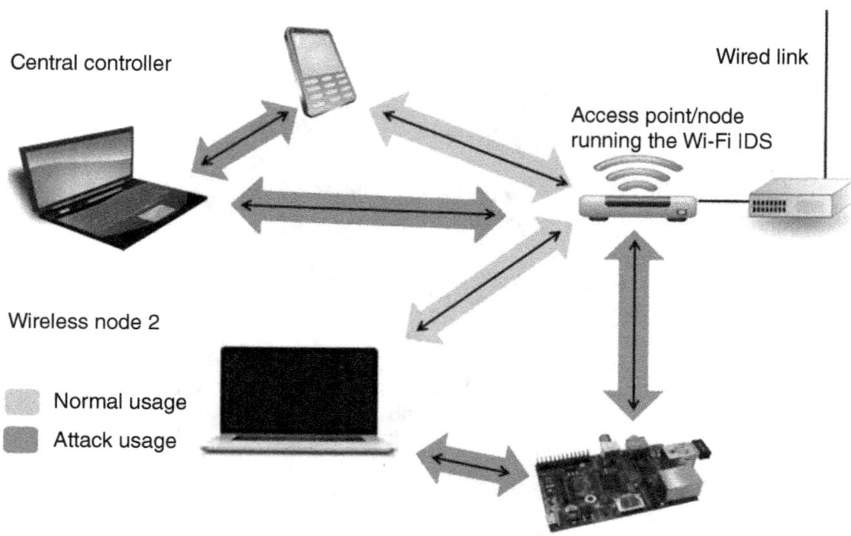

Figure 13.16 Wi-Fi testbed.

Table 13.6 Wireless attacks.

Attack type	Description
1. De-authentication attack	This attack involves the flooding of the target with de-authentication frames.
2. Fake authentication	This attack involves fake authenticating to an access point.
3. Minimal de-authentication	This attack is performed to de-authenticate a user from a network with just 15 frames.
4. Disassociation attack	This attack involves sending a disassociation frame to both the access point and the target.
5. Man in the middle attack	This attack results in a man in the middle on the layer two of the network.

13.4.4.2 Experimental Results

The performance of the IDS was measured through multiple experiments that tested the capability of the IDS to detect attacks accurately under different operating conditions.

- *Experiment 1*
  This experiment is designed to evaluate the percentage rise in the number of false positives given by the system due to the increase in the frame drop rate due to interference and wireless channel-related issues. On conclusion of this experiment, it was observed that no false positives were observed even at very high drop rates of 95–100%.
- *Experiment 2*
  This experiment has been designed to detect the percentage rise in the number of false positives given by the system as the frame drop rate of the system increases. The increase in the frame drop rate could be a result of interference and other channel noises. Thus, to simulate this interference, the received frames were randomly dropped by the packet sniffer unit. The percentage of the frames dropped was increased gradually over a period of time. It was observed that the IDS gives zero false positives till 80% frame drop rate. As the frame drop rate increases beyond 80–93%, there is a steady increase in the false positives till the system stops performing at 93% frame drop rate. This experiment shows that the designed IDS is resilient to signal interference and has low false negatives under normal Wi-Fi network operating conditions (Figure 13.17).
- *Experiment 3*
  In this experiment, the network monitored by the IDS was attacked with a modified de-authentication attack. This modified de-authentication attack had a

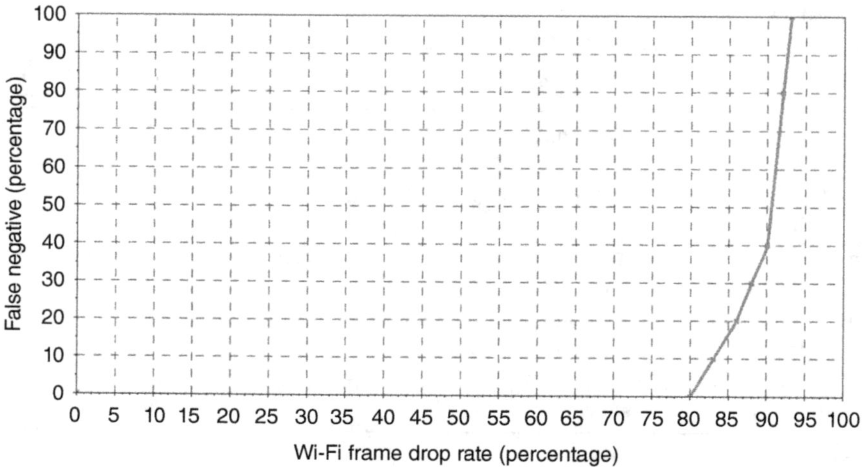

Figure 13.17 False negative versus frame drop rate.

very low foot print compared to the de-authentication attack available through the attack library aircrack-ng. This de-authentication attack was able to successful de-authenticate a user with seven to eight frames over a six second window. Our IDS was successfully able to detect this modified attack each time.

- *Experiment 4*

  This experiment demonstrates the differences between the four wireless flowscore values for the attacks listed in Table 13.6, which differ from normal WFlow values by a large margin. This large margin enables us to detect attacks accurately with low false alarms. This high sensitivity of our detection algorithm to the Wireless traffic can justify the results of Experiment 2 shown in Figure 13.6, where it is observed that the detection algorithm works even when the frame drop rate is as high as 80%. Table 13.7 shows the rules used to detect anomalous behavior due to Wi-Fi network attacks.

  The results of the experiment are shown in Figure 13.18. It also highlights the minimum difference between the condition for the anomalous behavior and the normal flowscore values.

- *Experiment 5*

  This experiment has been designed to evaluate the accuracy of attack classification algorithm as the frame drop rate increases. This increase in frame drop rate is due to the nature of the wireless physical layer being prone to error, interference, and noise. Thus, this experiment helps in identifying the maximum threshold beyond which the algorithm cannot effectively classify the attacks.

Table 13.7 Intrusion detection system rules.

Flowscore	Condition
1. Flowscore 1 (FS1)	Flowscore1 should be less than 97 **(FS1 < 97)**
2. Flowscore 2 (FS2)	Flowscore2 should be greater than 279 732 **(FS2 > 279 732)**
3. Flowscore 3 (FS3)	Flowscore3 should be greater than 1020 **(FS3 > 1020)**
4. Flowscore 4 (FS4)	For the current environment, the abnormal condition is independent of flowscore4 **(FS4 = NA)**

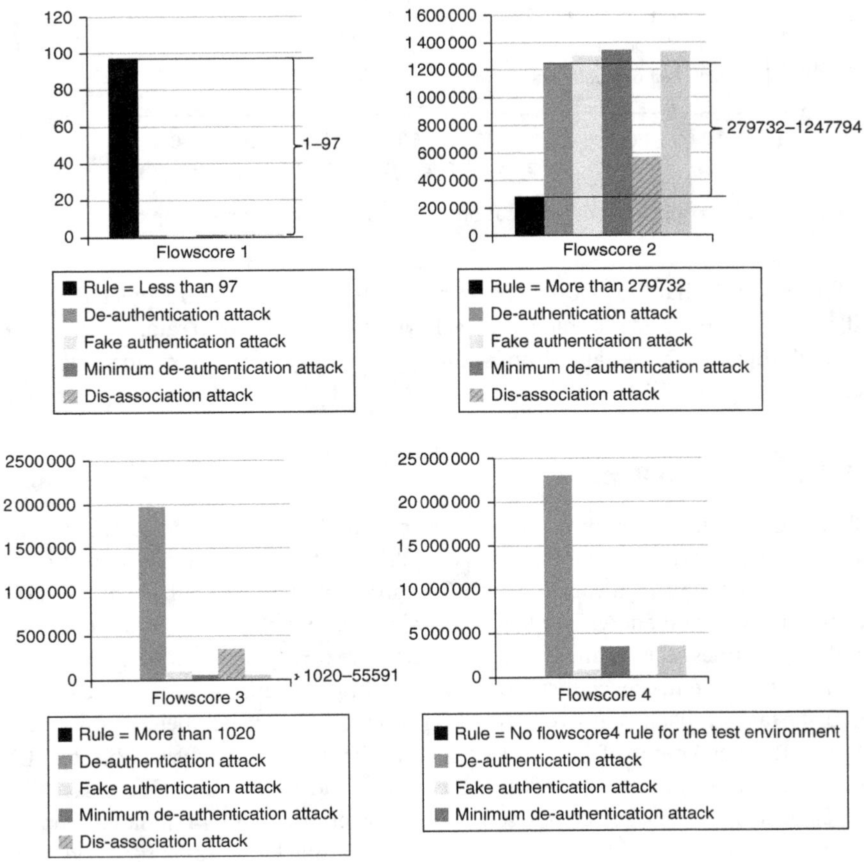

Figure 13.18 Flowscore value to rule comparison.

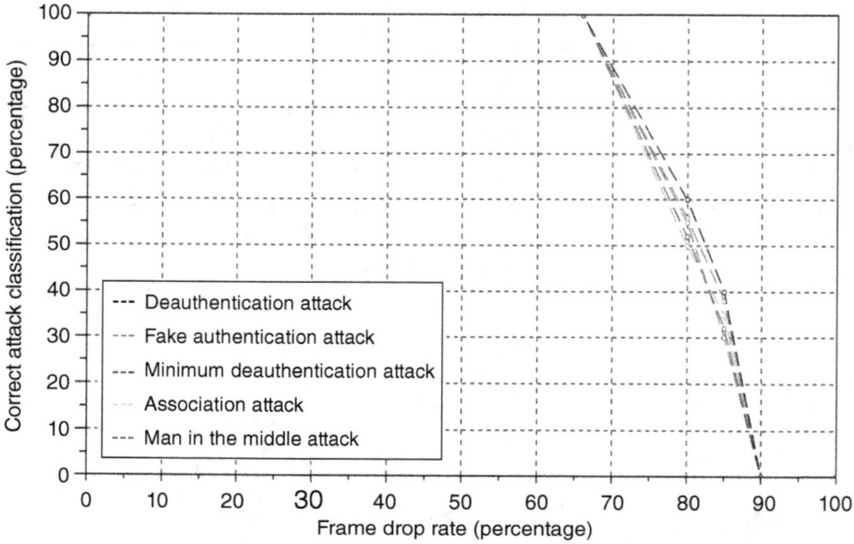

Figure 13.19 Correct attack classification (percentage) versus frame drop rate.

It is observed that the algorithm can classify all the attacks correctly at all times till the frame drop rate is as high as 66%. Beyond 66% frame drop rate, we observed a steady decline in the algorithm's ability to classify the attacks accurately as shown in Figure 13.19.

13.4.5 Bluetooth Protocol

Bluetooth is used for communication between closely connected devices, as a substitute for using cables for the data transfer [38]. Bluetooth by design is more secure than other wireless protocols like the Wi-Fi protocol. It depends on stealth, FHSS, authentication, and encryption for security [39]. Although the Bluetooth protocol by design defines security measures, these measures are typically not followed during the deployment, especially in IoT devices as implementation testing and device security takes a back seat over providing functionality. It has been shown that Bluetooth deployment of IoT devices such as smart cars are susceptible to PIN guessing attacks [40], thus allowing the attacker to send arbitrary messages to the car's network. IoT devices are sensitive to power utilization [41]. Battery draining attacks are possible on Bluetooth devices by the attacker sending repeated pairing requests or device information requests [39, 42, 43]. These battery draining attacks will result in the sensors running out of battery faster causing a DoS when used on Bluetooth sensor networks. Bluetooth protocol (IEEE 802.15.1) is

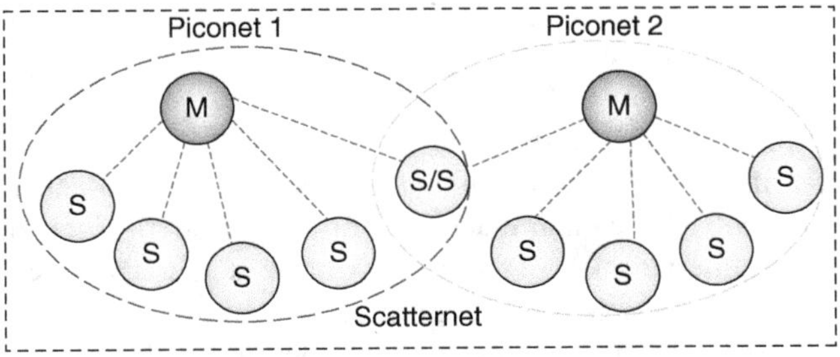

Figure 13.20 Bluetooth piconet.

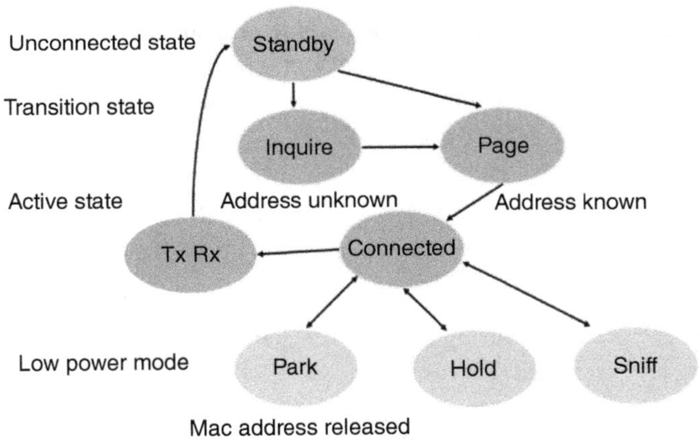

Figure 13.21 Bluetooth state machine.

a short-range wireless communication protocol (up to 100 m) designed to operate in 2.4 GHz ISM band [43, 44]. Bluetooth devices use FHSS over 79 frequencies and operate in a Master/Slave configuration. A master can connect up to seven slave devices to form a Piconet. A slave device can be part of multiple Piconets to form Scatternets as shown in Figure 13.20. Bluetooth devices in a Piconet are identified using unique physical address called the BD_ADDR. The Bluetooth protocol definition specifies the Bluetooth protocol and the Bluetooth state machine shown in Figure 13.21.

The state machine in Figure 13.21 can be used to find ASs for the Bluetooth protocol and perform threat modeling for the Bluetooth protocol. Based on the version of Bluetooth protocol used, the Bluetooth device may be exposed to different ASs. For instance, older versions of the Bluetooth protocol allowed the devices to operate in an unencrypted mode, requiring no authentication, while the newer Bluetooth devices force authentication and encryption before link establishment and use secure simple pairing to establish service-level pairing. Some attacks on Bluetooth devices are presented below:

- *Scanning for Bluetooth Addresses and Surveillance*
  Bluetooth address scanning, i.e. scanning for BD_ADDR, gives the attacker information on the Bluetooth device manufacturer, thus allowing him/her to target the device-specific vulnerabilities. BD_ADDR is transmitted unencrypted and hence can be sniffed easily. Tools like Blueprinting [45] and Bt_audit [46] can be used to easily fingerprint Bluetooth devices and then target them.
- *Pairing Attack*
  The pairing attack can be executed only during the pairing process. The pairing process which is used to exchange the link key and the initialization key is dependent on PIN code which in case of lot of Bluetooth devices is set to default, like in Bluetooth speakers or has a length of 4 (can be brute forced) like in cars. Tools like BTCrack [47] and btpincrack [48] can be used to brute force the PIN.
- *Bluejacking*
  Bluejacking [49] is used to send anonymous business cards to Bluetooth devices with offensive messages.
- *Backdoor Attack*
  Backdoor attack is executed by establishing a trusted relationship by pairing with the target devices. Once the pairing is done, attackers can perform attacks like stack smashing [50] on target devices.
- *Bluesnarfing Attack*
  Bluesnarfing [51] targets legacy devices, by giving the attacker control of the devices' phonebook, text messages, calendar data, business cards, and images.
- *Positioning and Tracking Attack*
  The sniffed BD_ADDR combined with the use of the received signal strength indicator (RSSI) can be used to track and locate the position of a Bluetooth device.
- *Replay Attacks*
  In the replay attack, the attacker records a Bluetooth conversation and replays the conversation, while hopping on correct Bluetooth frequencies.
- *Battery Draining Attacks*
  In Battery draining attacks, the attacker sends random packets like pairing request packets or device information request packet to the target device at a high rate. This results in the target device draining its battery at a higher rate.

Based on the above-mentioned ASs and attack vectors, features presented in Table 13.8 were selected to effectively characterize the normal and the abnormal behavior of the Bluetooth protocol.

Table 13.8 Bluetooth feature set.

Sr.no	Feature	Description
1.	Frame_epoch_time	Epoch time
2.	hci_h4_type	HCI packet type
3.	bthci_evt_code	Bluetooth HCI event code
4.	btchi_cmd_opcode	Bluetooth HCI command opcode
5.	btl2cap_scid	Bluetooth L2CAP protocol source CID
6.	btchi_acl_dst_bd_addr	Destination BD_ADDR
7.	bthci_acl_dst_name	Destination device name
8.	bthci_acl_dst_rolce	Destination device role
9.	bthci_acl_src_bd_addr	Source BD_ADDR
10.	bthci_acl_src_name	Source device name

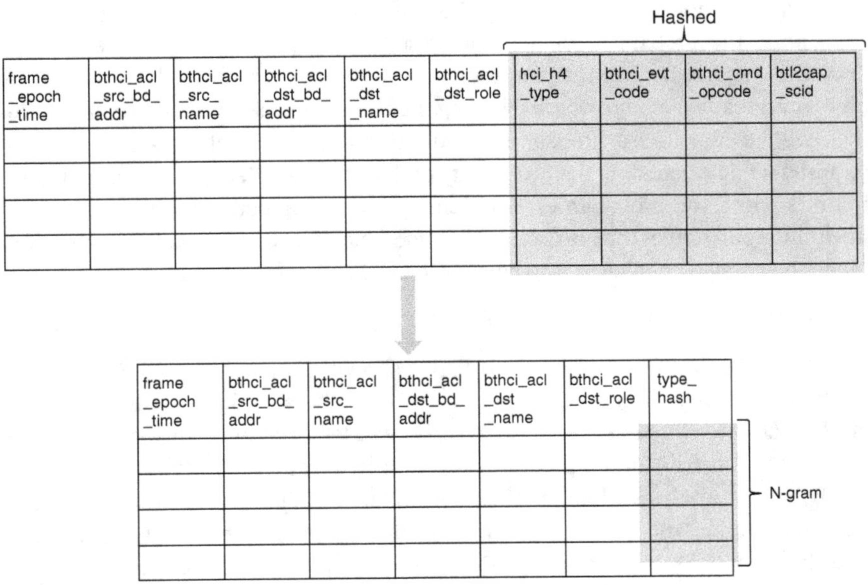

Figure 13.22 Feature set to n-grams.

Similar to the IDS designed for the Wi-Fi protocol, an n-gram and observation score data structure-based IDS is designed for the Bluetooth protocol. Once the feature set is collected, its mapping into n-grams is as shown in Figures 13.22 and 13.23 shows the Bluetooth observation-flow or the Bluetooth-flow-characteristics (BFC).

Probability of flow	C1	C2	C3	C4

Where, C1, C2, C3, and C4 are frame types.
C1 is HCL command frame
C2 is ACL data frame
C3 is SCO data frame
C4 is HCL data frame

Figure 13.23 Bluetooth observation-flow.

The probability of observation of an n-gram in an observation-flow (probability is obtained during the training phase based on heuristics of the training data set) "anded" with the probability of other n-grams in the observation-flow to get the "probability of flow." While calculating the probability of the flow, a count of number of HCI command frames, ACL data frames, SCO data frame, and HCI data frame is maintained. At the end of processing of each observation-flow, we obtain a data structure called the BFC as shown in Figure 13.23. The BFC as shown in Figure 13.23 is presented to the machine learning algorithms to classify the observation-flow as normal or abnormal. The "anding" of heuristic probabilities while calculating the probability of flow can result into numbers that are very small and tend to zero. During the runtime phase, although very unlikely, we may see an n-gram that we did not observe during the training phase. In this case, the "anding" will result in zero probability. Thus, to reduce these effects, smoothing is required while calculating the probability of flow. We used Jelinek-Mercer Smoothing to smooth the probabilities. The Jelinek-Mercer Smoothing method [7] uses linear interpolation of the maximum likelihood model with respect to the collection model by using a coefficient λ to control the influence of each model.

$$P(w|d) = \lambda P(w|C) + (1-\lambda)P(w|d) \tag{13.7}$$

The architecture for the Bluetooth IDS is shown in Figure 13.24 [37]. The IDS has two modules: Sniffer Module and Behavior Analysis Module. The Sniffer Module collects the Bluetooth frames from the wireless medium. On collecting the frames, the Sniffer Module processes the collected frames and passes the data to the Behavior Analysis Module. Host controller interface (HCI) protocol frame, SCO data frame, and HCI data frame are some of the frames that are collected. The Behavior Analysis Module analyzes the behavior of the received Bluetooth frames. Once the extracted data are presented to the Behavior Analysis Module, they are split into Observation-flows of size *t* seconds. The observation-flow is then converted to n-grams. The probability of that observed n-gram being either normal or abnormal is calculated using predetermined heuristics obtained during the training phase. These probabilities are then fed to the machine learning classifiers which classify the observation-flow as normal or abnormal. Figure 13.25 shows the test bed in the ACL at the University of Arizona, consisting of three devices, a

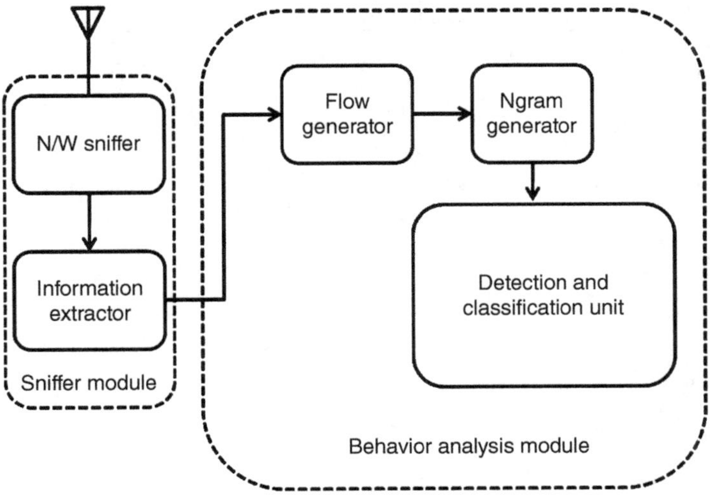

Figure 13.24 Bluetooth IDS architecture.

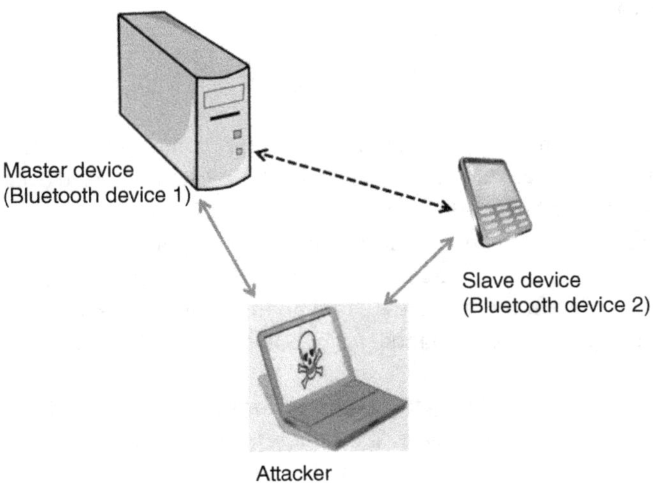

Figure 13.25 Bluetooth test bed.

Bluetooth master, and a Bluetooth slave. The third device is a Linux machine that behaves as an attacker.

Anomaly-based IDS are highly dependent on the complete understanding of the normal behavior of the system. The first two experiments help ensure complete understanding of the normal behavior of the system and help make the design

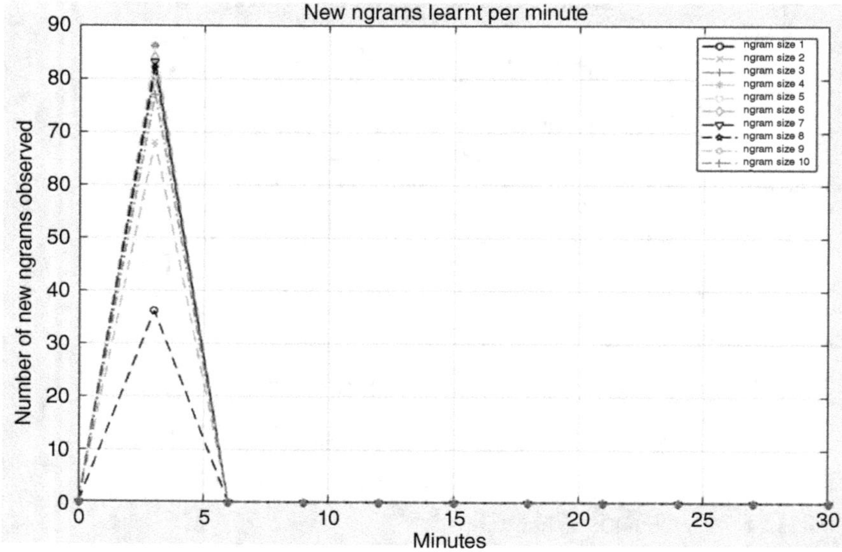

Figure 13.26 Rate of learning of unique n-grams.

decisions for the developers. In this experiment, the number of new n-grams collected with respect to time was calculated. The experiment used data collected over an hour, and a sweep was performed for n-gram sizes 1–10. Figure 13.26 shows that all the unique n-grams were learnt in first six minutes of the traffic collection. Figure 13.27 shows the total number of n-grams observed in the first 30 minutes of the traffic collection for different n-gram sizes. Based on the observations made in Figure 13.27, we use n-grams of size 3 to model the state transitions of the Bluetooth protocol.

The next experiment was designed to measure the performance of the machine learning models trained on the training data. The data set collected was split into 20 : 80 ratio where 20% of the data was used to train the models and 80% was used to measure the performance of the models, λ was set to 0.5 for the Jelinek-Mercer Smoothing, and 10-fold cross validation was used to measure the performance of the model. Machine learning classifiers like C4.5, Adaboost.m1, SVM, Naïve Bayes, RIPPER, and Bagging algorithms were used to build machine learning classifiers to detect abnormal behavior. Figures 13.28–13.30 show the Precision, Recall, and RocArea values for various classifiers, respectively. Our experimental results show that the classifiers can achieve a high precision of up to 99.6% and a recall of up to 99.6%.

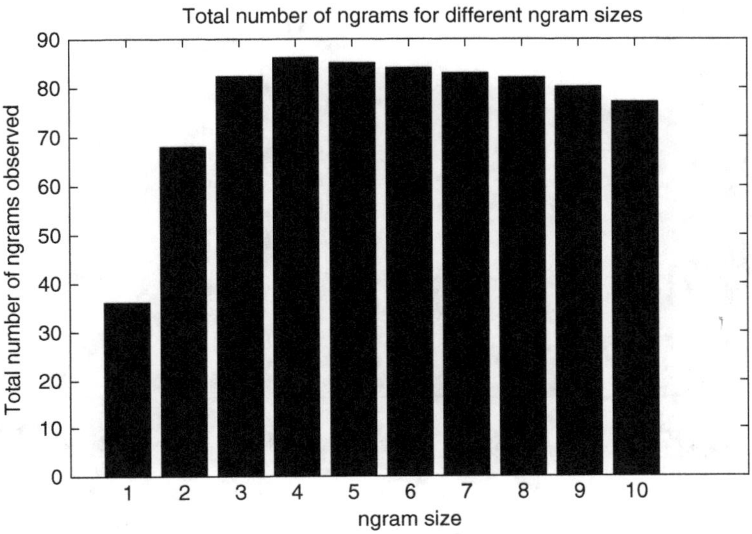

Figure 13.27 Unique n-grams for an n-gram size.

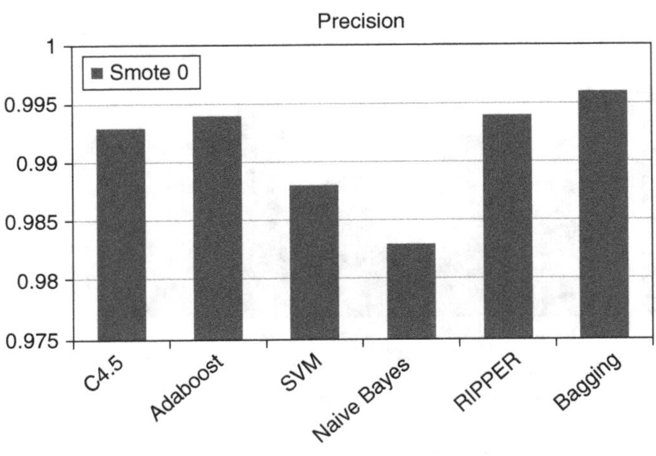

Figure 13.28 Precision for various classifiers.

13.5 Conclusion and Future Work

In this book chapter, we introduced the importance of IoT security and presented a systematic approach to perform threat modeling on IoT protocols. As an example for the approach, we applied the threat modeling to Wi-Fi protocol and the

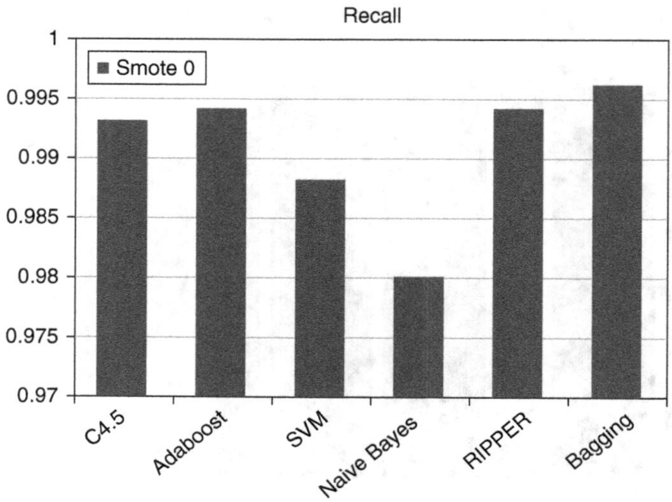

Figure 13.29 Recall for various classifiers.

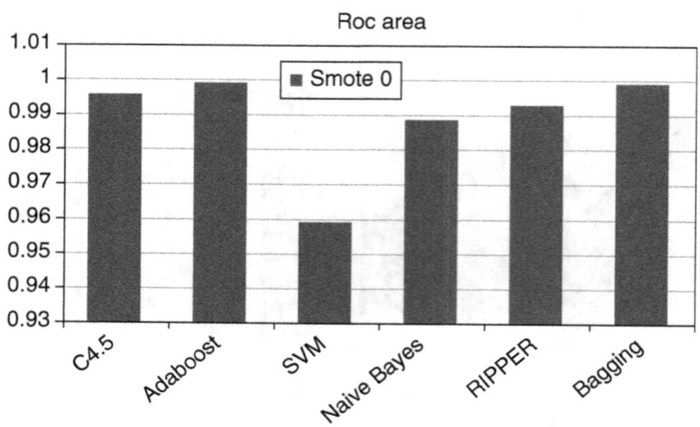

Figure 13.30 RoC area for various classifiers.

Bluetooth protocol. We present an IDS to detect attacks on a single access point Wi-Fi network using the ABA methodology. A simple conjunctive rule-based model was designed to model the normal behavior of the Wi-Fi protocol. The IDS designed was seen to have no false positives and negatives. It was also demonstrated through experimentation that the IDS is capable of detecting modified attacks. We followed the same approach to design an IDS to detect attacks on the Bluetooth protocol. The presented IDS used an `n-gram`-based approach to characterize the normal behavior of the Bluetooth protocol. Machine learning classifiers like C4.5, Adaboost.m1, SVM, Naïve Bayes, RIPPER, and Bagging algorithms were used to build machine learning classifiers to detect abnormal behavior. Our experimental results show that the classifiers can achieve a high precision of up to 99.6% and a recall of up to 99.6%.

Although the approach presented in this chapter is highly effective and has been applied by the authors to detect attacks on various other protocols and systems like DNS, HTML, HTTP, malicious system utilization by insiders and outsiders, etc., it has been observed that the normal behavior of the modeled systems drifts. Hence, unsupervised machine learning techniques need to be developed to enable continuous learning to capture the drift.

Once an IDS has detected threats on the target system, there is a need to develop frameworks that help users or system admins secure and patch their devices automatically based on the security updates and NIST best practices. Thus, future work will involve designing intelligent recommender systems that with little involvement of users will continuously monitor the IoT system configurations of all its layers, detect vulnerabilities, and implement the required mechanisms to remove them or change the deployed software modules to mitigate or tolerate these vulnerabilities if they were exploited.

References

1 Internet world stats. https://www.internetworldstats.com/stats.htm [accessed: October 2018].

2 Al-Fuqaha, Ala, Mohsen Guizani, Mehdi Mohammadi, Mohammed Aledhari, and Moussa Ayyash. "Internet of things: A survey on enabling technologies, protocols, and applications." *IEEE Communications Surveys & Tutorials* 17, no. 4 (2015): 2347–2376.

3 Whitmore, Andrew, Anurag Agarwal, and Li Da Xu. "The Internet of Things: A survey of topics and trends." *Information Systems Frontiers* 17, no. 2 (2015): 261–274.

4 Atzori, Luigi, Antonio Iera, and Giacomo Morabito. "The Internet of things: A survey." *Computer Networks* 54, no. 15 (2010): 2787–2805.

5 Lin, Jie, Wei Yu, Nan Zhang, Xinyu Yang, Hanlin Zhang, and Wei Zhao. "A survey on internet of things: Architecture, enabling technologies, security and privacy, and applications." *IEEE Internet of Things Journal* 4, no. 5 (2017): 1125–1142.

6 Yang, Nan, Lifeng Wang, Giovanni Geraci, Maged Elkashlan, Jinhong Yuan, and Marco Di Renzo. "Safeguarding 5G wireless communication networks using physical layer security." *IEEE Communications Magazine* 53, no. 4 (2015): 20–27.

7 Satam, Pratik. "An anomaly behavior analysis intrusion detection system for wireless networks." 2015.

8 Gaur, Padmini and Mohit P. Tahiliani. "Operating systems for IoT devices: A critical survey." In *Region 10 Symposium (TENSYMP), 2015 IEEE*, pp. 33–36. IEEE, 2015.

9 Speed the development of wearable devices with a solid, targeted platform. https://www.ept.ca/features/speed-development-wearable-devices-solid-targeted-platform [accessed: October 2018].

10 Google Assistant Speakers. https://assistant.google.com/platforms/speakers [accessed: October 2018].

11 Amazon Alexa Assistant. https://developer.amazon.com/alexa [accessed: October 2018].

12 212 Billion devices to make up the 'Internet of Things' by 2020. https://thejournal.com/articles/2013/10/07/212-billion-devices-to-make-up-the-internet-of-things-by-2020.aspx [accessed: October 2018].

13 Unlocking the potential of the Internet of Things. https://www.mckinsey.com/business-functions/digital-mckinsey/our-insights/the-internet-of-things-the-value-of-digitizing-the-physical-world [accessed: October 2018].

14 Paquet, Catherine. *Implementing Cisco IOS Network Security (IINS 640-554) Foundation Learning Guide*. Cisco Press, 2012.

15 Navetta, David, Boris Segalis, Erin Locker, and Andrew Hoffman. "Large Ransomware attack affects companies in over 70 countries," (May, 2017). [online] Available on: http://www.dataprotectionreport.com/2017/05/large-ransomware-attack-affects-companies-in-over-70-countries [accessed: October 2018].

16 Satam, Pratik, Jesus Pacheco, Salim Hariri, and Mohommad Horani. "Autoinfotainment security development framework (ASDF) for smart cars." In *2017 International Conference on Cloud and Autonomic Computing (ICCAC)*, pp. 153–159. IEEE, 2017.

17 Xu, Dianxiang, Manghui Tu, Michael Sanford, Lijo Thomas, Daniel Woodraska, and Weifeng Xu. "Automated security test generation with formal threat models." *IEEE Transactions on Dependable and Secure Computing* 9, no. 4 (2012): 526–540.

18 Schlegel, Roman, Sebastian Obermeier, and Johannes Schneider. "Structured system threat modeling and mitigation analysis for industrial automation systems." In *IEEE 13th International Conference on Industrial Informatics (INDIN)*, July 2015.

19 Jin, Jiong, Jayavardhana Gubbi, Slaven Marusic, and Marimuthu Palaniswami. "An information framework for creating a smart city through Internet of things." *IEEE Internet of Things Journal* 1, no. 2 (2014): 112–121.

20 Ferreira, Hiro Gabriel Cerqueira, Edna Dias Canedo, and Rafael Timóteo de Sousa Junior. " IoT architecture to enable intercommunication through REST API and UPnP using IP, ZigBee and arduino." In *2013 IEEE 9th international conference on*

wireless and mobile computing, networking and communications (WiMob), pp. 53–60. IEEE, 2013.

21 Soliman, Moataz, Tobi Abiodun, Tarek Hamouda, Jiehan Zhou1, and Chung-Horng Lung. "Smart home: Integrating Internet of Things with web services and cloud computing." In *IEEE 5th International Conference on Cloud Computing Technology and Science*, 2013.

22 Reddy, B. Raghavendhar and Erukala Mahender. "Speech to text conversion using android platform." *International Journal of Engineering Research and Applications (IJERA)* 3, no. 1 (2013): 253–258.

23 CMUSphinx Project. https://cmusphinx.github.io [accessed: October 2018].

24 Google Speech Recognition. https://cloud.google.com/speech-to-text [accessed: October 2018].

25 "IEEE Std 802.11-1997 information technology- telecommunications and information exchange between systems-local and metropolitan area networks-specific requirements-part 11: Wireless LAN medium access control (MAC) and physical layer (PHY) specifications." IEEE, 1997.

26 "802.11a-1999 high-speed physical layer in the 5 GHz band." IEEE, 1999.

27 "802.11b-1999 higher speed physical layer extension in the 2.4 GHz band." IEEE, 1999.

28 "IEEE 802.11g-2003: Further higher data rate extension in the 2.4 GHz band." IEEE, 2003.

29 "IEEE 802.11n-2009: Amendment 5: Enhancements for higher throughput." IEEE-SA, 29 October 2009. doi:10.1109/IEEESTD.2009.5307322.

30 "IEEE Std 802.11ac 2013: 22.5 Parameters for VHT-MCSs." IEEE, 2013.

31 Sivaraman, Vijay, Hassan Habibi Gharakheili, Arun Vishwanath, Roksana Boreli, and Olivier Mehani. "Network-level security and privacy control for smart-home IoT devices." In *2015 IEEE 11th International Conference on Wireless and Mobile Computing, Networking and Communications (WiMob)*, pp. 163–167. IEEE, 2015.

32 Satam, Pratik. "Cross layer Anomaly based intrusion detection system." In *2015 IEEE International Conference on Self-Adaptive and Self-Organizing Systems Workshops (SASOW)*, pp. 157–161. IEEE, 2015.

33 Satam, Pratik, Hamid Alipour, Youssif B. Al-Nashif, and Salim Hariri. "Anomaly behavior analysis of DNS protocol." *Journal of Internet Services and Information Security* 5, no. 4 (2015): 85–97.

34 Satam, Pratik, Douglas Kelly, and Salim Hariri. "Anomaly behavior analysis of website vulnerability and security." In *2016 IEEE/ACS 13th International Conference of Computer Systems and Applications (AICCSA)*, pp. 1–7. IEEE, 2016.

35 Alipour, Hamid, Youssif B. Al-Nashif, Pratik Satam, and Salim Hariri. "Wireless anomaly detection based on IEEE 802.11 behavior analysis." *IEEE Transactions on Information Forensics and Security* 10, no. 10 (2015): 2158–2170.

36 Satam, Pratik. *An Anomaly Behavior Analysis Intrusion Detection System for Wireless Networks*. The University of Arizona, 2015.

37 Satam, Pratik, Shalaka Satam, and Salim Hariri. "Bluetooth Intrusion Detection System (BIDS)." In *2018 IEEE/ACS 15th International Conference on Computer Systems and Applications (AICCSA)*, pp. 1–7. IEEE, 2018.

38 Ferro, Erina and Francesco Potorti. "Bluetooth and Wi-Fi wireless protocols: A survey and a comparison." *IEEE Wireless Communications* 12, no. 1 (2005): 12–26.

39 Dunning, John. "Taming the blue beast: A survey of Bluetooth based threats." *IEEE Security & Privacy* 8, no. 2 (2010): 20–27.

40 Oka, Dennis Kengo, Takahiro Furue, Lennart Langenhop, and Tomohiro Nishimura. "Survey of vehicle IoT bluetooth devices." In *2014 IEEE 7th International Conference on Service-Oriented Computing and Applications (SOCA)*, pp. 260–264. IEEE, 2014.

41 Martinez, Borja, Marius Monton, Ignasi Vilajosana, and Joan Daniel Prades. "The power of models: Modeling power consumption for IoT devices." *IEEE Sensors Journal* 15, no. 10 (2015): 5777–5789.

42 Moyers, Benjamin R., John P. Dunning, Randolph C. Marchany, and Joseph G. Tront. "Effects of Wi-Fi and Bluetooth battery exhaustion attacks on mobile devices." In *2010 43rd Hawaii International Conference on System Sciences (HICSS)*, pp. 1–9. IEEE, 2010.

43 Satam, Shalaka Chittaranjan. "Bluetooth Anomaly Based Intrusion Detection System." PhD diss., The University of Arizona, 2017.

44 Vainio, Juha T. "Bluetooth security." In *Proceedings of Helsinki University of Technology, Telecommunications Software and Multimedia Laboratory, Seminar on Internetworking: Ad Hoc Networking*, Spring, vol. 5, 2000.

45 Blueprinting. https://trifinite.org/trifinite_stuff_blueprinting.html [accessed: April 2017].

46 BT Audit. https://trifinite.org/trifinite_stuff_btaudit.html [accessed: April 2017].

47 BTCrack. https://github.com/mikeryan/btcrack.git [accessed: April 2017].

48 OpenCiphers. http://openciphers.sourceforge.net/oc/btpincrack.php [accessed: April 2017].

49 Thom-Santelli, Jennifer, Alex Ainslie, and Geri Gay. "Location, location, location: A study of bluejacking practices." In *CHI'07 Extended Abstracts on Human Factors in Computing Systems*, pp. 2693–2698. ACM, 2007.

50 Bluetooth Stack Smasher. http://www.securiteam.com/tools/5NP0220HPE.html [accessed: April 2017].

51 Bluesnarfer. http://git.kali.org/gitweb/?p=packages/bluesnarfer.git;a=summary [accessed: April 2017].

14

Dynamic Cyber Deception Using Partially Observable Monte-Carlo Planning Framework

Md Ali Reza Al Amin[1], Sachin Shetty[2], Laurent L. Njilla[3], Deepak K. Tosh[4], and Charles A. Kamhoua[5]

[1] *Computational Modeling and Simulation Engineering, Old Dominion University, Norfolk, VA, USA*
[2] *Virginia Modeling Analysis and Simulation Center, Old Dominion University, Norfolk, VA, USA*
[3] *Cyber Assurance Branch, US Air Force Research Laboratory, Rome, NY, USA*
[4] *Department of Computer Science, University of Texas at El Paso, El Paso, TX, USA*
[5] *US Army Research Laboratory, Adelphi, MD, USA*

14.1 Introduction

Cyber deception has attracted attention from security researchers and practitioners as an approach for designing secure cyber infrastructure. Cyber deception can provide twin advantages: (i) reduce likelihood of adversarial success and cost of cyber defense and (ii) provide insights into attacker's strategies, tactics, capabilities, and intent. Typically, attackers have a priori knowledge of the infrastructure they are targeting. In a cyber-deception approach, the defender can exploit this a priori knowledge to mislead the adversary in expending their resources and time using decoy networks thereby leading to attack paths that do not lead to a successful goal. The success of a cyber-deception strategy hinges on adversaries taking on more fake attack paths than actual attack paths that would increase the adversaries' resources in distinguishing between real and fake assets.

A significant number of physical devices are being connected to the Internet, increasing at an unprecedented rate which leads to realizing the idea of the Internet of Things (IoT). The adoption rate of IoT technologies in IT and OT sectors has resulted in exponential availability of various IoT solutions (Chapter 10). The secure mechanism technique not implemented in IoT devices leads to a higher risk of security. The real-world attacks can start with reconnaissance using insecure

Modeling and Design of Secure Internet of Things, First Edition. Edited by Charles A. Kamhoua, Laurent L. Njilla, Alexander Kott and Sachin Shetty.

Published 2020 by John Wiley & Sons, Inc.

IoT devices, followed by laterally moving through the environment to deepen access by spreading malicious code. The disadvantage with insecure IoT can be exploited to develop a proactive defense strategy, cyber deception, to benefit the defender by protecting the critical assets from vulnerabilities in IoT. Design and placement of network decoy are one of the significant steps in cyber deception where a defender can push the adversary to expend their resources and time to gather critically important information about the adversaries' strategies, tactics, capabilities, and intent. This approach will mitigate the information asymmetry that exists between adversaries and defenders by converting the defender's disadvantaged position to a position of strength.

Due to the static nature of organizations, IT/OT networks leading to adversaries perform reconnaissance activity and identify potential threats. The aim of the reconnaissance phase is collecting as much as critical information about the network including network topology, open ports and services running, and unpatched vulnerabilities. Having that critical information maximizes the intruder's chance of probability to penetrate a system and gaining a foothold successfully. Another essential factor to be noted is that patching a vulnerability will lower the possibility of being attacked where vulnerability patching depends upon discovery of a vulnerability and develop a patch for that specific weakness. Unfortunately, this period (vulnerability exposure window) often lasting more than five or six months [1]. This more extended period put the cyber network at higher risk. To address this, one needs an active form of a defense system to thwart cyberattacks while considering information in real time and providing with appropriate defense actions. But there is some issue with the development of such a system as the attacker considers series of exploit to deep dive into the network. Very few targeted cyberattacks consist of only a single vulnerability. It is always beneficial for the defender knowing how the intruder can infiltrate the network. To help with that, researchers develop such tools, e.g. attack graph/tree, dependency graph, to capture possible attack path. In [2], attack tree is first to introduce where it models the relationship between exploits and system states. But the problem with attack graph/tree is it can be enormous for a reasonably modest network [3]. To solve this issue, Ammann et al. [4] assumed the attacker's behavior states that the success of the previous exploit will not interfere with the success of future exploit. This assumption significantly reduces the amount of information in the attack graph while no loss of granularity.

To address the issue described earlier, several works have been done in the area with cyber deception. As by [5], traditional cyber defense systems (e.g. intrusion detection/prevention system, firewalls, antivirus system, malware scanner) are not adequate to defend critical cyber resources. Moreover, advanced persistent threat (APT) can infiltrate cyber networks using a combination of both social engineering and software vulnerability [6]. To defend against such types of attacks, we need proactive cyber defense system which engages with the intruder strategically and learns and influences his behavior (Chapter 2). Here comes cyber deception which is a vital component of dynamic cyber defense system. Cyber-deception technology can create and fortify the view of the network to an attacker by revealing or concealing the information to the intruder. By providing with a mix of true/false information to the attacker, we convert the static view to the dynamic view by introducing fake networks. To make the fake network looks like a real network, we take an approach described in [7]. By this way, an attacker cannot distinguish between real and fake system.

14.1.1 Research Goal and Contributions

In this chapter, we propose an approach to design and place network decoys while capturing attacker's lateral propagation. The lateral propagation is captured using exploit dependency graph [4, 8], where nodes in the graph represent possible security conditions and directed hyperedges represent exploits. The defender modifies the design and placement based on the history of security alerts and prior deployments of network decoys. The defender's decision to deploy network decoys balances the trade-off between security cost and availability cost. This decision is made by solving a modified version of partially observable Monte-Carlo planning (POMCP) algorithm [9], which provides the optimal action corresponding to the current belief vector comprising of the security state and prior actions.

By introducing fake networks, the overall system security state becomes substantial leading to a scalability issue. To solve the scalability and perform online deception algorithm, we adopt the approach described in [8], which is a modified version of POMCP algorithm [9]. The key contributions of the chapter are summarized below.

- Network decoy design in a dynamic environment manner where an attacker can distinguish the real network from the fake network with significantly low probability.
- Efficient and scalable deception approach that influences an attacker to take the path toward the fake network while capturing attacker progression using the dependency graph.
- Online deception algorithm that allows the defenders to select actions based on the attacker behavior in the real time.

14.1.2 Structure of this Chapter

The rest of this chapter is organized as follows. Section 14.2 discusses related work. Section 14.3 discusses the dynamic deception model. In Section 14.4, we present the defender's optimal action, Section 14.5 present the Online Deception Algorithm, and Section 14.6 present an IoT aware smart healthcare architecture. Finally, Section 14.7 reports the simulation results, and we conclude in Section 14.8.

14.2 Related Work

In recent publications [10], authors introduce systems that change the view of a cyber network by obscuring some system characteristics where [11] alters the system view by manipulating attackers' probe. Trassare et al. [12] use traceroute function to deceive the network reconnaissance attack. In Dunlop et al. [13], authors propose a mechanism where they hide IPv6 packets to achieve anonymity. They added a virtual network interface controller and shared a secret with all hosts. To defend against eavesdropping and denial of service (DoS) attack, Duan et al. [14] present a Random Route Mutation Technique to change the network's data flow. Reconnaissance tools such as Nmap or Xprobe2 collect critical network information like host OS or service by analyzing the packet received after probe [15]. Virtual DNS entries and software-defined networks can be used to do IP visualization to deceive the attacker shown in Jafarian et al. [16]. In Chapter 2, authors provide an overview of deception games in three different environments. The chapter introduces the baseline signaling game models to capture the information asymmetry, dynamic, and strategic behaviors of deceptions. The chapter first presents a leaky deception over a binary information space model based on signaling game framework with a detector that gives probabilistic evidence of the deception when the sender acts deceptively. The chapter later focuses on a class of continuous one-dimensional information space and use a signaling game model to characterize the deception by taking into account the cost of deception. Finally, the chapter explores the multistage incomplete-information Bayesian game model for defensive deception for advanced persistent threats (APTs).

All these approaches in cyber-deception area tend to change the network view from an attacker's point of view. But they all failed to answer the question what if the attacker enters the network where unpatched vulnerabilities are present, and patches are not released yet. A network administrator cannot just let the attacker compromise the system. As we mentioned earlier, the attacker always has time-advantage over unpatched vulnerability where vulnerability exposure window is high. A defender has to take defensive action while making a trade-off between availability and security. Our approach not only changes the network view but also influences the attacker to take the path toward fake networks while keeping availability and security at a satisfactory level.

In [8], authors use the dependency graph to provide solutions for the cyber defense system. The issue with that approach is network availability to the trusted user because the attacker always starts with the same static network and the defender has to take actions (system modification induces blocked vulnerabilities) which will have a more significant impact on availability. We solve this issue by introducing fake networks along with real networks. Our approach will help defender maintaining the network service availability and collect critical intelligence information about the attacker.

14.3 The Dynamic Deception Model

The proposed dynamic security model provides a basis for how a defender can deceive and prevent the infiltration of an attacker in a cyber network. Throughout the chapter, Figure 14.1 will be used to describe the model. Section 14.3 will firstly describe the attacker progression through the cyber network, later on, we will discuss how an exploit dependency graph used to deceive an attacker and forced him to choose a directed path toward a fake goal node.

14.3.1 Exploit Dependency Graph

It has been a long, exciting research topic that how to represent the steps that an attacker could take when compromising a system. To represent actual or possible attacks in a network through the graph is well established. Attack graph shows the relationship between network vulnerabilities which eventually lead to an understanding of the impact on cyber network security. Attack graphs are used in some areas of network security, which includes vulnerability analysis, intrusion alarm correlation, and attack response. It can be used in both an offensive or defensive context. There is another term called "Attack Tree," which is another form of attack representation diagram. It is a conceptual diagram showing how an asset or target might be attacked. Attack graph and attack trees are both developed

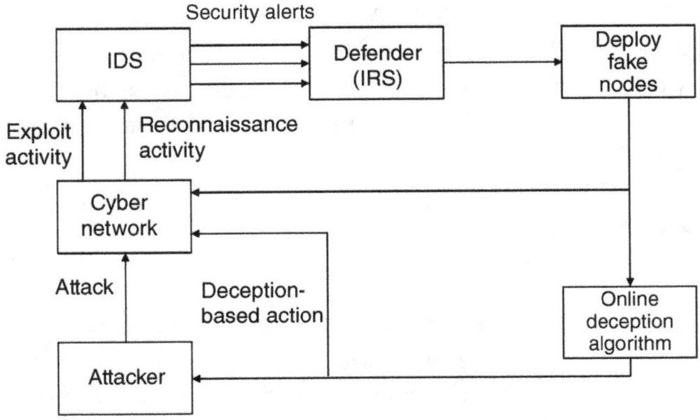

Figure 14.1 Dynamic deception model.

to allow one to study all possible sequence of exploits that an intruder can take to infiltrate a cyber network and reach its goal.

There are two components in attack graph, system states (nodes) and transition relations (edges), which connect each other via exploits. A significant amount of progress has been made in generating attack graphs automatically. The major issue with the attack graph is that it increases in dimension. For a minimal network with some vulnerabilities, the graph complexity from a usability standpoint has very little attention. To get rid of this, one assumption [4], monotonicity, regarding attacker behavior can improve the usability. The hypothesis states that there is no relationship between the success of an exploit and an attacker's ability to exploit future exploit. Under the assumption, one does not need to enumerate all the system states of the attack graph, instead can construct a dependency graph where the graph will show how the security alerts are related with exploits [4, 17]. The approach is taken by Ammann et al. [4], who described nodes as security condition and edges represent exploits, which is termed as exploit dependency graph. Security conditions are atomic facts (either true or false) that are related system attributes like exploiting related security conditions via precondition and postcondition [8]. Here, we took the slightly modified definition from [4] as described in [8] for security condition where it states that conditions not representing the typical network configuration is termed as the initial condition; instead, we assume that attack conditions consist of a set of security conditions.

14.3.2 Fake Nodes Deploying

We deploy fake nodes/networks along with the real nodes/networks whenever a reconnaissance alert is received from an intrusion detection system (IDS). An attacker can certainly differentiate real nodes from fake nodes by analyzing round trip times or measured bandwidth on the link. To render real and fake network indistinguishable, we introduce delay and bandwidth handler methods to ensure consistency of the network measurements collected during reconnaissance mission [7], which proves that the probability of distinguishing real from fake networks is very low.

14.3.3 POMDP Model

14.3.3.1 State Space

An exploit dependency graph [6], a directed acyclic hypergraph (H), consists of nodes and hyperedges where nodes represent a set of security conditions (c) and hyperedges render a set of exploits (e). The security condition of the hypergraph has two options: either true or false, where true means the attacker has the particular capability and false means the attacker does not pose the skill.

The node which the defender wants to protect termed as goal node is denoted by $N_r^g \subseteq N, N_f^g \subseteq N$, where N_r^g and N_f^g are real network goal node and fake network goal node, respectively. Defender's goal is to protect the goal nodes (real nodes) and drive the attacker toward the fake nodes. Each hyperedge has two conditions regarding exploits (e_i) termed as pre (N_i^-) and post (N_i^+). It is assumed that the attacker can exploit e_i if all preconditions $j \in N_i^-$ are enabled [8]. There will be entry points for the attacker to penetrate the network without having the prior capabilities termed as initial exploits. An attacker can increase the skill set by exploiting more vulnerabilities. Whenever an exploit's attempt is successful, all the postconditions of the exploit become successful. A security state, $s \subseteq N$, is called a feasible security state if for every condition $c_j \in s$ there exist at least one exploit $e_i = (N_i^-, N_i^+) \in E$ such that $c_j \in N_i^+$ and $N_i^-, N_i^+ \subseteq s$ [8] and set $S = \{s_1, \ldots, s_n\}$ represents the state space for this model.

14.3.3.2 Action

Defender's action influences an attacker to choose a network path. So, we assume that the defender can change its network configuration on the fly based on the attacker's action to prevent vertical movement. A simple mean of network configuration can be blocking a port from further communication, which will inhibit the attacker to progress. Another way to prevent attacker's progression is to apply countermeasure of any discovered exploit. The space of defender's available action set is represented by $U = \{u^0, u^1, \ldots, u^n\}$. Here, u^0 represents defender's null action which eventually means the defender will not block any exploit. The remaining actions from the set U signify the network changes which will induce a set of blocked exploits. Each action associated with the set of blocked exploits influences the attacker to traverse the available paths.

14.3.3.3 Threat Model

We construct a model based on a single attacker who is attempting to penetrate the network. An attacker can only increase its capability by exploiting more vulnerabilities, on the other hand, it also increases the chance of being detected. Defender's goal is to prevent exploitation on the real network by allowing the exploits to work on the fake network, which will not be blocked by the defender intentionally. The set of available exploits for real and fake network at a given state s can be defined as [8]:

$$E(s_t = s) = \{er_i = (N_i^-, N_i^+) \in | N_i^- \subset s, N_i^+ \not\subseteq s\} \tag{14.1}$$

$$E(s_t = s) = \{ef_i = (N_i^-, N_i^+) \in | N_i^- \subset s, N_i^+ \not\subseteq s\} \tag{14.2}$$

Two essential requirements must be satisfied for an exploit $e_i = \left(N_i^-, N_i^+\right)$ to be available: (i) $N_i^- \subset s$, i.e. all of the exploit's preconditions must be satisfied and (ii) $N_i^+ \not\subseteq$ s, i.e. the exploit's postconditions must not all be satisfied [8].

The attack probability which defines the attacker will attempt each real network exploits while security state s_t and defense action u_t for a given exploit $er_k \in E$ are given by

$$P_{er_k}(s_t, u_t) = \begin{cases} P_{er_k} \text{ when } er_k \in E(s_t) \setminus B(u_t) \\ P_{er_k} \text{ when } er_k \in E(s_t) \cap B(u_t) \\ 0 \text{ when } er_k \notin E(s_t) \end{cases} \tag{14.3}$$

similarly, for the fake network,

$$P_{ef_k}(s_t, u_t) = \begin{cases} P_{ef_k} \text{ when } ef_k \in E(s_t) \setminus B(u_t) \\ P_{ef_k} \text{ when } ef_k \in E(s_t) \cap B(u_t) \\ 0 \text{ when } ef_k \notin E(s_t) \end{cases} \tag{14.4}$$

In the above equations, P_{er_k} represents the probability of attack when there is no action and P_{-ef_k} defines the attack probability when defender's action block exploits.

To block any vulnerabilities, the defender will choose the action $u \in U$ from the set of accessible defense actions represented by U. Attackers always try to create a set of available initial exploits from reconnaissance state to penetrate the network. So, for any given exploit, er_k and ef_k, there is a probability of success,

$$\alpha_{er_k}(s_t, u_t) = \begin{cases} \overline{\alpha}_{er_k} \text{ when } er_k \notin B(u_t) \\ 0 \text{ when } er_k \in B(u_t) \end{cases} \tag{14.5}$$

similarly, for the fake network,

$$\alpha_{ef_k}(s_t, u_t) = \begin{cases} \overline{\alpha}_{ef_k} \text{ when } ef_k \notin B(u_t) \\ \underline{\alpha}_{ef_k} \text{ when } ef_k \in B(u_t) \end{cases} \tag{14.6}$$

Defenders lack information regarding the current security state and attacker's true strategy, which can be learned from noisy security alerts. In the next section, we describe how the defender uses that information to construct the belief by getting security alerts from the IDS. These security alerts are a mixture of false positive and false negative alerts. For defenders, it is important to differentiate those mixed alerts for better defense actions.

14.3.3.4 Defender's Observation

The IDS is a major component in this model because defender's certainty over the security state depends on security alert. The IDS generates security alert in a sequential form when an attacker attempts to exploit and progress through the network. Those security alerts are not free-form noise terms: false positive and false negative. Even sometimes there will be no alert for exploit activity which solely depends on the attacker capability (stealthiness) termed as a false negative. Similarly, it generates alert for legitimate user activity termed as false positive. It is critically important for the defender to know the exploit activity is going on. Based on the alert, the defender will choose his defensive action to drive the attacker toward deployed fake networks. Filtering out the noisy alert from true alert is an important factor to improve the efficiency of the defender when it turns in real time. In this work, we are considering only known vulnerabilities. There are several alert correlations with exploit activity techniques out there [18–20]. In this work, we are not focusing on alert correlation, rather we are assuming that the defender can do the alert correlation.

Let $Z = \{z_1, z_2, \ldots, z_n\}$ and $Z' = \{z'_1, z'_2, \ldots, z'_n\}$ represent the finite set of security alerts, real and fake network, respectively, generated by the IDS which is eventually the observation set for the defender. Each of the alert from real nodes set and fake nodes set can be generated by the IDS, given by the set $Z(e_{ri}) = \{z_{A_i(1)}, z_{A_i(2)}, \ldots, z_{A_i(ai)}\} \in P(Z)$ and $Z(e_{fi}) = \{z_{D_i(1)}, z_{D_i(2)}, \ldots, z_{D_i(di)}\} \in P'(Z')$, where $P(Z)$ and $P'(Z')$ are the power set of Z and Z'. Using these security alerts, the defender constructs a belief, $b_t \in \Delta S$, where ΔS is the space of probability distribution over security state. The belief state specifies the probability of being in each state given the history of action and observation experienced so far, starting from an initial belief b_0 and the belief update procedure is given in the next section.

14.3.3.5 Defender's Belief Update

For any defense action $u_t = u$ and observation $y_{t+1} = y_k$, the belief update is defined as $b_{t+1} = [T_j(b_t, y_k, u)]s_j \in S$, where $(j)'$ th is the update function, $T_j(b_t, y_k, u) = P(S_{t+1} = s_j \mid U_t = u, Y_{t+1} = y_k, B_t = b_t)$ is given by [8]

$$b^j_{t+1} = T_j(b_t, y_k, u) = \frac{p^u_j(b_t) r^u_{jk}(b_t)}{\rho(b_t, y_k, u)} \tag{14.7}$$

The above terms are defined as

$$p^u_j(b_t) = P(S_{t+1} = s_j \mid U_t, B_t) = \sum_{s_i \in S} b^i_t p^u_{ij} \tag{14.8}$$

$$r^u_{jk}(b_t) = P(Y_{t+1} \mid S_{t+1} = s_t, U_t, B_t) = \sum_{s_i \in S} b^i_t r^u_{ijk} \tag{14.9}$$

$$\rho(b_t, y_k, u) = P(Y_{t+1} | U_t, B_t) = \sum_{s_j \in S} r^u_{jk}(b_t) p^u_j(b_t) \tag{14.10}$$

where p^u_j is the transition probability from state s_i to s_j under defense action u, and $r^u_{jk}(b_t) = P(Y_{t+1} | S_{t+1} = s_t, U_t, B_t)$ is the probability that the IDS will generate observation vector y_k when transitioning from state s_i to s_j under a defense action u. Equation (14.8) defines the trajectory of beliefs based on security alerts termed as observations and series of actions. Under a defense action u, transition probability s_i to s_j is controlled by a set of exploit events. For the available set of exploits from Eq. (14.1), each event in the set of exploits is in binary form (successful and unsuccessful).

The belief update procedure is a controlled Markov Chain, where control is defender action [8]. The majority of partially observable Markov decision process (POMDP) planning methods operate under Bayes theorem [21]. For a large-scale cyber network, a single Bayes update procedure could be computationally infeasible. To plan efficiently for large-scale POMDP, we adopted the model described in [9] for the approximation of the belief state.

14.4 Defender's Action

As soon as the attacker progresses through the network, the defender will take action in real time to limit the attacker progression. Selection of action steps can be improved if the defender has some domain knowledge beforehand. To aid with the domain knowledge, we introduce utility function. Before taking any defensive action, it is also necessary to measure the impact on availability and security cost.

14.4.1 Utility Function

Attackers build an array of node utility function based on the base score metrics for exploiting vulnerabilities [22]. For every exploit, the attacker uses the metrics to justify the attack success probability which is illustrated in (14.13) and serves as the attacker's initial knowledge about the network and vulnerability. Defenders also create the same utility array. From [22], we borrow the impact (*I*) and exploitability (*V*) metrics to define the defender's utility.

$$I = 10.41 * (1 - (1 - CI) * (1 - II) * (1 - AI)) \tag{14.11}$$

$$V_i = 20 * AC * AI * AV \tag{14.12}$$

The above terms are defined as CI = ConfImpact, II = IntegImpact, AI = AvailImpact, I = Impact, Vi = Exploitability, AC = AccessComplexity, AI = Authentication, and AV = Accessvector. The utility array function is defined as

$$U_{a(r,f)} = I * V_i \tag{14.13}$$

Example 14.1 Consider a scenario where there are five nodes and the attacker sends scan queries to the neighbors of node 1. The defender needs to respond to the scan queries deceptively by mixing of true/false information at random. Here, 2, 3 are real nodes and 4, 5 are fake nodes having the following vulnerabilities: $vul(n_2)$, $vul(n_3)$, $vul(n_4)$, and $vul(n_5)$. The defender wants to drive the attacker toward nodes 4 and 5. We are assuming that using the above utility array equation, the defender comes up with the following values: $U_a(n_2) = 15$, $U_a(n_3) = 5$, $U_a(n_4) = 30$, and $U_a(n_5) = 50$. A true rational attacker will go after node 5.

14.4.2 Cost Function

In cyber deception, there is a possibility where you can leverage the availability cost over the security cost. There are two benefits when the attacker is in the fake network: (i) the defender can collect as much as intelligence information on the adversary which helps to derive the attacker's capability, intentions, targets, etc., and (ii) the defender can maximize the network availability to the trusted user during a cyberattack. An availability cost c_a for each action the defender takes to drive the adversary toward the fake network. For some defense action there will be no impact on the availability, and sometimes there will be a more significant impact. To formalize this notion, we represent the availability cost $c_a : U \rightarrow \mathbb{R}$ for each defense action taken by the defender, similarly for the security cost c_s : $S \times U \rightarrow \mathbb{R}$ to depict the cost while the system is in various security state under defense action u. Here, we are considering the availability of a node regarding end-to-end packet delay (considering IT system).

14.4.2.1 End-to-End Packet Delay

Let us assume that d_E and N represent total delay and number devices between a source and destination. The end-to-end delay defined in [23] is

$$d_E = N\left(d_{proc} + d_{trans} + d_{prop} + d_{queue}\right) + d_{proco} \tag{14.14}$$

The above equation's terms are defined as follows: d_{proc} = processing delay, d_{trans} = transmission delay, d_{prop} = propagation delay, d_{queue} = queuing delay, and d_{proco} = processing overhead because of authentication, integrity, and confidentiality. For an uncongested enterprise network, $d_{queue} \simeq 0$ and the distance between a source and destination node is very small so that $d_{prop} \simeq 0$. The processing delay, d_{proc}, is often negligible; however, it strongly influences a router's maximum throughput, which is the maximum rate at which a router can forward packets [24]. So that, (14.14) can be reduced to

$$d_E = N \times d_{trans} \tag{14.15}$$

where $d_{trans} = L/R$, L = packet size and R = transmission rate.

For every defense action, the defender will measure the total end-to-end packet delay. So, the availability cost in terms of delay is defined as $c_u = d_E$. We assign more cost to the goal conditions (attacker's target node) as defender's goal is to keep away the attacker from achieving the goal. The total cost regarding a security state and defense action is

$$c(s_t, u_t) = (1-f)c_s(s_t, u_t) + f*d_E(u_t) \tag{14.16}$$

Here, f is a weighted factor that determines which cost focused more ($f = 0$ represents the defender is concerned only with security cost, $f = 1$ means the defender is only concerned with availability cost). The proposed online deception algorithm is based on an existing online solver [9], computes optimal action from deception standpoint to deceive the attacker with the fake network while balancing availability and security cost.

Algorithm 14.1 Defender's belief update algorithm [8].

Initialize: n_k, $\mathfrak{B}_{t+1} = U_{a(r,f)}b$, numAdded = 0

procedure BELIEF UPDATE $(\mathfrak{B}_t, u_r, y_r)$
  while *numAdded* $< n_k$ **do**
    $(s)\tilde{\mathfrak{B}}_t$
    $(s', y, -) \sim G(s, u_r)$
    If $y^{Z(s)} = y_r^{Z(s)}$ **then** [If alerts Z(s) match]
      $\mathfrak{B}_{t+1} \leftarrow \mathfrak{B}_{t+1} \cup \{s'\}$
      numAdded $\leftarrow$ *numAdded* + 1

14.5 Online Deception Algorithm

Online defense algorithm is a heuristic search algorithm for determining defense actions in real time as the attacker progresses through the network and security alerts are generated where scalability is achieved via a sample-based, online defense algorithm that takes advantage of the structure of the security model to enable computation in large-scale domains. After employing defense actions (e.g. blocking vulnerability), the defender can evaluate the improvements by assessing the attacker's attacking path.

For a scalable network, computing optimal action while deceptively interacting with the attacker is a challenge. Off-line POMDP solver aims to compute the optimal action for each belief state before runtime. Although such solvers have improved their efficiency [25], capturing the optimal action can be intractable for large networks. To resolve this issue, Silver and Veness [9] developed an online algorithm termed as POMCP to handle large-scale network while computing optimal action. Online methods interleave the computation and execution (run-time) phases of policy [8], yielding a much more scalable approach than off-line methods.

POMCP algorithm is based on and makes use of POMDP [25]. There are two types of nodes in POMCP: belief nodes which represent a belief state and action nodes which are their children nodes that can be reached by doing an action. In this work, action selection procedure as same as POMCP algorithm is described in [9] and belief update procedure is based on [8] where it solves the large observation space problem. In POMCP, a belief state updates when a sample observation matches with real-world observation, but for large observation space, it barely matches with real-world observation. In the modified belief update procedure presented in Algorithm 14.1, check a statement whether each incoming alert $z_i \in Z$ matches with over a security state, $Z(s) = Z(e)$. The alerts are generated whenever an attacker attempts an exploit. Alerts not in $Z(s)$ cannot be generated by exploit activity for that security state. We are referring those alerts as false alarms for the defender.

An agent begins the simulation by calling a generative model provides a sample successor state, observation, and cost given a state and action, $(s', y, c) \sim G(s, u)$. The modified belief update procedure is given in Algorithm 14.1, where $\mathfrak{B}_t$ is a state–action pair named particles. History of search tree as shown in Figure 14.2 is constructed by calling the generative model and successive sampling from current belief. Monte-Carlo tree search (MCTS) uses Monte-Carlo simulation for assessing search tree nodes [26]. In the search tree, nodes represent histories and branches from the node in forwarding direction represent the possible future histories because of having partial observability of the fundamental process. A simpler version of MCTS uses greedy tree policy in the very beginning of the simulation, where it selects the action with the highest value. UCT algorithm [27] is used to improve the greedy action selection stage. In the search tree, each action selection is made using UCB1 [28], and the state is being viewed as multiarmed bandit rule to balance the exploration and exploitation. In the UCT algorithm, there is an option to use the domain knowledge [27] to initialize the new nodes. We use the utility array function $U_{a(r,f)}$ as our initial domain knowledge which is improved during more simulation runs. The optimum action for the defender while interacting with the attacker turns into a POMDP. Casting optimum action is defined as

$$V^{\pi}(b_0) = \sum_{t=0}^{\infty} \gamma^t c(b_t, u_t) = \sum_{t=0}^{\infty} \gamma^t E[c(s_t, u_t) \mid b_0, \pi] \tag{14.17}$$

where $0 < \gamma < 1$ is the discount factor, and $c(b_t, u_t)$ represents the cost for each belief state b_t when an action u_t is selected from the space of action where $c(b_t, u_t) = \sum_{s_i \in S} b_t^i c(s_t, u_t)$. For each belief state, defense action generates according

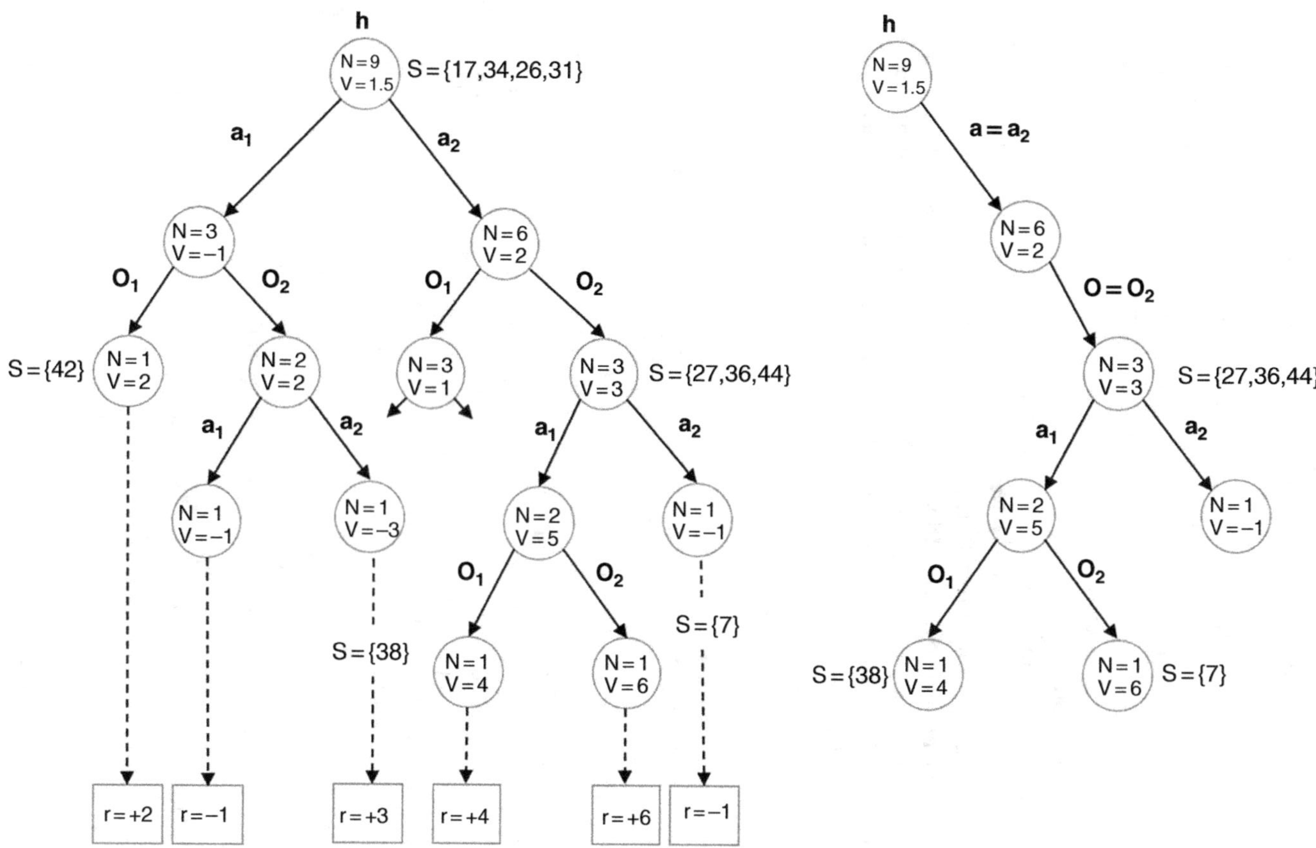

Figure 14.2 An illustration of POMCP in an environment with 2 actions, 2 observations, 50 states, and no intermediate rewards. The agent constructs a search tree from multiple simulations and evaluates each history by its mean return (left). The agent uses the search tree to select a real action *a* and observes a real observation *o* (middle).

to the policy function and belief update must follow the procedure defined in Eq. (14.7). The optimal policy π^* is obtained by optimizing the long-term cost which is

$$\pi^* = \arg\min_{\pi} V^{\pi}(b_0) \tag{14.18}$$

The optimal policy defined in Eq. (14.18) specifies the optimal action for each belief state $b_t \in \Delta S$ where the expected minimum expected cost is calculated over the infinite time horizon. The defender will choose the action where the cost makes the trade-off between availability and security cost.

14.6 IoT Aware Smart Healthcare Architecture

A secure thing and a security attack are the most commonly used terms in the scope of IoT. To make IoT secure, specific security requirements have to be met. To understand the secure thing, we have to follow the characteristics of security first. There are three terms in the security named as: (i) confidentiality, (ii) integrity, and (iii) availability referred to as CIA-triad. Confidentiality is a set of rules to limit unauthorized access to specific information. IoT devices contain sensitive information regarding the patient's health information, e.g. medical records and prescription. Integrity is an essential factor to be considered. It is essential to make sure that the communication between medical devices and collected information are legitimate. Compromising integrity may lead to serious threat sometimes life-threatening. Like, integrity attack against medical devices, e.g. an insulin pump or a pacemaker may lead to fatal outcomes. IoT devices need to available all the time to collect data and give an uninterrupted service to doctors and patients. To comply with CIA-triad and evaluate our approach with the real world, we propose an IoT aware Smart Healthcare System architecture with network decoys where a defender can redirect the attacker toward decoy networks.

The architecture shown in Figure 14.4 is composed of four parts: (i) the RFID-enhanced wireless sensor network (WSN); (ii) the IoT smart gateway; (iii) the user interface and data visualization; and (iv) network decoys. The first part which is also called edge devices is composed of an integrated RFID-WSN 6LoWPAN (low-power wireless personal area network) network composed of four typologies [29]: (i) 6LowPAN border router (6LBR); (ii) 6LowPAN routers (6LR); (iii) 6LowPAN router readers (6LRR); and (iv) 6LowPAN host tag (HT). Here, 6LoWPAN is responsible for connecting the network to the Internet by translating 6LowPAN packets into IPv6 packets and vice versa, and 6LR provides forwarding and routing capabilities [29]. In this architecture, we assumed that several 6LR are deployed throughout the hospital to collect data, such as temperature and pressure. IoT smart gateway is the core of smart healthcare system which is responsible for data collection and processing, system management, and service execution. IoT smart gateway composed of three components, such as two-way proxy, management application, and control DB, and secure access manager and user DB. Two-way proxy performs a transparent communication with Constrained Application Protocol (CoAP) devices. It is also responsible for translating HTTP request from a user interface and Management Application (MA) into CoAP messages and vice versa. The management application and control DB perform two certain tasks: (i) it allows the network operator to control hospital environment and (ii) monitor patients' health status and alerting doctor in case of emergency. The last component, secure access manager and user DB, ensures privacy and data protection. Compromising any of these components leads to a serious threat to patients. To make smart healthcare system more secure while vulnerabilities are present in the network and patches are not available, our proposed algorithm can play a significant role. Our model not only allows the cyber defender to redirect the attacker toward decoy networks but also gives the opportunity to gather critical information about the adversary.

The placement of network decoys in this architecture will be between edge devices and the rest of the corporate network, in this case, IoT smart gateway. For deception, network decoys mimic the configuration of the IoT smart gateway eventually to serve as a fake gateway. After getting the alarm from the IDS, the network administrator blocks the vulnerability from edge devices to push the cyber adversary toward the fake deployed network. Defender's ultimate goal is to secure the IoT smart gateway so that the attacker cannot access user DB and control DB where a massive amount of user information is stored.

14.7 Experimental Results and Discussion

In this simulation, we assume that an attacker is a rational attacker where his aggression, knowledge, and stealthiness are moderate, high, and high, respectively. Also, we assume that the defender can probabilistically correlate the alert from the IDS with exploit activity. Therefore, we create a probability table for alert detection with assumed attacker type where the column represents exploit activity and rows are triggered alert. The probability of detection table is not presented here due to the high volume of the data set. We previously considered that the

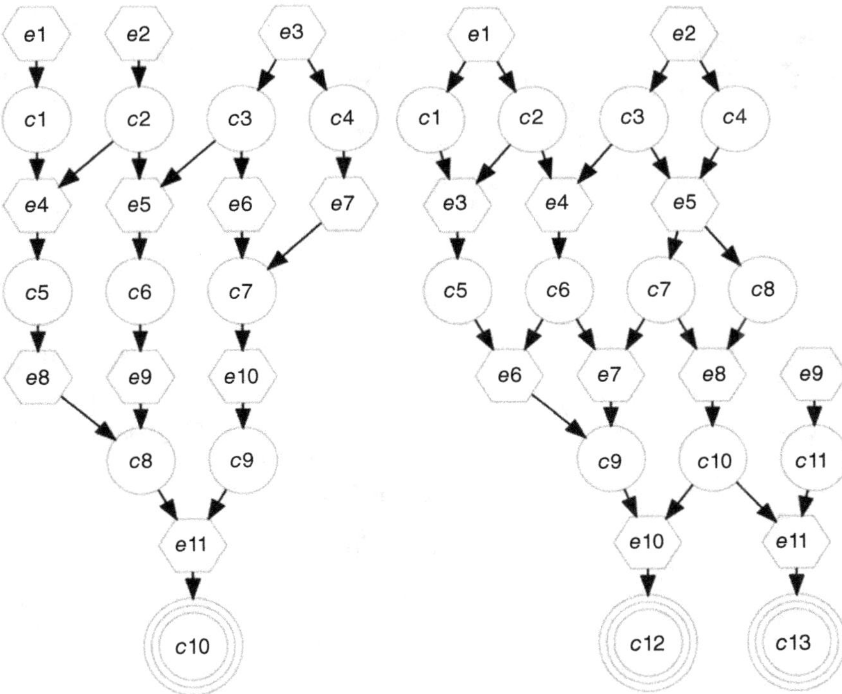

Figure 14.3 A sample exploit dependency graph with a real network (left) and fake network (right). The above dependency graph for real network $H = (N,E)$ consists of $n_{c_r} = 10$ security conditions and $n_{e_r} = 10$ exploits (in the form of hyperedges). Triple-encircled nodes are representing as goal conditions $N_r^g = \{c_{10}\}$ and $N_f^g = \{c_{12}, c_{13}\}$.

attacker can distinguish between fake nodes and real nodes with much less probability with delay and bandwidth handler approach. So, under null defender action, the probability of attacking real nodes and fake nodes are same. For this simulation, we assume that the exploit dependency graph is already generated using Topological Vulnerability Analysis (TVA) [30]. We use the POMDPy software package [31] to use the POMCP solver in our simulation and use python and Matlab to implement our model. Attacker progression depends on the defender's action, and we assume that the defender moves first with null action and wait for the attacker to proceed. In this simulation, we use the sample exploit dependency graph presented in Figure 14.3 to evaluate our approach and present our simulation results. Attack probabilities for each of the exploit under assumed true rational attacker type,

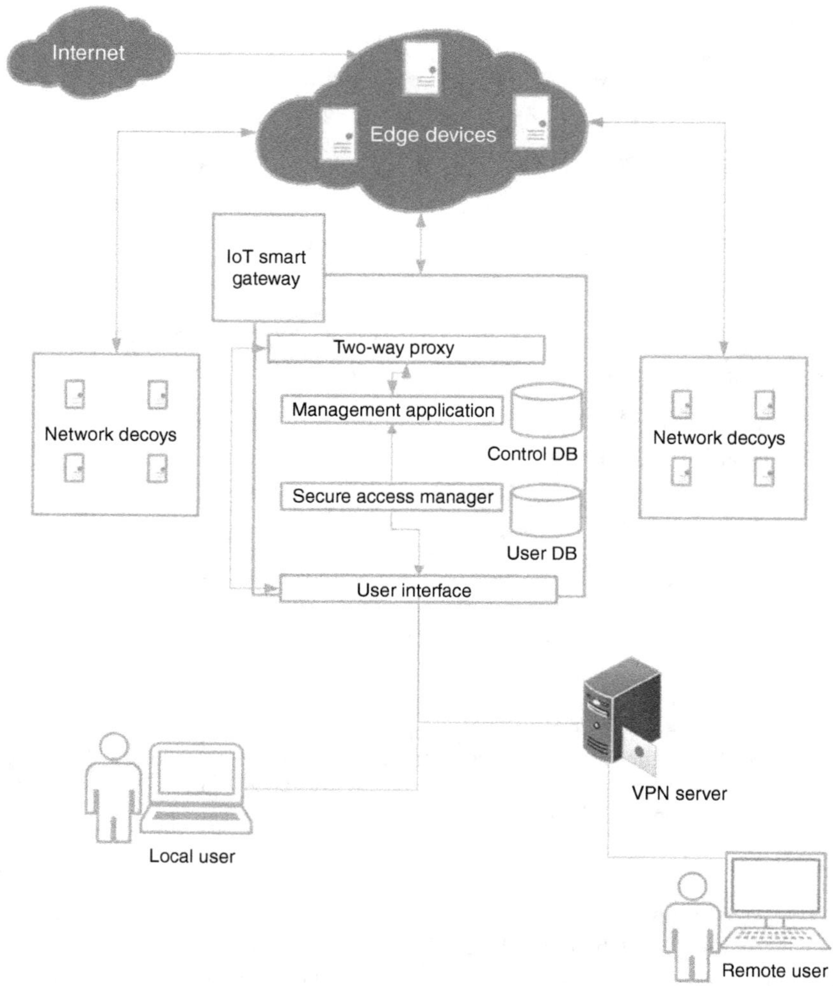

Figure 14.4 IoT aware Smart Healthcare System with network decoys.

$$\left(\bar{P}_{er_k}, \underline{P}_{er_k}\right) = (0.8, 0.3) \text{ for } er_k \in E_0$$

$$\left(\bar{P}_{ef_k}, \underline{P}_{ef_k}\right) = (0.8, 0.3) \text{ for } ef_k \in E_0$$

$$\left(\bar{P}_{er_k}, \underline{P}_{er_k}\right) = (0.7, 0.3) \text{ for } er_k \in \{e_4, e_5, e_6, e_8, e_9\}$$

$$\left(\bar{P}_{ef_k}, \underline{P}_{ef_k}\right) = (0.9, 0.7) \text{ for } er_k \in \{e_5, e_7\}$$

$$\left(\bar{P}_{er_k}, \underline{P}_{er_k}\right) = (0.7, 0.3) \text{ for } er_k \in \{e_7, e_{10}, e_{11}\}$$

$$\left(\bar{P}_{ef_k}, \underline{P}_{ef_k}\right) = (0.9, 0.7) \text{ for } er_k \in \{e_7, e_8\}$$

similarly, the probability of success,

$$\alpha_{er_k} = \begin{cases} 0.7 \text{ when } er_k \in E_0 \\ 0.5 \text{ when } er_k \in E \setminus E_0 \end{cases}$$

$$\alpha_{ef_k} = \begin{cases} 0.85 \text{ when } ef_k \in E_0 \\ 0.7 \text{ when } ef_k \in E \setminus E_0 \end{cases}$$

As we defined earlier, the space of actions is the power set of each defense action. In this simulation, we consider there are three actions for real network which induce a set of block exploits defined as $B(u^1) = \{e_1, e_2, e_3\}$, $B(u^2) = \{e_4, e_5, e_6, e_7, e_8\}$, $B(u^3) = \{e_9, e_{10}, e_{11}\}$. Similarly for the fake network, $B(u^1) = \{e_5, e_7\}$, $B(u^1) = \{e_7, e_8\}$, where the cost of each action is 0.30. The weight cost in Eq. (14.16) is 0.5 and the discount factor $\gamma = 0.95$. In total (real and fake), there are $n_s = 356$ security states and $n_z = 12$ security alerts leading to $2^{12} = 4096$ distinct observation vectors. To approximate the belief, all simulations use particles $n_z = 1500$. The sample evolution of computed deception policy when $N_{sim} = 5000$ is given in Figures 14.5 and 14.6. The computed deception policy is intuitive.

It is assumed that the security state starts from the empty state defined as $s_0 = \emptyset$. The defender uses utility array function to construct the initial belief which is defined in Eq. (14.13). We run the simulation 5000 times. The defender initially (from $t = 1$ to $t = 4$) does not take any action to save the availability cost. As the attacker progresses and enables more conditions, the defender belief gradually updates based on the received security alerts. Then the defender begins to deploy actions ($t = 5$) to block exploits. As we know from monotonicity assumption, once a security condition is enabled, it remains to enable all the time. Whenever the defender belief reflects that the attacker is close to goal conditions, will block the exploits to prevent the attacker reaching his goal.

As we can see from Figure 14.5, at time step $t = 8$ defender blocks exploit $\{e_8, e_9, e_{10}\}$ which prevents an attacker to move forward. From this point, the attacker will try to progress from another way as he received the response from the defender in the reconnaissance stage with a mix of true and false information. Then he moves toward the fake network, Figure 14.6, based on his available set of exploits dictated by Eq. (14.1). At this stage, the defender let the attacker move forward. From time

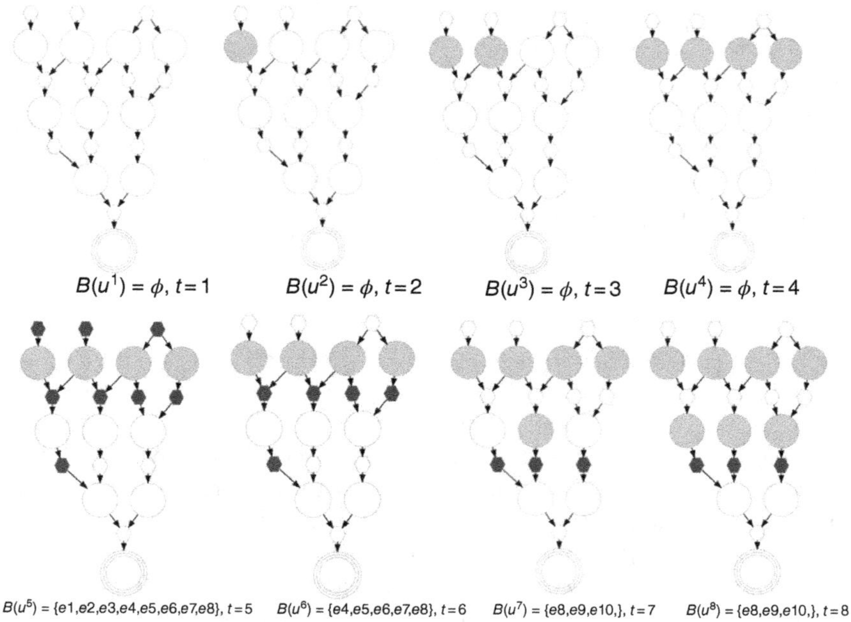

Figure 14.5 Sample evolution of deception policy when an attacker is in a real network. The security state is represented by the shaded node, and blocked exploits are represented by the red-shaped hyperedge.

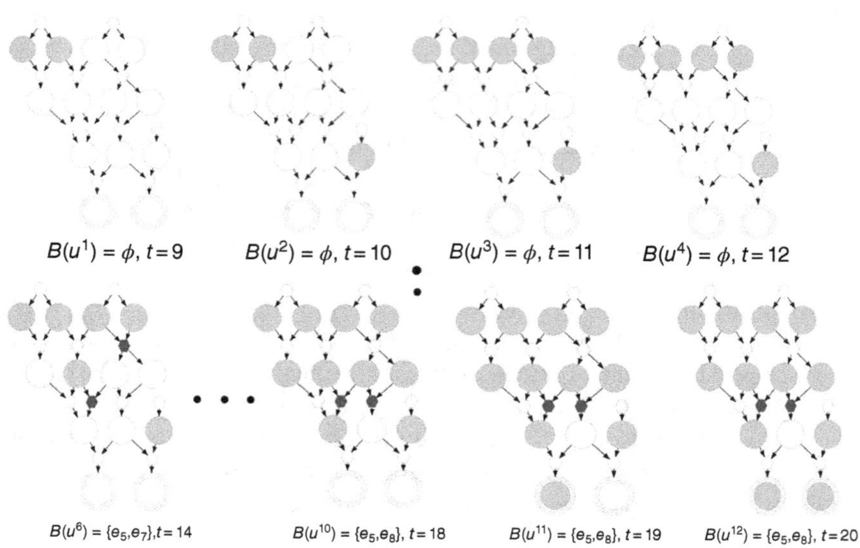

Figure 14.6 Sample evolution of deception policy when the attacker is in the fake network. The security state is represented by the shaded node, and blocked exploits are represented by red-shaped hyperedge.

step $t = 9$–13, defender action is null. As it (fake) is same as the real network from the attacker perspective, the defender will take action only when the attacker has an alternative way to reach the next security state (see time steps $t = 14$–20 in Figure 14.6). This way the attacker will have more confidence that he is in the right track and ultimately he gains nothing. On the other hand, the defender can save more availability cost and learn the attacker which will increase his capability.

From Figure 14.7, we can see that average 60% of the time attackers start with fake initial nodes and carry out the series of exploits to achieve the fake goal node. When $N_{sim} = 500$, out of 25 sample runs, 15 times attackers start with the real network because of poor quality of possible future histories' estimation. When the number of simulations increases and more possible future histories are taken into account, the action estimation quality increased as well as policy function (e.g. $N_{sim} = 5000$, 19 times attackers start with the fake network).

In Figure 14.8, we plot the discounted cost against each time step for 25 sample runs while attackers are in real network state. When $N_{sim} = 500$, 15 times attackers start with the real network where out of 15 times attackers reached the real goal state (node) 13 times. Trajectories which ended up with a circle represent the path where attackers reached the goal. Initially, for low simulation counts, e.g. $N_{sim} = 500$, the defender does not have much information about the attacker's strategy and capability. Because of this, the defender aggressively blocks exploit from the very beginning ($t = 0$) which eventually produces the low quality of estimation and ended up with less availability. For poor estimation, the attacker also reaches the goal node several times as shown in Figure 14.7 upper left corner. As

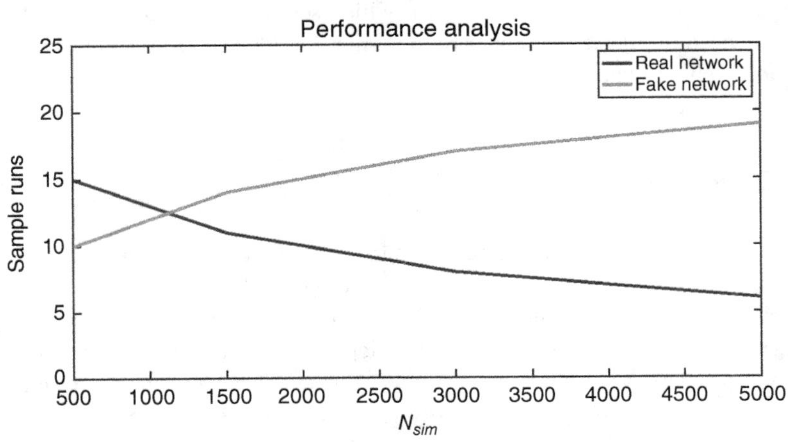

Figure 14.7 Performance evaluation.

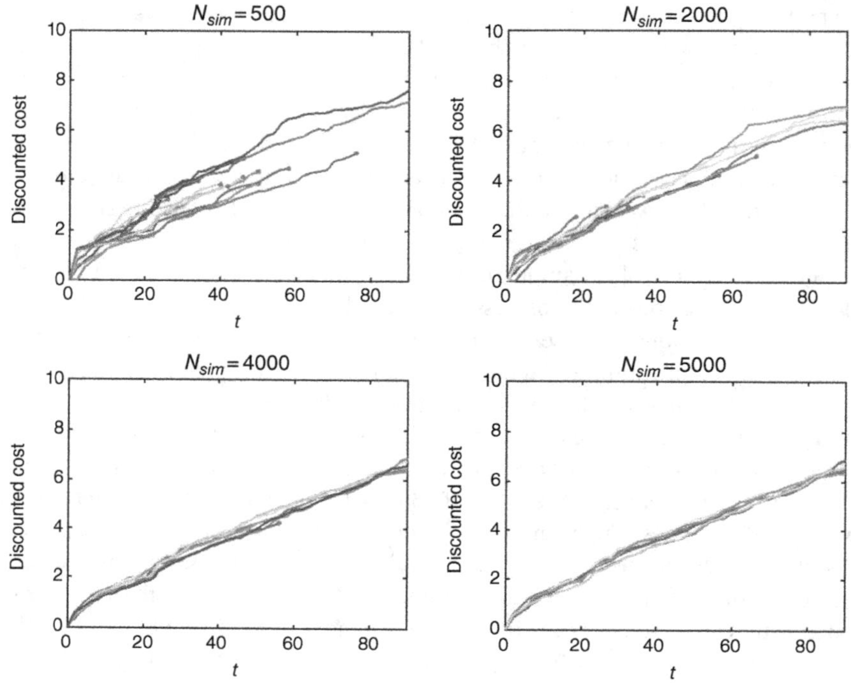

Figure 14.8 Discounted cost.

soon as simulation count increases, more possible future histories are included which results in high quality of estimation (which set of exploits to be blocked). As it is evidenced in Figure 14.8 lower right corner, for 5000 trials attackers could not able to reach any goal state.

14.8 Conclusion

In this chapter, we develop a technique to change the static view of the system to the dynamic view and a cyber-deception approach where defender's action influences an attacker to take different attack path while maintaining availability cost and security cost. To do so, we use an exploit dependency graph which describes attacker progression throughout the network. For every cyber defense system, there is a goal to maintain the availability to the trusted user while security is at a satisfactory level. Using our approach, the defender not only saves availability cost but also can learn the attacker while the attacker is in fake networks. This knowledge will help the defender to make better security planning in the future.

In our future work, we will try to improve the approach so that the attacker will always start with the fake network and achieve a fake goal state and we will design the network decoys in such a way so that it can detect ransomware activity and manage the deceptions in an optimal fashion. The defender can leverage the attack paths using the vulnerability dependency graph to place the decoys at strategic locations that will be most likely targeted.

References

1 R. J. Robles and M.-K. Choi, "Assessment of the vulnerabilities of SCADA, control systems and critical infrastructure systems," *Assessment*, vol. **2**, no. 2, pp. 27–34, 2009.

2 B. Schneier, "Attack trees," *Dr. Dobb's Journal*, vol. **24**, no. 12, pp. 21–29, 1999.

3 O. Sheyner, J. Haines, S. Jha, R. Lippmann, and J. M. Wing, "Automated generation and analysis of attack graphs," *Proceedings 2002 IEEE Symposium on Security and Privacy*, Berkley, CA, USA, 2002, pp. 273–284.

4 P. Ammann, D. Wijesekera, and S. Kaushik, "Scalable, graph-based network vulnerability analysis," in *Proceedings of the 9th ACM Conference on Computer and Communications Security*. ACM, 2002, pp. 217–224.

5 F. J. Stech, K. E. Heckman, and B. E. Strom, "Integrating cyber-D&D into adversary modeling for active cyber defense," in *Cyber Deception*. Springer, 2016, pp. 1–22.

6 K. Horák, Q. Zhu, and B. Bošanskỳ, "Manipulating adversary's belief: A dynamic game approach to deception by design for proactive network security," in *International Conference on Decision and Game Theory for Security*. Springer, 2017, pp. 273–294.

7 S. Achleitner, T. F. La Porta, P. McDaniel, S. Sugrim, S. V. Krishnamurthy, and R. Chadha, "Deceiving network reconnaissance using SDN based virtual topologies," *IEEE Transactions on Network and Service Management*, vol. **14**, no. 4, pp. 1098–1112, 2017.

8 E. Miehling, M. Rasouli, and D. Teneketzis, "A POMDP approach to the dynamic defense of large-scale cyber networks," *IEEE Transactions on Information Forensics and Security*, vol. **13**, no. 10, pp. 2490–2505, 2018.

9 D. Silver and J. Veness, "Monte-Carlo planning in large POMDPs," in *Advances in Neural Information Processing Systems*, Curran Associates, Inc., 2010, pp. 2164–2172.

10 A. Schlenker, O. Thakoor, H. Xu, F. Fang, M. Tambe, L. Tran-Thanh, P. Vayanos, and Y. Vorobeychik, "Deceiving cyber adversaries: A game theoretic approach," in *Proceedings of the 17th International Conference on Autonomous Agents and Multiagent Systems*. International Foundation for Autonomous Agents and Multiagent Systems, 2018, pp. 892–900.

11 M. Albanese, E. Battista, S. Jajodia, and V. Casola, "Manipulating the attacker's view of a system's attack surface," in *2014 IEEE Conference on Communications and Network Security (CNS)*. IEEE, 2014, pp. 472–480.

12 S. T. Trassare, R. Beverly, and D. Alderson, "A technique for network topology deception," in *MILCOM 2013–2013 IEEE Military Communications Conference*. IEEE, 2013, pp. 1795–1800.

13 M. Dunlop, S. Groat, R. Marchany, and J. Tront, "Implementing an IPv6 moving target defense on a live network," in *Moving Target Research Symposium 2012*. Cyber-Physical Systems Virtual Organization, 2012.

14 Q. Duan, E. Al-Shaer, and H. Jafarian, "Efficient random route mutation considering flow and network constraints," in *2013 IEEE Conference on Communications and Network Security (CNS)*. IEEE, 2013, pp. 260–268.

15 D. Watson, M. Smart, G. R. Malan, and F. Jahanian, "Protocol scrubbing: Network security through transparent flow modification," *IEEE/ACM Transactions on Networking*, vol. **12**, no. 2, pp. 261–273, 2004.

16 J. H. Jafarian, E. Al-Shaer, and Q. Duan, "Openflow random host mutation: Transparent moving target defense using software defined networking," in *Proceedings of the First Workshop on Hot Topics in Software Defined Networks*. ACM, 2012, pp. 127–132.

17 S. Noel and S. Jajodia, "Managing attack graph complexity through visual hierarchical aggregation," in *Proceedings of the 2004 ACM Workshop on Visualization and Data Mining for Computer Security*. ACM, 2004, pp. 109–118.

18 B. Morin, L. Mé, H. Debar, and M. Ducassé, "M2D2: A formal data model for ids alert correlation," in *International Workshop on Recent Advances in Intrusion Detection*. Springer, 2002, pp. 115–137.

19 R. Gula, "*Correlating Ids Alerts with Vulnerability Information*," Tenable Network Security, Inc., 2002.

20 F. Valeur, G. Vigna, C. Kruegel, and R. A. Kemmerer, "Comprehensive approach to intrusion detection alert correlation," *IEEE Transactions on Dependable and Secure Computing*, vol. **1**, no. 3, pp. 146–169, 2004.

21 S. Ross, J. Pineau, S. Paquet, and B. Chaib-Draa, "Online planning algorithms for POMDPs," *Journal of Artificial Intelligence Research*, vol. **32**, pp. 663–704, 2008.

22 P. Mell, K. Scarfone, and S. Romanosky, "A complete guide to the common vulnerability scoring system version 2.0," in *Published by FIRST-Forum of Incident Response and Security Teams*, National Institute of Standards and Technology, vol. **1**, 2007, p. 23.

23 K. Hasan, S. Shetty, A. Hassanzadeh, M. B. Salem, and J. Chen, "Modeling cost of countermeasures in software defined networking-enabled energy delivery systems," in *2018 IEEE Conference on Communications and Network Security (CNS)*. IEEE, 2018, pp. 1–9.

24 J. Kurose and K. Ross, "Computer networking: A top down approach," pp. 143–144, 2007.

25 H. Kurniawati, D. Hsu, and W. S. Lee, "SARSOP: Efficient point-based POMDP planning by approximating optimally reachable belief spaces," in *Robotics: Science and Systems*, MITP, Cambridge, MA, Vol. **2008**. Zurich, Switzerland, 2008.

26 R. Coulom, "Efficient selectivity and backup operators in Monte-Carlo tree search," in *International Conference on Computers and Games*. Springer, 2006, pp. 72–83.

27 L. Kocsis and C. Szepesvári, "Bandit based Monte-Carlo planning," in *European Conference on Machine Learning*. Springer, 2006, pp. 282–293.

28 P. Auer, N. Cesa-Bianchi, and P. Fischer, "Finite-time analysis of the multiarmed bandit problem," *Machine Learning*, vol. **47**, no. 2–3, pp. 235–256, 2002.

29 L. Catarinucci, D. De Donno, L. Mainetti, L. Palano, L. Patrono, M. L. Stefanizzi, and L. Tarricone, "An IoT-aware architecture for smart healthcare systems," *IEEE Internet of Things Journal*, vol. **2**, no. 6, pp. 515–526, 2015.

30 S. Jajodia and S. Noel, "Topological vulnerability analysis," in *Cyber Situational Awareness*. Springer, 2010, pp. 139–154.

31 P. Emami, A. J. Hamlet, and C. Crane, "POMDPy: An extensible framework for implementing POMDPs in python," 2015.

15

A Coding Theoretic View of Secure State Reconstruction

Suhas Diggavi and Paulo Tabuada

University of California, Los Angeles, Los Angeles, CA, USA

15.1 Introduction and Motivation

Every day, hundreds and thousands of sensors continuously transmit measurements to controllers that transform them into actuation commands regulating systems of many scales, from small consumer systems to large-scale critical infrastructure. Security of such cyber-physical systems (CPS), also known as Internet of Things (IoT), has been an afterthought and not an integral part of the original design considerations. However, security risks to these systems have recently escalated due to several factors: (i) standardization of equipment has led to the widespread use of equipment with well-known security flaws; (ii) increasingly these control systems are being interconnected through communication networks both for distributed control and sensing, as well as for remote monitoring and reconfiguration; (iii) there is widespread availability of information about the physical and control systems that can be exploited by hostile entities. Hence, modern society is at a critical juncture in that we rely more and more on CPS and IoT for basic needs and yet, these are increasingly exposed to security vulnerabilities both at the cyber as well as at the physical level. This led to the emergence of new security challenges that are distinct from those addressed by traditional cyber security.

At the center of any control algorithm for CPS and IoT is the reconstruction of the state of the physical system being controlled. In this context, it is crucial to understand the fundamental limits for state reconstruction, an integral aspect of CPS and IoT, in the presence of malicious attacks. With this motivation, we focus in this chapter on securely reconstructing the state of a linear dynamical

Modeling and Design of Secure Internet of Things, First Edition. Edited by Charles A. Kamhoua, Laurent L. Njilla, Alexander Kott and Sachin Shetty.

Published 2020 by John Wiley & Sons, Inc.

system from a set of maliciously corrupted sensor measurements and actuator commands. We restrict the attacks to be sparse in nature, i.e. an adversary can arbitrarily corrupt an unknown subset of sensors and actuators in the system but is restricted by an upper bound on the number of attacked sensors and actuators. In this chapter, we review our recent work on state reconstruction in the presence of adversarial manipulation of sensing and actuation. In particular, we will focus on some of the new approaches to security of CPS and IoT, which use physics/dynamics as a mechanism to defend against attacks, and complement information security approaches to CPS and IoT. On the one hand, physical dynamics offer a new attack surface against which existing cyber-security mechanisms have no defense, e.g. by spoofing physical signals before they reach the cyber domain. On the other hand, the physical dynamics offer additional degrees of redundancy that can be exploited to improve security. Therefore, one can turn the perceived vulnerabilities in physical systems into an advantage, as an adversary can attack the sensing, communication, or computation, but cannot alter physics. In this chapter, we take the perspective that physical dynamics provides redundancy in the form of an "error-correcting" code enabling state reconstruction despite attacks (adversarial errors). This perspective offers an alternative view of our recent work based on coding theory. Although connections between coding theory and systems theory have been established long ago [1], a detailed discussion of secure state reconstruction under the light of coding theory has not yet appeared in the literature with the exception of the abbreviated discussion in section VII of [2]. Moreover, as the purpose of this chapter is to present the authors' recent work under a new perspective, we ask the readers to consult the cited papers for a complete description of all the related literature.

15.2 Problem Definition and Model

15.2.1 Notation

We denote the sets of real and natural (including zero) numbers by $\mathbb{R}$ and $\mathbb{N}_0$. We represent vectors and real numbers by lowercase letters, such as $\mathbf{u}, \mathbf{x}, \mathbf{y}$, and matrices with uppercase letters, such as $\mathbf{A}$. If S is a set, $|S|$ is the cardinality of S. Given a vector $\mathbf{x} \in \mathbb{R}^n$ and a set $O \subseteq \{1, \ldots, n\}$, we use $\mathbf{x}|_O$ to denote the vector obtained from $\mathbf{x}$ by removing all elements except those indexed by the set O. Similarly, for a matrix $\mathbf{C} \in \mathbb{R}^{n_1 \times n_2}$, we use $\mathbf{C}|_{(O_1,O_2)}$ to denote the matrix obtained from $\mathbf{C}$ by eliminating all rows and columns except the ones indexed by O_1 and O_2, respectively, where $O_i \subseteq \{1, \ldots, n_i\}$ with $n_i \in \mathbb{N}$ for $i \in \{1, 2\}$. In order to simplify the notation, we use $\mathbf{C}|_{(\cdot,O_2)}:=\mathbf{C}|_{(\{1,\ldots,n_1\},O_2)}$ and $\mathbf{C}|_{(O_1,\cdot)}:=\mathbf{C}|_{(O_1,\{1,\ldots,n_2\})}$. We denote the complement of O by $\overline{O}:=\{1,\ldots,n\}\backslash O$.

We use the notation $\{\mathbf{x}(t)\}_{t=0}^{\ell}$ to denote the sequence $\mathbf{x}(0), \dots, x(\ell)$, and we drop the sub(super)scripts whenever it is clear from the context. The support set of a vector $\mathbf{x} \in \mathbb{R}^n$, denoted by supp($\mathbf{x}$), is the set of indices of its nonzero components. Similarly, we define the support of the sequence $\{\mathbf{x}(t)\}$ as $\text{supp}(\mathbf{x}(t)) := \cap_t \text{supp}(\mathbf{x}(t))$.

15.2.2 System Model

We consider a discrete-time linear time-invariant (LTI) system subject to sensor and actuator attacks as described by the following model:

$$\begin{aligned} \mathbf{x}(t+1) &= \mathbf{A}\mathbf{x}(t) + \mathbf{B}(\mathbf{u}(t) + \mathbf{e}_a(t)), \\ \mathbf{y}(t) &= \mathbf{C}\mathbf{x}(t) + \mathbf{e}_s(t), \end{aligned} \tag{15.1}$$

where $\mathbf{u}(t) \in \mathbb{R}^m$, $\mathbf{x}(t) \in \mathbb{R}^n$, and $\mathbf{y}(t) \in \mathbb{R}^p$ are the input, state, and output variables, respectively, at time $t \in \mathbb{N}_0$, $\mathbf{e}_a(t)$ and $\mathbf{e}_s(t)$ model actuator and sensor attacks, respectively. In Section 15.2.3, we describe in detail the attack model. The symbols **A**, **B**, and **C** denote system matrices with appropriate dimensions. We use (**A**, **B**, **C**) to denote the system described by (15.1). The order of an LTI system is defined as the dimension n of its state space. For an LTI system,

$$\mathcal{O} = \begin{bmatrix} C \\ CA \\ \vdots \\ CA^{n-1} \end{bmatrix}, \qquad \mathcal{N} = \begin{bmatrix} 0 & 0 & \dots & 0 \\ CB & 0 & \dots & 0 \\ CAB & CB & \dots & 0 \\ \vdots & & \ddots & \\ CA^{n-2}B & CA^{n-3}B & \dots & CB \end{bmatrix}, \tag{15.2}$$

are the *observability* and *invertibility* matrices, respectively.

15.2.3 Attack Model

In this section, we discuss the adversarial setup along with assumptions on the adversary, and provide a mathematical formulation of the secure state reconstruction problem considered in this chapter.

The adversary is able to arbitrarily change sensor measurements. Although this assumption may seem to endow the adversary with an unreasonable ability, we refer the interested readers to [3] for a concrete example on how sensor measurements can be arbitrarily manipulated by an adversary. Changes to sensor measurements **y** are modeled by the signal $\mathbf{e}_s$ injected by the adversary in (15.1). Similarly, actuator attacks are modeled by the signal $\mathbf{e}_a$ in (15.1). We make no assumptions on the signals $\mathbf{e}_s$ and $\mathbf{e}_a$ except for a bound on the number of sensors and actuators that can be attacked. In particular, we place no deterministic or stochastic assumptions on the magnitude or time evolution of the signals $\mathbf{e}_s$ and $\mathbf{e}_a$.

Assumption 15.1 (Bound on the Number of Attacks) The number of inputs and outputs under attack are bounded by r and s, respectively.

According to this assumption, an adversary can attack a subset of inputs and outputs denoted by $\Gamma_a \subseteq \{1, \ldots, m\}$ and $\overline{\Gamma}_s \subseteq \{1, \ldots, p\}$,[1] respectively, with $|\Gamma_a| \leq r$ and $|\overline{\Gamma}_s| \leq s$, such that $\text{supp}(\mathbf{e}_a(t)) \subseteq \Gamma_a$ and $\text{supp}(\mathbf{e}_s(t)) \subseteq \overline{\Gamma}_s$. Note that the sets Γ_a and $\overline{\Gamma}_s$ are not known to the controller and only upper bounds on their cardinality are known. Once the adversary chooses these sets, inputs and outputs outside these sets remain attack-free. This assumption is realistic when the time it takes for the adversarial agent to attack new inputs and outputs is large compared to the time scale of the algorithms employed for secure state reconstruction.

15.2.4 The Secure State Reconstruction Problem

We now precisely define the problems we address in this chapter. The first addresses attack detection.

Problem 15.1 (Attack Detection) For the LTI system defined by (15.1), under Assumption 16.1, what are necessary and sufficient conditions to detect an attack on its sensors and/or actuators based on the outputs $\mathbf{y}$ and inputs $\mathbf{u}$?

Once an attack is detected, the natural next question is to determine if it is still possible to reconstruct the state.

Problem 15.2 (Secure State Reconstruction) For the LTI system defined by (15.1), under Assumption 16.1, what are necessary and sufficient conditions under which the state $\mathbf{x}$ can be reconstructed from the outputs $\mathbf{y}$ and inputs $\mathbf{u}$?

15.3 Secure State Reconstruction as Error Correction

In this section, we show how the secure state reconstruction problem can be formulated as an error-correction problem.

Consider the following thought experiment. The plant seeks to send the state $\mathbf{x}(\tau)$ at time τ to the controller. The state $\mathbf{x}(\tau)$ is then the *real-valued message* to be transmitted. Since the only information available to the plant about the state is $\mathbf{Cx}$, it sends instead the sequence of measurements:

1 For ease of exposition, we use Γ_a to denote under-attack inputs while using Γ_s for the set of attack-free outputs, i.e. the set of under-attack outputs is represented by $\overline{\Gamma}_s := \{1, \ldots, p\} \setminus \Gamma_s$ in this chapter.

$$\{\mathbf{Cx}(t)\}_{t=\tau}^{\tau+n-1} = \mathbf{Cx}(\tau), \mathbf{Cx}(\tau+1), \dots, \mathbf{Cx}(\tau+n-1), \tag{15.3}$$

to the controller. We regard this sequence as the *encoded message*. The controller receives the following corrupted version of (15.3):

$$\{\mathbf{y}(t)\}_{t=\tau}^{\tau+n-1} = \{\mathbf{Cx}(t) + \mathbf{e}_s(t)\}_{t=\tau}^{\tau+n-1}. \tag{15.4}$$

We are thus regarding sensor attacks as *errors* in the received message. In order for the controller to extract the state from the received message, it must use:

1) The dynamics (15.1)
2) The knowledge of the inputs that were fed into the system

Hence, we can interpret the dynamics (15.1) as implicitly defining the code that will be used for error detection and correction. Here, "decoding" means extracting $\mathbf{x}(\tau)$ from $\{\mathbf{y}(t)\}_{t=\tau}^{\tau+n-1}$.

Since the dynamics is parameterized by the inputs, we can think of it as defining a family of codes, one for each sequence of used inputs. In the absence of actuator attacks, the controller knows exactly which inputs are used and thus knows the code. However, in the presence of actuator attacks, the inputs $\{\mathbf{u}(t) + \mathbf{e}_a(t)\}_{t=\tau}^{\tau+n-1}$ fed to the plant differ from the inputs $\{\mathbf{u}(t)\}_{t=\tau}^{\tau+n-1}$ sent by the controller and decoding also needs to account for input errors.

Let us first consider the nominal behavior that corresponds to the absence of attacks, i.e. $\mathbf{e}_a = 0 = \mathbf{e}_s$. The evolution of the output is then given by:

$$\mathbf{Y}_\tau^n(\mathbf{x}(\tau), \mathbf{U}_\tau^{n-1}) = \begin{bmatrix} \mathbf{y}(\tau) \\ \mathbf{y}(\tau+1) \\ \vdots \\ \mathbf{y}(\tau+n-1) \end{bmatrix} = \mathcal{O}\mathbf{x}(\tau) + \mathcal{N} \underbrace{\begin{bmatrix} \mathbf{u}(\tau) \\ \mathbf{u}(\tau+1) \\ \vdots \\ \mathbf{u}(\tau+n-2) \end{bmatrix}}_{\mathbf{U}_\tau^{n-1}} = \mathcal{O}\mathbf{x}(\tau) + \mathcal{N}\mathbf{U}_\tau^{n-1}, \tag{15.5}$$

where the observability matrix $\mathcal{O}$ is defined in (15.2). At time $\tau + 1$, we have:

$$\mathbf{Y}_{\tau+1}^n(\mathbf{x}(\tau+1), \mathbf{U}_{\tau+1}^{n-1}) = \mathcal{O}\underbrace{\{\mathbf{Ax}(\tau) + \mathbf{Bu}(\tau)\}}_{\mathbf{x}(\tau+1)} + \mathcal{N}\mathbf{U}_{\tau+1}^{n-1}. \tag{15.6}$$

We note that the matrix product $\mathcal{O}\mathbf{A}$ can be expressed as:

$$\mathcal{O}\mathbf{A} = \begin{bmatrix} CA \\ CA^2 \\ \vdots \\ CA^n \end{bmatrix} = \begin{bmatrix} 0 & I & 0 & \dots & 0 \\ 0 & 0 & I & \dots & 0 \\ \vdots & & & \ddots & \\ 0 & 0 & 0 & \dots & I \\ -c_0 I & -c_1 I & -c_2 I & \dots & -c_{n-1} I \end{bmatrix} \mathcal{O} = \mathcal{A}\mathcal{O},$$

where $c_0, \ldots, c_{n-1}$ are the coefficients of the characteristic polynomial of A and we used Cayley-Hamilton's theorem [4]. Using the equality $\mathcal{O}\mathbf{A} = \mathcal{A}\mathcal{O}$, we can rewrite (15.6) (suppressing the arguments for simplicity) as:

$$\begin{aligned}\mathbf{Y}^n_{\tau+1} &= \mathcal{A}\mathcal{O}\mathbf{x}(\tau) + \mathcal{O}\mathbf{B}\mathbf{u}(\tau) + \mathcal{N}\mathbf{U}^{n-1}_{\tau+1} \\ &= \mathcal{A}\mathbf{Y}^n_\tau + \mathcal{O}\mathbf{B}\mathbf{u}(\tau) + \mathcal{N}\mathbf{U}^{n-1}_{\tau+1}.\end{aligned}$$

This equality can be written in (affine) kernel form (see also the next section on affine codes):

$$\mathbf{L}\mathbf{Y}^{n+1}_\tau = \mathbf{M}\mathbf{U}^n_\tau. \tag{15.7}$$

The matrix $\mathbf{L}$ in (15.7) defines the code induced by the dynamics in the sense that the outputs $\mathbf{Y}^{n+1}_\tau$ and inputs $\mathbf{U}^n_\tau$ must live in the affine subspace defined by the kernel of $\mathbf{L}$, with an input-dependent shift $\mathbf{M}\mathbf{U}^n_\tau$. Error detection consists in checking if the affine code (15.7) is satisfied, i.e. if the sequence of received outputs $\mathbf{Y}^{n+1}_\tau$ and inputs $\mathbf{U}^n_\tau$ satisfies (15.7). Error correction asks for a suitable modification to the received outputs $\mathbf{Y}^{n+1}_\tau$ and inputs $\mathbf{U}^n_\tau$ so that they satisfy (15.7). If we denote this modification by $\mathbf{E}_s$, resulting in the *corrected message* $\mathbf{Y}^{n+1}_\tau - \mathbf{E}_s$, error correction can be formulated as the following optimization problem:

$$\begin{aligned}\text{minimize} \quad & |\overline{\Gamma_s}| + |\Gamma_a| \\ \text{subject to} \quad & \text{supp}(\{\mathbf{e}_s\}) \subseteq \overline{\Gamma_s}, \\ & \text{supp}(\{\mathbf{e}_a\}) \subseteq \Gamma_a, \\ & \mathbf{L}\left(\mathbf{Y}^{n+1}_\tau - \mathbf{E}_s(\{\mathbf{e}_s\}, \{\mathbf{e}_a\})\right) = \mathbf{M}\left(\mathbf{U}^n_\tau - \mathbf{E}_a(\mathbf{e}_a)\right).\end{aligned} \tag{15.8}$$

In other words, we are searching for the smallest number $|\overline{\Gamma_s}|$ of attacked sensors and the smallest number $|\Gamma_a|$ of attacked actuators for which there exist sequences of attacks $\{\mathbf{e}_s\}$ and $\{\mathbf{e}_a\}$, supported on $\overline{\Gamma_s}$ and Γ_a, respectively, leading to a corrected message $\mathbf{Y}^{n+1}_\tau - \mathbf{E}_s(\{\mathbf{e}_s\}, \{\mathbf{e}_a\})$ satisfying the code (15.7) with the corrected shift $\mathbf{M}\left(\mathbf{U}^n_\tau - \mathbf{E}_a(\mathbf{e}_a)\right)$.

Given the bounds r and s on sensor and actuator attacks, we now seek to determine if the code defined by (15.7) is strong enough to allow for attack detection and correction. In particular, the ability to perform correction implies that the optimization problem (15.8) has a unique solution. Before doing so, however, we review some well-established coding theoretic results in the next section.

15.4 Connections to Classical Coding Theory

In classical coding theory, the transmissions are over a finite field $\mathbb{F}_q$, and the goal is to construct a "codebook" $\mathcal{C}$, such that despite noisy observations of the codewords $\mathbf{c} \in \mathcal{C}$, one can correctly infer the transmitted message [5]. To make this more specific, we will review the ideas of classical coding theory in terms of real numbers and vectors, using a notation which is compatible with the previous section.

A *linear* code (m, k) linearly transforms an underlying message $\mathbf{x} \in \mathbb{R}^k$ to a "transmission" $\mathbf{y} = \mathbf{G}\mathbf{x}$, $\mathbf{y} \in \mathbb{R}^m$, where $\mathbf{G}$ is the "generator" matrix of the code. An alternative representation of the code is given by a parity check matrix where the valid codewords $\mathbf{z}$ are represented as the kernel of a parity check matrix $\mathbf{H}$, i.e.

$$\mathcal{C} = \{\mathbf{y} \in \mathbb{R}^m : \mathbf{H}\mathbf{y} = \mathbf{0}\}, \tag{15.9}$$

In a similar manner, we can define an *affine* code as

$$\mathcal{C} = \{\mathbf{y} \in \mathbb{R}^m : \mathbf{H}\mathbf{y} = \mathbf{z}\}, \tag{15.10}$$

where $\mathbf{z}$ is a fixed affine shift, similar to the representation of the code (15.7) induced by the dynamics.

We consider additive errors, i.e. we observe the transmission $\mathbf{w}$ corrupted by an additive error $\mathbf{e}$:

$$\mathbf{w} = \mathbf{y} + \mathbf{e}, \tag{15.11}$$

where $\mathbf{e}$ satisfies $\text{supp}(\mathbf{e}) \leq d$, i.e. at most d dimensions of the observations can be corrupted. One can then define the error-correcting capability of a code through the notion of minimum Hamming distance between valid codewords.

Definition 15.1 The Hamming distance between two codewords $\mathbf{y}, \mathbf{y}' \in \mathcal{C}$ is given by:

$$d_H(\mathbf{y}, \mathbf{y}') = \text{supp}(\mathbf{y} - \mathbf{y}'),$$

i.e. the number of co-ordinates in which $\mathbf{y}$ and $\mathbf{y}'$ differ.

From the error model (15.11), it is easy to see that:

$$d_H(\mathbf{y}, \mathbf{w}) = \text{supp}(\mathbf{e}) \leq d.$$

It is also clear that if the codebook $\mathcal{C}$ is constructed so that for all valid codewords $\mathbf{y}, \mathbf{y}' \in \mathcal{C}$ we have:

$$d_H(\mathbf{y}, \mathbf{y}') \geq 2d + 1,$$

then no two valid codewords in the codebook are confusable under the error model (15.11). This can be seen by noting that given $\mathbf{w}$ there always exists a codeword in $\mathcal{C}$ that is the closest (with respect to the Hamming distance) to $\mathbf{w}$. Hence, if we define the minimum distance of a code $\mathcal{C}$ as:

$$d_{\min}(\mathcal{C}) = \min_{\mathbf{y}, \mathbf{y}' \in \mathcal{C}, \mathbf{y} \neq \mathbf{y}'} d_H(\mathbf{y}, \mathbf{y}'),$$

then

$$d_{\min}(\mathcal{C}) \geq 2d + 1,$$

guarantees that any error with supp($\mathbf{e}$) $\leq d$ can be corrected by using the code $\mathcal{C}$. Moreover, correcting the effect of $\mathbf{e}$ consists of finding the nearest codeword (in terms of Hamming distance) to $\mathbf{w}$, i.e. by solving the problem:

$$\begin{aligned} \text{minimize} \quad & |\,\Gamma\,| \\ \text{subject to} \quad & \text{supp}(\mathbf{e}) \subseteq \Gamma \\ & \mathbf{H}(\mathbf{w} - \mathbf{e}) = \mathbf{y}, \end{aligned} \tag{15.12}$$

where Γ is the set of indices that are corrupted by the error $\mathbf{e}$. Although we can simplify this optimization formulation by taking Γ to be supp($\mathbf{e}$), this formulation highlights the connection with the optimization problem (15.8) in the previous section.

15.5 Sparse Strong Observability, Hamming Distance, and Secure State Reconstruction

In the absence of attacks, the ability to reconstruct the state of a system from its inputs and outputs is characterized by the notion of observability [4]. In the presence of sensor attacks only, reconstructing the state requires redundancy in the observations and this led to a redundant notion of observability that was first introduced in [6] and later refined and termed *sparse observability* in [7]. The same notion was independently introduced in [8] in the context of continuous-time LTI systems. When the inputs are also under attack, one must be able to reconstruct the state without their knowledge. The problem of reconstructing the state, in the absence of attacks, without the knowledge of the inputs has been studied in the literature and its solvability is characterized by the notion of strong observability [9].

Definition 15.2 (Strong Observability) An LTI system is called strongly observable if for any $\tau \in \mathbb{R}$, any state $\mathbf{x}(\tau) \in \mathbb{R}^n$, and any input sequence $\{\mathbf{u}(t)\}_{t=\tau}^{\infty}$ there exists an integer $\ell \in \mathbb{N}_0$ such that $x(\tau)$ can be uniquely recovered from $\{\mathbf{y}(t)\}_{t=\tau}^{\tau+\ell}$.

Note that $\ell + 1$ is always upper-bounded by the order of the system and thus we can replace ℓ with $n - 1$ in the preceding definition.

Although sparse observability provides redundancy against sensor attacks, it does not account for the possibility of input attacks. This suggests the need for a redundant notion of strong observability, termed *sparse strong observability*, that was introduced in [10].

Definition 15.3 ((*r*, *s*)-Sparse Strong Observability) An LTI system $(\mathbf{A}, \mathbf{B}, \mathbf{C})$ with m inputs and p outputs is (r, s)-sparse strongly observable if for any $\Gamma_a \subseteq \{1, \dots, m\}$ and $\Gamma_s \subseteq \{1, \dots, p\}$ with $|\Gamma_a| \leq r$ and $|\Gamma_s| \geq p - s$, the system $\left(\mathbf{A}, \mathbf{B}_{(\cdot,\Gamma_a)}, \mathbf{C}_{(\Gamma_s,\cdot)}\right)$ is strongly observable.

Note that the system $\left(\mathbf{A}, \mathbf{B}_{(\cdot,\Gamma_a)}, \mathbf{C}_{(\Gamma_s,\cdot)}\right)$ is the restriction of the original system in (15.1) to inputs from Γ_a and outputs from Γ_s. A coding theoretic interpretation of strong sparse observability is now provided by relating it to the Hamming distance of the code defined by the dynamics. We start by making precise the notion of Hamming distance for this code since its codewords are sequences of inputs and outputs. Let $T \in \mathbb{N}$ and consider the vectors $\mathbf{W}, \mathbf{W}' \in \mathbb{R}^{pT}$ given by:

$$\mathbf{W} = \begin{bmatrix} \mathbf{w}(\tau) \\ \mathbf{w}(\tau+1) \\ \vdots \\ \mathbf{w}(\tau+T-1) \end{bmatrix}, \quad \mathbf{W}' = \begin{bmatrix} \mathbf{w}'(\tau) \\ \mathbf{w}'(\tau+1) \\ \vdots \\ \mathbf{w}'(\tau+T-1) \end{bmatrix}, \tag{15.13}$$

where $\mathbf{w}(t), \mathbf{w}'(t) \in \mathbb{R}^p$, with:

$$\mathbf{w}(t) = \begin{bmatrix} w_1(t) \\ w_2(t) \\ \vdots \\ w_p(t) \end{bmatrix}, \quad \mathbf{w}'(t) = \begin{bmatrix} w_1'(t) \\ w_2'(t) \\ \vdots \\ w_p'(t) \end{bmatrix}.$$

We can think of $\mathbf{W}$ and $\mathbf{W}'$ as a vectorization of a sequence of T elements $\mathbf{w}(t) \in \mathbb{R}^p$ and $\mathbf{w}'(t) \in \mathbb{R}^p$, respectively. To define the Hamming distance between the vectors $\mathbf{W}$ and $\mathbf{W}'$, we use the vectors $\widetilde{\mathbf{w}}_i$ and $\widetilde{\mathbf{w}}_{i'}$ containing the subsequence of $\mathbf{W}$ and $\mathbf{W}'$ consisting of the ith entries in $\mathbf{w}$ and $\mathbf{w}'$, respectively:

$$\widetilde{\mathbf{w}}_i = \begin{bmatrix} w_i(\tau) \\ w_i(\tau+1) \\ \vdots \\ w_i(\tau+T-1) \end{bmatrix}, \quad \widetilde{\mathbf{w}}_i' = \begin{bmatrix} w_i'(\tau) \\ w_i'(\tau+1) \\ \vdots \\ w_i'(\tau+T-1) \end{bmatrix}. \tag{15.14}$$

Definition 15.4 (Hamming Distance and Hamming Weight) The Hamming distance between the vectors $\mathbf{W}, \mathbf{W}'$ defined in (15.13) is given by:

$$d_H(\mathbf{W}, \mathbf{W}') = \left|\left\{i \in \mathbb{N}_0 : \widetilde{\mathbf{w}}_i \neq \widetilde{\mathbf{w}}_i'\right\}\right|,$$

and the Hamming weight of vector $\mathbf{W}$ is $d_H(\mathbf{W}, \mathbf{0})$, where $\mathbf{0}$ is the all-zero vector. Note that $d_H(\mathbf{W}, \mathbf{W}')$ is also the Hamming weight of $\mathbf{W} - \mathbf{W}'$.

To define the Hamming distance of the affine code induced by the linear dynamics, we use the previous definition with $\mathbf{W}$ replaced with the received outputs $\mathbf{Y}$.

Definition 15.5 (Minimum Hamming Distance of the Affine Code (15.7)) The minimum Hamming distance of the affine code (15.7), induced by the LTI system (15.1), under an attack to at most α actuators is given by:

$$\min_{\mathbf{x},\mathbf{x}',\mathbf{U},\mathbf{E}_a,\mathbf{E}_{a'}} d_H\left(\mathbf{Y}_\tau^n(\mathbf{x}(\tau),\mathbf{U}+\mathbf{E}_a),\mathbf{Y}_\tau^n(\mathbf{x}'(\tau),\mathbf{U}+\mathbf{E}_{a'})\right),$$

with the minimization being performed over any two initial conditions $\mathbf{x}, \mathbf{x}' \in \mathbb{R}^n$, any input sequence $\mathbf{U} \in \mathbb{R}^{m\times(n-1)}$, and any two sequences of actuator attacks $\mathbf{E}_a, \mathbf{E}_{a'} \in \mathbb{R}^{m\times(n-1)}$ supported on (possibility different) sets of cardinality at most α.

Lemma 16.1 For *a* $(2\alpha, 2\beta)$-sparse strongly observable LTI system, the minimum Hamming distance of the code (15.7), induced by the LTI system (15.1), is at least $2\beta + 1$, if there are at most α actuators attacked. Conversely, if the minimum Hamming distance of the code (15.7), induced by the LTI system (15.1), is at least $2\beta + 1$, if there are at most α actuators attacked, then the LTI system (15.1) is $(2\alpha, 2\beta)$-sparse strongly observable.

Proof: For the sake of contradiction, assume that the system is $(2\alpha, 2\beta)$-sparse strongly observable, that there are at most α sensors attacked, and that the minimum Hamming distance of the code, as defined in Definition 16.5, is 2β or less. This implies the existence of initial conditions $\mathbf{x}, \mathbf{x}' \in \mathbb{R}^n$, an input sequence $\mathbf{U} \in \mathbb{R}^{m\times(n-1)}$, and two sequences of actuator attacks $\mathbf{E}_a, \mathbf{E}_{a'} \in \mathbb{R}^{m\times(n-1)}$ supported on sets of cardinality at most α, such that:

$$d_H\left(\mathbf{Y}_\tau^n(\mathbf{x},\mathbf{U}+\mathbf{E}_a),\mathbf{Y}_\tau^n(\mathbf{x}',\mathbf{U}+\mathbf{E}_{a'})\right) \le 2\beta. \tag{15.15}$$

Linearity of the dynamics implies:

$$\mathbf{Y}_\tau^n(\mathbf{x},\mathbf{U}+\mathbf{E}_a) - \mathbf{Y}_\tau^n(\mathbf{x}',\mathbf{U}+\mathbf{E}_{a'}) = \mathbf{Y}_\tau^n(\mathbf{x}-\mathbf{x}',\mathbf{E}_a-\mathbf{E}_{a'}),$$

and we conclude from (15.15) that $\mathbf{Y}_\tau^n(\mathbf{x}-\mathbf{x}',\mathbf{E}_a-\mathbf{E}_{a'})$ has at most Hamming weight 2β. Let Γ_s denote the set of outputs where $\mathbf{Y}_\tau^n(\mathbf{x},\mathbf{U}+\mathbf{E}_a) - \mathbf{Y}_\tau^n(\mathbf{x}',\mathbf{U}+\mathbf{E}_{a'}) = \mathbf{Y}_\tau^n(\mathbf{x}-\mathbf{x}',\mathbf{E}_a-\mathbf{E}_{a'})$ differ and note that $|\Gamma_s| \le 2\beta$. If we remove the sensors corresponding to the outputs in Γ_s, we conclude that the measurements provided by the remaining sensors from $\mathbf{Y}_\tau^n(\mathbf{x}-\mathbf{x}',\mathbf{E}_a-\mathbf{E}_{a'})$ are zero. This shows that from the initial condition $\mathbf{x}-\mathbf{x}'$ and the actuator attack $\mathbf{E}_a - \mathbf{E}_a{}'$ on 2α

actuators, we obtain a sequence of outputs that is identically zero. Since from the initial condition $\mathbf{0}$ and an identically zero actuator attack on the same actuators, we also obtain a sequence of outputs that is identically zero, we cannot distinguish between $\mathbf{x} - \mathbf{x}'$ and $\mathbf{0}$ which contradicts $(2\alpha, 2\beta)$-sparse strong observability. The converse is similarly proved. □

Combining this Lemma with the results in Section 16.4, we arrive at the solution of Problems 16.1 and 16.2.

Theorem 16.1 Let the number of attacked inputs and outputs be upper-bounded by r and s, respectively. An attack on the sensors and actuators of system (15.1) can be detected iff the system is $(2r, s)$-sparse strongly observable. Moreover, the state can be reconstructed from a sequence of inputs and outputs iff the system is $(2r, 2s)$-sparse strongly observable.

Proof: Lemma 16.1 shows that $(2\alpha, 2\beta)$-sparse strong observability is equivalent to the affine code induced by the linear system having Hamming distance $2\beta + 1$ when there are at most α sensors attacked. Sufficiency of the Hamming distance being $\beta + 1$ for error detection and being $2\beta + 1$ for error correction was explained in Section 16.4. Necessity, follows from classical coding theory results in [5]. □

Note that in classical coding theory, errors are usually only assumed to occur at the observations (outputs), and the case when there are input (or encoding) errors has not received much attention, except perhaps in [11].

15.6 Decoding Algorithms

As discussed in the previous sections, reconstructing the state, i.e. performing error correction, requires solving the optimization problem (15.8). This is a combinatorial problem as we must search over all the sets of attacked sensors Γ_s and actuators Γ_a in order to find the attack signals $\mathbf{e}_s$ and $\mathbf{e}_a$. The first efficient algorithm for the solution of this problem was proposed in [6] by leveraging compressed sensing ideas [12]. Intuitively, the minimization of cardinality is replaced with a minimization in the ℓ_1 norm thereby leading to convex problem that can be efficiently solved. In [6], however, conditions under which this relaxation (replacing cardinality with the ℓ_1 norm) is exact (i.e. leads to the same solution) were not discussed. Such conditions were introduced in [7] for a new optimization algorithm that was purposely designed for the problem of secure state reconstruction. Moreover, it was shown in [7] that the performance of this novel optimization algorithm was superior to using off-the-shelf algorithms to solve the relaxation proposed in [6]. To the writing of this chapter, the most efficient algorithm for solving the secure state reconstruction problem under sensor attacks is given in [13] and is based on a very different set of technical tools: satisfiability modulo theory (SMT) solving. In addition to being the most efficient algorithm, it also offers the advantage of always providing the correct solution to Problem 16.2 when the necessary sparse observability assumption is satisfied. This is in contrast with the algorithms in [6, 7] that may fail to produce the correct solution even when sparse observability holds. The results in [13] have recently been extended into three different directions: (i) to systems under sensor attacks and subject to plant and measurement noise [2]; (ii) to systems subject to sensor and actuator attacks [10], (iii) and by recognizing that the techniques in [13] lead to a methodology for efficiently solving a class of non-convex optimization problems that is now known as satisfiability modulo convex (SMC) programming [14].

References

1 J. Massey and M. Sain, Codes, automata, and continuous systems: Explicit interconnections, *IEEE Transactions on Automatic Control*, **12**, pp. 644–650, 1967.

2 S. Mishra, Y. Shoukry, N. Karamchandani, S. Diggavi, and P. Tabuada, Secure state estimation: Optimal guarantees against sensor attacks in the presence of noise, *IEEE Transactions on Control of Network Systems*, **4**, pp. 49–59, 2017.

3 Y. Shoukry, P. D. Martin, P. Tabuada, and M. B. Srivastava, Non-invasive spoofing attacks for anti-lock braking systems, in *Workshop on Cryptographic Hardware and Embedded Systems*, G. Bertoni and J.-S. Coron (Eds.): CHES 2013, LNCS 8086, International Association for Cryptologic Research, pp. 55–72, 2013.

4 P. J. Antsaklis and A. N. Michel, *Linear Systems*, Springer Science & Business Media, 2006.

5 R. Blahut, *Algebraic Codes for Data Transmission*, Cambridge University Press, 2003.

6 H. Fawzi, P. Tabuada, and S. Diggavi, Secure estimation and control for cyber-physical systems under adversarial attacks, *IEEE Transactions on Automatic Control*, **59**, pp. 1454–1467, 2014.

7 Y. Shoukry and P. Tabuada, Event-triggered state observers for sparse sensor noise/attacks, *IEEE Transactions on Automatic Control*, **61**, pp. 2079–2091, 2016.

8 M. S. Chong, M. Wakaiki, and J. P. Hespanha, Observability of linear systems under adversarial attacks, in American Control Conference (ACC), 2015.

9 M. Hautus, Strong detectability and observers, *Linear Algebra and Its Applications*, **50**, pp. 353–368, 1983.

10 M. Showkatbakhsh, Y. Shoukry, R. H. Chen, S. Diggavi, and P. Tabuada, An SMT-based approach to secure state estimation under sensor and actuator attacks, in IEEE 56th Conference on Decision and Control (CDC), IEEE, pp. 7177–7182, 2017.

11 J. Hachem, I.-H. Wang, C. Fragouli, and S. N. Diggavi, Coding with encoding uncertainty, in IEEE International Symposium on Information Theory (ISIT), pp. 276–280, June 2013.

12 E. Candes and T. Tao, Decoding by linear programming, *IEEE Transactions on Information Theory*, **51**, pp. 4203–4215, 2005.

13 Y. Shoukry, P. Nuzzo, A. Puggelli, A. L. Sangiovanni-Vincentelli, S. A. Seshia, and P. Tabuada, Secure state estimation for cyber physical systems under sensor attacks: A satisfiability modulo theory approach, *IEEE Transactions on Automatic Control*, **62**, pp. 4917–4932, 2017.

14 Y. Shoukry, P. Nuzzo, A. L. Sangiovanni-Vincentelli, S. A. Seshia, G. J. Pappas, and P. Tabuada, SMC: Satisfiability modulo convex programming, *Proceedings of the IEEE*, **106**, pp. 1655–1679, 2018.

16

Governance for the Internet of Things

Striving Toward Resilience

S. E. Galaitsi, Benjamin D. Trump, and Igor Linkov

US Army Engineer Research and Development Center, Vicksburg, MS, USA

16.1 Introduction

Accelerating innovations in the connectivity and remote operations of software-embedded household devices offer previously unimaginable improvements to everyday life. From digitally connected washing machines that alert cell phones upon cycle completion, onboard diagnostic systems in automobiles, to assorted household devices that can be powered cycled by a single universal remote, "smart" appliance development has increased users' centralized control over their lives via the Internet and computing devices [1]. The network of interconnected devices is known as The Internet of Things, or IoT, and is quickly becoming ingrained in much of Western culture, lifestyle, and economic activity [2].

Formally, IoT denotes consumer devices that are connected through the Internet, enabling continuous data collection and transfer between these devices [3]. Through wireless digital interconnectivity, a central computer or smartphone can operate and manage a wide array of devices which historically required independent or manual operation [4]. The trend toward digitalization and wireless operation is driven by market demand as much as technological progress – consumers want "smart" TVs, cars, appliances, and various other goods. Such arrangements increase levels of precision and control over much of daily life, but also introduce new risks from potential digital disruptions [5, 6].

Cyberattacks and hacking attempts, both from state-based and non-state-based actors, are constantly evolving to exploit new and unique vulnerabilities within complex information systems and digitally connected devices. These introduce

Modeling and Design of Secure Internet of Things, First Edition. Edited by Charles A. Kamhoua, Laurent L. Njilla, Alexander Kott and Sachin Shetty.

Published 2020 by John Wiley & Sons, Inc.

risks in (i) the safe management and use of information and data generated from the IoT use, and (ii) the safe operation of IoT devices in general. These risks include unwanted home surveillance through video baby monitors, access to unencrypted information revealing when a house is typically empty [7], and the ability to remotely remove brake or throttle control from automobile drivers [8]. These examples range from privacy violations to outright capabilities of physical harm and reflect the personalized connections that IoT devices foster.

International governance agencies and alliances have already expressed concern of such growing risks, such as with the Organisation for Economic Co-operation and Development (OECD), United Nations (UN), and the North Atlantic Treaty Organization (NATO), among others. In the United States, government agencies like the Department of Defense (DoD) and Department of Homeland Security (DHS), industrial players such as IBM, Facebook, Google, and General Electric, and insurance companies like the American International Group (AIG) have all reflected on the need to better address cybersecurity and digital risk-related concerns [9]. But efforts to protect or reduce a system's vulnerabilities from a given threat have generally been implemented only after initial disruptions or losses have occurred. While this is effective at mitigating future losses and disruption from the same threat, these strategies are limited in protecting IoT systems from evolving cyber disruptions.

Meeting the growing challenge of safe and efficient IoT development and use will require informed and adaptive governance strategies [10, 11]. Governance strategies differ across industries and between companies, but the critical question being asked of IoT developers and users centers upon how they intend to prevent, avoid, mitigate, manage, and recover from disruptions to IoT devices and interconnected information systems. This chapter explores the emerging scholarly and policy discussions of the digital economy and IoT governance and examines how such strategies might incorporate philosophical and methodological applications of resilience to enable efficient IoT device recovery and adaptation from an evolving cyber threat landscape.

16.2 Risk and Resilience

Comparatively evaluating emerging governance strategies for digitalization and IoT requires examining the philosophical and methodological options framing the countermeasures needed or available to prevent IoT disruption. The concepts of risk and resilience can support identifying, evaluating, and improving protective strategies.

The notion of risk, methodologically operationalized, imagines how a threat might affect a system. It is a function of (i) specific threats to a system, (ii) the vulnerabilities which a threat might exploit, and (iii) the consequences of the threat-vulnerability exploitation event. Risk assessment quantifies the likelihood and consequences of an event to identify critical components of a system vulnerable to specific threats, and to strengthen them to avoid losses. Statutory requirements for assessing infrastructural and environmental risks often apply this approach to characterize specific threats and their corresponding vulnerabilities in clearly defined situations with robust data or with straightforward methods for generating additional data [12]. However, for complex and rapidly evolving systems like IoT, risk assessments' benefits are limited to assessing the profile and impact of specific threats, rather than preparing systems for disruption from a variety of uncertain and ever-evolving threats. In a situation of unknown risks, the best strategy may not be preventing all risks, but minimizing their negative impacts [13].

The emerging concept of resilience refers to a system's response to an actualized threat. A 2012 National Academy of Sciences (NAS) report on "disaster resilience" defines resilience as the ability of a system to perform four functions with respect to adverse events: (i) planning and preparation, (ii) absorption, (iii) recovery, and (iv) adaptation [14]. Advancing the fundamental understanding and practical application of resilience requires greater attention to the development of resilience process metrics, as well as comparison of resilience approaches in multiple engineering contexts for the purpose of extracting generalizable principles.

Risks and resilience are inherently intertwined. Improving system resilience decreases the consequences of a threat-vulnerability exploitation event, which is the third component of the risk definition. Likewise, understanding the three components of risk will increase the efficacy of the planning and preparation inherent in resilience, and the subsequent success of its absorption, recovery, and adaptation. In contrast to risk, which evaluates specific threats, resilience examines a system's capability for recovery after any type of threat has materialized [15]. A resilient system has integral characteristics capable of responding to threats in a manner that is both generic and comprehensive to minimize their ability to damage the system. As a metaphor for resilience, we consider the horseshoe crab. Like any creature, a horseshoe crab risks infection from the ever-evolving world of bacteria. However, the crab's amoebocytes in its blood have the unique ability to coagulate around detected traces of bacterial presence. The resulting blood clots isolate harmful bacteria and prevent them from threatening the crab's health. The bacteria remains a risk, but the inherent resilience of the system it infiltrates can render it harmless. This denotes a highly resilient system.

The realm of IoT does not exist, however, within the confines of a single body. By design, its existence owes to its connection to the Internet and all the data, processes, and organizations operating in that system. Resilient IoT must address how a theoretical disruption to one component can percolate to disruption and degradation of other services. Resilience incorporates the properties of a system and a network, where it is imperative for systems planners to understand the complex and interconnected nature within which most individuals, organizations, and activities operate. Risk-based approaches can be helpful to understand how specific threats affect a system, yet often lack the necessary characteristic of reviewing how linkages and nested relationships with other systems leave one vulnerable to cascading failure and systemic threat. For example, a car thief who accesses both the "Go Home" navigation feature and the car's garage door opener has a much broader capacity for disruption. Chapter 12 examines these stepping-stone attacks digitally, where hacking one system can grant access to many others.

A resilience-based approach can examine how the structure and activities of systems influence one another and provide a method to understand and even quantify network interconnections and their potential for disruption via a cascading systemic threat [16]. This approach has earned prominence on the international stage as a method for determining whether a system can survive disruption and its subsequent condition and functionality in the disruption's aftermath.

In this way, a resilience-based approach requires considering the direct and indirect linkages between infrastructural, technological, social, informational, and environmental systems [17]. Alberts and Hayes [18], as well as Linkov and Trump [12] contend that interconnections between different domains of resilience can dampen or magnify the potential for cascading failure. For example, increased use of IoT to manage sensitive information by individuals with limited scientific literacy can increase the potential of a digital threat event – even if the devices were initially programmed or developed to be resistant to disruption or attack.

Resilience must extend beyond any single component of IoT cybersecurity – it requires a multifactorial effort to improve the capacity for threat identification, analysis, management, and communication within multiple stakeholder groups (IoT developers, manufacturers, consumers, insurance agencies, etc.). As the use of IoT grows, so too does the diversity of potential cyberattacks and disruptions that could compromise its systems. The IoT ecosystem must be structured for recovery with minimal losses in time, money, and device/information security. Several approaches to address this need are available, yet must be consistent with the unique institutional, cultural, and political values and norms held by the system's diverse stakeholders. These preconditions are essential to determine the appropriate strategy of IoT governance.

16.3 Strategies of IoT and Digital Economy Governance

Any emerging technology or complex system faces challenges over the governance of its activities and externalities. From a scholarly perspective, governance is the collection of formal and informal rights and responsibilities held by various stakeholders within a given activity or domain.

Formal governance stems from hard law and comprises the statutorily defined legal requirements for action within a given area. A violation of hard law often results in legally proscribed punishments. Such hard law arises as legislative law or proclamation, an executive order or rule, a judicial ruling and precedent, or similar action on the part of a recognized governmental institution. Likewise, informal governance activity is driven by soft law, which is no less important or impactful for the overall functions of policies or stakeholder groups. Soft law includes best practices, codes of conduct, multi-stakeholder agreements, or operating norms and procedures that lack the legal backing of a government agent for enforcement. However, soft law can be sufficiently pervasive within a given community that its violation can cause a significant loss in validity, respect, or collaboration opportunities with other community members or stakeholders. While hard law has the added benefit that compliance is legally required, soft law is often more flexible to emerging conditions and can be adapted or adjusted through the collective assent of pertinent stakeholders. Both hard and soft law present strengths and limitations in their capacity to identify, analyze, manage, and communicate the risks and benefits of a given activity.

The current governance status of the digital economy may be best characterized as "wild west" in terms of rules and expectations for best practices or safety-by-design requirements. Few governing norms or requirements are shared beyond small collections of stakeholders or companies, and large policymaking and scholarly bodies are just beginning to consider the implications and concerns that widespread digitalization might impose upon society. This is equally true for IoT, where requirements and even expectations for the capacity of digital devices to absorb and recover from risk remain relatively piecemeal and inconsistent across industries. This is a challenge for traditional risk assessment for cybersecurity or safe and proper use of IoT devices, let alone developing resilience-based approaches. Building resilience into IoT-governing strategies and practices requires simultaneously understanding the innovation horizon, the range of threats that innovation can introduce, as well as how any system governing IoT can complement, contribute to, and benefit from existing governance structures. These structures include hard law statutory requirements; soft law historical norms and cultural expectations of a given country, people, or discipline; and the development situation and stated development strategies of the country or community [19].

In April 2017, G20 ministers met to discuss strategies for maximizing contributions that digitization provides to economies [20]. Their declaration reflects previous international statements, namely that technology should support free expression and information access but should respect existing legal frameworks, both national and international, that safeguard privacy and intellectual property rights (see [21]). The declaration recognizes that balancing these objectives requires strengthening the trust and confidence in the digital economy through reliable and secure applications, sufficient digital literacy for all stakeholders, especially consumers, and collaborative environments. The G20 envisions that these collaborative efforts will develop risk-based technical standards, guidelines, and best practices to identify, assess, and manage security risks across the public–private sector divide and international borders. Any resulting cooperative development of international standards for technology should adhere to existing international rules, including WTO rules and principles [20–22].

Despite these agreements, the path to develop technical standards and guidelines to manage risks and regulate the digital economy is not yet clear. The OECD Ministerial Council Meeting of June 2017 heard three perspectives: (i) to promote favorable conditions for innovation and development while resisting pre-emptive digital regulation (Switzerland, Luxemburg), (ii) for central governments to actively shape the direction and soften negative consequences of digitalization (Portugal, Germany), and (iii) for a passive governance process where industry innovates and simultaneously builds the structures necessary to address any resulting social, economic, and environmental harms (United States, Mexico) [23]. These proposals represent three main governing processes which the Member States and Observers might adopt for governing IoT, corresponding to (i) a laissez-faire, industry-driven approach, (ii) a precautionary and pre-emptive strategy on the part of government, and (iii) a stewardship and "passive surveillance" approach by government agencies to reduce risks derived from digitalization while promoting private sector innovation.

A laissez-faire approach assumes the market will organize and protect itself from threats. With government intervention only in emergency circumstances, companies and individuals are responsible to identify their security interests and act to achieve them [24]. The virtue of this system is flexibility, which enables fast adaptations and responses to emerging threats [25]. However, without a centralized authority to enforce best practices and coherence with pre-existing legal frameworks and policy objectives, innovations can veer from established best practices that protect against security threats as well as encourage social inequities through uninhibited digitization and automation [26]. Furthermore, the G20 has emphasized the importance of enhancing digital literacy for consumers [20], which raises questions over whether consumers currently have enough fluency to advocate for appropriate IoT security in laissez-faire situations.

The second approach would enforce a pre-emptive strategy of government regulation to ensure adequate security measures in technological sectors, including IoT. A centralized authority would review available information about threats and develop regulatory requirements meant to avoid adverse consequences and protect any citizen populations made vulnerable by digital use. Governments could enforce specific requirements for locations and access to data servers, or computer system architectures, or features of open software. Such risk-averse governance strategies often seek to verify that new activities and technologies are safe before endorsing their use [27]. Critics of this approach argue that digitalization systems are highly dynamic and a rigid centralized governance structure lacks the ability to respond quickly to evolving threats as hackers scour each new system to identify and exploit vulnerabilities. The centrally imposed norms and practices may also limit innovations in defensive digitization and subsequently hinder realizing stated security objectives. Furthermore, a centralized governance approach may slow the speed of digital innovations in general and render the country less economically competitive against countries with more fluid governance systems.

The third digitization strategy advocates for stewardship on the part of the government, meaning that it passively monitors the paths taken by industry and consumers and intervenes when it identifies inconsistencies with existing processes and objectives. This governance system can also guide innovations through supporting the sustainability of operations in the face of future threats [24]. Governments can work with industrial and academic experts to articulate hard and soft strategies that support the secure access and data management of technological products while maximizing the derived technological benefits. Governments may create overseeing agencies, convene stakeholders, and conduct research to elucidate trade-offs. While the stewardship model still places constraints on industrial innovations, the government interventions are less determinant and more likely to be developed through a collaborative forum that favors profits while remaining informed about the relationship of imposed constraints to underlying governance policies and their motivations, which commonly protect the same populace that the technological innovations seek to serve. The development process remains dynamic, but risk components are minimized.

The three approaches require the different stakeholders – government, private users, and industry – to assume varying levels of responsibility for technological security. Of the three strategies, the stewardship strategy represents a compromise between innovation and caution and may be best poised to avoid the security risks potentially incurred under rigid governance structures or situations with unfettered innovation. Its inherent flexibility may also be best suited to forward the international cooperative goals expounded for the digital technology in the G20 summits. It is also the most amorphous strategy because it balances the difference between defining a strategy and abstaining from intervening. This invites broad flexibility in the interpretation of how a system based on guidance and constraints would manifest.

The G20, in their reports, recognized that any digital economy governance systems must account for the development situations, historical and cultural traditions, national legal system, and national development strategies of the implementing countries [20]. Constraints in local contexts, especially those that require politicians understanding technology and its characteristics, may play a role framing the international governance discussions. However, herein we argue that regardless of the governance strategy chosen, resilience can be built into the structure.

16.4 Applying Resilience to IoT Governance

Resilient governance under any system must be incentivized to conduct risk assessments regardless of the governance system. New information may provoke an adaptive response to emerging threats, including adjusting regulatory rules and best practices and, when necessary, reconsidering the risk–benefit analysis for various technology activities or availability [28, 29]. In IoT, these adjustments can be executed by a manufacturer who might add physical safeguards to a device, the developer who revises the software code to ensure digital security, the user who chooses to update their device or change their habits of use, or the government that refines its hard and soft policies to incorporate new insights [30]. These adaptions require a shift from knowledge to action. Each of the three government systems has the capacity for adaptation, whether through hard legislation (precautionary), voluntary changes among affected stakeholders (laissez-faire), and adjustments in the policy guidance surrounding cybersecurity (stewardship).

The incentives for executing such adaptations may vary between governance systems and their respective actors of responsibility. Mobilizing organizations and private citizens to commit specific actions to gain information and act upon it, even in their best interests, is a multifaceted process involving carrots and sticks. Motivators might include the policies of insurance agencies, legal structures for recourse following harm or damage, consumer responses as manifested in the market, and private or public enablement of citizen vigilance through education, whether acquired in schools, science museums, or through the support of research and dissemination by competitive grants. These components of the governance system all play a role in assuring its ability to incentivize adaptation in IoT device management.

Resilient governance recognizes that stakeholders must cooperate to monitor and characterize the uncertain and complex nature of emerging threats. Together, stakeholders from industry and academic and nongovernmental institutions can propose holistic strategies and best practices that minimize any externalities as they adjust to new circumstances. Stakeholders in each governance approach all benefit from ensuring that a proposed solution is relevant and useful to other stakeholders. This can assist the system as a whole absorb, recover from, and adapt to new threats as attackers shift their specific targets and objectives. The need for collaboration is a cornerstone of stewardship governance but should contribute to laissez-faire and precautionary systems as well.

Finally, following the establishment of a functioning security system within a given country, that system must maintain consistency with the rules and regulations of the international community that shares the network, as well support and contribute to broad communications about risk. Anyone connecting to the network assumes a duty to care for others [31]. In an interconnected system, vulnerabilities are universal, and international actors must be prepared to respond collectively to emerging threats, regardless of the nationality of the initially compromised target. International organizations have clearly stated goals of what they intend to support and what they need to protect in digital economies. A transparent process for operation in the realm of IoT will support trust, confidence, and reciprocity that can enable actors to collaborate quickly in times of actualized threats and to assist the system recovery.

16.5 Conclusions

The scope of IoT development and social adoption remains uncertain but is expected to significantly affect economic and social activities for consumers worldwide. As such, resilience-based strategies are needed to address disruption or attack to IoT within any governance system. As the United States, European Union, and OECD reflect upon the need for effective and forward-thinking digital governance practices, it is useful to examine the practices already extant and how to best use them to support structural resilience [32].

As the G20 has stated and reiterated, the path forward lies in transparency and the recognition and respect of norms, such as individual privacy, the free pursuit of objectives, and safety in person and possessions. Resilient governance structures can ensure the benefits and security risks are balanced, and that all actors are prepared to respond to new information emerging from the IoT threat landscape.

References

1 W. Li, H. Song, and F. Zeng, "Policy-based secure and trustworthy sensing for internet of things in smart cities," *IEEE Internet Things J.*, vol. 5, no. 2, pp. 716–723, 2018.
2 S. Greengard, *The Internet of Things*. MIT Press, 2015.
3 S. Madakam, R. Ramaswamy, and S. Tripathi, "Internet of Things (IoT): A literature review," *J. Comput. Commun.*, vol. 3, no. 5, p. 164, 2015.

4 A. Gaur, B. Scotney, G. Parr, and S. McClean, "Smart city architecture and its applications based on IoT," *Procedia Comput. Sci.*, vol. 52, pp. 1089–1094, 2015.

5 H. Margetts and P. Dunleavy, "The second wave of digital-era governance: A quasi-paradigm for government on the Web," *Philos. Trans. R. Soc. A*, vol. 371, no. 1987, p. 20120382, 2013.

6 R. Van Kranenburg and A. Bassi, "IoT challenges," *Commun. Mobile Comput.*, vol. 1, no. 1, p. 9, 2012.

7 W. Yakowicz, "Why Hackers Are Trying to Get Into Your Refrigerator," *Inc.com*, February 5, 2018.

8 E. Tegler, "Cyber threats targeting your car," *Autoweek*, October 2013.

9 J. R. Heft, "Top 10 writers of cybersecurity insurance," *Property Casualty 360*, 2017.

10 V. A. Almeida, D. Doneda, and M. Monteiro, "Governance challenges for the Internet of Things," *IEEE Internet Comput.*, vol. 19, no. 4, pp. 56–59, 2015.

11 L. Zheng, H. Zhang, W. Han, X. Zhou, J. He, and Z. Zhang, "Technologies, applications, and governance in the Internet of things, Internet of Things: Global Technological and Societal Trends," *Smart Environments and Spaces to Green ICT*, O. Vermesan and P. Friess (Eds.) pp. 143–178, River Publishers, 2011.

12 I. Linkov and B. Trump, *The Science and Practice of Resilience*. Springer International Publications, 2019.

13 S. Nastic, G. Copil, H.-L. Truong, and S. Dustdar, "Governing elastic IoT cloud systems under uncertainty," in 2015 IEEE 7th International Conference on Cloud Computing Technology and Science (CloudCom), pp. 131–138, 2015.

14 S. Larkin *et al.*, "Benchmarking agency and organizational practices in resilience decision making," *Environ. Syst. Decis.*, vol. 35, no. 2, pp. 185–195, 2015.

15 I. Linkov, B. D. Trump, and J. Keisler, "Risk and resilience must be independently managed," *Nature*, vol. 555, no. 7694, pp. 30–30, 2018.

16 M.-V. Florin *et al.*, "*Guidelines for the Governance of Systemic Risks*," ETH Zurich, 2018.

17 B. D. Trump, M.-V. Florin, and I. Linkov, "*IRGC Resource Guide on Resilience*," Vol. 2, International Risk Governance Center (IRGC), 2018.

18 D. S. Alberts and R. E. Hayes, "*Power to the Edge: Command... Control... in the Information Age*," Office of the Assistant Secretary of Defense, Washington DC Command and Control Research Program (CCRP), 2003.

19 I. Linkov, B. D. Trump, and K. Fox-Lent, "Resilience: Approaches to risk analysis and governance," IRGC Resource Guide on Resillience EPFL International Risk Governance Center, Lausanne V29-07-2016, https://www.irgc.org.riskgovernanceresilience, accessed October 10, 2016.

20 Bundesministerium für Wirtschaft und Energie, "G20 Digital Economy Ministerial Declaration – Shaping Digitisation for an Interconnected World," G20 Research Group at the University of Toronto, 2017.

21 G20 Research Group, "G20 Digital Economy Development and Cooperation Initiative," G20 Research Group at the University of Toronto, 2016.

22 G20 Research Group, "G20 Digital Economy Ministerial Declaration," G20 Research Group at the University of Toronto, 2018.

23 I. Linkov, B. D. Trump, K. Poinsatte-Jones, and M.-V. Florin, "Governance strategies for a sustainable digital world," *Sustainability*, vol. 10, no. 2, p. 440, 2018.

24 E. Ostrom, "Collective action and the evolution of social norms," *J. Econ. Perspect.*, vol. 14, no. 3, pp. 137–158, 2000.

25 A. A. Ganin, P. Quach, M. Panwar, Z. A. Collier, J. M. Keisler, D. Marchese, and I. Linkov, "Multicriteria decision framework for cybersecurity risk assessment and management," *Risk Anal.*, vol. 40, no. 1, pp. 183–199, 2017.

26 R. Böhme, N. Christin, B. Edelman, and T. Moore, "Bitcoin: Economics, technology, and governance," *J. Econ. Perspect.*, vol. 29, no. 2, pp. 213–38, 2015.

27 A. Stirling, "Precaution, foresight and sustainability: Reflection and reflexivity in the governance of science and technology," *Reflexive Governance for Sustainable Development*, J.-P. Voβ, D. Bauknecht, and R. Kemp (Eds.) p. 225, Edward Elgar Publishing Limited, 2006.

28 G. Baldini, T. Peirce, M. Handte, D. Rotondi, S. Gusmeroli, S. Piccione, B. Copigneaux, F. Le Gall, F. Melakessou, P. Smadja, A. Serbanati, and J. Stefa, "Internet of Things Privacy, Security, and Governance," *Internet of Things Security: Fundamentals, Techniques and Applications*, S. K. Shandilya, S. A. Chun, S. Shandilya, and E. Weippl (Eds.) pp. 19–36, River Publishers, 2018.

29 B. D. Trump, K. Poinsatte-Jones, T. Malloy, and I. Linkov, "9 Resilience and risk governance: Current discussion and future action," *Handbook on Resilience of Socio-Technical Systems*, M. Ruth and S. Goessling-Reisemann (Eds.) p. 136, Edward Elgar Publishing Limited, 2019.

30 R. H. Weber, "Internet of things–Governance quo Vadis?," *Comput. Law Secur. Rev.*, vol. 29, no. 4, pp. 341–347, 2013.

31 D. Rogers, "*Establishing Principles for Internet of Things Security*," Internet of Things Security Foundation.

32 I. Linkov, B. Trump, K. Poinsatte-Jones, P. Love, W. Hynes, and G. Ramos, "Resilience at OECD: Current state and future directions," *IEEE Eng. Manag. Rev.*, vol. 46, no. 4, pp. 128–135, 2018.

Part III

IoT Security Design

17

Secure and Resilient Control of IoT-Based 3D Printers

Zhiheng Xu and Quanyan Zhu

Department of Electrical and Computer Engineering, Tandon School of Engineering, New York University, Brooklyn, NY, USA

17.1 Introduction

Due to the fast development of prototyping and small-scale manufacturing, two main primary solutions have obtained increasing attention. The first solution is the Subtractive Manufacturing (SM), in which workers pour a mold or reduce a block of material via milling and turning. The second method is the Additive Manufacturing (AM), also known as 3D-printing technology, which creates part and prototypes of the objects by fusing layers of material [1]. Given the layer-by-layer property, 3D-printing technology can generate more complicated components. This critical merit makes the AM attractive for various company and agencies, such as NASA, ESA, and SpaceX. The benefits of AM enable a broad range of applications, e.g. printing of device components, replacement parts, and houses [2].

While producing high quality and precision devices and equipment, 3D-printing technology also incurs expansive costs, e.g. over $20 000 for high-quality plastics [3]. Besides, in different applications, each manufactured object needs to satisfy a group of rigorous physical requirements, e.g. resistance to mechanical or thermal stress. Therefore, to use the AM technology, we need to determine various parameters and coefficients, the pattern of heat source, i.e. the AM technology requires specialized knowledge, making it challenging for general users. Given these two reasons, there is a growing need to outsource the production to third parties, who specialize in 3D-printing technology [4].

Modeling and Design of Secure Internet of Things, First Edition. Edited by Charles A. Kamhoua, Laurent L. Njilla, Alexander Kott and Sachin Shetty.

Published 2020 by John Wiley & Sons, Inc.

In recent years, the Internet of Things (IoT) has provided a promising opportunity to facilitate the development of intelligent industrial systems and cyber manufacturing systems [5–7]. The rapid growth of the IoT can also promote the outsourcing process of 3D-printing production, creating an IoT-based 3D printer. However, the various applications of the outsourcing 3D-printing process produce an attractive target. There exist two significant threats in the 3D-printing system [2]: intellectual property theft and physical damage. To protect information and cyber operations, researchers have developed many conventional solutions, e.g. encryption and verification algorithms. However, these solutions are not sufficient to capture the cyber-physical nature of the IoT-based 3D printer. Since 3D printers are critical to many military and civil applications (e.g. creating weapons, vehicles, and other critical infrastructures), attackers and opponents have a growing desire to impair the 3D-printing systems. A real accident is that a 3D printer explosion occurred in November 2013 at Powderpart Inc. due to misoperation [8]. Besides, any other cyber-physical attack can also threaten IoT-based 3D printer, e.g. Stuxnet malware attacks Supervisory Control and Data Acquisition (SCADA) systems of Industrial Control Systems (ICSs) [9].

The security issues of the IoT have gained attention in recent years [10–12]. Among many attack models of IoT-based systems, we focus on data-integrity attack, where the attacker aims to deviate the IoT-based 3D printer from the original, damaging both the objects and systems. To develop defense strategies, we will depict the potential weaknesses of the outsourcing 3D-printing process. Figure 17.1 illustrates the steps and details of an outsourcing procedure. In an outsourcing process, customers design a computer-aided design (CAD) model,

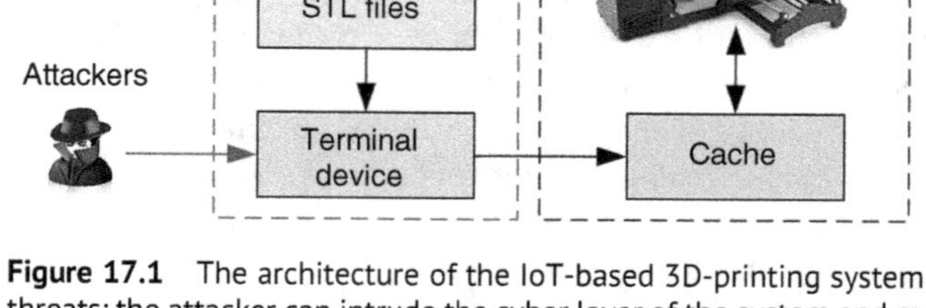

Figure 17.1 The architecture of the IoT-based 3D-printing system and the potential threats: the attacker can intrude the cyber layer of the system and modify the data stored in the terminal device; the fake data will cause severe damage on the physical part of the 3D-printing system.

convert it to a. STL file, and sends the data to the terminal device of the 3D-printing system through cyber networks. Based on this outsourcing process, an attacker can intrude the terminal device and modify the. STL file. By doing so, the attacker can change the forms of the objects in a low resolution, wasting the material and impairing the physical parts through mismanagement. Hence, the attacker can sabotage the physical system through the cyber layer. One possible attack can achieve this goal is the Advanced Persistent Threat (APT). In an APT, the attacker can entirely steal the secret keys in the terminal storage to access the. STL file [13]. In our previous work [14], we have shown how APT-type attacker intrudes the networked 3D-printing system.

To protect the IoT-based 3D printer, we use a game-theoretic approach to build game models, capturing the interactions among different players. Due to the cyber-physical nature of the 3D printer, we use different game models to describe their properties. In the physical layer, we use a zero-sum game to find the optimal controller to attenuate the disturbance. In the cyber layer, to mitigate the ATP attack, we formulate a *FlipIt* game to find the optimal defense strategies. To capture the relationship between these two layers, we use a Stackelberg game to emphasize the sequential architecture of the IoT-based 3D-printing system. Similar to the work in Part I of this book, we use game-theoretic tools to analyze the behavior of the attackers and evaluate the impact of the attack. Based on the game framework, we aim to find the optimal defense strategies. Finally, we present numerical simulation to show the performance of the proposed defense strategies.

The remainder of the chapter is organized as follows. Section 17.2 gives the problem formulation. Section 17.3 formulates the game-theoretical framework and analyzes the equilibrium. Section 17.4 provides the numerical experiments. Finally, Section 17.5 concludes the chapter.

17.2 Problem Formulation

In this section, we will present a cyber-physical structure of the 3D printer. We present a dynamical model and use the Markov jump system (MJS) to capture the variability of the parameters of the system. After that, we offer a cyberattack model. Given the cyber and physical problems, we formulate game models for different layers.

17.2.1 The Dynamical Model of a 3D-Printing System

Before presenting the attack model, we need to analyze the weakness of the IoT-based 3D printer. To this end, we identify the physical model of the printer. Generally, a 3D printer uses stepper motors to manipulate an extruder, which constructs the objects layer by layer. To simplify the problem, we only depict the one-dimensional model since the model of other dimensions are similar. In Figure 17.2, we present a simplified structure of the extruder moving in one dimension.

As illustrated in Figure 17.3, we use a linear model to describe the movements of the extruder with three assumptions: (i) we model rotor torque as a spring with a constant $k_2 > 0$; (ii) we model the elastic belt as a linear spring with another constant $k_1 > 0$; (iii) we assume that all damping forces are viscous with constants c_1 and c_2. Accordingly, we can write down kinetic equations:

$$m_1\ddot{x}_1 = -k_1(x_1 - x_2) - c_1\dot{x}_2,$$
$$m_2\ddot{x}_2 = -k_2(x_2 - u) + k_1(x_1 - x_2) - c_2\dot{x}_2,$$

where $x_1 > 0$ is the position of the extruder, and $x_2 > 0$ is the movement of the elastic belt; m_1 and m_2 are the modeling mass of the extruder and rotor. Then, we define a physical state $x := [x_1, \dot{x}_1, x_2, \dot{x_2}]^T$. Given the state and the above dynamical equations, we obtain the following continuous linear system, given by

$$\dot{x}(t) = A_c x(t) + B_c u(t) + D_c w(t), \qquad (17.1)$$
$$x(0) = x_0, \quad x_1 \in [0, L],$$

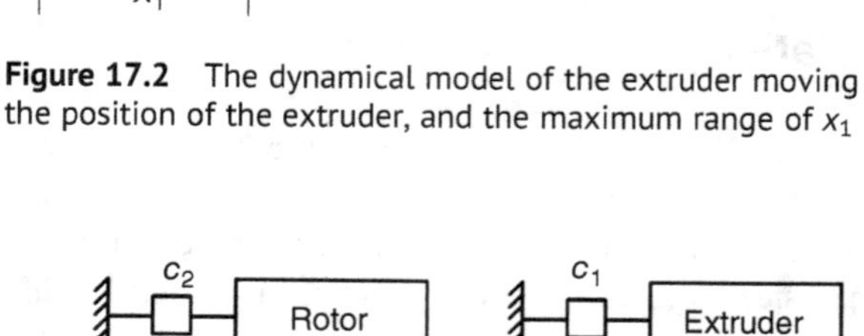

Figure 17.2 The dynamical model of the extruder moving in one dimension: the x_1 is the position of the extruder, and the maximum range of x_1 is L.

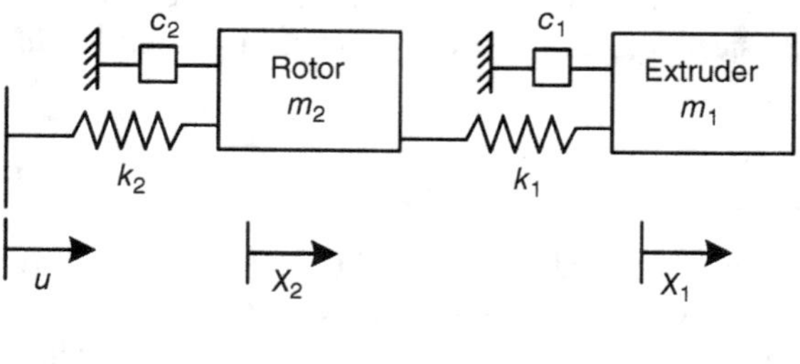

Figure 17.3 The linearized model of the rotor and extruder: we linearize the rotor torque and elastic belt as two strings with two constant coefficients k_1 and k_2; we assume that all the damping forces are viscous with constants c_1 and c_2.

where

$$A_c = \begin{bmatrix} 0 & 1 & 0 & 0 \\ -k_1/m_1 & -c_1/m_1 & k_1/m_1 & 0 \\ 0 & 0 & 0 & 1 \\ -k_1/m_2 & 0 & -(k_1+k_2)/m_2 & -c_2/m_2 \end{bmatrix}, B_c = \begin{bmatrix} 0 \\ 0 \\ 0 \\ k_2/m_2 \end{bmatrix},$$

and $E_c \in \mathbb{R}^{4\times 1}$ is a constant vector; $u(t) \in \mathbb{R}$ is the control input; $w(t) \in \mathbb{R}$ is the disturbance; $x_0 \in \mathbb{R}^{4\times 1}$ is a given initial condition.

The printing system is a digital system. We need to discretize the system with a sampling time T_s. Besides discretization, we also note that the motor speed will change the parameters in system matrices A_c, B_c, and D_c. Hence, we use a discrete-time MJS to capture this parameters' variation of the system [15]. The discrete-time MJS model is given by

$$x_{k+1} = A_{\theta_k} x_k + \mathrm{B}_{\theta_k} u_k + D_{\theta_k} w_k, \tag{17.2}$$

where $k \in \mathbb{Z}$, $x_k = x(kT_s)$, and $u_k = u(kT_s)$. The stochastic variable $\theta_k \in \Theta$ is a Markov process with a given initial condition and a given transition probability, i.e.

$$\lambda_{ij}(p_a) := Pr(\theta_{k+1} = j \mid \theta_k = i, \zeta = a)p_a + Pr(\theta_{k+1} = j \mid \theta_k = i, \zeta = d)(1 - p_a),$$

where p_a is the ratio of the period when the attacker takes over the terminal devices. Variable $\zeta \in \{a, d\}$ is the indicator of the player, where $\zeta = a$ represents that the attacker controls the devices, and $\zeta = d$ means the defender holds the system. We can obtain the probability $Pr(\theta_{k+1} \mid \theta, \zeta)$ by studying the historical data or using reinforcement learning. Specifically, probability $Pr(\theta_{k+1} \mid \theta, \zeta = a)$ depends on the attack model, while probability $Pr(\theta_{k+1} \mid \theta, \zeta = d)$ stands for the reliability. Designers can compute the reliability of the system through the reliability of its constituent devices. This information can be found in product manuals and specifications. In our application, we assume that the system has two modes, i.e. $\Theta := \{0, 1\}$, where $\theta = 0$ indicates that the extruder moves in a low-speed mode, and $\theta = 1$ means the high-speed mode. We can obtain matrices A_{θ_k}, B_{θ_k}, and D_{θ_k} by discretizing the system model (17.1) based on different speed modes.

17.2.2 The Zero-Sum Game Framework for the Control Problem

For each printing mission, the 3D printer aims to track a specific trajectory provided by the corresponding customers. The customer will send the. STL files to the printers, and the printer will convert the. STL files to the predefined trajectories for the extruder. Therefore, the control problem of the printer is a trajectory tracking problem.

To mitigate the impact noise, we consider developing a robust control problem, which can be viewed as a zero-sum game, denoted by G_Z, between the two players: the controller and the disturbance. We formulate the zero-sum game problem as follows:

$$\min_u \max_w J_d(x, u, w; \hat{x}, \hat{u}), \tag{17.3}$$

with the cost function $J(x, u, w; \hat{x}, \hat{u})$ defined by

$$J_d(x, u, w; \hat{x}, \hat{u}) := E_\theta\{L(x, u, w; \hat{x}, \hat{u}; \theta)\},$$

$$L(x, u, w; \hat{x}, \hat{u}; \theta) := \sum_{k=0}^{T-1} \ell(x_k, u_k, w_k; \hat{x}_k, \hat{u}_k; \theta_k) + q_f(x_T, \hat{x}_T)$$

$$\ell(x_k, u_k, w_k; \hat{x}_k, \hat{u}_k; \theta_k) := ||x_k - \hat{x}_k||^2_{Q_{\theta_k}} + ||u_k - \hat{u}_k||^2_{R_{\theta_k}} - \gamma^2 ||w_k||^2,$$

where $T \in \mathbb{Z}_+$ is the horizon length of the problem; $\|\cdot\|$ denotes the Euclidean norm; $Q \in \mathbb{R}^{4\times 4}$ and $R \in \mathbb{R}^{4\times 1}$ are two constant matrices; $\gamma > 0$ is a given threshold; and $q_f(x_T, \hat{x}_T)$ is a terminal cost; $(\hat{x}, \hat{u})$ are the desire trajectories sent by the customer.

In game G_Z, we define two strategies for the two players, where $\mu \in \mathbb{R}$ is the strategy of the controller, and $\nu \in \mathbb{R}$ is the strategy of the disturbance. To find the optimal controller against the worst-case disturbance, we need to find the equilibria (μ^*, ν^*), such that

$$J_d(x, \mu^*, \nu; \hat{x}, \hat{u}) \leq J_d(x, \mu^*, \nu^*; \hat{x}, \hat{u}) \leq J_d(x, \mu, \nu^*; \hat{x}, \hat{u}).$$

In the next section, we will use dynamic programming to obtain the equilibria.

17.2.3 The Attack Model of the IoT-Based 3D Printer

The feature of the IoT-based 3D printer inevitably introduces new threats to the physical layers. L. Sturm et al. have discussed several weaknesses of the cyber and physical layers in a networked 3D-printing system [4, 16]. As illustrated in Figure 17.1, the procedure of the outsourcing 3D-printing task provides many opportunities for the attacker to infiltrate the system. Especially, the attacker can intrude the terminal device that reserves the. STL files.

In our work, we consider an APT [17]. As stated in [18], an APT-type attacker can launch a zero-day exploration to pinpoint the vulnerabilities of the IoT-based 3D printer, e.g. stealing the password of the terminal device. After that, the attacker can completely gain control of the terminal device and modify the. STL files, which lead to fake trajectories for the extruder. While modifying the data, the APT attackers can remain undiscovered for a long period of time. The fake paths can deviate extruder to a fetal position, lowering the product resolution, wasting the materials, and damaging the system. We assume that the APT-type attacker will make unnoticeable changes in the data, and it will take a long period of time for the defender to discover the impact, similar to the work [19]. To mitigate this type of attacks, we cannot depend on monitoring the printing outcome. We develop an intelligent defense strategy to upgrade the cyber system according to a prescribed probability. We will discuss the design of the defense strategy in the following section.

Now, we analyze the impact of a successful attack. When the attacker takes over the terminal device, we assume that they will deviate the extruder to trajectory $(\widetilde{x}, \widetilde{u})$. Hence, the cost function becomes

$$J_a(x, u, w; \widetilde{x}, \widetilde{u}) := E_\theta\{L(x, u, w; \widetilde{x}, \widetilde{u}; \theta)\}.$$

Even though we might not predict fake trajectories $(\widetilde{x}, \widetilde{u})$, we assume that we can estimate the worst-case trajectories. Therefore, we can also estimate cost $J_a(x, \mu^*, \nu^*; \widetilde{x}, \widetilde{u})$ introduced by the attacker. We will use this cost in the following formulation of the cyber game.

17.2.4 The Cyber *FlipIt* Game

In this section, we will design the defense strategy. At the cyber part of the system, we aim to use a game framework to capture the interactions between the defender $(\zeta = d)$ and the attacker $(\zeta = a)$. The game model we use is called *FlipIt* game, denoted by G_F. In a *FlipIt* game, two players fight for a given resource, which, in our case, is the terminal device. In our framework, the 3D printer has N missions to process, where $N \in \mathbb{Z}_{++}$. During each mission, both the attacker and the defender aim to control the system. To compromise the cyber defense of the system, we assume that the attacker needs to pay a cost $\alpha_a > 0$. Similarly, the defender also needs to pay a cost $\alpha_d > 0$ to remove the impact of the attacker, e.g. the defender can upgrade the cyber defense strategy or change the password. Figure 17.4 presents the sequential movements of the cyber *FlipIt* game. At each time slot, both players can pay a cost to control the given resource.

At each time slot, the user can achieve a unite utility, denoted by $W_d > 0$, if the defender holds the system. Besides, the user also needs to pay a printing cost $J_d > 0$. We assume that $W_d - J_d > 0$. Hence, the total utility of the defender, denoted by U_d, is defined by

$$U_d := (W_d - J_d) \cdot l_d - \alpha_d \cdot h_d,$$

where $l_d \in \mathbb{Z}_+$ is the total length of period that the defender controls the system, and $h_d \in \mathbb{Z}_+$ is the total times that the defender upgrades the cyber defense system. Similarly, we can also define the attacker's utility function, given by

$$U_a := J_a \cdot l_a - \alpha_a \cdot h_a,$$

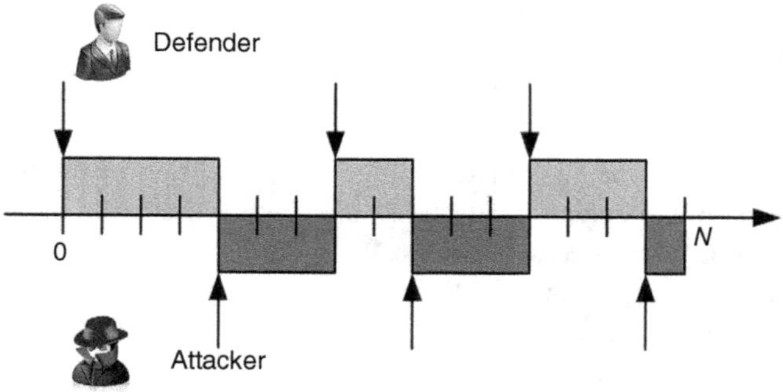

Figure 17.4 The sequential architecture of the *FlipIt* game: at each time slot, each player can pay a cost to gain control of the shared resource. In our case, the shared resource is the terminal device, which stores the. STL files for the 3D printer.

where $J_a > 0$ is the printing cost when the attacker controls the printer, $l_a \in \mathbb{Z}_+$ is the length that the attacker holds the system, and $h_a \in \mathbb{Z}_+$ is the total times that the attacker intrudes the system.

Suppose that total length N is sufficiently large. Then, we need to consider the players' average utility rates, defined by

$$\beta_d := U_d/N = (W_d - J_d) \cdot p_d - \alpha_d \cdot f_d, \tag{17.4}$$

$$\beta_a := U_a/N = J_a \cdot p_a - \alpha_a \cdot f_a, \tag{17.5}$$

where $p_\zeta := l_\zeta/N$ is the proportional time that the player $\zeta \in \{d, a\}$ holds the terminal system, and $f_\zeta := h_\zeta/N$ denotes the frequency that players takes their actions. Note that $l_d + l_a = N$. Hence, we have

$$p_d + p_a = \frac{l_d + l_a}{N} = 1.$$

In this cyber *FlipIt* game, denoted by G_F, both players aim to find their best policies (p_d^*, p_a^*) to maximize their utility average rates (β_d, β_a). In our work, we assume that both players' strategies are nonadaptive, i.e. both players cannot obtain the information that who controls the system until they take the actions. In next section, we will define the Nash equilibrium and present a theorem to find the optimal strategies.

17.2.5 The Cyber-Physical Stackelberg Game Model

In the previous sections, we defined the cyber and physical games for both cyber and physical layers. To describe the relationship between the cyber and physical layers, we define a Stackelberg game, denoted by G_S. In game G_S, the physical layer is the follower who takes actions after observing the actions of the leader, and the

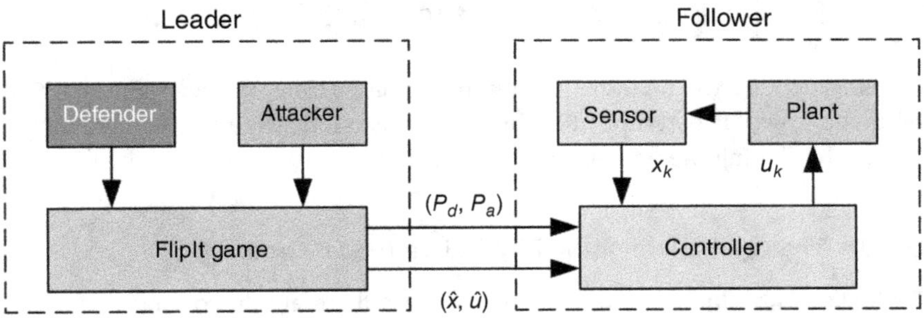

Figure 17.5 The structure of the Stackelberg game-theoretic framework: the physical part acts as a follower who waits for the command from the cyber space; the cyber space plays the role of leader who makes its optimal decisions by expecting the best respond of the follower.

cyber layer is the leader who takes its optimal strategies by anticipating the best strategies of the follower [20]. The main reasons for using the Stackelberg game are twofold. First, the interactions between the cyber and physical layers are sequential in nature, i.e. the physical printing system will work after receiving the message from the cyber space. The second reason is that the design of the cyber and physical policies is often asynchronized, i.e. the cyber operator does not design the physical controller, and vice versa. Hence, we aim to design the cyber-physical problem in a decentralized fashion.

Figure 17.5 illustrates the structure of the cyber-physical Stackelberg game G_S, where the cyber layer is the leader, and the physical layer is the follower. In game G_S, the physical layer has a pair of strategies (μ, ν), and the cyber layer has a pair of strategies (p_d, p_a). The following statement defines a Nash equilibrium of the cyber-physical Stackelberg game.

Definition 17.1 In cyber-physical game G_S, a Stackelberg equilibrium is a strategy pair $\{(p_d^*, p_a^*), (\mu^*, \nu^*)\}$ satisfying that

$$p_d^* \in \arg\max_{p_d} \beta_d(p_d, p_a^*, J_d(\mu^*, \nu^*), J_a(\mu^*, \nu^*)),$$
$$p_a^* \in \arg\max_{p_a} \beta_a\,(p_d^*, p_a, J_d(\mu^*, \nu^*), J_a(\mu^*, \nu^*)),$$

$$\overline{\mu}^*(p_d, p_a) \in \{\overline{\mu}^* \in \mathbb{R} \mid J_\zeta(\overline{\mu}^*, \overline{\nu}^*) \leq J_\zeta(\overline{\mu}, \overline{\nu}^*)\},$$
$$\overline{\nu}^*(p_d, p_a) \in \{\overline{\nu}^* \in \mathbb{R} \mid J_\zeta(\overline{\mu}^*, \overline{\nu}^*) \geq J_\zeta(\overline{\mu}^*, \overline{\nu})\},$$

where $\overline{\mu}$: $[0,1] \times [0,1] \to \mathbb{R}$ is the respond function of the controller, and $\overline{\nu}$: $[0,1] \times [0,1] \to \mathbb{R}$ is the respond function of the disturbance.

Given Definition 17.1, we will show the analysis of the cyber-physical Nash equilibrium in the next section.

17.3 Equilibrium Analysis of the Cyber-Physical Games

In this section, we will analyze the equilibrium of the cyber-physical games. Given the analysis, we will determine the optimal defense strategies against an intelligent attacker by finding the equilibrium of the games.

17.3.1 The Equilibrium of the Physical Zero-Sum Game

The first equilibrium is the one of game G_Z. We will present the solution concept of game G_Z defined in Section 17.2.2. To solve the problem, we define two tracking errors $\widehat{x}_k^e := x_k - \widehat{x}_k$ and $\widehat{u}_k^e := u_k - \widehat{u}_k$. Then, we use the dynamic programming approach to solve the problem. Let $\theta_k = i$ and $\theta_{k+1} = j$. We define a value function:

$$V^*\left(x_k^e, i, p_a\right) := \min_{u_k} \max_{w_k} \left\{\ell(x_k, u_k, w_k; \widehat{x}_k, \widehat{u}_k; i) + \sum\nolimits_{j\in\Theta} \lambda_{ij}(p_a) V^*\left(x_{k+1}^e, j, p_a\right)\right\}. \tag{17.6}$$

We let $V\left(x_k^e, i, p_a\right) = \left(x_k^e\right)^T P_{i,k}(p_a) x_k^e$. Hence, we obtain that

$$V\left(x_k^e, i, p_a\right) = \ell(x_k, u_k, w_k; \widehat{x}_k, \widehat{u}_k; i) + \left(x_{k+1}^e\right)^T S_{i,k+1}(p_a) x_{k+1}^e,$$

$$S_{i,k+1}(p_a) := \sum\nolimits_{j\in\Theta} \lambda_{ij}(p_a) P_{j,k+1}(p_a).$$

To solve problem (17.3), we need to ensure that (17.3) is convex in u and concave in w. To this end, we require the following inequality, given by

$$\gamma^2 I_4 - D_i^T S_{i,k+1}(p_a) D_i > 0, \tag{17.7}$$

where I_4 is the 4×4 identity matrix. Variable $\gamma > 0$ in (17.7) indicates the upper bound of the noise attenuation of the physical control system [21]. A small γ leads to a more stable system, i.e. the system uses more power to withstand the noise, i.e. the printing system will have a better control performance. In the numerical experiments, we will analyze the relationship between the attacker's strategies and parameter γ.

Given inequality (17.7), we consider the following first-order derivative conditions:

$$\frac{\partial V\left(x_k^e, i, p_a\right)}{\partial u_k} = 0 \quad \text{and} \quad \frac{\partial V\left(x_k^e, i, p_a\right)}{\partial w_k} = 0.$$

Thus, we obtain the optimal solutions, given by

$$\mu^*\left(x_k^e, \widehat{u}_k\right) := K(\theta_k, p_a) x_k^e + \widehat{u}_k, \quad \nu^*\left(x_k^e\right) := L(\theta_k, p_a) x_k^e, \tag{17.8}$$

where

$$K(i,p_a) := \{I_4 + B_i^T S_{i,k+1}(p_a)B_i - B_i^T S_{i,k+1}E_i H_{i,k+1}^{-1} E_i^T S_{i,k+1}B_i\}^{-1} \\ \times \{B_i^T S_{i,k+1}(p_a)E_i H_{i,k+1}^{-1} E_i^T S_{i,k+1}A_i - B_i^T S_{i,k+1}(p_a)A_i\}, \tag{17.9}$$

$$H_{i,k+1} := E_i^T S_{i,k+1}E_i - \gamma^2 I_4,$$

$$L(i,p_a) := \{E_i^T S_{i,k+1}E_i - \gamma^2 E_i^T S_{i,k+1}B_i G_{i,k+1}^{-1} B_i^T S_{i,k+1}E_i\}^{-1}, \\ \times \{E_i^T S_{i,k+1}(p_a)B_i G_{i,k+1}^{-1} B_i^T S_{i,k+1}(p_a)A_i - E_i^T S_{i,k+1}(p_a)A_i\}, \tag{17.10}$$

$$G_{i,k+1} := B_i^T S_{i,k+1}(p_a)B_i + I_4.$$

Note that $S_{i,k+1}(p_a) := \sum_{j\in\Theta}\lambda_{ij}(p_a)P_{j,k+1}(p_a)$. We update $P_{i,k}(p_a)$ by solving the coupled Riccati equations, given by

$$P_{i,k}(p_a) = \phi_i(S_{i,k+1}, Q_i) := Q_i + A_i^T S_{i,k+1}A_i - \left[A_i^T S_{i,k+1}(p_a)B_i \;\; A_i^T S_{i,k+1}(p_a)E_i\right] \\ \times \begin{bmatrix} I_4 + B_i^T S_{i,k+1}(p_a)B_i & B_i^T S_{i,k+1}(p_a)E_i \\ E_i^T S_{i,k+1}(p_a)B_i & E_i^T S_{i,k+1}(p_a)E_i - \gamma^2 I_4 \end{bmatrix} \begin{bmatrix} B_i^T S_{i,k+1}(p_a)A_i \\ E_i^T S_{i,k+1}(p_a)A_i \end{bmatrix}, i \in \Theta, \tag{17.11}$$

where $\phi : \mathbb{R}^{4\times4} \times \mathbb{R}^{4\times4} \to \mathbb{R}^{4\times4}$ is a function to update the coupled Riccati equations. We can simplify (17.11) if there is no disturbance or the value of attenuation is infinity, i.e. $\gamma \to \infty$. The simplified (17.11) is given by

$$P_{i,k}(p_a) = Q_i + A^T S_{i,k+1}A_i^T - A_i^T S_{i,k+1}B_i\left(I_4 + B_i^T S_{i,k+1}B_i\right)^{-1} B_i^T S_{i,k+1}A_i,$$

where $S_{i,k+1}$ is short for $S_{i,k+1}(p_a)$.

We present the following theorem to characterize a monotonic relationship between the coupled Riccati equations given by (17.11).

Theorem 17.1 Suppose that attack ratio p_a is given. The $P_{i,k}(p_a)$ is always monotonic in k, i.e. $P_{i,k}(p_a)$ converges to

$$P_{i,k}^*(p_a) = \phi_i\left(\sum_{j\in\Theta}\lambda_{ij}(p_a)P_{j,k+1}^*, Q_i\right), \forall i \in \Theta.$$

Given Theorem 17.1, we can see that the values of physical cost (J_d^*, J_a^*) depends on the references and ratio p_a. In our work, we assume that the normal references $(\widehat{x}, \widehat{u})$ and the malicious $(\widetilde{x}, \widetilde{u})$ are given. Hence, we define a map T_Z: $[0, 1] \to \mathbb{R}_+ \times \mathbb{R}_+$, i.e.

$$(J_d^*, J_a^*) = T_Z(p_a). \tag{17.12}$$

Given a p_a, we can use $T_Z(\cdot)$ to compute the physical costs. In the next section, we will propose the solutions of the cyber *FlipIt* game.

17.3.2 The Equilibrium of the Cyber *FlipIt* Game

Given the formulation of the *FlipIt* game, we present the following definition to characterize its Nash equilibrium.

Definition 17.2 In game G_F, a pair of Nash equilibria (p_d^*, p_a^*) satisfying that

$$p_d^* \in \arg\max_{p_d} \beta_d(p_d, p_a^*, J_d, J_a),$$
$$p_a^* \in \arg\max_{p_a} \beta_a(p_d^*, p_a, J_d, J_a),$$

where (J_d, J_a) are the physical costs of the defender and attacker, respectively.

Based on Definition 17.2, we present the following theorem to characterize the Nash equilibria of the cyber *FlipIt* game.

Theorem 17.2 (Equilibria of the Cyber Game) Suppose that (J_d^*, J_a^*) are given. When both players use periodic nonadaptive strategies, *FlipIt* game G_F admits the following Nash equilibria:

$$p_a^* = \begin{cases} \dfrac{1}{2}\dfrac{\alpha_d J_a^*}{\alpha_a(W_d - J_d^*)}, & \text{if } \dfrac{\alpha_d}{W_d - J_d^*} \le \dfrac{\alpha_a}{J_a^*}; \\ 1 - \dfrac{1}{2}\dfrac{\alpha_a(W_d - J_d^*)}{\alpha_d J_a^*}, & \text{if } \dfrac{\alpha_d}{W_d - J_d^*} > \dfrac{\alpha_a}{J_a^*}; \end{cases}$$
$$p_d^* = 1 - p_a^*.$$

Readers can find the proof of Theorem 17.2 in [22]. Given Theorem 17.2, we also define another mapping $T_F: \mathbb{R}_+ \times \mathbb{R}_+ \to [0, 1]$, such that

$$p_a^* = T_F(J_d, J_a). \tag{17.13}$$

In the following section, we will analyze the equilibria of the cyber-physical Stackelberg game G_S.

17.3.3 The Equilibria of the Cyber-Physical *FlipIt* Game

Now, we use the following Theorem 17.3 to characterize the equilibria.

Theorem 17.3 (Equilibria of the Cyber-Physical Game) Given game G_Z and G_F, the cyber-physical Stackelberg game admits the following equilibria:

$$p_a^* = T_F(J_d(\mu^*, \nu^*),\ J_a(\mu^*, \nu^*)),$$
$$\mu^*(x, u) = K(\theta, p_a^*)x + u,$$
$$\nu^*(x) = L(\theta, p_a^*)x,$$

where $K(\cdot)$ and $L(\cdot)$ are defined by (17.9) and (18.10), respectively.

Based on the mapping T_Z and T_F, we observe a fixed point p_a^*, i.e.

$$p_a^* = T_F \circ T_Z(p_a^*). \tag{17.14}$$

Hence, we can define a new mapping $T_S = T_F \circ T_Z$: $[0, 1] \rightarrow [0, 1]$. The following theorem shows the existence of the fixed point in mapping $T_S(\cdot)$.

Theorem 17.4 (Existence of the Fixed Point) In the cyber-physical Stackelberg game, mapping T_S admits at least one fixed point p_a^*, such that

$$p_a^* = T_S(p_a^*). \tag{17.15}$$

Proof: We first show that $T_S(\cdot)$ is continuous in $[0, 1]$. To this end, we verify that

1) $\lambda_{ij}(p_a)$ is continuous in p_a
2) $S_{i,k} = \sum_{j\in\Theta}\lambda_{ij}P_{j,k}$ is continuous in λ_{ij}
3) $\phi_i(S_{i,k}, Q_i)$ is continuous in $S_{i,k}$
4) $J_\varsigma(P_{i,k})$ is continuous in $P_{i,k}$
5) $T_F(J_d, J_a)$ is also continuous in (J_d, J_a), respectively

Note that $T_S(\cdot)$ is a combination of the above functions. Hence, $T_S(\cdot)$ is also continuous in $[0, 1]$. Based on Brouwer fixed-point theorem [23], mapping $T_S(\cdot)$ admits at least one fixed point in $[0, 1]$. ∎

Given Theorem 17.4, we can use iterative algorithms to find the fixed point. Later, in the numerical experiments, we demonstrate methods to compute the fixed point of $T_S(\cdot)$. Besides, given Theorems 17.3 and 17.4, we can quickly propose the following corollary.

Corollary 17.1 Given a fixed point p_a^* satisfying (17.15), we obtain the equilibria $\{(p_d^*, p_a^*), (\mu^*, \nu^*)\}$ of game G_S, i.e.

$$p_d^* = 1 - p_a^*, \quad \mu^*(x, u) = K(\theta, p_a^*)x + u, \quad \nu^*(x) = L(\theta, p_a^*)x.$$

Hence, if we have the fixed point p_a^*, then we can use Corollary 17.1 to compute the optimal defense strategies p_d^* and the optimal controller's policy μ^*. In the following section, we will present numerical experiments to evaluate the performance of the proposed defense strategies.

17.4 Simulation Results

In this section, we use numerical experiments to evaluate the performance of the proposed defense strategies. We use a mathematical model of a 3D-printing system. We have linearized the model at different equilibria due to the moving speed of the extruder. We illustrate the parameters in Table 17.1. The parameters come from the work [24].

Table 17.1 System parameters.

	x-Axis		y-Axis	
Parameters	**Low speed**	**High speed**	**Low speed**	**High speed**
m_1	0.13 kg	0.12 kg	0.13 kg	0.12 kg
m_2	0.45 kg	0.31 kg	1.07 kg	1.05 kg
k_1	110 kN/m	180 kN/m	110 kN/m	110 kN/m
k_2	355 kN/m	410 kN/m	392 kN/m	432 kN/m
c_1	112 Ns/m	130 Ns/m	71.5 Ns/m	86.5 Ns/m
c_2	1100 Ns/m	1350 Ns/m	411 Ns/m	670 Ns/m

Given the parameters in Table 17.1, we will linearize the system model with a sampling time $T_s = 0.1$ seconds. The transition matrices of the both players are given by

$$\Lambda_d = \begin{bmatrix} 0.99 & 0.01 \\ 0.98 & 0.02 \end{bmatrix}, \quad \Lambda_a = \begin{bmatrix} 0.20 & 0.80 \\ 0.10 & 0.90 \end{bmatrix}.$$

In the transition matrices, we note that the defender aims to maintain the extruder at low-speed mode to generate a stable object, increasing the resolutions. However, the attacker controls the extruder at high-speed mode to harm the system. We also define the weighting matrices as

$$Q_0 = Q_1 = \text{diag}[100, 5, 10, 4], \quad R_0 = R_1 = 0.9.$$

The cost ratio is given by $\alpha = \alpha_a/\alpha_d := 1$, i.e. both players have the same value of the move cost.

In the experiment, we let the printer to create a cylinder. To simplify the problem, we only plot the trajectories in a two-dimensional space. The reference trajectories at x-axis and y-axis of the normal user and attacker are given by

$$\begin{bmatrix} \hat{x}_1 \\ \dot{\hat{x}}_1 \\ \hat{x}_2 \\ \dot{\hat{x}}_2 \end{bmatrix} = \begin{bmatrix} 0.6\cos(0.2\pi t) \\ -0.12\sin(0.2\pi t) \\ 0.6\sin(0.2\pi t) \\ 0.12\cos(0.2\pi t) \end{bmatrix}, \quad \begin{bmatrix} \tilde{x}_1 \\ \dot{\tilde{x}}_1 \\ \tilde{x}_2 \\ \dot{\tilde{x}}_2 \end{bmatrix} = \begin{bmatrix} 0.6\cos(0.5\pi t) \\ -0.3\sin(0.5\pi t) \\ 0.6\sin(0.5\pi t) \\ 0.3\cos(0.5\pi t) \end{bmatrix}.$$

Given the above trajectories, we can see that the moving speed of the attacker's trajectories is much higher than that of the normal user. Hence, the malicious trajectories may harm the printing system, lowering the resolution of the product.

Figure 17.6a and b depicts the two-dimensional trajectories of the extruder in the normal case and the attack case. In the usual case, the extruder can track the

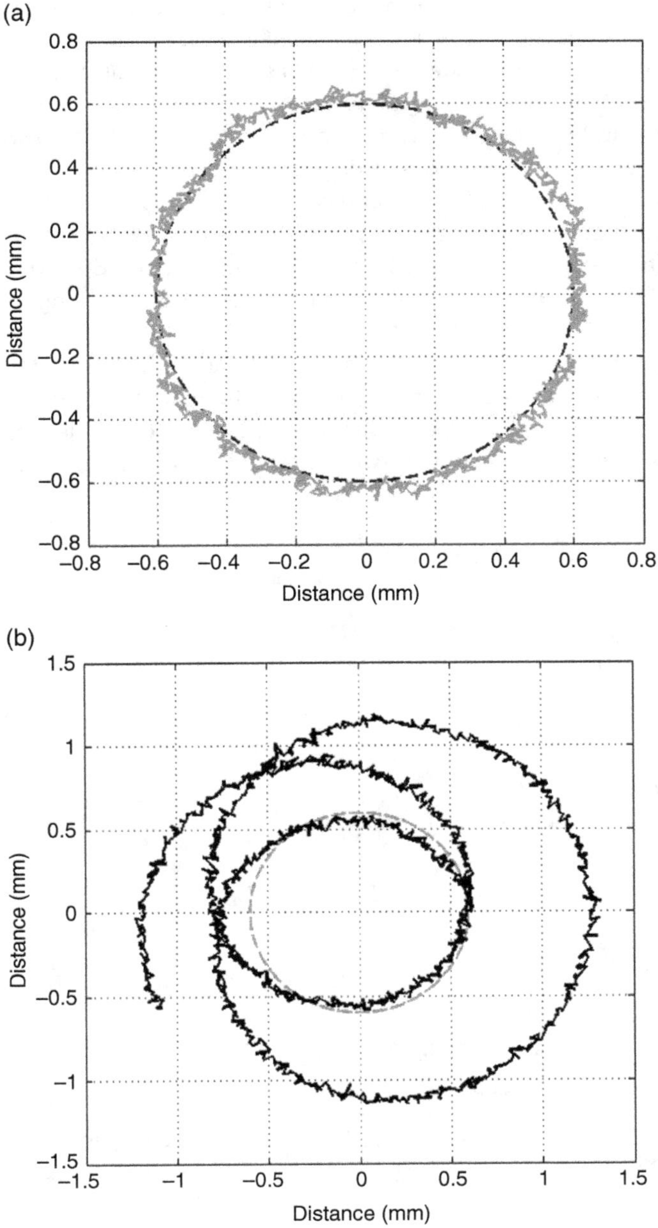

Figure 17.6 (a) The tracking performance under the control of the normal user: the extruder can track the trajectory smoothly. (b) The tracking performance under the control of the attacker: the extruder fails to track the trajectory, damaging the system.

trajectories smoothly, leading to a desirable performance. However, in the attack case, the extruder fails to follow the paths since the reference speed is too fast. The poor performance will waste the material as well as damaging the system. Figure 17.7 shows the control inputs in these two cases. We can see that the control input changes more sharply in the attack case than that in the normal one. The high value of the input also leads to poor control performance.

Now, in the next part, we evaluate the performance of the defense mechanism. Figure 17.8a shows the tracking performance without the protection mechanism, i.e. $p_a^* = 0$. We can see that the controller fails to track the trajectories. However, with defense strategies, i.e. $p_a^* = 0.6$, the extruder succeeds in tracking the trajectories, reducing the loss.

We are also interested in the relationship between the γ and p_a^*. In Figure 17.9, we see that p_a^* decreases when γ increases since the system becomes less valuable when γ increases. Recall that γ is the upper bound of the disturbance attenuation. Similarly, we can also see the relationship between the ratio α and p_a^*. When the

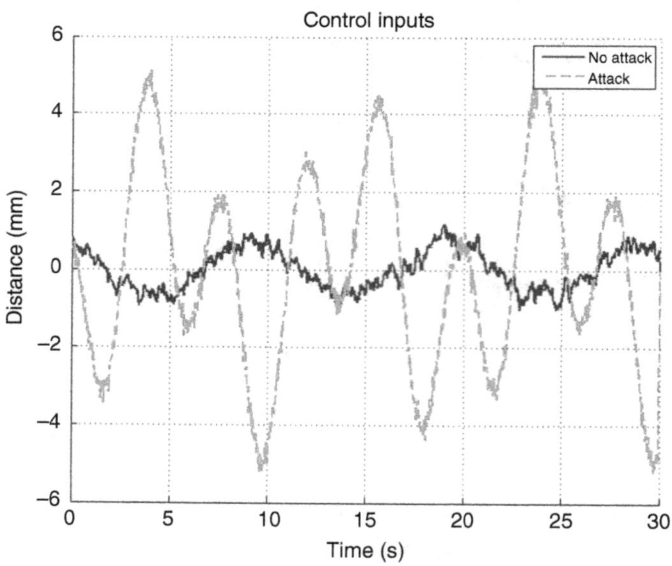

Figure 17.7 The input of the control in different cases: in the normal case, the input stays in a certain range; in the attack case, the input moves sharply, leading to a poor performance.

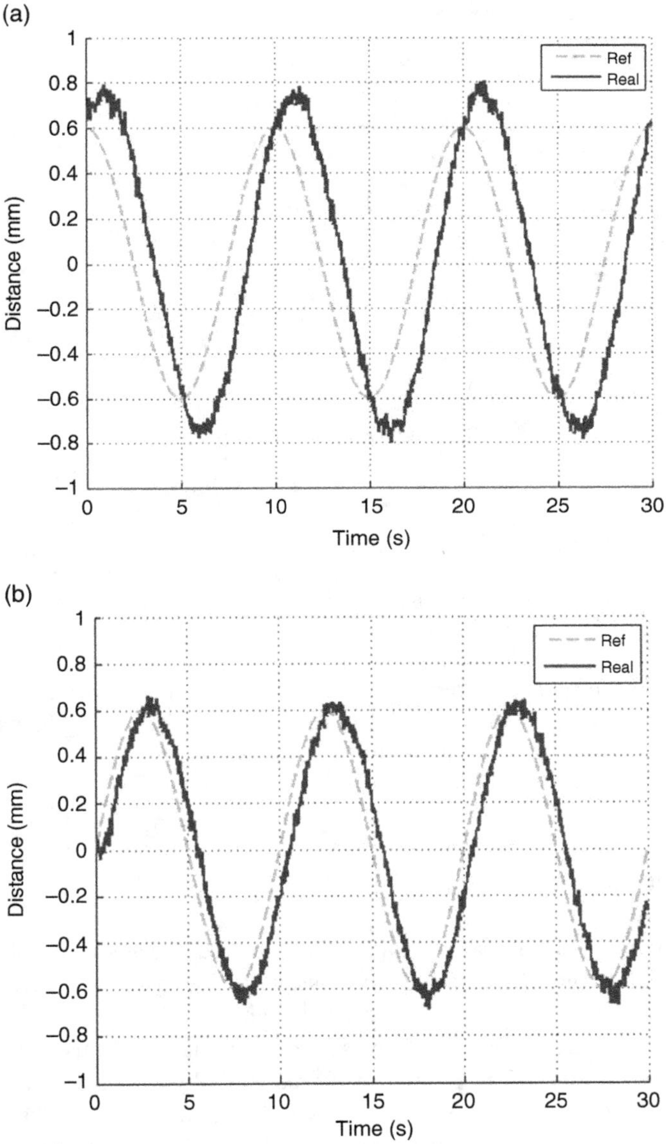

Figure 17.8 (a) Case 1 (No Defense Strategies): When there is no defense strategy (i.e. $p_a = 0$), the system fails to track the trajectories. (b) Case 2 (With Defense Strategies): With the defense strategies (i.e. $p_a^* = 0$), the extruder can track the given trajectories, minimizing the physical costs.

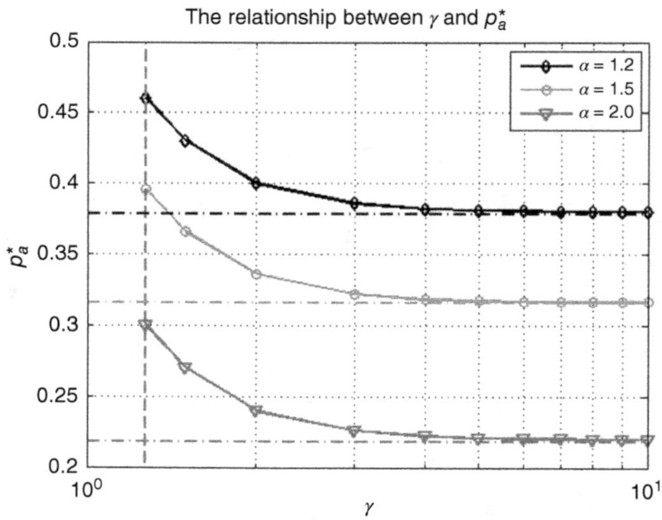

Figure 17.9 The relationship among α, γ, and p_a^*: (i) when we decrease γ, attack ratio p_c^* increases since the system's performance is better and more valuable to the attacker. (ii) When α decreases, p_a^* will increase since the move cost is cheaper for the attacker.

move cost is greater for the attacker, it reduces the attack ratio. Figure 17.10 shows that we can run $T_S(\cdot)$ iteratively. We can see that using $T_S(\cdot)$ can find the optimal solution in a few steps.

17.5 Conclusions

Integrated with networks, the IoT-based 3D printer constitutes a cyber-physical architecture, introducing new challenges to the system. In our work, we have investigated the weaknesses of the networked printer and developed a game-theoretic framework to protect the system from APT type of attacks. We use different game models to capture certain properties of cyber and physical layers. We have studied the equilibria of each game and proposed the optimal defense strategies for the defender. The proposed mechanism can enhance the system's resiliency since the strategies will upgrade the system security mechanism in an optimal period. In the end, we have presented numerical experiments to evaluate the performance of the proposed defense mechanism.

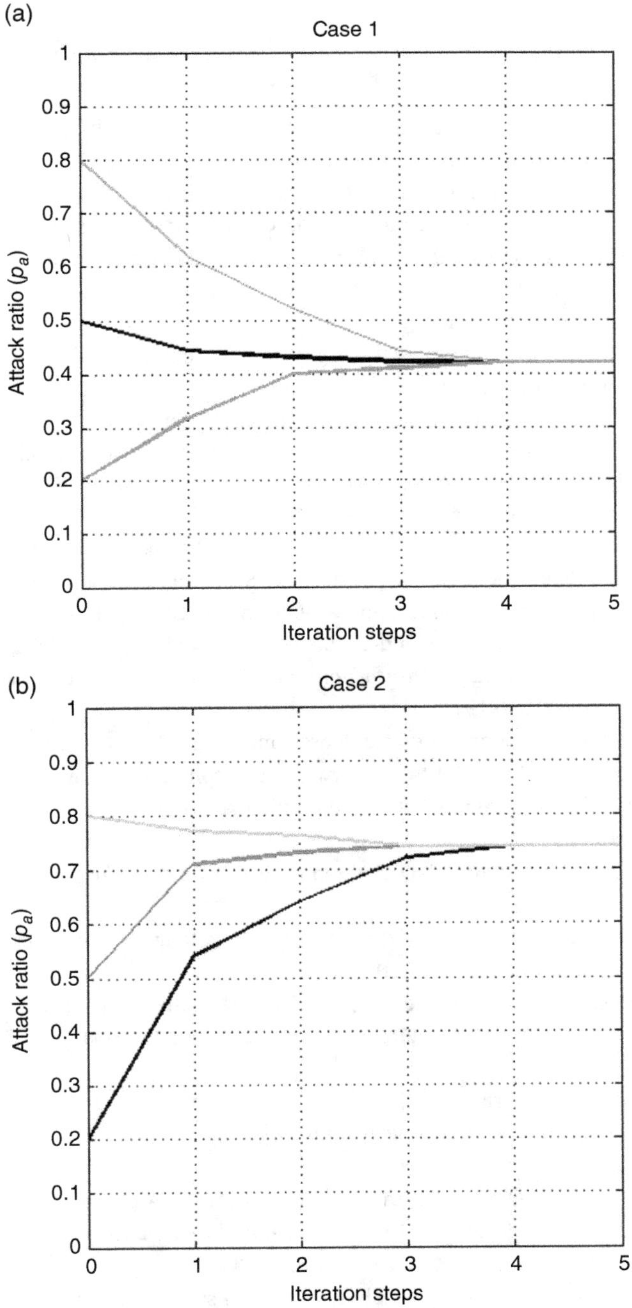

Figure 17.10 Given different move-cost ratio α, we use mapping $T_S(\cdot)$ to do iteration and obtain the optimal attack ratio p_a^*. Given difference initial conditions, the algorithm always converge to the optimal value.

References

1 Datum 3D. "Additive vs Subtractive manufacturing: Which approach is best?." *Datum 3D*, September 10, 2018. [Last accessed in October 29, 2018], https://www.datum3d.com/additive-manufacturing-vs-subtractive-manufacturing.

2 M. Yampolskiy, T. R. Andel, J. Todd McDonald, W. B. Glisson, and A. Yasinsac. "Intellectual property protection in additive layer manufacturing: Requirements for secure outsourcing." In *Proceedings of the 4th Program Protection and Reverse Engineering Workshop*, p. 7. ACM, 2014.

3 L. G. Cima and M. J. Cima. "Computer aided design; stereolithography, selective laser sintering, three-dimensional printing." 1996.

4 L. Sturm, C. Williams, J. Camelio, J. White, and R. Parker. "Cyber-physical vulnerabilities in additive manufacturing systems." *Context* 7, no. 2014 (2014): 8.

5 L. Da Xu, W. He, and S. Li. "Internet of things in industries: A survey." *IEEE Transactions on Industrial Informatics* 10, no. 4 (2014): 2233–2243.

6 S. Jeschke, C. Brecher, T. Meisen, D. Özdemir, and T. Eschert. "Industrial internet of things and cyber manufacturing systems." In: Jeschke S., Brecher C., Song H., Rawat D. (eds) *Industrial Internet of Things*, pp. 3–19. Springer, Cham, 2017.

7 J. Pawlick and Q. Zhu. "Strategic trust in cloud-enabled cyber-physical systems with an application to glucose control." *IEEE Transactions on Information Forensics and Security* 12, no. 12 (2017): 2906–2919.

8 A. Sternstein. "Things can go kaboom when a defense contractor's 3-D printer gets hacked." *Nextgov*, September 11, 2014. https://www.nextgov.com/cybersecurity/2014/09/heres-why-you-dont-want-your-3-d-printer-get-hacked/93923/. Last accessed: 01/07/2020.

9 B. Miller and D. C. Rowe. "A survey SCADA of and critical infrastructure incidents." *RIIT* 12 (2012): 51–56.

10 R. H. Weber. "Internet of Things: New security and privacy challenges." *Computer Law & Security Review* 26, no. 1 (2010): 23–30.

11 Q. Jing, A. V. Vasilakos, J. Wan, J. Lu, and D. Qiu. "Security of the Internet of Things: Perspectives and challenges." *Wireless Networks* 20, no. 8 (2014): 2481–2501.

12 G. Gan, Z. Lu, and J. Jiang. "Internet of things security analysis." In *2011 International Conference on Internet Technology and Applications*, pp. 1–4. IEEE, 2011.

13 F. Li, A. Lai, and D. Ddl. "Evidence of advanced persistent threat: A case study of malware for political espionage." In *2011 6th International Conference on Malicious and Unwanted Software*, pp. 102–109. IEEE, 2011.

14 Z. Xu and Q. Zhu. "Cross-layer secure cyber-physical control system design for networked 3D printers." In *2016 American Control Conference (ACC)*, pp. 1191–1196. IEEE, 2016.

15 O. L. V. Costa, M. D. Fragoso, and R. P. Marques. *Discrete-Time Markov Jump Linear Systems*. Springer Science & Business Media, New York, 2006.

16 C. Neuman. "Challenges in security for cyber-physical systems." In *DHS Workshop on Future Directions in Cyber-Physical Systems Security*, pp. 22–24. 2009.

17 M. K. Daly. "Advanced persistent threat." *Usenix* 4 (2009): 2013–2016.

18 S. Rass and Q. Zhu. "GADAPT: A sequential game-theoretic framework for designing defense-in-depth strategies against advanced persistent threats." In: Zhu Q., Alpcan T., Panaousis E., Tambe M., Casey W. (eds) *International Conference on Decision and Game Theory for Security*, pp. 314–326. Springer, Cham, 2016.

19 L. Huang and Q. Zhu. "Analysis and computation of adaptive defense strategies against advanced persistent threats for cyber-physical systems." In: Bushnell L., Poovendran R., Başar T. (eds) *International Conference on Decision and Game Theory for Security*, pp. 205–226. Springer, Cham, 2018.

20 T. Başar and G. J. Olsder. "*Dynamic Noncooperative Game Theory.*" SIAM, Philadelphia, vol. 23, 1999.

21 T. Başar and P. Bernhard. "*H-infinity Optimal Control and Related Minimax Design Problems: A Dynamic Game Approach.*" Springer Science & Business Media, New York, 2008.

22 M. Van Dijk, A. Juels, A. Oprea, and R. L. Rivest. "FlipIt: The game of 'stealthy takeover'." *Journal of Cryptology* 26, no. 4 (2013): 655–713.

23 A. Granas and J. Dugundji. "*Fixed Point Theory.*" Springer Science & Business Media, New York, 2013.

24 B. Weiss. "Closed-loop control of a 3D printer gantry." *Ph.D. dissertation*, University of Washington, 2014.

18

Proactive Defense Against Security Threats on IoT Hardware

Qiaoyan Yu[1], Zhiming Zhang[1], and Jaya Dofe[2]

[1] *Department of Electrical and Computer Engineering, University of New Hampshire, Durham, NH, USA*
[2] *Department of Computer Engineering, California State University, Fullerton, CA, USA*

18.1 Introduction

The Internet of Things (IoT) offers more advanced services than a single device or an isolated system, as IoT connects diverse components/subsystems through the Internet. Typical IoT devices include sensors, actuators, and microchips. One well-known IoT application is smart monitors on the power grid, which facilitate to optimize energy transmission. As more and more IoT applications are emerging, industries are optimistic with the IoT market. In 2011, Cisco predicted that 50 billion devices will be connected to the Internet by 2020 [1]. The McKinsey analysis in 2014 shows that 26–30 billion objects are expected to be connected to IoT by 2020 [2].

Networking a tremendous number of devices into the IoT framework potentially brings in new concerns on security and privacy. Since traditional isolated devices that lack security protection are connected in the IoT structure, attackers now have a bridge to invade the network through the low-end vulnerable components. As it is difficult to regulate the quality of all IoT devices, security management for IoT becomes much more difficult than before. Moreover, the wide coverage of IoT incurs the increasing number of end users to join hacker teams, either motivated by hobbies or driven by economic profits.

Although various threats challenge IoT security, the root of trust originates from the hardware security. For instance, attacks, like Bootkit malware that corrupts

Modeling and Design of Secure Internet of Things, First Edition. Edited by Charles A. Kamhoua, Laurent L. Njilla, Alexander Kott and Sachin Shetty.

Published 2020 by John Wiley & Sons, Inc.

the boot-up sequence to load and execute hostile codes, cannot be detected by software-level countermeasures. Without trustworthy and authenticated devices, high-level approaches such as data encryption and code authentication cannot thwart tampering. In this chapter, hardware security is the primary focus.

The outline of this chapter is as follows. We first review three threat models associated with the hardware security: side-channel attack (SCA), hardware tampering on IoT device, and hardware Trojan (HT) insertion via untrusted computer-aided design (CAD) software. Next, we will introduce the countermeasures against those security threats, respectively.

18.2 Hardware Security Threats

Under the current global business model, the IC supply chain typically involves multiple countries and companies that do not use the same regulation rules. Despite reducing the cost of fabrication, assembly, and test, the globalized IC supply chain has led to serious security concerns. Intellectual property (IP) piracy, IC overbuilding, reverse engineering attack, physical attack, counterfeiting chip, and hardware Trojans (HTs) are recognized as critical security threats on maintaining the integrity and trust of hardware.

The hardware security threats can be originated from various stages in the IC design flow. As shown in Figure 18.1, the potential IP piracy attackers could be designer, third-party IP (3PIP) vendor, and SoC integrator at design, synthesis, and verification stages. In the fabrication stage, an untrusted foundry may overbuild the IP cores and sell them under a different brand name to make profit. Reverse engineering an IC is a process of identifying its structure, design, and functionality. Traditionally, reverse engineering is a legal process as per US Semiconductor Chip Protection Act for teaching, analysis, and evaluation of techniques incorporated in mask work [3]. However, reverse engineering is a double-edged sword. One could use reverse engineering techniques to pirate ICs. Reverse engineering attacks can be done at various abstraction levels of supply chain depending upon the attackers' objectives.

18.2.1 Side-Channel Analysis Attack

SCA [4–8] in cryptography is an attack that uses the information extracted from the physical implementation of a crypto-system rather than theoretical analysis on the cryptographic algorithm. SCAs performed on electronic systems are noninvasive. This means attackers do not tamper the crypto-system. The success of SCAs relies on the correlation between the physical measurements taken during computations and the internal state of the processing device, which is related to the secret key. As IoT uses multiple sensors, the information leakage is the most common

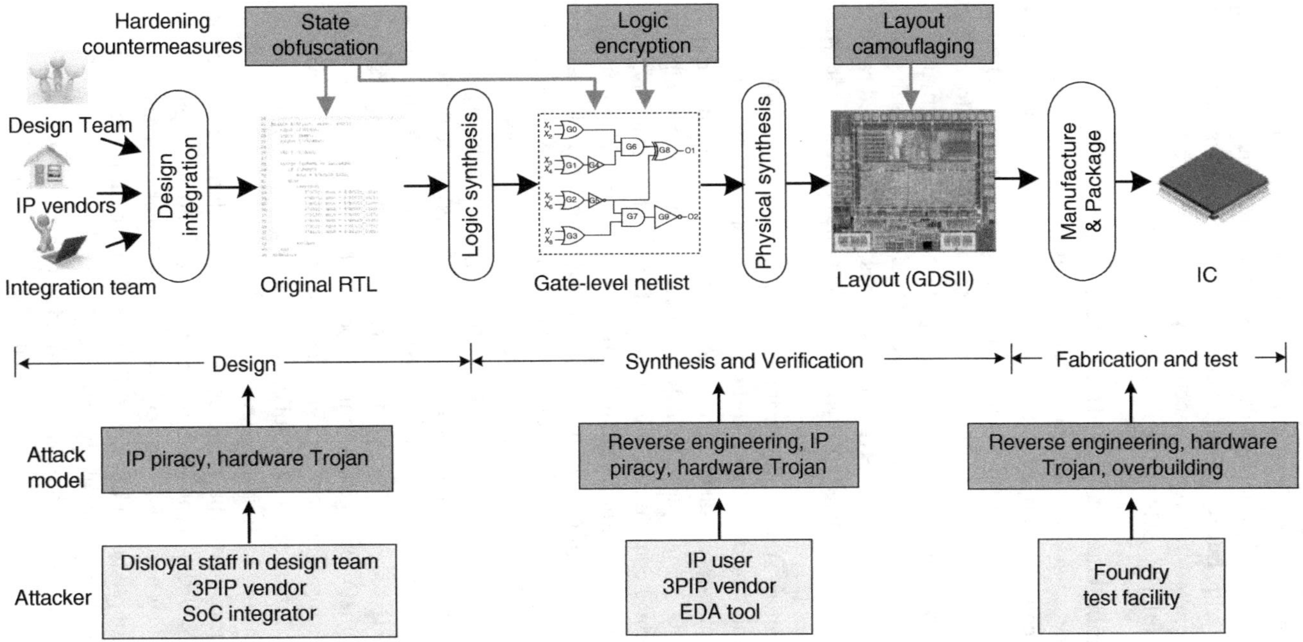

Figure 18.1 Hardware security threats from multiple stages of IC design flow.

sensor-based threat. Sensors used in IoT applications may reveal important information such as secret keys, passwords, and probably credit card information [9].

SCAs show interest in side-channel signals such as delay, electromagnetic leaks, power consumption, temperature, timing, and even sound. Such side-channel information can be exploited to reveal the secret key applied in the crypto-system. We briefly introduce different SCAs in the rest of this section.

a) *Timing Attacks:* Timing attacks are based on the total time that the system takes to perform the operations to retrieve the crypto-key. Similar to the data-dependent power consumption exploited by power analysis attacks, timing side-channel is also dependent on the internal operation of a crypto engine. In this way, attackers can recover the secret information of a cryptographic algorithm by measuring the time that crypto operation takes and then analyzing time variation due to different keys. The idea of timing attacks was first proposed by Kocher in 1996 [10]. This type of attacks has been developed rapidly since 1996 and researchers have implemented timing attacks in different types of computing devices [11, 12].
b) *Electromagnetic Analysis Attacks:* Electromagnetic analysis attacks exploit the electromagnetic emission of a device to breach the crypto engine. Attackers capture the electromagnet signal emanated from the device in which a cryptographic algorithm is being operated. After analyzing the captured signals, attackers could retrieve confidential information of the device. Basically, electromagnetic analysis exploits the fact that different types of logic switching generate different electromagnetic signals. Electromagnetic analysis can be adopted in different ways to form various attacks such as simple electromagnetic analysis (SEMA) attacks and differential electromagnetic analysis (DEMA) attacks. The attack targets fall in a wide range of computing devices including, for example, FPGAs and smart cards. The attacks can be implemented to break advanced encryption standard (AES) [13], or lightweight ciphers, like TWINE and PRINCE [14, 15], which is more suitable for IoT devices than AES. An electromagnetic leakage is a common occurrence for IoT devices. The adversary can eavesdrop the electromagnetic emission and make the system vulnerable to SCAs. Smart cards also radiate the electromagnetic waves, which can be detected by using radio frequency antenna [16].
c) *Power Analysis Attacks:* Power-based SCAs exploit the correlation between the power consumption of the crypto-system and the hypothetical crypto key to retrieve the secret key applied in the crypto-system. Power analysis attacks are categorized in simple power analysis (SPA) [8], differential power analysis (DPA) [8], and correlation power analysis (CPA) [17, 18]. The success of SPA is vulnerable to power noise. To address the limitation of SPA, DPA categorizes the measured power traces based on the predicted key bit until the average power for one of power trace groups stands out. In CPA, given a hypothesized

crypto key, attackers estimate the power consumption for the cryptographic module implemented in hardware. Typically, attackers use Hamming weight or Hamming distance power model [17, 19] to estimate the number of bit transitions happened in the crypto state registers and then predict the associated power change between two consecutive clock cycles. Next, the analytical power estimation is compared with the real measured power consumption. The Pearson correlation coefficient has been identified as an effective metric to assess how close the estimated power is to the real one. The highest Pearson correlation coefficient indicates the crypto key that leads to the corresponding estimated power is the key in use. Power analysis attack is one of the major threats to IoT devices, especially sensors. As demonstrated in [20], power analysis attacks can successfully recover the encryption key from nodes running an IEEE 802.15.4 stack. Note that many IoT communication protocols are built on top of IEEE 802.15.4.

18.2.2 Hardware Trojan

HTs are malicious hardware modifications on the original chip. Typically, the rarely switched internal signals from the original design are used to trigger the inserted HT and the Trojan output is fed back to influence the original design [21], as shown in Figure 18.2a. More specifically, an HT is composed of trigger circuit and payload circuit as depicted in Figure 18.2b and c, respectively.

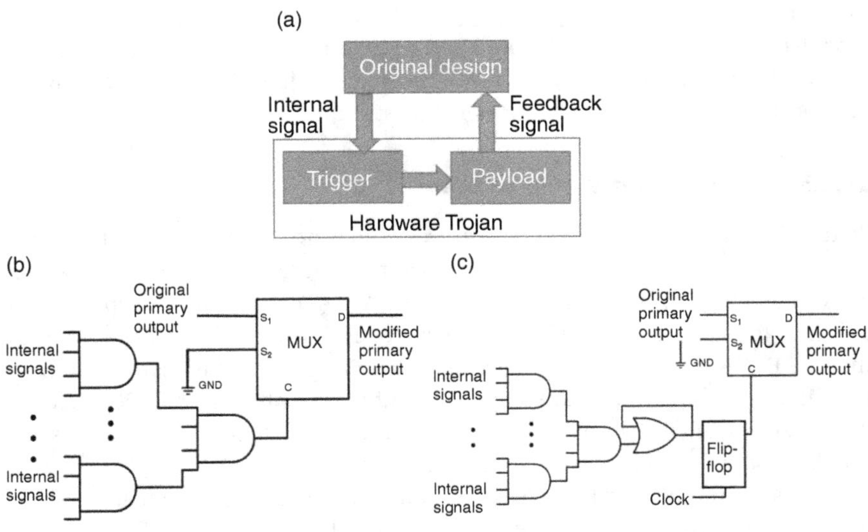

Figure 18.2 Hardware Trojan. (a) Structure of hardware Trojan, (b) combinational hardware Trojan, and (c) sequential hardware Trojan.

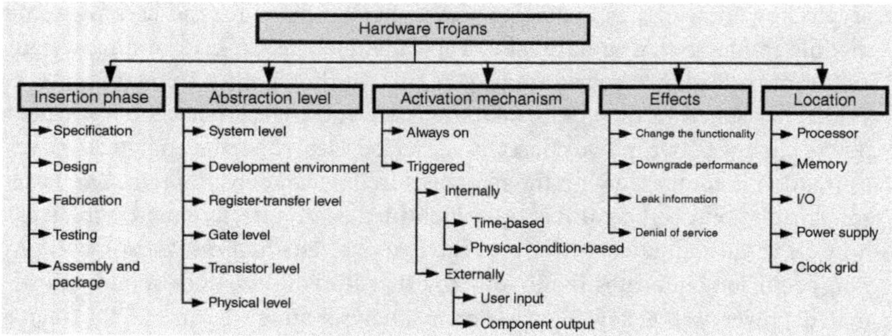

Figure 18.3 Hardware Trojan taxonomy [26].

The trigger circuit is used to examine the arrival of the trigger condition that the attacker specifies in the stage of Trojan insertion. The Trojan payload circuit could cause a denial of service (DoS) problem, alter a chip normal operations, or provide the adversary with the privilege to access a confidential memory space [22–24]. The Semiconductor Research Corporation (SRC) concluded that one of the most challenging hardware security areas is the lack of HT detection methods at and before the register-transfer level (RTL) [25].

HTs are designed to deliver attacks such as DoS, information leakage, data manipulation, and system degradation. During the design stage, HTs can be embedded at the RTL description [23], or straight into the gate-level netlist, leading to logical attacks on the system. In the fabrication stage, design layouts can be modified to include a HT, changing internal circuit characteristics such as delay. Figure 18.3 summarizes the taxonomy of HTs in detail [26]. For IoT devices, memory and I/O are the two hardware elements susceptible to Trojan insertion.

18.2.3 Security Threats from Software

Some IoT devices have configurable features, which are implemented with FPGAs. FPGA design software has been considered as potential attack surfaces [27], which are challenging IoT security. Untrusted FPGA CAD tools can be exploited by attackers to insert HTs [28, 29]. The attack model shown in Figure 18.4a assumes that the FPGA deployment engineers, in-house designs, the bitstream downloading channel, and procedure are trusted. The untrusted phase is the FPGA configuration, especially the design mapping, place, and route stages. The attacks are originated from malicious software mounted on top of the original FPGA design suite, as shown in Figure 18.4b. The FPGA design suite may not be malicious initially, but advanced attackers could exploit the vulnerability of the FPGA design suite to implant malicious software in the original suite through software

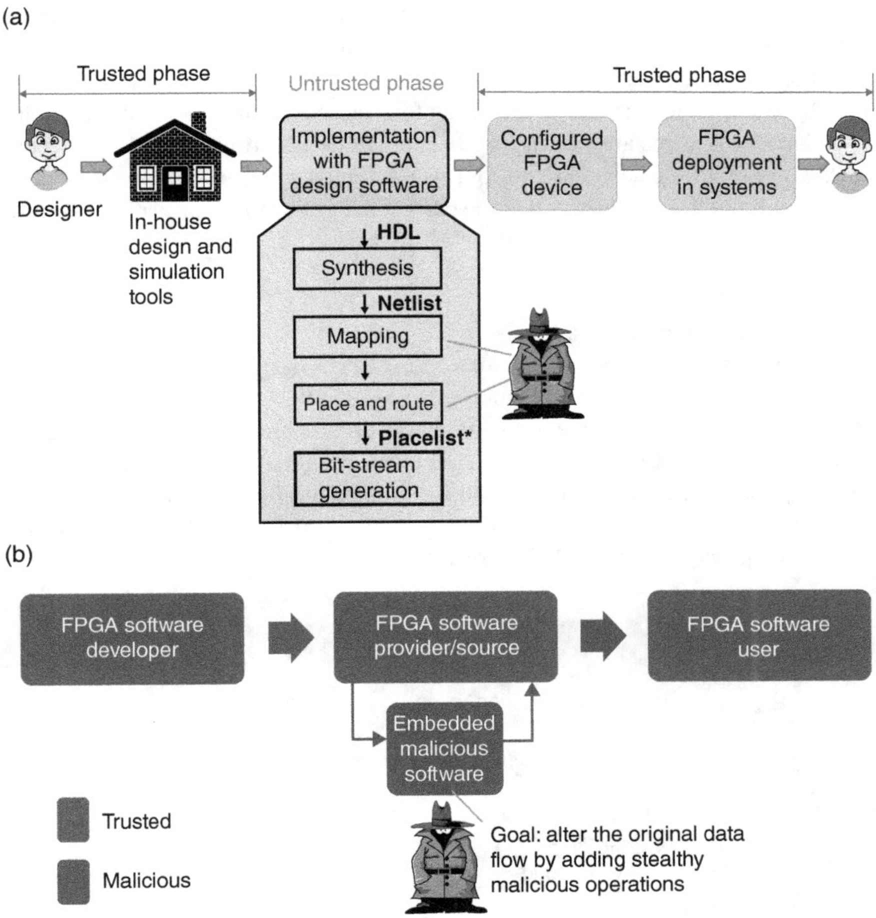

Figure 18.4 Contaminated FPGA design suite leading to a stealthy modification on the placelist for an FPGA device. (a) Software compromising stage, and (b) malicious software add-on in the supply chain of FPGA tools.

upgrading. We argue that the FPGA design suite will be propagated through computer network or retailers, so the integrity of the software may be sabotaged by advanced attackers.

As the malicious program is mounted on top of the original FPGA CAD tool before FPGA users utilize the FPGA tool and development board, it is reasonable to assume that the attacker does not know what exact design will be mapped to the FPGA die. Depending on the attacker's capability, we classify the attacks into two levels.

- *L-1 Attacks:* Based on his/her experiences, the attacker places HTs in the most popular FPGA die area. At this level, the attacker does not have any knowledge of the design to be configured on the FPGA.
- *L-2 Attacks:* The attacker is capable of extracting information, for example, which FPGA slices are utilized by the current design from the FPGA placelist (the output after placement and routing). Although the attack at this level does not analyze the exact function of the design, the exploration space of L-2 attacks is significantly smaller than that of L-1 attacks.

18.2.3.1 Attacks on Xilinx ISE

Figure 18.5 depicts the design flow for a Xilinx FPGA design suite. There are three potential attack surfaces for maliciously implanted FPGA tools to land on. We use Xilinx ISE 14.1 as an example in the following discussion. In the step of mapping, an attacker could introduce additional I/O pins, exchange the existing I/O pin connection, and modify the slew rate and the voltage level of I/O pins. As the tampered mapping output *map.ncd** is not readable (unless the FPGA design suite provides a program like *ncd2xdl* to read back the native circuit description file),

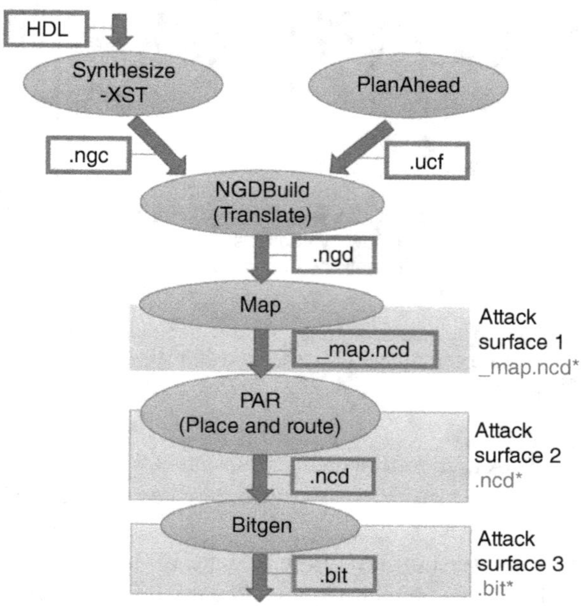

Figure 18.5 Attack surfaces on the Xilinx FPGA design flow. The rectangles represent the output file from each step. The file with the symbol of * is an output file modified by the malicious FPGA software.

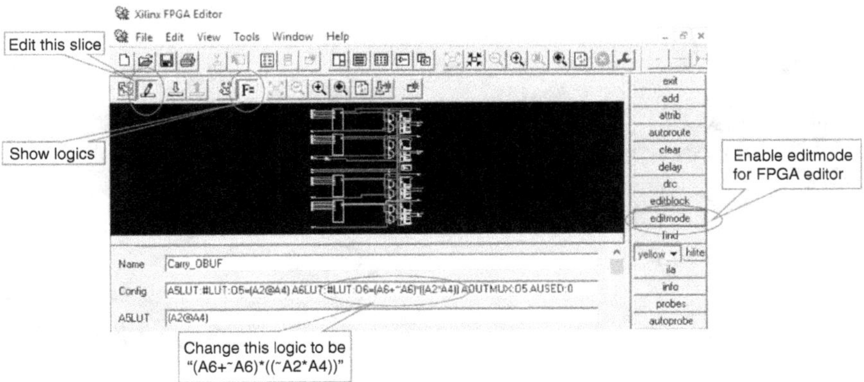

Figure 18.6 An example of practical attack performed through the FPGA editor tool available in the Xilinx ISE 14.1 design suite.

it is not easy to notice the modification performed by the malicious FPGA software. More tampering on the FPGA configuration can be done in the step of Place and Route (PAR) than in the mapping stage because all the LUTs, flip-flops, SRAM blocks, and interconnects are specifically designated on the FPGA die. The attack on the stage of bitstream generation is mainly for the purpose of IP piracy, which is out of the scope in this work. For interested readers, many existing literatures [30–32] have extensive discussion on this issue. Our work focuses on the first two attack surfaces shown in Figure 18.5.

We successfully modified the configuration of the target slice through the *FPGA editor* tool from Xilinx. Figure 18.6 shows the graphic interface. In the edit mode of the FPGA editor, we changed the logic configuration after the PAR stage, and then re-did bitstream generation. The attack can also be performed via XDL file editing followed by the command *xdl2ndc*. All attack actions here can be implemented in a malicious FPGA software implanted in the original FPGA design suite.

18.2.3.2 Attacks on Altera Quartus

The Altera FPGA design suite, *Quartus*, leaves similar back doors for attackers to insert HTs. The security vulnerability of Quartus is in the process of placement and routing *Fitter*, like PAR in the Xilinix ISE. Attackers can, in theory, manipulate the entire FPGA configuration if they control *Fitter* or access and alter the design file that the tool *Fitter* is dealing with. As shown in Figure 18.7, attackers can change buffer slew rate, I/O standard, or logic function of the design via the Quartus built-in tool *Chip Planner*. The malicious changes can be done after design compilation and *no* recompilation process is needed to save the changes. The attacks performed through *Chip Planner* are stealthy because they do not disturb the

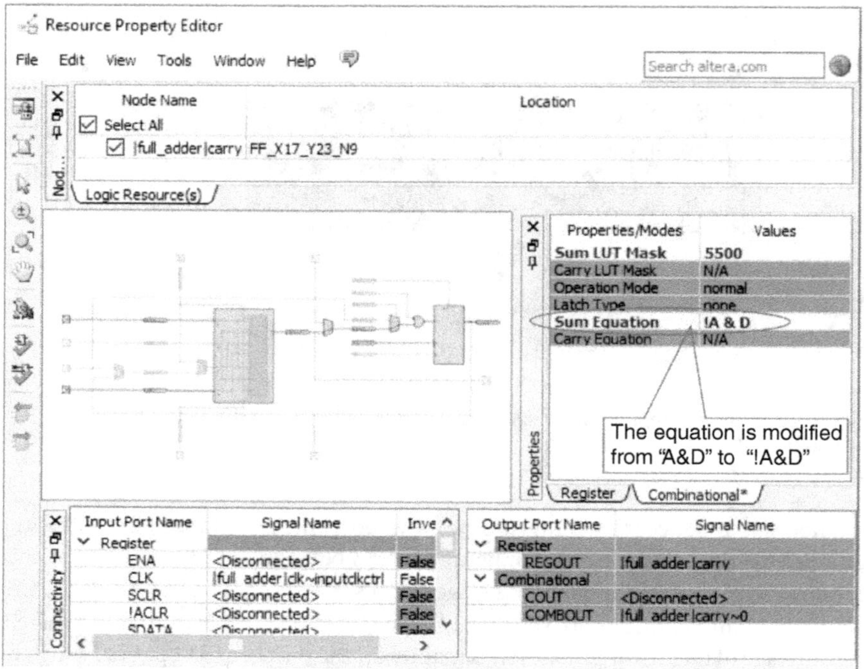

Figure 18.7 An example of practical attack performed through Quartus Chip Planner.

functional module in a format of the hardware description language and the constraint settings.

18.3 Countermeasure Against SCAs

As IoT devices typically have limited computation power and need to be energy efficient, sophisticated cryptographic algorithms and authentication protocols are not suitable for every IoT device. In an IoT application, sensors, actuators, and simple data collection devices are connected through the Internet. Among those connected devices, at least one data computation chip is needed to analyze the massive amount of data and make a corresponding decision. Due to the important role that the computation chip plays, we assume that some cryptographic algorithm (at least a lightweight one) will be applied to that chip. We propose to apply a dynamic permutation method to protect the processing unit from hardware attacks, more precisely, HT insertion and SCAs simultaneously.

18.3.1 Dynamic Masking and Error Deflection Method for Protection Against Combined Attacks

As mentioned earlier, IoT devices are constrained power budget, and hence it is imperative to design the unified countermeasures that can address the multiple attacks simultaneously. As single fault detection method cannot thwart the CPA and fault analysis (FA) attacks simultaneously, we propose to integrate a dynamic masking technique with an error control code (ECC)-based error deflection mechanism. The main principle of this method is highlighted in Figure 18.8 [18]. This method is designed for the AES, but can be easily implemented for the other ciphers too.

Unlike the fault detection in the previous work encoding the S-Box input [33], we encode the intermediate state value before it reaches the state registers. We argue that it makes more sense to manipulate a register value than modify the S-Box logic to achieve a better correlation of the measured power consumption

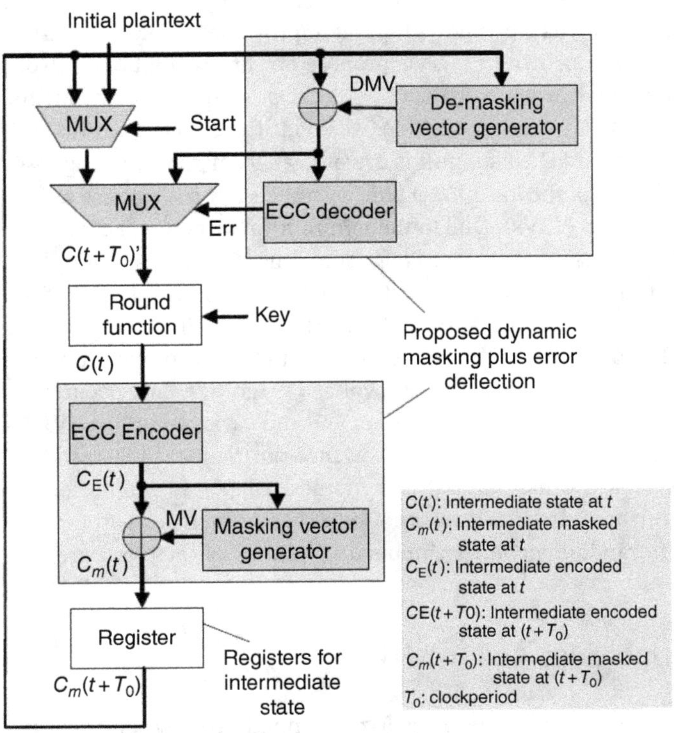

Figure 18.8 Dynamic masking plus error deflection method for thwarting CPA and FA attacks [18].

and the hypothetical key for CPA attacks. Another innovation of our method is the dynamic masking. Instead of using a fixed masking vector, we propose to generate the masking vector at runtime using the masking vector generator. The masking vector is a modification of the intermediate state value. As the intermediate state register changes the value over the time, the (de)masking vector is not consistent. Consequently, the power model modification according to a guessed masking vector will fail. Our FPGA-based experimental results show that the proposed method can successfully thwart the CPA attack for 100 000 power traces and reduce the FA success rate by 54 and 90% over the masking only and the joint of masking and cyclic redundancy check codes methods.

18.3.2 PDN Noise-Based Countermeasure Against CPA

IoT devices perform different functionalities and use wide range of sensors and actuators to carry out operations like processing, computation, monitoring, and communication. Heterogeneous integration of different technologies in a single die using 3D or 2.5D integration can be beneficial for the IoT devices to perform the mentioned diverse tasks as it can provide great improvement in system performance and efficiency. In our work [34], we propose power distribution network (PDN) noise-based countermeasure against CPA attacks in 3D ICs to thwart the timing and power-based SCAs. In our method, we divide the crypto unit into multiple subunits (e.g. four) and each subunit is driven by a local supply voltage. We utilize a crossbar to connect the local VDD pins with the PDN nodes close to four power through-silicon vias (TSVs). Due to the nonuniform switching activities in all 3D planes, *each TSV passes a unique voltage* from other 3D planes to the plane carrying the crypto unit. The effect of parasitic resistance and capacitance of the metal wire between the power grid and the local VDD pin further increases the variance of the VDDs connected to the crypto unit. From the power measurement of each TSV, we observed that both the peak voltage magnitude and occurrence timing for the four VDDs are not exactly same, and the variance on four VDDs over baseline are noticeable. We exploit the variance on the local VDDs to blur the correlation between the measured and hypothesized power profile for the crypto module. To further introduce power noise, we can bring in a small random number generator to randomly connect inputs and outputs of the crossbar.

18.4 Countermeasure Against HTs

Existing efforts on HT detection can be categorized as nondestructive and destructive detection methods [26, 28, 35]. The nondestructive methods include functional testing [36, 37], accelerated Trojan activation followed by testing [38], and analysis on side-channel signals (e.g. power, delay, temperature, or electromagnetic profiles) [39–41]. The destructive method category is mainly attributed to destructive reverse engineering, such as chemical metal polishing followed by scanning electron microscope (SEM) image reconstruction and analysis [42], and fingerprint examination [43]. As HT designs are becoming more intricate and the HT target system integrates more transistors, HT detection in a very-large-scale integrated system is comparable to finding a needle in a haystack. Consequently, HT detection at testing time (or static time) cannot guarantee a HT detection rate of 100%. The residual HTs on chips will further harm the systems with those malicious circuits.

We propose to harden a hardware design at the design time. In this section, we use network-on-chip (NoC) as a case study to show how to strengthen the design resilience against HT insertion. The proposed HT mitigation method aims to detect and mitigate the HT attacks that (i) modify the flit type, (ii) change the legal packet destination address to an unauthorized one, and (iii) sabotage the integrity of a packet. The main consequence of the HT targeted in this work is the NoC bandwidth depletion. We assume that the links between NoC routers are trusted (ensured by using the technique such as in [44]). Due to the regularity of implementation and large area consumption of the FIFOs, we assume that FIFOs are more likely and easier to be compromised by the adversary than the other router submodules (such as flit processing unit).

As shown in the left side of Figure 18.9, a generic mesh-NoC is composed of links, network interfaces (NIs), and routers with five ports (north, east, south, west, and local input/output ports). We use one pair of input and output ports (shadowed area in Figure 18.9) as an example to present the novelty of our router design. The zoomed-in picture in Figure 18.9 depicts the high-level view of the proposed HT countermeasures (HTC) applied in the south input port and the east output port.

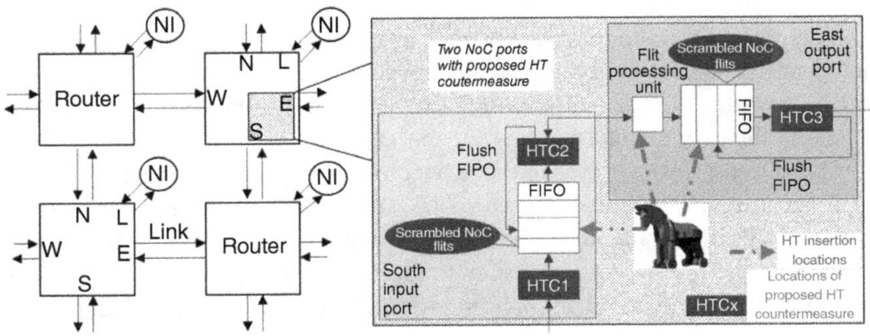

Figure 18.9 High-level overview of the proposed countermeasure.

In the input port, the HTC1 module dynamically permutes the incoming NoC flit bits before reaching the FIFO. As a result, the NoC flit saved in the input FIFO is scrambled. Due to our randomness and dynamicity of the permutation patterns, it is much harder for an attacker to change the flit content into something meaningful through HTs in our proactively defensed router than in a router with static protection. The goal of proposed permutation is to reduce the probability of a HT inserted in the FIFO successfully modifying the critical bits in the NoC packet. The HTC2 module is used to examine the integrity of flits, which could be sabotaged by the HTs placed in the input FIFO. The main function of the HTC2 is to prevent the malicious flits from entering the rest of the network by dropping the flit and flush the input FIFO if necessary. HTs could also be located in the flit processing unit and the FIFO in the east output port. Hence, we place a HTC3 module after the FIFO to stop the malicious flit from leaving the current router and then recover the scrambled flit back to the original bit order. The three HTC modules proposed in Figure 18.9 cooperatively provide proactive defense to thwart the potential HTs that harm the NoC flit integrity and thus deplete the NoC bandwidth.

The HTC modules in Figure 18.9 are detailed in Figure 18.10. The highlighted solid blocks in Figure 18.9 are the changes over the generic router. To guarantee a packet reach its original destination, the critical fields (i.e. source/destination address and flit type bits) in a NoC packet cannot be tampered with. To maintain the flit integrity in despite of the potential HTs in the router, we propose to first encode the critical flit fields with an ECC. Next, we apply a dynamic flit permutation (DFP) technique to the packet buffers, so that critical information bits in a NoC flit is "randomly" reordered before that flit is sent to the input FIFO. If the attacker could not precisely know the dynamic permutation pattern at each specific moment, the HTs he/she inserted in the router will not be triggered as expected or execute the malicious mission as designed. Symmetrically, the dynamic flit de-permutation (DFDeP) is applied at the output port. After de-permutation, an ECC decoder (DeECC) facilitates the router to detect and correct all 1-bit errors on the critical flit content.

The permutation randomness is achieved by a pattern selection vector, which is dynamically generated by the local random-vector generator. Each router has its own random number generator, which provides the selection vector to choose one of the flit permutation patterns for the DFP and DFDeP units. We exploit the unique routing history saved in the Arbiter to generate a random vector through a PUF structure implemented in each router. Due to the uniqueness of each PUF and the routing history, the dynamic permutation pattern for DFP and DFDeP varies over time and is different from router to router. As the packet routing history in each router depends on the NoC application at deployment stage, the output of local random-vector generator is unpredictable at the NoC design stage. Thus, the adversary cannot easily place HTs to perform successful attacks at the design time.

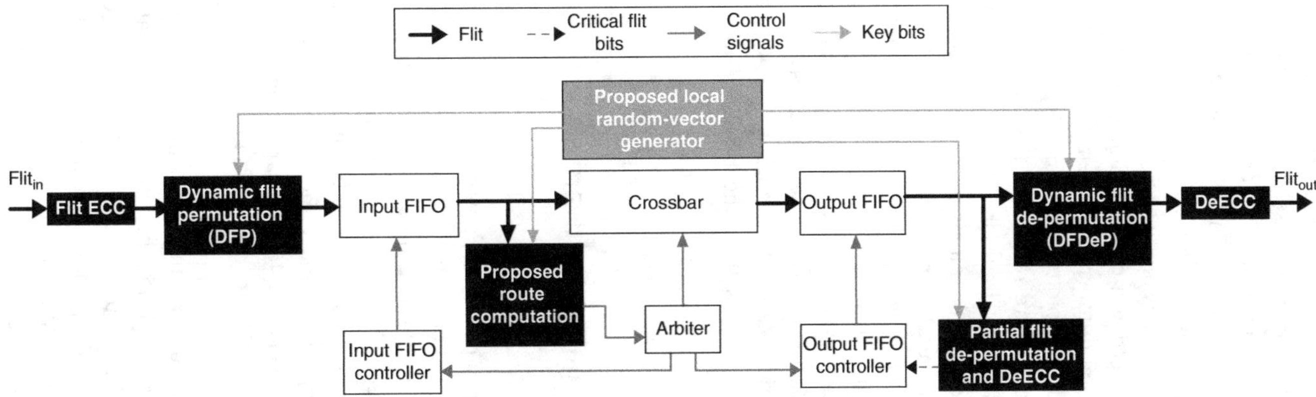

Figure 18.10 Router architecture used in this work. One proposed input and output port pair for hardware Trojan detection and mitigation. Solid blocks highlight our innovations over the generic router architecture.

The Proposed Route Computation unit is designed to interpret multiple versions of the permutated flit content into a request to use the correct output port for the next routing hop. To reduce the hardware cost, we propose to use partial flit depermutation and DeECC to correctly interpret important internal and external signals such as write request and FIFO buffer status signals.

We monitored the arrival activity of new flits at each output port of the routers to estimate the switching activity (thus power consumption) of each router throughout the entire simulation. The total number of different flits transferred through each router is accumulated and then plotted in a node-type plot, as shown in Figure 18.11. Each square in the figure represents the switching activities of one router, with the corresponding coordinates in the physical NoC implementation. The color of each square indicates the intensity of the number of output port switching operations occur over the entire simulation period. The tile plots of each method were compared to the baseline NoC without HTs in the network, which is subplot (a) in the figure. The tile plots for the HEAD HT case are shown in Figure 18.11. As all of the methods, excluding the proposed method, cannot prevent the effects of the HEAD HTs, those methods will have less link usage at the affected routers. This is an indication that the R-Term, Baseline, R-AddrFilter, and NI-Term methods receive fewer packets than the baseline NoC without the HT case.

18.5 Countermeasure Against Software Security Threats

Defenders must protect every entry point from potential security threats. In contrast, an adversary only needs to find one way to breach the attack surface; moreover, the attacker may even have unlimited time to perform attacks. The main motivation of applying the concept of moving target defense (MTD) to a system is to reduce, if not completely eliminate, the imbalanced advantage that an attacker could have. MTD techniques can make the system less predictable and thus the attack surface is changed over time [45]. The early concept of MTD is illustrated in [46] and the application of MTD is observed in the domain of cyber security [47].

We exploit the principle of MTD as a mean to proactively address the security threats from malicious FPGA software. Different with the traditional MTD methods applied in the domain of cyber security, our FOMTD method explores the unpredictability of the way that a hardware design is configured on FPGAs to deter attackers from precisely inserting HTs. More specifically, the key idea of FOMTD is to make the output of FPGA placement and routing unpredictable, such that

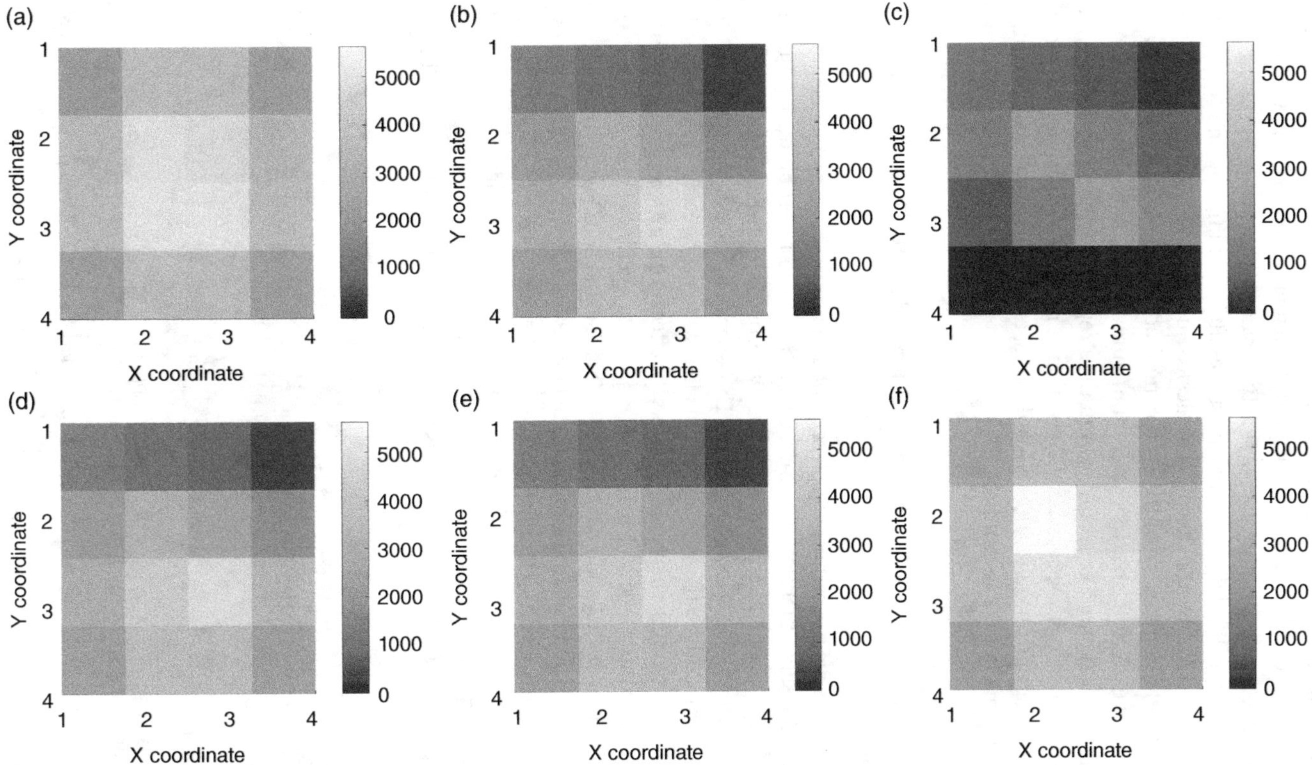

Figure 18.11 Traffic hotspot migration and bandwidth depletion induced by HTs. (a) Baseline No HT. (b) Baseline HEAD HT. (c) NI-Term HEAD HT. (d) R-AddrFilter HEAD HT. (e) R-Term HEAD HT. (f) Proposed HEAD HT.

attackers who mount a malicious program on the original FPGA design suite cannot easily and successfully alter the original implementation on an FPGA. Note, our method does not guarantee to completely prevent all hardware intrusions but it will increase the difficulty of a Trojan successfully landing on one (or more) FPGA slices occupied by the design. The desired unpredictabilities are achieved by the three defense lines provided by our method. In the domain of hardware (i.e. FPGA), we exploit the following configuration resources to realize the FOMTD method: (i) the availability of multiple replicas of the intended design, (ii) random selection of one replica for operation at runtime, (iii) random designation of FPGA slice positions for the selected LUTs, and (iv) hot-swappable submodules for runtime design assembling.

18.5.1 Defense Line 1 (DFL1): Slice Position Selection Through User Constraints File

The use of FPGA default settings for placement and route will make the location of occupied FPGA slices predictable, which eases the Trojan insertion through malicious FPGA CAD tools. To address this issue, we propose to specify some slice locations for the selected LUT configurations. This specification can be performed by appending commands to the user constraints file, which is typically used to specify pin and timing constraints. Figure 18.12 shows the effect of the proposed defense line 1 (DFL1). As can be seen, the entire design is mapped to a different area of the FPGA grid, thanks to the reallocation of three LUTs (black squares in Figure 18.12).

The selection of slice positions is conducted by FPGA users at the FPGA deployment stage, which is after the implementation of the malicious FPGA software. Hence, it is not easy for the attackers (malicious software designers) to ensure the injected HTs to successfully alter every unknown user designs. Here, we assume that attackers do not have access to the user constraints file applied after

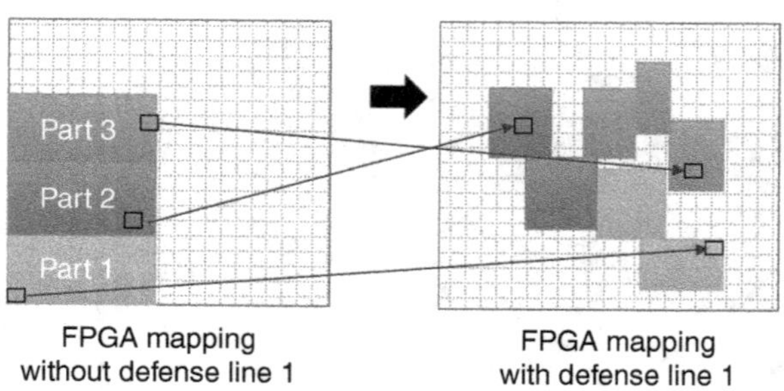

Figure 18.12 FPGA mapping modified by proposed defense line 1. Three parts in different colors represent three partitions of the intended design. Black squares are three LUT configurations. Proposed defense line 1 alters the default LUT mapping on the FPGA grid.

the FPGA CAD tool is delivered to the FPGA user. A blindly inserted Trojan may not effectively impact the design on the FPGA.

18.5.2 Defense Line 2 (DFL2): Pseudorandom Replica Selection

FPGA has a nature of reconfiguration and redundancy. We exploit this nature to implement the principle of MTD on FPGAs. Assume a design is composed of multiple parts (however, design portioning is not always necessary). We duplicate the entire design (as a single unit) n times. Only one of the replicas will be active at a time, and the rest of the replicas are inactive by using input gating technique. The replica selection and input gating are controlled by a pseudorandom selector, which is *not* a true random number generator. Because we only have a limited number of replicas on the FPGA, the range of the random number is not large. A user-defined arbitrary logic function and a set of external inputs are good enough to pseudorandomly choose one of the replicas and further prevent attackers from searching the typical random number generator circuit to nullify the countermeasure in advance. Figure 18.13 shows the concept of our defense line 2 (DFL2), in which we do not have a comparison logic to examine the consistency among the n replicas for the purpose of power saving. As the fact that which replica will be active is determined after the FPGA configuration, an attacker (at L-1) needs to blindly place the HT to the entire FPGA die to make a successful attack.

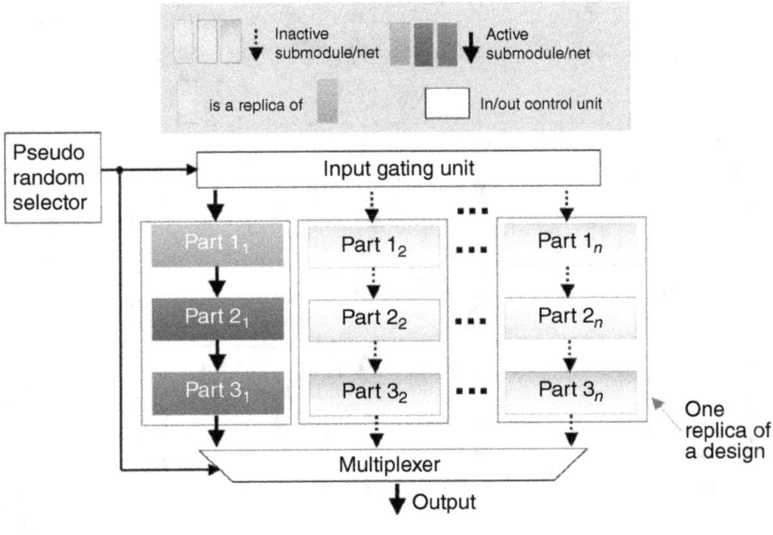

Figure 18.13 Schematic diagram of proposed defense line 2.

18.5.3 Defense Line 3 (DFL3): Runtime Design Assembling

Our defense line 3 (DFL3) is the hot-swappable submodule assembling technique, as shown in Figure 18.14. We partition the original design into m submodules and each submodule is duplicated by n times. Only one replica of each submodule will be assembled into a complete design. The pseudorandom selector is utilized to determine which replica to be chosen at runtime. After a period of time, the selection of submodule replicas will be changed without stopping the normal operation (i.e. *hot-swappable assembling*). The maximum number of design configurations is n^m. This large number of configurations further increases the difficulty for the attacker to recognize the entire design for attack. The hot-swappable assembling technique is shown in Figure 18.14.

In the following experiments, we used the Xilinx ISE 14.1 design suite to synthesize, place, and route the netlist of ISCAS'85 and ISCAS'89 benchmark circuits, and the Amber 23 processor core (hereafter, *a23*) and the communication controller Ethernet MAC (hereafter, *ethmac*) downloaded from the OpenCores website. The ISCAS circuits were configured for a Xilinx Spartan-6 XC6SLX16 FPGA, and the large-scale *a23* and *ethmac* circuit were mapped to a Xilinx XC6SLX75 FPGA. The detailed slice utilization of each circuit was analyzed by our Python script to extract the occupied FPGA slice positions. We used MATLAB to insert HTs blindly or purposely (depending on the experimental goal) in the extracted placelists to mimic the Trojan injection in the FPGA mapping and Place&Route stages, and then measured the HT hit rate. We assume that only the Trojans having payloads in the FPGA slices occupied by the design under protection will lead to a Trojan hit. The attack resilience of the baseline and our method are compared. Two attack levels mentioned previously are considered in the following assessment.

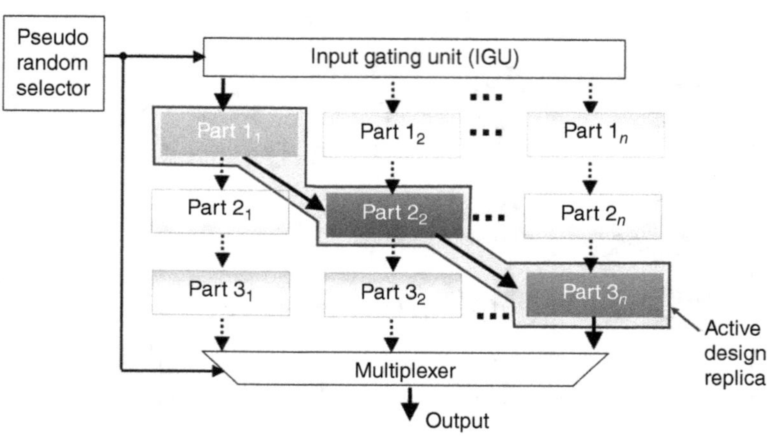

Figure 18.14 Schematic diagram of the hot-swappable submodule assembling technique provided by proposed defense line 3.

18.5.3.1 HT Hit Rate for L-1 Attacks

Recall that attackers who execute L-1 attacks do not know the locations of all occupied slices for the design of interest. We varied the range of attack exploration space from 5 to 50% of the entire FPGA die in the following experiments. Figure 18.15

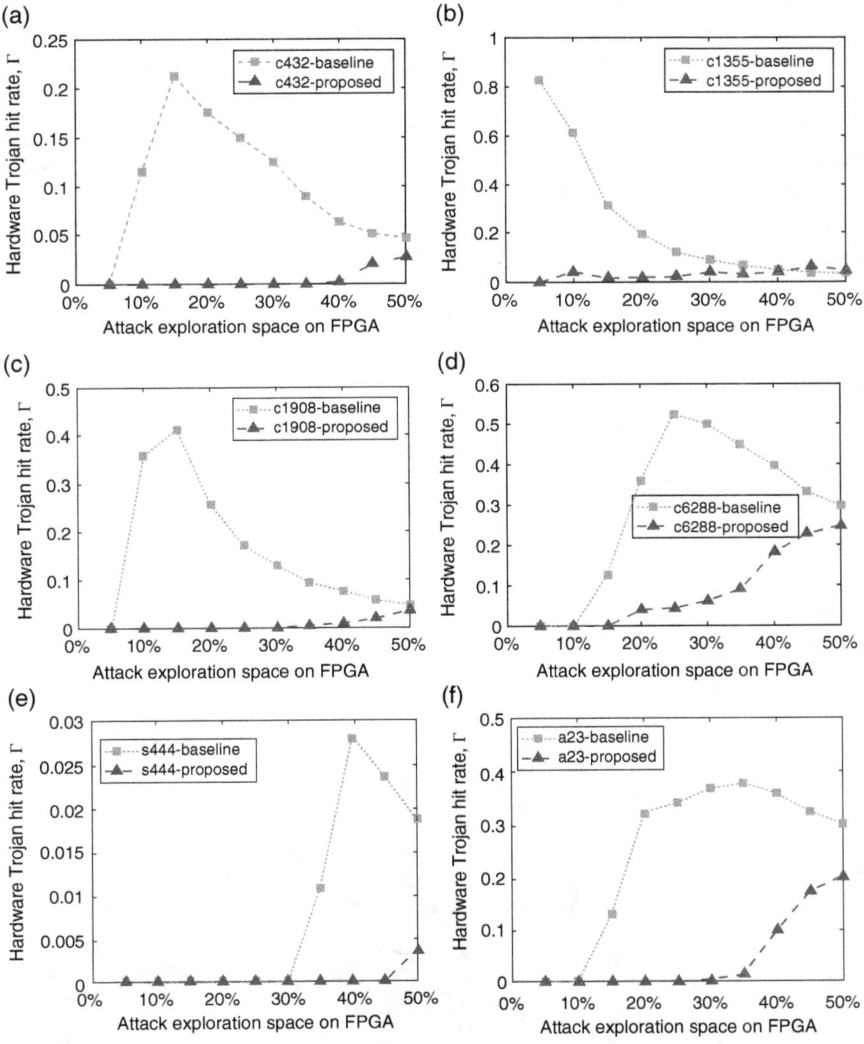

Figure 18.15 Hardware Trojan hit rate reduction by proposed defense line 1 applied in the benchmark circuit (a) c432, (b) c1355, (c) c1808, (d) c6288, (e) s444, and (f) a23 in the scenario of L-1 attacks.

shows that the proposed method achieves a lower HT hit rate Γ than the baseline in a wide range of the attack exploration space. When the attack exploration space is large enough to cover the entire design placed on the FPGA die, the Trojan hit rate of the proposed method will approach to the Trojan hit rate of the baseline eventually.

18.5.3.2 HT Hit Rate for L-2 Attack

Different from L-1 attacks, L-2 attacks are able to retrieve the exact locations of the occupied slices. Consequently, the baseline design does not have any resilience against L-2 attacks. The proposed DFL2 activates one complete design replica according to the pseudorandom selection and the DFL3 assembles the hot-swappable submodules at runtime. Here, we used two design replicas and each replica composed of four submodules. Our simulation indicates that DFL2 and DFL3 further increase the unpredictability of the truly activated design copy and achieve a lower Trojan hit rate over the baseline. As shown in Figure 18.16a, the baseline

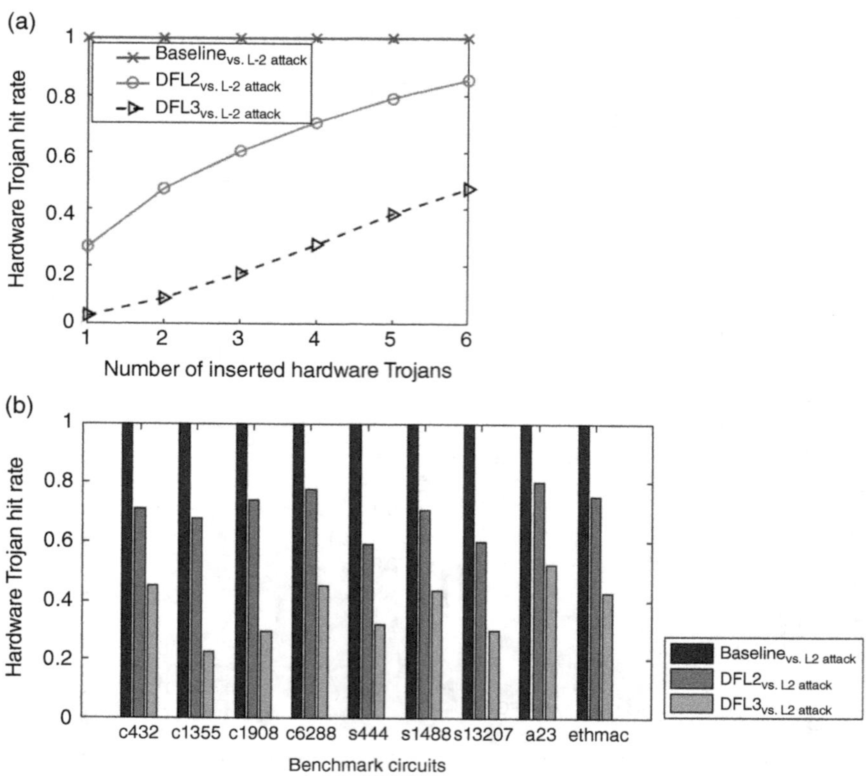

Figure 18.16 Hardware Trojan hit rate for (a) c432, and (b) nine benchmark circuits suffering from four hardware Trojans inserted via L-2 attacks.

yields a HT hit rate of 1, which means Trojans are always injected to the occupied slices. In contrast, our DFL2 and DFL3 significantly reduce the Trojan hit rate over the baseline, especially for the small number of injected Trojans. When more Trojans are placed in the utilized FPGA slices, our Trojan hit rate eventually increases due to the limited number of replicas available in the design. We examined the Trojan hit rate for nine benchmark circuits, which suffer from four Trojan insertions via L-2 attacks. Each HT hit rate was obtained from 10 000 test cases. The average Trojan hit rate of DFL2 (DFL3) is 71% (38%). As shown in Figure 18.16b, the DFL2 reduces the hit rate by up to 40% over the baseline. The reduction on the Trojan hit rate can be further improved by up to 91% with DFL3.

Many security mechanisms for FPGA-based systems have been investigated to prevent the system from IP theft and bitstream reverse engineering. However, there is limited literature available that studies the security threats originated from the untrusted FPGA CAD tools. This work fills the gap. First, we demonstrate two practical attacks through Xilinx and Altera FPGA design suites. Next, we classified two Trojan attack levels, depending on the attacker's prior FPGA experience and access to the FPGA software. Then, we propose a FOMTD method to mitigate the HTs induced by the malicious CAD tools. Our method offers three defense lines, each offering a different degree of unpredictability from the malicious FPGA software designer's point of view. As our unpredictability is formed after the untrusted CAD tool is delivered to FPGA users, our method facilitates the FPGA user to thwart the two levels of Trojan insertion attacks during the process of FPGA configuration. To the best of our knowledge, our research effort is the first work that investigates the FPGA-based MTD for SRAM FPGAs.

18.6 Conclusion

Hardware is the root of trust. Assurance of hardware integrity and security is increasingly important due to the globalized semiconductor supply chain. This chapter presents three typical hardware security threats: side-channel analysis attacks, hardware tampering, and HT insertion from untrusted CAD tools. A dynamic masking and error deflection method is introduced to address the fault attack. To thwart the CPA attack, the noise originated from the power distribution network is exploited to obscure the correlation between real power consumption and the input/key-dependent power estimation. On-chip network router is hardened by dynamic permutation and error control coding to reduce the success of HT insertion in application-specific integrated circuits. The principle of MTD is employed to the Trojan placement in the FPGA platform. Despite the emergence of various security threats, dynamicity of design parameters, system inherent variables, and configurable architecture have great potential to be applied to address hardware security issues.

References

1 D. Evans, "The Internet of Things – How the next evolution of the Internet is changing everything." 2011. https://www.cisco.com/c/dam/en_us/about/ac79/docs/innov/IoT_IBSG_0411FINAL.pdf.

2 H. Bauer, M. Patel, and J. Veira, "The Internet of Things: Sizing up the opportunity." 2014. https://www.mckinsey.com/industries/semiconductors/our-insights/the-internet-of-things-sizing-up-the-opportunity.

3 "Federal statutory protection for mask works," United States Copyright Office, 1996. http://www.copyright.gov/circs/circ100.pdf.

4 Y. Tsunoo, T. Saito, T. Suzaki, M. Shigeri, and H. Miyauchi, *Cryptanalysis of DES Implemented on Computers with Cache*, pp. 62–76. Berlin, Heidelberg: Springer, 2003.

5 Y. Zhou and D. Feng, "Side-channel attacks: Ten years after its publication and the impacts on cryptographic module security testing," 2005.zyb@is.iscas.ac.cn 13083 received 27 October 2005.

6 M. Hutter and J.-M. Schmidt, *The Temperature Side Channel and Heating Fault Attacks*, pp. 219–235. Cham: Springer International Publishing, 2014.

7 E. Brier, C. Clavier, and F. Olivier, *Correlation Power Analysis with a Leakage Model*, pp. 16–29. Berlin, Heidelberg: Springer, 2004.

8 P. Kocher, J. Jaffe, B. Jun, and P. Rohatgi, "Introduction to differential power analysis," *Journal of Cryptographic Engineering*, vol. 1, no. 1, pp. 5–27, 2011.

9 A.K. Sikder, G. Petracca, H. Aksu, T. Jaeger, and A.S. Uluagac, "A survey on sensor-based threats to Internet-of-Things (IoT) devices and applications," arXiv preprint arXiv:1802.02041, 2018.

10 P.C. Kocher, "Timing attacks on implementations of Diffie-Hellman, RSA, DSS, and other systems," in Annual International Cryptology Conference. Berlin, Heidelberg: Springer, 1996.

11 C. Reinbrecht, A. Susin, L. Bossuet, G. Sigl, and J. Sepúlveda, "Side channel attack on NoC-based MPSoCs are practical: NoC Prime+Probe attack," in 2016 29th Symposium on Integrated Circuits and Systems Design (SBCCI), Belo Horizonte, pp. 1–6, 2016.

12 P. Kaushik and R. Majumdar, "Timing attack analysis on AES on modern processors," in 2017 6th International Conference on Reliability, Infocom Technologies and Optimization (Trends and Future Directions) (ICRITO), Noida, pp. 462–465, 2017.

13 Z.L. Ding, X. Chang, and Q. Zhao, "Differential electromagnetic analysis on AES cryptographic system," in 2009 Second Pacific-Asia Conference on Web Mining and Web-based Application, Wuhan, pp. 120–123, 2009.

14 M. Yoshikawa, Y. Nozaki, and K. Asahi, "Electromagnetic analysis attack for a lightweight block cipher TWINE," in 2016 IEEE/ACES International Conference

on Wireless Information Technology and Systems (ICWITS) and Applied Computational Electromagnetics (ACES), Honolulu, HI, pp. 1–2, 2016.

15 M. Yoshikawa and Y. Nozaki, "Electromagnetic analysis attack for a lightweight cipher PRINCE," in 2016 IEEE International Conference on Cybercrime and Computer Forensic (ICCCF), Vancouver, BC, pp. 1–6, 2016.

16 Y. Ren and L. Wu, "Power analysis attacks on wireless sensor nodes using CPU smart card," in 2013 22nd Wireless and Optical Communication Conference, Chongqing, pp. 665–670, 2013. doi:https://doi.org/10.1109/WOCC.2013.6676458.

17 E. Brier, C. Clavier, and F. Olivier, "Correlation power analysis with a leakage model," in Cryptographic Hardware and Embedded Systems -CHES 2004, pp. 16–29. Berlin Heidelberg: Springer, 2004.

18 J. Dofe, H. Pahlevanzadeh, and Q. Yu, "A comprehensive FPGA-based assessment on fault-resistant AES against correlation power analysis attack," *Journal of Electronic Testing: Theory and Applications*, vol. 32, no. 5, pp. 611–624, 2016.

19 J. Dofe, J. Frey, and Q. Yu, "Hardware security assurance in emerging IoT applications," in International Symposium on Circuits and Systems (ISCAS'16), pp. 2050–2053, May 2016.

20 C. O'Flynn and Z. Chen, "Power analysis attacks against IEEE 802.15.4 Nodes," IACR Cryptology ePrint Archive 2015, p. 529, 2015.

21 T. Inoue, K. Hasegawa, M. Yanagisawa, and N. Togawa, "Designing hardware Trojans and their detection based on a SVM-based approach," in 2017 IEEE 12th International Conference on ASIC (ASICON), Guiyang, pp. 811–814, 2017.

22 J.-P. Diguet, S. Evain, R. Vaslin, G. Gogniat, and E. Juin, "NOC-centric security of reconfigurable SoC," in Proceedings of NOCS'07, pp. 223–232, May 2007.

23 Y. Jin, N. Kupp, and Y. Makris, "Experiences in hardware Trojan design and implementation," in Proceedings of HOST'09, pp. 50–57, July 2009.

24 S. Adee, The hunt for the kill switch, https://spectrum.ieee.org/semiconductors/design/the-hunt-for-the-kill-switch..

25 Semiconductor Research Corporation, "Reliable needs for secure, trustworthy, and reliable semiconductors." 2013. https://www.src.org/calendar/e004965/sa-ts-workshop-report-final.pdf.

26 R. Karri, J. Rajendran, K. Rosenfeld, and M. Tehranipoor, "Trustworthy hardware: Identifying and classifying hardware Trojans," *IEEE Computer*, vol. 43, no. 10, pp. 39–46, 2010.

27 "SoC FPGA Hardware Security Requirements and Roadmap." https://www.intel.com/content/dam/www/programmable/us/en/pdfs/education/events/northamerica/isdf/SoC-FPGA-Hardware-Security.pdf.

28 M. Tehranipoor and F. Koushanfar, "A survey of hardware Trojan taxonomy and detection," *IEEE Design Test of Computers*, vol. 27, no. 1, pp. 10–25, 2016.

29 J.A. Roy, F. Koushanfar, and I.L. Markov, "Extended abstract: Circuit cad tools as a security threat," in Proceedings of HOST'08, pp. 65–66, June 2008.

30 J. Zhang, Y. Lin, Y. Lyu, and G. Qu, "A PUF-FSM binding scheme for FPGA IP protection and pay-per-device licensing," *IEEE Transactions on Information Forensics and Security*, vol. 10, pp. 1137–1150, 2015.

31 R. Karam, T. Hoque, S. Ray, M. Tehranipoor, and S. Bhunia, "MUTARCH: Architectural diversity for FPGA device and IP security," in Proceedings of ASPDAC'17, pp. 611–616, January 2017.

32 S.M. Trimberger and J.J. Moore, "FPGA security: Motivations, features, and applications," *Proceedings of the IEEE*, vol. 102, pp. 1248–1265, 2014.

33 P. Maistri and R. Leveugle, "Double-data-rate computation as a countermeasure against fault analysis," *IEEE Transactions of Computers*, vol. 57, no. 11, pp. 1528–1539, 2008.

34 J. Dofe and Q. Yu, "Exploiting PDN noise to Thwart correlation power analysis attacks in 3D ICs," in Proceedings of the 20th IEEE/ACM International Workshop on System-Level Interconnect Prediction (SLIP), pp. 1–5, June 2018.

35 H. Salmani, M. Tehranipoor, and J. Plusquellic, "New design strategy for improving hardware Trojan detection and reducing Trojan activation time," in Proceedings of HOST'09, pp. 66–73, 2009.

36 S. Bhunia, M.S. Hsiao, M. Banga, and S. Narasimhan, "Hardware Trojan attacks: Threat analysis and countermeasures," *Proceedings of the IEEE*, vol. 102, no. 8, pp. 1229–1247, 2014.

37 J. Rajendran, O. Sinanoglu, and R. Karri, "Regaining trust in VLSI design: Design-for-trust techniques," *Proceedings of the IEEE*, vol. 102, no. 8, pp. 1266–1282, 2014.

38 M. Banga and M. Hsiao, "VITAMIN: Voltage inversion technique to ascertain malicious insertions in ICs," in Proceedings of HOST'09, pp. 104–107, 2009.

39 S. Narasimhan, W. Yueh, S. Mukhopadhyay, X. Wang, and S. Bhunia, "Improving IC security against Trojan attacks through integration of security monitors," *IEEE Design & Test of Computers*, vol. 29, no. 5, pp.37–46, 2012.

40 Y. Jin and Y. Makris, "Hardware Trojan detection using path delay fingerprint," in Proceedings of HOST'08, pp. 51–57, 2008.

41 L. Lin, W. Burleson, and C. Paar, "Moles: Malicious off-chip leakage enabled by sidechannels," in Proceedings of ICCAD'09, pp. 117–122, 2009.

42 R. Torrance and D. James, "The state-of-the-art in IC reverse engineering," in Proceedings of CHES, vol. 5747 of LNCS, pp. 363–381, September 2009.

43 D. Agrawal, S. Baktir, D. Karakoyunlu, P. Rohatgi, and B. Sunar, "Trojan detection using IC fingerprinting," in Proceedings of the IEEE Symposium on Security and Privacy, pp. 296–310, 2007.

44 Q. Yu and J. Frey, "Exploiting error control approaches for hardware Trojans on network-on-chip links," in Proceedings of the 16th IEEE Symposium Defect and Fault Tolerance in VLSI and Nanotechnology Systems, pp. 266–271, October 2013.

45 D. Last, D. Myers, M. Heffernan, M. Caiazzo, and C. N. Paltzer, "Command and control of proactive defense," *Journal of Cyber Security and Information Systems*, vol. 4, no. 1, pp. 8–13, 2015.

46 "Moving Target Defense." https://www.dhs.gov/science-and-technology/csd-mtd.

47 R. Zhuang, S. A. DeLoach, and X. Ou, "Towards a theory of moving target defense," in Proceedings of the First ACM Workshop on Moving Target Defense, pp. 31–40, 2014.

19

IoT Device Attestation

From a Cross-Layer Perspective

Orlando Arias[1], *Fahim Rahman*[2], *Mark Tehranipoor*[2], *and Yier Jin*[2]

[1] *University of Central Florida, Orlando, FL, USA*
[2] *University of Florida, Gainesville, FL, USA*

19.1 Introduction

Under the umbrella terms of Internet of Things (IoT) and cyber-physical systems (CPS), millions of low-power devices have become entrenched in our lives. Reports state that 15 billion IoT devices are currently deployed [1], and deployment is expected to reach 50 billion by the year 2020 [2]. The number of CPS has also risen, with the advent of the smart grid [3, 4] and autonomous vehicles [5].

This massive deployment of devices has led to significant security concerns. Various attacks have shown weaknesses in IoT and CPS infrastructure, with a swarm of light bulbs potentially leaving a city in darkness [6], rogue devices attacking infrastructure [7], to attacks in critical infrastructure [8, 9].

Device attestation has risen as a promising solution to the security demands of embedded devices. In this chapter, we discuss the requirements of an attestation scheme, as well as classify attestation schemes by their functionality and coverage. We then summarize representative device attestation approaches in the different categories. We present a discussion of the approaches as well as possible pitfalls and limitations that we have encountered in our survey of the literature.

19.2 Background on Attestation

In this section, we introduce a brief background to the concepts behind attestation and its requirements.

Modeling and Design of Secure Internet of Things, First Edition. Edited by Charles A. Kamhoua, Laurent L. Njilla, Alexander Kott and Sachin Shetty.

Published 2020 by John Wiley & Sons, Inc.

19.2.1 Common Terminology

Before proceeding any further, we would like to introduce some terminology and notation that will be used throughout this chapter. We first present a few definitions.

Definition 19.1 (State) A *state* is an abstract representation of the behavior of a device at a specific point in time.

The *state* of a device may be composed of, but not limited to, the contents of memory of the device, the CPU registers, the temperature of the components, the brand and model of the components, and any serial numbers associated with the device. In general, any identifiable information from the device can be considered part of the state.

Definition 19.2 (Measure) A *measure* is the act of collecting the operational *state* of a device by a *prover*.

By its nature, the *measure* of the device may include a plethora of information. For this reason, it is often useful for the prover to compress the measure in some meaningful way before it is obtained by the verifier. We discuss the characteristics of the measure in the next section.

Definition 19.3 (Prover) A *prover* is a trusted entity which *measures* the state of a device.

A device's *prover* is a mechanism which is tasked with measuring the state of the device at any arbitrary time. The prover, by definition, must be tightly coupled with the device as to be able to access the state of the device. We discuss the characteristics of the prover in the next section.

Definition 19.4 (Verifier) A *verifier* is a trusted entity which decides whether the state reported by a *prover* corresponds to a valid state of a device.

A *verifier* in an attestation scheme need not be the device's operator or owner. It may be the manufacturer, vendor, or any other third party. Furthermore, a verifier need not be a person; it could be a machine or a component inside a device.

We also use the nomenclature in Table 19.1 to refer to the different portions of the attestation mechanism throughout this chapter.

Table 19.1 Common symbols and terminology used in this chapter.

Symbol	Meaning
$\mathcal{V}$	The *verifier* in an attestation mechanism
$\mathcal{P}$	The *prover* in an attestation mechanism
$\mathcal{M}$	A *measure* computed by a prover
$\mathcal{S}$	A particular *state* in the device being attested

19.2.2 Problem Definition

In its most basic form, device attestation is defined as making a claim about the properties of a target [10]. Two mutually exclusive parties are involved in an attestation scheme: a *verifier* V and a *prover* P. Attestation is performed using a challenge–response mechanism upon V's request. During the servicing of an attestation, request P does a *measure* M of the device. V receives M and then determines whether M represents a valid device state.

Formally, let $\mathbb{A}$ represent the set of possible responses from V. Since V replies with a yes or no answer

$$\mathbb{A} = \{0, 1\}. \tag{19.1}$$

Where 1 represents the device attested successfully and 0 represents an attestation failure. Let $\mathbb{M}$ be the set of measures returned by P. The operation of V is then given by the mapping of $\mathbb{M}$ into $\mathbb{A}$, that is

$$V_{k,n} : \mathbb{D} \rightarrow \mathbb{A} \tag{19.2}$$

P generates $M \in \mathbb{M}$ by performing a computation over the state S of the device. S can be any identifiable quality of the device, such as the code it runs, the contents of its memory, or an intrinsic characteristic of the system on chip (SoC) it utilizes. We allow $\mathbb{S}$ to represent the set of possible S for the device, which may include illegal ones. Then, P performs the operation

$$P_{k,n} : \mathbb{S} \rightarrow \mathbb{M}. \tag{19.3}$$

We also say that $M \in \mathbb{M}$ is the result of a measure performed by P. With this, we are now free to define the operation of attestation.

Definition 19.5 (Device Attestation) Let S be a state of a device. Let $P_{\{k,n\}} : \mathbb{S} \rightarrow \mathbb{M}$ be the prover operation of mapping an S to a M. Let $V_{\{k,n\}} : \mathbb{M} \rightarrow \mathbb{A}$ be the verifier operation of determining whether M corresponds to a valid state of the device. Then, we define *device attestation* to be the operation of computing V_{k, n}(M).

We further define an *operational device* if and only if

$$\forall S \in \mathbb{S}, M = P_{k,n} \in \mathbb{M}, V_{k,n}(M) = 1. \tag{19.4}$$

Conversely, a device is deemed *compromised* or *inoperable* if

$$\exists S \in \mathbb{S}, M = P_{k,n}(S) \in \mathbb{M}, V_{k,n} = 0. \tag{19.5}$$

19.2.3 Attacker Capabilities

Although attestation can be passively used to monitor the operational state of a device under regular operation, its main objective is to detect whether a device is operating outside specified bounds due to an attack. We now discuss the objectives and assumed capabilities of an attacker under an attestation model.

Attestation assumes an attacker wishes to compromise a device to alter its functionality. For example, an attacker may wish to change the values reported by a current meter to underpay for energy utilization; or an attacker may wish to tamper with a network of connected traffic lights to cause infrastructure damage. An attacker may choose to follow one or many of the following avenues.

19.2.3.1 Change the Software

The attacker may change the code being run by the device with malicious code. This can be performed by injecting new code into the device [11, 12], overwriting existing code through open programming interfaces [13] or a vulnerable firmware update process [14], or glitching the memory bus to change fetched data [15].

19.2.3.2 Alter Software Behavior

The attacker may alter the behavior of the software run by means such as code-reuse attacks [16–19]. This only requires the corruption control-flow information through memory errors to execute arbitrary sequences of code. This attack methodology is meant to bypass attestation approaches that only check code memory.

19.2.3.3 Tamper with Communications

The attacker may also attempt to tamper with the communications between devices in a swarm, or between P and a remote V. Tampering may consist of inserting messages, removing messages, or changing the contents of messages sent between devices or P and V. The attacker's objective is to imitate or leak secrets from a swarm to conceal a compromised device.

19.2.3.4 Compromising the Hardware Root of Trust

Hardware attacks use semi-invasive and invasive techniques, such as probing the device's circuit board to expose backdoors. To facilitate the process, an attacker may be an insider in a fabrication house where a counterfeit printed circuit board is used to manufacture the device allowing for the extraction of cryptographic secrets. Attackers with physical access to the device may extract secrets held in nonvolatile memories to facilitate the intrusion into larger systems or to steal the device's unique identity in order to launch different types of attacks against the system, e.g. relay and replay attacks [20].

19.2.4 Scheme Requirements

For the results of Eq. (20.4) to be valid, a few considerations about the attestation scheme need to be in place when considering the threat of an attacker.

19.2.4.1 Trustworthiness of *P*

Since the *P* oversees collecting measures on the device, it must be devised in such a way that it is tamper resistant. That is, *P* must be provably trustworthy. Trustworthiness of *P* may be established by means of a root of trust [21–23]; isolation [24, 25] by adding it as an intrinsic part of the platform [15]; or by utilizing properties intrinsic to hardware, such as the effect of process variations [26, 27].

Trust in *P* relies not only on its proper operational state but also on its semantic approach at performing the event of measuring the device. *P* may be in a proper operational state, but its implementation does not gather sufficient information to construct $S_i \in \mathbb{S}$, the current state of the device being attested. The trustworthiness of *P* has a direct effect in the trustworthiness of *M*. However, a trustworthy *P* does not necessarily reflect a trustworthy *M*. That is to say, the trustworthiness of *P* is a necessary but not sufficient condition to determine the trustworthiness of *M*.

19.2.4.2 Liveness of *M*

Any *M* computed by *P* must reflect the current *S* of the system. Any requests by *V* must be answered by *P* by performing a fresh collection of state S_0 in order to compute *M* while still considering all previously measured *S* that have not been used in computation to determine a given *M*. This ensures an attacker is unable to alter the behavior of the device (a state) in a way that *P* will not utilize it.

19.2.4.3 Unforgeability of *M*

An attacker must not be able to reproduce any *M* computation done by *P*; that is, forge *M*. Two attributes are necessary for *M* to meet this condition: *non-replayability* and *non-reconstructability*. We discuss the need for these approaches below.

Consider the case where a trustworthy *P* has collected states in $\mathbb{S}_\mathbb{C} = \{S_0, S_1, S_2, S_3\}$. *P* has also computed measures in $\mathbb{M}_\mathbb{C} = \{M_0, M_1, M_2, M_3\}$, respectively. Allow for the condition expressed in Eq. (20.4) to be met under these conditions. Without loss of generality, suppose that during the next attestation request *P* obtains state S_0. If *P* reports M_0 as the computed measure, then an attacker is able to present M_0, or any other previously computed $M \in M$ as a response to an attestation request.

Consequently, given a particular $S_i, S_j, S_l, \ldots \in \mathbb{S}$ collected at different times, with $S_i = S_j = S_l = \cdots$, Eq. (20.3) must return $M_i, M_j, M_l, \ldots \in \mathbb{M}$ so that $M_i \neq M_j \neq M_l \neq \cdots$. This is to say, any *M* reported to *V* by *P* must be nonreplayable. For this particular purpose, an agreed upon nonce *n* is used by *P* and *V* during each attestation request to generate *M* and determine the functional status of the device, respectively.

Furthermore, an attestation approach must guarantee that any *M* is the result of *P* collecting state data on the device. That is, under the assumption that an attacker has knowledge of the function and the agreed upon nonce *n* used by *P* to compute

M, M cannot be computed by an entity other than P. For this purpose, a secret shared by P and V is utilized. The secret can be a preshared key or negotiated session key k. In this instance, standard key management procedures must be followed.

19.2.5 Types of Attestation

Device attestation may take many forms. We first define two methods P may utilize to collect S.

Definition 19.6 (Static and Dynamic Attestation) An attestation mechanism is *static* if P halts normal device operation to collect state information on the device.

An attestation mechanism is *dynamic* if P collects state information without interrupting normal device operation.

Attestation may also be local or remote. In a local attestation setting, both V and P reside within the device. In the event of an attestation failure, V must take a reporting action to record the failure and possibly set the device into a known stable failsafe mode. Under a remote attestation setting, V is a far away trusted entity which sends attestation requests to P. In this case, requests and responses must be authenticated and properly secured from tampering.

We further categorize attestation approaches based on the portion of the device which is measured by P.

19.2.6 Software Attestation

In a *software attestation* model, P checks the functionality of the code in the device. Here, S may consist of code segments on the device, control-flow metadata, and the contents of RAM. P computes M based on this type of state data, for example, by using a cryptographic hash function, then sends M to V.

19.2.6.1 Swarm Attestation

A *swarm attestation* model concentrates on checking the trustworthiness and proper functionality of multiple interconnected devices. In a swarm, devices may not have the same hardware and software configuration. Devices that are part of a swarm are only able to communicate with their direct neighbors. The protocol used for swarm attestation must be designed so that a compromised device in the swarm does not tamper with any measure performed in the system. Each device in the swarm may contain its own V and P. P collects S, computes M, which then sends to V in a neighboring node to complete the attestation request.

19.2.6.2 Hardware Attestation

Hardware attestation techniques center on validating the authenticity and trustworthiness of the underlying hardware. This may be printed circuit boards, SoC, onboard sensors, or communication channels. Hardware attestation schemes require a robust design in *P*, often utilizing intrinsic hardware properties such as physical unclonable functions (PUF) [26, 27], to compute *M*.

19.3 Notable Approaches

We now describe a few notable attestation approaches and their workings. We briefly explore each approach by target category starting with software, then swarm, and finish with hardware attestation. Table 19.2 summarizes the approaches we will talk about.

19.3.1 Software Attestation

We begin our discussion of attestation approaches with those that attempt to safeguard software.

Table 19.2 Comparison of different approaches.

	Attestation parameters						
Approach	**T**	**L**	**R**	**SW**	**SM**	**B**	**SoC**
Samsung KNOX [28]	S	•	•	•			
SMART [24]	S		•	•			
Control-flow attestation (C-FLAT) [25]	S		•	•			
ATRIUM [15]	D		•	•			
Scalable Embedded Device Attestation (SEDA) [29]	S		•	•	•		
DARPA [30]	D		•	•	•		
PUF for device authentication [31]	S	•	•				•
BoardPUF [32]	S	•	•			•	

Attestation mechanisms flagged as static will halt normal device operation during execution of *P*. Attestation mechanisms flagged as dynamic allow device operation to remain uninterrupted while *P* performs operations to compute *M*. In the table, the following key is used: T, type; L, local; R, remote; SW, software; SM, swarm; B, board; SoC, system on chip. Furthermore, types have the following subcategories: S, static; D, dynamic.

19.3.1.1 Samsung KNOX

Samsung introduced KNOX in 2013 for some of their Android-based devices as an enterprise-grade security solution [28]. KNOX builds upon an ARM Trust-Zone [33] environment while adding new features to an ARM-based SoC. KNOX-enabled devices provide a root of trust as part of the secure world software, as well as operating system safeguards and verification mechanisms to ensure a safe execution environment. Prior to launching security critical tasks, such as enterprise applications, the KNOX *P* reads software configuration as *S* and computes a signature as *M*. *V* deems some variations from the original configuration as a compromise. Security-sensitive data, held in KNOX Containers, are only accessible if the security of the device is not deemed as compromised. For example, in the event the boot chain of the device is found to have been tampered with, an e-fuse internal to the SoC is blown branding the unit as untrusted. From this point KNOX Containers can no longer be created and any secured data become inaccessible.

Although Samsung KNOX has been certified by the United States government and other countries as safe for use [34], there are known attacks that are capable of bypassing some versions of the system [35]. Furthermore, KNOX is limited to a few Samsung-made devices and not available for other vendors to use and implement.

19.3.1.2 SMART

Eldefrawy et al. propose a small root of trust for embedded devices in SMART [24], providing a static remote attestation solution. It incorporates *P* into a small on-chip ROM. *P* utilizes a range of the device's application code as *S* and computes a HMAC [36] over it as *M*. The hash is sent to a remote *V* to affirm its correctness. A preshared attestation key is stored in the device and gated so that only ROM code is capable of accessing it. To ensure that no leakage occurs, a safe memory erasure is performed every time the device reboots and whenever the ROM code finishes executing. Further precautions are taken to avoid indirect leakage by controlling code execution within the ROM. ROM code is only allowed to be executed from a fixed entry point to a fixed return point as to ensure that no code-reuse attacks attempt to leak the secret key.

We show the basic flow of SMART in Figure 19.1. When *P* receives an attestation request from the remote *V* ❶, *P* suspends the currently running task and hashes the requested portion of code memory as *M*, ❷. The result of the hash operation is sent to *V*, ❸. Using the returned *M*, *V* can determine whether program code on the device has been mutated. SMART was implemented and tested in an OpenMSP430 [37] core, with a reported 9.2% area overhead. HMAC computation was reported to range from 48 to 287 ms on a block size ranging from 32 B to 1 kB when run at a clock frequency of 8 MHz.

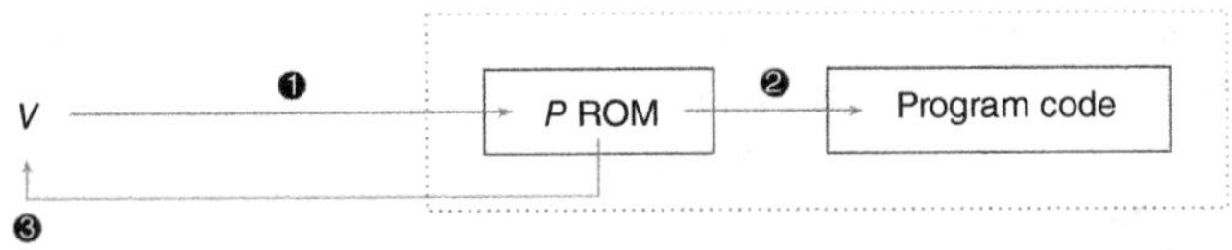

Figure 19.1 Flow of operations in SMART. A remote *V* sends a request to a device, to which *P* responds by securely hashing the requested portion of program code using a pre-shared key.

19.3.1.3 C-FLAT

Abera et al. demonstrated in [25] that it is insufficient to use only device code as *S* in order to compute *M*. Code-reuse attacks are able to bypass attestation mechanisms that use this type of system. As a result, they proposed control-flow attestation (C-FLAT) [25]. C-FLAT is a remote attestation mechanism that statically aggregates the execution path of a running program, including branches and function returns. In this approach, *P* collects control-flow information as *S* and hashes it to compute *M*. On an attestation request, *V* receives the hashed control-flow information and computes the expected *M* on its end. The device is said to attest correctly if both *M* match.

C-FLAT was implemented and tested in a Raspberry Pi 3 single-board computer. We show a simplified view of the implementation in Figure 19.2. Code is instrumented so that control-flow instructions target a trampoline section which belongs to a runtime tracer. The branch instruction in question targets the trampoline area ❶. The trampolines allow software to transition to a BLAKE2 [38] Measurement Engine which resides in a TrustZone environment ❷. The Measurement Engine

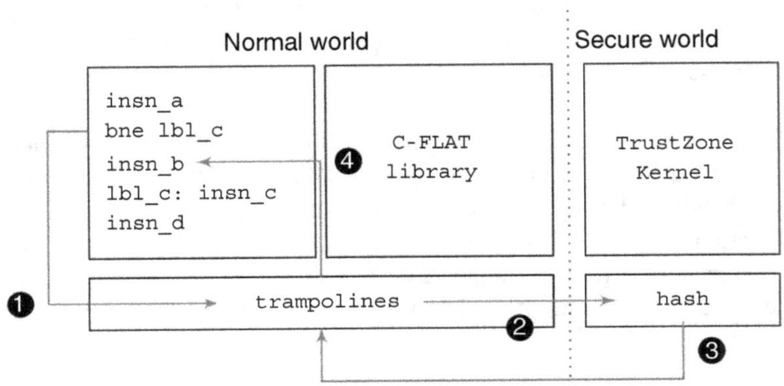

Figure 19.2 C-FLAT test implementation model. Trampolines are used to enter the secure world component and perform control-flow attestation.

is part of P and is responsible for hashing control-flow. When the hash engine finishes executing, it returns control to the trampolines, ❸, which ultimately returns control-flow to the program ❹. When V sends an attestation request to the device, P replies with the collected information. The authors report that as the number of control-flow events increase, overhead increases linearly. When testing against dedicated functions in Open Syringe Pump, a relatively large overhead 72–80% was reported. The authors further caution that a limiting factor in their target application is that of external sensors and actuators. Thus, the perceived overhead may be reduced to a mere 0.03–0.13%.

19.3.1.4 ATRIUM

Zeitouni et al. showed how a series of time of check to time of use (TOCTOU) attacks in previous static attestation schemes that can compromise the way S is collected, and consequently manipulate the results of P computing M. The authors then proposed ATRIUM as a solution in [15]. ATRIUM is a dynamic remote attestation mechanism that borrows concepts from C-FLAT [25] and SMART [24], in that it utilizes control-flow and code that is executed as S. However, unlike C-FLAT and SMART, ATRIUM adds an instruction filter as an extension to the execution stage of a pipelined processor. All executed instructions are collected in an instruction buffer. The instruction filter checks if the currently executing instruction is a control-flow instruction and sends a signal to a loop encoder. Loop information as well as function call information is kept as states in the ATRIUM core. When a full basic block is executed, the collected instructions are hashed using a hardware BLAKE2b implementation and the results are stored in a dedicated memory. If the instruction buffer fills while instruction hashing is taking place, a signal is sent to the CPU to halt execution until hashing has completed. The ATRIUM core is what allows P to dynamically collect S to compute M. When an attestation request arrives, the stored hashes are sent as a response to the remote V for checking. Attestation concludes in a fashion similar to C-FLAT.

A RISC-V PULPino core [39] was extended to include ATRIUM. The modified core was synthesized and tested targeting a Virtex-7 XC7Z020 FPGA with minimal hardware overhead. Under the configured conditions, the authors report a total resource utilization of 15% of the total slice registers, 20% of slice LUTs, and 18 kbit of BRAM. Performance overhead ranged from 1.7 to 22.69%, depending on the amount and frequency of control-flow instructions in the tested algorithm.

19.3.2 Swarm Attestation

We now discuss different swarm attestation approaches.

19.3.2.1 SEDA

Asokan et al. propose Scalable Embedded Device Attestation (SEDA) to attest a swarm of devices in [29]. The SEDA protocol has two phases, an off-line device initialization and registration phase, and an online device attestation and swarm attestation phase. During the off-line phase, each device in the swarm is initialized by a trusted operator and given a signing key pair alongside an identity certificate. The device is then registered into the swarm, executing the join action of the SEDA protocol. This produces a handshake between the device and its neighbors, where it learns and verifies the public parameters of the neighbors. Furthermore, an attestation key is established during the join operation between the device and its neighbors in the swarm. During the online phase, devices can request attestation from its neighbors using an attdev request. The key generated during the off-line phase is used to attest the neighboring nodes. Swarm attestation starts with V sending an attestation request attest to a random device in the swarm. This device becomes the *initiator* of the swarm attestation. By recursively sending attdev messages to neighboring devices, a spanning tree is built, with devices that have not received the message being added. Eventually, V receives a reply containing the results of the swarm attestation.

SEDA was tested with SMART-based [24] and TrustLite-based [23] implementations. The protocol is disjoint from the details regarding P and the state information it uses to compute M. The protocol was evaluated using *computational cost, communication cost, memory cost, runtime cost, and energy cost* as metrics under a simulation using up to 106 devices using different spanning tree configurations. Performance was shown to fluctuate depending on the number of devices in the network and the number of child nodes in the spanning tree. Energy consumption was shown to increase linearly with respect to the number of neighboring devices, and to be evenly distributed by all members of the swarm. Only the *initiator* exhibits higher energy consumption due to the extra number of computations it must perform; however, the authors state that this energy cost can be amortized throughout the swarm by choosing a different *initiator* on every swarm attestation request.

19.3.2.2 DARPA

DARPA [30] is a lightweight collective attestation protocol aimed at defending unattended networks of embedded systems from physical attacks. The types of physical attacks considered are focused upon those that require disabling or disconnecting the device from the network. In considering such an adversary, DARPA leverages the non-negligible time required during disconnection to identify a device's absence using a *heartbeat*. The *heartbeat* identifies a device in the network uniquely, and is intermittently probed at regular intervals by its immediate neighbors to detect liveness. If at any time a device heartbeat fails to be detected, then the offending device can be considered compromised and appropriately quarantined.

19.3.3 Hardware Attestation

Lastly, we discuss two proposed hardware attestation approaches.

19.3.3.1 PUF IC Attestation

Suh et al. propose utilizing a PUF to authenticate integrated circuits and generate cryptographic keys in [31]. A PUF is a hardware function which utilizes process variations as a source of entropy to generate random numbers based on a challenge–response setup [26, 27]. Because the operation of a PUF is subject to environmental factors, the authors add an error-correction mechanism to ensure reliability on response (key) generation. Chip attestation works as follows: *V* applies a challenge (attestation request) to the chip. *P* utilizes the PUF circuit as *S* to compute *M*. *V* uses a previously populated challenge–response database to finish the attestation request. Authors discuss that the proposed method has a 2.1×10^{-21} at misidentifying integrated circuits, and a probability of less than 5×10^{-11} at failing to identify the integrated circuit. Although the authors claim that the method can be used in very low-power environments, such as RFID, no metrics are provided regarding energy consumption. Also, no formal protocol description is provided.

19.3.3.2 BoardPUF

Wei et al. propose BoardPUF in [32]. BoardPUF provides an attestation mechanism for circuit boards using capacitance variations during the manufacturing process. The approach has two components, a source of variations fabricated on the board, and a chip to perform measurement and authentication. The variation units are designed as capacitor structures in the internal layers of the circuit board. Protecting them from noise sources are uninterrupted ground planes in the outer layers of the circuit board. Burr edges resulting from the chemical etching process of the boards, as well as the misalignment of layers while laminating, and changes in the thickness of boards are used as variations that alter the capacitance of the structures on the board. The chip component of the mechanism serves as *P*, capturing *S* as a series of frequencies generated by an oscillator circuit using the onboard capacitances. *M* is computed by utilizing the captured frequencies as clock sources for sequential logic. The final logic state is error-corrected and used by *V* to determine the authenticity of the board.

The proposed mechanism was manufactured and tested under different environmental conditions. Authors report a variation between 130 and 158 kHz in the measured frequencies. Results claim that boards are incorrectly rejected with a ratio of 5.89×10^{-6}, and incorrectly accepted with a ratio of 3.01×10^{-11}. However, no discussions on board area overhead and power are given.

19.4 Discussion and Future Directions

Throughout the literature, we have seen a race with constantly increasing stakes between attacks and defenses. As attackers use more sophisticated methods to tamper with devices, attestation defenses have become more complex. We argue that a successful device attestation scheme must not only meet the requirements stated in Section 19.2, they must also meet deployability criteria. Attestation defenses should be readily implementable by vendors wishing to add security to their devices. Furthermore, attestation defenses should concern themselves with the power, memory, and computational constrains of embedded devices.

For example, SMART [24] and ATRIUM [15] require the processor to be modified in invasive ways. This is impossible for device manufacturers under the current Intellectual Property licensing model used in industry, or limited to a select few such as with the case of Samsung KNOX [28]. This limits the deployment of any proposed security mechanism. Although the implementation in C-FLAT [25] requires no hardware modifications, it was shown to be vulnerable attack [15], which makes it undesirable as a security solution.

Of the discussed approaches, we note that static attestation approaches (those where *P* interrupts device operation) may be detrimental to timing sensitive systems, where a small delay can result in the stability of the system being lost. Furthermore, Zeitouni et al. demonstrated in [15] that the switch to a *P* results in a timing channel that can be exploited by an attacker to give *P* a false *S*. Our survey points to dynamic approaches result in more accurate collection of *S* by *P* to compute *M*. Dynamic data collection also results in reduced performance overhead during execution. As such, we argue that future attestation mechanisms should be dynamic in nature.

We also note the difference in the way performance overhead is reported. SMART [24] reports the latency in execution caused by the HMAC function, whereas CFLAT [25] reports overhead in partially instrumented code in a particular application (Open Syringe Pump [40]). We find that computational overhead is a function of any instrumentation required by *P*, any delays introduced by *P* including its cryptographic requirements, and any delays caused by hardware modifications, such as in the case with ATRIUM [15]. Furthermore, of the discussed approaches, only SEDA [29] reports energy measurements. This is an important aspect for power-constrained devices, as it directly affects the field life of the units. Much like computing performance can be tested with a set of industry standard benchmarks such as SPEC CPUINT2006 [41], a common embedded test suite should be used to test performance and energy requirements of proposed device attestation solutions. The test suite should consist of representative applications in IoT and CPS devices. For example, Open Syringe Pump [40] and OpenPLC [42] can serve as baselines for testing new attestation methods for CPS, while demonstration programs targeting sensor boards can be used as a baseline for IoT devices. We urge researchers in this area to take this as a consideration when presenting new attestation solutions.

Hardware attestation approaches are limited by area overhead and energy needs of the devised solution. PUF-based approaches like the ones presented suffer from accuracy issues and thus need extra circuitry to correct errors, adding to area and power overhead. Also, the PUF-based security mechanism presents scalability issues. As the number of devices increases, the number of challenge–response pairs that must be evaluated during factory time increase as well, elevating the manufacturing cost. Further research in this area should address these limitations. Furthermore, we argue that these approaches should report area and energy metrics as to facilitate comparison and view applicability.

Acknowledgments

This work is partially supported by National Science Foundation (DGE-1802701), Semiconductor Research Corporation (2016-TS-2724), Florida Center for Cybersecurity (FC2), and Cisco. Mr. Orlando Arias is also supported by the National Science Foundation Graduate Research Fellowship Program under Grant No. 1144246.

References

1 D. Evans, "The internet of things: how the next evolution of the internet is changing everything," *White Paper*. Cisco Internet Business Solutions Group (IBSG), 2011.

2 A. Nordrum, "Popular internet of things forecast of 50 billion devices by 2020 is outdated," *IEEE Spectrum*, 2016.

3 S. M. Amin and B. F. Wollenberg, "Toward a smart grid: Power delivery for the 21st century," *IEEE Power and Energy Magazine*, vol. 3, no. 5, pp. 34–41, 2005.

4 H. Farhangi, "The path of the smart grid," *IEEE Power and Energy Magazine*, vol. 8, no. 1, pp. 18–28, 2010.

5 L. D. Burns, "Sustainable mobility: A vision of our transport future," *Nature*, vol. 497, no. 7448, pp. 181–182, 2013.

6 E. Ronen, A. Shamir, A.-O. Weingarten, and C. O'Flynn, "IoT goes nuclear: Creating a Zigbee chain reaction," in *2017 IEEE Symposium on Security and Privacy (SP)*, IEEE, pp. 195–212, 2017.

7 M. Antonakakis, T. April, M. Bailey, M. Bernhard, E. Bursztein, J. Cochran, Z. Durumeric, J. A. Halderman, L. Invernizzi, M. Kallitsis, D. Kumar, C. Lever, Z. Ma,

J. Mason, D. Menscher, C. Seaman, N. Sullivan, K. Thomas, and Y. Zhou, "Understanding the Mirai botnet," in *26th USENIX Security Symposium (USENIX Security 17)*, pp. 1093–1110, 2017.

8 M. Abrams and J. Weiss, "*Malicious Control System Cyber Security Attack Case Study–Maroochy Water Services, Australia*," McLean, VA: The MITRE Corporation, 2008.

9 R. Langner, "Stuxnet: Dissecting a cyberwarfare weapon," *IEEE Security and Privacy*, vol. 9, no. 3, pp. 49–51, 2011.

10 G. Coker, J. Guttman, P. Loscocco, A. Herzog, J. Millen, B. O'Hanlon, J. Ramsdell, A. Segall, J. Sheehy, and B. Sniffen, "Principles of remote attestation," *International Journal of Information Security*, vol. 10, no. 2, pp. 63–81, 2011.

11 A. Francillon and C. Castelluccia, "Code injection attacks on Harvard architecture devices," in *Proceedings of the 15th ACM Conference on Computer and Communications Security*. ACM, pp. 15–26, 2008.

12 A. One, "Smashing the stack for fun and profit (1996)," *Phrack Magazine*, 2007.

13 L. Garcia, F. Brasser, M. H. Cintuglu, A.-R. Sadeghi, O. Mohammed, and S. A. Zonouz, "Hey, my malware knows physics attacking PLCS with physical model aware rootkit," in *Proceedings of the Network and Distributed System Security Symposium*, San Diego, CA, pp. 26–28, 2017.

14 A. Cui, M. Costello, and S. J. Stolfo, "When firmware modifications attack: A case study of embedded exploitation," in *NDSS Symposium*, 2013.

15 S. Zeitouni, G. Dessouky, O. Arias, D. Sullivan, A. Ibrahim, Y. Jin, and A.R. Sadeghi, "Atrium: Runtime attestation resilient under memory attacks," in *International Conference on Computer Aided Design (ICCAD)*, 2017.

16 F. Schuster, T. Tendyck, C. Liebchen, L. Davi, A.-R. Sadeghi, and T. Holz, "Counterfeit object-oriented programming: On the difficulty of preventing code reuse attacks in C++ applications," in *2015 IEEE Symposium on Security and Privacy (SP)*, pp. 745–762, 2015.

17 T. Bletsch, X. Jiang, V. W. Freeh, and Z. Liang, "Jump-oriented programming: A new class of code-reuse attack," in *Proceedings of the 6th ACM Symposium on Information, Computer and Communications Security*. ACM, pp. 30–40, 2011.

18 M. Prandini and M. Ramilli, "Return-oriented programming," *IEEE Security and Privacy*, vol. 10, no. 6, pp. 84–87, 2012.

19 S. Checkoway, L. Davi, A. Dmitrienko, A.-R. Sadeghi, H. Shacham, and M. Winandy, "Return-oriented programming without returns," in *Proceedings of the 17th ACM conference on Computer and Communications Security*. ACM, pp. 559–572, 2010.

20 B. Krebs, "Cyber incident blamed for nuclear power plant shutdown," *Washington Post*, June 5, 2008.

21 W. A. Arbaugh, D. J. Farber, and J. M. Smith, "A secure and reliable bootstrap architecture," in *1997 IEEE Symposium on Security and Privacy*. IEEE, pp. 65–71, 1997.

22 F. Brasser, B. El Mahjoub, A.-R. Sadeghi, C. Wachsmann, and P. Koeberl, "Tytan: Tiny trust anchor for tiny devices," in *2015 52nd ACM/EDAC/IEEE Design Automation Conference (DAC)*, pp. 1–6, 2015.

23 P. Koeberl, S. Schulz, A.-R. Sadeghi, and V. Varadharajan, "Trustlite: A security architecture for tiny embedded devices," in *Proceedings of the Ninth European Conference on Computer Systems*. ACM, 2014.

24 K. Eldefrawy, G. Tsudik, A. Francillon, and D. Perito, "Smart: Secure and minimal architecture for (establishing dynamic) root of trust," in *NDSS Symposium*, vol. 12, pp. 1–15, 2012.

25 T. Abera, N. Asokan, L. Davi, J.-E. Ekberg, T. Nyman, A. Paverd, A.-R. Sadeghi, and G. Tsudik, "C-flat: Control-flow attestation for embedded systems software," in *Proceedings of the 2016 ACM SIGSAC Conference on Computer and Communications Security*. ACM, pp. 743–754, 2016.

26 B. Gassend, D. Clarke, M. Van Dijk, and S. Devadas, "Silicon physical random functions," in *Proceedings of the 9th ACM Conference on Computer and Communications Security*. ACM, pp. 148–160, 2002.

27 R. Pappu, B. Recht, J. Taylor, and N. Gershenfeld, "Physical one-way functions," *Science*, vol. 297, no. 5589, pp. 2026–2030, 2002.

28 Samsung Knox, "White paper: An overview of Samsung Knox," xhttps://image-us.samsung.com/SamsungUS/samsungbusiness/solutions/topics/iot/081717/Samsung_KNOX_whitepaper_June-0.pdf. 2013.

29 N. Asokan, F. Brasser, A. Ibrahim, A.-R. Sadeghi, M. Schunter, G. Tsudik, and C. Wachsmann, "SEDA: Scalable embedded device attestation," in *Proceedings of the 22nd ACM SIGSAC Conference on Computer and Communications Security*, pp. 964–975, 2015.

30 A. Ibrahim, A.-R. Sadeghi, G. Tsudik, and S. Zeitouni, "DARPA: Device attestation resilient to physical attacks," in *Proceedings of the 9th ACM Conference on Security and Privacy in Wireless and Mobile Networks*, pp. 171–182, 2016.

31 G. E. Suh and S. Devadas, "Physical unclonable functions for device authentication and secret key generation," in *Proceedings of the 44th Annual Design Automation Conference*, pp. 9–14, 2007.

32 L. Wei, C. Song, Y. Liu, J. Zhang, F. Yuan, and Q. Xu, "BoardPUF: Physical unclonable functions for printed circuit board authentication," in *IEEE/ACM International Conference on Computer Aided Design (ICCAD)*. IEEE, pp. 152–158, 2015.

33 A. ARM, "Security technology building a secure system using TrustZone technology (white paper)." ARM Limited, 2009.

34 Samsung Knox, "Is Knox Certified by the US Government?" https://support.samsungknox.com/hc/en-us/articles/115013733108-Is-Knoxcertified-by-the-US-government, 2017.

35 U. Kanonov and A. Wool, "Secure containers in android: The Samsung Knox case study," in *Proceedings of the 6th Workshop on Security and Privacy in Smartphones and Mobile Devices*. ACM, pp. 3–12, 2016.

36 M. Bellare, R. Canetti, and H. Krawczyk, "Message authentication using hash functions: The HMAC construction," *RSA Laboratories' CryptoBytes*, vol. 2, no. 1, pp. 12–15, 1996.

37 OpenCores, "Openmsp430 project," http://opencores.org, 2019.

38 J.-P. Aumasson, S. Neves, Z. Wilcox-O'Hearn, and C. Winnerlein, "Blake2: Simpler, smaller, fast as MD5," in *International Conference on Applied Cryptography and Network Security*. Springer, pp. 119–135, 2013.

39 ETH Zurich and University of Bologna, "PULP Platform," http://www.pulp-platform.org, 2019.

40 B. Wijnen, E. J. Hunt, G. C. Anzalone, and J. M. Pearce, "Open-source syringe pump library," *PLoS One*, vol. 9, no. 9, p. e107216, 2014.

41 J. L. Henning, "Spec cpu2006 benchmark descriptions," *ACM SIGARCH Computer Architecture News*, vol. 34, no. 4, pp. 1–17, 2006.

42 T. R. Alves, M. Buratto, F. M. de Souza, and T. V. Rodrigues, "OpenPLC: An open source alternative to automation," in *IEEE Global Humanitarian Technology Conference (GHTC)*, pp. 585–589, 2014.

20

Software-Defined Networking for Cyber Resilience in Industrial Internet of Things (IIoT)

Kamrul Hasan[1], Sachin Shetty[1], Amin Hassanzadeh[2], Malek Ben Salem[2], and Jay Chen[2]

[1] *Virginia Modeling Analysis and Simulation Center, Old Dominion University, Norfolk, VA, USA*
[2] *Accenture Technology Lab, Arlington, VA, USA*

20.1 Introduction

Industrial Internet of Things (IIoT) is a new communication paradigm that combines information and technologies to enable real-time monitoring and provide controlling for industrial domains [1]. In the context of energy delivery systems (EDS), the vision of deploying IIoT relies on using low-power communication technologies by employing standard Internet Protocols (IP) [2] to enable end-to-end communication between IIoT-based control center and remote access devices. EDS needs improved efficiency, reliability, and safety [3]. Researchers have proposed techniques to detect and mitigate threats to the EDS infrastructure. However, these techniques typically ignore the impact of a security solution on the ability to conduct a normal operation in EDS. There is a need to ensure operation resilience for EDS while implementing security solutions. Enforcing security countermeasures should not come at the expense of quality of service (QoS), such as an end-to-end delay, network throughput, and available network bandwidth (BW). In this chapter, end-to-end delay or one-way delay (OWD) refers to the time taken for a packet to be transmitted across the Supervisory Control and Data Acquisition (SCADA) communication network from SCADA master (source) to regional substation's Regional Transmit Unit/Intelligent Electronic Device (RTU/IED) (destination). The network throughput is the rate of successful

Modeling and Design of Secure Internet of Things, First Edition. Edited by Charles A. Kamhoua, Laurent L. Njilla, Alexander Kott and Sachin Shetty.

Published 2020 by John Wiley & Sons, Inc.

message delivery over the SCADA communication channel. The available network BW represents the maximum message passing ability at a certain time over the SCADA communication link. The static, nonadaptive design of current grid communication networks does not support network reconfiguration to mitigate the impact of zero-day attacks [4]. Several studies advocate [5] adopting software-defined networking (SDN) to enrich functionality and improve the resilience of IIoT-based EDS communication networks by leveraging SDN's run-time configurability. System resilience in EDS is the ability of a system to recover and maintain critical services despite accidental failures and malicious attacks, such as Stuxnet [6] and Dragonfly [7]. However, SDN in EDS should not only support network orchestration but also provide the ability to balance the trade-off between enforcing security and operator resilience while managing cyber risks. For example, requirements like an end-to-end delay of the control commands from SCADA to any substation's Remote Terminal Unit (RTU) must be less than or equal to 100 ms [8, 9]. In this chapter, we propose a genetic algorithm (GA)-based multi-objective optimization approach to ensure QoS is maintained during deployment of security countermeasures in SDN-enabled IIoT-based EDS.

20.1.1 Contributions

In summary, we make the following contributions:

1) First, we analyze and evaluate the security risk level and the QoS class service metrics in EDS.
2) Then, we propose an efficient and dynamic optimization model that determines a combination of optimal security services settings, satisfying QoS requirement of each class of SCADA communication service using an elitist non-dominated sorting genetic algorithm (NSGA-II). We are confident that this optimization scheme is the first attempt in IIoT-based SCADA communication network which optimizes between security aspects and QoS parameters.
3) Finally, we evaluate the correctness of the optimization model for a set of network and optimization model parameters.

20.1.2 Organization

In Section 20.2, we review the related research. In Section 20.3, we cover the system model. In Section 20.4, we demonstrate the security risk and QoS framework. In Section 20.5, we discuss the optimization problem of security risk and QoS. Analysis and simulation results are discussed in Section 20.6, and we conclude in Section 20.7.

20.2 Related Work

Researchers have analyzed EDS resilience in the context of static failures [10–12] with the traditional IP network. These efforts evaluated the performance of the power grid when a certain subset of the system elements fails. Researchers have analyzed the resilience of the power grid by removing a sizable group of system elements. However, these efforts do not model the dynamic behavior. Because of a conventional communication network, at run time, it is often tedious, cumbersome, and even impossible to reconfigure a network to react in time to accidental and malicious events that undermine grid efficiency and safety. In that case, researchers proposed that SDN paradigm can mitigate those challenges and can be employed for system resilience, communication network management, and security.

In the SDN context, Song et al. [13] developed an SDN-based architecture for autonomous attack containment which dynamically modifies access control rules based on configurable trust levels. Xinshu et al. [4] demonstrated how SDN could enhance the resilience of typical EDS in the presence of malicious attacks. Al-Rubaye et al. [5] discuss how SDN can improve the resiliency in IIoT-based EDS. Li et al. [14] showed how the scheduling of services-oriented IoT works is based on network QoS. Dong Wei et al. [8] emphasize different network QoS constraints in EDS to protect cyberattack. Aujla et al. [15] demonstrate SDN-enabled multi-attribute-based secure communication for smart grid in IIoT environment. In all those work, the researchers introduced SDN but did not mention the effect of countermeasure selection to network QoS for system resilience and cybersecurity threat mitigation scheme.

On the other hand, Chen et al. [16] showed how security selection impacts network QoS in a traditional IP network. In this work, we notify all the challenges mentioned above and incorporate the effect of countermeasure selection to network QoS and show how it is possible to balance between network security level and QoS parameters selection in SDN-enabled power grid communication system.

20.3 System Model

In this section, we present the system model that consists of a system framework and optimization model. Both together ensure assurance of QoS while enforcing security countermeasures in the event of cyberattack. Before going to systems model, we discuss a little bit about EDS and the function of SCADA communication network in EDS.

20.3.1 EDS and SCADA Communication Network

National Institute of Standards and Technology (NIST) adopted the approach of dividing the smart grid into seven domains: Customers, Markets, Service Providers, Operations, Bulk Generations, Transmission, and Distribution. The transmission domain predominantly refers to the EDS (Figure 20.1). Like other domain, SCADA also monitors the status and sends supervisory commands to substation through the wide area network (WAN) in EDS.

There are different types of cyberattack in WANs and mostly categorized into attacks from outside the WAN communications flow and inside attacks that use existing critical WAN paths to target critical assets and systems (see Figure 20.2). External attacks typically originate from the same medium through which the network traffic flows. For example, if WAN data are traveling over a wireless medium, an attacker may use another radio to try to monitor or manipulate that link. Internal attacks use the existing critical asset WAN transport mechanisms and communications devices to spread an attack beyond the original exploitation point (either at the centralized control point or the remote outstation). An example of this is an attacker who has compromised the corporate network of a critical asset owner and then uses the WAN infrastructure to gain access to remote critical sites.

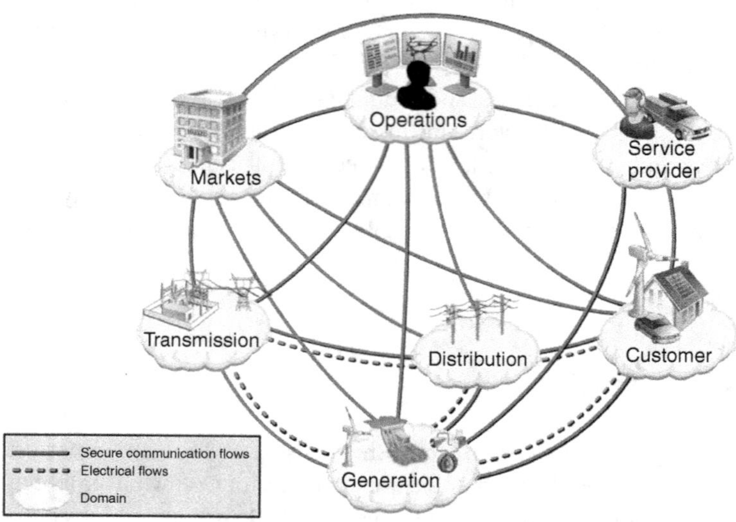

Figure 20.1 Interaction of roles in different EDS domains through secure communication. Source: Updated NIST Smart Grid Framework 3.0 Feb 2014.

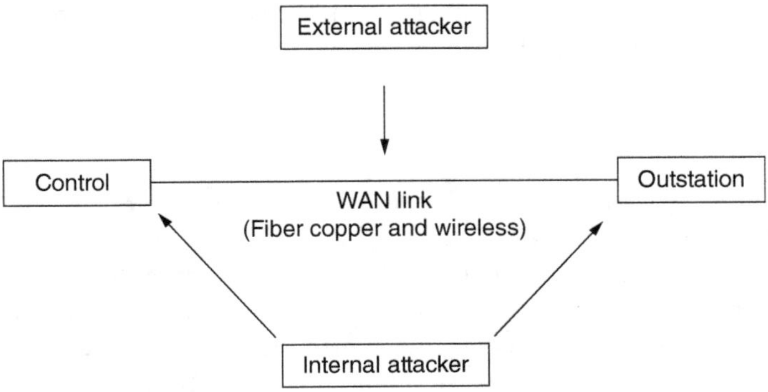

Figure 20.2 Internal and external WAN attacks.

External attacks of WAN are the main concern of this book chapter. External attacks are further dissected into three attack types: confidentiality, integrity, and availability. Confidentiality attacks encompass all attacks that compromise the secrecy of the data traveling over the WAN. Attackers use well-known sniffing techniques to gather the actual communications data traveling over copper, fiber, and especially wireless network. Not even modern cellular networks are immune to these techniques. Integrity attacks are attempts to manipulate data traveling over the WAN link. A malicious entity can seek to disrupt the authenticity of information by manipulating existing data or injecting new data. Integrity attacks are especially dangerous because they undermine the trust of even the most robust communications networks and possibly cause adverse operations on critical equipment, including tripping of the breaker on the substation bus system. The attack against availability simply seeks to disrupt normal WAN services by rendering them unavailable for normal operation. Availability attacks range from the purposeful disruption of traffic flows and the malicious cutting of communications cables to mass distributed denial of service (DDOS) attacks against EDS network.

One of the best methods to mitigate external WAN attacks is to use encryption and authentication of data on the WAN link between the control center and outstations or between outstations themselves. Encryption, the process by which data are scrambled so adversaries cannot analyze the data flowing on the WAN, helps mitigate confidentiality attacks. Authentication, which helps prevent injection or manipulation of data, is excellent mitigation of integrity attacks. However, this adoption of encryption of data and authentication of users cost extra processing delay in EDS devices and transmission latency in EDS WAN. Most of the researchers did not quantify this cost concerning a tolerable security level of this SCADA communication networks of EDS. In this chapter, we consider this extra cost and formulate it in an EDS context and optimize this cost concerning EDS security level. The successive systems model and other sections depict the formulation of this cost, security level, and how optimization scheme works between them.

20.3.2 Framework

The proposed framework illustrated in Figure 20.3 is influenced by the software-defined infrastructure proposed by Song et al. [13]. The difference between those two architectures is the optimization framework in the SDN controller to achieve the balance between QoS and deployment of security countermeasures. We assume the EDS consists of one SCADA master (control center) connected with regional SCADA substation slaves via SDN-enabled communication network. We also

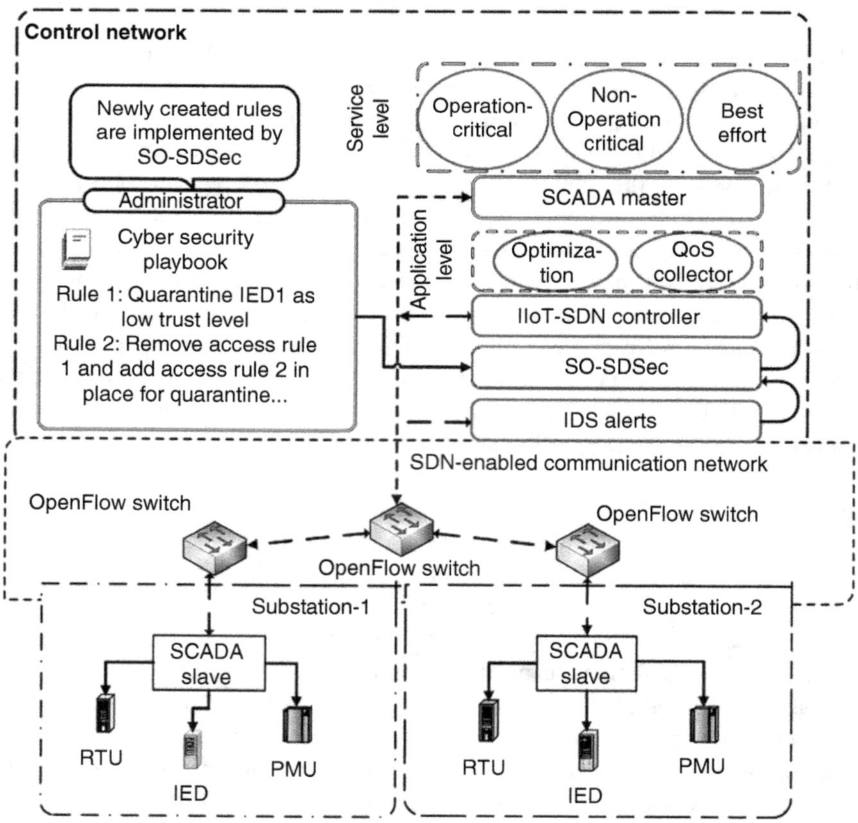

Figure 20.3 Security and QoS framework for SDN-enabled EDS.

assume that the communication protocol between master and slaves is DNP3 whose Ethernet packet size is 1500 bytes and the shared link capacity is 10 Mbps.

- **SCADA master** is responsible for providing grid substation operation-critical service (configuration and setting updating, etc.), non-operation critical service (power quality monitoring, etc.), and best effort service (exchanging historical data for mid-term and long-term planning, etc.) [8]. SCADA master collects measurement data and transmits control commands from/to SCADA slaves in the grid via the SDN-enabled communication network.
- **SCADA slave** interacts with Intelligent Electronic Devices (IEDs) and remote terminal units (RTUs), sensors (traditional meters, phasor measurement unit [PMUs]) and issues commands to actuators, e.g. circuit breakers, relays, and tap changers.
- **SDN controller** supports applications such as *optimization* and *QoS collector* through a northbound interface (NBI). The controller communicates with *OpenFlow* switch through the southbound interface (SBI) with *OpenFlow* protocol. It also includes IIoT connectivity adapter or software [5] element is responsible for interconnection between a smart sensor or actuators and future smart application mechanisms. This adapter employs special protocol agents to support the connectivity interface to the IIoT protocol. The adapter can support front/back end transmissions to allow collected data for delivering to the above layers for further processing.
- **The SDN-enabled communication network** is reconfigured by the controller to optimize between QoS and security thereby ensuring resilience support.
- **Playbook** contains operational plans provided by security administrators should take actions and include alerts from security controls, network monitors, server programs, or any other sensors. The actions in playbook specify service-related operations such as changing the key size of communication service for mutual authentication and messages authentication code (MAC) size for data integrity, updating a security control's service descriptions, or modifying a security service binding.
- **Service-oriented software-defined security (SO-SDSec)** converts security functions into abstract security services, with security appliances serving as service providers and power/IT assets represent service consumers [13]. In the event of intrusion detection system (IDS) alerts, SO-SDSec enforces rules provided by the playbook. SO-SDSec orchestrator monitors the IDS alerts and automatically requests for security services. The security services will interact with optimization service so that a trade-off between security parameters and QoS constraints settings is enforced. Next, the SDN controller communicates with service consumer and applies the optimized security setting parameters.

- **QoS collector:** This application uses a mix of passive and active monitoring techniques to measure different network metrics, like BW usage, residual capacity, and number of dropped packets, at different aggregation levels, e.g. per flow, per switch port/link, and per user. It also measures per flow latency, error rate, and jitter by inserting packet probes in the network.

20.3.3 Optimization Model

The objective of the proposed optimization model is to ensure security services while guaranteeing QoS. Figure 20.3 illustrates the optimization module which can communicate with SCADA master and other applications of SDN for coordinated actions. The optimization model consists of four inputs, an optimization module, and one output. The first input represents IDS alerts which result in the selection of an effective security countermeasure. The second input is communication services which can be classified according to its type (operation-critical, non-operation-critical, and best effort services); the third input is network QoS, which represent the network performance assessment, such as throughput, delay, and so on. The fourth input is security service settings, which includes the parameters of each security service (the message authentication code (MAC) length, the encryption key in our case). For attack alerts rèported by IDS, the playbook rule instructs that, for service being attacked, the SO-SDSec must modify the device security service requirements by changing MAC length or key length. After the rule is applied, the SO-SDSec orchestrator starts monitoring the IDS alerts and automatically requests for the security parameters settings to optimization module through IIoT-SDN controller. The optimization module executes the GA for each class of service of SCADA control center and assesses the resource availability regarding network performance parameters. If they are sufficient to implement the security service settings having the highest security level, then they will be directly sent to the output. Otherwise, the optimization module trades-off between the security level and QoS and calculates the optimal result of security service settings, QoS requirement, and network performance. Then, SDN controller applies the security parameters through SCADA master for that particular security countermeasure.

20.4 Security Risk Levels and Impact on QoS

Security risk and QoS are two opposite parameters in a communication system. If you want to increase one part, you must sacrifice in another part. For this reason, proper optimization is needed between these two parameters during run time which is quite impossible to do in traditional IP-enabled SCADA communication network. But SDN-enabled network can easily balance between them. In the following sections, we discuss the security risk levels and the impact of those security metrics to network QoS.

20.4.1 Security Risk Levels

To provide a secure data communication in SDN-enabled IIoT-based EDS, we use the combination of three security risk levels (end-to-end authentication level, integrity level, and confidentiality level). The end-to-end integrity level is provided by a Hash-based Message Authentication Code (HMAC) function based on a shared key between the SCADA master server and regional RTUs. The integrity level ensures no one in between SCADA master and RTU can temper the transmitted messages. On the other hand, the confidentiality level provided by the shared key length between these two entities ensures that transmitted information is only disclosed to parties for which it is intended. So, key management plays an important role in maintaining SCADA communication security levels. As per the standard of DNP3 protocol [17], the SCADA master only has the authority to assign keys to SCADA master and outstation (SCADA slave, RTUs, IEDs, PMUs, etc.). There are three types of communication in a SCADA system: unicast, multicast, and broadcast. In this work, we only consider unicast also known as point-to-point communication that is utilized to communicate between a SCADA master and an RTU or IED, a SCADA slave and an RTU or IED, and an RTU or IED to another RTU or IED. SCADA master usually generates two types of keys known as Update Key and Session Key for all SCADA communication network devices. By default, Update Keys are pre-shared to the master and outstation and must be changed by symmetric cryptography or asymmetric (public key) cryptography. Such a mechanism must ensure that the Update Key is kept secret and cannot be obtained by eavesdropping in transit. Usually, Update Key changes in a month to year intervals. On the other hand, the master initializes the Session Keys immediately after communications are established and regularly changes the Session Keys. This practice of periodically changing the Session Keys protects them from being compromised through analysis of the communications link. This Session Key maintains the message authenticity by frequently changing the session key, integrity by applying encryption algorithms (SHA-1-HMAC, SHA-256-HMAC, AES-GMAC) to create HMAC, and confidentiality by keeping key length large enough (at least 128 bits) [17]. Thus, security service settings of each security risk level vary according to the SDN-enabled network performance parameters, the class of network service from SCADA master to RTU, and the amount of resource available at a given time. Maximizing the security level implies to maximize the network services (more details in section VI). For each security service settings, we evaluate the following security risk levels:

20.4.1.1 Security Risk at Authentication (AU) Level

The authentication risk level depends on session key update rate r_{sk}, which is the arrival rate of authentication requests. During every authentication procedure, devices frequently share session keys in a short period of time. The higher the frequency of session key update rate, the stronger the security of the service. When session key update rate increases, the authentication risk level also increases. The authentication risk level is defined as [17]

$$AU_l = c_1 \frac{r_{sk}}{r_{sk} + c_2} \tag{20.1}$$

where both c_1 and c_2 are constants and can be determined by the range of r_{sk} and authentication risk level settings. In our case, the authentication risk level is scaled from 1 to 4 and according to DNP3 standard, session key rate varies from 0.067 to 1 per minute [17]. We can observe that AU_l grows steadily along with r_{sk}. When r_{sk} approaches the maximum values, AU_l slowly approaches the maximum value that represents the highest security level of authentication.

20.4.1.2 Security Risk at Integrity (*I*) Level

Data integrity is ensured by checksums generated using cryptographic hash functions with strong collision resistance property. The probability to generate the same hash code for two different messages is higher in lower hash values. The maximum integrity level is defined as [18]:

$$I_l = \left(2^{\frac{\mathrm{MAC}}{K_{min}}} - 1\right) \frac{c_3}{2^{\frac{\mathrm{MAC}}{K_{min}}} + c_4} \tag{20.2}$$

where MAC represents the length of the checksum digest from a hash function, $\frac{c_3}{2^{\frac{\mathrm{MAC}}{K_{min}}} + c_4}$ is the impact factor. Both c_3 and c_4 are determined from the range of MAC (16–512 bits) and from the range of integrity risk level (1–4). Eq. (20.2) can assure that the security level reaches the maximum as four when the checksum approaches infinity. The K_{min} indicates the minimum key length is 128 bits.

20.4.1.3 Security Risk at Confidentiality (*C*) Level

Confidentiality is determined by the length of the key and the encryption algorithm. The confidentiality level of security is defined as [18]

$$C_l = \frac{1}{2}\left[\left(2^{\frac{K_L}{K_{min}}} - 1\right) \frac{8 - c_5}{2^{\frac{K_L}{K_{min}}} + c_6} + c_5\right] \tag{20.3}$$

where the term $\frac{8 - c_5}{\frac{K_L}{2^{K_{min}}} + c_6} + c_5$ indicates the impact factor, K_L represents the length of the key. The value of the c_5 is determined by the specific algorithm and c_5 is decided by range of K_L (128–512 bits) and range of confidentiality risk level (1–4) in practical applications.

20.4.1.4 Total Security Risk Level (SL)

The security level SL can be defined as

$$SL = w_1 AU_l + w_2 I_l + w_3 C_l \tag{20.4}$$

where the weights of the security features denoted by w_1, w_2, and w_3 are configurable by the security administrator.

20.4.2 The Impact of Security on QoS Metrics

Authentication, data integrity, and confidentiality all bring extra overhead on packet delay, throughput, and packet discard probability. In this section, we present the QoS metrics for EDS.

20.4.2.1 End-to-End Packet Delay

Let us consider d_E and N represent the total delay and forward devices between source and destination, respectively. The end-to-end delay is defined as [19]

$$d_E = N\left(d_{proc} + d_{trans} + d_{prop} + d_{queue}\right) + d_{proco} \tag{20.5}$$

In (20.5), the term d_{queue}, d_{prop}, d_{proc}, and d_{trans} refers to the queuing, propagation, processing, and transmission delay, respectively. Queuing delay of a packet is the waiting time in the output buffers of the forwarding devices to be forwarded. Propagation delay is the time that a transmitted packet needs to travel from one end of a link (SCADA master) to the other end (regional RTU). Processing delay of a packet includes time to look up the routing table and to move the packet over the switch fabric, and lastly, transmission delay is the time it takes to transmit a packet on a link. If the network is not congested ($d_{queue} \approx 0$) and the distance between the source node and destination node is very small ($d_{prop} \approx 0$). The processing delay, d_{proc}, is often negligible; however, it strongly influences a forwarding device's maximum throughput, which is the maximum rate at which a router can forward packets [19]. In the presence of uncongested network, (20.5) reduces to

$$d_E = N \times d_{trans} + d_{proco} \tag{20.6}$$

where d_{proco} indicates the processing overhead because of authentication, integrity, confidentiality, and firewalls rules checking and $d_{trans} = \frac{L}{R}$, where L = packet size (1500 bytes for DNP3 Ethernet packet), R = transmission rate out of each forwarding device (bits/s). The d_{proco} can be defined as: $d_{proco} = 2 \times d_{AU} + 2 \times d_I + 2 \times d_C$. The terms d_{AU}, d_I, and d_C refer to authentication, data integrity, and data confidentiality delay, respectively.

20.4.2.2 Authentication Delay (d_{AU})

The authentication delay is proportional to authentication rate r_{sk} and can be defined as [17]:

$$d_{AU} = c_7 r_{sk} + c_8 \tag{20.7}$$

where c_7 and c_8 are constants that are determined by concrete network status.

20.4.2.3 Data Integrity Delay (d_I)

The data integrity delay d_I increases linearly with the check sum length and can be defined as [18]:

$$d_I = c_9 \text{MAC} + c_{10} \tag{20.8}$$

where MAC denotes the length of the checksum, c_9 and c_{10} are determined by different computers and integrity algorithms.

20.4.2.4 Data Confidentiality Delay (d_C)

Confidentiality delay d_C is proportional to the encryption key length and can be defined as [18]

$$d_C = c_{11} K_L + c_{12} \tag{20.9}$$

where K_L represents the length of the key used and c_{11}, c_{12} are determined by different environment. Obviously, the longer the key length, the more the overhead.

20.4.2.5 Total End-to-End Packet Delay (d_E)

Based on the previous analysis, the total end-to-end delay includes transmission delay, authentication delay, data integrity delay, confidentiality delay. Therefore, the total end-to-end delay looks like

$$d_E = (N \times d_{trans}) + (2 \times d_{AU}) + (2 \times d_I) + (2 \times d_C) \tag{20.10}$$

The constraints of an end-to-end delay for individual network services from SCADA master to RTU refer to the service availability timing for respective service.

20.4.2.6 Throughput

We define throughput as [20]:

$$Thr = \text{efficiency} \times \text{bitrate} = \frac{1500}{1538 + \text{MAC}} \times \text{bitrate} \tag{20.11}$$

since the Ethernet packet size is maximum 1500 octet payload +8 octet preamble +14 octet header +4 octet trailer + minimum inter-packet gap corresponding to 12 octets = 1538 octets. The maximum efficiency is $\frac{1500}{1538} = 97.53\%$ and the physical layer net bit rate depends on the Ethernet physical layer standard and maybe 10 Mbit/s, 100 Mbit/s, 1 Gbit/s, or 10 Gbit/s.

20.5 Optimal Security Countermeasure Selection Problem Formulation

IIoT-based EDS has several network services that can be classified based on the specifications for delay, integrity, and confidentiality. In this work, we focus on three types of services, namely operation-critical service, noncritical services, and best effort services. An example of operation-critical service is transmitting a control signal from SCADA to the controlled device. It has a strong delay constraint; data confidentiality can be ignored. An example of noncritical service is customers' metering data. It requires high throughput and strong data integrity but allows relatively large delay. The problem of achieving the trade-off among security risk, delay, and throughput can be cast as a multi-objective optimization problem as follows:

$$F(x) = (f_1(x), f_2(x), \dots, f_k(x)) \tag{20.12}$$

where $f_1, \dots, f_k$ are the k objective functions to optimize and $x = (x_1, x_2, \dots, x_n)$ is a vector of n decision variables. The goal is to find the vector x that optimizes the k objective functions. In our case, we formulate three objectives:

- Maximize the security level: $SL(x)$
- Minimize the end-to-end delay: $d_E(x)$
- Maximize the Throughput: $Thr(x)$

Subject to,

$$d_E \leq 100\text{ ms}$$

$$Thr \geq 97\%$$

$$x \in 16, 32, 64, 128, 256, 512$$

where $x = (\text{MAC}, K_L)$ is the vector of the decision variables which represents the security services settings. By varying this vector, we will determine the optimum or even near-optimum solution of the objective functions.

In the case of a multi-objective optimization problem, there is usually no single solution that is optimum concerning all objectives and constraints. There are a set of optimal or near-optimal solutions known as Pareto-optimal solutions or Pareto

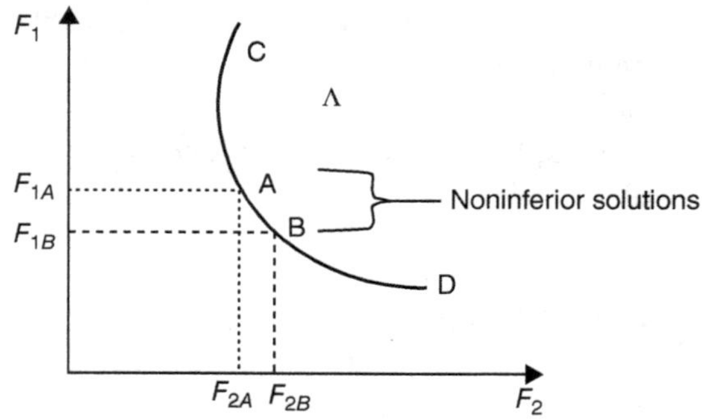

Figure 20.4 Pareto front.

front. A Pareto front is a set of points in parameter space (the space of decision variables) that have non-inferior fitness function values. In other words, for each point on the Pareto front, you can improve one fitness function only by degrading another [21].

In Figure 20.4, A and B are non-inferior solution points because of an improvement in one objective, F_1 requires a degradation in the other objective, F_2, i.e. $F_{1B} < F_{1A}$; $F_{2B} > F_{2A}$.

Since any point in Λ that is an inferior point represents a point in which improvement can be attained in all the objectives, it is clear that such a point is of no value. Multi-objective optimization is, therefore, concerned with the generation and selection of non-inferior solution points. Non-inferior solutions are also called Pareto optima. A general goal in multi-objective optimization is to construct the Pareto optima [22].

In our problem context, this translates into simultaneously optimizing the objectives: maximize security level, minimize end-to-end delay, and maximize throughput and constraints: an end-to-end delay ≤100 ms for operation-critical services and throughput ≥97% for non-operation-critical services [8, 9]. In search of Pareto-optimal front, many multi-objective optimization methods have been proposed. In the next section, we present approaches to solve multi-objective optimization method and our rationale to select a GA.

20.5.1 Multi-Objective Optimization

In traditional multi-objective optimization approach, the objectives are aggregated together to form a single (scalar) fitness function, which is solved by using classical techniques such as multiple objective linear programming (MOLP), multiple attribute utility theory (MAUT), random search, simulated annealing, etc. [23]. The optimization of the single objective may guarantee a Pareto-optimal solution but results in a single-point solution. In real-world situations, decision makers often need to evaluate several alternatives during decision-making. Moreover, the techniques above are not effective if some of the objectives are noisy or have discontinuous variable space. Some of these techniques are also expensive as they require knowledge of the individual optimum before vector optimization. Another drawback is the sensitivity to weights or demand levels [24]. The decision maker must have a thorough knowledge of the priority of each objective. The solutions obtained largely depend on the underlying weight vector or demand level. Thus, for different situations, different weight vectors need to be used, and the optimization process needs to be repeated several times. A more effective technique would be one that can find multiple Pareto-optimal solutions simultaneously so that decision makers may be able to choose the most appropriate solution for a given optimization scenario. The knowledge of many Pareto-optimal solutions is also useful for later use, particularly when the given scenario has changed, and an updated solution is required to be implemented. Since GA deals with a population of several points instead of one-point, multiple Pareto-optimal solutions can be captured in the population, in a single run. To speed up the performance of the GA significantly, elitism keeps track of good solutions already encountered during optimization. Therefore, using elitist method is attractive to reduce the delay and increase the SL through the optimization process in the SCADA communication network. Furthermore, to choose the best optimization method, we need to search for the Pareto front to select the optimal security setting solution. We also need to ensure a good distribution on the Pareto front with the aim to discover security setting solutions while ensuring the desired trade-offs among the three objectives functions. A comparative study of elitist multi-objective optimization methods is provided to evaluate the performance of each one in [23]. Results show that elitist non-dominated sorting genetic algorithm (NSGA-II) (a fast and elitist multi-objective GA) is the best-suited method to maintain a better spread of solutions and converges better in the obtained non-dominated front. Therefore, we choose the fastest GA (NSGA-II) to optimize our multi-objective problem, whose time-complexity is $O(MN^2)$, where M is the number of objective functions and N is the population size. In this work, we implement the NSGA-II in *Matlab* using the *gamultiobj* function. NSGA-II maintains the diversity of population for convergence to an optimal Pareto front. Diversity is maintained by controlling the elite members of the population as the algorithm progresses. Two options, first non-domination sorting *ParetoFraction* and crowded distance estimation procedure *DistanceFcn*, control the elitism. *ParetoFraction* limits the number of individuals on the Pareto front (elite members). The *DistanceFcn* helps maintain diversity on a front by favoring individuals that are relatively far away on the front. The algorithm stops if the spread, a measure of the movement of the Pareto front, is small [22]. When NSGA-II is adopted to solve our multi-objective problem, an individual will symbolize a possible solution, therefore, the security settings combination of the two security services. Thus, the vector of variables (MAC; K_l) is considered as an individual. To get optimal security settings, the NSGA-II implementation illustrated in Figure 20.5 is applied. The step-by-step procedure shows that the NSGA-II algorithm is simple and straightforward [25]. At first, an initial population P_t of S possible solutions of the security services settings $(x_1, x_2, ..., x_s)$ is created randomly, where an individual x_i = (MAC; K_L) represents the combination of the two security services parameters.

We assign each x_i with the three objective functions, security services level, throughput, and end-to-end delay. The usual binary tournament selection, recombination, and mutation operators are used to create a new population of possible security services settings called child population Q_t of size S. Thereafter, a combined population $R_t = P_t \cup Q_t$ is formed. The population R_t is of size $2S$. Then, the population R_t is sorted according to non-domination. Since all previous and current population members are included in R_t, elitism is ensured. Then, the total population R_t is sorted according to non-domination and non-dominated fronts F_1, F_2, ..., F_l are obtained. Now, solutions belonging to the best non-dominated set F_1 are of best solutions in the combined population and must be emphasized more than any other solution in the combined population. If the size of F_1 is smaller than S, we definitely choose all members of the set F_1 for the new population P_{t+1}. The

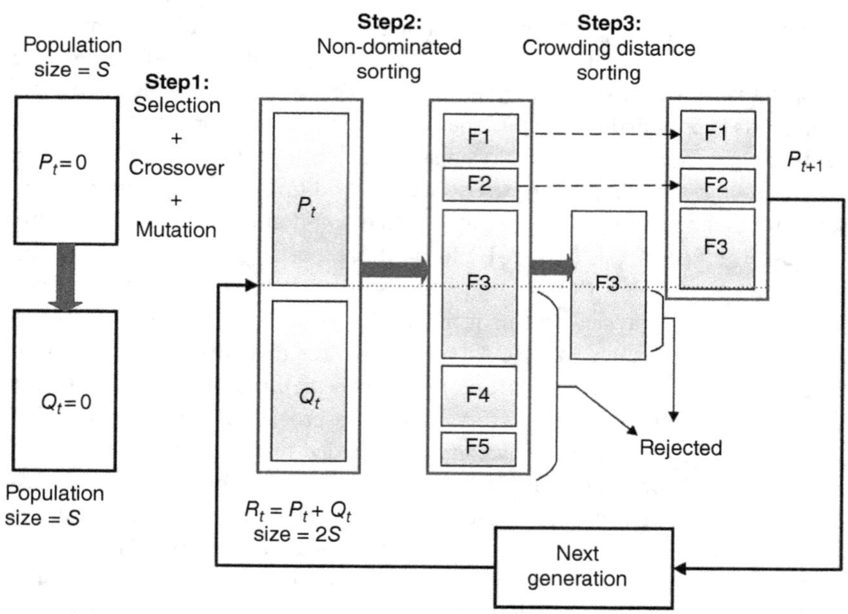

Figure 20.5 NSGA-II procedure.

remaining members of the population P_{t+1} are chosen from subsequent non-dominated fronts in the order of their ranking. Thus, solutions from the set F_2 are chosen next, followed by solutions from the set F_3, and so on. This procedure is continued until no more sets can be accommodated. Say that the set F_l is the last non-dominated set beyond which no other set can be accommodated. In general, the count of solutions in all sets from F_1 to F_l would be larger than the population size. To choose exactly population members S, we sort the solutions of the last front F_l using the crowded distance operator in descending order and choose the best solutions needed to fill all population slots. The new population P_{t+1} of size S is now used for selection, crossover, and mutation to create a new population P_{t+1} of size S that will be used in the same way to generate the next generation. This procedure will be applying until finding the optimal solution and then optimal security configuration.

20.6 Simulation Results

We developed the simulation environment in MATLAB 2016a on a windows 7 Intel(R) Core (TM) i7-6820 HQ CPU of 2.67 GHz with parameters of NSGA-II for all simulations are as following:

Initial population	**50**
Maximum generation	150
String length in binary code (n)	32
The probability of crossover	0.8
The probability of mutation	$\frac{1}{n}$

The assigned values to initial population size, maximum generation, and crossover rate ensure the diversity as well as less processing time to get the result. If we select the initial population and maximum generation are high, it gives more accurate diversified solutions to converge but time complexity increases. In EDS operation, time is also a critical factor for which we keep value to those parameters in such a way that, they maintain high diversity as well as low processing time to converge. The value to the crossover rate also ensures the better diversity to converge. After that, the program is allowed to iterate over several generations, and the final optimized security settings values of the non-dominated solutions resulting from this run are noted.

Good performance of any service playing between SCADA and regional substation depends on settings of security level, delay time, and throughput. The trade-off among those objective functions is a multi-objective problem. The parameters

Table 20.1 Range of basic parameters for security level.

Name	MAC (bits)	K_L (bits)	r_{sk} (/min)	$d_{trans}(ms)$
Value	[16, 512]	[128, 512]	[0.067, 1.0]	12

Table 20.2 Parameters for evaluating security level.

Name	c_7	c_8	c_9	c_{10}	c_{11}	c_{12}
Value	5.29	0.22	4.07	0.72	8	1.2

Table 20.3 Parameters for evaluating delay and throughput.

Name	c_7	c_8	c_9	c_{10}	c_{11}	c_{12}
Value	0.6	0.21	0.2	8.1	0.2	65

for the simulation of all services are given in detail in Tables 20.1–20.3. In Table 20.1, the range of basic parameters for security is provided. The checksum MAC varies from 16 to 512 bits. The key length K_l also ranges from 128 to 512 bits. The session key refresh rate r_{sk} changes continuously from 0.067 to 1.0 per minute [17]. The end-to-end transmission time is 12 ms for the context of 1500 bytes packet size, 10 Mbps end-to-end shared link, and 10 intermediate forwarding devices. Based on the range of Table 20.1, the values of parameters in Table 20.2 reflect the normal requirement from Eqs. (20.1) to (21.3) and the setting of three security levels. The parameters of Table 20.3 are calculated based on the ranges of delay factor imposing from Eqs. (20.7) to (21.9) to packet's end-to-end delay in an SDN-enabled SCADA communication network.

20.6.1 Operation-Critical Services

For the operation-critical services (i.e. a control command from SCADA master to RTU/IED), the authentication function operates in a short time, mostly needs the data integrity and smaller part of data confidentiality. Hence, weights are: $w_1 = 0.3$, $w_2 = 0.6$, $w_3 = 0.1$ and Eqs. (20.4) and (21.10) are given as below:

$$SL = 0.3AU_l + 0.6I_l + 0.1C_l \tag{20.13}$$

$$d_E = \left(N \times \frac{L}{R}\right) + 2 \times 0.3 \times d_{AU} + 2 \times 0.6 \times d_I + 2 \times 0.1 \times d_C \tag{20.14}$$

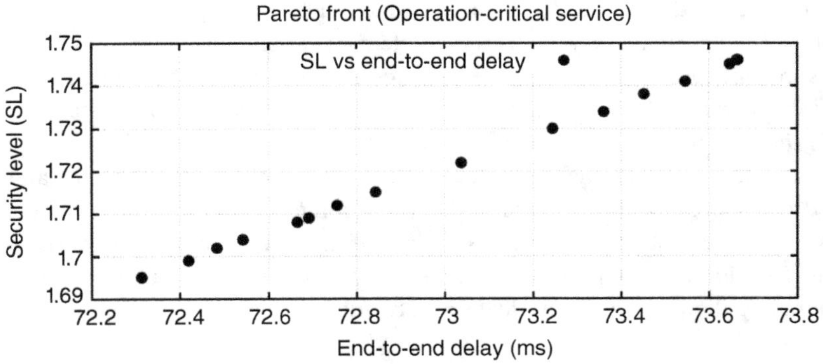

Figure 20.6 Pareto front to maintain delay ≤100 ms when N = 10.

Figure 20.6 shows that the trend of the delay increases with the increase of security level up to 100 ms for operation-critical services. The delay is determined by the session key update rate, some intermediate nodes, the communication link BW, the MAC size, and key length. In our case, the session key update rate of intermediate nodes and BW keep constant, but MAC is varied from 16 to 512 bits, key length is varied from 128 to 512 bits. All 15 points of Pareto front are optimized points. Among these points, one point may give better SL and another point may ensure a better end-to-end delay. Like the leftmost point whose co-ordinates when $(SL, d_E) = (1.69, 72.3\text{ ms})$ when $(\text{MAC}, K_L) = (16, 128)$, the rightmost point is $(SL, d_E) = (1.75, 73.7\text{ ms})$ when $(\text{MAC}, K_L) = (32, 128)$ [8]. The leftmost point ensures better delay constraint at smaller MAC length than the rightmost point where better SL constraint is ensured at the cost of bigger MAC length. So, both points are important for SDN-aware EDS in different situations as per the need of respective operation-critical service but slight increment of SL (from 1.69 to 1.75) from left point to right point, the system has to increase MAC length from 16 bits to 32 bits which actually increases end-to-end delay from 72.3 to 73.7 ms.

According to the time complexity of NSGA-II which is big O mentioned earlier, total 7500 functions are to be evaluated in every simulation because of our problem domain consists of three objective functions and initial population of 50. The calculated time requirement for every function evaluation in our simulation environment is 10^{-4} seconds. So, the total time overhead added due to this optimization solver to the SDN controller is 0.75 seconds. But this time delay is incurred one time and is only applicable when the attacker attacks SCADA communication link and the optimization module identifies the optimized security settings parameters to mitigate that attack. After application of new security settings, there will be no additional latency due to the optimization solver until another threat has been detected to the SCADA communication networks. In the absence of the optimization solver, the time taken to identify the problem and apply the appropriate countermeasure will consume considerable time and may not satisfy the timeliness of delivering operational critical messages from SCADA master to substations. Though the countermeasure was applied, the fact that the end-to-end delay was not met would result in the substation operating in pulsating condition and thereby responsible for an unstable grid. By adopting our optimization module, we can implement optimized security settings within 0.75 seconds which ensures that the operation-critical messages are exchanged in a timely fashion and ensures operational resilience of communication network between SCADA master and substation's RTU.

20.6.2 Non-Operation-Critical Services

For a defined network condition, the throughput has an inverse relationship with the delay. We can say that the network delay multiplying the throughput is a constant in the known network. Non-operation-critical services deem authenticity, integrity, and confidentiality at equal share [8]. Hence, the weights are: $w_1 = 0.33$, $w_2 = 0.33$, $w_3 = 0.33$ and Eqs. (20.4) and (21.10) are as given below:

$$SL = 0.33AU_l + 0.33I_l + 0.33C_l \tag{20.15}$$

$$d_E = \left(N \times \frac{L}{R}\right) + 2 \times 0.33 \times d_{AU} + 2 \times 0.33 \times d_I + 2 \times 0.33 \times d_C \tag{20.16}$$

Figures 20.7 and 20.8 together indicate that when the security level increases, the throughput decreases and also end-to-end delay increases. There are 29 optimized

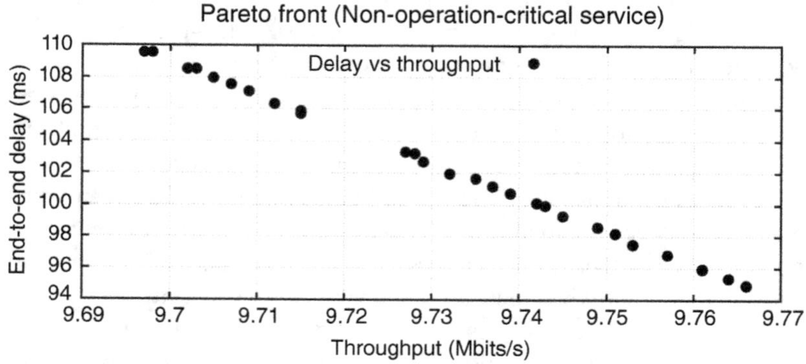

Figure 20.7 Pareto front with constraint $Thr \geq 97\%$ when $N = 14$.

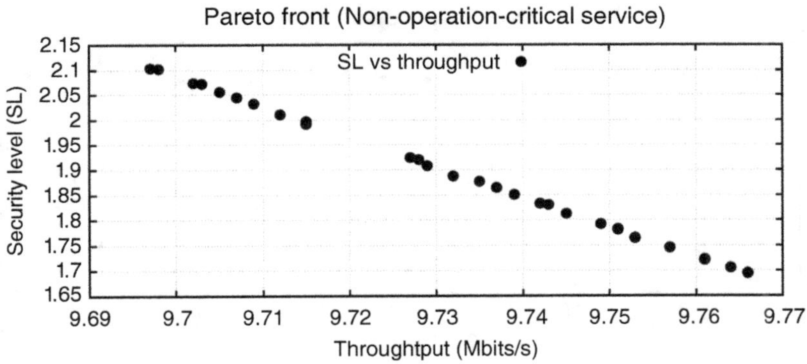

Figure 20.8 Pareto front with constraint $Thr \geq 97\%$ when $N = 14$.

points in each Pareto front. If we consider two points from Figure 20.8, the leftmost point is (SL, Thr) = (2.1, 97.10%) when (MAC, K_L) = (128, 128) and the rightmost point is (SL, Thr) = (1.7, 97.76%) when (MAC, K_L) = (64, 128). Both points are important in SDN-aware EDS at different situations and cost. The leftmost point is important for those non-operation-critical services (power quality monitoring, customer metering data) where data throughput is important as well as SL but delay factor can be considerable up to a certain level [8]. But in that case, there is an extra processing cost that adds with the system to improve SL because of selecting bigger MAC. On the other hand, the rightmost point is applicable to those non-operation-critical services, where SL can be considerable up to a certain level, but data throughput is more important than SL because of huge data has to collect in short time. Collecting remote customers' metering data can be treated as this type of service where MAC is smaller which also indicates less processing time.

20.6.3 Best Effort Services

Best effort service has no specific requirements regarding security features as well as network QoS. In that case, network specialist can run the optimization solver and set security parameters, and QoS parameters depend on the service requirements. Figure 20.9 shows the Pareto front of optimization output when no constraint is set as input. In that case, the security level increases with an increasing trend of an end-to-end delay. Like the leftmost point, where (SL, d_E) = (1.75, 75 ms) when (MAC, K_L) = (16, 128) is perfectly applicable to collecting data for automation system engineering [9] and troubleshooting-type service in SDN-aware EDS where there are no hard and fast constraints, but the delay factor is slightly important. On the other hand, the rightmost point) (SL, d_E) = (4, 325 ms) when (MAC, K_L) = (512, 256) is applicable to exchanging historical data

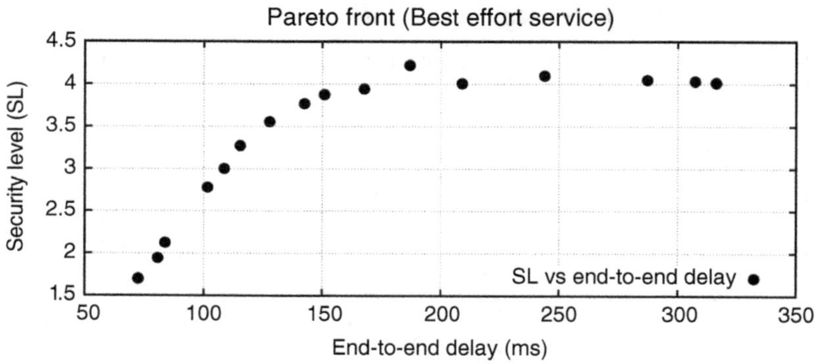

Figure 20.9 Pareto front with no constraint when N = 14.

for midterm and long-term planning where data integrity is more important than end-to-end delay.

20.6.4 Performance Evaluation of the Optimized Security Settings

To evaluate the performance of the obtained security setting solutions, different performance metric has been proposed [26], such as error ratio. The error ratio is used to evaluate the percentage of the convergence to the known Pareto-optimal front. It is defined as $E = \frac{sum_{i=1}^{n} e_i}{n}$ where n is the total number of security setting solutions, $e_i = 0$, if a solution is a member of the Pareto-optimal front, otherwise 1. Like from Figure 20.9, there is one point so far little bit out of Pareto front, so the error $E = \frac{1}{29} = 0.034$. That means 96.6% of the security setting solutions are close to the Pareto front.

20.7 Conclusion

In this chapter, we presented an SDN-enabled IIoT-based EDS architecture that provides the ability to enforce security countermeasures to reduce the risk of cyberattack and ensure QoS. We proposed a GA-based multi-objective optimization approach to aid in selecting the optimal security countermeasure which balances the reduction of security risk and maintaining QoS, which thereby ensures EDS resilience. Simulation results indicate that the proposed approach can provide resiliency by balancing the trade-off between reducing security risk and QoS guarantees. For future work, we will extend the optimal countermeasure selection as to enable cyber security deception in an SDN-enabled IIoT (Chapter 21).

Acknowledgment

This material is based upon work supported by the Department of Energy under Award Number DE-OE0000780 and Office of the Assistant Secretary of Defense for Research and Engineering agreement FA8750-15-2-0120.

Disclaimer

This report was prepared as an account of work sponsored by an agency of the United States Government. Neither the United States Government nor any agency thereof, nor any of their employees, makes any warranty, express or implied, or assumes any legal liability or responsibility for the accuracy, completeness, or usefulness of any information, apparatus, product, or process disclosed, or represents that its use would not infringe privately owned rights. Reference herein to any specific commercial product, process, or service by trade name, trademark, manufacturer, or otherwise does not necessarily constitute or imply its endorsement, recommendation, or favoring by the United States Government or any agency thereof. The views and opinions of authors expressed herein do not necessarily state or reflect those of the United States Government or any agency thereof.

References

1 A. Zanella, N. Bui, A. Castellani, L. Vangelista, and M. Zorzi, "Internet of things for smart cities," *IEEE Internet of Things Journal*, vol. 1, no. 1, pp. 22–32, 2014.

2 S. Mumtaz, A. Alsohaily, Z. Pang, A. Rayes, K. F. Tsang, and J. Rodriguez, "Massive internet of things for industrial applications: addressing wireless IIoT connectivity challenges and ecosystem fragmentation," *IEEE Industrial Electronics Magazine*, vol. 11, no. 1, pp. 28–33, 2017.

3 C. Cecati, C. Citro, A. Piccolo, and P. Siano, "Smart operation of wind turbines and diesel generators according to economic criteria," *IEEE Transactions on Industrial Electronics*, vol. 58, no. 10, pp. 4514–4525, 2011.

4 X. Dong, H. Lin, R. Tan, R. K. Iyer, and Z. Kalbarczyk, "Software-defined networking for smart grid resilience: opportunities and challenges," in Proceedings of the 1st ACM Workshop on Cyber-Physical System Security. ACM, 2015, pp. 61–68.

5 S. Al-Rubaye, E. Kadhum, Q. Ni, and A. Anpalagan, "Industrial internet of things driven by SDN platform for smart grid resiliency," *IEEE Internet of Things Journal*, 6(1), pp. 267–277. 2017.

6 R. Langner, "Stuxnet: dissecting a cyberwarfare weapon," *IEEE Security & Privacy*, vol. 9, no. 3, pp. 49–51, 2011.

7 D. Deka, R. Baldick, and S. Vishwanath, "Optimal data attacks on power grids: leveraging detection & measurement jamming," in 2015 IEEE International Conference on Smart Grid Communications (SmartGridComm). IEEE, 2015, pp. 392–397.

8 D. Wei, Y. Lu, M. Jafari, P. Skare, and K. Rohde, "An integrated security system of protecting smart grid against cyber attacks," in Innovative Smart Grid Technologies (ISGT), 2010. IEEE, 2010, pp. 1–7.

9 IEEE Power Engineering Society, "IEEE Standard for SCADA and Automation Systems," vol. IEEE Std C, 37.

10 Y. Zhu, J. Yan, Y. Tang, Y. L. Sun, and H. He, "Resilience analysis of power grids under the sequential attack," *IEEE Transactions on Information Forensics and Security*, vol. 9, no. 12, pp. 2340–2354, 2014.

11 J. Yan, Y. Tang, B. Tang, H. He, and Y. Sun, "Power grid resilience against false data injection attacks," in Power and Energy Society General Meeting (PESGM), 2016. IEEE, 2016, pp. 1–5.

12 R. Kinney, P. Crucitti, R. Albert, and V. Latora, "Modeling cascading failures in the North American power grid," *The European Physical Journal B-Condensed Matter and Complex Systems*, vol. 46, no. 1, pp. 101–107, 2005.

13 S. Luo and M. B. Salem, "Orchestration of software-defined security services," in 2016 IEEE International Conference on Communications Workshops (ICC). IEEE, 2016, pp. 436–441.

14 L. Li, S. Li, and S. Zhao, "QoS-aware scheduling of services-oriented internet of things," *IEEE Transactions on Industrial Informatics*, vol. 10, no. 2, pp. 1497–1505, 2014.

15 G. S. Aujla, R. Chaudhary, S. Garg, N. Kumar, and J. J. Rodrigues, "SDN-enabled multi-attribute-based secure communication for smart grid in IIoT environment," *IEEE Transactions on Industrial Informatics*, 14(6), pp. 2629–2640. 2018.

16 J. Chen, H. Zeng, C. Hu, and Z. Ji, "Optimization between security and delay of quality-of-service," *Journal of Network and Computer Applications*, vol. 34, no. 2, pp. 603–608, 2011.

17 "IEEE Std 1815–2012 for electric power systems communications distributed network protocol (dnp3)," pp. 171–265, June 2012.

18 P. Blomgren and S.M Kotronx, "Cryptographic protection of SCADA communications part 1: background, policies and test plan," American Gas Association (AGA), Draft, 4, 2006.

19 J. F. Kurose and K. W. Ross, *Computer Networking: A top down approach featuring the internet*. Peorsoim Addison Wesley. 2010.

20 "Ethernet frame," https://en.wikipedia.org/wiki/Ethernet_frame, accessed: 2018-07-01.

21 K. Deb, S. Agrawal, A. Pratap, and T. Meyarivan, "A fast elitist non-dominated sorting genetic algorithm for multi-objective optimization: NSGA-II," in International Conference on Parallel Problem Solving from Nature. Springer, 2000, pp. 849–858.

22 D. Zheng, M. Gen, and R. Cheng, Multiobjective optimization using genetic algorithms. Engineering Valuation and Cost Analysis, 2(1), pp. 303–310, 1999.

23 V. Khare, X. Yao, and K. Deb, "Performance scaling of multi-objective evolutionary algorithms," in International Conference on Evolutionary Multi-Criterion Optimization. Springer, 2003, pp. 376–390.

24 N. Srinivas and K. Deb, "Muiltiobjective optimization using nondominated sorting in genetic algorithms," *Evolutionary Computation*, vol. 2, no. 3, pp. 221–248, 1994.

25 K. Deb, A. Pratap, S. Agarwal, and T. Meyarivan, "A fast and elitist multiobjective genetic algorithm: NSGA-II," *IEEE Transactions on Evolutionary Computation*, vol. 6, no. 2, pp. 182–197, 2002.

26 D. A. Van Veldhuizen, "Multiobjective evolutionary algorithms: classifications, analyses, and new innovations," DTIC Document, Technical Report, 1999.

21

Leverage SDN for Cyber-Security Deception in Internet of Things

Yaoqing Liu[1], Garegin Grigoryan[2], Charles A. Kamhoua[3], and Laurent L. Njilla[4]

[1] *Computer Science, Fairleigh Dickinson University, Teaneck, NJ, USA*
[2] *Computing and Information Sciences, Rochester Institute of Technology, Rochester, NY, USA*
[3] *US Army Research Laboratory, Adelphi, MD, USA*
[4] *Cyber Assurance Branch, US Air Force Research Laboratory, Rome, NY, USA*

21.1 Introduction

The Internet of Things (IoT) is becoming an increasingly attractive target for cybercriminals. We observe that many attacks on the IoTs are launched in a collusive way, such as brute-force hacking of usernames and passwords, to target a particular victim. However, most of the time our mechanisms that defend against such attacks operate individually and independently, which leads to ineffective and weak defense. To this end, we propose to leverage Software Defined Networks (SDN) to enable cyber-security deception and cooperative security for legacy Internet Protocol (IP)-based IoT devices. SDN decouples control plane and data plane and can help bridge the knowledge divided between the application and network layers. In this chapter, we discuss IoT security problems and challenges and propose to leverage cyber-security deception techniques and an SDN-based architecture to enable IoT security in a cooperative manner. Furthermore, we implemented a platform that can simulate cyber-security deception techniques, quickly share the attack information with peer controllers, and block the attacks. We carried out our experiments in both virtual and physical SDN environments with OpenFlow switches. Our evaluation shows that both environments can scale well to handle attacks, but hardware implementation is much more efficient than a virtual one.

Modeling and Design of Secure Internet of Things, First Edition. Edited by Charles A. Kamhoua, Laurent L. Njilla, Alexander Kott and Sachin Shetty.

Published 2020 by John Wiley & Sons, Inc.

We observe that normally IoT systems can be both static and mobile targets and are expected to be attacked because many of them have low resource capabilities. They are not equipped with dynamic anti-attack mechanisms to fight against collusive attacks from dedicated attackers. They also work individually and independently to defend against the attacks, which makes them easily compromised and particularly vulnerable when the same attacks are applied to similar systems having the same vulnerabilities. It is important to note that many illegitimate access attempts are neglected by the current IoT software. For instance, we are able to collect a large amount of attack information from the system logs in an IoT system when many attackers use a brute-force approach to guess usernames and passwords of the IoT devices. Even when these attacks fail, they could potentially threaten to compromise the system sooner or later. However, most of the current practice ignores the failed attack's traces instead of taking advantage of them. One reason is that such IoT devices do not have enough computing capabilities to run custom programs to handle such kind of attack information. Another, and more important, reason turns out to be the knowledge gap between the application and network layers. The application layer can easily detect application-layer attacks, such as brute-force password guessing, which are transparent to the network layer because of the original design of the Transmission Control Protocol (TCP) IP protocol stack that naturally separates application-layer and network-layer identifiers.

Considering the problem of IoT devices' insufficient computing capabilities for running custom programs to mitigate attacks, cyber-security deception techniques (e.g. camouflage and decoys) are potential solutions enabling planned actions that mislead and/or confuse attackers, causing them to take (or not take) specific actions that aid computer-security defenses. The essential parts of cyber deception are carefully crafted, misleading information by the defender and wrong actions taken by the adversary as a result of the deception. To achieve this, we leverage SDN, an emerging and promising technology, to make it happen. SDN decouples the control plane from the data plane. An SDN controller can take input from end systems' applications, such as honeypots and decoys, and make decisions to the data plane about what traffic can go through. SDN can potentially benefit the security of IoT systems in at least three aspects. First, it can help form a feedback-control loop from end IoT systems to the SDN controller, which further controls one or more programmable switches. With the loop, attacks to a victim or a honeypot can be blocked by the intermediate network device, such as an OpenFlow switch or another programmable gateway. Second, similar attacks to other victims from the same source in the same network can be blocked and thus benefit the whole IoT network in a collaborative way. Third, the attack information can be shared among multiple peering controllers that oversee and control different networks. Other controllers that obtain the attack information can take the same actions to block the malicious activities before they break into these peering networks. To this end, we designed and built an SDN-based cyber-deception platform that enables cooperative security for IoT systems. We make the following contributions in this chapter, as follows:

1) *Literature Review:* We reviewed the state-of-the-art cyber-deception research in a variety of applications – wireless network, cloud computing, virtual networks, and game theory – that applied different cyber-deception approaches to protecting networking resources and defending against attackers.
2) *Protocol Design:* We articulate the technical challenges to implement SDN-based solutions in a mobile IoT environment with multiple distributed SDN controllers; also, define the interfaces, functions, and protocols to enable an efficient SDN-based cyber-deception communication model among honeypots, victims, OpenFlow switches, and the controllers.
3) *Platform Implementation:* We implemented the platform using Ryu controllers and OpenFlow switches in the Global Environment for Network Innovations (GENI). We also tested the platform in a real hardware environment using an OpenFlow switch and a Ryu-based controller.
4) *Performance Evaluation:* The platform was evaluated with its effectiveness and efficiency. We measured various times, such as attack-detection time, controller-response time, controller-to-controller sharing time, and flow-entry installation time. The evaluation shows that the platform can efficiently share attack information with trustworthy peers and stop attacks in a timely manner. In addition, hardware-based implementation can handle attacks more efficiently.
5) *SDN for Distributed Networks:* We argue that it is very challenging to deploy SDN-based cyber-deception techniques in mobile and distributed networks with the traditional TCP/IP architecture. We propose to combine SDN and Named Data Networks (NDN) to implement cyber deception in a mobile IoT environment. We discuss the benefits with examples as well as the challenges for this new research direction.

21.2 Literature Review

There are generally three techniques to enable cyber deception: camouflage, decoys, and misrepresentation. Camouflage is an act, means, or result of obscuring things to confuse an adversary by painting or screening objects so that they are not recognizable from a distance. Decoys are lures that emulate real systems and are deployed at multiple locations of a network so the real assets can be kept away from the attackers. Honeypots are a good example of decoys. Misrepresentation is the network system's generation of false responses to the attacker so that they are not able to obtain the accurate information about the network to start an attack.

21.2.1 Reconnaissance Deception System (RDS)

RDS [1] aims to give the attacker a false view of network features, including topological location of hosts, number of hosts, addresses of hosts, and connectivity among hosts. These goals are achieved by the following:

- *Dynamic Address Translation:* On-the-fly packet-header rewriting hides real addresses and makes the network appear larger than it is, causing the search space to increase for the attacker; further, this technique allows addresses of nodes in the virtual network to be changed regularly.
- *Route Mutation:* Using virtual routers, the system simulates virtual paths over multiple hops, allowing RDS to alter the topology of different network views; thus, the scanner cannot accurately see the network topology.
- *Vulnerable Host Placement:* This places real hosts in the underlying network in virtual subnets to increase the time needed to identify them.
- *Honeypot Placement:* RDS uses dynamic address translation to make one honeypot appear as multiple servers, making the network look much larger than it is and creating many pitfalls for an attacker.
- *Dynamic Detection of Malicious Flows:* By using the statistics of the flow rules in the SDN switches, RDS can detect malicious flows before the attacking scanner can find vulnerable hosts. The system also involves several components to enable its operation.
- *Virtual Network View Generator:* This part of the system generates a machine-readable topology description for the controller that contains virtual network-view node specification, port and address information of the deception server, IP and Media Access Control (MAC) addresses of visible real hosts and honeypots, deceptive IP addresses of real hosts and honeypots as they appear to the view node, and virtual path information to the network nodes. These topologies are used by the controller and can be changed in seconds to redefine the network view for the node.
- *SDN Controller:* This uses the virtual network-view descriptions to generate flow rules to construct the virtual network views; moreover, it watches the statistics of the current rules to detect malicious traffic, handles routing packets to and from honeypots, and handles Dynamic Host Configuration Protocol (DHCP) requests and Address Resolution Protocol (ARP) traffic.
- *Deception Server:* This server manipulates network traffic and simulates certain resources like ARP, Internet Control Message Protocol (ICMP), DHCP, and DNS. This works in tandem with the SDN Controller and Network View Generator to enable the system to fool scanning attempts.

21.2.2 Decoys and Cybermoat

Most cyber-deception systems use reactive strategies with decoys for attack detection and observation [2]. These systems are limited by poor scalability and static decoy configurations. Detection of a data breach takes, on average, 82 days. Specially crafted decoys can shorten the detection and prevention cycle. Most decoys are based on the honeypot technique for information gathering and attack detection, which makes it easy for attackers to find and avoid decoys. This creates a trade-off between the number of decoys and the fidelity of the decoys: higher-fidelity decoys take more resources, and deployment of a large number of decoys also requires a large amount of resources; however, low-fidelity decoys are easily identifiable by attackers. Cybermoat addresses this trade-off by using a pool of very light proxies, numbering in the hundreds or even thousands, that can redirect traffic between the Internet and back-end decoy servers or protected servers. Furthermore, advanced attackers are able to identify decoy servers using fingerprinting and timing techniques to compare the discovered servers. Decoys will typically be running with fewer resources, giving them slower response times than the real servers. More specifically, Cybermoat aims to prevent targeted remote attacks, and does not address "remote insiders" that have gotten into the network by compromising legitimate users. Cybermoat consists of four major components: control plane, proxy pool, decoy server farm, and the Cybermoat gateway.

The control plane monitors and regulates traffic to the decoy proxies, shuffles and manages the proxies, and manages the back-end decoy servers. Also, the control plane has five core functions: (i) The authentication module verifies legitimate clients to white-list them through the proxy pool. (ii) The traffic monitor looks at traffic going to the proxies to detect large-scale attacks and alert the controller so it can react to the attacks; upon detection of scanning, it will initiate proxy shuffling to invalidate any information gathered from the scanning. (iii) The proxy manager creates proxies as needed to adapt to changing attack conditions and the topology of the proxy pool. (iv) The randomization controller coordinates mutation of the proxy configuration based on alarm signals from the traffic monitor; also, it randomizes the addresses based on an algorithm to determine a frequency to randomize them. (v) The decoy manager creates decoys as needed and monitors existing decoys; further, it recycles compromised decoy servers to create new ones as needed.

The system uses ClickOS Virtual Machines (VMs) for the proxy handlers. These VMs can run in an environment with only a few MBs of memory and can boot in tens of milliseconds. This allows for quickly deployable, highly scalable proxies. These proxies transparently redirect network traffic between attackers and users and back-end decoys or servers, making the network appear to have many more end hosts than it truly does. The decoy server farm contains a small number of high-fidelity decoy servers that process the malicious requests diverted from front-end proxies. The gateway enforces the forwarding rules configured by the control plane and helps with connection migration during proxy shuffling.

21.2.3 Probabilistic Logic

This work proposes that in response to a malicious scan, the network could generate true and false responses [3]. If an attacker believes all scan results are true, they will be on the incorrect path to their attack. However, if they believe that some scan results are false, they will have to separate the true results from the false results, costing time and effort and minimizing the potential damage from an attack. This technique can exponentially increase the difficulty of gathering useful data from a scanning attempt, making the attacker unsure of exactly which features the network really has. The technique uses a bulk offline computation to create a scan table to answer scan queries, allowing for computational costs to only be incurred upon a network change.

21.2.4 Deceptive Routing in Relay Networks

Wireless networks are vulnerable to jamming attacks, which prevent packets from being properly decoded [4]. This is made worse when the adversary can exploit weaknesses in the physical or MAC protocols in the target network. Current mechanisms in the physical layer that defend against jamming involve beamforming, spread-spectrum, and directional antennas. MAC layer protocols can use channel surfing for defense. When multihop routing is used, sources can decrease the flow on paths with high packet loss to correct for issues; however, this is a reactive technique and suffers loss until the attack is detected. This work puts forth a proactive defense against jamming in multihop relay networks, where one or more network sources use a deceptive network flow along a disjoint network path. This mechanism uses strategic jamming behaviors, making it so the attacker has to target deceptive flows, reducing the impact on real network traffic. These flows consist of randomly generated dummy packets that flow on a path that is not regularly used by normal traffic. This is modeled using a two-stage game to find strategies that are at Stackelberg equilibrium for selfish and altruistic nodes. The first stage consists of the sources playing a noncooperative game to select the real and deceptive flow locations. The second stage has the adversary observing the total flow allocation of each source and selecting a jamming strategy to maximize the decrease in throughput. This game is studied under two source behaviors: one with selfish sources that wants to maximize its own throughput at the cost of other nodes and another altruistic version that incorporates the delays of other sources when deciding flow rates. The results of this study are illustrated through simulation. The simulation results show that altruistic behavior improves the utility of the sources.

21.2.5 Game Theory

Since threats and tactics are constantly evolving, a reciprocally diverse set of defense tactics need to be developed [5, 6]. A technique for this is using the mathematical foundations provided by game theory to investigate the fundamental needs for securing a system. Game theory is useful for studying these problems because it provides a framework for simulating interactions between multiple agents with different goals. Furthermore, it provides a platform for reasoning about uncertainty, randomization, and information manipulation in these contexts. The work by Carroll and Grosu [7] proposes three game-theoretic models that capture aspects of how honeypots can be used in network security. These models show different strategies for the deployments of honeypots and how they can be deployed to the greatest effect. Also, it contains a basic honeypot-selection game, an extended version of that game, allowing for probing actions from the attacker, and lastly, a version involving attack strategies represented using attack graphs. In addition, attackers can detect honeypots using clues like slow input/output and other time delays, unusual system calls, obvious file names, and the type/amount of data in memory. However, a real system can be disguised as a honeypot, creating new difficulty for an attacker, because now they have to check if a honeypot is truly a honeypot. The research by Kiekintveld et al. [8] uses game theory to investigate interactions between attackers and defenders in a network. The defender can camouflage either a normal system as a honeypot or a honeypot as a normal system. These interactions are modeled as a signaling game, a noncooperative two-player dynamic game with incomplete information. The work focused on finding which strategies create perfect Bayesian equilibriums, a refined Nash equilibrium where neither the defender nor attacker will deviate from their chosen strategy.

21.2.6 Dolus

Cloud computing is the backbone of many online services [9]. Cloud computing allows for on-demand elasticity in the back end and other benefits, which have lead many operations to rely heavily on the infrastructure. SDN also allows for the creation of Software-Defined Internet Exchange Points (SDXs) to allow multiple SDN domains or autonomous systems to communicate, enabling app-specific peering, sharing of threat information, and other collaboration. These SDX clouds are starting to mature, but are still at risk of distributed denial of service (DDoS) attacks bringing services down at least temporarily, causing financial impact to the business operating the service. To compound this issue, traditional DDoS-mitigation techniques only block the attack at the destination, not the neighboring networks, causing the target to be safe while the nearby paths are still suffering from the increased traffic flow. The Dolus system leverages a two-stage ensemble learning scheme to analyze the attack features, using foundations from pretense theory in child play. This allows the system to use "quarantine virtual machines" and SDxI coordination across network domains to give the attacker(s) a false sense of success. This method fools the attacker into believing they are attacking the target system long enough for the target system's operators to decide on, and put in place, an appropriate countermeasure. Dolus uses Frenetic to monitor traffic to detect attacks; then, once an attack is detected, it uses statistical methods to classify the attacks. Outlier classification is used to differentiate attacks from infrastructure failure, such as a downed router. Once an attack is detected, Dolus sets up a quarantine VM using elastic provisioning. This quarantine VM responds with spoofed packets to pretend to be the targeted VM. Depending on the nature and volume of the attack, either the pretense is upheld or the traffic is dropped. The pretense has the quarantine VM respond as if it is the target host for a calculated time to simulate the attacked machine eventually failing. This pretense method allows the operator to use the collected information to black-list the attacker very accurately. This method is repeated should the attack characteristics change, allowing for agile responses to changing circumstances. Evaluation of the system shows very high accuracy for multiple flows to one server, but multiple flows to multiple servers cause a drop in accuracy. However, during a simulated attack Dolus was able to keep a service up and running throughout, which is an improvement over an MTD-based defense, which causes the target to be down for about six seconds.

21.2.7 Distributed SDN Architectures

A variety of IoT networks have been deployed, such as cellular, Wi-Fi, ZigBee, and LoRa, for various applications [10–13]. These networks bear multiple-access technologies with various routing protocols but do not have a seamless communication platform to interconnect them. From their research, Qin et al. [14] proposed to design a software-defined approach for IoT to dynamically achieve differentiated quality levels in a heterogeneous wireless network. The developed IoT controller can incorporate and support flexible scheduling over multihop and heterogeneous ad hoc wireless paths. The preliminary simulation results show that the extended system can support efficient exploitation of different IoT capabilities. There are many other research efforts in the flexibility and scalability of the distributed SDN architecture. The research by Kotronis et al. [15] considered the feasibility of deploying SDN-based solutions incrementally to current BGP-based Internet. To optimize network-routing decisions, the routing of control planes' functionalities from multiple individual domains is outsourced to a logically centralized, multi-AS (Autonomous System) routing-control platform. The platform can make efficient routing decisions, detect policy conflicts, and troubleshoot routing problems. In addition, ONOS [12] and HyperFlow [13] are two SDN systems that realize logically centralized but physically distributed architecture. Specifically, ONOS is designed to operate as a group of identical nodes and thus can withstand failure of individual nodes from disruptions of the whole network control and operation. The platform provides a number of high-level abstractions for its applications that can learn about the network state and thus control the traffic flow through the network. The network graph abstraction summarizes the information about the network topology and structure. The Intent abstraction allows programmers and end users to specify what they wish to forward in the traffic instead of how to do it.

21.2.8 SDN in MANET

SDN's premise – decouple the control plane from the data plane – can facilitate the deployment of new network services and protocols because it enables a paradigm shift from traditional fixed network functionality to programmable network management. SDN can bring the programmability not only in a traditional wired-backbone network but also in MANETs, where network operations are carried out without infrastructure. The work by Poularakis et al. [16] proposed a set of new architecture designs for SDN-enabled mobile ad hoc networks in the tactical field. The work discussed how to leverage SDN to design tactical ad hoc networks: where to place and how to organize the SDN control logic in the network. Poularakis et al. concluded that SDN-based MANETs are feasible but have certain limitations and overheads. The study by Bellavista et al. [17] proposed a model and an architecture that loosely federate MANET and FiWi (fiber-wireless). The main idea is to deploy two types of controllers, FiWi controller and MANET controllers, each in charge of managing its own network but both cooperate with each other for critical information sharing. The implementation results show the potential to make better decisions based on requirements specified by applications.

21.3 Design

21.3.1 Threat Model

A successful cyberattack normally consists of several stages: reconnaissance, scanning, compromising, and escalation. The detection of an attack at its early stages, such as network scanning to discover open ports, can significantly decrease the chances of the attack escalation. IoT devices are of particular interest for the attacks that use network scanning because they have certain ports open all of the time [7, 8]. In our threat model, attackers attempt to compromise multiple IoT devices simultaneously in different networks that are connected by a few programmable switches such as OpenFlow switches. Instead of using IoT devices directly, honeypots are used to emulate them. The switches are further controlled by many logically centralized controllers. We also assume the honeypots that are under attack have not been compromised yet. Namely, we primarily focus on stopping attacks when they are still in scanning and compromising stages. We also posit that the honeypots in the networks are able to detect attacks happening through the information collected by their applications (e.g. system logs). They can extract attackers' identity information (e.g. source IP addresses or port numbers). Also, these honeypots can run custom programs, such as smart SDN agent, to communicate with the SDN controller. Normally, such devices are designed to be powerful enough to run our SDN agent, which requires relatively small central processing unit (CPU) resources. Note that our design does not require other real IoT devices to run a user program to identify attacks due to the low resource capabilities of the small devices. However, in our proposed topology those devices will be protected, as well, thanks to the predeployed honeypots in the network.

21.3.2 Overall Design

In the design, we leverage SDN to build a scalable framework that enables cyber-security deception and cooperative IoT security. Figure 21.1 illustrates the high-level architecture. When an attacker attempts to attack multiple honeypots located in different networks, assume one of them – an emulated mobile phone – can detect such attacks. The phone runs a user program to compose a message that contains the attacker's identity (e.g. IP address). The message then is shared with the SDN controller over User Datagram Protocol (UDP) or TCP. After confirming the victim is a registered IoT system, an SDN application running on the controller is used to compose an OpenFlow switch message. The message is then issued to a programmable forwarding device coupled with the controller (in this example). Once the command takes effect, the attack will be stopped and other IoT systems in the network no longer will not receive similar malicious attempts. Meanwhile, the attack information from the victim is shared with other controllers that cover different networks. The other controllers perform the same tasks as the first controller did when they receive the information: issuing controlling commands to their programmable gateways to filter out malicious traffic. As we can see from Figure 21.1, an attack to the smart watch in the second network is blocked because

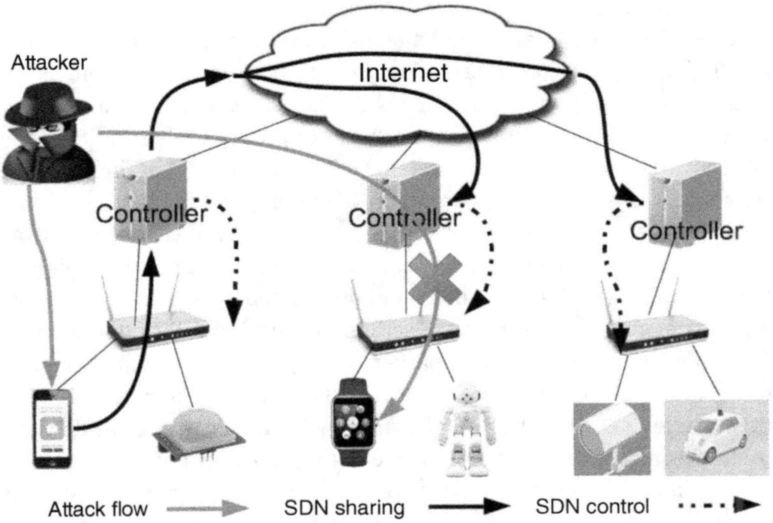

Figure 21.1 Architecture of cooperative IoT security.

of the cooperation. The following sections present the detailed designs for corresponding interfaces, functions, and protocols.

21.3.3 Protocol Design

A controller needs to handle many events; therefore, when we design the protocol between controllers and victims a few modules are considered in the controller, as follows:

- *Initialization:* During boot-up, the controller needs to initialize itself and the programmable gateway (e.g. an OpenFlow switch), where a default table-miss flow entry needs to be installed. Because initially the flow table is empty and no data packets can be forwarded between end hosts, the default flow entry can direct all received packets to the controller, which subsequently processes these packets based on their types and content.
- *Registration:* A controller needs to ensure the information it receives is trustworthy; thus, whoever shares attack information with it should register it at the controller. The IoT systems that have good resource capabilities and can detect attacks need to be registered. Otherwise, a malicious user or system can fool the controller by sharing arbitrary false attack information. Other controllers that want to be shared with and are willing to share attack information need

to be registered, as well, to benefit their networks. The controller uses a specific port to receive messages from them and maintains a list of registered IoT systems and controllers that can share information with each other. In addition, as an acknowledgment message, the controller sends a list of required parameters such as passcode, timestamp, source IP address, and port number to the registrars. When an attack occurs, the registered victim needs to fill out the required parameters in a data packet and share them with the controller.

- *Normal Packet Handling:* Two types of normal packets are handled by the controller: ARP and IP. Upon receiving an ARP request message, the controller replies to the requester with the MAC address of the unknown IP address. For an IP packet, which contains source and destination IP addresses, the controller finds the corresponding output port and forwards the packet to that port. The controller does not do this for every packet; rather, the controller installs a flow entry in the flow table of the programmable gateway. Subsequently, an IP packet that has the same destination IP address can be routed by the flow rule in the flow table to its next hop.
- *Attack-Information Message Handling:* One of the most important jobs for the controller is to handle messages from the registered IoT systems, which as victims under attack may send the attack information to the controller for help. Upon receiving such a message, the controller needs to confirm the sender is a registered and trustworthy host. It is easy to confirm a registered host since the host sends to the controller the passcode assigned during the registration. In addition, the host's IP address must be found in the list of IP addresses of all registered hosts that the controller maintains. It is relatively tricky to confirm a trustworthy host; the challenge is for the controller to tell whether the host is faking a source IP address when asking the controller to block it. In other words, the controller needs to see certain evidence to prove the reported suspicious traffic exists. Recall that any IP packet with a previously unknown source or destination IP address needs to be forwarded to the controller by the default table-miss rule installed in the programmable gateway; thus, the controller keeps a record of prior communication of the reported attack if it exists. If a victim reports an attack, the controller searches a relevant record; if such a record exists, the controller trusts the attack is ongoing and takes actions to block the attack. Otherwise, the controller temporarily releases the traffic and keeps monitoring. This approach may not be able to fully stop a compromised host from reporting a false alarm, but it is a good level of safeguard to ensure the host does not misuse its privileges. On the other hand, if the message is from a neighbor controller, the controller will conduct similar checks. It is possible the neighbor network has not yet been under attack by the same source, so the neighbor controller will not install the flow entry immediately to the flow table in the programmable gateway. However, it will keep the attacker's information in its repository. When a new data packet, originating from the same source, appears

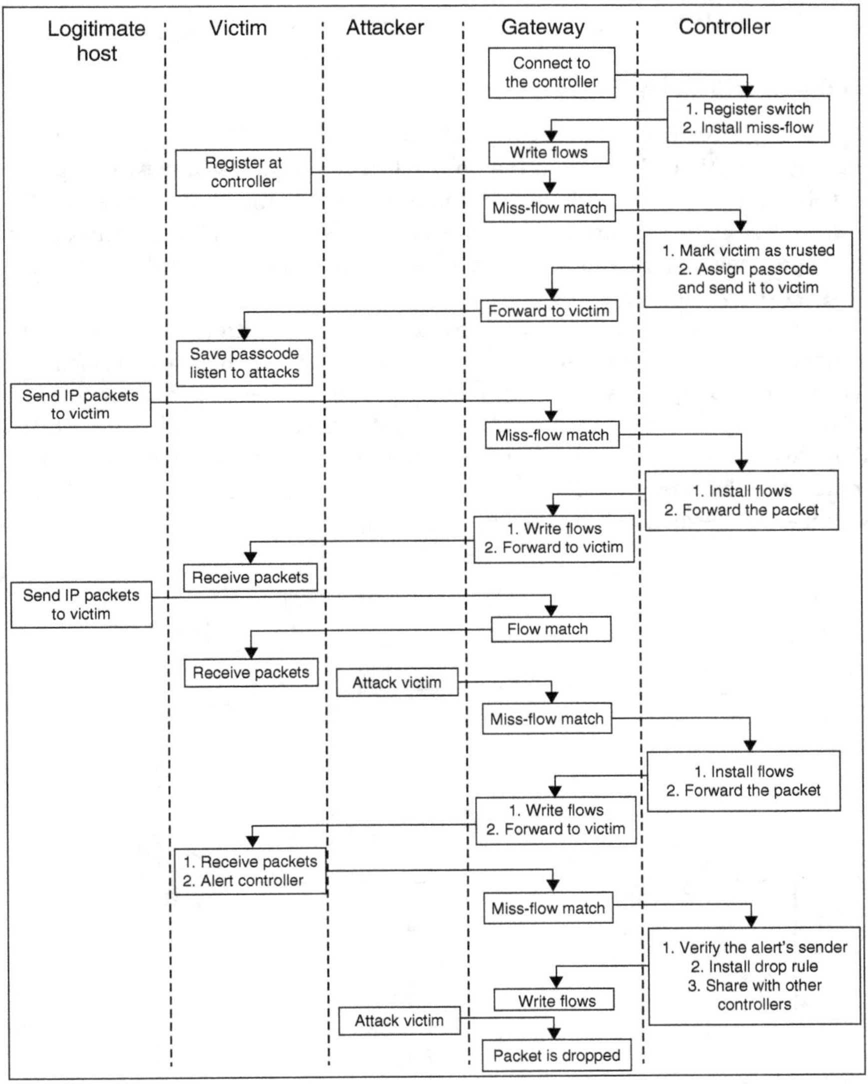

Figure 21.2 SDN workflow.

at the controller, it will install the flow rules to the programmable gateway to stop the attack. As a result, both the entire victim and neighbor networks can benefit from such information sharing and collaboration.

Figure 21.2 shows the workflow among different components.

21.4 Evaluation

21.4.1 Methodology

We built up an experimental environment in a GENI testbed [14] to emulate cyber-security deception. GENI is an open infrastructure for at-scale networking and distributed systems research and education that spans the United States. In the testbed, two controllers, two programmable gateways (OpenFlow switches), one honeypot, one victim, and one attacker are emulated and located on different VMs, as shown in Figure 21.3.

The attacker is connected to two networks, which are controlled by the two controllers. Each of the controllers operates one OpenFlow switch on which Open Virtual Switch (OVS) software is running. We set up the experiments as follows: initially, both controllers are initialized and install miss-flow entries in their OpenFlow switches. After that, the honeypot and victim communicate with their controllers and the two controllers talk to each other to register them. Eventually, the victim, the honeypot, and the peer controllers are added to the list of trusted hosts.

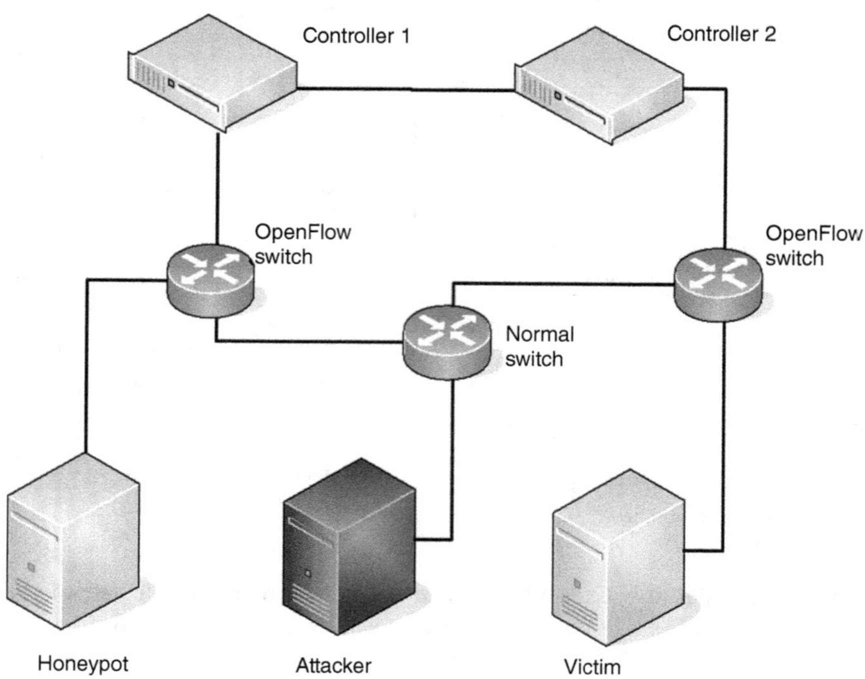

Figure 21.3 Experimental topology.

Subsequently, the attacker launches the attack on the honeypot and the first packet is redirected to the first controller. The controller installs corresponding flow entries to the first OVS and lets the following packets from the same source go through. When the attacker's packets arrive at the honeypot, it immediately discovers it is an attack. Instantly, the victim extracts the attacker's identity information and sends it to the first controller. After verifying the honeypot as "trusted" and that the attacker's data trace was recorded previously, the controller issues OpenFlow commands to stop the attack and meanwhile shares the attack information with the peer controller, which takes similar actions to block the attack. Based on this scenario, we measured (i) the time when the attacker's packet reached the controller, (ii) the time when the honeypot detected the attack, (iii) the time when the first controller obtained the attack information, (iv) the time when the flow entries were installed in OVS-1, (v) the time when the second controller received the attack information, and (vi) the time when the flow entries were installed in OVS-2. We used the iPerf client tool [15] to simulate the attack.

The results from the GENI environment:

1) *Attack Alert Time:* Figure 21.4a shows the attack alert time from the victim to the controller. It is the duration from when the attack was discovered by the honeypot to when the controller received the alert message. On average, it takes 520 ms to alert on a new attack for the 10 experiments.
2) *Flow-Installation Time on OpenFlow Switches 1 and 2:* It is challenging to obtain the flow-installation time because it requires highly precise time synchronization between the switch and the controller. The controller can record the time when the rule was issued out to the flow queue, where rules are waiting to be installed in the flow table, but it cannot keep track of when the flow entry was successfully installed in the switch's flow table. In our experiment, we addressed this problem by the following steps. First, we set a hard timeout value for the flow entry and issue the installation command to the OpenFlow switch. When the flow entry expires, it generates an event that is handled by the controller. The timestamps of the flow issuing and the event-handling procedure call can be captured at the controller. Thus, we can calculate the total round-trip time from when the flow rule was issued to when the expiration message was received. In order to calculate the flow-installation time, we need to subtract the hard timeout value and the single trip time from the OpenFlow switch to the controller. The single trip time can be obtained by halving the round-trip time of a packet traveling from the controller to the switch and then back to the controller. Finally, Figure 21.4b demonstrates the flow-installation time from Controller 1 to OVS-1. On average, it takes about 46 ms for the 10-round experiments to install a flow rule in OVS-1. Figure 21.5b shows similar results in OVS-2.

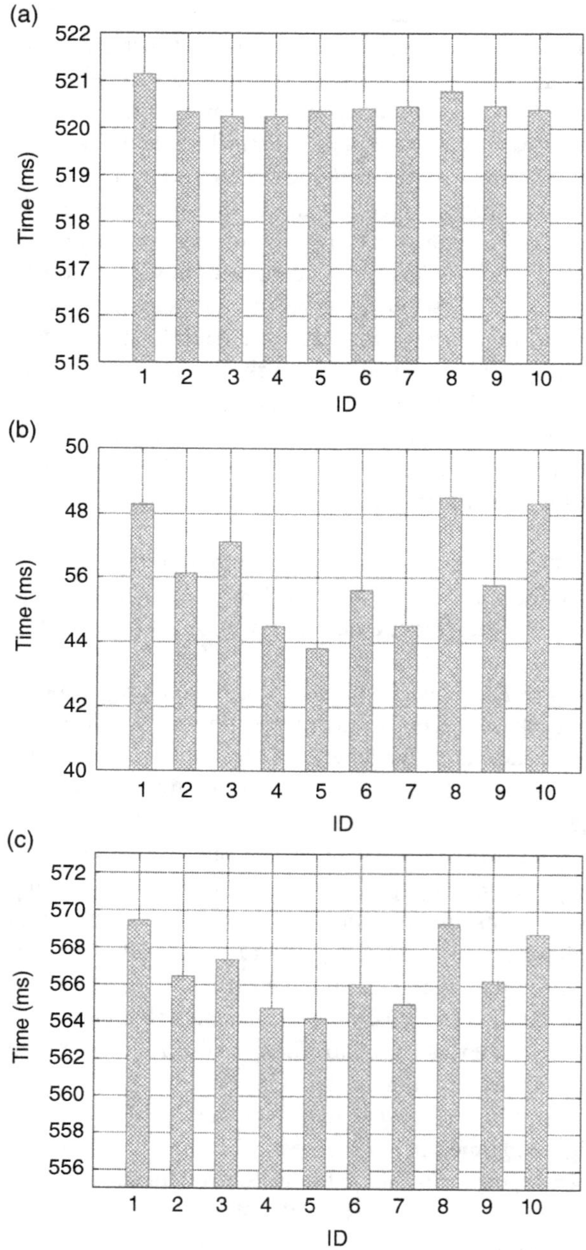

Figure 21.4 GENI experiments (Controller 1's network). (a) Attack alert time. (b) Flow installation time on OVS-1. (c) Total time to defend the network.

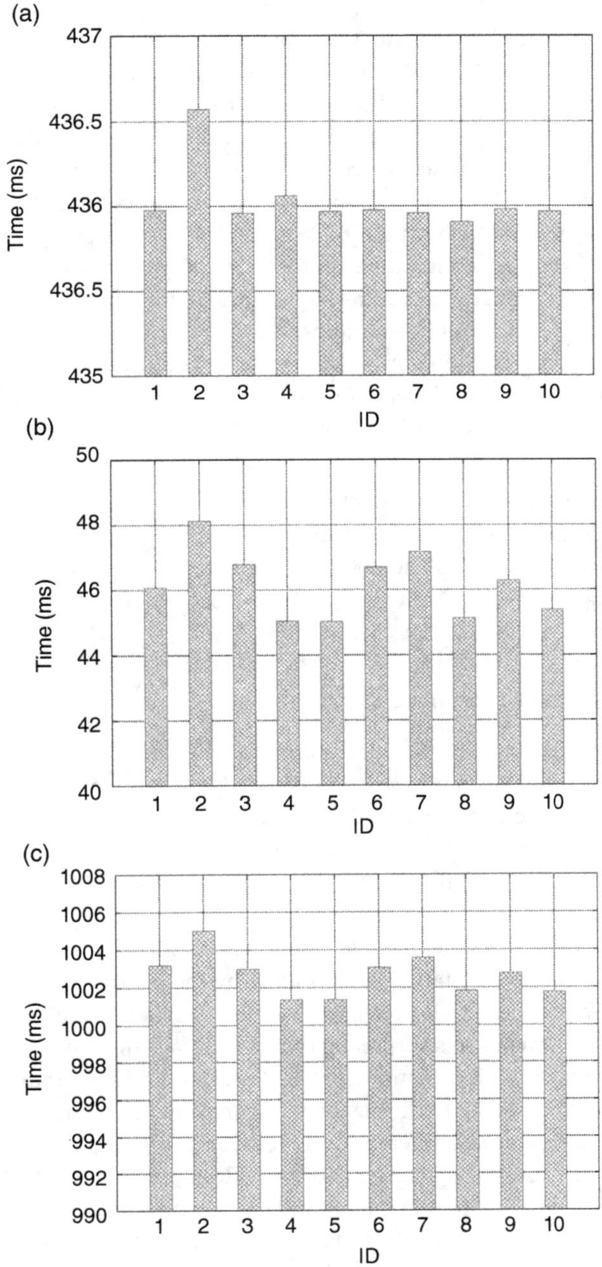

Figure 21.5 GENI experiments (Controller 2's network). (a) Information sharing time between two controllers. (b) Flow installation time on OVS-2. Total time to defend the second network.

3) *Attack-Information Sharing Time:* It can also be called propagation or delay time from one controller to another. Figure 21.5a shows the time period from when Controller 1 starts sharing the attack information to when Controller 2 accepts the information. The average sharing time is around 436 ms. When there are multiple controllers who want to share the same attack information, the delay determines the overall performance of the system. The delay also depends on the type of network topology being used. For example, the efficiency of a ring topology will be higher than that of a straight-line topology.
4) *Total Time:* Figure 21.4c shows the total time consumed to stop the malicious traffic for a single-controller scenario, including alert time and flow-installation time. On average, it takes 566 ms to block the attack in a virtual environment on GENI. Figure 21.5c shows the total time in a two-controller scenario, where it takes about 1003 ms to finish the whole loop. From the results, we can observe that the system is scalable, particularly when multiple controllers are well connected to each other.

In addition to the GENI experiment, we conducted similar experiments in a real hardware environment that included a Pica8 OpenFlow switch [18], a controller with Intel(R) Core(TM)2 Duo CPU E8400 at 3.00-Ghz machine, and an attacker and a honeypot on two Raspberry Pis [19] with ARMv7 Processor (rev 4 (v71)). We performed 10 rounds of experiments for each setting.

21.4.2 Results in Hardware Environment

1) *Attack Alert Time:* Figure 21.6a shows the duration from when the attack was detected by the honeypot to the time when the controller learns about the attack from the honeypot. The average alert time is 306 ms, which is 40% less than in the virtual environment. This is because the hardware implemented a flow-matching mechanism in Pica8 that operates significantly faster than the algorithm implemented in OVS.
2) *Flow-Installation Time on Pica8 OpenFlow Switch:* Figure 21.6b shows the interval between issuance of the flow-installation command by the controller and the actual installation on the Pica8 switch. We used similar calculation techniques as in the GENI environment. It is surprising that the average time of flow installation on the hardware switch is 50 ms, which is slower by almost 9% than that in the OVS in GENI. The slower performance in the real environment can be explained by two main reasons: (i) different from flow-matching operations, writing on hardware such as ternary content-addressable memory (TCAM) chips is relatively slow because in order to guarantee fast matching, the entries on TCAM need to be sorted properly [16, 17]; and (ii) resources obtained in the GENI experiment were located on the same hardware, but in the real environment the controller and the OpenFlow switch are connected via a physical link.

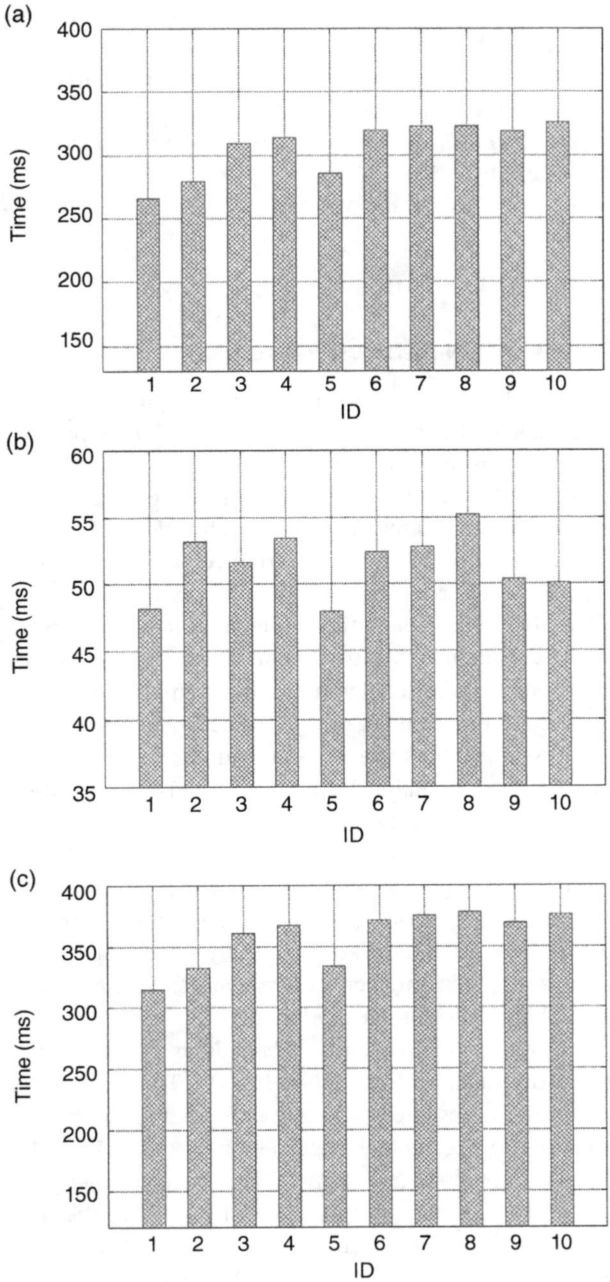

Figure 21.6 Hardware experiments. (a) Attack alert time. (b) Flow installation time. (c) Total time.

3) *Total Time:* Figure 21.6c shows the total time needed to detect the attack and install the rules in the Pica8 switch so the switch will drop all subsequent packets from the attacker. The average total time is 358 ms. The same measurement in the GENI environment is about 566 ms, as shown in Figure 21.4c. Overall, the performance of our solution in the hardware environment outperforms the one in the virtual environment by 37%.

21.5 SDN for Distributed Networks

21.5.1 Technical Challenges of SDN Distributed Networks

SDN is an emerging network architecture featuring programmable network management, flexible configuration, and efficient resource allocation. One important attribute of SDN, compared with conventional networks, is the separation of control plane and data plane. In particular, for the best decision-making, all information is collected to the controller, which can realize global optimization of network performance. However, a single-controller network architecture suffers from severe scalability issues. As a network size increases, the management and operational complexity as well as the volume of traffic to the centralized controller grow significantly, and eventually the network performance will degrade due to the high computational and communication overhead by a single controller. Therefore, distributed SDN is proposed to mitigate the scalability issues. In this new framework, multiple SDN controllers are present and each controller is responsible for a set of subnetworks or domains. In this distributed-SDN architecture, controllers exchange network-status information through proactive probing and passive listening. On the one hand, complete synchronization among all controllers may incur high costs in terms of traffic load and status synchronization; on the other hand, IoT devices are mobile and heterogeneous. The challenge is the interoperability among those devices when there are different types of data format, protocols, and resources. The varying performance requirements such as throughput, latency, and sensitivity enhance the complexity for such architecture.

Our previous experiments have shown the potential of leveraging SDN to enable cyber-security deceptions in a traditional infrastructure-based IoT environment. However, there still are many challenges to extending the programmable control and cyber-deception capabilities to the mobile edge networks (e.g. MANET). Firstly, there is no fixed network topology in a MANET environment, network links may be up and down, and nodes may join and leave frequently. The first key question is where to place and how to organize the network control in a MANET network. If a controller is placed too far away from forwarding nodes, they may not be able to receive timely instructions on network operations from the controller. If placed too close, a large number of controllers may be needed to accommodate communication needs and may also introduce scalability issues. Secondly, even if we can figure out where to install SDN controllers, another challenge is how to enforce cyber-security deceptions; that is where to place decoys, since network nodes themselves may be able to communicate with each other in a multihop manner and malicious traffic may still be able to penetrate the MANET environment without the coordination from the SDN controller. Thirdly, the SDN controller should have the global view of the network, such as link status and topological information. A challenge is obtaining such information in an efficient way. If network state changes (e.g. link failure), how can the controller make the best routing decision for traffic forwarding without causing network congestion and introducing high communication overhead? A potential solution is to let each data-plane node intelligently and autonomously react to network changes without much communication overhead with the controller, but the controller is still aware of what routing decision has been made by individual forwarding node. Finally, real-time network traffic monitoring in a MANET is important but also extremely challenging. With accurate traffic activities of the network, SDN controllers can acquire full knowledge of the network status so as to take proper management decisions when an emergency event occurs in the network. The close cooperation among multiple distributed SDN controllers may be needed to make this happen.

21.5.2 Potential Solutions with Information Centric Network (ICN) and SDN

Named Data Networking (NDN) [20] has been playing an important role as one of the Information Centric Networking architectures. In NDN, objects and nodes are defined by names instead of addresses (e.g. IP addresses). When retrieving content, one does not need to know "where" to obtain the data by specifying destination IP addresses. Instead, a consumer only needs to express an interest with "what" to retrieve. The NDN network will forward the interest to the corresponding producers to satisfy the request. NDN features named data, in-network caching, multipath routing, and built-in security. Different from the conventional TCP/IP-based approach, NDN-enabled routers have a content store where pass-through content can be cached temporarily. When there are multiple requests with the same name, only one copy of data is adequate to satisfy all interests. Another important feature of NDN is its flexible forwarding strategies at the data plane. A forwarding strategy can define where and how to forward an interest with a single path or multiple paths (interfaces). The feature is extremely important and effective over a MANET environment with many heterogeneous wireless links available. Cyber-deception techniques, such as camouflaging and decoy, are easier to deploy. In an information-centric network, nodes or devices are not identified by IP addresses; camouflage can be more stealthy and deceitful because the paths toward a particular node are harder to discover and identify. In addition, since there is no destination or location concept in the network, any node can be a potential data producer; decoying techniques can be more effective. In terms of handling network dynamics, NDN can define smart-forwarding strategies that can directly control how traffic is forwarded toward which interface without maintaining a global routing or forwarding table. SDN capabilities can be easily extended in this type of network as well. In NDN, the data plane can work independently and adaptively without the involvement of a control plane [21], which indicates a control plane can be flexibly placed in a needed place when possible but, meanwhile, has a global overview of the network state. Therefore, a promising solution for cyber deception is to deploy it in an SDN–NDN-mixed environment [22].

We use both decoy and camouflage examples to illustrate that the combination of SDN and NDN may lead to effective cyber-deception architectures for distributed and mobile networks. For a decoy example, we can build content providers with diverse content names in a NDN network. We assume both legitimate users and malicious users each have a name or ID (e.g. public key), so that the network knows what traffic has been carried or requested by which name. Also, we assume legitimate users know the correct content names to retrieve but malicious users must guess the right names to access the content providers to obtain the right content. In this case, we can deploy many decoy nodes that provide fake content names. Since they do not provide services for legitimate users, any traffic to these decoy nodes is potentially malicious. Once receiving such requests, the decoy nodes can report them to the SDN controller, which can eventually block the malicious traffic from entering the network or redirect it to a dedicated intrusion-detection system for further inspection. For a camouflage example, in a traditional TCP/IP network a popular practice is to change the target IP addresses so the attackers are not able to figure out the real locations of the targets. In NDN scenarios, instead of changing IP addresses, we can easily change the name-mapping to the content at different times so that attackers may not be able to use the same names to retrieve the desirable content. For legitimate users, they can coordinate with the SDN controller about the naming change so as to retrieve the right content accordingly. However, this is not to say there are no challenges for SDN–NDN-combined architecture. In this new architecture, attackers may hide themselves easier through fake names so that the network may not easily block them. Due to the use of names, malicious users may come from inside the network and obtain the name exchanges silently, then eventually figure out the right names for attacks. Name confidentiality is a big challenge to enable secure information exchange in an information-centric network.

21.6 Conclusion

In this chapter, we conducted a literature review in cyber-security deception areas and introduced a trustworthy, cooperative, and scalable architecture to enable cyber-security deception and IoT security among multiple networks. The architecture is powered by SDN technologies, where the controller application can take input about malicious activities from its honeypots and translate their requirements to the network-level flow rules to stop attacks quickly. The architecture does not only benefit the victims under attack but any other potential targets in the networks, as well. In addition, we tried to mitigate the trust problem through double-checking the traces of the malicious traffic. Meanwhile, we measured the time spent in each phase in both virtual and real environments. The results show the overall hardware implementation outperforms the same implementation in a virtual environment. Finally, we discussed the challenges and potential solutions to enable cyber-security deceptions in an SDN-based MANET environment.

References

1 "New Threat Report: A New IoT Botnet Is Spreading over HTTP 81 on a Large Scale," http://blog.netlab.360.com/a-new-threat-an-iot-botnet-scanning-internet-on-port-81-en.

2 "IoT Security News Issues with Authentication and Open Ports," https://www.securerf.com/iot-security-news-issues-with-authentication-and-open-ports.

3 "What Is GENI?" http://www.geni.net/about-geni/what-is-geni.

4 "iPerf—The Ultimate Speed Test Tool for TCP, UDP and SCTP," https://iperf.fr.

5 "PicOS—Pica8," http://www.pica8.com/products/picos.

6 "Raspberry Pi," https://www.raspberrypi.org/about.

7 Carroll, Thomas E., and Daniel Grosu. "A game theoretic investigation of deception in network security." *Security and Communication Networks*, vol. 4, no. 10, pp. 1162–1172, 2011.

8 Kiekintveld, Christopher, Viliam Lisý, and Radek Píbil. "Game-theoretic foundations for the strategic use of honeypots in network security." In *Cyber Warfare*, pp. 81–101. Springer, Cham, 2015.

9 Bifulco, Roberto, and Anton Matsiuk. "Towards scalable SDN switches: Enabling faster flow table entries installation." *ACM SIGCOMM Computer Communication Review*, vol. 45, no. 4, pp. 343–344, 2015.

10 Wen, Xitao, Bo Yang, Yan Chen, Li Erran Li, Kai Bu, Peng Zheng, Yang Yang, and Chengchen Hu. "RuleTris: Minimizing rule update latency for TCAM-based SDN switches." In *2016 IEEE 36th International Conference on Distributed Computing Systems (ICDCS)*, pp. 179–188. IEEE, 2016.

11 Achleitner, Stefan, Thomas La Porta, Patrick McDaniel, Shridatt Sugrim, Srikanth V. Krishnamurthy, and Ritu Chadha. "Cyber deception: Virtual networks to defend insider reconnaissance." In *Proceedings of the 8th ACM CCS International Workshop on Managing Insider Security Threats*, pp. 57–68. ACM, 2016.

12 Neupane, Roshan Lal, Travis Neely, Nishant Chettri, Mark Vassell, Yuanxun Zhang, Prasad Calyam, and Ramakrishnan Durairajan. "Dolus: Cyber defense using pretense against DDoS attacks in cloud platforms." In *Proceedings of the 19th International Conference on Distributed Computing and Networking*, p. 30. ACM, 2018.

13 Clark, Andrew, Quanyan Zhu, Radha Poovendran, and Tamer Başar. "Deceptive routing in relay networks." In *International Conference on Decision and Game Theory for Security*, pp. 171–185. Springer, Berlin, Heidelberg, 2012.

14 Qin, Zhijing, Grit Denker, Carlo Giannelli, Paolo Bellavista, and Nalini Venkatasubramanian. "A software defined networking architecture for the internet-of-things." In *2014 IEEE Network Operations and Management Symposium (NOMS)*, pp. 1–9. IEEE, 2014.

15 Kotronis, Vasileios, Xenofontas Dimitropoulos, and Bernhard Ager. "Outsourcing the routing control logic: Better internet routing based on SDN principles." In *Proceedings of the 11th ACM Workshop on Hot Topics in Networks*, pp. 55–60. ACM, 2012.

16 Poularakis, Konstantinos, George Iosifidis, and Leandros Tassiulas. "SDN-enabled tactical ad hoc networks: Extending programmable control to the edge." *arXiv Preprint* arXiv:1801.02909, 2018.

17 Bellavista, Paolo, Carlo Giannelli, Thomas Lagkas, and Panagiotis Sarigiannidis. "Multi-domain SDN controller federation in hybrid FiWi-MANET networks." *EURASIP Journal on Wireless Communications and Networking*, vol. 2018, no. 1, p. 103, 2018.

18 Berde, Pankaj, Matteo Gerola, Jonathan Hart, Yuta Higuchi, Masayoshi Kobayashi, Toshio Koide, Bob Lantz, et al. "ONOS: Towards an open, distributed SDN OS." In *Proceedings of the Third Workshop on Hot Topics in Software Defined Networking*, pp. 1–6. ACM, 2014.

19 Tootoonchian, Amin, and Yashar Ganjali. "Hyperflow: A distributed control plane for OpenFlow." In *Proceedings of the 2010 Internet Network Management Conference on Research on Enterprise Networking (INM/WREN'10)*, pp. 3–3. USENIX Association, Berkeley, CA, USA, 2010.

20 Zhang, Lixia, Alexander Afanasyev, Jeffrey Burke, Van Jacobson, Patrick Crowley, Christos Papadopoulos, Lan Wang, and Beichuan Zhang. "Named data

networking." *ACM SIGCOMM Computer Communication Review*, vol. 44, no. 3, pp. 66–73, 2014.

21 Yi, Cheng, Alexander Afanasyev, Lan Wang, Beichuan Zhang, and Lixia Zhang. "Adaptive forwarding in named data networking." *ACM SIGCOMM Computer Communication Review*, vol. 42, no. 3, pp. 62–67, 2012.

22 Mishra, Vinod. *Coexistence of Named Data Networking (NDN) and Software-Defined Networking (SDN)*. Army Research Laboratory (US), Adelphi, MD, 2017. Technical Report.

23 Sun, Jianhua, Kun Sun, and Qi Li. "CyberMoat: Camouflaging critical server infrastructures with large scale decoy farms." In *2017 IEEE Conference on Communications and Network Security (CNS)*, pp. 1–9. IEEE, 2017.

24 Jajodia, S., N. Park, F. Pierazzi, A. Pugliese, E. Serra, G. I. Simari, and V. S. Subrahmanian. "A probabilistic logic of cyber deception." *IEEE Transactions on Information Forensics and Security*, vol. 12, pp. 2532–2544, 2017.

22

Decentralized Access Control for IoT Based on Blockchain and Smart Contract

Ronghua Xu[1], *Yu Chen*[1], *and Erik Blasch*[2]

[1] *Department of Electrical and Computer Engineering, Binghamton University, SUNY, Binghamton, NY, USA*
[2] *US Air Force Research Laboratory, Rome, NY, USA*

22.1 Introduction

Internet of Things (IoT) technology significantly transforms our daily life by ubiquitously providing data rich applications and services covering transportation, healthcare, industrial automation, emergency response, and so on [1]. These "smart things" include everything from resource-constrained devices, like temperature measuring sensors; to powerful things, such as smart phones, smart meters, etc. Through the existing Internet infrastructure and communication standards, intelligent devices see, hear, and touch each other to perform cooperative tasks like collecting data, sharing information, and coordinating decisions. These capabilities enable context awareness, of which smart things sense and analyze surrounding environment, and interact with peers according to situational context changes. In another words, IoT stands for "a world-wide network of interconnected objects uniquely addressable, based on standard communication protocols" [2].

IoT has demonstrated a great market potential with an increasing growth rate in recent years. In 2014, results estimate that there were 1.5 billion Internet-enabled computers and one billion Internet-enabled mobile phones. By 2020, the number of Internet connected IoT devices will be somewhere between 50 and 100 billion [3]. The global market for sensors will total $283.4 billion in 2023 increasing from $152.2 billion in 2018 at a compound annual growth rate of 13.2% from 2018 through 2023 [4]. The large volumes of smart things are

Modeling and Design of Secure Internet of Things, First Edition. Edited by Charles A. Kamhoua, Laurent L. Njilla, Alexander Kott and Sachin Shetty.

Published 2020 by John Wiley & Sons, Inc.

utilized by many service-oriented platforms, such as clouds, to provide IoT oriented services and applications. Those IoT oriented services support many different application domains, such as on-line business, smart-grid energy, smart-home control, and intelligent-city coordination. Overall, IoT oriented services can be categorized into four classes [1]:

- *Identity-Related Services:* provide basic and important identity services, such as biometric authentication, and recognition. The logistics applications that utilize radio-frequency identification (RFID) technology belong to this field.
- *Information Aggregation Services:* collect and summarize raw sensory measurements processed by the IoT application. For example, the smart grid aims to improve and enhance the energy consumption of houses and buildings by utilizing the IoT technology.
- *Collaborative-Aware Services:* use the obtained data from Information Aggregation Services (IAS) to perform high-level operations, like environmental sensing, feature extraction, and decision-making. A smart home is one of the typical applications to improve the quality of people's life by remotely monitoring and operating home devices and systems (e.g., air conditioner, heating systems, energy consumption meters, etc.).
- *Ubiquitous Services:* aim to provide Collaborative-Aware Services (CAS) anytime, anywhere, and to anyone. Smart cities is considered as one application of ubiquitous services, which aims to improve the quality of life in the city by making it easier and more convenient for the residents to find information of interest [5, 6].

While benefiting from the large-scale applications like Smart Gird and Smart Cities, highly connected smart IoT devices with heterogeneous platforms and insufficient security enforcement incur more concerns on safety and privacy. Security issues, such as confidentiality, authentication, access control, availability, non-repudiation, integrity, and privacy, are the main challenges that these IoT-based services and applications are facing [7]. Conventional access control approaches, like Access Control List (ACL), Role-based Access Control (RBAC), and Attribute-based Access Control (ABAC) have been widely used in information technology (IT) systems. However, they are not able to provide a manageable and efficient mechanism to meet the requirements raised by IoT networks:

- *Scalability:* Access control strategies must handle the connection problem resulting from a fast growing number of devices and services in distributed IoT networks.
- *Heterogeneity:* Access control mechanisms should be flexible to provide complex arrangements and flexible configurations for IoT systems, which normally

integrate heterogeneous cyber physical objects with variant underlying technologies or in different application domains.

- *Causality:* The IoT world is mainly characterized by short-lived, often causal and/or spontaneous interactions, which require access control schemes to deal with dynamic challenges based on the changing situational awareness.
- *Lightweight:* IoT devices are usually resource-constrained and communicate to each other under a low power in a noisy network environment. Consequently, access control protocols should be lightweight and not impose significant overhead on devices and communication networks.

The extraordinary large number of devices with heterogeneity and dynamicity necessitate more scalable, flexible and lightweight access control mechanisms for IoT networks. In addition, the majority of access control solutions rely on a centralized architecture, which incur many problems, such as performance bottlenecks and single point of failures. IoT networks need a new access control framework that provides decentralized authentication and authorization in trustless application network environments, such that intelligence diffuses among a large number of distributed edge devices. *Blockchain*, which is the fundamental technology of Bitcoin [8], demonstrates potential to revolutionize the fundamentals of IT technology because of its key features such as decentralization and anonymity. A *smart contract* [9] based on a blockchain protocol achieves agreement among parties, as opposed to relying on third parties for maintaining a trust relationship. By encapsulating operational logic as a bytecode and performing Turing-complete computation on distributed miners, a smart contract allows a user to transcode more complex business models as new types of transactions on a blockchain network. The smart contract provides a promising solution to implement more flexible and fine-grained access control models on blockchain networks, but requires adaptation for IoT applications.

This chapter provides readers a decentralized framework for IoT security based on blockchain and smart contract methods. Particularly, the chapter details access control mechanisms for distributed IoT system development for untrusted network environments. The rest of this chapter is organized as the follows. Section 22.2 briefly discusses the state of art in access control solutions for IoT systems, followed by an introduction to blockchain and smart contract in Section 22.3. Section 22.4 presents current research on blockchain-based security solutions to IoT, and lists the main challenges by evaluating the pros and cons of each of the proposed approaches. Section 22.5 provides a case study of decentralized access control mechanism for IoT with experimental results for more insight. Finally, Section 22.6 summarizes the chapter with a brief discussion on the future research directions in IoT security.

22.2 Access Control in IoT

An *Access Control* (*AC*) mechanism, which specifies admittance to certain resources or services, contributes to the protection, security, and privacy for IoT devices. As a fundamental mechanism to enable security in computer systems, AC is the process that decides who is authorized to have what communication rights on which objects with respect to some security models and policies [10]. An effective AC system is designed to satisfy the main security requirements, such as confidentiality, integrity, and availability. In general, a comprehensive access control system addresses three main security issues: authentication, authorization, and accountability [11]. This chapter focuses on authorization, which is considered as a narrow conception of AC. *Authorization* involves the following phases: defining a security policy (set of rules), selecting an access control model to encapsulate the defined policy, implementing the model, and enforcing the access rules [11]. The authorization process incorporates two aspects: the authorization model and architecture.

22.2.1 Authorization Model in IoT

Various AC methods and solutions with different objectives exist to address IoT security challenges. Several most popular AC schemes used in IoT system are introduced here and the pros and cons of each model are discussed.

22.2.1.1 Role-Based Access Control Model

The *Role-Based Access Control* (*RBAC*) [12] model provides a framework that authorizes user's access to resources based on functions. In a RBAC model, permissions are assigned to each role according to organizational functionality definition, and access rights are indirectly granted by associating a user with certain specified role. The role acts as an intermediary to bring users and permissions together. The RBAC model supports principles, such as least privilege, partition of administrative functions, and separation of duties [13]; and has been widely used in computer systems. The RBAC model implemented in IoT networks adopts a Web of Things (WoTs) framework [14, 15], which is a service-oriented approach, to enforce AC policies on the smart things via a web service application. To enable environment awareness by using context constraints in AC decisions, an extended RBAC model with context information collected from the environment of the physical object enhances the security for web services applications [16].

The RBAC based solutions using the web service technology provide great interoperability between heterogeneous devices. However, they are not able to address the key issues of implementing RBAC in a highly distributed network environment, such as self-management to handle the explosion of roles in complex and ambiguous IoT scenarios and the autonomy for the physical objects through device-to-device communication.

22.2.1.2 Attribute-Based Access Control Model

Compared to the RBAC model, the *Attribute-Based Access Control* (*ABAC*) [17, 18] model defines permissions based on any security relevant characteristics, known as attributes. In the ABAC, AC policies are defined through directly associating pre-defined attributes with subjects, resources, and conditions, respectively. Given all the attributes assignments, a *Policy Rule*, which is a Boolean function, decides whether to grant the subject's access to the resource under specific conditions. The ABAC model eliminates the definition and management of static roles used in the RBAC model; and hence, it also eliminates the need for the administrative tasks for user-to-role assignment and permission-to-role assignment [17]. To address the weaknesses of the RBAC model in a highly distributed network environment, the ABAC model introduced to IoT networks reduces the number of rules resulting from role explosion. Some ABAC extensions to the AWS-IoTAC model enhance the flexibility of access control in cloud-enabled IoT platform [19]. An efficient Elliptic Curve Cryptography (ECC)-based authentication and the ABAC policy together solve the resource-constrained problem of perception layer of the IoT [20].

The ABAC is more manageable and scalable than the RBAC by providing finer-grained AC policies that involve multiple subject and object attributes [17]. However, specifying a consistent definition of the attributes within a single domain or across multiple domains could significantly increase the effort and complexity of policy management as the number of devices grow, and a user-driven and delegation strategy are not supported by ABAC model. Hence, the attribute-based proposal is not suitable for large-scale distributed IoT scenarios.

22.2.1.3 Capability-Based Access Control Model

Capability-based Access Control (*CapAC*) utilizes the concept of capability that contains rights granted to the entity holding it [11]. The capability is defined as tokens, tickets, or keys that give the possessor permission to access an entity or object in a computer system [21]. Two implementations of CapAC are (i) the Access Control Matrix (ACM) model, which provides a framework for describing Discretionary Access Control (DAC) [13], and (ii) the Access Control List (ACL) and capability are widely used in authorization systems.

The ACL and capability are two permission relationships in the ACM model, and Figure 22.1 shows the elements and relationships. In the ACL model, each object is associated with an access control list that records the subjects and their access rights for the objects. Meanwhile, in the CapAC model, each subject is associated with a capability list that represents its access rights to all concerned objects.

ACL

Subject \ Object	Photo and camera	Contact	Gmail	Facebook
Alice Capability	RW	RW	RW	RW
Bob	R	R	#	R
Carol	RW	#	R	#

Figure 22.1 Capability-based access control model.

The CapAC has been implemented in many large-scale IoT-based projects, like IoT@Work [22]. However, the direct application of the original concept of CapAC model in a distributed networks environment has raised several issues, like capability propagation and revocation. To tackle these challenges, a Secure Identity-Based Capability (SICAP) System [10] was proposed to provide a prospective capability-based AC mechanism in distributed IoT-based networks. By using an exception list, the SICAP enables monitoring, mediating, and recording capability propagations to enforce security policies as well as achieving rapid revocation capability [10]. By introducing a delegation mechanism for the capability generation and propagation process, a Capability-based Context-Aware Access Control (CCAAC) model was proposed to enable contextual awareness in federated IoT devices [23]. The federated delegation mechanism enables the CCAAC model advantages in terms of addressing scalability and heterogeneity issues in IoT networks.

The CapAC is a user-driven AC model and supports machine-to-machine (M2M) communication. It presents great scalability and flexibility to deal with spontaneous and dynamic changes in distributed IoT networks. However, management of capability propagation becomes difficult for a proper delegation and revocation mechanism.

22.2.2 Authorization Architecture for IoT

At the architecture level, the IoT authorization solutions are categorized as either the centralized or the decentralized approach.

22.2.2.1 Centralized Architecture

In a centralized AC architecture, the key components, like authorization policy management and policy decision making, employ a centralized authority. Outsourcing computationally intensive tasks to a back-end cloud or a gateway relieves smart device from the burden of handling access control-related functionalities. The approaches in [16, 19, 20, 23] are centralized methods. The centralized AC solutions present many advantages:

- *Easy to Adapt Existing Access Control Model:* convenient to reuse traditional access control models and implement security standard technologies and protocols on a centralized cloud server, which has much more computation power and storage space.
- *Simple Security Policy Management:* Since all identity information and access control policies are stored on a fully trusted central authority, it is easy to define and manage authentication and authorization rules on the centralized framework.

However, enforcing access control on the centralized architecture also suffers from several drawbacks:

- *Single Point of Failure:* Since only one single centralized entity stores and manages all the data from devices, any vulnerability of the server might lead to compromise all sensitive information and make the whole system unavailable.
- *Performance Bottleneck:* Owing to the fact that all data have to be transferred and processed on a centralized authority, it becomes the performance bottleneck to ensure Quality of Service (QoS) for the user.
- *Data Management and Privacy:* Since AC policies and data management are located in a central entity, the user does not have a full control over his/her own data. Therefore, data usage and privacy protection rely on a third-party authority to maintain trust relationship between the users and the central entity.

22.2.2.2 Decentralized Architecture

To address the drawbacks in the centralized access control solutions, designing a decentralized AC mechanism remains a popular research topic. A *distributed Capability-based Access Control (DCapAC)* mechanism was proposed that directly deploys access control on resource-constrained devices [24, 25]. The DCapAC allows smart devices to autonomously make decisions on access rights based on an authorization policy, and it has advantages in scalability and interoperability. Macaroons is another approach that uses nested and chained message authentication code (MAC) to address decentralized delegation problems in IoT systems [26]. Through cryptographic operations, new restrictions on an authorization certificates are hashed to hash-based MAC (HMAC) and chained in a nested fashion. Macaroons support the attenuation, delegation, and contextual confinement for cloud-based applications. To address challenges like scalability, granularity, and dynamicity in AC strategies for IoT; a Federated Capability-based Access Control model (FedCAC) is proposed to enable an effective AC mechanism to devices, services, and information in large scale IoT systems [27]. Migrating part of the centralized authorization validation tasks to local devices helps the FedCAC to be lighter and context-awareness enabled. The decentralized access control approach presents the following advantages:

- *User-Driven Security Mechanism:* Deploying authorization policy definition and verification logic on the edge of network allows user to have more intelligence to control the private data usage and granularity of their own access control policies. It significantly reduces the risk of data misuse and privacy leakage by centralized authorities.
- *Offline Mode:* Since a security mechanism is implemented among autonomous entities, part of failures, such as connection lost or malfunction on devices, have limited influences on the whole network, and the remaining edge devices still carry on-site intelligence to maintain the core functionalities of the system.
- *No Centralized Trust Authority:* In a decentralized approach, the edge of the network defines AC policies, and distributed entities establish trust relationships instead of relying on any central authority.

Nonetheless, the decentralized approach also brings many other issues:

- *Complex Access Control Mechanism:* Owing to the fact that various security policies are enforced on large volumes of smart things, it is enviable to increase complexity of AC mechanism based on heterogeneous security scenarios.
- *Overhead on Edge Networks:* Implementing authorization mechanisms on the edge devices might not only introduce computation and storage overhead on those resource-constrained devices, but also incur latency over network communication.

According to the discussions on above proposed AC solutions for IoT, designing more scalable, decentralized, and lightweight AC strategies addresses limitations with the state-of-the-art. To introduce decentralization, anonymity, and auditability; blockchain and smart contract provide a prospective solution, which is introduced in the next section.

22.3 Blockchain and Smart Contract

22.3.1 Blockchain

A blockchain is a replicated public database (Ledger) that records, stores, and updates all data as chained blocks. The blockchain is essentially a complex system that consists of core technologies:

- *Asymmetric Cryptography:* Blockchain uses asymmetric cryptography mechanism for encryption and decryption on data transmission, and the blockchain address derives from the public key. The use of asymmetric cryptography by blockchain brings authentication, integrity, and nonrepudiation into the network.

- *Digital Signature:* Users employ their private key to sign their own transactions. The authenticity of digital transactions and data verifies the digital signatures singed by issuers.
- *Hash Function:* A one-way hash function maps any size of transaction data to a fixed-size string given specified hash algorithm, and those hashed values could be appended to blockchain and verified by miners.
- *Consensus Mechanism:* The blockchain uses a consensus algorithm to maintain the sanctity of the data recorded on the blocks. The prevalent consensus algorithms are Proof-of-Work (PoW), Proof-of-Stake (PoS), and Practical Byzantine Fault Tolerance (PBFT).
- *Peer-to-Peer (P2P) Network:* Using a P2P network enables blockchain a decentralized architecture that does not rely on a centralized authority.

In a public blockchain network, any node can join a network and send transactions to interact with other nodes. The participating nodes include two types: active node and passive node. The passive nodes only read data in a blockchain and do not perform consensus algorithms to verify blocks in a chain. However, the active nodes, which are also referred to as *miners*, perform mining task to verify blocks and append new valid blocks to the blockchain in a chronological order. The miners that execute consensus algorithm contribute to maintain trust relationship among distributed nodes and receive rewards through validating transaction data in the blockchain. Figure 22.2 outlines the processing steps in blockchain.

Figure 22.2 Work process in blockchain.

- *Generate Block:* A hash value records the transactions, which can be of any type such as monetary transactions, system logs, health information, etc., and saved to a block. The miner, who is the first to solve the difficulty of the mathematical challenge, adds a new block to the blockchain and publishes it to network for miners' verification.
- *Verify and Consensus:* After receiving the latest blockchain, miners not only verify all transactions recorded in the block given pre-defined rules, but also perform a consensus mechanism to validate the new appended block. Only a verified blockchain could be accepted as the main chain among miners.
- *Extend Blockchain:* Since the majority of miners have verified the new block, the valid transactions are stored in the approved block, and the blockchain could be extended by appending the valid block, which is accepted by the whole network. The miner who first generated and validated the block could receive an award.

Thanks to the "trustless" proof mechanism enforced through mining tasks on miners across networks, users can trust the system of the public ledger stored worldwide on many different decentralized nodes maintained by "miner-accountants," as opposed to having to establish and maintain trust with the transaction counter-party or a third-party intermediary [28]. Because of the attractive characteristics, such as decentralization, auditability, and anonymity; the blockchain is an ideal architecture to ensure distributed transactions between all participants in a trustless environment.

22.3.2 Smart Contract

The blockchain has shown its success in the decentralization of currency and payments like Bitcoin. Currently, the design of programmable money and contracts that support a variety of customized transaction types has become a trend to extend blockchain applications beyond the cryptocurrency domain. The *Smart Contract*, which has emerged from the smart property, is a method of using a blockchain to achieve agreement among parties, as opposed to relying on third parties to maintain a trust relationship.

A *smart contract* is an automatable and enforceable agreement. Automation by a computer, although some parts may require human input and control, enhances efficiency of transactions. Enforceable either by legal enforcement of rights and obligations or via tamper-proof execution of computer code [29], ensures transactions are effective. By using cryptographic and other security mechanisms, smart contracts combine protocols with user interfaces to formalize and secure relationships over computer networks [30]. Since smart contracts are developed as scripts and stored on the blockchain, where each smart contract has a unique address that resides on the blockchain. The smart contract based on a blockchain network has the following properties:

- *Automation:* The business logic can be transcoded to a collection of pre-defined instructions and data resides at a specific address of a blockchain. Each smart contract executes independently and automatically in a prescribed manner and the encapsulated legal agreement in smart contract performs on distributed computers or devices without human interference.
- *Hierarchical Structure:* Smart contracts use Merkle hash trees, a constructed bottom-to-up binary tree data structure, to record all data and transactions generated in the blockchain network. All transaction data are hashed and saved as leaf nodes, and then the Merkle tree stores successive children node hashes from leaf-to-root upon any changes to the tree.
- *Auditability:* The Merkle tree structure demonstrates a global state indicated by the root hash value, and any attempt to change smart contract data in the Merkle tree will generate a different root hash value. Thus, the root hash value of the Merkle tree allows the auditability of large sets of data. After verified by a consensus mechanism running on a large volume of miners in the blockchain network, the data changed by executing smart contract can be accepted by the entire network through generating a new block and appending to the blockchain. Since a smart contract resides on the blockchain and is transparent to all nodes, its code can be inspected by every network participant.
- *Interoperability:* A smart contract is triggered by messages/transactions sent to its address. Through exposing business operation as public functions or application binary interfaces (ABI), a user could easily interact with a deployed smart contract to perform pre-defined business logic or contract agreement.

Through encapsulating operational logic as a bytecode and performing Turing complete computation on distributed miners, a smart contract allows the user to transcode more complex business models as new types of transactions on a blockchain network. A smart contract provides a promising solution to allow the implementation of more flexible and complex applications far beyond cryptocurrencies, such as financial products and services, supply chain management, data provenance, digital identity authentication, resource sharing and access control, etc.

22.4 Blockchain-Based Security Solutions to IoT

Because of many attractive characteristics shown by blockchain technology, numerous works investigate whether blockchain and smart contacts can address security issues in IoT ecosystems. This section reviews works that intend to integrate blockchain technology into IoT systems.

Researchers have examined whether blockchains and smart contracts can be integrated in the IoT sector [31]. Both the advantages and limits of blockchain usage in IoT are discussed. The blockchain-IoT combination, based on blockchain and smart contracts, not only facilitates the sharing of IoT services and resources; but also allows automations in a cryptographically verifiable manner of several existing, time-consuming workflows. Certain issues to consider before the deployment of a blockchain network in an IoT setting, from transactional privacy to the expected value of the digitized assets traded on the network.

Blockchain-based solutions in an IoT system provide decentralized security and privacy; however, limitations in public blockchain, such as energy consumption, delay, computational overhead and transaction costs, make it is not suitable for IoT scenarios that include large scales of resource-constrained IoT devices. A lightweight instantiation of a blockchain is presented [32], particularly geared for use in IoT by eliminating the Proof of Work (PoW) and the concept of coins. The proposed ideas exemplify in the context of a smart home. Relying on hierarchical structure and distributed trust to maintain the blockchain security and privacy makes the proposed framework more suitable for the specific requirement of IoT.

To improve throughput of transactions in blockchain, an hybrid blockchain architecture for Internet of Things, called Hybrid-IoT is proposed [33]. In Hybrid-IoT, subgroups of IoT devices form PoW blockchains to achieve distributed consensus among many IoT devices. Then, the connections among the PoW sub-blockchains employ a Byzantine fault tolerance (BFT) inter-connector framework to maintain a relative high throughput with only few nodes. Exploiting both *PoW blockchains* and *BFT protocols* allows Hybrid-IoT to process a high throughput of transactions and improve scalability of blockchain in IoT scenarios.

To offer a decentralized AC scheme in trustless network environments, a blockchain-based AC was proposed for the publication of AC policy and to allow distributed transfer access rights among users on Bitcoin networks [34]. The proposal allows distributed auditability, preventing a third party from fraudulently denying the rights granted by an enforceable policy. Based on the blockchain technology, FairAccess offers a fully decentralized pseudonymous and privacy-preserving authorization management framework for IoT devices [35]. In FairAccess, the AC policies enclose new types of transactions to grant, get, delegate, and revoke access. However, the scripting language used in Bitcoin allows the users to transcode two types of AC policies, so the proposed framework cannot support more complex and granular AC models.

To provide a decentralized, scalable, fine-grained, and lightweight AC solution for IoTs, BlendCAC [36, 37] provides a smart contract enabled decentralized capability-based AC mechanism to protect smart devices, services, and information in IoT networks. By enforcing a federated capability-based delegation model (FCDM) for an AC strategy, a hierarchical and multi-hop delegation mechanism enables support to complex AC policies. The AC policies are transcoded to the smart contract for registration, propagation, and revocation of the access authorization. The experimental results on a proof-of-concept prototype demonstrate feasibility and effectiveness of proposed BlendCAC.

22.5 BlendCAC: A Case Study

In this section, a case study presents the effectiveness and feasibility of blockchain enabled AC solution to secure IoT systems. This section discusses the experimental results and evaluations.

22.5.1 BlendCAC: A BLockchain-ENabled Decentralized CapAC

As one of the most actively investigated topics in the smart cities community, smart surveillance enables a broad spectrum of promising applications. However, most of the IoT-based smart surveillance systems rely on a centralized architecture, like a cloud server. A centralized method is not scalable for large IoT networks, and latencies are not tolerable to many mission-critical applications. To address the challenges in smart surveillance systems running on a cloud-based architecture, a *fog/edge computing* presents as a promising approach that enables the smart surveillance system to meet the delay-sensitive, mission-critical requirements [38–40]. A *Federated Capability-based Access Control system* (*FedCAC*) [27] design provides a scalable and fine-grained AC strategy for IoT system. However, it is essentially still a centralized AC scheme that suffers from a single point of failure and performance bottleneck.

Inspired by the smart contract and blockchain technology, a decentralized capability-based AC framework for IoTs, called BlendCAC, is proposed. Figure 22.3 illustrates the BlendCAC system architecture, which intends to function in a scenario including two isolated IoT-based surveillance domains without a pre-established trust relationship. In each domain, the domain owner, who has the ownership of devices, enforces predefined security policies to manage domain related devices and services. Operation and communication modes are as follows:

- *Registration:* In a blockchain network, each entity must create at least one main account defined by a pair of public key and private key to join the blockchain network. Each account uniquely indexed by its address derives from his/her own public key. Given the assumption that the authentication process is ensured by a blockchain network, a blockchain account address is ideal for identity

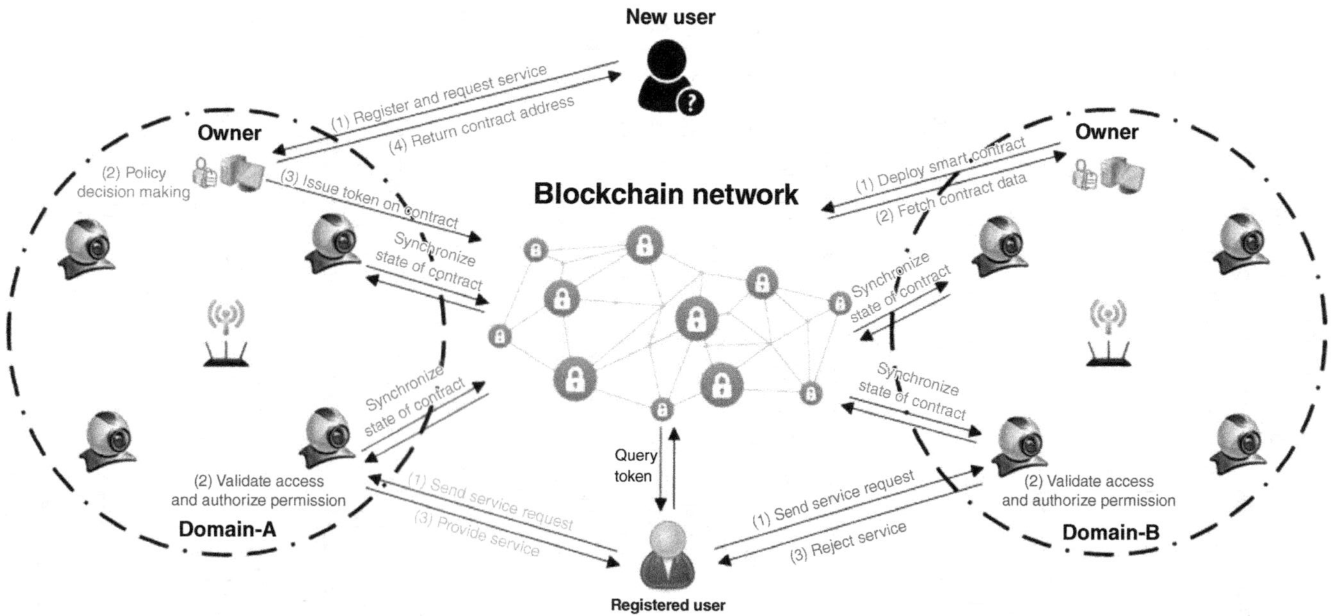

Figure 22.3 BlendCAC system architecture.

authentication in the BlendCAC system. In this scenario, identity authentication and management take advantage of the unique account address as Virtual Identity (VID), which resides in a local profile database maintained by the domain owner. New users could send registration requests to the domain owner. Once the identity information related to users or IoT devices is verified, the profile of each registered entity is created by using his/her address for authentication process when an access right request happens. As a result, the domain owners are able to define authorization policies and perform decision-making to directly control their own devices or resources instead of depending on third parties.

- *Smart Contract Deployment:* A domain transcodes the capability token management policy to a smart contract and deploys it on the blockchain network. After a smart contract deploys successfully on the blockchain network, it becomes visible to all nodes in the network so that any participants in the blockchain network can access transactions and smart contracts recorded in the chain data. Thanks to cryptographic and security mechanisms provided by a blockchain network, smart contracts secure any algorithmically specifiable protocols and relationships from malicious interference by third parties in a trustless network environment. After synchronizing the blockchain data, all nodes could access all transactions and recent state of each smart contract by referring local chain data. Each node interacts with the smart contract through the provided contract address and the public Remote Procedure Calls (RPC) interfaces.
- *Capability Propagation:* To successfully access services, (e.g., video surveillance at cameras), entity service providers in the system, initially send an access right request to the domain owner to get a capability token. Given the registered entity information established in the profile database, a policy decision-making module evaluates the access request by enforcing the predefined authorization policies. If the access request is granted, the domain owner issues the capability token encoded the access right, then launches a transaction to update the token data in the smart contract. After the transaction has been approved and recorded in a new block, the domain owner notifies the entity with a smart contract address and ABI for the querying token data. Otherwise, the access right request is rejected.
- *Authorization Validation:* Local service providers (e.g., cameras) perform the authorization validation process by receiving a service request from the entity. Through regularly synchronizing the local chain data with the blockchain network, a service provider just simply checks the current state of the contract in the local chain to get a capability token associated with the entity's address. Considering the capability token validation and access authorization process results, if the access right policies and conditional constraints are satisfied, the service provider grants the access request and offers services to the requester. Otherwise, the service request is denied.

22.5.2 Prototype Design

A concept-proof prototype system demonstrates the BlendCACon a real private Ethereum blockchain network environment. Compared with other open blockchain platforms, like Bitcoin and Hyperledger, Ethereum has a more matured ecosystem design to be more adaptable and flexible for development of smart contracts and business logic [41]. The BlendCAC access control model has been transcoded to a smart contract using Solidity [42], which is a contract-oriented, high-level language for implementing smart contracts. Thanks to Truffle [43], which is a world-class development and testing environment and asset pipeline for smart contract in Ethereum, source codes are easily compiled to Ethereum Virtual Machine (EVM) bytecode and migrated to the Ethereum blockchain network.

To enforce access authorization and validation policy on participants, a web service application based on the Flask [44] framework processes access requests between different entities. The lightweight and extensible micro architectures make the Flask a preferable web solution on resource constrained IoT devices. The capability token data structure is represented in the JSON [45] format to introduce some overhead over network communication and computations on resource-constrained IoT devices. The profiles and policy rule management use an embedded structured query language (SQL) database engine, called SQLite [46]. The lower memory and computation cost make the SQLite an ideal database solution for resource-constrained systems like Raspberry Pi.

22.5.3 Experiment and Evaluation

In the private blockchain network, mining tasks perform only on systems with stronger computing power, like a laptop or a desktop. In the experiment, two miners were deployed on a laptop, of which the configuration was as follows: the processor was 2.3 GHz Intel Core i7 (8 cores), the RAM memory was 16 GB and the operating system was Ubuntu 16.04. Another four miners were distributed to four desktops which were empowered with the Ubuntu 16.04 OS, 3 GHz Intel Core TM (2 cores) processor, and 4 GB memory. Each miner uses two CPU cores for mining. The edge devices are two Raspberry PI 3 Model B(s) with the configuration as follows: 1.2 GHz 64 bit quad-core ARMv8 CPU, the memory is 1 GB LPDDR2-900 SDRAM and the operation system is Raspbian based on the Linux kernel. Owing to fact that Raspberry PI is not powerful enough to perform high computational mining tasks, two Raspberry Pi boards worked as nodes to join the private blockchain without mining. All devices use Go-Ethereum [47] as the client application to interact with the blockchain network.

To verify the feasibility of the BlendCAC approach against unauthorized access requests, a service access experiment on a real network environment provides a verification test. In the test scenario, one Raspberry Pi 3 device worked as the client and another worked as the service provider. In the access right verification process, when any of the steps in the authorization procedure failed, the running process immediately aborts instead of continuing to step through all the authorization stages. As shown by Figure 22.4b, the server stopped the authorization process due to the failure to verify the granted actions or the conditional constraints specified in the access right list. In contrast, Figure 22.4a demonstrates a successful data request example, in which the whole authorization process was accomplished at the server side without any error, allowing the client to successfully retrieve the data from the service provider.

To measure the general cost incurred by the proposed BlendCAC scheme, including both on the IoT devices' processing time and the network communication delay, 50 test runs have been conducted based on the proposed test scenario, in which the client sends a data query request to the server for an access permission. Given an assumption that the requester has a valid capability token when it performs an action on service provider; all steps of authorization validation on the server side result in the maximum latency value.

The average total delay time (again over 50 runs) required by the BlendCAC operation of retrieving data from the client to server is 243 ms. The total delay time includes the Round Trip Time (RTT), time for querying capability data, and access right validation. As the most computing intensive stage, the execution

(a)

Client

```
{'project': {'time': '12:00 am', 'date': '4-29-2017', 'title': 'GET', 'description': 'test GET APIs'
, 'id': 2}, 'result': 'Succeed'}
Execution time is:0.231032
```

Server

```
Execution time of get_token is:0.198435
Execution time of is_token_valid is:0.000101
Execution time of is_access_valid is:0.000655
128.226.79.127 - - [22/Mar/2018 17:45:19] "GET /test/api/v1.0/dt/project?project id=2 HTTP/1.1" 201 -
```

(b)

Client

```
{'result': 'Failed', 'message': 'Authorization fail, deny access'}
Execution time is:0.243724
```

Server

```
Execution time of get_token is:0.212900
Execution time of is_token_valid is:0.000105
access valid fail
128.226.79.127 - - [22/Mar/2018 17:49:32] "GET /test/api/v1.0/dt HTTP/1.1" 401 -
```

Figure 22.4 Experimental results of BlendCAC: (a) access authorization succeed and (b) access authorization failed.

time of token processing is about 210 ms, which is accounted for almost 86% of the entire process time. The average time of the authorization process is about 0.86 ms.

To reduce high overhead introduced by querying token data from the smart contract, a *token data caching solution* is introduced by regularly synchronizing cached token data on service provider with smart contract status. As the client sends a service request to the server, the service side just extracts the cached token data from the local storage to validate the authorization instead of querying data from smart contract. The token synchronization is consistently executed by separate service processes with block generation time, which is about 15 seconds in the Ethereum blockchain network. Apart from the first service request scenario, when a long delay is observed; the service provider just uses the local cached token data for access authorization validation. Therefore, computation overhead on edge device becomes stable at a low level (0.86 ms).

To evaluate overall network latency incurred by BlendCAC, the benchmark scenario in which no access control is enforced on access request provides a baseline comparison. The benchmark without access control enforcement takes an average of 31 ms for fetching requested data versus that the BlendCAC consumes on average of 36 ms. To measure general network latency of inter-domain communication, HTTP is executed on the same test network environment by simulating a regular transaction, like connects, which sends a request and retrieves the reply. Compared with calculated average network latency that is about 300 ms, the tradeoff in the proposed BlendCAC is acceptable for the network environments by only incurring 5 ms latency (no more than 2%).

The authorized capability token validates transactions to the smart contract approved by miners and recorded to new blocks. The average block generation time was calculated to show impact of the number of miners on the transaction rate, which is proportional to the block generation time. Initially, only two miners ran the consensus algorithm and the block generation time is 16.07 seconds. As more miners performed the proof of work (PoW), the block generation time reduced and finally, became stable about 7.8 seconds given six miners. In the Ethereum network, gas pays for the computing resources consumed by miners. After calculating 100 transactions that assign capability token on blockchain network, the average gas cost for each transaction is 169 576.15 Wei (in Ethereum, 1 ether = 1.0×10^{18} Wei), which is about $1.018 given the gas price in the public Ethereum market.

Given the above evaluations, although the transaction rate constricted by the block-generation time introduces latency into the capability token generation and revocation, the financial cost is enviable to pay miners for transaction verification. BlendCAC demonstrates good scalability in the distributed IoT network environment with minimal overhead and consumption time.

22.5.4 Discussions

The experimental results demonstrate that the BlendCAC access control strategy is an effective and efficient method for protecting IoT devices from unauthorized access requests. Compared to the centralized AC approaches, the BlendCAC scheme has the following advantages:

- *Decentralized Authorization:* By leveraging the blockchain technique, the BlendCAC scheme allows users to control their devices and resources instead of relying on a third centralized authority to maintain the trust relationship with unknown nodes; therefore, BlendCAC can effectively mitigate risks resulting from centralized architecture, such as single point of failure and a performance bottleneck.
- *Edge Computing-Driven Intelligence:* Thanks to the blockchain technology, the BlendCAC framework provides a user-driven AC strategy that is suitable for the distributed nature of the IoT environment. Through transferring power and intelligence from the centralized cloud server to the edge of network, smart things are capable of protecting their own resources, enforcing privacy, and securing user-defined data content, which is meaningful to distributive, scalable, heterogeneous, and dynamic IoT-based applications.
- *Fine Granularity:* In the BlendCAC system, each entity uses their unique blockchain address for identity authentication, and a capability token authorizes the authenticated entity. Thus, an attacker cannot use fake identities to access service. A fine-grained AC model with a lease-privilege access principle prevents privilege escalation, even if an attacker steals the capability token.
- *Lightweight Design:* Compared to the XML-based language for AC, such as XACML, JSON is a lightweight technology that is suitable for the resource-constrained platforms. Given the experimental results, the JSON based capability token structure introduces small overhead on the computation and network latency.

Although the BlendCAC mechanism has demonstrated above attractive merits, using a blockchain to enforce the AC policy in IoT systems also incurs new challenges in performance and security:

- *Low Transaction Rate:* The transaction is associated with the confirmation time of the blockchain data, which depends on the block size and the difficult level for generating new blocks; thus, the latency for transaction validation by miners may not be able to meet the requirement in real-time application scenarios.
- *High Storage Overhead:* As data and transactions increases, the blockchain data become large. The continuously growing chain data requires more storage to save the public ledger data, and extracting transaction data in a larger chain data

need more computing resources on the host. It greatly increases overhead on each client, especially for resource-constrained IoT devices.

- *Majority Attacks:* The consensus algorithms used by blockchain, either PoW or PoS, are susceptible to majority attacks (also known as 51% attacks). If an attacker takes over 51% computing power of the network by colluding selfish miners, they could control the whole blockchain network and reverse the transactions.
- *Privacy Leakage:* Since the blockchain data is open to all participates joined to the blockchain network, and each node could have a backup data of whole chain, such a property of transparency inevitably brings privacy leakage concerns. Although using multiple addresses to interact with the network could prevent certain information leakage, attackers could reveal user's information by linking transactions data to user.

22.6 Conclusions

This chapter focuses on the access control (AC) mechanism for an IoT system. After a review of current AC solutions for IoT system, a thorough discussion on their pros and cons demonstrates tradeoffs to consider. As a prospective approach to enforce scalable and decentralized security policy on a distributed IoT network, the chapter describes blockchain and smart contract technologies as security solutions to IoT networks. A blockchain enabled decentralized AC model, called *BlendCAC*, provides a case study. A comparison experimental study deployed it on a physical network environment highlights the BlendCAC smart contract improvement over baseline methods.

Although the blockchain enabled security mechanism has shown significant potential, there is still a long way ahead towards achieving a complete decentralized security solution for an IoT system. Further exploration of blockchain-based AC schemes in real-world applications needs to address issues resulting from blockchain framework, such as low transaction rate and higher data storage. Furthermore, current consensus algorithms used by blockchain require high computation, and ever-growing blockchain data needs large memory to save public ledger on local host, from which these weaknesses prohibit performing a mining task on resource-constrained IoT devices. Thus, designing a lightweight consensus mechanism for IoT scenarios is mandatory to apply blockchain enabled security approaches on distributed IoT systems without introducing significant overhead.

References

1 A. Al-Fuqaha, M. Guizani, M. Mohammadi, M. Aledhari, and M. Ayyash, "Internet of things: A survey on enabling technologies, protocols, and applications," *IEEE Communications Surveys & Tutorials*, vol. 17, no. 4, pp. 2347–2376, 2015.

2 A. Bassi and G. Horn, "Internet of things in 2020: A roadmap for thefuture," *European Commission: Information Society and Media*, vol. 22, pp. 97–114, 2008.

3 C. Perera, A. Zaslavsky, M. Compton, P. Christen, and D. Georgakopoulos, "Semantic-driven configuration of internet of things middleware," in *IEEE International Conference on Semantics, Knowledge and Grids (SKG)*, pp. 66–73, 2013.

4 S. Rajaram, "Sensors: Technologies and markets to 2023," *BCC Research*, 2018. Available: https://www.bccresearch.com/market-research/instrumentation-and-sensors/sensors-technologies-markets-report-ias006j.html.

5 T. Gea, J. Paradells, M. Lamarca, and D. Roldan, "Smart cities as an application of internet of things: Experiences and lessons learnt in Barcelona," in *IEEE Seventh International Conference on Innovative Mobile and Internet Services in Ubiquitous Computing (IMIS)*, pp. 552–557, 2013.

6 J. Jin, J. Gubbi, S. Marusic, and M. Palaniswami, "An information framework for creating a smart city through internet of things," *IEEE Internet of Things Journal*, vol. 1, no. 2, pp. 112–121, 2014.

7 F. A. Alaba, M. Othman, I. A. T. Hashem, and F. Alotaibi, "Internet of things security: A survey," *Journal of Network and Computer Applications*, vol. 88, pp. 10–28, 2017.

8 S. Nakamoto, "*Bitcoin: A Peer-to-Peer Electronic Cash System*," Manubot, Tech. Rep., 2019.

9 V. Buterin et al., "A next-generation smart contract and decentralized application platform," White paper, 37, 2014.

10 L. Gong, "A secure identity-based capability system," in *IEEE Symposium on Security and Privacy*, pp. 56–63, 1989.

11 A. Ouaddah, H. Mousannif, A. A. Elkalam, and A. A. Ouahman, "Accesscontrol in the internet of things: big challenges and new opportunities," *Computer Networks*, vol. 112, pp. 237–262, 2017.

12 R. S. Sandhu, E. J. Coyne, H. L. Feinstein, and C. E. Youman, "Role-based access control models," *Computer*, vol. 29, no. 2, pp. 38–47, 1996.

13 P. Samarati and S. C. de Vimercati, "Access control: Policies, models, and mechanisms," in *International School on Foundations of Security Analysis and Design*. Springer, pp. 137–196, 2000.

14 L. M. S. De Souza, P. Spiess, D. Guinard, M. Kohler, S. Karnouskos, and D. Savio, "SOCRADES: A web service based shop floor integration infrastructure," in *The Internet of Things*. Springer, pp. 50–67, 2008.

15 P. Spiess, S. Karnouskos, D. Guinard, D. Savio, O. Baecker, L. M. S. De Souza, and V. Trifa, "SOA-based integration of the internet of things in enterprise services," in *IEEE International Conference on Web Services (ICWS)*, pp. 968–975, 2009.

16 G. Zhang and J. Tian, "An extended role based access control model for the internet of things," in *International Conference on Information Networking and Automation (ICINA)*, vol. 1, p. V1-319, 2010.

17 E. Yuan and J. Tong, "Attributed based access control (ABAC) for web services," in *IEEE International Conference on Web Services (ICWS)*, pp. 561–569, 2005.

18 W. W. Smari, P. Clemente, and J.-F. Lalande, "An extended attribute based access control model with trust and privacy: Application to a collaborative crisis management system," *Future Generation Computer Systems*, vol. 31, pp. 147–168, 2014.

19 S. Bhatt, F. Patwa, and R. Sandhu, "Access control model for AWS internet of things," in *International Conference on Network and System Security*. Springer, pp. 721–736, 2017.

20 N. Ye, Y. Zhu, R.-c. Wang, R. Malekian, and L. Qiao-min, "An efficient authentication and access control scheme for perception layer of internet of things," *Applied Mathematics & Information Sciences*, vol. 8, no. 4, p. 1617, 2014.

21 J. B. Dennis and E. C. Van Horn, "Programming semantics for multi-programmed computations," *Communications of the ACM*, vol. 9, no. 3, pp. 143–155, 1966.

22 S. Gusmeroli, S. Piccione, and D. Rotondi, "IoT@Work automation middleware system design and architecture," in *IEEE Conference on Emerging Technologies & Factory Automation (ETFA)*, pp. 1–8, 2012.

23 B. Anggorojati, P. N. Mahalle, N. R. Prasad, and R. Prasad, "Capability-based access control delegation model on the federated IoT network," in *IEEE International Symposium on Wireless Personal Multimedia Communications (WPMC)*, pp. 604–608, 2012.

24 J. L. Hernandez-Ramos, A. J. Jara, L. Marin, and A. F. Skarmeta, "Distributed capability-based access control for the internet of things," *Journal of Internet Services and Information Security (JISIS)*, vol. 3, no. 3/4, pp. 1–16, 2013.

25 J. L. Hernandez-Ramos, A. J. Jara, L. Marın, and A. F. Skarmeta Gomez, "DCapBAC: Embedding authorization logic into smart things through ECC optimizations," *International Journal of Computer Mathematics*, vol. 93, no. 2, pp. 345–366, 2016.

26 A. Birgisson, J. G. Politz, U. Erlingsson, A. Taly, M. Vrable, and M. Lentczner, "Macaroons: Cookies with contextual caveats for decentralized authorization in the cloud," in *Internet Society Network and Distributed Systems Security (NDSS) Symposium*, pp. 1–16, 2014.

27 R. Xu, Y. Chen, E. Blasch, and G. Chen, "A federated capability-based access control mechanism for internet of things (IoTs)," in *The SPIE Defense & Commercial*

Sensing 2018 (DCS), Conference on Sensors and Systems for Space Applications. SPIE, Vol. 10641, p. 106410U, 2018.

28 M. Swan, *Blockchain: Blueprint for a New Economy*. O'Reilly Media, Inc., 2015.

29 C. D. Clack, V. A. Bakshi, and L. Braine, "Smart contract templates: Foundations, design landscape and research directions," arXiv preprint arXiv:1608.00771, 2016.

30 N. Szabo, "Formalizing and securing relationships on public networks," *First Monday*, vol. 2, no. 9, 1997.

31 K. Christidis and M. Devetsikiotis, "Blockchains and smart contracts for the internet of things," *IEEE Access*, vol. 4, pp. 2292–2303, 2016.

32 A. Dorri, S. S. Kanhere, R. Jurdak, and P. Gauravaram, "Blockchain for IoT security and privacy: The case study of a smart home," in *IEEE International Conference on Pervasive Computing and Communications Workshops (PerCom Workshops)*, pp. 618–623, 2017.

33 G. Sagirlar, B. Carminati, E. Ferrari, J. D. Sheehan, and E. Ragnoli, "Hybrid-IoT: Hybrid blockchain architecture for internet of things-PoW sub-blockchains," arXiv preprint arXiv:1804.03903, 2018.

34 D. D. F. Maesa, P. Mori, and L. Ricci, "Blockchain based access control," in *IFIP International Conference on Distributed Applications and Interoperable Systems*. Springer, pp. 206–220, 2017.

35 A. Ouaddah, A. Abou Elkalam, and A. Ait Ouahman, "Fairaccess: A new blockchain-based access control framework for the internet of things," *Security and Communication Networks*, vol. 9, no. 18, pp. 5943–5964, 2016.

36 R. Xu, Y. Chen, E. Blasch, and G. Chen, "BlendCAC: A blockchain-enabled decentralized capability-based access control for IoTs," in *The IEEE International Conference on Blockchain, Selected Areas in IoT and Blockchain*, pp. 1027–1034, 2018.

37 R. Xu, Y. Chen, E. Blasch, and G. Chen, "BlendCAC: A smart contract enabled decentralized capability-based access control mechanism for the IoT," *Computers*, vol. 7, no. (3), p. 39, 2018.

38 N. Chen, Y. Chen, E. Blasch, H. Ling, Y. You, and X. Ye, "Enabling smart urban surveillance at the edge," in *IEEE International Conference on Smart Cloud (SmartCloud)*, pp. 109–119, 2017.

39 S. Y. Nikouei, Y. Chen, S. Song, R. Xu, B.-Y. Choi, and T. Faughnan, "Real-time human detection as an edge service enabled by a lightweight CNN," in *The IEEE International Conference on Edge Computing*, 2018.

40 R. Xu, S. Y. Nikouei, Y. Chen, S. Song, A. Polunchenko, C. Deng, and T. Faughnan, "Real-time human object tracking for smart surveillance at the edge," in *The IEEE International Conference on Communications, Selected Areas in Communications Symposium Smart Cities Track*, pp. 1–6, 2018.

41 "Ethereum Homestead Documentation," http://www.ethdocs.org/en/latest/index.html. (Accessed on Jan. 7th, 2020)

42 "Solidity," http://solidity.readthedocs.io/en/latest/. (Accessed on Jan. 7th, 2020)

43 "Truffle," http://truffleframework.com/docs/. (Accessed on Jan. 7th, 2020)

44 "Flask: A Pyhon Microframework," http://flask.pocoo.org/. (Accessed on Jan. 7th, 2020)

45 D. Crockford, "RFC 4627: The application/JSON media type for JavaScript object notation (JSON)," July 2006, Status: INFORMATIONAL.

46 "SQLite," https://www.sqlite.org/index.html. (Accessed on Jan. 7th, 2020)

47 "Go-ethereum," https://ethereum.github.io/go-ethereum/. (Accessed on Jan. 7th, 2020)

23

Intent as a Secure Design Primitive

Prashant Anantharaman[1], J. Peter Brady[1], Ira Ray Jenkins[1], Vijay H. Kothari[1], Michael C. Millian[1], Kartik Palani[2], Kirti V. Rathore[3], Jason Reeves[4], Rebecca Shapiro[5], Syed H. Tanveer[1], Sergey Bratus[1], and Sean W. Smith[1]

[1] *Department of Computer Science, Dartmouth College, Hanover, NH, USA*
[2] *Dartmouth College, Hanover, NH, USA*
[3] *Department of Electrical and Computer Engineering, University of Illinois at Urbana-Champaign, Champaign, IL, USA*
[4] *VMWare, Inc., Palo Alto, CA, USA*
[5] *Champlain College, Burlington, VT, USA*

23.1 Introduction

At the core of design lies *intent*, or the primary goals and objectives that drive the creation and usage of a technology. However, there are often multiple actors involved with a technology's production, all with their own ideas about intent.

- The designer has some idea of what their product should do and how it should do it.
- The developer produces code to realize their own interpretation of the designer's intent.
- The user uses the product according to their own intent, which may or may not match those of the designer and developer.

This chapter examines the role of intent when it comes to *security*. The security of a technology stems from how closely the intents of its designers, developers, and users align, as well as whether these intents reflect reality. However, ensuring that

Modeling and Design of Secure Internet of Things, First Edition. Edited by Charles A. Kamhoua, Laurent L. Njilla, Alexander Kott and Sachin Shetty.

Published 2020 by John Wiley & Sons, Inc.

these intents align and match reality can be difficult or even impossible. Designers tend to explicitly define the functionality of their technologies, but their security intent is often implicit and conceptualized in vague terms like "only legitimate users will be able to authenticate." This oversight is not intentional; rather, the designer lacks the proper tools to think through the security implications of their design decisions. Developers must then translate these vague notions into code, and their decisions may not align with the designer's mental model. Finally, users have their own goals and intentions; their focus is often on accomplishing their primary task rather than ensuring the system remains secure. As they are the final arbiters of how the system operates, their decisions have the most impact on the system's security, and their usability-centered choices can override the designer's and developer's security intentions.

Intent mismatches have caused many of the most severe and prevalent vulnerabilities, and this trend is continuing in the Internet of Things (IoT) era. The Mirai botnet [1], which utilized buffer overflows to infect routers, demonstrates one such vulnerability. Inadequate input validation on routers violated the designer's assumption that only properly formatted input would be accepted. Another example involves the first variant of the Spectre vulnerability [2], which demonstrates a mismatch between the intents of the processor designer and the operating system developer. Speculative execution of program instructions violated the developer's assumption that certain code branches would never be executed and should therefore leave no trace in the system hardware. Unchanged default passwords are a mismatch between developer and user intent; developers intend for users to create secure passwords once they start using the system, whereas users may not do so as they are reluctant to change anything once the system is running – and in some application scenarios, users even depend on well-known default passwords for system availability during crisis situations.

The unique nature of the IoT exacerbates the security ramifications of these intent mismatches. Almost by definition, the IoT brings computing to physical-world devices that have historically lacked such computational capabilities, thus merging previously separate design goals. Inexperienced designers who lack adequate security background or training may be thrust into the role of bringing their revolutionary vision to fruition. A lack of standards and conventions means many developers will develop ad-hoc protocols or put together makeshift devices that satisfy the demands of their use case without giving security adequate consideration. The unprecedented scale of the IoT presents an additional challenge. For example, when users do not change their default passwords, millions of unsecure devices could be compromised to create a massive botnet. These challenges render traditional security assumptions and models obsolete, requiring us to create new approaches suitable to the current landscape.

In this chapter, we present two emerging security paradigms we have helped develop to preserve designer and developer intent.[1] The growing field of *language-theoretic security* (*LangSec*) provides a theoretical foundation, as well as the machinery, to ensure that devices only accept developer-intended inputs, thus protecting these devices against a myriad of input-validation bugs. LangSec posits that input recognizers must be built from the formal specification of a protocol, and the recognizer must adopt a language-based approach to accept valid input while rejecting all invalid input without performing additional computation. LangSec compels the developer to contemplate the input language of the protocol being designed and implemented. Separately, *ELF-based access control* (*ELFbac*) empowers the developer to specify access control policies for their programs at a natural level of granularity, the *application binary interface* (*ABI*). ELFbac allows the developer to codify their intent by creating policies for intra-process memory isolation. Together, these techniques help to reduce the incidence (the purview of LangSec) and consequences (the purview of ELFbac) of zero-days, which are more critical in the IoT where patching is harder and the consequences of compromise may be so wide reaching. By restricting the input a program can accept and the memory it can access at any given time, these techniques effectively constrain and/or prevent undesirable behavior, forcing the program to conform to the intent of the developer (see Figure 23.1).

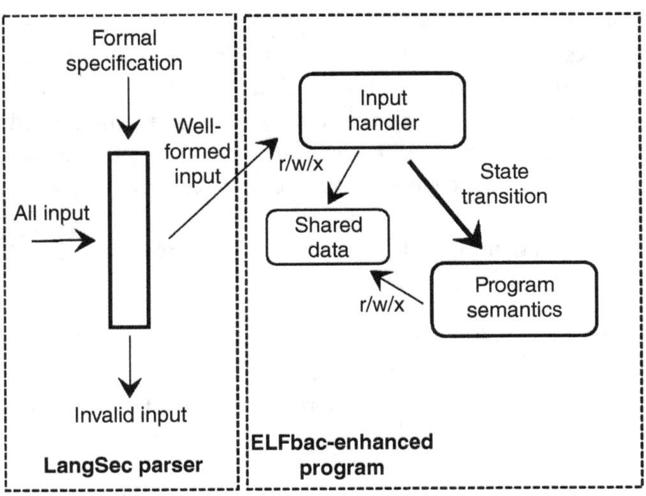

Figure 23.1 ELFbac and LangSec both enforce human intent by codifying the specifications.

---

1 Preserving user intent is also important, but it is not a focus of this chapter.

Section 23.2 presents LangSec. Section 23.3 presents ELFbac. Section 23.4 presents an IoT application area where we use the two techniques together. Section 23.5 discusses how we evaluate implementations, and Section 23.6 concludes.

23.2 A LangSec Primer

The field of *language-theoretic security* (*LangSec*) posits that since exploits are input-driven computations which are unintended by and unknown to designers and developers, defending against exploits requires taking a principled approach to input recognition that is rooted in *formal language theory* and *computability theory*. (Security based on *programming languages* is an orthogonal topic.)

LangSec examines both theoretic and applied challenges with input sanitization and input recognition that manifest in real-world exploits; it suggests parser design and implementation best practices, and it even provides a tool that facilitates the principled development of parsers – all with the goal of preserving designer and developer security intent.[2]

23.2.1 What Is LangSec?

The set of acceptable inputs to a system is often not explicitly defined by the developer, nor is it tested extensively. This leads to invalid input being processed instead of being rejected, which in turn leads to major bugs such as Heartbleed [3], Shellshock [4], and Android Master Key [5]. When a system receives an unanticipated input, the processing code may drive the system to a state that is unaccounted for by the developer. LangSec seeks to prevent such vulnerabilities driven by poor input handling.

At its core, LangSec is the idea that input-driven security exploits can be avoided by ensuring that the acceptable inputs to a program:

- are well-defined by a language grammar,
- are as simple as possible within the Chomsky hierarchy, and
- can be fully validated by a dedicated parser of appropriate power as defined by the Chomsky hierarchy.

The idea with LangSec-hardened parsers is to separate the input validation code of the program from the rest of the code, so that the main program never acts on input that has not been validated. The set of acceptable inputs is used to define a

2 LangSec offers many other contributions as well, such as identifying anti-patterns. Our focus in this section, however, is to give the reader an overview of the field.

grammar for a language. A LangSec parser must simply follow this grammar by rejecting any non-conforming input and operating correctly on well-formed input.

23.2.2 Exploits Shatter Intent

The designer and the developer often have some notion of what they want their software to do and not to do. Correct behavior may be explicitly expressed at different points within the design and development process – and it may even be enforced in the final product. However, it may also be an amorphous, ill-defined, high-level notion that is assumed, rarely thought about, and never explicitly stated. Many times, we see something in between, where attempts to express intent are occasionally made throughout the design and development process, but this intent cannot be correctly enforced because the designer and developer lack the requisite knowledge or tools to ensure the security principles they want are achieved in the software they produce. This is perhaps best exemplified by the anti-pattern of the shotgun parser [6], which comprises segments of code scattered throughout a program that collectively aim to sanitize the input but instead exhibit security vulnerabilities because input-driven computation occurs before the input is deemed legitimate or because the parser code is executed at different times and may not act on a single input as the developer expects. These shotgun parsers demonstrate that their developer not only intends to protect against bad input, but also that they expend great effort in pursuing this endeavor; however, despite this effort, shotgun parsers frequently fail to safeguard the program from malicious-input-driven exploits, as evidenced by the many exploits seen in the wild that leverage unintended computation enabled by these poorly designed parsers.

Indeed, exploits shatter designer and developer intent – and they do so by design. A computer exploit is input that is crafted and submitted to a target program to produce input-driven computation unintended by and unbeknownst to the designer and the developer [7]. First, the exploit programmer seeks to uncover unintended computational capabilities offered by the target program. Next, the exploit programmer identifies and distills these behaviors by determining simple input constructs that reliably produce them. Finally, the exploit programmer chains together these input constructs to create the exploit.

We stress that programming an exploit is programming. The basic unit or building block of an exploit is the "weird instruction," a gadget or a collection of sequentially executed instructions that collectively performs a useful operation for the exploit programmer, one that should not be allowed as neither the developer nor designer intended for such computational capabilities. Once the exploit programmer discovers and distills unintended behaviors into these weird instructions, they systematically piece these weird instructions together to create the exploit, just as, say, an assembly programmer pieces together assembly instructions to create an assembly program. The exploit runs on – and therefore serves as an attestation to the existence of – the "weird machine," a programmable machine harbored by the target program that offers the requisite weird instructions to construct the exploit.[3]

The exploit will not exist if the weird machine upon which it runs does not exist. LangSec attempts to prevent the emergence of weird machines.

23.2.3 A Brief Detour into the Theory of Computation

LangSec builds upon the Theory of Computation. Here, we give a very brief primer of the theory of computation and its branches to provide the requisite language and machinery to understand the theoretical underpinnings of modern-day exploits. For the reader who seeks a more complete treatment, we highly recommend Sipser's book [8].

In formal language theory, an *alphabet* is a non-empty finite set of *symbols*. A *string* over an alphabet is a finite sequence of symbols belonging to that alphabet. For example, consider what is colloquially considered the English alphabet: $\Sigma = \{a, b, \ldots, z\}$. Here, $a \in \Sigma$ is a symbol belonging to the alphabet and the word $cat \in \Sigma^*$ is a string over the alphabet. A *language* is defined in relation to an alphabet as a set of strings over that alphabet. A natural way to express a language is to give a grammar. A *grammar* specifies rules, each of which is a mapping from a variable to a sequence of variables and symbols. A grammar comprises a start symbol along with a sequence of rules. The language of the grammar is simply the language comprising all strings generated by the start symbol. These language-theoretic notions form the building blocks of automata theory and computability theory – the connection being what computation is required to "capture" languages. These fields, in turn, lay the foundation for language-theoretic security.

As suggested by its name, the central focus of automata theory is *automata* – or mathematical models of computation – and their capabilities. Core to automata is the notion of *state*, which roughly refers to a condition that may dictate a subsequent course of action. Automata maintain state, often have some form of memory, and perform computation on input in its pursuit of some task, such as generating output or determining whether the supplied input is well-formed.[4] The task that we are concerned with in this chapter, which is perhaps the most common task, is to accept or reject input. The automaton begins in a start state. The automaton (or perhaps its operator) then repeatedly examines the state that the automaton is in. Based on that state, it will read symbols from the input and/or the memory that is part of the automaton, such as a stack or a tape, it may

3 For further discussion of exploit programming, weird instructions, and weird machines, we recommend reading previous work by Bratus et al. [7].

4 The standard concepts of Deterministic Finite Automata (DFA) and Non-deterministic Finite Automata (NFA) are just special cases of these machines.

perform an action such as writing a symbol to memory, and it will transition to the next state. When supplied with input, the automaton may either run indefinitely or it may terminate when some condition is met, such as there being no symbols left to process. If the computation terminates, the state that the automaton is in during termination determines whether the input is to be accepted or rejected. That is, the automata we are interested in take as input a string and either *accept* or *reject* the string. An automaton *recognizes* a language if the automaton accepts only those input strings that belong to that language. And it *decides* a language if it accepts those strings that belong to the language and rejects those that do not (instead of sometimes merely running forever).

Central to discussion of languages, grammars, and automata is the notion of *expressiveness*, which enables us to meaningfully differentiate classes of a given type. A language L is considered more *expressive* than a language L' if L is a strict superset of L'. Similarly, a class of languages $\mathcal{L}$ is more expressive than a class of language $\mathcal{L}'$ if $\mathcal{L}$ is a strict superset of $\mathcal{L}'$. Similar notions of expressiveness exist for classes of grammars and automata, grounded in the languages associated with these grammars and automata. Indeed, this notion of expressiveness intimately links classes of languages, grammars, and automata, e.g. the grammars corresponding to finite state automata are regular grammars and the corresponding languages are regular languages. Moreover, it enables us to develop nested classification schemes such as the well-known Chomsky hierarchy. We find, for example, that in the Chomsky hierarchy, finite state machines (and the corresponding grammar and language classes of regular grammars and regular languages) are much less expressive than Turing machines (and the corresponding grammar and language classes of unrestricted grammars and recursively enumerable languages).

Computability theory and computational complexity theory provide us with a foundation for understanding the limitations of automata in regard to what they can do and how efficiently they can do it. Computability theory, in particular, provides us valuable negative results that inform LangSec.

23.2.3.1 The Halting Problem, Undecidability, and the Equivalence Problems

A particularly well-known, expressive class of automata is Turing machines. Certain classes of languages are not recognizable using a Turing machine and there are still more that are undecidable. The classic example of a Turing-undecidable language comes from the Halting Problem, which involves designing a Turing machine that accepts all (M, I_M) pairs where M is a Turing machine and I_M is an input for which M halts when run on I_M and rejects all other Turing machine, input pairs.

It turns out that while the language of the halting problem, L_{TM}, is not decidable, it is recognizable. However, Turing-unrecognizable languages also exist; one such example is the complement of L_{TM} [8]. Recursive languages are exactly those languages that are decidable; notably, the slightly less expressive context-sensitive languages in the Chomsky hierarchy are also decidable.

Undecidability is relevant to LangSec because if an input language is computationally undecidable, then it *cannot* be validated – a validation layer will always let some crafted attack input through. It is also relevant because of the *equivalence problem* – determining whether the language of one grammar is equivalent to the language of another. In terms of LangSec: Do two input validation layers accept the same input? While this problem is undecidable in the general case and for many context-free languages, it is decidable for deterministic context-free languages [9].

23.2.3.2 An Extension to the Chomsky Hierarchy

The well-known Chomsky hierarchy provides a containment classification for grammars, as well as their corresponding languages and automata, based on their expressiveness. With LangSec, we are largely concerned with the questions of decidability that we presented earlier. Hence, we extend the Chomsky hierarchy in Figure 23.2 by differentiating between non-deterministic pushdown automata and deterministic pushdown automata, where a crucial barrier lies with regard to the decidability of parser/grammar equivalence. We also note the barrier of decidability at linear-bounded automata within the hierarchy.

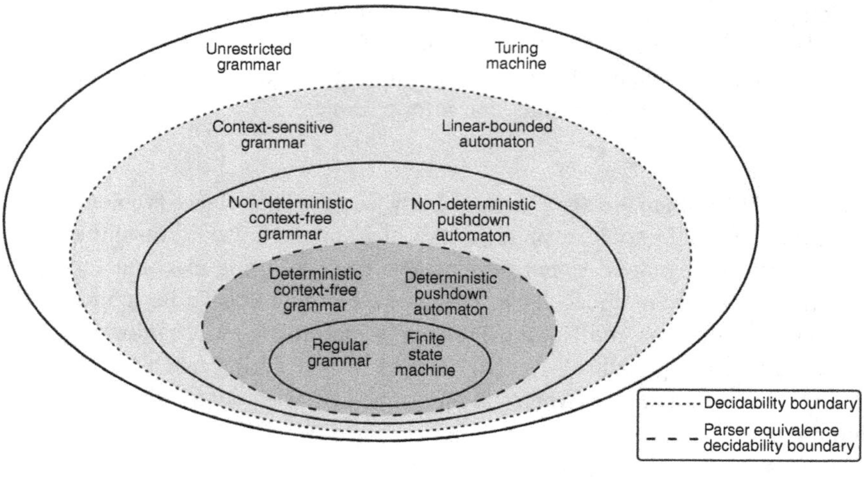

Figure 23.2 Chomsky hierarchy extended with LangSec boundaries.

23.2.4 Input Recognition

Sassaman et al. [7] state that "a system's security is largely defined by what computations can and cannot occur in it under all possible inputs." A program's input drives its control and data flow, and input-driven computation that does not conform to the intent of the designer and developer suggests the existence of a weird machine that an adversary can operationalize to program an exploit. This raises the question: How do we produce programs that match the intent of the designer and developer with the end goal of preventing such exploitation? LangSec, in its pursuit of an answer to this very question, advocates treating the input as a formal language, making this language as simple as possible, and ensuring that the input is correctly recognized in accordance with the design specification before permitting input-driven computation.

Ensuring the absence of unintended input-driven computation is intimately linked with the objective of secure composition. Sassaman et al. [7] argue that, as composition is integral to the construction of complex systems, secure composition has become a colossal security challenge. Modern systems comprise many components that perform a variety of tasks across different layers, and these components must be able to talk to one another. The challenge of secure composition lies not only in securing each of the component parts, but also securing the interfaces, i.e. the glue that enables communication between these parts. The vulnerabilities underpinning modern-day exploits often stem from improper input handling at these interfaces.

To secure programs and the interfaces between them, it is imperative that we harden the *parser*, the code responsible for input handling. A primary goal of input handling is to – or, rather, if we are to produce secure code, it *should* be to – *recognize* the input language. That is, we want an input handling routine, a *parser*, that only accepts strings that belong to the input language. However, this alone is insufficient; ensuring recognizability makes no guarantees about whether the parser will halt on all inputs – and we truly want our parsers to halt on all inputs! Ergo, what we really want is for the parser to *decide* the input language, that is, to accept strings that belong to the input language and to reject strings that do not. We may wish for additional properties from the input language and the parser, e.g. we may want to satisfy timing constraints, space constraints, or other objectives, but at a bare minimum, the input language must be decidable.

23.2.4.1 Full Recognition Before Processing

LangSec advocates recognizing the input language – or more precisely, deciding the input language – before the program performs any (non-parser) computation on that input. That is, the parser should be separated from the rest of the program and the parser should reject all invalid input immediately, only allowing valid input to be passed on to the program internals. This approach is expressed in Figure 23.3.

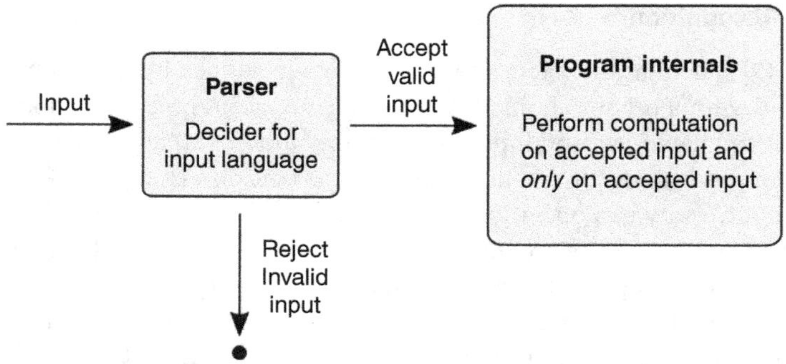

Figure 23.3 Full recognition before computation.

23.2.4.2 Principle of Least Expressiveness

The *principle of least privilege* roughly states that an entity within a system, such as a user or a process, should operate with the least privilege required to perform the task at hand [10]. LangSec advocates a similar principle for the design of protocol specifications or grammars and their parser implementations, the *principle of least expressiveness*:

> *One should always use the least expressive grammar, language, or automaton that achieves the task.*

This principle has the following two implications:

- Protocols should be designed to be minimally expressive.
- Parsers should be no more expressive than is required to decide the languages specified by the protocols they obey.

23.2.4.3 Principle of Parser Computational Equivalence

As mentioned earlier, secure composition is one of the leading security challenges. As such, it's vital to the security of interfaces. Given two parsers that act on the same input, those parsers must wholly agree, i.e. for every possible input, either they both accept the input or they both reject it. The *principle of parser equivalence* states:

> *Secure composition requires any two parsers that should decide the same language, e.g., due to sharing a component–component boundary, do decide that language.*

Recall that the equivalence problem is not decidable for non-deterministic context-free languages and more expressive languages, whereas it is decidable for deterministic context-free languages. When constructing parsers involved in secure composition, one should, therefore, aim to ensure they are no more expressive than deterministic context-free.

23.2.4.4 Updating Postel's Robustness Principle

Jon Postel's robustness principle states, in the context of TCP implementation: "[B] e conservative in what you do, be liberal in what you accept from others." [11]. While this may be sage advice in many situations, its misinterpretation has led to many Internet bugs. In particular, the advice does not account for the active adversary. While the principle, when read in context, displays understanding of bugs caused by weak input handling and also states that such a principle must be applied at every layer of the network stack, protocol implementers and the implementations, by extension – often mistakenly assume that the input that one layer passes to another is well-formed and any malformed input is filtered out at the layer boundary.

Previous work by Sassaman et al. [12] proposes an update to the robustness principle, which is particularly pertinent to the new network protocols being developed for the IoT:

- Be definite in what you accept. Clearly stating what language your machine understands is critical to the security of the device.
- Treat all input as a language that is parsed by an automaton of matching computational power. Also, the recognizer must be generated from the grammar of the language. Try to keep the language regular or at most context-free.

23.2.5 Incorporating LangSec into System Design and Development

Given the importance of specifying an input language formally at the protocol design phase, we want to make it as easy as possible for the parser developers – those people writing production code – to translate the language specification correctly. This job involves taking a grammar – the description of a language as a state machine or in prose – and writing code to implement the grammar. In the traditional workflow, every parser is written by hand using standard code blocks. In practice, it is difficult to determine that a series of *if* statements corresponds to the intended grammar. Any misunderstanding or oversight in reading the grammar on the part of the developer may result in an implementation that does not match the specification. We see this kind of mistake happen in many real-world parser bugs like Heartbleed [3] and Android Master Key [5]. In a formal sense, even a small difference changes the implementation to one for a completely different language. Additionally, writing a parser as a series of code blocks involves performing the entire translation process from scratch every time.

A LangSec solution to this problem is to use a *parser combinator tool*. The building blocks of grammars are combinators like concatenation, union, and Kleene star. A parser combinator tool is just a library that provides these combinators. In this way, the developer ends up with code that visually looks like the grammar. Verifying that the implementation matches the specification is trivial and can be done by inspection. Furthermore, the correctness of each of the combinators can be verified independently, meaning the developer does not have to worry about translating the combinators into code correctly – only calling the combinators as the grammar specifies.

23.2.5.1 Hammer

Hammer is a parser combinator tool written in C with bindings for several other languages, including C++, Java, Python, Ruby, Pearl, PHP, and .NET. Hammer provides implementations of all necessary combinators for context-free grammars and then some!

Future directions include adding a utility to create fuzz data based on the parser implementation and adding a utility to take a specification such as a state machine as input and automatically generate a parser (see Table 23.1).

Table 23.1 Syntax and usage of Hammer.

Syntax	Usage	Semantics
`h.ch`	`h.ch('a')`	Matches a single specified character token.
`h.ch_range`	`h.ch_range('a','z')`	Matches a single token in the specified character range.
`h.uint8`	`h.uint8()`	Matches a single integer token.
`h.int_range`	`h.int_range(1,16)`	Matches a single integer token in the specified range.
`h.sequence`	`h.sequence(h.ch('a'), h.ch('a'))`	Performs the concatenation operation.
`h.choice`	`h.choice(h.ch('a'), h.ch('b'))`	Performs the Boolean "or" operation.
`h.many`	`h.many(h.ch('a')`	Performs the Kleene Star operation.
`h.optional`	`h.optional(h.ch('a'))`	Specifies that matching this token is optional.

23.2.6 Summary

LangSec provides a theoretical foundation, a body of research, and usable tools for the design and implementation of protocols and parsers to avert many of the worst exploits of the modern day. The IoT and the incentive structure that drive both fledgling and established companies to rush to get broad market coverage have resulted in more component–component interactions with less focus on secure input handling. LangSec provides the fix: separate the parser from the remainder of the program and ensure it matches the specification; ensure parsing is done in full by the parser and only pass on valid input to the program internals; use the least expressive computation power necessary; and ensure parser equivalence. LangSec also delivers a tool in Hammer to facilitate development of parsers that match their underlying specifications.

23.3 An ELFbac Primer

In this section, we discuss *ELF-based access control* (*ELFbac*), a technique for intra-process memory isolation that can be used to mediate what code can operate on which data at what times. By allowing the user to express their intent through policy, ELFbac can protect programs from intent-violating exploits. We provide a tutorial on designing and enforcing ELFbac policies, and we show how ELFbac can mitigate existing high-profile vulnerabilities such as the SSH roaming bug and variant 1 of the Spectre processor bug.

23.3.1 What Is ELFbac?

The *principle of least privilege* states that the components of a system should have the most restrictive set of permissions to accomplish their tasks [10]. This level of isolation limits the exposure of vulnerabilities within a system. For example, a web browser should not have default access to critical operating system files. In this way, vulnerabilities in one component do not impact the overall operation of the system.

With ELFbac, we apply this principle all the way down to the individual code and data elements within a single process address space. A bug in one library, the browser uses could trigger functions from other libraries.

Privilege separation is critical to preserving developer intent. However, existing access control mechanisms do not address such intent. File system, process, or system call level policies are not granular enough for processes which may keep their objects in memory and never trigger an offending operation until it is too late. Mitigations for memory corruption vulnerabilities such as control flow integrity operate at finer granularities but do not address a "Trojan horse" attack where a malicious third-party library mapped to the process address space can read and manipulate process memory it should not have access to.

To this end, ELFbac preserves developer intent within the address space. We frame the problem of unintended, intra-process memory accesses involving buffer overflows or use-after-free as an out-of-type reference problem. We provide a technique for constructing a type-state system consistent with the existing ABI to allow developers to express intent.

23.3.2 Why ELFbac?

While several similar memory protection schemes have been proposed in the past, they do not achieve the granularity of protection that we desire here:

- Standard sandboxing is too coarse-grained for our purposes, as it requires us to "spawn extra processes or re-engineer code" [13] to achieve the code/data-level separation we need.
- SELinux [14] extends traditional Unix file protections to define how programs can interact with other objects, but cannot "protect a process' data from the process itself" [13].
- Mondrian [15] allows users to set the read/write/execute permissions to individual memory segments at the word level, but doing so required additional hardware mechanisms (more registers, a "permissions lookaside buffer") that are not currently present in commodity systems.[5]

23.3.3 Relationships Between Code and Data

A process has code chunks and data chunks. Standard OS controls provide little protection on what they can do to each other – but the developer certainly has intended behavior. We would like to have the program development process automatically give us info about what these chunks are and what the developer intends.

Systems already have standard ways to have an "executable" program file express the code and data for a given executable, as well as the metadata necessary for the creation of a process address space. In our work, we focus on the *Executable and Linkable Format (ELF)*, standard in the UNIX family (although our approach can easily generalize to other formats). Inherent within the ELF format are specific rules and conventions that are expected to be followed – for example, code in the `.init` section operates only on data in the `.ctors` section. ELFbac's key insight is that ELF sections can be leveraged to encode *explicit* rules beyond standard ELF

5 Later papers used the Dutch spelling *Mondriaan* for this project.

conventions, allowing users to create new sections that contain (i) specific pieces of code or data, and (ii) policies that govern the interaction between other sections [13].

To enforce these permissions, however, we must overcome the issue of the "forgetful loader" [13]. When a program is loaded into memory to be run, its layout is organized by *segments* rather than sections. Segments are grouped together based solely on their read/write/execute permissions, causing section-level permissions to be discarded and leaving the door open for programmer intent to be ignored. To get around the forgetful loader, we need a way to maintain section-level permissions all the way through program execution.

23.3.4 Design

ELFbac utilizes the existing linking and loading process to define policies (via common linker scripts) for intra-process memory isolation. These policies, as defined by the developer, consist of a set of rules codifying relationships between code and data, which are specifically the access controls (i.e. read, write, and execute permissions). ELFbac code within the Linux kernel then enforces these policies, via a finite-state machine where each state has a separate virtual memory context. The operating system kernel utilizes the information within the ELF file to link, load, and ultimately construct a runtime process.

Enforcement of an ELFbac policy happens in the kernel page fault handler. Memory accesses that happen outside the current state trigger page faults, which are then handled to verify if the memory access was valid at the current state or lead to an error state if the access was invalid.

We treat code and data units equally, and place fine-grained permissions over these units. We abstract the security-relevant phases of execution of a program into a finite-state machine (FSM). Permissions to code and data units are specified for each state, and the methods that trigger transitions are also specified. The permissions can be a combination of read, write, or execute. Sensitive data is placed in its own memory region, with read access only given when the data needs to be read. At all other states in the state machine, this permission is rescinded.

23.3.5 Implementation

Our team has built reference implementations for Intel x86 and ARM-based Linux systems. We believe that migrating ELFbac to other hardware platforms or similar operating systems should be made easier as portability is built into the current implementation. Wrappers for all the key kernel primitives were created to speed the porting process, so only a minimal set of files need to be changed.

Checking the policy by trapping the page fault on every memory access would cause an unacceptable drop in performance. To prevent this, ELFbac provides its own cache on top of the usual caching layers. To explain this, we need to first review how paging operates.

On the x86 platform, mapping is done through the page directory (PD) and the page table (PT), with each table containing 1024 4-byte entries. Each page directory entry (PDE) points to a page table, while each page table entry (PTE) points to a physical address; additionally, any page table lookup is cached by either the instruction or data translation lookaside buffer (TLB). Since the appropriate PDE and PTEs are created as needed by the running process and any system calls it makes, the page table entries serve as a proxy cache for the overarching policy in force in that memory space.

We add an additional layer of shadow memory where different code sections get different views of memory based on the policy FSM state. Adding additional contexts has the danger of "churning" the TLB, which would reduce performance; we counter this by filling the shadow contexts in a "lazy" fashion – every page starts empty and only when we have a valid access do we fill the shadow memory.

In practice, when the running process accesses an unmapped address, the Memory Management Unit (MMU) raises a page fault that ELFbac intercepts. ELFbac checks that the policy allows access and does not cause an FSM state change, and if true the shadow contexts are loaded so future access in the same state will not cause a fault. If the access causes a state transition, shadow contexts for the new state are loaded. The old memory context is unloaded so it can be validated against the policy on the next access.

Verification of the current state is done at each system call; the policy state can allow or disallow the call.

23.3.6 Using Mithril to Build Policies

ELFbac policies are written in a Ruby domain-specific language. Mithril embeds these policies in a separate `.elfbac` section in the binary, which is then enforced by the memory management system of the ELFbac enhanced kernel. We implement a policy for a parser using Mithril.

First, we build a state machine for the entire program. Figure 23.4 shows the state machine for our program. In the *start* state, all globals need to be given access since they are initialized. Following the initialization, we change state to the *parser* by invoking the *do_parse* method. The *libc* state handles access to the globals that are needed by the libraries. The variable being input by the user, the bytes received on the socket, the parse-tree built by the parser library, all need to be given access

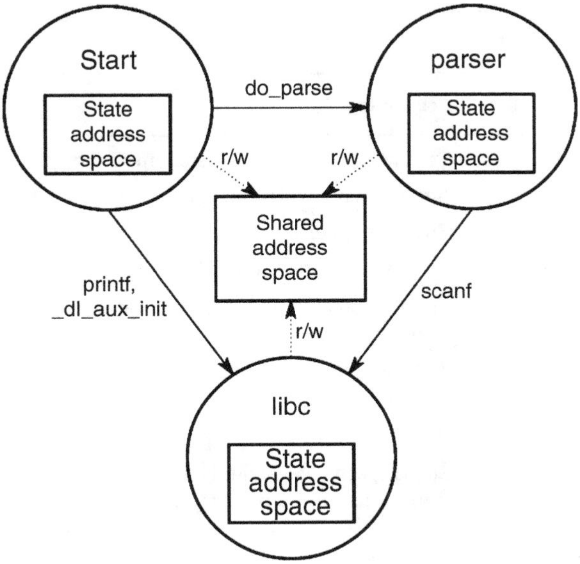

Figure 23.4 Building the state machine using Mithril.

in the *libc* state. This forces the developer to decide on the intentions for all the global variables and restrict their accesses between address spaces.

Second, we implement the state machine in the ruby-based domain specific language to reflect the origin state machine created. For each state, we specify the code sections that need to be given read, write, or executable access, and do the same for all the data sections. The function calls that trigger state transitions are also specified in the policy.

Third, we invoke Mithril to add the ELFbac section in the binary. We inspect the ELF binary to make sure the section was created. We trigger the code-paths that would violate our ELFbac permissions and cause *segfaults*. Exhaustive validation of all the possible state-transitions tells us that the policy is not over-permissive.

The syntax of Mithril involves keywords like: `tag`, `state`, `start`, `to`, `exec`, `readwrite` and `call`. Table 23.2 describes of each of these keywords.

23.3.7 How to Use ELFbac

While ELFbac requires some modifications to the OS kernel to enable policy enforcement, using ELFbac to preserve program intent is a straightforward process. Much of the work involved in creating an ELFbac policy revolves around modeling the security-relevant phases of execution of a program as a FSM and defining how the program transitions between these states at runtime. Here, we

Table 23.2 Syntax and usage for the keywords used in the Mithril Domain-specific language.

Syntax	Usage	Semantics
`start`	`start :main`	Set the start state for the state machine.
`exec`	`exec :code`	Specify that a code unit is only executable.
`readwrite`	`readwrite :data`	Specify the permissions for a data section.
`to`	`to :main`	Specify the state to transition to.
`call`	`call 'do_main'`	Specify a method that triggers a state transition.

`tag`	`tag :data do`
	`section ".datasec"`
	`end`
		Tag a section with a keyword. The tag can be used interchangeably with the section header.

`state`	`state :main do`
	`readwrite :data`
	`to :libc do`
	`call 'scanf'`
	`end`
	`end`
		States specify code and data units and the transitions from those states.

give a tutorial on how to design and enforce ELFbac policies that align with developer intent to prevent vulnerabilities.

ELFbac is most easily applied during the software development process, as the developer can create the policy based on their existing knowledge of the domain and the programmer's mental model [16]. At this point, the designer and developer can work together to determine which pieces of code pose security risks and formulate a policy that isolates these pieces from the rest of the program. While it is possible to explicitly separate every symbol, function, or library from one another, not all of these interactions present a security risk. Proper identification of these risks helps reduce the burden on the policy creator and prevents unnecessary policy complexity.

Incorporating ELFbac into legacy programs is possible, but may pose more of a challenge if the original designer and developer are unavailable, as "familiarity with a codebase is essential to understanding potential areas of vulnerability" [16]. Therefore, extra care must be taken to understand the code and identify the pieces of code that need to be isolated [16].

Once the pieces of the program that require isolation are identified, we can define our FSM with the appropriate states our policy will need to achieve our isolation goals. Most programs already include this sort of structure implicitly: Most programs already include this structure implicitly: sections of code handling encryption, network communications, user input, etc. ELFbac supports state labeling at a number of different granularities, ranging from individual symbols to function boundaries to entire libraries [13], so states can be as small or as large as required by the user. The key is to ensure that any code and data that needs to be isolated is placed inside its own state within the FSM so that ELFbac can protect it from unauthorized accesses.

Once this policy is defined, the user can use an ELFbac-instrumented compiler to produce a policy-aware binary. The controls explained in Section 23.3.5 that ELFbac uses to isolate the requested program pieces are mostly hidden to the user, until an attempted policy violation causes the program to halt.

23.3.8 ELFbac in Action

We highlight two examples of how ELFbac works to mitigate intent-violating vulnerabilities:

- The OpenSSH roaming bug is an example of such a mismatch of intent. A malicious server can trick an SSH client into potentially sharing private keys and other data. We will demonstrate how ELFbac uses existing memory isolation techniques to enforce the principle of least privilege over separate code and data.
- Spectre variant 1 exhibits yet another mismatch of intents: The processor designer's intent for performance (via speculative execution) unexpectedly conflicts with the developer's intent to preserve secret data. We will also demonstrate how ELFbac naturally mitigates Spectre variant 1 via policy.

23.3.8.1 SSH Roaming Bug

In 2010, developers working on the OpenSSH client (an open-source program used to securely connect to other machines using the SSH protocol) added an experimental "roaming" feature to their software, aiming to allow clients to reconnect to a server in the case of an unexpected network loss [16]. To accomplish this goal, the OpenSSH client would store any messages that could not be sent to the server in a buffer, which the server could then request the contents of when the connection was reestablished. However, there was a major flaw: The server was not bound by the buffer's actual content and could request an arbitrary amount of data from it upon reconnection even if the client had not filled that space.

In 2016, CVE-2016-0777 described how a malicious SSH server could steal sensitive information from an OpenSSH client using this flaw in the roaming feature [17]. If the memory allocated for the roaming buffer had previously contained private SSH keys or other secrets and had not yet been erased or overwritten by the client, it could be stolen by the server via a malicious buffer request [16]. While the risk of this vulnerability was limited to connections to compromised SSH servers, its existence demonstrated a clear violation of programmer intent: The feature was not intended for production release; the developers did not intend to give the server control over what data could be sent, and the developers certainly did not intend for sensitive information to be exposed via the roaming buffer.

To mitigate this flaw, we need to isolate the code and data involved in cryptographic operations from that which handles network operations, thus ensuring that secret keys cannot be leaked via the roaming buffer [16]. Since this is an existing codebase, we first needed to perform a thorough review to identify the pieces of the program that are involved in these tasks and thus need to be isolated.

Our resulting FSM consists of three states: A *crypto* state for tasks involving private and secret SSH keys, a *network* state that deals with network communications, and a *default* state that encompasses all other operations. (For the sake of simplicity, we assume that there are no other code/data interactions that pose a security risk; however, our existing states could be further split into smaller ones if the need arose.) Each state maintains its own copy of the process address space, but also has access to a shared address space in case data needs to be passed between states. Figure 23.5 depicts our full state machine to apply ELFbac to the OpenSSH Roaming bug.

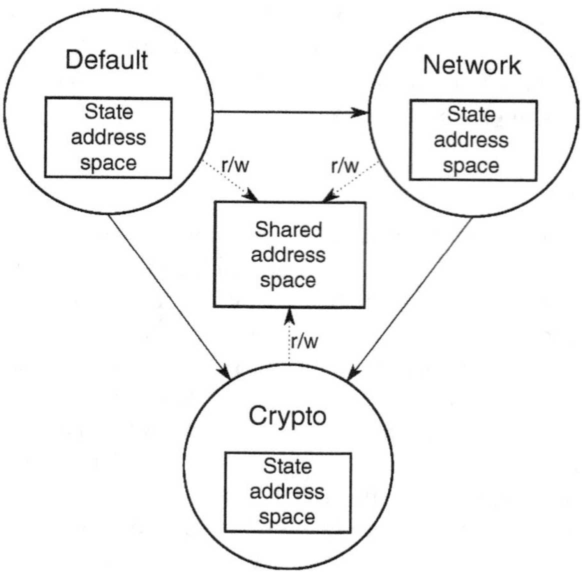

Figure 23.5 The finite state machine represented by our OpenSSH ELFbac policy. The memory allocated for private keys and the roaming buffer will come from different heaps within different address spaces, keeping the keys from being leaked to a malicious server.

The key insight here is that each state will maintain a private heap within its address space, which it uses for allocating memory for its own data items [16]. This separation of heaps means that even if a private key is not removed from memory after use, it will never overlap with the program's roaming buffer because the two items do not share the same memory space. The server can request as much of the buffer as they want, but any data used for cryptographic operations remains out of reach, thus mitigating the roaming bug.

23.3.8.2 Spectre Variant 1

In 2018, researchers revealed two major security flaws in the architecture of nearly every processor chip used in the last couple decades. One of these vulnerabilities was dubbed Spectre [2]. To achieve greater efficiency, modern processors are highly parallel in their operation. This parallelism inside instruction execution is often termed *pipelining*, and as an optimization technique, it allows the various execution and memory units of a CPU to operate simultaneously. A challenge for pipelining is conditional branching within a program. What code and data should be loaded into the pipeline when a branch is encountered? If the wrong decision is made, the pipeline must be flushed (or cleared away) so that the correct code can properly execute. Branch prediction is an additional optimization technique useful in limiting the amount of pipeline flushes that occur during execution. In essence, through various heuristics, a processor guesses which direction the conditional branch will take. Speculative execution is the pipelining and execution of instructions, after a conditional branch, based on a processor's guess about how the conditional will evaluate.

Spectre attacks take advantage of this branch prediction and the latent architectural effects of speculative execution. Kocher et al. show that "... speculative execution implementations violate the security assumptions underpinning numerous software security mechanisms, including operating system process separation, containerization, just-in-time (JIT) compilation, and countermeasures to cache timing and side-channel attacks" [2]. Here we find the violation of intents:

- Processor designers did not intend for there to be any noticeable, latent architectural effects.
- Operating system developers assumed their understanding of the architectural pipeline flushing mechanisms.
- Software developers intended for their program's secrets to be unaffected by spurious conditional branches.

To see this in action, Kocher et al. provide a proof-of-concept in C [2] that demonstrates an adversarial training of the branch predictor and its subsequent exploitation to reveal "secret" data that is never actually read during normal program operation. A snippet of the lines of interest follows in Listing 23.1.

The code begins with the creation of several regions of memory, e.g. `array1` and `array2`, including a `secret`. Additionally, there is a `victim_function` with a key conditional branch. To train the branch predictor, the PoC calls this function with valid values of input parameter `x`. These values touch areas of memory in `array1` and `array2`. After training, the `victim_function` is called with a malicious `x`. This malicious `x` should cause the conditional branch to evaluate false. Because of the prior training, the code is executed speculatively, which loads the page table entry that contains the secret string, thus caching the secret. After the secret is cached, the branch conditional is evaluated, and the CPU unrolls the speculatively executed instructions. However, this leaves behind the secret in the cache. Via some side-channel timing methods, this secret can be extracted.

A key observation is that the secret is never touched during normal execution of the program. The time when it is loaded is essentially by proxy, grouped along with other data within a page table entry. However, the latent effects of the architecture, i.e. the caching mechanisms, allow this data to be extracted. We draw the analogy of breadcrumbs being left behind during program execution.

So, what can ELFbac do to bridge this intent gap? ELFbac's primary means of policy separation is using page table differentiation. So, naively, the solution to this particular PoC, from the ELFbac perspective, is to place the secret on a separate page table.

Listing 23.1 Spectre PoC from Kocher et al.

```
/******************************
Victim Code
******************************/
uint8_t array1[16] =
{1,2,3,4,5,6,7,8,9,10,11,12,13,14,15,16};
uint8_t unused2[64];
uint8_t array2[256 * 512];

char * secret = "The Magic Words are Squeamish
Ossifrage.";

uint8_t temp = 0; /* Used so compiler won't optimize
out victim_function() */
void victim_function(size_t x) {
    if (x < array1_size) {
      temp &= array2[array1[x] * 512];
      }
}
```

(a)

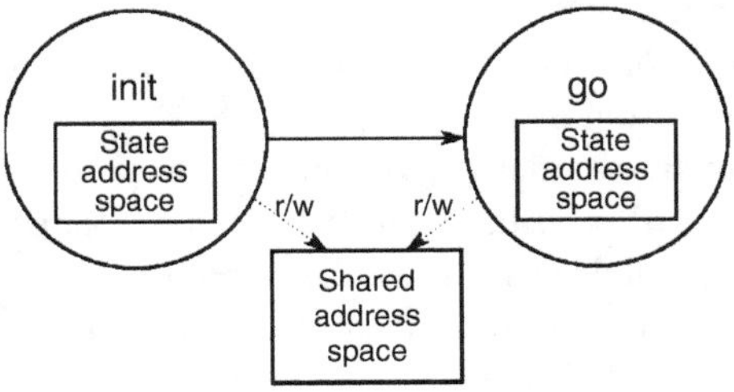

(b)

```
Elf::Policy.inject_symbols (file)
    x = Elf::Policy. build do
      state 'init_state' do
        readwrite :default
          exec :default
        readwrite :secret
         to 'go state' do
            call 'go'
       end
    end
    state 'go_state' do
       readwrite :default
       exec : default
       exec :go
     end
     Start 'init_state'
  end
  x.inject(file)
```

Figure 23.6 (a) State machine. (b) Mithril policy that mitigates Spectre variant 1.

As seen in Figure 23.6, two states are modeled. The `init` state represents the entire program with access to the `secret` data. Any access to the `secret` variable requires a state-transition to the `init` state. Outside of `init`, the `secret` is not accessible. As seen in part (b), only the `init_state` has `readwrite : secret`.

The key insight here is that page-table permissions are respected by the architecture. When the branch predictor is being trained, the `secret` variable is never "gobbled" by the paging mechanisms and thus cached. **When the speculative instructions are pipelined and executed, a page fault is raised when access to `secret` is requested in the wrong state.** This causes the pipeline to be stalled, voided, and eventually cleared before the caching mechanisms are invoked.

23.3.9 Summary

As IoT devices scale into the billions, it is absolutely imperative that they do not fall victim to intent violations stemming from vulnerabilities such as those described above. ELFbac is an effective, powerful policy solution that provides a critical layer of security against attackers looking to subvert designer and developer intent. It can be applied to both new and existing programs, and, for long-lived IoT devices where traditional security solutions (such as regular patching) are not feasible, ELFbac can still help protect against zero-day vulnerabilities and keep them from being exploited.

23.4 Building a Secure Implementation of AMQP

As an example of using LangSec and ELFbac together for the IoT, we consider hardening of AMQP protocol implementations.

Demonstrating that an implementation of a network protocol is secure from crafted-packet vulnerabilities includes ensuring that the parser rejects every message that does not conform to the grammar. Another implementation property to ensure: If a vulnerability that can be exploited exists in the parser, then it is isolated and placed in a separate memory region, so it does not affect the rest of the memory. (This example applies LangSec *and* ELFbac to the parser. Another family of use cases would apply LangSec to the parser and then apply ELFbac to the main program to contain damage from any attacks that get through this input validation component.)

The Advanced Message Queuing Protocol (AMQP) is a networking protocol that clients can use to communicate with each other through the server. AMQP requires at least one server (also known as a broker), and more than one client. The server consists of various components: an exchange, a message queue, and a binding. The exchange receives messages and decides which message queue must receive the message. Message queues store the messages sent if the messages haven't been consumed yet. Bindings specify which received message is put in which message queue. Traditionally, the clients *subscribe* to channels on the broker. Producers publish messages that are received by consumers.

The AMQP protocol is not very suitable for devices with very limited memory because of the size of the optional fields, but it is very suited for industrial IoT use cases such as SCADA systems with reliable network connectivity and bandwidth. Having an asynchronous queuing system for data means that the data will always be eventually processed, despite traffic spikes or network connectivity issues. AMQP supports better security protocols than its predecessor MQTT and also supports federation of various AMQP servers. Support for such delegation and a resilient security architecture make AMQP one of the most popular IoT protocols. A search for the AMQP port 5672 on Shodan shows that there are over 900 000 AMQP brokers operating on the Internet worldwide [18]. More than half of these AMQP brokers are operating in the United States alone. We also studied the number of reported vulnerabilities in implementations of the AMQP protocol. Table 23.3 shows that a significant portion of the number of reported vulnerabilities comprise of parser bugs which could be avoided using LangSec and isolated with the use of ELFbac.

In this section, we discuss the application of two tools: First, we use the Hammer parser-construction toolkit to demonstrate its efficacy on the AMQP protocol. Second, we implement and inject an ELFbac policy using Mithril.

Table 23.3 Number of vulnerabilities reported by the common vulnerabilities and exposures database maintained by Mitre for the AMQP protocol.

Year	Number of input-handling vulnerabilities	Total number of vulnerabilities
2015	2	3
2016	2	3
2017	4	5
2018	3	5

23.4.1 A Deeper Understanding of the AMQP Protocol

The various messages spoken by the client and the server in the AMQP protocol are demonstrated in Figure 23.7. The connection is initiated by the client, whereas the server is waiting to receive a communication.

The messages received by the broker are: Connection Start, Connection Tune, Connection Open, Channel Open, Queue Declare, Basic Publish, Basic Get, Channel Close, and Connection Close. The messages sent by the broker are: Connection Start OK, Connection Tune OK, Connection Open OK, Channel Open OK, Queue Declare OK, Basic Publish OK, Basic Get OK, Channel Close OK, and Connection Close OK. The receiving parser on the client needs to be able to recognize all the messages that are sent by the broker, and the parser on the broker needs to be able to recognize all the messages sent by the client. The state machines of the client and the servers are shown in Figures 23.8 and 23.9.

As a part of the LangSec approach to implementing parsers we analyze individual packet formats for the various messages supported by the protocol and extract the syntactic features used by the protocol. As seen in Figure 23.10, the packet format uses a length field that must be parsed to correctly parse the rest of the packet. The length of the rest of the packet must be equal to the length field in the packet.

The *payload* of an AMQP packet can be of various types: method frame, content-header frame, content-body frame, and heartbeat frame. The method frames are used the most, and the syntax structure of the method frame can be found in Figure 23.10.

As we mentioned earlier, LangSec argues protocols should be either regular or deterministic context-free. We consider all the structural requirements for the payload fields of the various packet formats in the AMQP protocol. We do not impose syntactical restrictions only on the headers but the entire packet as a part of the LangSec methodology. The AMQP specification includes a generic grammar for the AMQP packet format. Although the header usually remains similar across

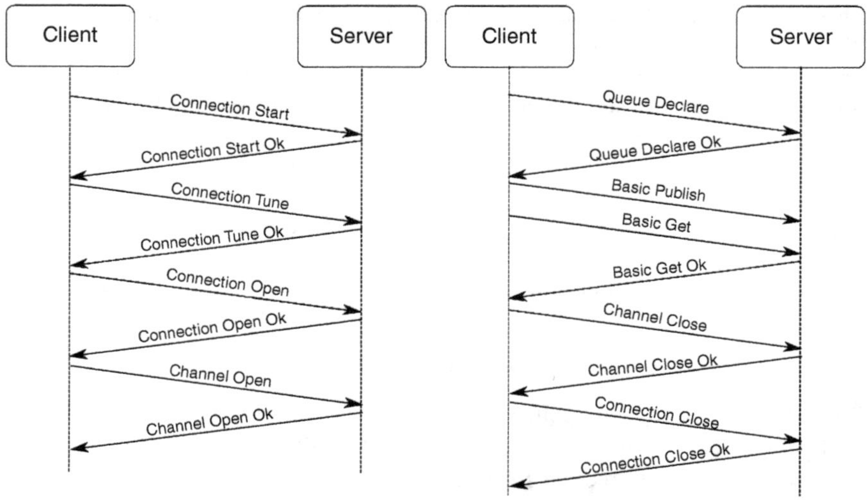

Figure 23.7 Flow of messages in the AMQP protocol.

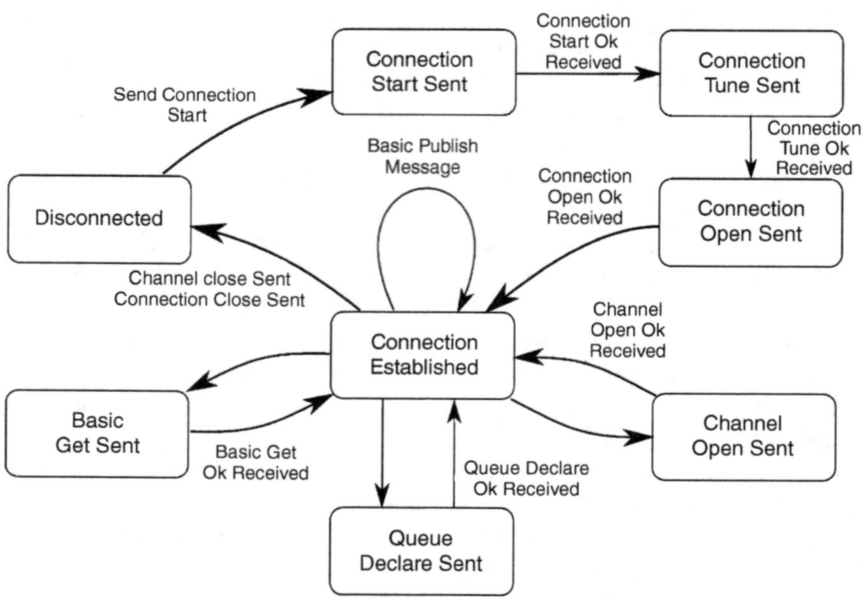

Figure 23.8 AMQP Client state machine. The client initiates the connection with the broker by sending a Connection Start message.

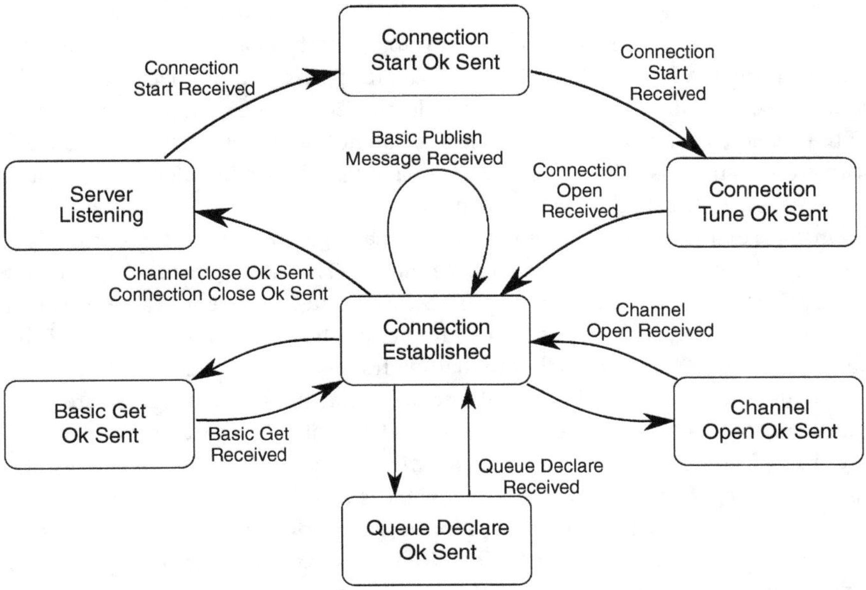

Figure 23.9 AMQP Broker State Machine diagram showing the various states and transitions. The server responds to connection initiation requests from clients.

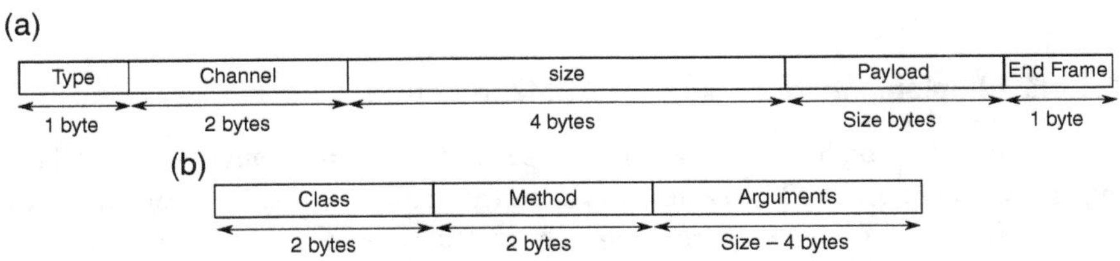

Figure 23.10 (a) Packet format of a generic AMQP packet. The number of bytes in the payload depends on the "size" bytes. (b) Format of the method frame used in AMQP payloads.

the different packet formats of AMQP, having separate grammars for the various packet formats would improve readability and make the parsers more fine-grained.

23.4.2 Hardening AMQP Parsers with Hammer

In the previous section, we detailed the AMQP protocol and its various message formats. Although the sequence of messages is important for any protocol filter, the contents of the messages aren't completely related, i.e. a parser built for the

Connection Tune message format does not rely on fields from its previous message, which was *Connection Start*. In case of protocols with reliance on data from previous message, we advocate performing stateful parsing, after checking for *well-formedness*. A well-formedness check would only ensure the structural correctness of the header, without checking for the equivalence of the reliant fields. This well-formedness check is followed by stateful parsing, where the fields are validated with respect to the data previously received.

Once we extract the syntax for the various messages in a particular protocol, we begin implementing our parsers for the protocol. The AMQP header begins with the `type` field, which is 8 bits long. There are a limited number of values that can comprise of the `type` message. For example, the field can only comprise of the values 1 through 4 corresponding to method frames, content header frames, content body frames, and heartbeats. We witness that the method frames are used most frequently in our data traces of the AMQP protocol. Apart from examining the sizes of each of the fields, we are also imposing constraints on the data itself – ranges of integers, ranges of characters, choices of strings, etc.

The syntaxes for each of the message formats are converted to specific parsers for each of the states in the state machine. We use `scapy` to extract the application layer of the packets and pass it on to the AMQP parser. The AMQP parser is implemented as a Docker container that receives raw messages and validates the syntax of the messages. The design of our parser implementation is in Figures 23.11 and 23.12.

23.4.3 Implementing a Policy for AQMP in Mithril

We implement a policy for the AMQP parser in the Mithril domain-specific language, already described in Table 23.2. First, the state machine is extracted for an implementation of the AMQP protocol. The states for the program as a whole are: start, parser, and libc. The globals are initialized in the start state, the message is received in the buffer. The buffer is then passed on to the parser. The parser

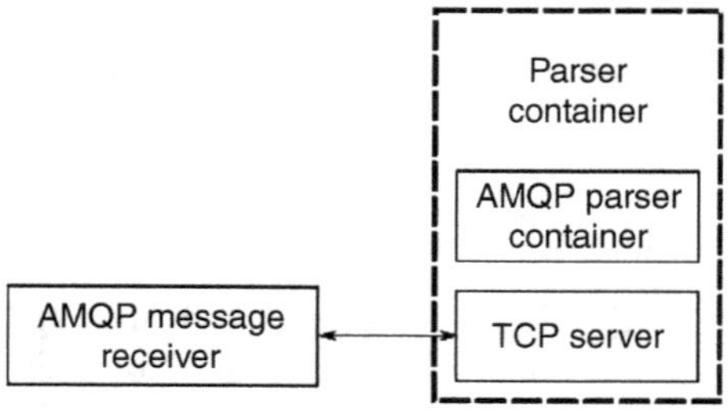

Figure 23.11 Design of our AMQP parser implementation.

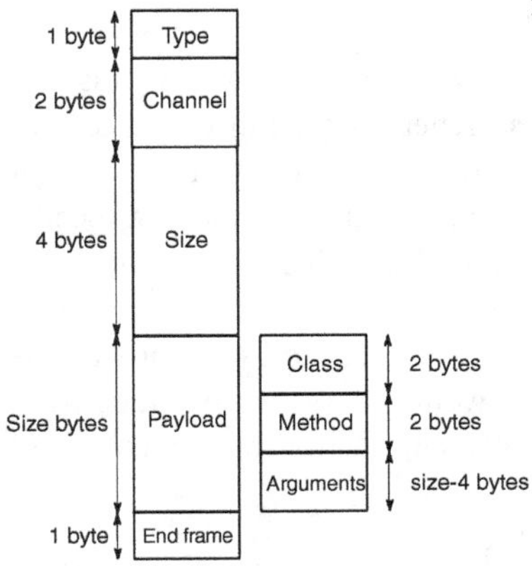


```
h_sequence(
   h_uint8(),    # Type
   h_uint16(),   # channel

    h_and(h_length_value(
        h_uint32(),
        h_ch_range('\x00', '\xff'))),
    h_uint32(),       # Size
    h_sequence(
        h_uint16(), # class
        h_uint16(), # Method
        h_optional(arguments)), # Arguments
    h_ch('\xce'),   # End Frame
    h_end_p(), NULL);
```

Figure 23.12 Conversion of a syntax structure for a packet format into Hammer parser-combinator code.

performs the parsing operation and hence has to have access to the necessary globals. In the libc state, the program performs operations such as `scanf` and `printf`.

Second, this state machine is then converted into a Mithril policy. Each state is specified with a fine-grained policy specifying all the global variables that need to be accessible and the transitions that would lead to that state. Our policy for AMQP starts with the `start` state and then transitions to the parser when the `do_parse` method is called. All globals need to be given `readwrite` access to the `start` state since the globals need to be initialized. Since Mithril makes use of dynamic loading, we transition to `libc` whenever `_dl_runtime_resolve` is called. Since in our implementation we do not make use of any other libraries, this method would only be executed when functions from `libc` are called.

The state machine and the corresponding Mithril policy code can be found in Figure 23.4.

23.5 Evaluation Techniques

Thus far in this chapter, we have studied existing research to enforce the programmer's intent with techniques such as LangSec and ELFbac. In this section, we will survey some of the techniques widely used to test software built on the principles of LangSec and ELFbac.

23.5.1 Coverage-Guided Fuzzing with AFL

American Fuzzy Lop (AFL) is a state-of-the-art coverage-guided fuzzer that is protocol agnostic. It takes a set of samples and runs genetic algorithms on it to generate more samples to run through a program [19]. Fuzzers run over their target many times with different inputs. AFL does not require that we extract the specific target methods into separate files and then compile those files as fuzz targets.

Other techniques, such as control-flow analysis and data-flow analysis, demonstrate interesting ways to understand how a binary operates and what operations the binary performs. However, they do not scale well for large applications. AFL can be given a single binary, and it can perform brute-force fuzzing that covers all edges of a program's control flow.

```
afl-fuzz -i testcase.dir -o findings.dir /path/to/program @@
```

Binaries are fuzzed with the script above. AFL provides a text-based GUI to display the progress, number of paths discovered, number of crashes and hangs, and the run time. The inputs that lead to crashes and hangs are all logged in separate folders.

23.5.2 Static Analysis Using Infer

Infer is a static-analysis tool that can be run on C code to detect certain categories of errors commonly found. Namely, Infer can catch null de-references, memory leaks, premature nil termination arguments, and resource leaks [20]. Static analysis provides another layer of assurance that the program doesn't have any vulnerabilities.

Static analysis requires access to the source code. These tools analyze all code paths of a program and check for paths that lead to crashes, memory violations, memory corruption, memory leaks, etc. They can also be programmed to detect style violations. In our implementations, we use [20] and make sure there are no memory violations.

Memory violations usually indicate poor handling of input, which could lead to vulnerabilities. Capturing programmer's intent with LangSec and ELFbac can alleviate these memory violations and isolate them.

23.6 Conclusions

In this chapter, we introduced the notion of intent as a secure design primitive and we discussed the importance of preserving designer and developer intent for system security. We presented two security paradigms to safeguard against adversaries who wish to subvert system security by violating designer and developer intent: *LangSec*, which constrains program inputs to safe options as defined by the protocol specification and the Chomsky hierarchy, and *ELFbac*, which isolates unrelated code and data and enforces boundaries between program pieces as defined by designer/developer policies. Finally, we showed how a simple program such as an AMQP parser could be hardened using tools delivered by these paradigms, and we introduced various ways to test and validate such hardened programs.

While LangSec and ELFbac provide some mechanisms to preserve designer and developer intent, work remains. For instance, we aim to extend LangSec's functionality beyond just validating packets against a protocol specification; we intend to start examining the data within the packet to ensure that it makes sense within the program's context. For example, can we examine the metadata around a piece of data (where it came from, what values we have seen in the past, etc.) to make a decision about how trustworthy it is? With ELFbac, we are looking for ways to formally prove specific security properties of ELFbac-enhanced programs, similar to projects such as Low* [21].

We also intend for our tools to extend beyond merely enforcing intent. We aim to help the designer and the developer think critically about their intentions and what they should or should not do to maintain system safety. Currently, LangSec and ELFbac blindly follow the instructions of the designer and developer and blindly ensure that whatever intent they express is not violated. However, if the original intent of the system is misguided (for example, if the desired protocol is too powerful to be verified as safe), our tools should alert the designer/developer to this issue and help them produce a better alternative that balances safety with functionality. At the very least, we want LangSec and ELFbac to make designers and developers explicitly consider their intent and ask themselves if they are asking too much of their system.

Finally, we must also consider preservation and refinement of user intent, which neither LangSec nor ELFbac currently address. As we noted before, users are ultimately the ones who determine how a system operates, and thus their actions can supersede the intent of the designer and developer. As a community, we must pay special attention to the *usability* of systems to ensure that the user uses the system the way the designer and developer intended, instead of violating designer and developer intent to achieve their primary task.

In sum, securing the IoT requires realizing the intent of each stakeholder. LangSec and ELFbac make great strides toward addressing this grand challenge, but more work must be done.

23.7 Further Reading

For an in-depth discussion of the theoretical aspects of LangSec, we recommend the reader consult Sassaman et al. [22]. For Linux kernel implementation details of ELFbac, we recommend consulting Bangert et al. [13].

The yearly LangSec Workshop at the IEEE Symposium on Security and Privacy provides a venue to discuss LangSec research ideas. For case studies of LangSec and ELFbac, the reader may consult Anantharaman et al. [23], Bratus et al. [24], and Jenkins et al. [16].

Acknowledgments

This material is based upon work supported by the United States Air Force and DARPA under Contract No. FA8750-16-C-0179, ONR under grant N00014-18-1-2022 and Department of Energy under Award Number DE-OE0000780.

Any opinions, findings and conclusions or recommendations expressed in this material are those of the authors and do not necessarily reflect the views of the United States Air Force, DARPA, ONR, United States Government, or any agency thereof.

References

1 M. Antonakakis, T. April, M. Bailey, M. Bernhard, E. Bursztein, J. Cochran, Z. Durumeric, J. A. Halderman, L. Invernizzi, M. Kallitsis et al., "Understanding the Mirai Botnet," *in Proceedings of the 26th USENIX Security Symposium,* USENIX Association, 2017, pp. 1093–1110.

2 P. Kocher, D. Genkin, D. Gruss, W. Haas, M. Hamburg, M. Lipp, S. Mangard, T. Prescher, M. Schwarz, and Y. Yarom, "Spectre Attacks: Exploiting Speculative Execution," In Proceedings of the *IEEE Symposium on Security and Privacy (SP),* San Francisco, CA, USA, IEEE Computer Society, 2019, pp. 1–19.

3 Z. Durumeric, F. Li, J. Kasten, J. Amann, J. Beekman, M. Payer, N. Weaver, D. Adrian, V. Paxson, M. Bailey, and J. Alex Halderman. "The Matter of Heartbleed," In *Proceedings of the Conference on Internet Measurement Conference (IMC '14),* Association for Computing Machinery, New York, NY, USA, 2014, pp. 475–488.

4 T. Fox-Brewster, "What Is the Shellshock Bug? Is It Worse than Heartbleed," *The Guardian.* https://www.theguardian.com/technology/2014/sep/25/shellshock-bug-heartbleed.

5 P. Ducklin, "Anatomy of a Security Hole – Google's "Android Master Key" Debacle Explained," *Naked Security,* Sophos. https://nakedsecurity.sophos.com/2013/07/10/anatomy-of-a-security-hole-googles-android-master-key-debacle-explained/.

6 F. Momot, S. Bratus, S. M. Hallberg, and M. L. Patterson, "The Seven Turrets of Babel: A Taxonomy of LangSec Errors and How to Expunge Them," in *Proceedings of the 1st IEEE Cybersecurity Development (SecDev),* IEEE Computer Society, 2016, pp. 45–52.

7 S. Bratus, M. Locasto, M. Patterson, L. Sassaman, and A. Shubina, "Exploit Programming: From Buffer Overflows to Weird Machines and Theory of Computation," *USENIX; login:,* 2011, vol. 36, no. 6, pp.13–21.

8 M. Sipser, *Introduction to the Theory of Computation*, Thomson Course Technology, Boston, 2006.

9 G. Sénizergues, "The Equivalence Problem for Deterministic Pushdown Automata is Decidable," in *Proceedings of the 24th International Colloquium on Automata, Languages, and Programming*, Springer-Verlag, Berlin, 1997, pp. 671–681.

10 J. H. Saltzer and M. D. Schroeder, "The Protection of Information in Computer Systems," In *Proceedings of the IEEE*, IEEE Computer Society, 1975, vol. 63, no. 9, pp. 1278–1308.

11 J. Postel, "DoD Standard Transmission Control Protocol," *ACM SIGCOMM Computer Communication Review*, ACM, 1980, vol. 10, no. 4, pp. 52–132.

12 L. Sassaman, M. L. Patterson, and S. Bratus, "A Patch for Postel's Robustness Principle," *IEEE Security & Privacy*, IEEE Computer Society, 2012, vol. 10, no. 2, pp. 87–91.

13 J. Bangert, S. Bratus, R. Shapiro, M. E. Locasto, J. Reeves, S. W. Smith, and A. Shubina, "ELFbac: Using the Loader Format for Intent-Level Semantics and Fine-Grained Protection," Dartmouth College, Department of Computer Science, Technical Report 727, June 2013.

14 P. A. Loscocco and S. D. Smalley, "Meeting Critical Security Objectives with Security-Enhanced Linux," in *Proceedings of the Ottawa Linux Symposium*, 2001, pp. 115–134.

15 E. Witchel, J. Cates, and K. Asanović, "Mondrian Memory Protection," in *10th International Conference on Architectural Support for Programming Languages and Operating Systems, ser. ASPLOS '02*, ACM, 2002, pp. 304–316.

16 I. R. Jenkins, S. Bratus, S. Smith, and M. Koo, "Reinventing the Privilege Drop: How Principled Preservation of Programmer Intent Would Prevent Security Bugs," in *Proceedings of the 5th Annual Symposium and Bootcamp on Hot Topics in the Science of Security, ser. HoTSoS '18*, ACM, 2018, pp. 3:1–3:9.

17 CVE-2016-0777. MITRE, January 2016. https://cve.mitre.org/cgi-bin/cvename.cgi?name=cve-2016-0777.

18 Shodan. "Shodan: The Search Engine for the Internet of Things." https://www.shodan.io/.

19 M. Zalewski, "American Fuzzy Lop," 2015.

20 C. Calcagno, D. Distefano, J. Dubreil, D. Gabi, P. Hooimeijer, M. Luca, P. O'Hearn, I. Papakonstantinou, J. Purbrick, and D. Rodriguez, "Moving Fast with Software Verification," in *Proceedings of the 7th International Symposium NASA Formal Methods Symposium*. Springer International Publishing, 2015, pp. 3–11.

21 J. Protzenko, J.-K. Zinzindohoué, A. Rastogi, T. Ramananandro, P. Wang, S. Zanella-Béguelin, A. Delignat-Lavaud, C. Hriţcu, K. Bhargavan, C. Fournet, and N. Swamy, "Verified Low-Level Programming Embedded in F∗," in *Proceedings of the ACM on Programming Languages*, ACM, 2017, vol. 1, no. 17, pp. 1–29.

22 L. Sassaman, M. L. Patterson, S. Bratus, M. E. Locasto, and A. Shubina, "Security Applications of Formal Language Theory," *IEEE Systems Journal*, IEEE Computer Society, 2013, vol. 7, no. 3, pp. 489–500.

23 P. Anantharaman, M. Locasto, G. F. Ciocarlie, and U. Lindqvist, "Building Hardened Internet-of-Things Clients with Language-Theoretic Security," in *Proceedings of IEEE Security and Privacy Workshops (SPW)*, IEEE Computer Society, 2017, pp. 120–126.

24 S. Bratus, A. J. Crain, S. M. Hallberg, D. P. Hirsch, M. L. Patterson, M. Koo, and S. W. Smith, "Implementing a Vertically Hardened DNP3 Control Stack for Power Applications," in *Proceedings of the 2nd Annual Industrial Control System Security Workshop*, ACM, 2016, pp. 45–53.

24

A Review of Moving Target Defense Mechanisms for Internet of Things Applications

Nico Saputro[1,2], Samet Tonyali[3], Abdullah Aydeger[1], Kemal Akkaya[1], Mohammad A. Rahman[1], and Selcuk Uluagac[1]

[1] *Department of Electrical and Computer Engineering, Florida International University, Miami, FL, USA*
[2] *Department of Electrical Engineering, Parahyangan Catholic University, Bandung, Indonesia*
[3] *Department of Electrical and Computer Engineering, Abdullah Gul University, Kayseri, Turkey*

24.1 Introduction

In recent years, we have witnessed an exponential growth in the number of Internet of Things (IoT) devices around us [1]. IoT offers promising solutions in many application domains. Today, IoT devices and application are transforming the operations and role of many existing industrial systems from transportation to manufacturing to healthcare service systems. For example, in the transportation systems, IoT is used to realize Intelligent Transportation System (ITS) implementations [2]. There are also other on-going efforts to use IoT in critical infrastructures (CIs) for a nation such as in the transformation of the existing electric grid into the Smart Grid [3]. Similarly, their applications in the battlefield domain are also evolving [4]. Besides being used on the areas such as unmanned units (aerial or land or underwater) and troop health [5], there is a need to use IoT to support and augment critical-missions battlefield operations such as in the Intelligence, Surveillance, and Reconnaissance operations [6].

Security is a supreme importance for these applications. On the one hand, we can reap the benefits of the implemented IoT systems, yet on the other hand, they can pose significant security threats. However, securing them towards attacks from the

Modeling and Design of Secure Internet of Things, First Edition. Edited by Charles A. Kamhoua, Laurent L. Njilla, Alexander Kott and Sachin Shetty.

Published 2020 by John Wiley & Sons, Inc.

adversaries is very challenging. Typically, the applications and the used IoT devices are developed by a wide variety of organizations, either from for-profit organizations (e.g. start-up companies, small and medium enterprises, large corporations) or from non-profit organizations such as academic research institutions. Besides the interoperability issue between applications and devices that come from multiple developers, to keep pace with competitors, the developers often develop an IoT device only with the necessary network capabilities without any strong implementation of network security features in it. Furthermore, the security threats are augmented by the lack of physical security and upgradeability since these devices can be placed in any non-traditional locations such as in appliances, automobile, streets, or any remote locations. When the remote locations are inaccessible or very difficult to reach, upgrading is nearly impossible or too costly and thus the IoT devices are designed to operate for years with no means of long-term supports.

To secure an IoT system, one approach that can be done before the deployment is by incorporating security measures that address as many of the security vulnerabilities as possible at the design phase (a.k.a *security by design*). However, this approach may not enough. The lessons learned from securing the digital infrastructures, which have been studied for years before the imminent arrival of IoT, indicate that it is a never-ending battle between the defenders and attackers. Once an exploited vulnerability can be identified by the defenders and a patch and/or a defense mechanism against an attack that exploits this vulnerability is developed, attackers are able to find new vulnerabilities or increase their attack sophistication to outsmart the newly developed defense mechanism. This situation is driven by the fact that the existing network configurations of any digital infrastructures (e.g. network address, operating system, software, hardware, etc.) mostly remain static all the time. Thus, attackers have no time constraints of finding any vulnerabilities in the systems and exploiting these vulnerabilities at any time. Moreover, since other similar systems may use the same network components (e.g. operating system, vulnerable software), attackers can easily replicate their attacks to these systems.

Given these attackers' advantages, the defender's traditional approaches [7] that strive to directly deal with the vulnerability are no longer effective since they always fall-behind the attackers. Thus, complementary proactive self-defending approaches to break these advantages by introducing uncertainty through randomization of system attributes have been initiated in 2003. These types of approaches are later classified under the name of Moving Target Defense (MTD). MTD dynamically shifts the systems attack surface (e.g. IP addresses and ports, software/hardware vulnerability, memory location to store instructions and data, etc.) by leveraging redundancy (e.g. adding extra components with the similar functions) and diversity in the systems to make the attack surface unpredictable for attackers. Since the coined MTD in 2009, a variety of MTD techniques have been proposed and can be classified into five major categories based on their implementation on the software stack model [8]: (i) dynamic data; (ii) dynamic software; (iii) dynamic runtime environment; (iv) dynamic platform; and (v) dynamic network.

MTD techniques have been promising in many of the traditional network domains and thus such success has given rise to bring this experience to the IoT domain. Thus, in recent years, we have started to see studies that apply a variety of MTD techniques to numerous IoT environments (e.g. see Chapters 10 and 18 in this book) while there is still a lot of challenges yet to be addressed. Given the wide variety of MTD techniques, and the IoT characteristics, limitation, and the security and privacy issues in IoT, it is imperative for professionals and researchers working in IoT security to have a firm grasp on the available MTD techniques that are appropriate for the IoT systems depending on the application's needs.

Therefore, this chapter aims to provide a review of MTD approaches that revolve around IoT applications and then investigate the feasibility and potential of specific MTD approaches that will benefit some of the IoT applications the most. Our goal is to not only provide a categorization of various MTD approaches but to also lay down new research directions by advocating certain MTD techniques that can be used in conjunction with some of the emerging networking paradigms such as

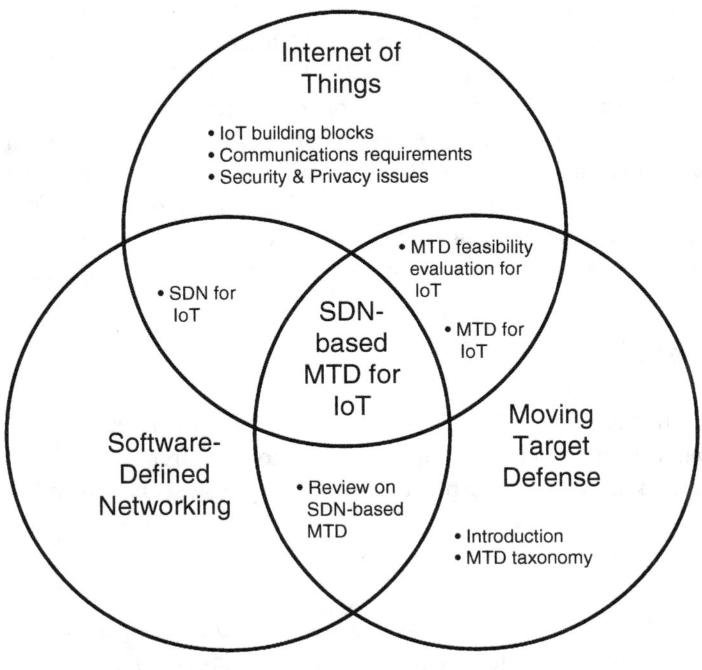

Figure 24.1 Three-different domains – IoT, SDN, and MTD.

Software Defined Networking (SDN) [9]. This creates a rich intersection of multiple concepts as shown in Figure 24.1 that would benefit IoT security and defense.

The chapter begins the discussion by first laying out the foundation of MTD in the digital infrastructures. The discussion includes the motivation behind MTD and a brief overview of the existing MTD classifications. Then the security and privacy challenges for IoT as well as its characteristics and limitation are discussed. Based on that, we provide a brief evaluation of the feasibility of implementing each of the five major MTD categories, which are basically intended for the enterprise network, for IoT. For example, the dynamic platform that provides diversity by utilizing multiple operating systems, different processor architecture, storage systems, etc. may not be applicable in IoT system since it may be too costly to provide multiple processor architectures or storage systems when the role of the IoT device is merely for a simple task. Based on the feasibility analysis discussion, we then focus on the use of the dynamic network option for IoT.

In the dynamic network domain, the primary attempt for diversity is by modifying the network properties. In this chapter, we will first propose a taxonomy for the MTD techniques in the dynamic network domain. This categorization will be based on certain criteria in terms of techniques, network architecture, and the types of network attacks that these techniques strive to deal with. We then emphasize the MTD techniques that are supported by SDN paradigm which provides flexibility in terms of network management and control. A brief introduction on SDN will be provided prior to the discussion of the SDN-based MTDs.

After the discussion on the SDN-based MTD for the general digital infrastructures, a review on the existing works on MTD for IoT, either SDN-based or non-SDN based, is provided. The review follows a similar structure, i.e. it includes the techniques, network architecture, and the type of attacks. Finally, we provide future research directions and challenges that relate to SDN-based MTD for IoT. The chapter is finalized with a conclusion that summarizes the contributions.

24.2 Internet of Things

In this section, we first explain the IoT and the Industrial Internet of Things (IIoT) and its components. Then, we explain the communication requirements to build IoT and IIoT networks. Finally, we explain security and privacy concerns in IoT and IIoT applications.

24.2.1 Overview

The term IoT was initially referred to inter-operable objects with radio frequency identification (RFID) technology [10]. Then, its span expanded so as to cover other "things" that are capable of computing and communication as the market delivered low-cost, mobile computation and communication technologies. Today, the IoT is envisioned as a self-established global network infrastructure comprised of interconnected and uniquely identifiable physical artifacts, services, and humans that can be accessed from anywhere in the world through the Internet using standard communication protocols [11–15]. The IoT network aims to make these three parties communicate, and exchange data and information without human intervention as far as possible to fulfill a common need in different application domains [16, 17]. To this end, it enables Human-to-Thing, Thing-to-Thing, and Thing-to-Things communications [18]. The term of "thing" in the IoT concept mostly refers to compact smart devices [19] such as smartphones, tablets, digital cameras, automation systems, and controllers rather than typical computational platforms such as personal computers and workstations. Moreover, the things that are already a part of our environments such as home appliances, light bulbs, consumer electronics equipped with sensors, actuators that have communication interface, our cars, and offices and some other devices equipped with, say RFID tags, connected to the Internet through a gateway take part in the IoT.

Over the past decade, the number of connected devices has exceeded 15 billion. If this trend continues, it is expected that an estimate of 50 billion devices will be connected by the year 2020 [20]. Considering the drastically increasing number of connected devices, it can be deduced that the IoT has initiated a technological revolution in ubiquitous connectivity, computing, and communications.

The variety in the "things" that can take part in the IoT leads to a large number of application domains for the IoT. The domains are including but not limited to transportation, agriculture, food processing, healthcare, entertainment, home automation, industrial automation, surveillance and military, smart energy monitoring and management, and smart cities [12, 16, 21].

The IoT applications facilitate monitoring, controlling, and thereby interacting with the surrounding environment to make it more plausible. For example, smart cars, roadside units, and smart traffic signals make the driving experience safer and more convenient [16]. Another IoT-enabled technology is the ongoing Smart Grid initiative which enables high-frequency data collection compared to existing metering systems from the consumers, distribution substations, and transmission lines [22–26]. Such industrial systems that interconnect production systems and integrate them into conventional business information technology (IT) systems by incorporating the IoT technologies are called Industrial IoT (IIoT) [27].

The IIoT is a gateway towards digitizing the entire industry. Roughly speaking, this is achieved by collecting sensor data from industrial systems and transmitting them to more powerful computation and storage units over the Internet to analyze the system state and take convenient actions thanks to intelligent machine applications. This paves the way for more efficient production systems/manufacturing processes which are a key to the growth and competitiveness in a global economy [28] since the IIoT has the potential to improve quality control, supply chain management, sustainable services, and maintenance services for the user in the loop [29].

24.2.2 IoT Building Blocks

In this subsection, we explain the fundamental components of an IoT infrastructure.

An IoT ecosystem is typically composed of physical and digital worlds and the communication network that connects these worlds to each other [30]. As shown in Figure 24.2, the IoT architecture is a four-layer model comprised of IoT devices and gateway, communication network, cloud server, and IoT application.

IoT devices are equipped with sensors, actuators, controllers/processors, network interfaces, and short- and long-term memory. An IoT device runs its hardware through a firmware or an operating system that resides on its long-term memory. The device senses its surrounding environment through its sensors, stores the digitized data in its short-term memory to process the obtained data through its controller or processor, and communicates with an IoT application running on a cloud server which can be accessed over the Internet. The application processes the data collected from the IoT device and provides feedback to the device. Accordingly, the device takes an appropriate action ranging from displaying the result of a computation to switching a relay if it is a part of a cyber-physical system.

The IoT devices exchange data with other devices in the network or cloud servers to fulfill an IoT application's objective. The communication technology may differ while conveying the data towards the target device/cloud server which requires a protocol conversion because the communication network of an IoT ecosystem is the typical internet network having different layers (physical, link, network, transport, and application) with different protocols operating at each layer. In such cases, a gateway is installed where the protocol conversion is needed.

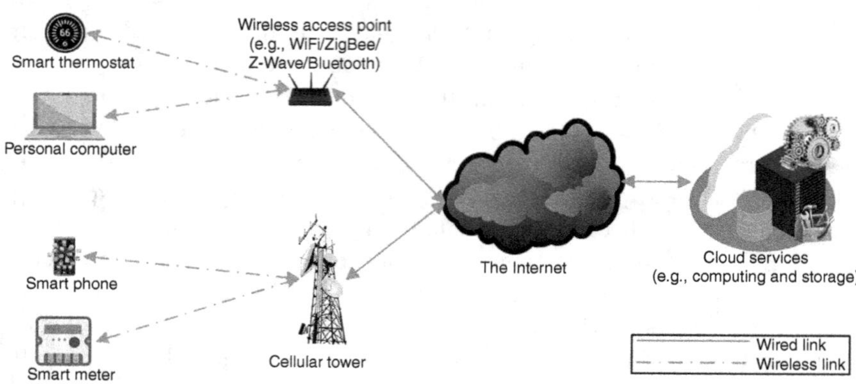

Figure 24.2 Essential building blocks of an IoT environment.

A gateway is supposed to be able to perform a bidirectional packet format conversion. For example, smart home appliances typically have a ZigBee [31] interface (Z-Wave [32] is also another promising protocol stack for smart home applications) and thereby communicate over the ZigBee protocol. To report the collected data from home appliances to the cloud server, the devices need a gateway that converts ZigBee packets into Long-Term Evolution (LTE) [33] or TCP/IP [34] packets and communicate with the cloud server by following the relevant protocol. Similarly, when the gateway receives some packets from the cloud server for the home appliances, it converts LTE or TCP/IP packets into ZigBee packets and follows the ZigBee protocol.

As mentioned above, the IoT devices communicate with a cloud server to store the collected data. These systems reside at the edge of the IoT system and have abundant storage resources and powerful processors. Thus, the stored data is used in data mining and analysis to make useful inferences for an IoT application. They also monitor the connected devices and manage device-to-device communication. They operate and synchronize different IoT devices and enable IoT applications. In addition, they communicate with other private and public servers or cloud services when needed for an IoT application.

An IoT application is the essential and indispensable part of an IoT ecosystem. It is a piece of software running on the cloud server that pre-processes, mines, and analyzes the stored data to derive useful insights and to manipulate the targeted IoT device(s) securely based on these insights. For instance, an IoT application designed for smart home automation can process data received from pressure sensors when households are not home, send commands from the cloud to lock the doors and windows and report the situation to the households and police department.

24.2.3 Communication Requirements of IoT

According to the statistics given in [35], there are approximately 25 billion IoT devices connected worldwide. The great majority of these devices are mobile, thereby should be equipped with required technology to be able to access the Internet whenever and wherever needed. Hence, it is inevitable to take advantage of wireless communication technologies because wireless solutions require far less cabling work and lower the infrastructure, deployment, and maintenance costs. Some of the wireless radio technologies currently used for IoT include 802.16 (WiMAX) [36], 802.11 (Wi-Fi) [37], and 802.15 (Bluetooth [38], ZigBee, and Z-Wave). These technologies differ from each other in the underlying PHY/MAC implementation which specifies the data rate, bandwidth, communication range, and so on. Hence, each of these technologies should be employed in convenient environments in which they can operate efficiently.

As we mentioned in Section 24.2.1, there will be more than 50 billion IoT devices connected by 2020. According to the survey conducted by SANS Institute [39], 72% of the survey participants who are industrial organizations base their communication infrastructure on Internet Protocol Suites where IP addresses are used to identify each device in IoT networks. Considering that there are fewer than 4.3 billion Internet Protocol version 4 (IPv4) [40] addresses (4 294 967 296 IPv4 addresses [41]) which are about to be exhausted, we can deduce that we will need a new mechanism to address the additional devices. Fortunately, Internet Protocol version 6 (IPv6) [42] has a potential of almost unlimited address space for more than trillions of devices [12]. Therefore, network stack of IoT devices is designed to support IPv6 routing protocol.

The IPv6 introduces an overhead such that the header occupies more space than the payload in a data fragment, which is not convenient for low-power and resource-constrained devices because this increases the number of datagrams to be transmitted. To overcome this problem, IPv6 over Low-Power Wireless Personal Area Network (6LoWPAN) protocol [43] was developed. 6LoWPAN is a low power communication network which connects resource-constrained wireless devices using compressed IPv6. It defines IPv6 header compression and how packets are routed through the IEEE 802.15.4 [44] links. It also defines fragmentation of IPv6 datagrams when the size is more than the IEEE 802.15.4 Maximum Transmission Unit (MTU) [45] which is 127 bytes.

An advantage of 6LoWPAN networks is that they support multihop communication (mesh network) where the nodes between a source and a destination node can forward data packets towards the destination node on behalf of the source node. Another advantage is on energy consumption. Instead of idle listening which is mostly used by power supplied devices, 6LoWPAN networks follow duty cycling paradigm which means that the radio is turned on only for a very short time for listening.

Currently used connection-oriented web protocols such as Hypertext Transfer Protocol (HTTP) [46] or HTTP Secure (HTTPs) [47] are designed to be used over Transmission Control Protocol (TCP) [48] which is not feasible for low-power and lossy links where it is hard to maintain a continuous connection between devices. Therefore, the Constrained Application Protocol (CoAP) [49] running on top of the connectionless User Datagram Protocol (UDP) [50] was developed for the IoT communication [12].

The protocols that were developed for the IoT communications can be listed as follows. Some of the most commonly used application layer protocols are CoAP, COAP Secure (COAPs) [49], Representational State Transfer (REST) [51], Message Queuing Telemetry Transport for Sensor Networks (MQTT-SN) [52], Live Long and Process (LLAP) [53], Extensible Messaging and Presence Protocol Internet of Things (XMPP-IoT) [54], and Advanced Message Queuing Protocol (AMQP)

[55]. TCP, UDP, Datagram Transport Layer Security (DTLS) [56], and Transport Layer Security (TLS) [57] are some examples of transport layer protocols. As mentioned above, IPv4 and IPv6 are the two most common network layer protocols. In addition, the Routing Protocol for Low-Power and Lossy Networks (RPL) [58] is a standardized routing protocol for the IoT. It is primarily designed for 6LoWPAN networks which are low-power and lossy networks. There is a wide variety of physical or link layer protocols that can be used for IoT. Low-Rate Wireless Personal Area Network (LR-WPAN) [59], Bluetooth/Bluetooth Low Energy (BLE) [60], 802.15.4, LTE, General Packet Radio Services (GPRS) [61], Collision Detect Multiple Access (CDMA) [62], Carrier Sense Multiple Access (CSMA) [63], CSMA Collision Detection (CSMA/CD) [64], CSMA Collision Notification (CSMA/CN) [65], Zigbee, 802.11 Wi-Fi, Near Field Communication (NFC) [66], Wireless Highway Addressable Remote Transducer (WirelessHART) [67], Z-Wave, Sigfox [68], DASH7 Alliance Protocol (D7A) [69], Long Range Wide Area Network (LoRaWAN) [70], Thread [71], and INSTEON [72] are some of the most known physical/link layer protocols [30]. In Table 24.1, we provide the IoT technologies developed for each OSI layer and their implementations in the Contiki operating system [12].

When it comes to IIoT both wired and wireless communication technologies can be used. In addition to those used by IoT, Ethernet, Profinet [81], and ModBus [82] are some examples to the protocols developed particularly for IIoT. MQTT was designed for IIoT networks although it is also used by IoT devices. It is a light-weight publish–subscribe messaging protocol which enables two-way machine-to-machine communications [14, 83]. Another publish–subscribe-based machine-to-machine messaging protocol is Data Distribution Service [84] which has a wide variety of application domains such as autonomous vehicles, air-traffic control, smart grid, robotics, medical devices, and so on.

Table 24.1 IoT technologies developed for each OSI layer and their implementations in Contiki OS.

OSI layer	IoT technology	Contiki implementation
Application	CoAP, CoAPs	Erbium [73]
Session	DTLS	TinyDTLS [74]
Transport	UDP	*micro*IP [75]
Network	IPv6, RPL, IPSec [76], 6LoWPAN	*micro*IP, ContikiRPL [77], IPSec in Contiki, SICSLoWPAN [78]
Data link	802.15.4 MAC	ContikiMAC [79]
Physical	802.15.4 PHY	Contiki 802.15.4 [80]

24.2.4 Security and Privacy Issues in IoT

Security has been an indispensable component of computers and computer networks. In a similar manner, IoT devices and IoT networks are expected to meet some security requirements. Hence, IoT security is an emerging research topic that is attracting attention both in academics and industrial sector. There are plenty of international organizations and companies that focus on the design and development of IoT-based systems. However, commercialization of IoT resulted in an increase in public security concerns such as privacy issues, cyber-attacks, and organized crime [19].

The integration of smart devices into the Internet introduces several security problems due to two major reasons. The first reason is the heterogeneity in the IoT devices. This causes some inconsistencies among the devices especially in security mechanisms employed because some of the devices are resource-constrained devices while the others are powerful hosts. The second reason is that most of the Internet technologies and communication protocols were designed for traditional web communications, but not to support IoT. This has led to the development and standardization of IoT-specific security protocols (e.g. lightweight compressed IPsec, lightweight DTLS, IEEE 802.15.4 link-layer security, etc.) that ensure end-to-end message integrity and confidentiality [16].

IoT devices are exposed to cyber threats from the Internet as well as from the local network they are connected to. Thanks to Shodan [85], it has been very easy to find some specific types of IoT devices linked to the Internet such as IP cameras, routers, servers, and so on. Shodan helps to find the public IP address of these kinds of devices. These devices are delivered with default username and password to access their management interface. Most of the owners do not change the username and password, and this enables cyber attackers to compromise the devices and use them for their cruel purposes. Mirai botnet attack [86] is one of such attacks taken place recently. Therefore, to thwart botnet attacks, California signed a new bill into law called "Security of connected devices." According to the law, device manufacturers are going to include either a pre-programmed password unique to each device or a mechanism that enables the user to generate a new means of authentication before being granted access to the device for the first time [87].

Security goals can be implemented with some policies guided by the confidentiality, integrity, and availability (CIA) triad [88]. Confidentiality is the first thing that comes to mind when it comes to security because it is not reasonable to envision a secure system in which sensitive data is stored in plaintext. Some data perturbation methods such as data obfuscation, encryption, and secret sharing are used to conceal users' sensitive data. Maintaining integrity of data at rest, code running on the device or a network packet is another essential item in the triad.

For instance, it is extremely important to ensure that a packet is not tampered with during its journey from the source to the destination. Message authentication codes and digital signatures are the two most commonly used techniques to prevent adversaries from breaking the integrity of data or message. Moreover, the data at rest local/remote or a web service should be reliably accessible by authorized parties whenever they attempt to access. This mostly depends on the performance of storage devices, so maintenance of hardware and design of software that makes the data/services available. A robust and resilient system avoids creating bottlenecks, provides adequate communication bandwidth and fast and adaptive disaster recovery.

As might be expected, IIoT also faces similar security challenges. One of the most significant security challenges in IIoT is to assure CIA-compliant but at the same time traceable and transparent data exchange between multiple stakeholders [89]. To comply with CIA features, certificates can be maintained, but it is time-consuming to use them due to the significant management overhead for certificate storage, distribution, verification, and revocation in traditional PKI [15]. Therefore, most of the IT/operational technology (OT) department managers would prefer network segmentation using firewalls, data diodes (unidirectional gateways), and IT/OT gateways to protect manufacturing systems against the risks imposed by existing and new IIoT devices [39].

The introduction of Internet technologies into the industrial domain drastically enlarges the potential attack surface and, therefore, poses some security risks. In a probable attack scenario, remote attackers can intrude themselves into an industrial communication network and gain privileged access to the equipment's processors although they do not have physical access to the hardware of the devices. They can get access to the authentication mechanisms and extract authentication keys after a successful attack. Moreover, remote attackers may have local partners inside the facilities which are called insider attacker. Insider attackers have physical access to the equipment and make it easy to obtain the credentials from the equipment's memory [90].

To reduce attack surface and protect their networks, industrial IT/OT security specialists have isolated their systems in cyber-domain. However, such an isolation cannot thwart zero-day vulnerability exploits and insider attacks. The attacks against power and nuclear plants performed by Stuxnet worm, Havex and BlackEnergy trojans, EternalBlue and WannaCry attacks, and Dragonfly group are major examples that demonstrate that isolation is not a panacea for security concerns of industrial control systems [91].

We have focused on security concerns so far. In addition to security concerns, frequent data collection from IoT devices puts the users' privacy in risk as it can help expose the users' location and daily habits to the service providers. For example, collection and storage of fine-grained energy consumption data raise the issue of privacy for the consumers who must use smart meters daily. Specifically, the collected consumption data can be analyzed using load monitoring techniques to infer activities of the consumers. Hence, typical privacy threats include, but not limited to: (i) Determining personal behavior patterns (can be used by marketers, government); (ii) Determining specific appliances used (can be used by insurance companies); (iii) Performing real-time surveillance (can be used by law enforcement and press); (iv) Target home invasions (can be used by criminals); and (v) Location tracking based on electric vehicle usage patterns (can be used by law enforcement). Secure in-network data aggregation can be used to both preserve consumers' privacy and reduce the packet traffic due to high-frequency metering data. The privacy can be provided by performing the aggregation on concealed metering data. Fully homomorphic encryption and secure multiparty computation are the other two systems that enable performing multiple operations on concealed data [24].

24.3 Software Defined Networking Background

IoT devices are both connected to each other and to the Internet, which forms a large-scale network. The communication between each other and to the Internet is established through the routers/switches deployed on the network. These devices are usually expensive and run the complex proprietary software. The software running on these devices is also vendor specific and each device should be configured individually by the network admin. Considering IoT devices and massive network communication between them, managing the network devices individually is very challenging since there can be too many of them in a small environment, such as in a smart home.

Meanwhile, researchers proposed SDN recently which is an emerging technology providing great flexibility and cost-effective solution to the control and management of the networks [9]. SDN proposes a separation of data and control planes and that is the main difference from the traditional network environment as shown in Figure 24.3. On the one hand, the data plane consists of switches which are not capable of any routing/blocking decisions by themselves. On the other hand, the control plane (also known as SDN Controller) is the brain of the network and is responsible for all critical operations in the network. SDN provides a flexible and cost-effective solution for network management by enriching network devices with programmability feature and a centralized controller can be used by the network admin to configure the network devices. Even though network programming has been investigated for a long time, SDN has been improved and named since a couple of years now [9]. With the introduction of SDN, it is made possible to do

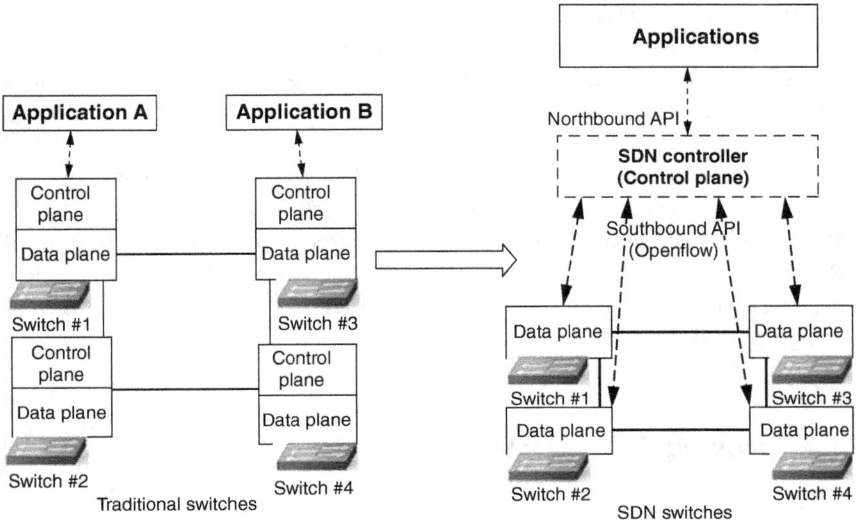

Figure 24.3 Traditional network switches versus SDN-based switches.

dynamic traffic engineering, drop packets, reconfigure the links in case of failures and enforce certain policies which make network management convenient and more flexible.

Traditional network devices (switch, router, etc.) require all the software on the hardware to run protocols before it can be installed in the network. Meanwhile, SDN-based switches are basic devices and can be updated easily even after the installation. However, there is a need for a new communication protocol for SDN Controller to communicate to SDN switches. Even though there is not a standardized protocol yet, OpenFlow is the widely-used protocol by both researchers and industry [92]. SDN-based switches are also called OpenFlow switches and these switches have flow tables that will provide the knowledge of the rules on what to do with each packet. *Flow table*, *OpenFlow Protocol* and *secure channel* between SDN Controller to OpenFlow switches are the main parts of OpenFlow switches. In addition to the mentioned OpenFlow switches and the protocol requirements, there is also a need to run an SDN Controller which network administrators can use to access the network devices remotely. For example, FloodLight [93] and OpenDayLight [94] are publicly available and preferred by network admins.

The SDN Controller typically should be running on a different machine in a centralized location and only network administrator(s) should have permission to access it. The SDN Controller can do many operations including but not limited to enforce security rules (block some packets, etc.) and decide forwarding tables. By exposing SDN-based solution to networks, the high cost of network devices and complexities of maintenance of such devices can be minimized. The network administrator can configure and update some parts of the SDN Controller to manipulate the network or s/he can implement some applications on top of SDN Controller to apply his/her own rules. The latter is more common and requires less knowledge of SDN Controller's source code. Northbound API of SDN Controller is required to be used for this purpose and mostly used protocol is REST interfaces [95]. REST interface is turning network devices to Web applications. These applications send REST inquiries to get information about the current situation or to update some flow tables by using SDN Controller interfaces.

24.4 Moving Target Defense

24.4.1 Introduction

In 2009, five new game-changing directions have been introduced to address some intractable problems in the digital infrastructures [96]. These directions strive to solve these problems from a different perspective instead of the typical traditional approaches that tackle them directly. One of these directions called MTD, was driven by the fact that the digital infrastructure settings are relatively static for a long period of time. For example, once the computer that we use every day has been installed with an operating system (OS) and assigned an IP address, these settings are barely changed. These conditions enable attackers to have unlimited time to perform any stage of the five-phase attack kill chain [8]. MTD enables defenders to build proactive self-defense mechanisms that dynamically change the digital system attributes while still ensuring the system accessibility for legitimate users. These self-defense mechanisms introduce the redundancy and diversity in the system to make the attack surface unpredictable for attackers. For instance, instead of using a single OS platform all the time, a computer can dynamically rotate its OS to a different OS platform at a random time interval. This way, it will make it harder for attackers to perform their attacks since their insight of the digital system from their previous attack attempt may become obsolete since the OS platform has changed.

24.4.2 A Brief Review on MTD Classifications

A variety of MTD techniques have been proposed since the MTD is coined in 2009. They can be classified into five top-level categories [8]: (i) *dynamic data*, which covers MTD techniques that dynamically change the data representation properties (e.g. format, syntax, encoding, etc.); (ii) *dynamic software*, which covers MTD techniques that utilize multiple equivalent functions to provide an application diversity by dynamically changing the application codes; (iii) *dynamic runtime environment*, which covers MTD techniques that strive to modify the operating system (OS) to provide the diversity, either through address space randomization (ASR) or instruction set randomization (ISR); (iv) *dynamic platforms*, which covers MTD techniques that utilize multiple unmodified OSes and other platform characteristics (e.g. processor architectures, virtual memory, storage systems, etc.) to enable the platform diversity by dynamically migrating between platforms; and (v) *dynamic network*, which covers MTD techniques that dynamically modify the network properties (e.g. addresses, ports, routing, etc.). Each category attempts to address a different attack phase of the five-phase attack kill chain (i.e. reconnaissance, access, development, launch, and persistence phases). The dynamic data, software, and runtime environment domains mainly focus on the *development* and *launch* phases. The dynamic platforms domain attempts to disrupt any attacks at the *access* and *persistence* phases while the dynamic network domain attempt to address the *reconnaissance* and *launch* phases.

Specifically in the dynamic network domain, the existing MTD techniques can be further classified into four categories: (i) *dynamic network-identity*, which covers any attempts that strive to introduce the dynamic to the network identity, such as the physical address, logical address, and port number [97–103]; (ii) *dynamic network-path*, which covers any attempts to provide the dynamic paths with the intention to thwart any network attacks [104,105]; (iii) *dynamic network-infrastructure*, which covers any attempts to introduce the dynamic to the network components such as by dynamically changing the proxies [106–108]; and (iv) *dynamic network traffic configuration*, which covers any attempts that strive to provide the dynamicity to the network traffic such as the dynamic time schedule for periodic traffic, data size, and the dynamic protocol information with the aim of obfuscating the attackers [109, 110]. Figure 24.4 shows our proposed MTD classification.

Among these categories, most of the research efforts fall into the dynamic network-identity category since it can be done on the legacy network with the minimum to moderate legacy network modifications. However, this is not the case for the dynamic network-path on the legacy network. Typically, network paths are static when there is no performance of failures issues [105]. Moreover, orchestrating dynamic paths among network devices is very challenging since most routing protocols are distributed in nature. Every network node operates based on the local knowledge to determine the path for a traffic while the dynamic network-path MTD requires a coordinated effort that needs a global knowledge to provide the path diversity. Therefore, a completely new dynamic routing protocol may be needed to support the dynamic network paths randomization. For the last two categories, the dynamic network-infrastructure, and dynamic network traffic

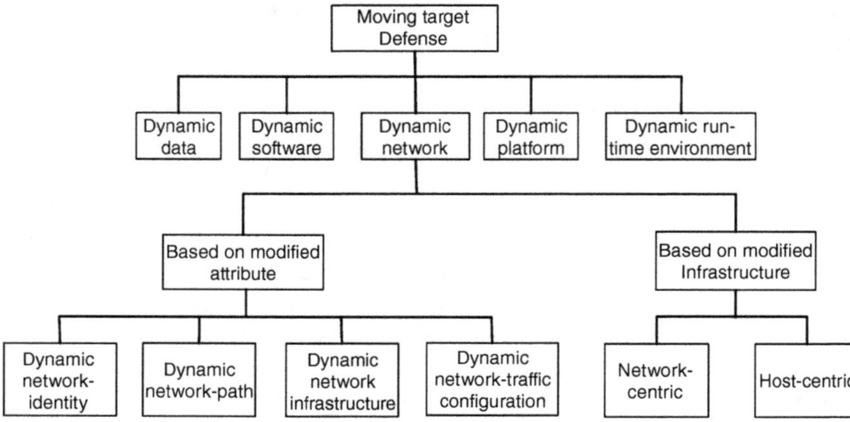

Figure 24.4 Proposed MTD classification for the dynamic network. This classification is extended from the MTD classification in [8].

configuration, on the other hand, (even though they can be implemented on the legacy networks), the possible moving options are somewhat limited and application-dependent (e.g. smart grid [109, 110]).

Based on which network infrastructure will be modified/added to support the MTD based network-identity randomization operations, an MTD approach can be classified as a host-centric [97, 102] or a network-centric approach [98–101]. In the host-centric MTD approach such as the TAP-based port and address hopping (TPAH) [97], the communicating end-hosts need to be modified to support the MTD operations. TPAH approach requires a hopping engine that needs to be added between the TAP virtual-network kernel driver and the physical network adapter as depicted in Figure 24.5. This hopping engine consists of two modules: (i) a port and address hopping module that handles port and address mutation and mapping; and (ii) an access control module that performs traffic monitoring, access control, and port and address mapping. The TAP driver acts as tunnel between a user-space process and the Operating System network stack. Operating System sends packet to a user-space process via a TAP driver and a user-space process can sends packets to the TAP device, which then injects these packets to the Operating System network. In the network-centric approach, network infrastructure needs to be modified such as by adding varying number of new gateways [98–101]. These gateways are typically located at the boundaries of the physical subnet to ensure the address translation between the real address and the short-lived temporary addresses. Figures 24.6 and 24.7 show an example of the network-centric architecture with three and two gateways respectively.

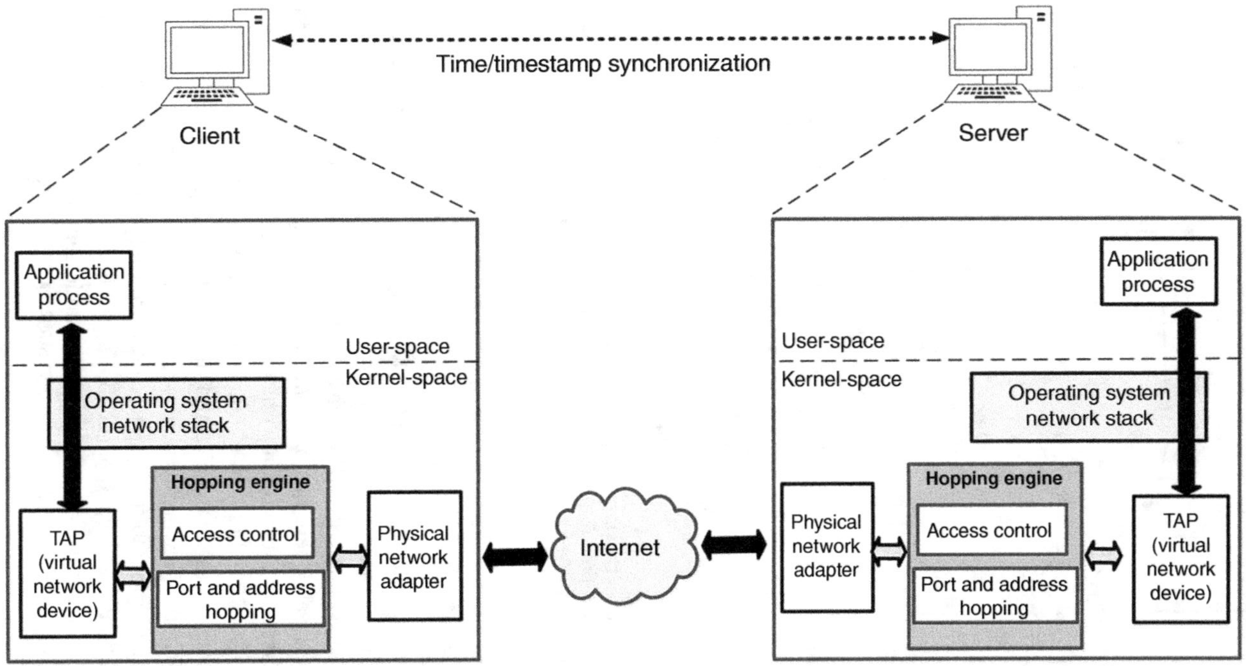

Figure 24.5 An example of host-centric approach: TAP-based port and address hopping (TPAH) architecture

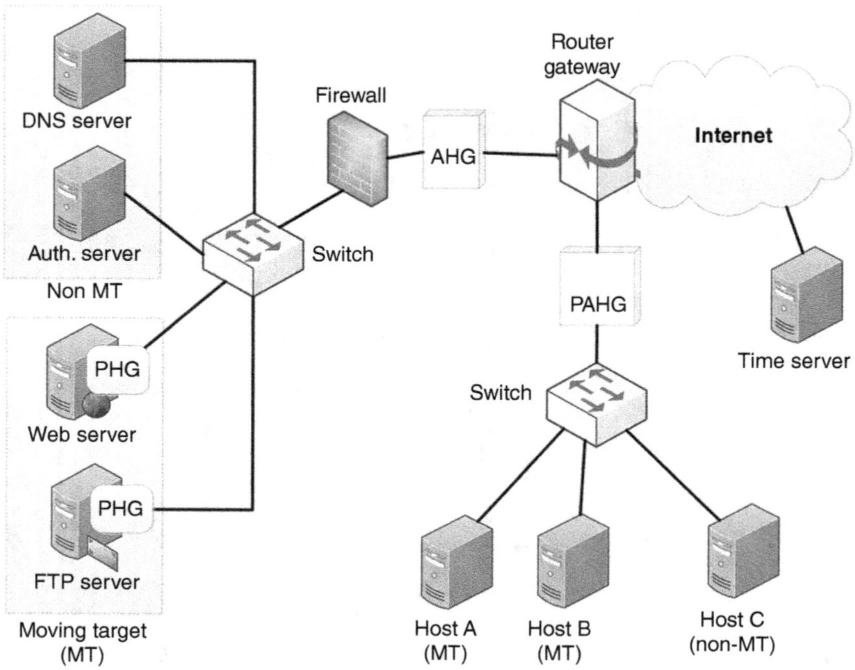

Figure 24.6 An example of network-centric architecture with three additional gateways, to support RPAH operations, adapted from [98]: Port Hopping engine Gateway (PHG), Address Hopping Gateway (AHG), and Port and Address Hopping Gateway (PAHG).

24.4.3 SDN-Based MTD Overview

The presence of emerging SDN technology that offers the separation of data and control plane, which in turn improves the network visibility and policy enforcement capability when compared to the legacy network, has brought the MTD research to the next level. There has been an increasing number of SDN-based MTD studies in the recent years, not only related to the MTD in the dynamic network domain [104, 111–124], but also to other areas such as for cloud networks [125, 126]. An overview of the SDN roles and the MTD techniques used to introduce the dynamicity in the networks is reviewed in the following subsections. The review is organized according to the dynamic network category (if applicable) in Section 24.4.2. Additionally, a separate section is provided for the review of the hybrid approaches, which utilize more than one dynamic network category collaboratively. Figure 24.8 shows the proposed SDN-based MTD classification.

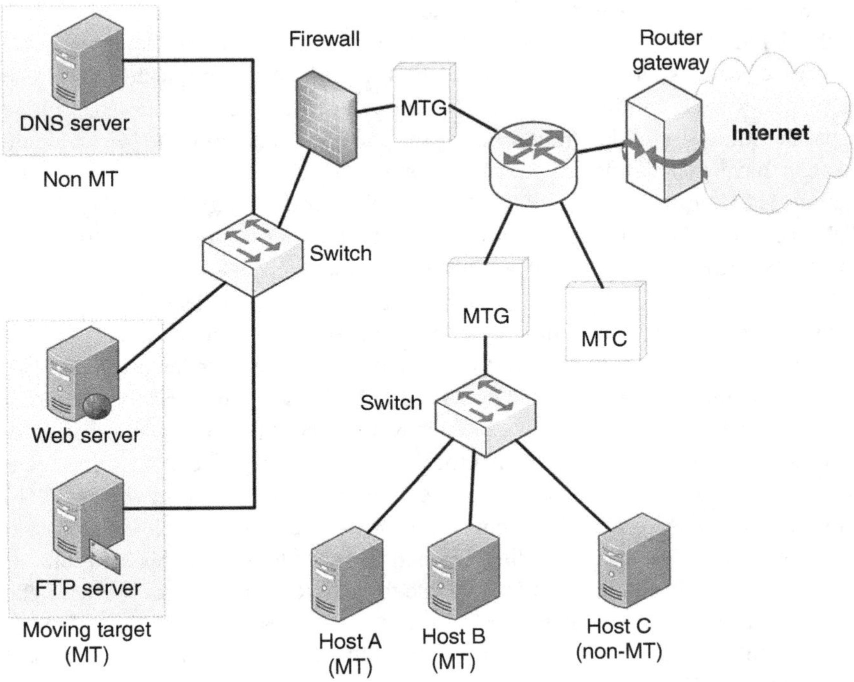

Figure 24.7 An example of network-centric architecture with two additional gateways to support the RHM operations: Moving Target Gateways (MTGs) and Moving Target central Controller (MTC).

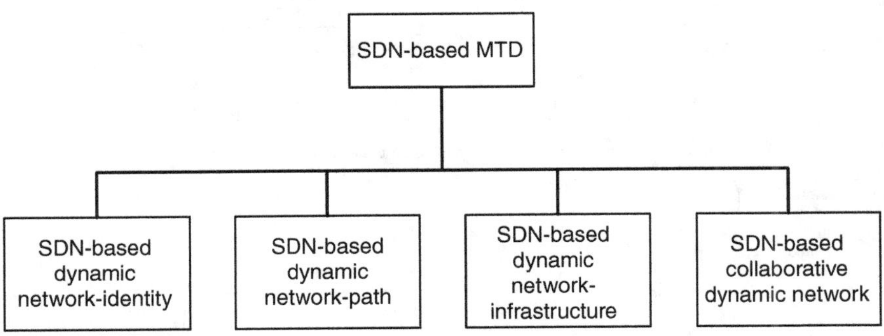

Figure 24.8 Proposed SDN-based MTD classification. This classification is derived from the existing works in the SDN-based MTD.

24.4.3.1 SDN-Based Dynamic Network-Identity

In the dynamic network-identity domain, instead of shuffling the real identity of a network device, a short-lived virtual identity is used in the communications between network devices to minimize the reconfiguration overhead on network hosts and to make the address mutation transparent to the end host. This short-lived virtual identity can have many different names depending on the approach, e.g. synthetic address [112], virtual address [99, 111], or ephemeral address [100, 101]. When SDN is employed in this domain, SDN can be used either in the host-centric based approach [112] or in the network-centric based approach [99–101, 111, 113].

The SDN shuffle technique [112] is a host-centric approach to defend against malicious reconnaissance. This technique enables both MTD clients and non MTD clients to access MTD (i.e. protected) servers by utilizing the SDN controller as the network address generator for the synthetic MAC and IP addresses. When the clients (either MTD or non-MTD clients) request a Domain Name Service (DNS) resolution of a protected server, the DNS server forwards the protected server synthetic IP address generated by the SDN controller to the client with a very low time-to-live (TTL) information to ensure that the client will reissue a new DNS resolution for each new connection. The corresponding synthetic MAC address for this synthetic IP address will be sent to the client in response to the ARP resolution of the synthetic IP address. Hence, both synthetic physical and logical addresses of the protected server are different for different clients. Figure 24.9 illustrates the SDN shuffle protocol operations.

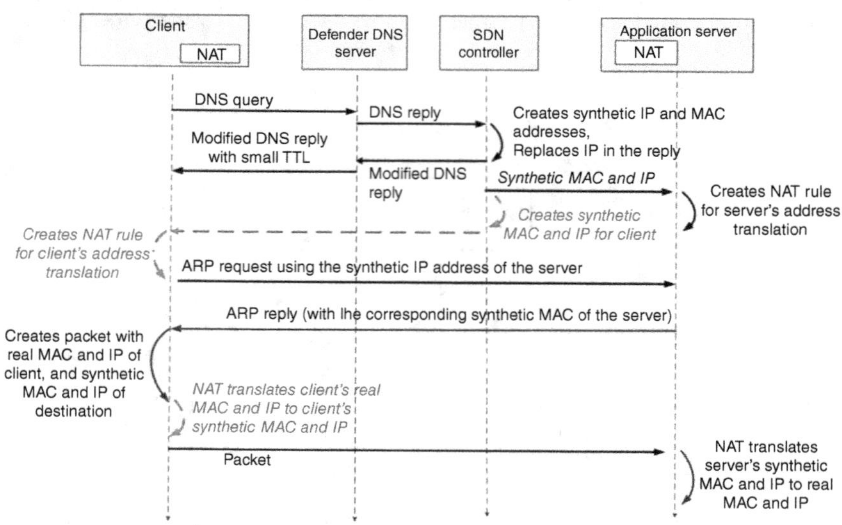

Figure 24.9 SDN shuffle protocol operations. The dotted lines are optional protocol operations when a client also employs the synthetic addresses.

In the network-centric approaches, SDN is used as the alternative replacement for the role of the MTD gateways in the legacy networks. For example, Figure 24.10 illustrates the RHM operations when it is used in the legacy networks and in the SDN-based network. SDN switch performs the address translation mechanism in the SDN-based network.

Packet Header Randomization (PHEAR) Technique [113], is a network-centric anonymous SDN-based communications system for enterprise networks that provide traffic unlinkability from its communicating hosts. It creates an identifier free traffic by removing both the explicit (e.g. MAC and IP address) and implicit identifiers (e.g. the IP initial TTL and TCP initial window size information may reveal

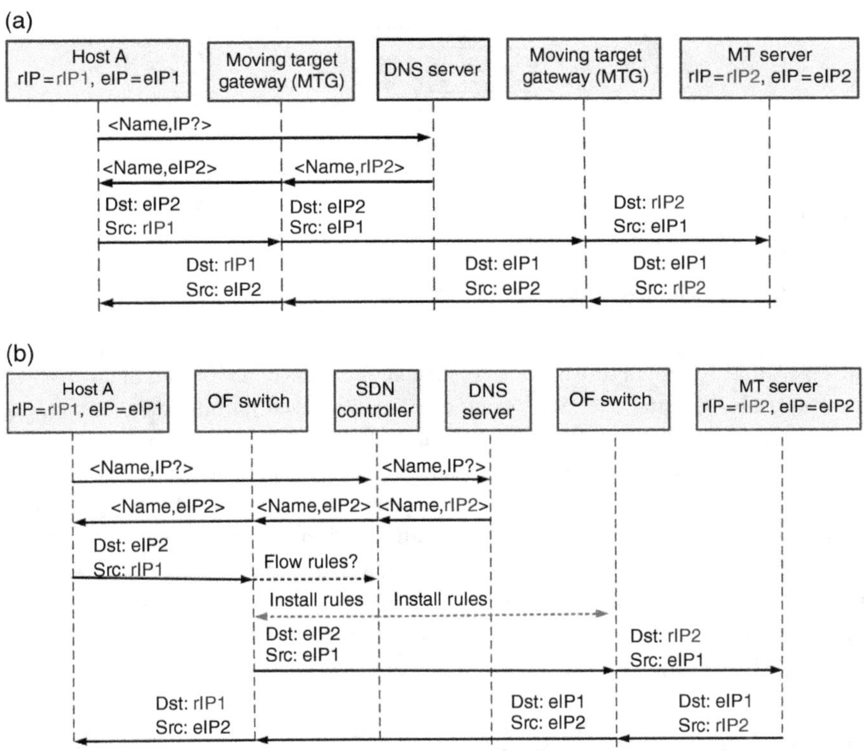

Figure 24.10 RHM operations comparison when a client tries to access the server using the domain name. In (a), gateways are used on the legacy network while in (b), an SDN-based network is used. When the MTG and SDN controller intercepts the DNS response on the domain name request, besides translating the real IP (*rIP*) to the ephemeral IP (*eIP*), they also change the time-to-live (TTL) to ensure that network hosts will perform DNS request again for each new connection.

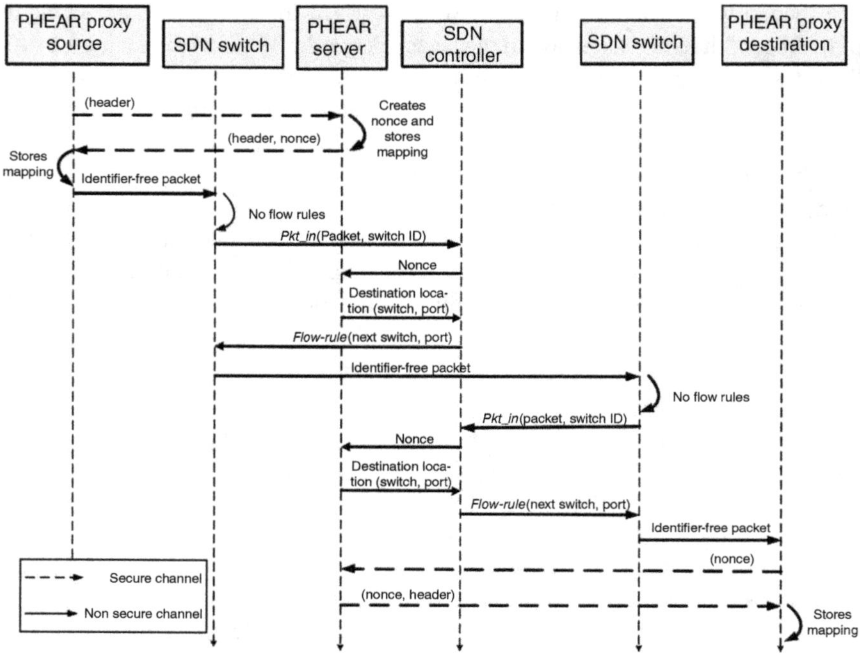

Figure 24.11 PHEAR per-switch basis operations.

the Operating system platform) from the packet headers while hiding the transport layer and above using the IPSec protocol. A short-lived 64-bit nonce is used to replace the source and destination IP address fields in the packet headers.

PHEAR leverages the SDN for the identifier-free packet forwarding using this short-lived nonce. Two types of network components need to be added to the SDN network to support the PHEAR operations: (i) a PHEAR server that creates a unique nonce for each packet header using a collision-resistant hash function and maintains its mapping; and (ii) a PHEAR proxy that resides on end-hosts and is responsible for the translation. Figure 24.11 illustrates the PHEAR per-switch basis operations.

24.4.3.2 SDN-Based Dynamic Network-Path

The SDN programmable feature and centralized control through the separation of data and control plane, has enabled proactive defense MTD against reconnaissance, eavesdropping, and DoS attacks by dynamically changing the network path [104, 119–123]. SDN eliminates the tedious tasks that must be done when implementing dynamic network-path MTD in the legacy network with respect to automatic routing table update to support any route changes without interrupting the on-going traffic or violating any security requirements. With SDN, changing routes can be done as a series of flow table updates in the SDN switches. In contrast, updating a routing table in the legacy network can be done by issuing a static route command manually [104] and/or by using a dynamic routing protocol. Obviously, issuing static route command is not an option due to a scalability issue. As previously mentioned in Section 24.4.2, a new routing protocol may be needed to ensure a timely update that will not interrupt the on-going traffic. One way to solve this problem is by utilizing MPLS (multiprotocol label switching) [127], which requires MPLS routers to carry out the operations. MPLS utilizes labels to identify virtual paths. Data packets are assigned labels and the forwarding decisions are made solely based on the label. MPLS, however, requires establishing the virtual path through the bandwidth reservation mechanism.

In the SDN-based network, the SDN controller acts as the central coordinator of the route mutation. The difference between approaches is typically on the route selection criteria. The random route mutation (RRM) [104] selects the route that satisfies the capacity, overlap, and quality of service constraints. When more than one eligible routes are found, one route will be randomly selected. In [119], a security constraint, which considers any previously used access control policies, is added into RRM route selection criteria. The route mutation in [120] selects the route based on the overlap constraint criterion to defend against the reconnaissance phase of the Crossfire attack, an indirect distributed DOS (DDoS) attack. As the network size increases, the efficiency of the route mutation becomes an issue since the bigger the network, the longer the processing delay of the route mutation to find the route(s) that satisfy the constraint(s) [121]. To address this issue, an effective and faster route mutation approach called Area-dividing Random Route Mutation (ARRM) [121] approach is proposed. The idea is by dividing the entire network into sub-areas and a backbone area. The communication between areas is enabled through the backbone. This way, when the internal link states in an area are changed, the ARRM approach only needs to calculate the route mutation for that area.

24.4.3.3 SDN-Based Dynamic Network *Infrastructure*

In an SDN-based network, the SDN controller plays a key role in network operations. Yet, it also can be the single point of failure of the network. Thus, the SDN controller becomes the ultimate attack target to disable the network. Considering a new type of DDoS attack, called the Blind DDoS attack, an MTD approach that provides diversity through dynamically changing the active SDN controller from a pool of SDN controllers has been proposed [116]. This type of DDoS attack attempts to attack the SDN controller by flooding the SDN switch with many packets that cannot be processed. In this case, the switch will forward them to the controller. Hence, the controller's capability will be degraded when packets from multiple SDN switches are forwarded to the controller.

24.4.3.4 SDN-Based Collaborative Dynamic Network

Recently, a hybrid MTD approach that combines different dynamic network categories to provide a proactive defense has been proposed. This collaborative approach typically combines the *Dynamic Network-identity* with the *Dynamic Network-path* [122–124]. This approach is also known as the *double hopping* approach. The differences between approaches are related to the attack models and the considered constraints in the approaches. The considered attack model in the Path Hopping SDN Network Defense (PH-SND) [122] is the traffic analysis attack that attempts to intercept and examine traffic to deduce information from the traffic patterns. Besides selecting a route based on the capacity and overlap constraints, PH-SND modifies the address and port information of the source and destination nodes during the forwarding process to further confuse the traffic analysis. Similarly, the Double Hopping Communication (DHC) [123] also modifies the address and port information and selects a route based on the path-length and overlap constraints to address the sniffer attack.

Another double hopping approach, a self-adaptive End-point Hopping Technique (SEHT) [124] is proposed to address the lack of ability from the existing approaches to adapt to different attacks strategies. Only a few existing approaches are the adversary-aware approach, which considers any potential attackers' actions. For example, Jafarian et al. [101] studied the use of a fast and accurate hypothesis testing for characterizing the adversarial scanning strategies in the legacy networks and adapts accordingly. Similarly, SEHT approach also considers the scanning attacks as its attack model and utilizes an analysis engine to perceive and analyze the adversary attack strategy. Four constraints are considered in SEHT: capacity, reachability, forwarding path delay, and hopping space selection constraints.

24.5 Moving Target Defense for IoT

In this section, we first discuss the feasibility of MDT for IoT environments and then move on to review the existing works under various categories.

24.5.1 A Brief Evaluation of MTD Feasibility for IoT

Besides the typical unconstrained nodes (e.g. servers, desktop computers, and powerful mobile devices such as smartphones), a plethora of IoT end-nodes are constrained nodes that have limited resources such as limited computation power,

Table 24.2 Constrained device categories (RFC 7228).

Class	Memory	Storage	Remark
Class 0	<<10 KiB	<<100 KiB	Typically, battery-operated, not enough space for the full IP stack and security implementations, typically preconfigured with a very small data set, assisting by gateways for Internet communications
Class 1	~10 KiB	~100 KiB	Enough space for optimized protocols such as CoAP (Constrained Application Protocol), it does not need any gateways assistance for Internet communications
Class 2	~50 KiB	~250 KiB	Enough to support full IP stack implementation and can be fully integrated into IP networks

limited memory, limited storage capacity, and limited power capacity (e.g. battery-operated devices). The Internet Engineering Task Force (IETF) has defined three constrained node categories in RFC 7228 as presented in Table 24.2.

The inherent limitations in each class eventually will limit the applicability of any MTD domains in the IoT. The class-0 IoT devices are severely constrained nodes with very limited memory and storage capacities since they only need to do very simple tasks (e.g. send an on/off or other basic health indicators). These capacities are not even enough for the IP stack and security implementations. Hence, any MTD domains cannot be implemented on the class-0 devices due to these resource limitations. Instead, since the IoT devices in this class utilize gateways, which are typically unconstrained nodes, for Internet communications, MTD can be implemented at the gateways level. For the other two less-constrained classes (i.e. class-1 and class-2), the MTD techniques need to be designed and optimized by considering these limitations.

Among the five top-level categories of MTD techniques, the dynamic data domain may not be affected by the resource-constrained limitation, unless encryption mechanisms are involved in the data encoding. Most of the existing works are in the dynamic network domain as will be explained in the next section. The works on the other domains, however, are limited.

The maneuverability of the dynamic software and dynamic run-time environment domains to introduce diversity will be limited by the remaining available memory and storage spaces. To overcome these limitations, Mahmood and Shila [128] proposed a context-aware code partitioning and code diversity approach for IoT. IoT devices only store a minimal trusted code and depend on the context, another trusted code can be downloaded from a secure source (e.g. cloud), which can then be removed and replaced by other trusted code after this code is no longer needed. To further obfuscate the attacker's view, code diversification is used in conjunction with the context-aware. Different versions of the same code can be run at different times.

In the dynamic platform domain, while providing diversity through hardware diversity (e.g. dual processor architectures) may not be technically (i.e. in terms of the form factor of the sensors) nor economically feasible (i.e. in terms of cost), providing software diversity for the dynamic platform can be a viable option. Casola et al. [129] proposed an MTD approach based on fine-grained reconfiguration at two different architectural layers of WSN, namely the security layer and the physical layer. In the first reconfiguration scheme, the security layer protocol is dynamically changed between two or more cryptosystem implementations [130]. Similarly, the second reconfiguration scheme enables each node to swap to a new image that was stored on an external flash memory. In [130], the decision to change can be based on the rules such as reaching maximum execution time of application or non-reachability of a node.

24.5.2 MTD for IoT in the Dynamic Network Domain

Several MTD approaches that have been proposed for IoT in the recent years are mainly in the dynamic network domain [131–137]. In the following sections, a review of each work is organized into two categories, MTD based on Identity and non-Identity randomization. Figure 24.12 shows the proposed MTD for IoT classification.

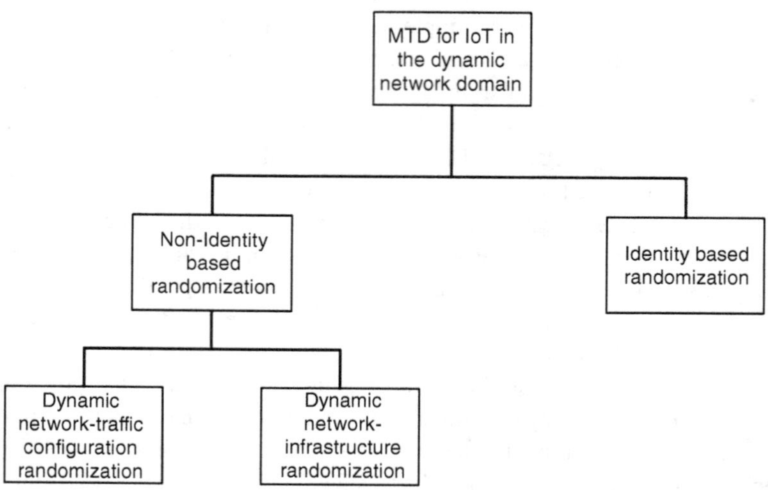

Figure 24.12 Proposed MTD for IoT classification. The classification is derived from the existing works in the dynamic network domain specifically for IoT.

24.5.2.1 MTD Based on Identity Randomization

Depending on the IoT application's needs, MTD has started to be used in various IoT environments for more effective defense. Obviously, the explicit identities such as the physical and logical addresses of IoT devices become the main focus for randomization even though another identity may also exist [138]. With respect to the explicit addressing, network address shuffling is one of the first MTD methods comes to mind. It is basically changing IPv4 or IPv6 address of the devices in a network periodically. Thus, the attackers reach the wrong target or way better, a non-existent target. Judmayer et al. [135] assess the overhead and the effects of periodically changing network addresses and ports under various scenarios. In their experiments, they used Linux-based IoT systems such as Raspberry Pi (RasPi) [139], RasPi 2 [140], RasPi 3 [141], and Carambola 2 [142]. They measured the number of address change operations per second for each device. As expected, the more advanced one among these devices, which is RasPi3, outperformed the others. What is surprising is that Carambola 2 outperformed RasPi although RasPi has a clock rate of 700 MHz while Carambola 2 has a clock rate of 400 MHz. Also, they tested using multiple IP addresses simultaneously. They observed that it takes more time to add new addresses as the number of addresses already in use increases. Based on their findings, they concluded that it is feasible to shuffle network addresses in IoT environments.

24.5.2.1.1 MTD-Based IPv6 for IoT

Due to the huge amount of IoT devices, IPv6 is considered as the viable addressing scheme for IoT. However, a privacy issue may arise from the IPv6 stateless address autoconfiguration (SLAAC) feature that enables a host to self-assign a unique address. Typically, the 64 bits interface identifier (IID) portion of the IPv6 address is derived from the host MAC address in the form of 64-bit extended unique identifier (EUI-64) format. By utilizing this static IID, a third party can perform address tracking or traffic correlation. Additionally, with this static IID, targeted network attacks such as the address-based DoS and Man-in-the-middle (MITM) attacks become a security concern.

The Moving Target IPv6 Defense (MT6D) [102] is a non-deterministic IPv6 dynamic addressing that continually modifies the IP and port addresses of the sender and receiver of two communicating end-devices without breaking the existing connections. The aims are to preserve user privacy and protect against DoS and MITM attacks. Due to these features and the immense IPv6 address space, MT6D is a potential solution for Smart Grid [143] and for the low-powered, resource-constrained WSN running on IPv6 over Low-Powered Wireless Personal Area Network (6LoWPAN) [133, 134]. MT6D, however, is designed for full-scale systems and devices, and thus it needs to be adapted to be used for IoT constrained devices.

MT6D uses a hash function **H** to create a $vIID_x$, an obscured IID of a host x, from the first 64-bits of the hash value in Eq. (24.1). The input of **H** is the concatenation of (i) the real IID_x for a host x, (ii) a shared symmetric key K_s, and (iii) the time t_i at instance i. The symmetric key is distributed between two communicating hosts through an out-of-band key exchanged. Besides for obscured IID generation, this key is also used for encryption.

$$vIID_{x(i)} = \mathbf{H}[IID_x \| K_s \| t_i]_{0\to 63} \tag{25.1}$$

$$vPort_{x(i)} = \mathbf{H}[IID_x \| K_s \| t_i]_{64\to 79} \tag{25.2}$$

The next 16-bits remaining unused bits of the hash value can be used as the obscured port number $vPort_x$ as in Eq. (24.2). The 128-bits MT6D IPv6 address is constructed by concatenating the global routing prefix (48 bits), subnet identifier (16 bits), and the obscured IID (64 bits). The minimum IID rotation time is at least twice of the single-trip time (STT) of a packet sent between a sender and receiver. Figure 24.13 illustrates the timeline of an obscured IPv6 address, from the creation to deletion.

When a new obscured IPv6 address is created, a host utilizes the Neighbor Discovery Protocol (NDP) to verify that there is no conflict with the existing addresses on the subnet and ensure that the routers have this new address in their routing table prior the usage of this address. This way, each host only stores two obscured IPv6 addresses, the current address and the next address, while routers maintain multiple obscured IPv6 addresses that refer to the same host. The deleted IPv6 addresses from hosts will then be removed from the router's routing table as well through the IPv6 internal mechanism. This new obscured IPv6 address, however, does not need to be disseminated to the other end of the communicating partner. Once the communicating partner knows the real IPv6 address of a host, it can

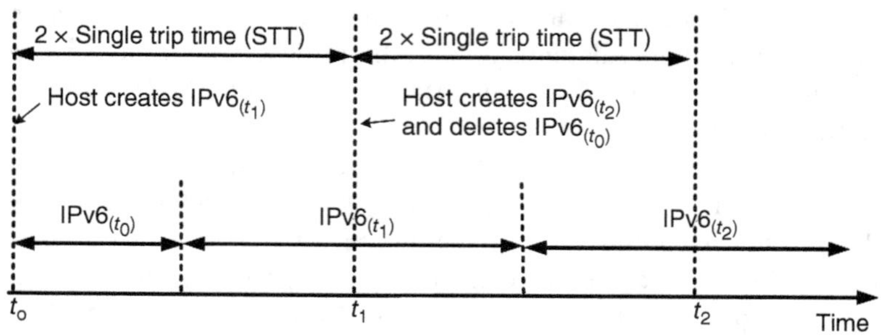

Figure 24.13 The creation, usage, and deletion of an obscured IPv6 address timeline. A packet sent within one STT of the next rotation will use the obscured IPv6 address from the next rotation cycle to ensure that there is no additional overhead of connection reestablishment or breakdown.

calculate the obscured IPv6 address of its partner using the same hash function since it shares the same symmetric key. However, to find the correct obscured IPv6 address of the communicating partner, time synchronization between these two communicating devices is mandatory.

Instead of using the obscured IPv6 addresses to replace the address information in the original packet, MT6D creates a packet that consists of an MT6D header and the anonymized version of the original packet where both IP and MAC addresses of the source and destination are overwritten to make it anonymous and disable address tracking while keeping all other protocol information in place. The MT6D packet is sent from a source to a destination, either through un-encrypted or encrypted UDP tunnel. UDP is used rather than TCP for the tunneling to avoid any TCP connection establishment and termination each time MT6D address changes. The encrypted UDP tunnel prevents traffic correlation since the anonymized original packet is encrypted using the shared symmetric key and thus a third party cannot get any information from it. The MT6D's overhead is 62 bytes that come from 40 bytes of MT6D header, 8 bytes of UDP header, and 14 bytes of Ethernet frame header. MT6D can be implemented either as an embedded software on a host or as a stand-alone gateway device.

Several efforts to adapt MT6D for IoT has been recently performed [131–134]. These efforts strive to implement the dynamically changing IPv6 address scheme on the 6LoWPAN protocol. Three different locations (from the most constrained node to the least constrained node) have been identified as the candidate for the scheme operations [131]: (i) on the 6LoWPAN mote (i.e. sensor), (ii) the 6LoWPAN border router, and (iii) the IPv6 Gateway. However, only the first two locations are further studied [132, 133]. While 6LoWPAN mote is a resource-constrained device, the border router in a typical 6LoWPAN network is usually attached to an external power source and/or greater processing power, which enables it to connect to the Internet and acts as a bridge to the 6LoWPAN network. Thus, implementing the MT6D on the border router, on the one hand, offers a simpler design [132]. However, control of the border router is not always guaranteed since it can belong to a third party. Implementing the MT6D on a resource-constrained 6LoWPAN mote (i.e. the end-point), as MT6D was originally intended, ensures that an attacker cannot effectively capture the IPv6 traffic to the mote [132]. However, it requires some adjustments. For example, when MT6D is implemented on TMote Skye motes [144], a class-1 constrained node with 10 kB of RAM and 48 kB of ROM, using a full IPv6 stack is too expensive and thus, a lightweight version of the network stack (e.g. Rime stack from Contiki OS [145]) can be used [132].

The Micro-Moving Target IPv6 Defense (μ-MT6D) [133, 134] was designed to operate over 6LoWPAN with both modes of operation: a *host-based* mode and *border-based* mode. In the former, μ-MT6D is implemented at the end-point (i.e.

sensor), while in the latter, μ-MT6D is implemented at the border router. Initially, MT6D uses Secure Hash Algorithm 256 (SHA-256) that requires around 80 kB of storage. Thus, besides utilizing a lightweight operating system (e.g. Contiki OS), another effort that can be done to adapt MT6D for the resource-constrained node is by optimizing the SHA-256 or utilizing more lightweight cryptographic hash algorithms.

24.5.2.1.2 MTD-Based Identity Virtualization for MANET

Albanese et al. [138] proposed an MTD based Identity Virtualization mechanism for mobile ad-hoc networks (MANET), a persistent self-organizing infrastructure less network of mobile wireless nodes that allows each mobile node to join and leave the network at any time, with the aims of protecting the identity of mobile nodes and obstructing the reconnaissance phase from external attackers by using a dynamically changing virtual identity, instead of the mobile node real identity, for communications with other legitimate mobile nodes. This virtual identity is selected from a pool of virtual identities. A validity interval determines how long a virtual identity can be used before it must be replaced with a new virtual identity from the pool. This way, the virtual identity is dynamically changing, and the real identity of the mobile node is never publicly used.

Each mobile node generates a pool of N virtual identities using a hash chain, a method that can create N virtual identities from a single input value x by successively applying a cryptographic one-way hash function F to the input value N times. For example, $F^4(x) \equiv F(F(F(F(x))))$ represents a hash chain of length $N = 4$. To create a pool using the hash chain, each mobile node uses two input values for the hash function: (i) a mobile node generates a random initial seed value x_i, which is generated whenever a mobile node i needs to create a pool such as when a new pool is needed since the existing pool is exhausted; and (ii) a shared secret seed s, which is known by all legitimate mobile nodes. This shared secret seed is used to change the argument of the hash function at each recursive step using Eq. (24.3):

$$F^k(x_i) \equiv \begin{cases} x_i & \text{for } k = 0, \\ F\left(F^{(k-1)}\right)(x_i), s) & \forall k \in [1, N]. \end{cases} \tag{25.3}$$

The goal of using a shared secret seed is to prevent an attacker, who has the knowledge of the hash function, to perform a brute-force attack by using every initial seed value in the seed space to generate the entire hash chains. The virtual identity vID of a mobile node i is then selected from the hash chain in the reverse order of the hash chain generation (i.e. starting from the last element of the hash chain) as in Eq. (24.4):

$$vID_i(t) = F^{N-(t-1)}(x_i) \ \ \forall t \in [1, N]. \tag{25.4}$$

Thus, the first virtual identity will be $vID_i(1) = F^N(x_i)$, which is called the commitment of the chain, and the last virtual identity will be $vID_i(N) = F(x_i)$. For each $vID_i(t)$, a validity interval $T_i(t)$ is randomly selected from the validity range $[Tmin, Tmax]$.

To ensure that mobile nodes can communicate in this new dynamically changing identity environment without synchronization, the authors proposed to modify the Internet layer of TCP/IP protocol stack with the following mechanisms: (i) a translation service for identity mapping using a translation table; and (ii) an update protocol for the virtual identity dissemination and periodically updating the translation table. Each entry in the translation table stores the real identity of a legitimate mobile node, its current virtual identity, and the hash-index corresponding to the current virtual identity in the hash chain. When a sender node wants to send a message to a destination node, the translation service at the network layer translates the source and destination real identities in the message to their currently corresponding virtual identities before the message is broadcast. At any intermediate node located on the route to the destination, when this intermediate node receives this message, the translation service translates the virtual identities to their real identities to find the correct route to the destination. After the correct route is found, the message, which still using the virtual identities, is forwarded again to the next hop. Figure 24.14 illustrates the translation service operations at the network layer when a message is sent from the sender to the destination.

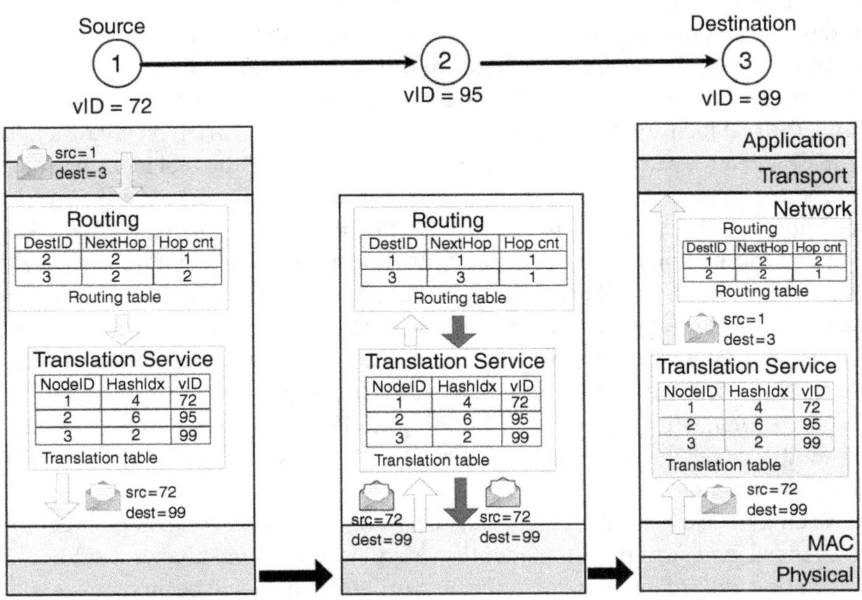

Figure 24.14 The translation service operations at the network layer.

When a mobile node i replaces its current virtual identity $vID_i(t)$ with a new $vID_i(t+1)$ since $T_i(t)$ expires, it sends a broadcast *Update* message that uses the new $vID_i(t+1)$ as the source identity and contains the hash index $N-(t-1)$ of that new identity in the mobile node i's hash chain. On receiving the broadcast *Update* from i, each receiver must verify the authenticity of the sender i before the receiver can update its translation table. The verification is performed to the source virtual identity in the received *Update* message by checking that the equality in Eq. (24.5) holds.

$$F\left(F^{N-1}(x_i), s\right) \equiv F^N(x_i). \tag{25.5}$$

For example, when the sender j receives an *Update* message with a $vID(k)$ as the source identity and contains hash index k in the message, it will check for every entry i in its translation table whether the stored $HashIndex(i) > k$ and $F^{HashIndex(i)-k}(vID(k)) \equiv$ the currently stored vID in the translation table. If that is the case, node i is the originator of the *Update* message and the corresponding entry in the translation table is updated.

In addition, the authors also proposed the join and leave mechanisms since a mobile node can join or leave the network at any time. To join a network, a legitimate mobile node must have the shared secret k to encrypt its messages. The node first selects two values, a real identity i and an initial random seed x_i; generates a hash chain of N virtual identities; composes an encrypted *join request* message that contains the real identity i and a random number r_i in its payload and uses the commitment of the chain as the source address. The random number r_i is used by the node i to distinguish its own request from others in case two or more *join request* messages that are issued at the same time have either the same the real identity, the commitment of the chain, or both. On receiving a *join request* message, when a receiver recognizes that either of the real identity, commitment, or both is the same as its own, a *join response* message is broadcast to indicate a duplicate. Otherwise, this *join request* message is stored and marked as pending until a timer expires. When the timer expires and no *join response* message from other node related to that *join request* message is received, the receiver assumes the request is valid and stores the identity and commitment in its translation table. Otherwise, when a *join response* message is received, the *join request* message is discarded. For the sender, when a *join response* message is received for its *join request*, which indicates that either the identity or commitment is owned by another node, the sender must choose a new identity and secret key s and redo the join process.

When a node leaves the network, its network information (e.g. node identity, hash index, and current virtual identity) is still available and always valid in the translation table of other nodes in the network since there will be no more *update* message from the leaving node. For this case, a valid timer associated with each entry in the translation table is introduced. When this timer expires, which indicates that no more *update* messages are received within the valid time interval, any additional data packet using this identity is no longer considered as a legitimate packet.

24.5.2.1.3 MTD-Based DDoS Resistant Multicast (DRM)

Andrea et al. [137] modified the RPL, a routing protocol developed for low power and lossy networks and presented the Simple Agile RPL multiCAST (SARCAST) protocol against DDoS attacks. The protocol is based on an address agility technique called DDoS Resistant Multicast (DRM). Address agility technique prevents hostile actors from performing meaningful reconnaissance or an attack against a previously discovered target by shuffling destination address in any message. SARCAST benefits from the DRM concept and mitigates the hostile multicast traffic that targets the interiors of a deployed network. The agile addressing mechanism is embedded into the destination address field of the packet. Each system on the network periodically updates that field to the current valid address, and any packets with an expired or invalid address are rejected. The SARCAST was tested on a working IoT testbed, and results showed that the SARCAST can protect IoT systems against DDoS attacks with almost no adverse impact on overall performance and that the size of agile address history has a significant effect on packet delivery ratios.

24.5.2.2 MTD Based on Non-Identity Randomization

In this section, the MTD-based on the non-Identity randomization is presented and organized according to the three dynamic-network sub-classifications (if applicable): the dynamic path, dynamic network infrastructure, and dynamic network traffic configuration.

24.5.2.2.1 MTD Based Network Infrastructure Randomization

Tracking and localization services developed for IoT applications pose some risks to the location security of base-stations serving for WSNs in case of a compromised node in the network. This is because damaging a base-station may result in catastrophic consequences. To avoid such disasters, Chin and Xiong [136] proposed moving proximity base station defense (MPBSD) to conceal the location of a base station. MPBSD complicates localization techniques based on received signal-Strength-Indicator (RSSI). It presumes that multiple base-stations exist in the WSN, and one of them is elected as the active base station for a specific period. In the meantime, inactive base-stations transmit deceptive beacons to mask the location of the active base-station by thwarting localization methods. The real-world testbed experiments demonstrated that MPBSD is an effective MTD approach to provide obscurity in the location information of the base-station on duty in term of end-to-end delay.

24.5.2.2.2 MTD Based Network Traffic Configuration Randomization

Some recent study focused on Smart Grid domain where IoT-based smart meters with wireless transmission capabilities are deployed. Advanced Metering Infrastructure (AMI) is one of the Smart Grid applications that utilizes a smart meter to enable two-way communications between the household and utility company. AMI has enabled the utility company to perform its operations remotely such as energy consumption data collection, outage detection, and diagnostics. Due to its critical role and its predictable and deterministic behavior (e.g. periodic data collection schedule, same size data collection, and same route to convey data from the smart meter to the AMI headend systems), AMI has been the target of various attacks, called mimicry attacks. These types of attacks mimic the AMI behavior and follow the protocol. Due to the resource-constrained devices in AMI networks, deep packet inspection may not be possible and thus, these attacks can go undetected. Ali et al. [109], studied how MTD can be an effective proactive defense against the attacks by randomizing three configuration parameters of AMI: (i) report size, (ii) report interval, and (iii) relaying nodes.

Algin et al. [110] studied how MTD can be an effective means to eliminate selective jamming attacks for Smart Grid AMI networks. In this setup, jamming could be an important attack to block a smart meter's transmission if the schedule of the transmission is learned. To mitigate this issue, the authors proposed randomizing the schedules in each round of data transmissions so that the attackers will not be able to determine the exact transmission times. The randomization of the transmissions follows the idea of MTD, basically changing the times randomly for each smart meter. To this end, they employ the Fisher–Yates shuffle algorithm that has been shown to provide secure randomness. This algorithm guarantees that a smart meter will transmit in a different time slot than its previous slot. The experiment results indicate that the proposed MTD approach does not bring any significant overhead. On the contrary, the authors demonstrate that in some cases MTD helps to improve packet deliver ratios as the randomness help in accessing the wireless channel more efficiently. In the same lines, there was another very recent study which explored MTD for vehicular environments [146]. The idea is transmitting data across dynamic multi paths relayed through vehicles traveling on a multi-lane road. Again, the study demonstrated that jamming a targeted data stream would be very difficult.

24.5.3 SDN and MTD for IoT

SDN architecture has been considered a cost-effective solution that can ease the network management and dynamic configuration for the IoT devices that are in many cases not easily accessible. Once SDN is employed, it can pave the way to also apply MTD techniques for IoT applications. In this subsection, we first explore the research efforts which strive to apply SDN to IoT environments and then make a case for SDN-based MTD for certain IoT applications.

24.5.3.1 SDN for IoT Environments

Besides for cyber resilience in IIoT and cybersecurity deception in IoT as presented in Chapters 20 and 21 respectively, there are some recent works on possible use-cases, architecture, implementation, and even security of SDN integration for IoT environments.

For instance, the authors in [147] discuss two use-cases of SDN in IoT environments. In the first one, they suggest using OpenFlow-enabled switches in gateways for smart homes and claim that it will decrease the cost of the management and maintenance of the network. Their proposal framework is shown in Figure 24.15.

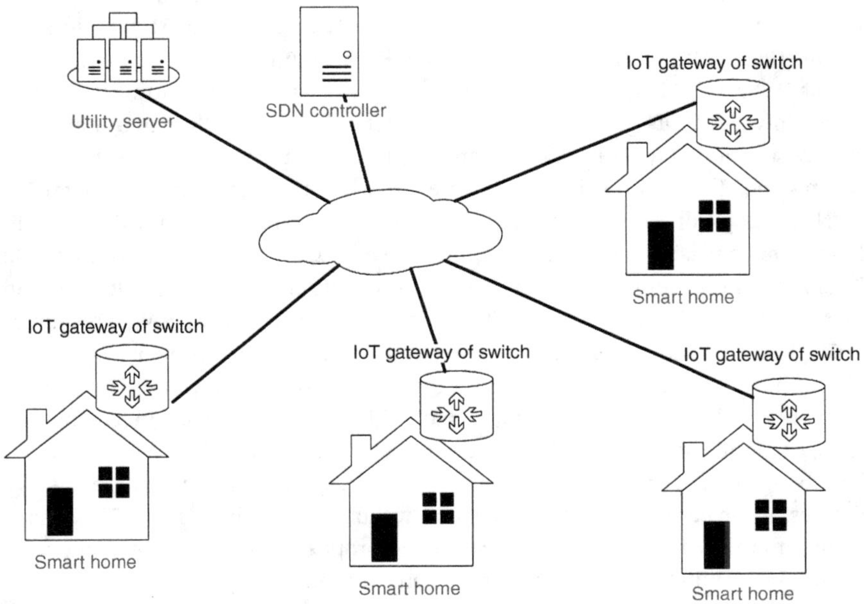

Figure 24.15 SDN-based switches for Smart Home.

As a second use-case, the authors consider SDN-enabled Evolved Packet Core (EPC) in LTE networks in order to separate and handle efficiently the traffic that must pass through EPC (e.g. voice traffic), and traffic that can be offloaded to the Internet (e.g. video traffic). This simple solution reduces the load on EPC and the operating expenditures for the service provider. While the second example relates mostly to EPC, the first use-case is a perfect fit solely for IoT environments. In another study [148], researchers demonstrate the scenarios and standards for virtualization and SDN in IoT environments. They argue that SDN would provide a flexible and interoperable environment which is critical for IoT communications. They show SDN on Wireless Mesh Network (WMN) by connecting SDN Controller to Wireless Mesh Routers (WMRs). In addition, they also mention that virtualization cannot be directly applied to wireless environments due to the dynamic nature of air interface.

There are a couple of architectural frameworks for SDN-based IoT that researchers have proposed. For instance, Zhijing et al. proposed a layered controller design [149]. He showed that their proposed generic algorithm for flow scheduling has better performance in terms of throughput, delay, and jitter compared to two common scheduling algorithms, namely bin packing and load balancing. Alejandro et al. introduced Software Defined Wireless Sensor Network (SDWSN) where SDN Controller is considered in the base station [150]. They follow the flow table idea of OpenFlow and try to solve the compatibility issue of other SDN nodes. Different studied also proposed to change OpenFlow to support Wireless Sensor Networks (WSNs) [151, 152]. The approaches in these works can control the flow of data. However, SDN Controller could also be used for controlling and managing the sensor hardware as it would change sensors' status depending on the need. Meanwhile, the authors in [153] came up with a framework for integrating SDN and Fog Computing in the IoT domain. Smart transportation, video surveillance, and precision agriculture are presented as use-cases of such framework. There is also an effort on providing communication between IoT devices and SDN Controller. They propose REST protocol to turn IoT devices to Web resources [154]. REST interface is used as Northbound API from SDN Controller.

Another noteworthy IoT application domain for SDN deployment is vehicular networks. SDN has started to be applied to Vehicular environments in recent years [155]. Among these efforts, there were a few works which focused on routing, bandwidth management, QoS, etc. among vehicles which are considered as IoT devices. As an earlier work in this subject, authors in [156] introduced SDN-based communication for vehicular devices. They propose LTE-based connection for control plane and Wi-Fi for data plane. Their solution provides better packet delivery ratio compared to Adhoc on Demand Distance Vector Routing (AODV) [157], On-demand Link State Routing (OLSR) [158], and Destination Sequence Distance Vector (DSDV) [159] routing protocols considering quick response mechanism of

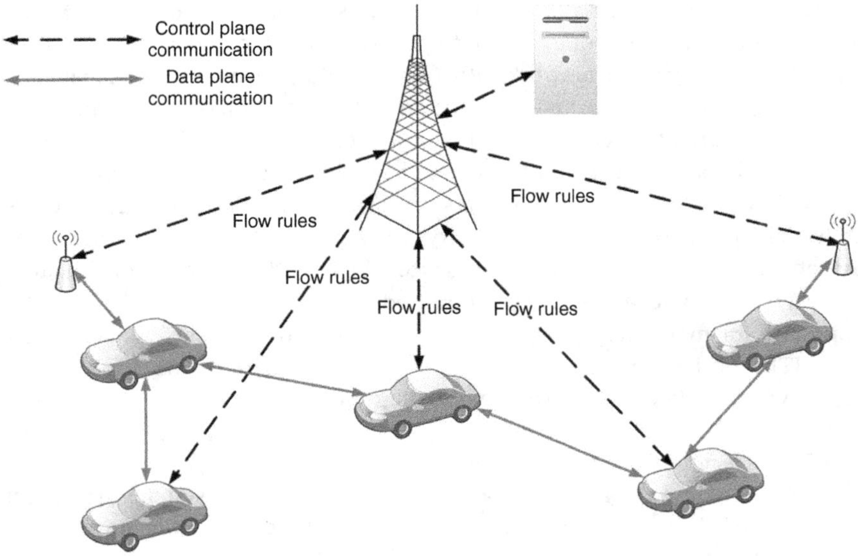

Figure 24.16 SDN-based vehicular network.

SDN Controller in case of topology changes. As a different work in this area, Baihong et al. [160] introduced SDN-based on-demand routing protocol (SDOV) with two level design (two-level controller); one for deciding which path to forward by utilizing vehicle information, and the other one to select forwarding vehicles according to the decision made in the upper layer (see Figure 24.16). Their simulation demonstrates better results on the impact of vehicle density, speed on data transmission rate and average packet delay compared to common ad-hoc routing protocols namely OLSR, Dynamic Source Routing (DSR) [161], DSDV, etc.

24.5.3.2 SDN-Based MTD for IoT

As was shown in the previous studies, applying SDN and MTD at the same time to IoT environments may not always be a feasible operation due to resource constraints and operational needs of the applications. In this section, we advocate two specific applications of IoT where SDN-based MTD can be an effective means for secure and dependable operations.

24.5.3.2.1 Internet of Battlefield Things

One of such applications is the Internet of Battlefield Things (IoBT) which focuses on the military applications where various IoT devices (e.g. sensors, munitions, weapons, vehicles, robots, and human-wearable devices) and human warfighters need to communicate in real-time through wireless communications [162]. Mobility is the inherent nature of this ad hoc network. The command and control messages flow through the network either in a multi-hop manner or single-hop depending on the priorities and needs. As critical information is shared within this network, securing the communication infrastructure is essential. While sophisticated security approaches and tools can be employed, the resource-constrained "things" (IoT devices) do not allow for such solutions. In addition, fault-tolerance or availability of IoT devices are also very critical due to the adverse nature of the battlefield environments. While these devices will have the capability to act autonomously with some AI features, centralized management capabilities are also important for more effective defense and strategy spreading. In this vein, SDN and MTD can be a perfect fit with their centralized nature which would also in-line with the hierarchical aspects of military infrastructure and information.

24.5.3.2.2 Industrial Internet of Things

The other potential application is the IIoT environments. IIoT systems are perfect examples of CPS and they are often identified as CIs for their national importance to national economy and security. Security of these IIoT systems is essential, especially in the wake of cyber warfare, including Advanced Persistent Threats (APT) and nation-state attacks. MTD, in addition to the static security measures, can provide an edge to fight against incessant and highly-capable state-backed cybercriminals, who mostly exploits zero-day vulnerabilities. IIoT systems are often equipped with communication infrastructures for control channels, which easily allow the deployment of SDN to employ MTD. However, these communication systems are often large, widespread, with legacy devices and delicate industrial protocols. Hence, the deployment of SDN, considering technical difficulties, limited resources, MTD measures, and security requirements, is challenging, creating an interesting decision-making problem.

In an IIoT environment, the cyberinfrastructure is highly integrated with physical systems through control routines. MTD traditionally considers random and/or periodic, often proactive, moves of cyber components, protocols, or behaviors. The physical moves, if feasible, can open a new dimension to MTD. A successful attack against an IIoT system often needs the knowledge of the physical properties, physical agility can thwart such attacks. A pioneer work of this kind has been proposed in a smart grid scenario [163]. An integration of physical agility with cyber moves will provide a powerful MTD technique (Figure 24.17a). The design of the MTD technique (particularly, the selection and timing of MTD moves/actions) for an IIoT system must consider the IIoT control structure. An MTD move can easily deteriorate the system's operation. While the impact of a move needs to be within the optimal operational requirement, the move must affect the adversary's capability. In the case of the integration of cyber and physical MTD moves, both types

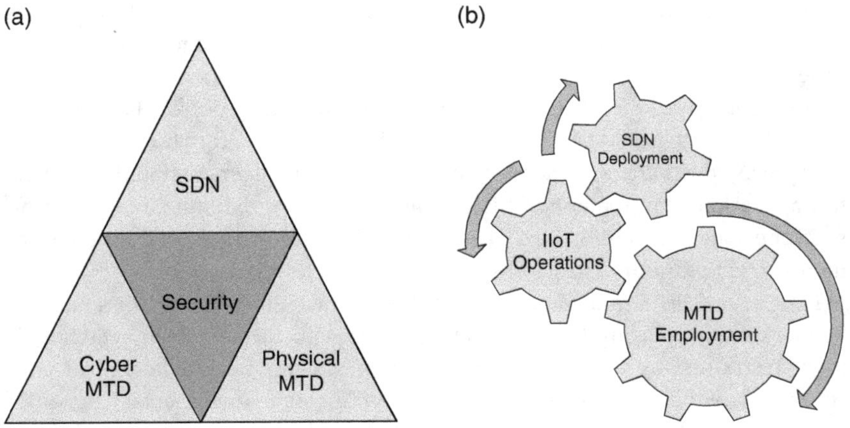

Figure 24.17 (a) MTD for IIoT systems and (b) optimal design of an MTD technique.

of actions must comply with each other such that one kind should not negatively influence the other, rather collectively offer a larger impact. An SDN-based control structure can be leveraged to govern these MTD actions for an effective impact. The optimal design of an MTD technique, as shown in Figure 24.17b, needs to consider the MTD moves, IIoT operational requirements, and SDN deployment factors.

24.6 Future Research Challenges

A full application of MTD techniques for IoT specifically for IoBT and IIoT comes with additional challenges that necessitate new research regardless of the use of SDN or not. In addition, any other IoT application might benefit from MTD while still there are challenges regarding the availability of resources. In this section, we elaborate on these research challenges as listed below:

- *SDN Control Channel in IoBT:* IoBT requires constant communication among a possibly large number of mobile nodes. Any attack on this network will thus affect almost all the nodes in terms of information gathering, command and control, and resources. Therefore, a fast access to these devices from the SDN controller is needed. Given that this channel needs to be wireless, reliability and availability of the communications will be a challenge. Security and robustness would be a must for this communication. Therefore, new wide area control protocols are needed. The emerging 5G technologies have the potential to address this challenge with its reliability and long-distance coverage. However,

5G may not be available everywhere and thus its features such as D2D as well as technologies from 802.11 families such as IEEE 802.11ah need to be investigated in terms of QoS they can guarantee.

- *Local Intelligence in IoBT:* Given the challenges with the control channel reliability, the nodes should be given some local intelligence so that they can operate when SDN controller is not accessible. The level of intelligence should be subject to application requirements. This intelligence can be dynamic and based on machine learning technologies where the devices can learn from their environments to act without instructions.
- *MTD-Aware SDN Deployment in IIoT:* The deployment of SDN in an existing network (e.g. replacing the traditional routers with SDN-enabled switches) is often restrained by limited resources, legacy systems, and/or technical constraints. Therefore, the challenge of deploying (often incremental) the SDN architecture within the limits while optimally achieving the security/defense objective should be explored under different parameters.
- *Integrated Cyber and Physical MTD Measures in IIoT:* While the cyber and physical moves together can bring a larger MTD capability, these actions must comply with one another, without declining the optimal operation of the IIoT system. Therefore, an important research direction for IIoT security is to explore the feasible physical moves and their MTD effectiveness, as well as to study the optimal integration of physical agility measures with the traditional cyber moves in the SDN-based MTD framework.
- *MTD Games:* Given the resource constraints and operational requirements, MTD in IoBT/IIoT may not be able to arbitrate many properties extensively. Therefore, game-theoretical approaches are needed to deceive the attackers based on their communication and action patterns. This will introduce a trade-off between the resources, service constraints, technical/communication feasibility, and the attacker's assumed capabilities. Another interesting piece of this challenge is to be able to model the behavior of the attackers in certain conditions so that MTD moves could be better arranged.
- *Smart Moving Target Defense for IoT:* Most proposed MTD approaches are typically not an attacker-aware and the decision to select the move is based on the pre-defined constraints. IoT, on the other hand, produces a huge amount of data, which is known as the big data. The IoT data analytics is much powerful than the traditional data analytics since the IoT data analytics are intended to do the real-time processing before the data becomes irrelevant or obsolete. Typically, the IoT data analytics can be placed at the edge (i.e. fog/edge computing) or at the cloud. Therefore, the IoT data analytics can be exploited to support a smart attacker-aware MTD as opposed to traditional networks.
- *Collaborative Multi-Attribute Moving Target Defense:* In this chapter, we have shown that most MTD approaches concentrate only on a single attribute (e.g.

address, path, configuration, etc.). Only a few approaches attempt to utilize more than one attribute for the proactive defense. Therefore, a collaborative multi-attribute MTD can be a potential future research direction.

24.7 Conclusion

In this chapter, we provided a survey of existing MTD mechanisms geared for IoT environments. We first provided an overview of MTD and present a classification of MTD mechanisms in general before we make the transition to IoT environments. With respect to IoT, we first discussed how MTD could be fit to IoT for security and defense purposes. We demonstrated that despite the overview of MTD, there are many applications of MTD for IoT security that may provide defense against a diverse set of attacks including zero-days. Our specific focus on the topic was IoBT and IIoT which we claim to welcome SDN-based MTD for a comprehensive security that will be flexible and cost-effective. We laid off several future research challenges that will be of value to new researchers and industry. Overall, we advocated that for IoBT and IIoT, MTD can be an effective defense when it is integrated with SDN.

References

1 Rajkumar Buyya and Amir Vahid Dastjerdi. *Internet of Things: Principles and Paradigms*. Morgan Kaufmann Publishers Inc., San Francisco, CA, USA, 1st edition, 2016.

2 J. A. Guerrero-ibanez, S. Zeadally, and J. Contreras-Castillo. Integration challenges of intelligent transportation systems with connected vehicle, cloud computing, and internet of things technologies. *IEEE Wireless Communications*, 22(6):122–128, Dec 2015.

3 N. Bui, A. P. Castellani, P. Casari, and M. Zorzi. The internet of energy: a web-enabled smart grid system. *IEEE Network*, 26(4):39–45, July 2012.

4 N. Suri, M. Tortonesi, J. Michaelis, P. Budulas, G. Benincasa, S. Russell, C. Stefanelli, and R. Winkler. Analyzing the applicability of internet of things to the battlefield environment. In *2016 International Conference on Military Communications and Information Systems (ICMCIS)*, pages 1–8, May 2016.

5 D. Singh, G. Tripathi, A. M. Alberti, and A. Jara. Semantic edge computingand IoT architecture for military health services in battlefield. In *2017 14th IEEE Annual Consumer Communications Networking Conference (CCNC)*, pages 185–190, Jan 2017.

6 F. T. Johnsen, Z. Zieliski, K. Wrona, N. Suri, C. Fuchs, M. Pradhan, J. Furtak, B. Vasilache, V. Pellegrini, M. Dyk, M. Marks, and M. Krzyszto. Application of IoT in military operations in a smart city. In *2018 International Conference on Military Communications and Information Systems (ICMCIS)*, pages 1–8, May 2018.

7 Jonathan Lampe. *IoT Security: Locking Down the Internet of Things*. Auerbach Publications, Boston, MA, USA, 1st edition, 2017.

8 H. Okhravi, T. Hobson, D. Bigelow, and W. Streilein. Finding focus in the blur of moving-target techniques. *IEEE Security Privacy*, 12(2):16–26, Mar 2014.

9 Nick Feamster, Jennifer Rexford, and Ellen Zegura. The road to sdn: An intellectual history of programmable networks. *ACM SIGCOMM Computer Communication Review*, 44(2):87–98, 2014.

10 Kevin Ashton et al. That internet of things thing. *RFID Journal*, 22 (7):97–114, 2009.

11 D INFSO. Networked enterprise & rfid infso g. 2 micro & nanosystems. In *Internet of Things in Cooperation with the Working Group RFID of the ETP EPOSS,* 2020, 4.

12 Linus Wallgren, Shahid Raza, and Thiemo Voigt. Routing attacks and countermeasures in the rpl-based internet of things. *International Journal of Distributed Sensor Networks*, 9(8):794326, 2013.

13 Li Da Xu, Wu He, and Shancang Li. Internet of things in industries: A survey. *IEEE Transactions on Industrial Informatics*, 10(4):2233–2243, 2014.

14 Lane Thames and Dirk Schaefer. Industry 4.0: An overview of key benefits, technologies, and challenges. In *Cybersecurity for Industry 4.0*, pages 1–33. Springer, 2017.

15 Arijit Karati, SK Hafizul Islam, and Marimuthu Karuppiah. Provably secure and lightweight certificateless signature scheme for IIoT environments. *IEEE Transactions on Industrial Informatics*, vol. 14, no. 8, pages 3701–3711, Aug. 2018.

16 Rwan Mahmoud, Tasneem Yousuf, Fadi Aloul, and Imran Zualkernan. Internet of things (IoT) security: Current status, challenges and prospective measures. In *2015 10th International Conference for Internet Technology and Secured Transactions (ICITST),* pages 336–341. IEEE, 2015.

17 Shahid Mumtaz, Ahmed Alsohaily, Zhibo Pang, Ammar Rayes, Kim Fung Tsang, and Jonathan Rodriguez. Massive internet of things for industrial applications: Addressing wireless IIoT connectivity challenges and ecosystem fragmentation. *IEEE Industrial Electronics Magazine*, 11(1):28–33, 2017.

18 Dhananjay Singh, Gaurav Tripathi, and Antonio J. Jara. A survey of internet-of things: Future vision, architecture, challenges and services. In *2014 IEEE world forum on Internet of things (WF-IoT),* pages 287–292. IEEE, 2014.

19 Arsalan Mosenia and Niraj K. Jha. A comprehensive study of security of internet of-things. *IEEE Transactions on Emerging Topics in Computing*, 5 (4):586–602, 2017.

20 Dave Evans. The internet of things: How the next evolution of the internet is changing everything, 2011. URL: https://www.cisco.com/c/dam/en_us/about/ac79/docs/innov/IoT_IBSG_0411FINAL.pdf

21 Shuo Feng, Peyman Setoodeh, and Simon Haykin. Smart home: Cognitive interactive people-centric internet of things. *IEEE Communications Magazine*, 55 (2):34–39, 2017.

22 Samet Tonyali, Ozan Cakmak, Kemal Akkaya, Mohamed MEA Mahmoud, and Ismail Guvenc. Secure data obfuscation scheme to enable privacy-preserving state estimation in smart grid AMI networks. *IEEE Internet of Things Journal*, 3(5):709–719, 2016.

23 Hawzhin Mohammed, Samet Tonyali, Khaled Rabieh, Mohamed Mahmoud, and Kemal Akkaya. Efficient privacy-preserving data collection scheme for smart grid ami networks. In *2016 IEEE Global Communications Conference (GLOBECOM)*, pages 1–6. IEEE, 2016.

24 Samet Tonyali, Kemal Akkaya, Nico Saputro, A. Selcuk Uluagac, and Mehrdad Nojoumian. Privacy-preserving protocols for secure and reliable data aggregation in IoT-enabled smart metering systems. *Future Generation Computer Systems*, 78:547–557, 2018.

25 Samet Tonyali. *Privacy-Preserving Protocols for IEEE 802.11s-based Smart Grid Advanced Metering Infrastructure Networks*. PhD thesis, Florida International University, 2018.

26 Ahmad Alsharif, Mahmoud Nabil, Samet Tonyali, Hawzhin Mohammed, Mohamed Mahmoud, and Kemal Akkaya. Epic: Efficient privacy-preserving scheme with e2e data integrity and authenticity for ami networks. *arXiv preprint arXiv:1810.01851*, 2018.

27 Ahmad-Reza Sadeghi, Christian Wachsmann, and Michael Waidner. Security and privacy challenges in industrial internet of things. In *2015 52nd ACM/EDAC/IEEE Design Automation Conference (DAC)*, pages 1–6. IEEE, 2015.

28 Amin Hassanzadeh, Shimon Modi, and Shaan Mulchandani. Towards effective security control assignment in the industrial internet of things. In *2015 IEEE 2nd World Forum on Internet of Things (WF-IoT)*, pages 795–800. IEEE, 2015.

29 Michele Albano, Jos´e Bruno Silva, and Luis Lino Ferreira. The industrial internet of things. *22º Seminário da Rede Temática de Comunicações Móveis*, 2017.

30 Priya. IoT building blocks and architecture: IoT part 2, 2012. URL: https://www.engineersgarage.com/Articles/Internet-of-Things-Architecture

31 ZigBee Alliance. IEEE 802.15. 4, Zigbee standard, 2009.

32 Z-Wave Alliance. About z-wave technology. Z-Wave Alliance, Technical Report, 2013.

33 Yasir Zaki. Long term evolution (lte). In *Future Mobile Communications*, pages 13–33. Springer, 2013.

34 Behrouz A. Forouzan and Sophia Chung Fegan. *TCP/IP Protocol Suite*. McGrawHill Higher Education, 2002.

35 Statista. Internet of things (IoT) connected devices installed base worldwide from 2015 to 2025 (in billions), 2018. https://www.statista.com/statistics/471264/iot-number-of-connected-devices-worldwide/

36 Arunabha Ghosh, David R. Wolter, Jeffrey G. Andrews, and Runhua Chen. Broadband wireless access with wimax/802.16: Current performance benchmarks and future potential. *IEEE Communications Magazine*, 43(2):129–136, 2005.

37 Brian P. Crow, Indra Widjaja, Jeong Geun Kim, and Prescott T. Sakai. IEEE 802.11 wireless local area networks. *IEEE Communications Magazine*, 35(9):116–126, 1997.

38 SIG Bluetooth. Bluetooth specification, 2003.

39 SANS Institute InfoSec Reading Room. *The 2018 Sans Industrial IoT Security Survey: Shaping IIoT Security Concerns*. SANS Institute, 2018.

40 Jon Postel et al. Rfc 791: Internet protocol, 1981.

41 RIPE NCC. Understanding ip addressing and cidr charts, 2016.

42 Steve Deering and Robert Hinden. Internet protocol, version 6 (ipv6) specification. Technical report, 2017.

43 Geoff Mulligan. The 6lowpan architecture. In *Proceedings of the 4th Workshop on Embedded Networked Sensors*, pages 78–82. ACM, 2007.

44 Jose A. Gutierrez, Marco Naeve, Ed Callaway, Monique Bourgeois, Vinay Mitter, and Bob Heile. IEEE 802.15. 4: A developing standard for low-power low-cost wireless personal area networks. *IEEE Network*, 15(5):12–19, 2001.

45 Gabriel Montenegro, Nandakishore Kushalnagar, Jonathan Hui, and David Culler. Transmission of ipv6 packets over IEEE 802.15. 4 networks. Technical report, 2007.

46 Roy Fielding, Jim Gettys, Jeffrey Mogul, Henrik Frystyk, Larry Masinter, Paul Leach, and Tim Berners-Lee. Hypertext transfer protocol–http/1.1. Technical report, 1999.

47 Eric Rescorla. Http over tls. Technical report, 2000.

48 Jon Postel. Transmission control protocol. Technical report, 1981.

49 Zach Shelby, Klaus Hartke, and Carsten Bormann. The constrained application protocol (coap). Technical report, 2014.

50 Jon Postel. User datagram protocol. Technical report, 1980.

51 Roy Fielding. Representational state transfer. In *Architectural Styles and the Design of Network-based Software Architecture*, pages 76–85, 2000. https://www.ics.uci.edu/~fielding/pubs/dissertation/fielding_dissertation.pdf

52 Urs Hunkeler, Hong Linh Truong, and Andy Stanford-Clark. Mqtt-sa publish/subscribe protocol for wireless sensor networks. In *3rd International Conference on Communication Systems Software and Middleware and Workshops, 2008. comsware 2008*, pages 791–798. IEEE, 2008.

53 Apache Hive. Live long and process (llap), 2018. URL: https://gerardnico.com/db/hive/llap

54 Peter Saint-Andre. Extensible messaging and presence protocol (xmpp): Core. Technical report, 2011.

55 Steve Vinoski. Advanced message queuing protocol. *IEEE Internet Computing*, Vol. 10, no. 6, pp. 87–89, Nov.–Dec. 2006.

56 Eric Rescorla and Nagendra Modadugu. Datagram transport layer security version 1.2. Technical report, 2012.

57 Sean Turner. Transport layer security. *IEEE Internet Computing*, 18 (6):60–63, 2014.

58 T. Winter. Routing protocol for low-power and lossy networks. Technical report, rfc 6550, 6551, 6552. IETF, 2012.

59 Jose A. Gutierrez, Edgar H. Callaway, and Raymond L. Barrett. *Low-Rate Wireless Personal Area Networks: Enabling Wireless Sensors with IEEE 802.15. 4*. IEEE Standards Association, 2004.

60 Carles Gomez, Joaquim Oller, and Josep Paradells. Overview and evaluation of Bluetooth low energy: An emerging low-power wireless technology. *Sensors*, 12(9):11734–11753, 2012.

61 Regis J. Bates. *GPRS: General Packet Radio Service*. McGraw-Hill Professional, 2001.

62 Ahmad Jalali, Witold Krzymien, and Paul Mermelstein. Medium access control scheme for data transmission on code division multiple access (cdma) wireless systems, Oct 27, 1998. US Patent 5,828,662.

63 Simon S. Lam. A carrier sense multiple access protocol for local networks. *Computer Networks*, 4(1):21–32, 1980.

64 Fouad A. Tobagi and V. Bruce Hunt. Performance analysis of carrier sense multiple access with collision detection. *Computer Networks (1976)*, 4 (5):245–259, 1980.

65 Souvik Sen, Romit Roy Choudhury, and Srihari Nelakuditi. Csma/cn: Carrier sense multiple access with collision notification. *IEEE/ACM Transactions on Networking (ToN)*, 20(2):544–556, 2012.

66 Roy Want. Near field communication. *IEEE Pervasive Computing*, 10 (3), pp. 4–7, July–September 2011.

67 Ivan Muller, Joao Cesar Netto, and Carlos Eduardo Pereira. Wirelesshart field devices. *IEEE Instrumentation & Measurement Magazine*, 14 (6), pp. 20–25, December 2011.

68 Juan Carlos Zuniga and Benoit Ponsard. Sigfox system description. LPWAN@ IETF97, Nov. 14, 2016.

69 Maarten Weyn, Glenn Ergeerts, Rafael Berkvens, Bartosz Wojciechowski, and Yordan Tabakov. Dash7 alliance protocol 1.0: Low-power, mid-range sensor and actuator communication. In *2015 IEEE Conference on Standards for Communications and Networking (CSCN)*, pages 54–59. IEEE, 2015.

70 Ferran Adelantado, Xavier Vilajosana, Pere Tuset-Peiro, Borja Martinez, Joan Melia-Segui, and Thomas Watteyne. Understanding the limits of lorawan. *IEEE Communications Magazine*, 55(9):34–40, 2017.

71 Wikipedia. Thread (network protocol), 2018.

72 Paul Darbee. Insteon the details, smarthouse. Inc., Aug, 11:68, 2005.

73 Matthias Kovatsch, Simon Duquennoy, and Adam Dunkels. Erbium (er) rest engine and coap implementation for contiki, 2014.

74 Olaf Bergmann. Tinydtls. https://projects.eclipse.org/projects/iot.tinydtls, pages 2–15, 2013.

75 D. Gothberg. Micro-ip for embedded systems. *Computer Club West*, 2005. https://tools.ietf.org/html/draft-gothberg-micro-ip-00

76 Naganand Doraswamy and Dan Harkins. *IPSec: The New Security Standard for the Internet, Intranets, and Virtual Private Networks*. Prentice Hall Professional, 2003.

77 Nicolas Tsiftes, Joakim Eriksson, and Adam Dunkels. Low-power wireless ipv6 routing with Contiki rpl. In *Proceedings of the 9th ACM/IEEE International Conference on Information Processing in Sensor Networks*, pages 406–407. ACM, 2010.

78 A. Dunkels. Sicslowpan-internet-connectivity for low-power radio systems. *SICS*, 2008. https://internetstiftelsen.se/docs/SICS_Lowpan-report.pdf

79 Adam Dunkels. The Contiki mac radio duty cycling protocol, 2011. URL: http://dunkels.com/adam/dunkels11contikimac.pdf

80 Muhammad Omer Farooq and Thomas Kunz. Contiki-based IEEE 802.15. 4 node's throughput and wireless channel utilization analysis. In *Wireless Days (WD), 2012 IFIP*, pages 1–3. IEEE, 2012.

81 Joachim Feld. Profinet-scalable factory communication for all applications. In *2004 IEEE International Workshop on Factory Communication Systems, 2004. Proceedings*, pages 33–38. IEEE, 2004.

82 IDA Modbus. Modbus application protocol specification v1. 1a. *North Grafton, Massachusetts*. URL: www.modbus.org/specs.php, 2004.

83 Thomas Ulz, Thomas Pieber, Christian Steger, Sarah Haas, Holger Bock, and Rainer Matischek. Bring your own key for the industrial internet of things. In *2017 IEEE International Conference on Industrial Technology (ICIT)*, pages 1430–1435. IEEE, 2017.

84 Gerardo Pardo-Castellote. Omg data-distribution service: Architectural overview. In *Proceedings. 23rd International Conference on Distributed Computing Systems Workshops, 2003*, pages 200–206. IEEE, 2003.

85 John Matherly. Shodan search engine. URL: https://www.shodan.io, 2009.

86 Constantinos Kolias, Georgios Kambourakis, Angelos Stavrou, and Jeffrey Voas. DDoS in the IoT: Mirai and other botnets. *Computer*, 50(7):80–84, 2017.

87 Kieren McCarthy. California cracks down on internet of crap passwords with new law to stop the botnets, 2018. https://www.theregister.co.uk/2018/10/04/california_iot_password/

88 Chad Perrin. The cia triad, 2008. URL: https://www.techrepublic.com/blog/it-security/the-cia-triad/

89 Christian Lesjak, Holger Bock, Daniel Hein, and Martin Maritsch. Hardware secured and transparent multi-stakeholder data exchange for industrial IoT. In *2016 IEEE 14th International Conference on Industrial Informatics (INDIN)*, pages 706–713. IEEE, 2016.

90 Christian Lesjak, Daniel Hein, and Johannes Winter. Hardware-security technologies for industrial IoT: Trustzone and security controller. In *Industrial Electronics Society, IECON 2015-41st Annual Conference of the IEEE*, pages 002589–002595. IEEE, 2015.

91 Sreejaya Viswanathan, Rui Tan, and David KY Yau. Exploiting power grid for accurate and secure clock synchronization in industrial IoT. In *2016 IEEE Real-Time Systems Symposium (RTSS)*, pages 146–156. IEEE, 2016.

92 Nick McKeown, Tom Anderson, Hari Balakrishnan, Guru Parulkar, Larry Peterson, Jennifer Rexford, Scott Shenker, and Jonathan Turner. Openflow: Enabling innovation in campus networks. *ACM SIGCOMM Computer Communication Review*, 38(2):69–74, 2008.

93 Floodlight Project. Floodlight controller, 2014. URL: http://www.projectfloodlight.org/floodlight/

94 Opendaylight. Opendaylight, 2018. URL: https://www.opendaylight.org/

95 Mark Masse. *REST API Design Rulebook: Designing Consistent RESTful Web Service Interfaces*. O'Reilly Media, Inc., 2011.

96 NITRD. National cyber leap year summit 2009, co-chairs report. Technical Report, Federal Networking and Information Technology Research and Development (NITRD) Program, 2009.

97 Yue-Bin Luo, Bao-Sheng Wang, Xiao-Feng Wang, Xiao-Feng Hu, and Gui-Lin Cai. TPAH: A universal and multi-platform deployable port and address hopping mechanism. In *2015 International Conference on Information and Communications Technologies (ICT 2015)*, pages 1–6, April 2015.

98 Y. B. Luo, B. S. Wang, X. F. Wang, X. F. Hu, G. L. Cai, and H. Sun. RPAH: Random port and address hopping for thwarting internal and external adversaries. In *Trustcom/BigDataSE/ISPA, 2015 IEEE*, volume 1, pages 263–270, Aug 2015.

99 Ehab Al-Shaer, Qi Duan, and Jafar Haadi Jafarian. *Random Host Mutation for Moving Target Defense*, pages 310–327. Springer Berlin Heidelberg, Berlin, Heidelberg, 2013.

100 Jafar Haadi H. Jafarian, Ehab Al-Shaer, and Qi Duan. Spatio-temporal address mutation for proactive cyber agility against sophisticated attackers. In *Proceedings of the First ACM Workshop on Moving Target Defense*, MTD '14, pages 69–78, New York, NY, USA, 2014. ACM.

101 J. H. Jafarian, E. Al-Shaer, and Q. Duan. Adversary-aware ip address randomization for proactive agility against sophisticated attackers. In *2015 IEEE Conference on Computer Communications (INFOCOM)*, pages 738–746, April 2015.

102 M. Dunlop, S. Groat, W. Urbanski, R. Marchany, and J. Tront. The blind man's bluff approach to security using IPv6. *IEEE Security Privacy*, 10(4):35–43, July 2012.

103 S. Yan, X. Huang, M. Ma, P. Zhang, and Y. Ma. A novel efficient address mutation scheme for ipv6 networks. *IEEE Access*, 5:7724–7736, 2017.

104 Qi Duan, Ehab Al-Shaer, and Haadi Jafarian. Efficient random route mutation considering flow and network constraints. In *2013 IEEE Conference on Communications and Network Security (CNS)*, pages 260–268. IEEE, 2013.

105 Usman Rauf, Fida Gillani, Ehab Al-Shaer, Mahantesh Halappanavar, Samrat Chatterjee, and Christopher Oehmen. Formal approach for resilient reachability based on end-system route agility. In *Proceedings of the 2016 ACM Workshop on Moving Target Defense*, MTD'16, pages 117–127, New York, NY, USA, 2016. ACM.

106 Huangxin Wang, Quan Jia, Dan Fleck, Walter Powell, Fei Li, and Angelos Stavrou. A moving target ddos defense mechanism. *Computer Communications*, 46:10–21, 2014.

107 P. Wood, C. Gutierrez, and S. Bagchi. Denial of service elusion (dose): Keeping clients connected for less. In *2015 IEEE 34th Symposium on Reliable Distributed Systems (SRDS)*, pages 94–103, Sept 2015.

108 S. Venkatesan, M. Albanese, K. Amin, S. Jajodia, and M. Wright. A moving target defense approach to mitigate ddos attacks against proxy-based architectures. In *2016 IEEE Conference on Communications and Network Security (CNS)*, pages 198–206, Oct 2016.

109 M. Q. Ali, E. Al-Shaer, and Q. Duan. Randomizing AMI configuration for proactive defense in smart grid. In *2013 IEEE International Conference on Smart Grid Communications (SmartGridComm)*, pages 618–623, Oct 2013.

110 Ramazan Algin, Huseyin O. Tan, and Kemal Akkaya. Mitigating selective jamming attacks in smart meter data collection using moving target defense. In *Proceedings of the 13th ACM Symposium on QoS and Security for Wireless and Mobile Networks*, Q2SWinet'17, pages 1–8, New York, NY, USA, 2017. ACM.

111 Jafar Haadi Jafarian, Ehab Al-Shaer, and Qi Duan. Openflow random host mutation: Transparent moving target defense using software defined networking. In *Proceedings of the First Workshop on Hot Topics in Software Defined Networks*, HotSDN'12, pages 127–132, New York, NY, USA, 2012. ACM.

112 Douglas C. MacFarland and Craig A. Shue. The SDN shuffle: Creating a moving target defense using host-based software-defined networking. In *Proceedings of the Second ACM Workshop on Moving Target Defense*, pages 37–41. ACM, 2015.

113 Richard Skowyra, Kevin Bauer, Veer Dedhia, and Hamed Okhravi. Have no PHEAR: Networks without identifiers. In *Proceedings of the 2016 ACM Workshop on Moving Target Defense*, MTD'16, pages 3–14, New York, NY, USA, 2016. ACM.

114 S. Y. Chang, Y. Park, and A. Muralidharan. Fast address hopping at the switches: Securing access for packet forwarding in SDN. In *NOMS 2016–2016 IEEE/IFIP Network Operations and Management Symposium*, pages 454–460, April 2016.

115 H. Zhou, C. Wu, M. Jiang, B. Zhou, W. Gao, T. Pan, and M. Huang. Evolving defense mechanism for future network security. *IEEE Communications Magazine*, 53(4):45–51, April 2015.

116 Duohe Ma, Zhen Xu, and Dongdai Lin. *Defending Blind DDoS Attack on SDN Based on Moving Target Defense*, pages 463–480. Springer International Publishing, Cham, 2015.

117 Ankur Chowdhary, Adel Alshamrani, Dijiang Huang, and Hongbin Liang. Mtd analysis and evaluation framework in software defined network (mason). In *Proceedings of the 2018 ACM International Workshop on Security in Software Defined Networks & Network Function Virtualization*, SDN-NFV Sec'18, pages 43–48, New York, NY, USA, 2018. ACM.

118 J. B. Hong, S. Yoon, H. Lim, and D. S. Kim. Optimal network reconfiguration for software defined networks using shuffle-based online mtd. In *2017 IEEE 36th Symposium on Reliable Distributed Systems (SRDS)*, pages 234–243, Sept 2017.

119 Jafar Haadi Jafarian, Ehab Al-Shaer, and Qi Duan. *Formal Approach for Route Agility against Persistent Attackers*, pages 237–254. Springer Berlin Heidelberg, Berlin, Heidelberg, 2013.

120 A. Aydeger, N. Saputro, K. Akkaya, and M. Rahman. Mitigating crossfire attacks using sdn-based moving target defense. In *2016 IEEE 41st Conference on Local Computer Networks (LCN)*, pages 627–630, Nov 2016.

121 Huiting Tan, Chaojing Tang, Chen Zhang, and Shaolei Wang. Area-dividing route mutation in moving target defense based on sdn. In Zheng Yan, Refik Molva, Wojciech Mazurczyk, and Raimo Kantola, editors, *Network and System Security*, pages 565–574. Springer International Publishing, Cham, 2017.

122 L. Zhang, Q. Wei, K. Gu, and H. Yuwen. Path hopping based SDN network defense technology. In *2016 12th International Conference on Natural Computation, Fuzzy Systems and Knowledge Discovery (ICNC-FSKD)*, pages 2058–2063, Aug 2016.

123 Zheng Zhao, Daofu Gong, Bin Lu, Fenlin Liu, and Chuanhao Zhang. SDN-based Double Hopping Communication against sniffer attack. *Mathematical Problems in Engineering*, vol. 2016 (13), 2016.

124 Duohe Ma, Cheng Lei, Liming Wang, Hongqi Zhang, Zhen Xu, and Meng Li. *A Self-adaptive Hopping Approach of Moving Target Defense to thwart Scanning Attacks*, pages 39–53. Springer International Publishing, Cham, 2016.

125 Ankur Chowdhary, Sandeep Pisharody, and Dijiang Huang. SDN based scalable MTD solution in cloud network. In *Proceedings of the 2016 ACM Workshop on Moving Target Defense*, MTD'16, pages 27–36, New York, NY, USA, 2016. ACM.

126 S. Debroy, P. Calyam, M. Nguyen, A. Stage, and V. Georgiev. Frequency-minimal moving target defense using software-defined networking. In *2016 International Conference on Computing, Networking and Communications (ICNC)*, pages 1–6, Feb 2016.

127 R. Callon E. Rosen, A. Viswanathan. Multiprotocol Label Switching Architecture. RFC 3031, RFC Editor, Jan 2001.

128 Kaleel Mahmood and Devu Manikantan Shila. Moving target defense for internet of things using context aware code partitioning and code diversification. In *2016 IEEE 3rd World Forum on Internet of Things (WF-IoT)*, pages 329–330. IEEE, 2016.

129 Valentina Casola, Alessandra De Benedictis, and Massimiliano Albanese. *A Multi-Layer Moving Target Defense Approach for Protecting Resource-Constrained Distributed Devices*, pages 299–324. Springer International Publishing, Cham, 2014.

130 Ermanno Battista, Valentina Casola, Antonino Mazzeo, and Nicola Mazzocca. Siren: A feasible moving target defence framework for securing resource constrained embedded nodes. *International Journal of Critical Computer-Based Systems*, 4(4):374–392, 2013.

131 Matthew Sherburne, Randy Marchany, and Joseph Tront. Implementing moving target ipv6 defense to secure 6lowpan in the internet of things and smart grid. In *Proceedings of the 9th Annual Cyber and Information Security Research Conference*, CISR'14, pages 37–40, New York, NY, USA, 2014. ACM.

132 T. Preiss, M. Sherburne, R. Marchany, and J. Tront. Implementing dynamic address changes in Contiki OS. In *International Conference on Information Society (i-Society 2014)*, pages 222–227, Nov 2014.

133 K. Zeitz, M. Cantrell, R. Marchany, and J. Tront. Designing a micro-moving target ipv6 defense for the internet of things. In *2017 IEEE/ACM Second International Conference on Internet-of-Things Design and Implementation (IoTDI)*, pages 179–184, April 2017.

134 K. Zeitz, M. Cantrell, R. Marchany, and J. Tront. Changing the game: A micro moving target ipv6 defense for the internet of things. *IEEE Wireless Communications Letters*, vol. 7(4), pp. 578–581, Aug. 2018.

135 Aljosha Judmayer, Georg Merzdovnik, Johanna Ullrich, Artemios G. Voyiatzis, and Edgar Weippl. A performance assessment of network address shuffling in IoT systems. In Roberto Moreno-D´ıaz, Franz Pichler, and Alexis Quesada-Arencibia, editors, *Computer Aided Systems Theory – EUROCAST 2017*, pages 197–204, Cham, 2018. Springer International Publishing.

136 Tommy Chin and Kaiqi Xiong. Mpbsd: A moving target defense approach for base station security in wireless sensor networks. In Qing Yang, Wei Yu, and Yacine Challal, editors, *Wireless Algorithms, Systems, and Applications*, pages 487–498, Cham, 2016. Springer International Publishing.

137 K. Andrea, A. Gumusalan, R. Simon, and H. Harney. The design and implementation of a multicast address moving target defensive system for internet-of-things applications. In *MILCOM 2017–2017 IEEE Military Communications Conference (MILCOM)*, pages 531–538, Oct 2017.

138 Massimiliano Albanese, Alessandra De Benedictis, Sushil Jajodia, and Kun Sun. A moving target defense mechanism for manets based on identity virtualization. In *2013 IEEE Conference on Communications and Network Security (CNS)*, pages 278–286. IEEE, 2013.

139 Matt Richardson and Shawn Wallace. *Getting Started with Raspberry PI*. O'Reilly Media, Inc., 2012.

140 Raspberry Pi. Model b. 2015. URL: http://www.alliedelec.com/raspberry-piraspberry-pi-2-model-b/70465426, 2.

141 Raspberry Pi. Model b. *Raspberrypi. org. Saatavissa*. URL: https://www.raspberrypi.org/products/raspberry-pi-3-model-b/. Hakup¨aiva¨, 6:2018, 3.

142 8 devices. Carambola 2, 2018. URL: https://www.8devices.com/products/carambola-2

143 S. Groat, M. Dunlop, W. Urbanksi, R. Marchany, and J. Tront. Using an IPv6 moving target defense to protect the smart grid. In *2012 IEEE PES Innovative Smart Grid Technologies (ISGT)*, pages 1–7, Jan 2012.

144 J. Polastre, R. Szewczyk, and D. Culler. Telos: Enabling ultra-low power wireless research. In *IPSN 2005. Fourth International Symposium on Information Processing in Sensor Networks, 2005*, pages 364–369, April 2005.

145 A. Dunkels, B. Gronvall, and T. Voigt. Contiki – A lightweight and flexible operating system for tiny networked sensors. In *29th Annual IEEE International Conference on Local Computer Networks*, pages 455–462, Nov 2004.

146 Esraa M. Ghourab, Effat Samir, Mohamed Azab, and Mohamed Eltoweissy. Diversity-based moving-target defense for secure wireless vehicular communications. In *2018 IEEE Security and Privacy Workshops (SPW)*, pages 287–292. IEEE, 2018.

147 Vishwapathi Rao Tadinada. Software defined networking: Redefining the future of internet in IoT and cloud era. In *2014 International Conference on Future Internet of Things and Cloud (FiCloud)*, pages 296–301. IEEE, 2014.

148 Fabrizio Granelli, Anteneh A. Gebremariam, Muhammad Usman, Filippo Cugini, Veroniki Stamati, Marios Alitska, and Periklis Chatzimisios. Software defined and virtualized wireless access in future wireless networks: Scenarios and standards. *IEEE Communications Magazine*, 53(6):26–34, 2015.

149 Zhijing Qin, Grit Denker, Carlo Giannelli, Paolo Bellavista, and Nalini Venkatasubramanian. A software defined networking architecture for the internet-of things. In *Network Operations and Management Symposium (NOMS), 2014 IEEE*, pages 1–9. IEEE, 2014.

150 Alejandro De Gante, Mohamed Aslan, and Ashraf Matrawy. Smart wireless sensor network management based on software-defined networking. In *2014 27th Biennial Symposium on Communications (QBSC)*, pages 71–75. IEEE, 2014.

151 Tie Luo, Hwee-Pink Tan, and Tony QS Quek. Sensor openflow: Enabling software defined wireless sensor networks. *IEEE Communications Letters*, 16(11):1896–1899, 2012.

152 Amr El-Mougy, Mohamed Ibnkahla, and Lobna Hegazy. Software-defined wireless network architectures for the internet-of-things. In *2015 IEEE 40th Local Computer Networks Conference Workshops (LCN Workshops)*, pages 804–811. IEEE, 2015.

153 Slavica Tomovic, Kenji Yoshigoe, Ivo Maljevic, and Igor Radusinovic. Software defined fog network architecture for IoT. *Wireless Personal Communications*, 92(1):181–196, 2017.

154 Zhigang Wen, Xiaoqing Liu, Yicheng Xu, and Junwei Zou. A restful framework for internet of things based on software defined network in modern manufacturing. *The International Journal of Advanced Manufacturing Technology*, 84 (1–4):361–369, 2016.

155 Manisha Chahal, Sandeep Harit, Krishn K. Mishra, Arun Kumar Sangaiah, and Zhigao Zheng. A survey on software-defined networking in vehicular ad hoc networks: Challenges, applications and use cases. *Sustainable Cities and Society*, Vol. 35, Pages 830–840. 2017.

156 Ian Ku, You Lu, Mario Gerla, Rafael L. Gomes, Francesco Ongaro, and Eduardo Cerqueira. Towards software-defined vanet: Architecture and services. In *2014 13th annual Mediterranean ad hoc networking workshop (MED-HOC-NET)*, pages 103–110. IEEE, 2014.

157 Charles Perkins, Elizabeth Belding-Royer, and Samir Das. Ad hoc on-demand distance vector (aodv) routing. Technical report, 2003.

158 Thomas Clausen and Philippe Jacquet. Optimized link state routing protocol(olsr). Technical report, 2003.

159 Charles E. Perkins and Pravin Bhagwat. Highly dynamic destination-sequenced distance-vector routing (dsdv) for mobile computers. *In ACM SIGCOMM Computer Communication Review*, volume 24, pages 234–244. ACM, 1994.

160 Baihong Dong, Weigang Wu, Zhiwei Yang, and Junjie Li. Software defined networking based on-demand routing protocol in vehicle ad hoc networks. In *2016 12th International Conference on Mobile Ad-Hoc and Sensor Networks (MSN)*, pages 207–213. IEEE, 2016.

161 David B. Johnson, David A. Maltz, Josh Broch, et al. Dsr: The dynamic source routing protocol for multi-hop wireless ad hoc networks. *Ad Hoc Networking*, 5:139–172, 2001.

162 A. Kott, A. Swami, and B. J. West. The internet of battle things. *Computer*, 49 (12):70–75, Dec. 2016.

163 Mohammad Ashiqur Rahman, Ehab Al-Shaer, and Rakesh B. Bobba. Moving target defense for hardening the security of the power system state estimation. In *ACM Workshop on Moving Target Defense (MTD)*, pages 59–68, 2014.

25

Toward Robust Outlier Detector for Internet of Things Applications

Raj Mani Shukla and Shamik Sengupta

Department of Computer Science and Engineering, University of Nevada, Reno, Reno, NV, USA

25.1 Introduction

Internet of Things (IoT) envisions the presence of ubiquitous devices and sensors that gather environment parameters and transfers them to either nearby device like mobile phones or to a centralized unit like edge and cloud for processing [1, 2]. At the centralized unit, the data are analyzed and processed to extract useful information and assist in efficient IoT-enabled applications' performance. For example, the sensors, loop detectors, and cameras installed throughout the city collect traffic information, and the processed information helps in making better travel decision or traffic management [3]. Similarly, a plethora of IoT based applications exist that enable smart services in connected communities like intelligent transportation system, smart home automation, or smart grid [4, 5].

Often the data gathered from sensors has time-series patterns with periodic behavior and follows certain trends [6]. For example, the health monitoring sensor measuring the ECG signal collects periodic patterns [7]. Similarly, the roadside traffic data have periodic behaviors with periodicity in the granularity of day, week, and season. The household electricity consumption also has specific daily and seasonal trends. The time-series data are leveraged to make optimal decisions by various IoT enabled applications. However, the patterns in the data are nondeterministic and affected by endogenic or exogenic factors. Therefore, the anomaly or outliers are common in time-series data. A frequent occurrence of outliers

Modeling and Design of Secure Internet of Things, First Edition. Edited by Charles A. Kamhoua, Laurent L. Njilla, Alexander Kott and Sachin Shetty.

Published 2020 by John Wiley & Sons, Inc.

reduces the effectiveness of intelligent and automated decision making of applications due to bad data quality.

The anomaly in time-series data may occur due to several possible factors. First, a time-series can deviate from the probable value due to changes in external conditions. For instance, some event in a city may affect the traffic congestion and traffic count values at certain locations. Similarly, a person's heartbeat variation from the average value may be due to either physical activity or health conditions. The anomaly may also arise due to the poor quality of sensors. For example, a sensor may be poorly calibrated thus sending incorrect information. The degradation in sensor quality may be due to sensors being physically tampered. If large number of sensors are compromised, it may affect the data analytic decisions by applications.

Another kind of vulnerability that may occur in time-series data is due to Data falsification attacks. Data falsification attack is a kind of vulnerability where an adversary may intentionally alter data values [8]. The corresponding adversary may have enough resources that can bypass cryptographic security mechanisms [9]. The adversary may intelligently attack the spatially and temporally correlated sensors to affect the data quality. Furthermore, an adversary may inject attacks in such a way that the statistical properties of the data like mean, mode, or median values are preserved, but cumulative anomaly affects the data analysis to a more considerable extent. Additionally, the adversary may also optimize the attacking noise such that a minimal noise can cause a substantial error in application performance. The resultant anomaly is hard to detect since the actual data and attacked data patterns differ by a tiny amount [10].

Since, the IoT based applications make automated decisions based on the processed data, the outlier detection is a significant concern that needs to be thoroughly investigated. Further, because an anomaly can arise due to several factors, it is also essential to determine the original cause of an anomaly. That is to detect whether an anomaly is due to any environmental parameters or *Data-falsification attack*. Moreover, if the detection method finds an adversary is tampering with the data, the type of the attack needs to be identified so that the signal processing can subsidize its effect.

The problem of determining anomaly in time-series data is complicated because of several reasons. The amount of data from the sensors is of enormous quantity such that manually analyzing the data is an infeasible task. Additionally, the possible types of attack that an attacker can introduce are generally not known in advance. An attacker can intelligently attack a group of sensors without affecting the statistical properties of the data points to a considerable extent. An adversary may also come with new kind of attacks models that current detection technique may not handle. Orchestrated or adversarial attacks are hard to detect even with some of the best known current state of the art techniques [11].

Moreover, time-series data are affected by external factors like climatic conditions, events, construction, or emergency situations. These conditions may cause the time-series to behave abnormally. For example, snow fall in a certain region can cause road traffic to differ from normal value. However, during normal conditions the traffic can be still be different than normal value since the sensors may have been attacked by an adversary. Therefore, distinguishing whether the abnormal number is due to external factors or *Data-falsification attack* complexifies the detection of such attacks.

Although, the anomaly detection problem has been studied by the researchers, it does not meet the rising demand of data generated due to IoT devices. For example, the data set often has diverse nature and thus simple statistical analysis cannot be used due to varied statistical properties. Supervised learning techniques require huge set of known anomaly patterns which may not be available. Since in the high dimensional data the different dimensions may not be orthogonal, the techniques like Principal Component Analysis (PCA) cannot be efficiently utilized.

To address the problem of anomaly detection for IoT data, we propose a novel detection architecture shown in Figure 25.1. The proposed architecture leverages the diversity in time-series data to better detect outliers. The proposed method consists of different modules which complement each other. Furthermore, the method complements existing techniques by finding the relation between different time-series data. To explain in detail our method we begin by describing some of the existing techniques used by the researchers. The main contributions of this work are as follows:

- We describe the various anomaly detection techniques used by the researchers and discuss their limitations.
- We present a novel outlier detector and discuss how it can be implemented by fusing different modules.

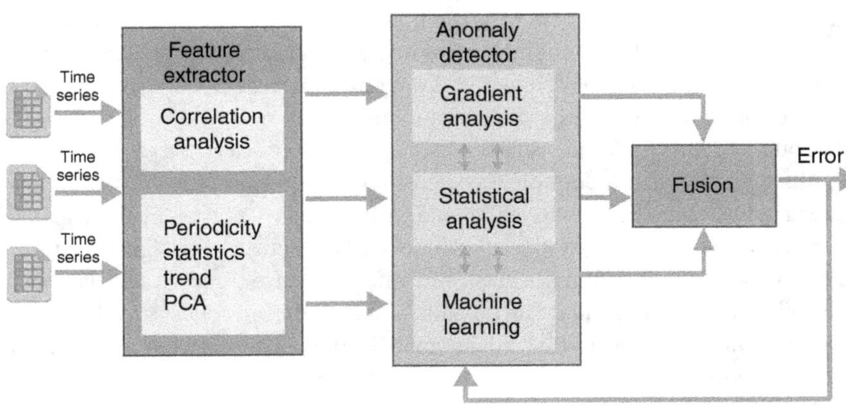

Figure 25.1 Proposed anomaly detector.

- We provide some of the open research issues that need to be tackled in the context of the proposed outlier detector.

The rest of the chapter is organized as follows:

Following this section, we discuss the state-of-the-art anomaly detection methods in Section 25.2. Section 25.3 describes the proposed outlier detector. In Section 25.4, we discuss future research directions and Section 25.5 concludes this chapter.

25.2 State of the Art Anomaly Detection Techniques

There have been several works in the literature on anomaly detection of time-series data. Chandola et al. [12], Gupta et al. [13], Akoglu et al. [14], and Hodge and Austin [15] have comprehensively studied the anomaly detection techniques in diverse contexts. This section complements the above survey papers by describing some of the outlier detection techniques relevant to the IoT.

In IoT domain, there are two approaches to process the data for outlier detection. First, the distributed approach where anomaly detection task is distributed between different IoT nodes and the centralized unit. Second, a centralized approach where outlier detection responsibility is solely put on the centralized unit such as cloud. We cover some of the prominent works for both distributed and centralized methods for anomaly detection of IoT data.

25.2.1 Distributed Approach

Thanigaivelan et al. has presented a distributed approach form anomaly detection for IoT in [16]. In this work, both nodes and routers are responsible for anomaly detection and task is distributed between them. The nodes observe the properties like data rate and packet size for detecting any anomalous value. The edge router collects the information about anomaly from different nodes and based on correlated information it classifies a node as anomalous.

The outlier detection for IoT sensor data has been proposed by Yu et al. [17]. The proposed method has a hierarchical system consisting cluster of IoT sensors, cluster head, and cloud. The mentioned scheme groups the IoT sensors by their spatial location using clustering algorithm like k-means. Every cluster is assigned a cluster head having strong computational power. The IoT sensors record the data and transfer it to the cluster head for processing. The cluster head collects the data from corresponding IoT nodes to find the anomaly in the cluster using recursive-PCA (R-PCA) algorithm. The R-PCA is a modified version of PCA that iteratively updates the basis of PCA transformation based on the changes in IoT system.

Li et al. has proposed a distributed method for anomaly detection in IoT sensors and its trust management in [18]. In the given method first IoT nodes verify for any anomaly in other IoT nodes within its transmission range. Based on the information each IoT node informs its neighborhood about the possible outliers. Then the data are fused according to the Dempster–Shafer Theory (DST) [19]. To verify the trustworthiness of the collected data the proposed method checks if a sensor data is in consistent with its neighborhood. If not, then it is further checks if abnormal value is due to environmental factors or any adversary.

Although above approaches have found to be efficient in terms of anomaly detection, the distributed approaches often pose the problems due to computing power of the sensor nodes. The IoT sensors are low powered device and as the data size grows, they cannot efficiently process it due to their limited computing facilities. Therefore, the IoT systems are centered around cloud and edge such that the nodes can be easily configured, and their computing power does not limit its processing capability. Therefore, we describe below some of the centralized approaches for anomaly detection of IoT data.

25.2.2 Centralized Approach

Due to the popularity of cloud computing in IoT there are a plethora of works that use centralized approach. In centralized approach, the sensor nodes are considered as dumb devices and the intelligence is shifted to the cloud due to its computing capabilities. Following are some of the notable works that use centralized outlier detector.

Netflix has recently proposed an outlier detection method (RAD) for their big data applications. The method uses a modified version of PCA, Robust PCA (RPCA), an algorithm for efficient anomaly detection in their cloud applications [20]. The RPCA recovers the principal components of the data matrix even if a fraction of the data is corrupt [21]. Netflix has used the method in two of its major applications. Netflix involves many daily banking transactions. It tracks those transactions both in real-time and batch mode and detects the failure (anomaly) in the transaction to assist the user. Every day many peoples sign in the Netflix in their browser. Netflix determines anomaly in account sign-in process by utilizing the combination of country, language, browsers, etc., that is used in the login process.

Data falsification by an organized adversary poses a serious threat to business models. Such adversaries may inject anomaly by launching different kind of attacks through altering data values but preserving statistical measures. To avoid this problem, Bhattacharjee et al. has proposed a semi-supervised anomaly detection technique in [22]. The paper presented a robust statistical measure namely ratio of Harmonic Mean to Arithmetic Mean (HM–AM ratio) to infer the attack.

HM–AM ratio is resilient against changes in mean value. The HM–AM ratio combined with HM and AM values finds the attack and its type efficiently.

The time-series Twitter data have seasonal trends and often contain anomalies due to either internal or external factor. To detect anomalies in the cloud data, Vallis et al. have proposed a novel detector method [23] that is used by Twitter Inc. The paper discussed two statistical techniques that automatically detect anomalies in the cloud. The described method decomposes time-series data by filtering seasonal components. Then it applies Extreme Studentized Deviate (ESD) on the filtered time-series for anomaly detection. The mean and median in ESD are found to be highly sensitive to many anomalies. Therefore, the paper further describes the use of Seasonal-Hybrid-ESD (S-H-ESD) for anomaly detection. S-H-ESD uses median and Median Absolute Deviation (MAD) to detect anomalies. S-H-ESD performs effectively even if 50% of data values are perturbed. The method found to be efficient for time-series data like tweets per second or CPU utilization per second.

Ouyang et al. have described the anomaly detection in power consumption data using machine learning method in [24]. The data are collected in the centralized cloud where it is detected for any anomaly. The work has used time-series feature engineering where time-series data are represented in multiple components using Hierarchical Time-Series Feature Extraction (HTSF) method. The features are then fed to the three multi-stage multi-view stacking ensemble (TMSE) machine learning model. The TMSE classifies the data based on supervised learning by finding any abnormal distribution.

Xie and Chen. have explained the anomaly detection in IoT Big Data in [25]. The given method works in two stages that are rough detection and careful detection. In rough detection, the features of the data are extracted using PCA and if features deviate by a substantial amount at a certain point then it is considered as anomalous. Rough detection checks only the anomaly in gathered data stream. The careful detection finds the specific data stream that is causing anomaly. For this purpose, careful detection uses the Bayesian network that leverages the fact that certain amount of dependency exists between data collected from different nodes.

The use of the neural network for anomaly detection using clustering for the cloud data has been described by Pandeeswari and Kumar. [26]. Pandeeswari and Kumar have presented a fuzzy clustering-based technique for anomaly detection in cloud data. The given method uses a combination of Fuzzy C-Means and Artificial Neural Network (FCM–ANN) methods that improve the detection accuracy as compared to the techniques like Naïve Bayes classifier and simple ANN algorithm. The hybrid FCM–ANN method is effective even in the presence of low-frequency outliers and performs with a low false alarm rate.

The above techniques using centralized approach uses machine learning approaches like PCA, Neural Network, Bayesian Network, and SVM or are based on statistical methods like HM–AM ratio, S-ESD, and S-H-ESD. Both the machine learning and statistical techniques have their inherent disadvantages. Machine learning provides probabilistic output with certain possible error. Thus, it can tell only about the expected value plus or minus the error of the trained model. Thus, it cannot provide an accurate representation of time-series data.

Second, statistical techniques assume that all the data points are generated from specific distribution and follow a pattern. The presumption does not always hold. Especially for high dimensional data, the statistical techniques cannot be efficiently applied due to the above assumption. Moreover, there are different kinds of statistical methods and choice of specific statistical measure is difficult [12]. Therefore, relying solely on statistical method or machine learning method reduces the effectiveness of anomaly detector. To address this issue, we describe an outlier detection method that fuses machine learning and statistical model and combines it with correlation and gradient analysis to better detect anomaly in time-series data.

25.3 Problem Statement and Proposed Outlier Detector

We describe the problem that need to be solved and the proposed outlier detector. For our description we use the time-series road-side traffic data obtained from PeMS system [27]. PeMS measures and collects the freeway traffic from California highways using loop detectors installed along the roads. The traffic count data from different locations are transferred to the centralized unit after every five minutes. We obtained the traffic data for freeway I-680 at 136 locations for the month of October 2017 in our analysis.

25.3.1 Problem Statement

As we have seen in Section 25.2, the well-known anomaly detection methods are not enough to meet the demand of IoT data. Additionally, there is another problem that exists with the methods mentioned above and need to be considered. An anomaly detector should also inform whether outlier is due to environmental factor or an attack. To consider its importance, suppose the case that traffic congestion occurs due to bad road-side conditions. In that case, the anomaly detectors will classify the traffic at that point as anomalous. However, the applications using the traffic data require true values which are the congested traffic to find optimal solution. Therefore, the outlier detector should not only tell about if the data does not follow normal pattern but also find if abnormal values are due to attack scenario or external environmental conditions.

Moreover, the diversity in time series data is growing at a very high rate. Same type of data has different statistical properties and therefore it is difficult to

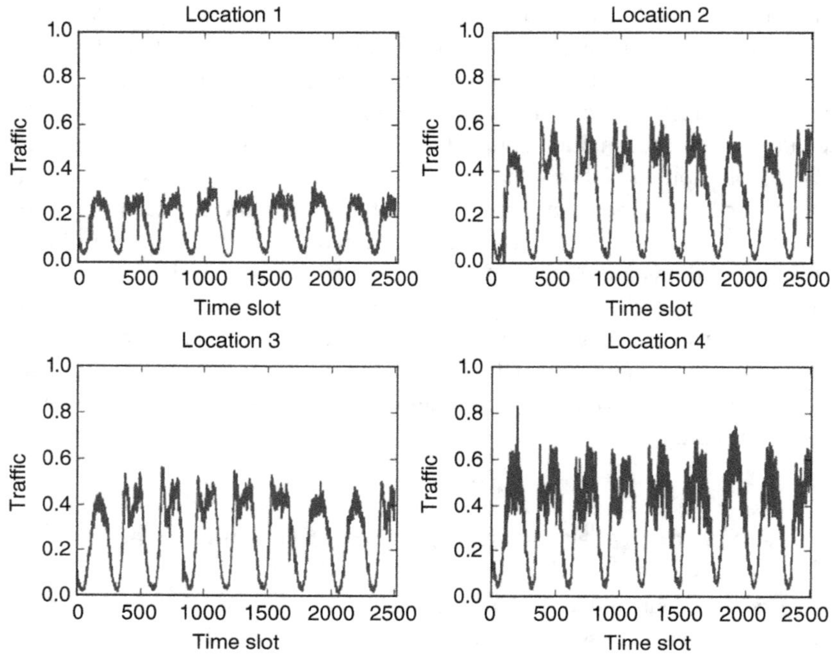

Figure 25.2 Time series traffic data at different locations.

determine parameters of the outlier detector. To explain this, consider Figure 25.2 that shows the time-series traffic at four different locations in the same highway segment. The time-series data have different statistical properties, and their variation is also significant for four locations. Using simple statistical methods as a benchmark for such data coming from different sources cannot be accurate due to diverse measured values.

Also, due to this diversity the anomaly detection requires a scalable solution. For example, the outlier detection technique used by the Twitter collects the data from all the users and finds the anomaly in aggregated data values [23]. Therefore, the parameters of the detection method in those methods need to be set only once which is an easy task. If the same technique is followed to determine the anomaly in individual sensors, then the parameters must be found for individual sensors which are a cumbersome task due to large number of sensors. Therefore, the methods proposed in the literature are not scalable and cannot be applied for IoT in general due to large number of devices generating time-series data. Thus, the anomaly detection solutions for the IoT devices need scalable solutions. To address above issue, we describe below the possible techniques that can be applied to impart scalability and robustness in outlier detector for IoT.

25.3.2 Proposed Outlier Detector

The proposed system consists of IoT sensors that record the time-series data and the aggregate of the data over a suitable time interval τ is sent to the edge server. At the edge node, spatial and temporal data patterns are collected and then based on that the optimization decisions are made. The adversary may tamper with the data at the sensor. This affects the optimization decision as it is based on the spatiotemporal data from sensors thus affecting the performance of the corresponding IoT-enabled application. Adversary can incrementally alter the data points to cause change in behavior of applications. Incremental change can be either adding small values to the measured data or decreasing the measured values by some quantity. An adversary may also use different techniques to evade state of the art detection techniques. The technique may include attacking a group of sensors providing correlated information. For example, altering data values in sensors that are spatially close to each other for measuring road-side traffic. The data points in the group of sensors can be changed such that the statistical properties over the group are not altered. In this case, the attacker may increase the measured value in some sensors and decrease values in others.

The proposed outlier detector is placed at the edge node to find such attacks. The detector filters the time-series data obtained from the sensor and filtered data is sent to the optimization applications. Figure 25.1 in Section 25.1 shows the broad-view architecture of the proposed outlier detector. The proposed architecture takes different time-series as its input. The time-series data may be coming from the varied range of sensor nodes. For example, different time-series can be traffic data, bandwidth requirement, social media posts, and weather conditions. We divide the proposed detector into two portions namely; (i) Feature extractor and (ii) Anomaly detector. The feature extractor performs the offline processing of the known correct data-set values. Outlier detector works in two phases. In the first phase, we set its parameters like threshold values for statistical analysis and machine learning hyper-parameters. The parameters can be set by injecting the possible types of attacks in the correct data values by developing attack models. Once the anomaly detector is trained, the time-series data is fed to the detector, and it will output possible outliers. We describe below the feature extractor and anomaly detector modules in detail.

25.3.3 Feature Extractor

To develop an outlier detection model first the properties of the time-series data need to be determined in terms of its mean, mode, periodicity, dominant components, etc. The feature extractor module should determine the characteristics of the time-series data that are gathered from diverse sources. Feature extractor itself consists of two parts viz. Pattern analysis module and Correlation module. Pattern analysis is the very important first step in any outlier detection method since it helps in finding the benchmark values of the data points to distinguish between true points and anomaly. Correlation analysis assists in finding whether anomaly is due to natural factors or an attack.

Pattern analysis module determines the statistical properties and principal components of the individual time-series data. Ratio of HM and AM is an important characteristic that can be used as one of the feature [22]. However, the ratio approaches to the value of 1 if data points are close to each other. Since, in that case $AM \cong HM$. The AM to HM ratio can be useful for finding point as well as group anomalies.

Median is another important property that is robust measure of the data sample. While mean value of a time-series are changed by changing a single value of the sample, median is robust again 50% of data to be changed. That is to change median 50% valued above or below it need to be altered. Therefore, median and median absolute deviation (MAD) is supposed to be one of the robust measures of the data. However, for the normal data samples the median and MAD are less efficient than the mean. Therefore, another technique is to use M-estimator. M-estimator is defined as the solution of the Eq. (25.1).

$$\sum_{i=1}^{n} \varphi\left(\frac{x_i - \widehat{\theta}}{\widehat{\sigma}}\right) \tag{26.1}$$

Here, the number of samples are n, x_i is a particular sample, φ is a real function and the denominator $\widehat{\sigma}$ is an initial estimate such as MAD [28]. The solution $\widehat{\theta}$ of the Eq. (25.1) is an robust measure of the scale.

Interquartile range (IQR) is also a robust measure that can be used to determine anomalies. IQR is defined as the difference between third and first quartile. It has the breakdown of 25% and thus can be used as a robust measure [28]. The variability of the data is defined by the IQR since it divides the data into quartiles.

The statistical features can also be determined for different time window size. That is the statistical analysis is done for different time granularity. Here, the problem is to determine the size of the window under consideration. Furthermore, regarding application processing, all the features of the data may not be significant. Data-set may also contain large number of components. Therefore, finding dominant features reduces the complexity of further analysis. The neural network comes in handy such a case. For large number of features Convolution Neural Network (CNN) can be helpful as they are used in image processing with many features. In contrast to the traditional PCA based methods, neural networks are not based on the assumptions of linearly correlated dimensions or the requirement of orthogonal components.

25.3.4 Anomaly Type Detection

Correlation analysis assists in determining whether anomaly is due to natural factors or due to an attack. Often the changes in one time-series data affect some other time-series data. For example, traffic deviation at a certain location also affects the traffic change in a correlated location. Similarly, the traffic volume and air quality can also be correlated and high traffic count results in higher air quality. Likewise, the traffic congestion at place also affects the mobile bandwidth requirement, and there may also be changes in social media posts.

To explain this further, we determine the correlation between traffic data at 136 locations of freeway I-680 in Figure 25.3. The corresponding traffic at 4 of such locations is shown in Figure 25.2. As we can see from both the figures that the actual traffic data follows a varied pattern but the correlation at all 136 locations is around 0.90. Thus, correlation information can be leveraged to determine anomaly in the time-series patterns. Thus, if a traffic at a certain point changes then traffic at the correlated point should change. Similarly, if there is traffic congestion at a certain point then air quality at that point must be poorer. An abnormal value in one time-series and normal value in other signifies the presence of outliers due to attacks. Therefore, the different time-series can complement each other for outlier type detection. Thus, if anomaly is observed in different time-series then it may be due to natural factors. However, if anomaly occurs in

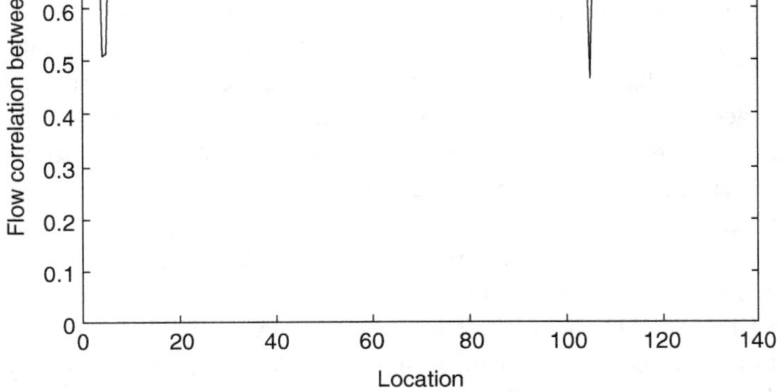

Figure 25.3 Spatial correlation of traffic flow between different positions.

one-time series not in correlated time-series then the possibility of attack is there. For this purpose, it is required to find the extent by which two time series are correlated. If one time series can be approximated as a function of other correlated time-series, then the function can be used to determine the outlier type. Since diverse time-series help each other in finding anomaly the model is scalable. This also makes an attacker's task difficult since he may have to alter data values not in a single time-series but across multiple time-series to evade the anomaly detector. Once the time-series characteristics are obtained the outlier-detector need to find the anomalies in the time-series data if the patterns deviate from the expected values. The next section describes the proposed anomaly detector that uses fusion of machine learning, statistical analysis, and gradient analysis modules.

25.3.5 Anomaly Detector

The anomaly detector must adhere to three rules. First, at a certain time it should follow expected pattern based on available values. For example, it is expected that the roadside traffic in weekdays and weekend will be different but somewhat similar over different weekdays and for different weekends. Any departure from the probabilistic data point raises the possibility of an anomaly. Supervised learning is a very powerful prediction tool once it is trained with the known input and output patterns. The supervised learning using neural networks can be trained to predict the expected values of a data point based on the available historical values. This property of the supervised learning can be leveraged to check if the time-series is following an expected pattern. If not, then the possibility of the attack is there.

Even if the time-series values follow expected pattern there is possibility that anomalies are introduced in it. For example, the traffic count before office hours is expected to be higher. However, the actual values of the traffic can be increased or decreased to evade the supervised learning method since it gives a probabilistic output with certain error. Therefore, outlier detector should check whether data points follow statistical characteristic or not. If it does not, then the possibility is that either an attack has occurred, or the deviation is due to the external environmental conditions. Third, the rate of change in time-series is also an important property. The actual value of the time-series data at a location can be large or small depending upon different conditions but the sudden change of data points is unexpected. For instance, during some event in the city the traffic count may be high and certain roads may be congested. But after the event traffic cannot subsidies immediately and it takes certain amount of time to follow usual behavior. Any sudden deviation in the data points raises an alarm about the presence of adversary. Therefore, gradient changes along with supervised learning and statistical analysis provide an insight about the time-series data. Thus, fusion of three techniques has a potential to provide efficient method for outlier detection. We describe below some of the techniques that can be used for different modules.

25.3.5.1 Machine Learning

Neural network is a very powerful tool that can be used for predicting expected values in the time-series data. The multi-layer perceptron models can extract deep features in time-series data. Outlier detector can determine the expected values of the traffic count using neural network based on historical values in spatially correlated sensors. To verify this, we trained the neural network on PeMS traffic data and used it to determine expected traffic count at a certain location. The input to the network is spatiotemporal data and output is the expected traffic at a certain time. Figure 25.4 compares the expected traffic obtained using neural network and actual traffic count at a certain location. As we can see from the figure the expected and true values are very close each other. The RMSE error between the two time-series is around 2%. The network can be used in real-time to compare the measured values with expected values and possible prediction error. If the value deviates to a large amount, then it will signify possibility of the attack.

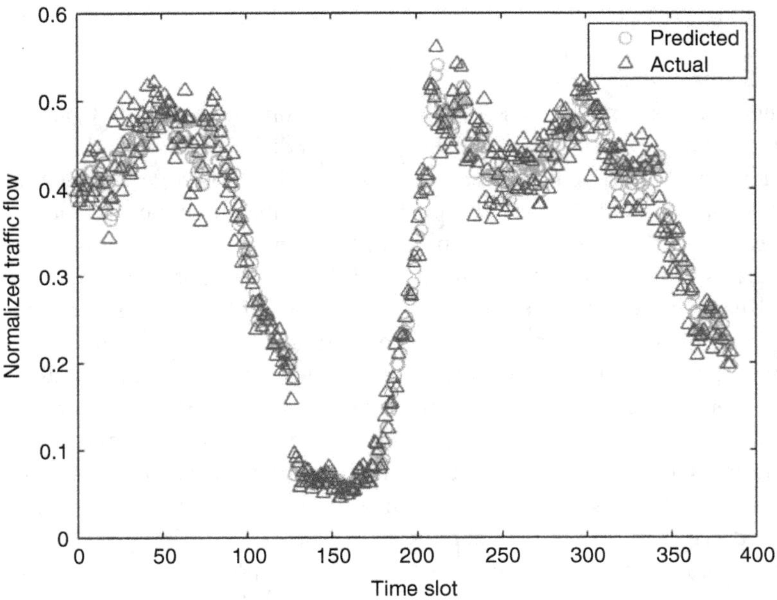

Figure 25.4 Actual and predicted traffic count values.

25.3.5.2 Statistical Analysis

To enable on-line and accurate detection of anomaly its extension with statistical and gradient analysis module is needed as explained earlier. In statistical model, the time-series pattern is observed against any deviation in pattern. Several possible statistical properties can be used for comparison. For example, a 3-sigma rule has been often used. Here, if the patterns deviate by more than three times of standard deviation with respect to mean value then value is anomalous. Recently, Seasonal Extreme Studentized deviate (S-ESD) or Seasonal Hybrid ESD (S-H-ESD) has been found effective in statistical techniques. Also, the Median Absolute Deviation (MAD) is found to be robust statistical measure [29]. While absolute mean values are distorted if few data points are corrupt, the MAD is found to be robust against about 50% values are anomalous. Another feature to use is the ratio of HM to AM which has been used in [22]. However, the paper has used the statistics for the power usage when the ratio of the HM to AM is quite high. High HM–AM ratio signify that the HM $\approx$ AM. However, this is possible only when the values of the data are close to each other. The effect of HM–AM ratio under the scattered data values is yet to be investigated. In statistical technique, the changes in properties need to be verified for different time-series. For example, it needs to be examined whether multiple correlated time-series simultaneously shows change in pattern. If pattern changes in multiple time-series, then it may be due to environmental factors.

25.3.5.3 Gradient Analysis

The gradient analysis provides useful information about the anomaly. Gradient is the change in aggregate time-series data values over different time-window size. The gradients at the different time can be used as a threshold to compare with respect to actual gradient changes. If there is a remarkable difference in gradient than the expected value, the points can be considered anomalous. Figure 25.5 presents the average gradients for different window size of the time of traffic data. The window size l is of the order five minutes. Thus, a value of $l = 2$ means the time-window size is of 10 minutes and the change in traffic after every 10 minutes is determined. Here, it should be noted that a lower window size provides less values of gradient, but the graph is not smooth. For high window size, the curve is smooth, but gradient values are higher. Selecting low window size may result in large number of false alarms since variation in actual data points may be counted as anomaly. A large window size may result in an anomalous value to be computed as true value due to higher threshold. Therefore, the intelligent determination of window size to select threshold value is an important problem. One of the approaches that we have used in our recent work is to dynamically select window size at runtime and based on that discover any anomalous data [30]. The given

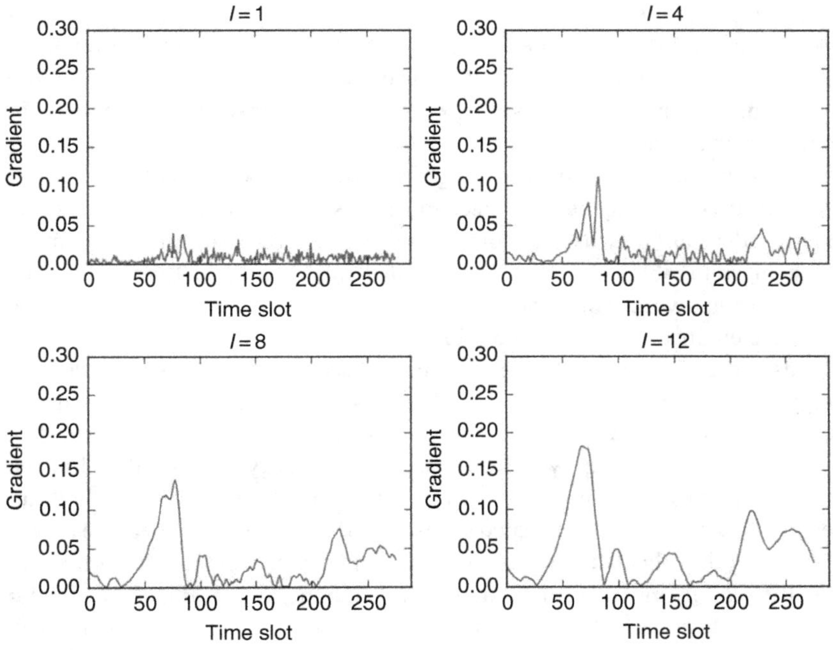

Figure 25.5 Time-series traffic gradient for different window size.

approach provides good accuracy but at the cost of detection time such that anomalies are not detected at real-time.

25.4 Future Research Directions

There are some of the open research issues that need to be tackled for creating robust outlier detector for IoT data. Concerning the above proposed model, we discuss two of such issues that need to be handled for its efficient operation.

25.4.1 Tailored Attack Model

IoT involves large number of things, devices, and application to be interfaced. Often the devices and applications are distributed by different vendors. In the IoT paradigm, the new devices and applications are frequently added or removed. This requires IoT based solutions to be open source rather than customized. Open source software, packages, and operating system enable easy to reconfigure, add, or remove devices and applications. Also, it assists in easy communication between devices; otherwise few of them may go defunct due to unavailability of supported formats [31, 32]. Due to open source software and well-documented devices, an adversary may have a different degree of knowledge of the defense mechanism. The adversary can use this knowledge of defense mechanism to tailor the attack model that can bypass security mechanisms.

An adversary may not know anything about the defense mechanism. He may also not aware that if any defense mechanism exists in the system. In such a scenario, the adversary may have to come up with a powerful attack model to make the system vulnerable. Although, it is pointed in [33] that such an obfuscated method does not work. This is because there is always a possibility that someone can reverse engineer the implemented mechanism.

It is also possible that the attacker may have a perfect knowledge of the defense strategy. For example, adversary not only knows the algorithm used but also is aware of associated parameters like threshold values. Having perfect knowledge, an adversary may come up with *white-box attack* strategy that is tailored to the defense mechanism. For example, in the context of the proposed method, an adversary may be aware of correlated time-series that are used to detect the anomaly. Knowing this, the data falsification attack may be injected into all of them to bypass the security mechanism. Similarly, knowing the statistical method enables the adversary to control the attack amount that can bypass the security method. One of the defense strategies against *white-box attack* may be to select different parameters during run-time dynamically. However, the impact of *white-box attack* strategies on the performance of the IoT applications still needs to be investigated. Further, the strategies to defend it are yet to be explored.

Another possibility is that the adversary may have some knowledge of the defense mechanism not all. The attack method to be employed in such a case is *black-box attack*. For example, the attacker may know the statistical technique but may not know threshold values. The attack model is intermediate between zero knowledge and perfect knowledge. That is if zero attack model attack succeeds to bypass security mechanism then black box attack will also be successful. In contrast, if perfect knowledge attack fails then black box attack will also fail.

The adversary may utilize the open source software and methodologies in IoT system to come up with strong attack model that can bypass security mechanisms. The comprehensive analysis of the attack model and investigation of defense strategies for the tailored attack model still need to be explored.

25.4.2 Computation Requirement

The anomaly detection method needs to process a massive amount of time-series data in real-time. Many of the algorithms mentioned above like SVM or E-H-ESD are computationally complex. The defense mechanism is also bundled with the applications that host different services. Both defense method and applications may need to perform in real-time. Therefore, their combined processing requirement may need significant computation resources.

Thus, the proposed method requires an efficient technique for its implementation. One of the ways may be to parallelize operations within the algorithm. Techniques like Map-Reduce enable efficient processing of large volumes of data and can become handy to parallelize the processing of different time-series. Also, since the proposed method constitutes different modules, they can be parallelized to speed up the performance. The various modules may also be implemented at different edge locations. The parallel and distributed operations of different modules require them to be synchronized so that modules are aware of their status and decisions.

Further, application placement also needs to be discussed in detail. There is a possibility that defense strategy and IoT applications are placed at different edge nodes. Thus, one edge server takes the responsibility of defense and then it transfers the data to other edge servers where different services may be hosted. However, this strategy also opens security vulnerability while clean data is transferred from one edge to another edge.

Computation of anomaly and processing of applications needs to be efficiently implemented in parallelized and distributed manner. Therefore, research is required to parallelize, synchronize, and distribute computing tasks to meet the stringent real-time requirement of applications. Although there has been some work done on improving computation performance for anomaly detection by Angiulli et al. [34] and Matthews and St. Leger [35], a detailed exploration of the computation method still needed to be investigated.

25.5 Conclusions

This chapter has discussed the importance of a robust outlier detector for effective processing of IoT applications. We discussed several state-of-the-art methods of anomaly detection and described their limitation for anomaly detection in time-series data. We presented a robust outlier detector that has potential to handle the anomaly detection problem in IoT data. The proposed method uses different time-series data and has different modules to extract outliers in time-series data. Further, we discussed some of the challenges associated with the proposed method and pointed the need to explore them so that proposed method can be implemented in a practical framework.

References

1 R. Shukla, S. Sengupta, and M. Chatterjee, "Sofware-defined network and cloud-edge collaboration for smart and connected vehicles," in *Proceedings of ACM International Conference on Distributed Computing and Networking (ICDCN)*, Varanasi, India, January 2018.

2 R. Shukla and A. Munir, "A computation offloading scheme leveraging parameter tuning for real-time IoT devices," in *Proceedings of 2016 IEEE International Symposium on Nanoelectronic and Information Systems (iNIS)*, Gwalior, India, December 2016.

3 R. Shukla and S. Sengupta, "A novel software-defined network based approach for charging station allocation to plugged-in electric vehicles," in *Proceedings of 2017 IEEE 16th International Symposium on Network Computing and Applications (NCA)*, Boston, MA, USA, November 2017.

4 I. Lee and K. Lee, "The Internet of Things (IoT): Applications, investments, and challenges for enterprises," *Business Horizons*, vol. 58, no. 4, pp. 431–440, July 2015.

5 R. Shukla, P. Kansakar, and A. Munir, "A neural network-based appliance scheduling methodology for smart homes and buildings with multiple power sources," in *Proceedings of 2016 IEEE International Symposium on Nanoelectronic and Information Systems (iNIS),* Gwalior, India, December 2017.

6 E. Siow, T. Tiropanis, X. Wang, and W. Hall, "TritanDB: Time-series Rapid Internet of Things Analytics," arXiv preprint arXiv: 1801.07947, January 2018.

7 M. Hossain and G. Muhammad, "Cloud-assisted Industrial Internet of Things (IIoT) – Enabled framework for health monitoring," *Elsevier Compuer Networks*, vol. 101, pp. 192–202, June 2016.

8 B. Kailkhura, S. Brahma, and P. K. Varshney, "Data falsification attacks on consensus-based detection systems," *IEEE Transaction on Signal and Information Processing over Networks*, vol. 3, no. 1, pp. 145–158, March 2017.

9 AMI-SEC task force, "AMI System Security Requirements" (December 2008). [Online]. Available: https://www.energy.gov/sites/prod/files/oeprod/DocumentsandMedia/14-AMI_System_Security_Requirements_updated.pdf.

10 D. Hendrycks and K. Gimpel, "Early methods for detecting adversarial images," arXiv Preprint arXiv: 1608.00530, August 2016.

11 N. Carlini and D. Wagner, "Adversarial examples are not easily detected: Bypassing ten detection methods," in *Proceedings of the 10th ACM Workshop on Artificial Intelligence and Security (AISec)*, New York, NY, USA, November 2017.

12 V. Chandola, A. Banerjee, and V. Kumar, "Anomaly detection: A survey," *ACM Computing Surveys*, vol. 41, no. 3, p. 15, July 2009.

13 M. Gupta, J. Gao, C. Aggarwal, and J. Han, "Outlier detection for temporal data: A survey," *IEEE Transanction on Knowledge and Data Engineering*, vol. 26, no. 9, pp. 2250–2267, September 2014.

14 L. Akoglu, H. Tong, and D. Koutra, "Graph based anomaly detection and description: A survey," *ACM Journal of Data Mining and Knowledge Discovery*, vol. 29, no. 3, pp. 626–688, May 2015.

15 V. Hodge and J. Austin, "A survey of outlier detection methodologies," *Journal of Artificial Intelligence Reviews*, vol. 22, no. 2, pp. 85–126, October 2004.

16 N. Thanigaivelan, E. Nigussie, S. Virtanen, and J. Isoaho, "Hybrid internal anomaly detection system for IoT: Reactive nodes with cross-layer operation," *Security and Communication Networks*, vol. 2018 August 2018.

17 T. Yu, X. Wang, and A. Shami, "Recursive principal component analysis-based data outlier detection and sensor data aggregation in IoT systems," *IEEE Transaction on Internet of Things*, vol. 4, no. 6, pp. 2207–2216, December 2017.

18 W. Li, H. Song, and F. Zeng, "Policy-based secure and trustworthy sensing for Internet of Things in smart cities," *IEEE Transaction on Internet of Things*, vol. 5, no. 2, pp. 716–723, April 2018.

19 G. Shafer, *A Mathematical Theory of Evidence*, vol. 42. Princeton University Press, April 1976.

20 Netflix Technology Blog, "RAD — Outlier detection on Big Data" (February, 2015) [Online]. Available: https://medium.com/netflix-techblog/rad-outlier-detection-on-big-data-d6b0494371cc.

21 E. Candès, X. Li, Y. Ma, and J. Wright, "Robust principal component analysis?," *Journal of ACM*, vol. 58, no. 3, p. 11, May 2011.

22 S. Bhattacharjee, A. Thakur, and S. Das, "Towards fast and semi-supervised identification of smart meters launching data falsification attacks," in *Proceedings of the 2018 on Asia Conference on Computer and Communications Security(ASIACCS)*, Incheon, Korea, June 2018.

23 O. Vallis, J. Hochenbaum, and A. Kejariwal, "A novel technique for long-term anomaly detection in the cloud," in *Proceedings of the 6th {USENIX} Workshop on Hot Topics in Cloud Computing (HotCloud)*, Philadelphia, PA, USA, June 2014.

24 Z. Ouyang, X. Sun, J. Chen, D. Yue, and T. Zhang, "Multi-view stacking ensemble for power consumption anomaly detection in the context of industrial Internet of Things," *IEEE Access*, vol. 6, pp. 9623–9631, February 2018.

25 S. Xie and Z. Chen, "Anomaly detection and redundancy elimination of big sensor data in Internet of Things," arXiv prepint arXiv:1703.03225, March 2017.

26 N. Pandeeswari and G. Kumar, "Anomaly detection system in cloud environment using fuzzy clustering based ANN," *Journal of Mobile Networks and Applications*, vol. 21, no. 3, pp. 494–505, June 2016.

27 Performance Measurement System, "California Department of Transportation" [Online]. Available: http://pems.dot.ca.gov/.

28 P. Rousseeuw and M. Hubert, "Robust statistics for outlier detection," *Wiley Interdisciplinary Reviews: Data Mining and Knowledge Discovery*, vol. 1, no. 1, pp. 73–79, January 2011.

29 J. Hochenbaum, O. Vallis, and A. Kejariwal, "Automatic anomaly detection in the cloud via statistical learning," arXiv prepint arXiv: 1704.07706, April 2017.

30 R. Shukla and S. Sengupta, "Analysis and detection of anomaly due to data falsification attacks on traffic prediction applications," in *Proceedings of the 9th IEEE Annual Ubiquitous Computing, Electronics & Mobile Communication Conference (UEMCON)*, New York, NY, USA, November 2018.

31 Brian Buntz, "Open source IoT is growing in importance" (May, 2018) [Online]. Available: https://www.iotworldtoday.com/2018/05/24/open-source-iot-growing-importance/.

32 Vivek Ratan, "The role of open source in IoT" (July, 2017) [Online] Available: https://opensourceforu.com/2017/07/open-source-role-in-iot/.

33 Glenn Longley, "Security considerations for the IIoT challenge" (April, 2016) [Online]. Available: https://www.iotworldtoday.com/2016/04/02/security-considerations-iiot-challenge/.

34 F. Angiulli, S. Basta, S. Lodi, and C. Sartori, "GPU strategies for distance-based outlier detection," *IEEE Transanction on Parallel and Distributed Systems*, vol. 27, no. 11, pp. 3256–3268, November 2016.

35 S. Matthews and A. St Leger, "Leveraging mapReduce and synchrophasors for real-time anomaly detection in the smart grid," *IEEE Transanctions on Emerging Topics in Computing*, vol. 7, p. 1–1, April 2017.

26

Summary and Future Work

Charles A. Kamhoua[1], Laurent L. Njilla[2], Alexander Kott[1], and Sachin Shetty[3]

[1] *US Army Research Laboratory, Adelphi, MD, USA*
[2] *Cyber Assurance Branch, US Air Force Research Laboratory, Rome, NY, USA*
[3] *Virginia Modeling Analysis and Simulation Center, Old Dominion University, Norfolk, VA, USA*

Internet of Things (IoT) is a ubiquitous part of our lives with the presence of billions of Internet-connected devices in commercial, government, and military domains across the entire planet. IoT has realized technologies such as the smart home, self-driving car, smart grid, smart city, and intelligent transportation. In addition, IoT devices have also realized the Internet of Battlefield Things (IoBT) platform that has sensors, wearable devices, robots, drones, and autonomous vehicles, facilitating the intelligence, surveillance, and reconnaissance to command and control and battlefield services.

IoT device manufacturers have not implemented security mechanisms, making IoT devices vulnerable when connected to the Internet. For example, real-world attacks can now start with reconnaissance using insecure IoT devices and follow with lateral movements through the environment to deepen access by spreading ransomware and malicious code. There are several reasons for the need for IoT security. First, IoT devices are mass-produced rapidly to be low-cost commodity items without security protection in their original design. Second, IoT devices are highly dynamic, mobile, and heterogeneous without common standards. Third, it is imperative to understand the natural world, the physical process(es) under IoT control, and how these real-world processes can be compromised before recommending any relevant security

Modeling and Design of Secure Internet of Things, First Edition. Edited by Charles A. Kamhoua, Laurent L. Njilla, Alexander Kott and Sachin Shetty.

Published 2020 by John Wiley & Sons, Inc.

countermeasure. As a result, those systems are the frequent targets of sophisticated cyber attacks that aim to disrupt mission effectiveness. Moreover, unprotected IoT devices can be used as "stepping stones" by attackers to launch more sophisticated attacks such as advanced persistent threats (APTs). These challenges and the high risk and consequence of IoT attacks in the battlefield drive the need to accelerate basic research on IoT security.

The preceding chapters in the book have addressed the issues in modeling and designing secure IoT to provide flexible, low-cost, and secure means to provide resilient critical infrastructure, reduce risks of exploitable attack surfaces, and improve survivability of physical processes. The chapters provide a defense-in-depth approach to layered security, with each layer representing cyber deception, modeling, and design applied to information processing, networking, and IoT subdomains. Each chapter presents techniques to secure the IoT devices and methodologies to realize a secure interconnection between the IoT devices and requisite distributed information-processing techniques required to achieve scalable and robust protection. In addition to the broad theme of modeling techniques to secure IoT in military and commercial domains and empirical validation of IoT platforms, the technical content in this book encompasses several themes, including game-theoretic models, cyber deception models, moving target defense models, and adversarial machine learning models. The chapters include cutting-edge research findings that promise to attract strong interest (on topics including IoBT, APTs, and cyber deception). The contributions address design issues in developing secure IoT such as secure software-defined networking (SDN)-based network orchestration, networked device identity management, tactical battlefield settings, and smart cities.

26.1 Summary

26.1.1 Game Theory, Cyber Deception, and IoT

The disadvantage with insecure IoT can be exploited to develop a proactive defense strategy, cyber deception, to benefit the defender by protecting mission-critical assets from vulnerabilities in IoT. The premise of the cyber deception approach is to design a network of decoy assets and camouflage with the aim to expend adversaries' resources and time and gather information about the adversaries' strategies, tactics, capabilities, and intent. The cyber deception approach will mitigate the information asymmetry that exists between adversaries and defenders by converting the defender's disadvantageous position to a position of strength. However, in order to deploy cyber deception to realize a resilient cyber infrastructure, several research challenges need to addressed: (i) design of network decoys,

(ii) placement of network decoys, (iii) maximize information uncertainty for attacker, (iv) anticipate adversarial attack strategy and nodes targeted, and (v) quantify the performance of the deception mechanism to assess the level of asymmetry between attacker and defender.

Each of the six chapters in Part I provides insight into how game-theoretic techniques can be used as a tool to design cyber deception schemes to protect IoT devices, adapting in adversarial cyber settings, trust management, understanding the interactions between defender and attacker by analyzing the impact of configuring honeynets, and characterizing the information that the defender can use to understand adversaries' strategies, tactics, and impact of attack vectors.

We kicked off with Chapter 2, focusing on an overview of game-theoretic analysis for cyber deception in three different environments. The chapter introduces the baseline signaling game models to capture the information asymmetry, dynamic, and strategic behaviors of deceptions. To enrich the game models with exogenous factors, evidence-based strategies through side-channel knowledge acquisition were also included. The chapter discusses several theoretical results that provide insights into the effectiveness of cyber deception as a risk mitigation strategy for IoT.

In Chapter 3, we present a hyper-game-based cyber deception strategy to mitigate a targeted attack where an adversary compromises an edge device, and through exploitation of vulnerabilities of connected IoBT devices, launches a targeted attack on a critical device on the network. Using the two-player hyper-game model, the chapter provides quantitative metrics such as recommended times to disable all the links to a target device, reboot the network devices to clean up all potentially compromised devices, and reset the network topology.

In Chapter 4, we focus on a resource-starving attack due to non-availability of dedicated spectrum channels to IoT devices. The chapter proposes a decentralized game-theoretic trust-management framework to enable cooperative spectrum leasing to potentially selfish IoT devices. The premise of the framework is to allow unlicensed users to borrow the spectrum access from licensed users, and in return, they provide cooperative service to enhance information secrecy of licensed users via adding intentional jamming to protect them from potential eavesdroppers, and enhance the quality of communication through cooperative relaying. The proposed framework improves the secrecy rate of the primary users and reduces the possible attacks from selfish IoT devices.

In Chapter 5, we analyze deception in adversarial cyber operations. With the rise in reinforcement learning that has increased the level of autonomy of computing systems, the benefits have not necessarily translated in combating adversarial cyber operations. With the increasing adoption of reinforcement learning techniques in computing, it is imperative to review the challenges with using reinforcement learning in protecting cyber operations.

In Chapter 6, we conduct a formal characterization of the interactions between attacker and defenders within the context of cyber deception. The chapter focuses on increasing the total space of the attacker's options and ensuring there are more deceptive options than real ones, which will ensure the workload of the attacker in detecting the difference will increase. Specifically, the chapter presents a game-theoretic model and Nash-equilibrium-based deception tactics to analyze attack–defense scenarios. The scenarios use fake nodes (computing devices) for deception under consideration of the system deploying defense resources to protect individual nodes in a cost-effective manner applicable to IoBT.

Finally in Chapter 7, we summarize the section of cyber deception by providing a status of state-of-the art work challenges involved in realizing cyber deception for IoT and the opportunities of an effective cyber-deception-based risk management scheme for IoT. The chapter provides specific advantages for using deception techniques for cybersecurity over traditional security defenses where the system could learn and adapt the security parameters on the fly to combat malicious cyber operations. The chapter also summarizes the ongoing game-theoretic deception approach for realizing cyber deception.

26.1.2 IoT Security Modeling and Analysis

In Part II, we provide modeling formalisms and formal analytic techniques to characterize threats to IoT and secure IoT solutions. Given the plethora of attack surfaces in IoT, the first step to developing a secure IoT solution is to characterize the attack surfaces and compute the security risk posed by exploiting the attack surfaces. Once the attack surfaces are formally characterized, one can identify the countermeasure that mitigates the risk. However, prior to applying the countermeasure, it is also critical to model and analyze the effectiveness of the proposed countermeasure and assess if the countermeasure can indeed address the security risk.

We kick off in Chapter 8 by conducting an analysis of vulnerabilities in IoT applications in a smart home domain. In the chapter, we focus on physical layer vulnerabilities in sensors, communication links, and devices used for video monitoring, storage, and heating, ventilating, and air-conditioning control. The chapter does a deep dive into an attack involving energy injection to electronic hardware that is widely used in IoT devices. Using a multi-modeling approach that represents the geometric structure of the hardware, the analysis in the chapter shows low-energy electromagnetic (EM) pathways for corrupting system behavior and argues the importance of EM injection analysis at behavior-critical inputs and outputs.

In Chapter 9, we scale the attack surface by moving to techniques for realizing security and privacy in a complex system-of-systems such as a smart city. With IoT potentially playing a big role in building the infrastructure for future smart cities, the chapter focuses on approaches to monitor the behavior of IoT devices and protocols. The chapter proposes modeling techniques to analyze IoT device function and security for a large and complex system like a smart city. The chapter addresses the issues of composition as a key challenge in realizing the secure IoT solution for a smart city because security challenges will arise in integration of disparate system components. We address issues with balancing the tradeoff between interoperability of disparate components and reliability, safety, security, and privacy.

In Chapter 10, we switch gears to modeling and analysis of software-defined IoT network based on cyber deception technology (i.e. a decoy system) and moving target defense (MTD). We propose a graphical security model-based evaluation framework using a hierarchical attack representation model to evaluate the effectiveness of the proposed MTD over the cyber deception strategy. The analysis in the chapter aids in quantifying the system lifetime (i.e. mean time to security failure), the number of decoy nodes used in attack paths, and the defense cost introduced by the proposed MTD over the cyber deception technique.

In Chapter 11, we focus on techniques that can augment existing intrusion detection to expand the ability to detect threats in IoT. The detection of threats in IoT would require analyzing information acquired across layers in the network and software stack. This chapter discusses a technique to assist the intrusion detection of APTs in the home network by proposing Security Layer for Smart Home (SLaSH), a cross-layer security mechanism that incorporates capabilities and features of the various layers to help secure the entire smart home network. The cross layer approach involves analyzing the attack surface of each of the three layers: device, network, and service. The chapter proposes a machine-learning-based model to realize the cross-layer intrusion detection that uses rule-based and behavior-based features.

In Chapter 12, we move from polymorphic threats in the IoT device to similar threats in the network of IoT devices. Specifically, this chapter uses a vulnerability graph model to analyze the exploitability of stepping stones in a network of IoT devices. Unlike traditional vulnerability graphs that are static and do not account for dynamic defense, in this chapter we propose biased min-consensus technique for dynamic graphs with switching topology as a distributed technique to determine the attack paths with more probable stepping-stones.

In Chapter 13, following the same vein as understanding threats in IoT networks, we conduct a deep dive into analysis of the security of IoT communication protocols. Specifically, we develop an Anomaly Behavior Analysis (ABA)-based intrusion detection system that uses machine learning models to analyze IoT communication protocols. If the normal behavior of the IoT communication protocols is understood, the ABA technique not only helps detect known attacks but also detects new and modified attacks. The chapter describes a detailed threat for the Wi-Fi and Bluetooth protocols and applies our ABA methodology to detect attacks on Wi-Fi and Bluetooth protocol.

In Chapter 14, we model the effectiveness of using a Partially Observable Monte-Carlo Planning (POMCP) model for cyber deception in IoT. We use the model to quantify the amount of resources expended by the adversary and time taken to gather critical information. The chapter also looks into scalability aspects in deploying network decoys among real network infrastructure. The POMCP algorithm provides insights into where to deploy the network decoys within the network infrastructure. The POMCP algorithm presents an approach that can influence an attacker to take the path toward the fake network while capturing attacker progression using the vulnerability dependency graph.

In Chapter 15, we focus on modeling and analyzing the effectiveness of state estimation in cyber physical systems as a technique to detect malicious attacks. The chapter presents secure state-estimation and control mechanisms in the presence of sensing and actuating attacks, extending classical observers and Kalman estimation methods to the case when there are adversaries in IoT networks. The chapter describes techniques to securely estimate the state of a linear dynamical system from a set of noisy and maliciously corrupted sensor measurements and actuator commands.

In Chapter 16, we shift the perspective from technical to governance to address the challenge of maintaining safe and efficient IoT systems. The chapter presents governance strategies to prevent, avoid, mitigate, manage, and recover from disruptions to IoT devices and interconnected information systems. This chapter explores the emerging scholarly and policy discussions of the digital economy and IoT governance, and examines how such strategies might onboard philosophical and methodological applications of resilience to enable efficient IoT device recovery and adaptation from an evolving cyber threat landscape.

26.1.3 IoT Security Design

In Part III, we present several IoT security design techniques to protect information systems, network, and things. We present nine diverse security design techniques that not only comprise techniques that are applicable to the broad family of IoT environments but also to specific domains and, we provide insights into the effectiveness of the design techniques.

In Chapter 17, we present a defensive strategy against an APT threat impacting IoT-based 3D printers. This APT threat can intrude an IoT-based

3D printer, modify the critical files, and damage the physical components. The chapter presents game-theoretic framework to capture the properties of the ATP attack. We also use other games to describe the behaviors of the cyber and physical layers. We use the equilibrium of the game to develop optimal defense strategies.

In Chapter 18, we present MTD as a strategy to defend against typical hardware security threats, such as side-channel analysis attacks, hardware tampering, and hardware Trojan insertion from untrusted computer-aided design tools. The chapter presents MTD techniques such as dynamic masking, error deflection, dynamic permutation, and error control coding. In order to combat emergence of various security threats, the chapter presents the need to introduce dynamicity of design parameters, system inherent variables, and configurable architecture have great potential to be applied to address hardware security issues.

In Chapter 19, we present device attestation as a solution to meet the security demands of embedded devices. We discuss the requirements of an attestation scheme, as well as classify attestation schemes by their functionality and coverage. The chapter provides a survey of device attestation approaches in the different categories and the possible pitfalls and limitations of existing device attestation approaches.

In Chapter 20, we present a software-defined security design for IoT wherein we use SDN to quarantine the attack from spreading across the network and present an optimal countermeasure selection solution to ensure the balance between resilience and security risk. The design described in the chapter will aid in the selection of security countermeasures dynamically in an SDN platform based on Industrial IoT (IIoT)-technology-enabled Energy Delivery Systems (EDS) and achieving tradeoff between providing security and quality of service (QoS).

In Chapter 21, we present an integrated SDN and cyber deception technique to address the security of IoT network. The SDN controller collects inputs about malicious activities from honeypots and translates their requirements to the network-level flow rules to quarantine attacks. The chapter presents experimental results that provide the evaluation of a hardware implementation.

In Chapter 22, we present a decentralized design framework based on blockchain and smart contract methods to address security issues in IoT networks. Current research on blockchain-based security solutions to IoT are introduced, and the main challenges are highlighted. To demonstrate the feasibility of a blockchain-enabled security mechanism, we present a case study of a decentralized access control mechanism for IoT. The access control logic is captured as a smart contract deployed on a blockchain network that allows users to control their devices and resources instead of relying on a centralized third authority to maintain the trust relationship in a trustless network environment.

In Chapter 23, we tackled the challenge of capturing intent of designer, developer, and user of a software system as the security design primitive. The presence of this primitive would address security failures that arise due a disconnect between what an actor intends for a technology to do and what it actually does. The techniques presented in the chapter are based on the language-theoretic security (LangSec) and the intent enforcement mechanism ELF-based access control (ELFbac). This chapter discusses the importance of preserving intents, provides primers on both LangSec and ELFbac, demonstrates how these two paradigms can be used to construct a secure implementation of the Advanced Message Queuing Protocol, and examines evaluation techniques.

In Chapter 24, we present an SDN-based MTD defense technique to protect IoBT and IIoT. As MTD-based techniques have been applied in non-IoT domains, the chapter provides a discussion of the feasibility of applying MTD to resource-constrained IoT devices. The chapter presents the challenges of merely adopting existing MTD schemes to protect IoT. Due to the resource constraints in IoT devices, an SDN-based network infrastructure can provide an effective means to manage the devices that will facilitate more effective defense approaches through MTD.

In Chapter 25, we present an anomaly detection technique to protect against attacks where an adversary alters the data value and induces outliers in data that result in data-falsification attacks in IoT devices. The anomaly detection technique fuses several modules to detect the outliers in IoT data. We propose to use correlation modules to find how diverse time series are related to each other. Correlation modules can be used to verify whether an anomaly is due to external conditions or an attack. We use a combination of neural network, statistical analysis, and gradient analysis to determine the outliers in time-series data. Neural network can estimate the expected data and assist in a rough estimation of an anomaly. Statistical and gradient analysis can provide more accurate information about the outliers.

26.2 The Future

26.2.1 Game Theory, Cyber Deception, and IoT Security

In the first part, several chapters focused on game-theoretic-based cyber deception approaches for IoT security. There are several challenges and possible opportunities for cyber-deception-based cyber security. One of the research challenges and opportunities worth exploring and developing for future use in cybersecurity is *mimetic deception*, which offers advantages over static deception and cryptic deception. Most of the studies have focused primarily on static games but there are research challenges to be addressed by considering *evolution* or *dynamic games* for cyber deception. Thus, the application of the game theory in cybersecurity and cyber deception is in its early stage. Some of the techniques such as game theory are studied well in research but there is an urgent need to realize them in practical scenarios. Furthermore, a lack of collaboration between cyber security analysts and game theorists hinder the implementation of game theory for cyber defense.

Deception is regarded as a multidisciplinary topic that has been used in a diverse set of applications to reach domain-specific goals such as deception in criminology, economics, and trading business. *Interdisciplinary teamwork* is needed to successfully implement cyber deception to combat attacks targeted to vulnerable cyber systems. The ultimate goal of any cyber defense solution in an organization, including cyber deception, is to protect sensitive information from habitual attacks and ensure that the potential attacks are prevented and the cyber system is not vulnerable to outside attacks. Among other approaches, the game-theoretic approach can help to model the interactions between victims and attackers for cyber deception.

26.2.2 IoT Security Modeling and Analysis

The open architecture, physical elements, and scale render modeling and analysis of the attack surfaces in IoT systems an open challenge. The sheer volume of information required for managing devices makes centralized solutions all but impossible in many current and future IoT applications. Susceptibility to physical layer attacks increases the attack surfaces of IoT devices and brings in new dimensions to vulnerability analysis and risk mitigation. This susceptibility increases the importance to research targeting the development and deployment of IoT composition platforms that enforce the use of security frameworks in all potentially safety-critical IoT applications.

In order to conduct comprehensive and accurate vulnerability analysis, one needs to have detailed information of designs, which is often proprietary. Constructing surrogate models based on circuit technologies will assist in the electrical/physical analysis. The approaches proposed here involve large search problems, across multiple domains. Creating efficient solutions that minimize computational time is highly desirable. Furthermore, reasoning about the temporal and sequencing impacts will make vulnerability assessment more accurate.

For future work, applying the modeling and analysis schemes in Part II to large-scale IoT networks by showing high scalability will be necessary. Also necessary will be conducting sensitivity analysis for the proposed models by varying the values of other key design parameters (e.g. weights to consider each system objective, increasing the number of decoy nodes [for cyber deception schemes] deployed in the network, increasing the number of attackers, and/or system security vulnerability thresholds).

26.2.3 IoT Security Design

We have presented a combination of MTD- and SDN-based techniques to protect IoT security, which will require future work in the following areas to realize these techniques.

- *SDN Control Channel in IoBT:* IoBT requires constant communication among a possibly large number of mobile nodes. Any attack on this network will thus affect almost all the nodes in terms of information gathering, command and control, and resources. Therefore, a fast access to these devices from the SDN controller is needed. Given that this channel needs to be wireless, reliability and availability of the communications will be a challenge. Security and robustness would be a must for this communication. Therefore, new wide-area control protocols are needed. The emerging 5G technology has the potential to address this challenge with its reliability and long-distance coverage. However, 5G may not be available everywhere and thus its features, such as D2D as well as technologies from 802.11 families such as IEEE 802.11ah, need to be investigated in terms of QoS they can guarantee.
- *Local Intelligence in IoBT:* Given the challenges with the control channel reliability, the nodes should be given some local intelligence so that they can operate when the SDN controller is not accessible. The level of intelligence should be subject to application requirements. This intelligence can be dynamic and based on machine learning technologies where the devices can learn from their environments to act without instructions.
- *MTD-Aware SDN Deployment in IIoT:* The deployment of SDN in an existing network (e.g. replacing the traditional routers with SDN-enabled switches) is often restrained by limited resources, legacy systems, and/or technical constraints. Therefore, the challenge of deploying (often incremental) the SDN architecture within the limits while optimally achieving the security/defense objective should be explored under different parameters.
- *Integrated Cyber and Physical MTD Measures in IIoT:* While the cyber and physical moves together can bring a larger MTD capability, these actions must comply with one another, without declining the optimal operation of the IIoT system. Therefore, an important research direction for IIoT security is to explore the feasible physical moves and their MTD effectiveness, as well as to study the optimal integration of physical agility measures with the traditional cyber moves in the SDN-based MTD framework.
- *MTD Games:* Given the resource constraints and operational requirements, MTD in IoBT/IIoT may not be able to arbitrate many properties extensively. Therefore, game-theoretical approaches are needed to deceive the attackers based on their communication and action patterns. This will introduce a trade-off between the resources, service constraints, technical/communication

feasibility, and the attacker's assumed capabilities. Another interesting piece of this challenge is to be able to model the behavior of the attackers in certain conditions so that MTD moves could be better arranged.

- *Smart Moving Target Defense for IoT:* Most proposed MTD approaches are typically not an attacker-aware and the decision to select the move is based on the pre-defined constraints. IoT, on the other hand, produces a huge amount of data, which is known as big data. The IoT data analytics is much more powerful than the traditional data analytics since the IoT data analytics are intended to do the real-time processing before the data become irrelevant or obsolete. Typically, the IoT data analytics can be placed at the edge (i.e. fog/edge computing) or at the cloud. Therefore, the IoT data analytics can be exploited to support a smart attacker-aware MTD as opposed to traditional networks.
- *Collaborative Multi-Attribute Moving Target Defense:* In this chapter, we have shown that most MTD approaches concentrate only on a single attribute (e.g. address, path, and configuration). Only a few approaches attempt to utilize more than one attribute for the proactive defense. Therefore, a collaborative multi-attribute MTD can be a potential future research direction.

Index

a

access control (AC) approach, IoT 506
 authorization architecture 510–512
 authorization model 508–510
 description 508
active identification methods 190
adaptive shuffling (AS) 228–229, 234
adaptive strategic cyber defense for APT
 case study 51–52
 equilibrium concept 50–51
 multistage dynamic Bayesian game model 48–50
additive manufacturing (AM), benefits of 385
address space layout randomization (ASLR) 220
advanced encryption standard (AES) 410, 417
advanced message queuing protocol (AMQP) 22, 552–557
advanced metering infrastructure (AMI) 596
advanced persistent threat attacks 62
advanced persistent threats (APT) 31, 48, 332, 387, 636, 640–641
 attack surface analysis 252–253
  attacks due to devices 253–255
  attacks on network 254–256
  attacks on service 256–257
 intrusion detection system 249
 security layer
  cross-layer security 263–269
  design 257–258
  functions and capabilities 258–263
 SLaSH 249
 system model
  device layer 250–251
  network layer 251, 252
  service layer 252
adversarial cyber operations
 CGT 111
 game models and analyses 111
 key aspects of recent game theory successes 112–114
 military and security domains 111
 paradoxes and paradoxes resolved 117–118
 real-world adversarial situations 114–117
adversarial engagements 112
adversarial modeling 114–117
adversary's perceived game 67–68
adversary's strategies 65–67
aggressive detector 39, 40

Modeling and Design of Secure Internet of Things, First Edition. Edited by Charles A. Kamhoua, Laurent L. Njilla, Alexander Kott and Sachin Shetty.

Published 2020 by John Wiley & Sons, Inc.

algebraic graph 281
Altera Quartus, attacks on 415–416
American Fuzzy Lop (AFL) 558
amplify-and-forward (AF) relaying 86
anomaly behavior analysis-based intrusion detection system (ABA-IDS) 16, 308, 639–640
anomaly behavior analysis of IoT protocols
  attack sophistication *vs.* intruder technical knowledge 297–298
  Bluetooth protocol 318–326
  cyber attackers 297
  growth of IoT 296–297
  IoT architecture 299–300
  IoT threat modeling methodology 301–303
  rise of wearables and future wearable technology 295–296
  sensing technologies 296
  timeline of cyber-threads 297–298
  WannaCry attack 297
  Wi-Fi protocol 303–306
    data-link layer behavior of the 306–308
    experimental results and evaluation 314–318
    Wi-Fi IDS to detect attacks 308–314
anomaly behavior analysis of the Wi-Fi protocol 308
anomaly detection 642
  black-box attack strategy 630
  centralized approach 619–621
  computation requirement 630–631
  correlation analysis 624, 625
  distributed approach 618–619
  feature extractor 623–624
  gradient analysis 628–629
  interquartile range 624
  machine learning 627
  M-estimator 624
  pattern analysis module 624
  problem statement 621–622
  proposed anomaly detector 617
  proposed outlier detector 623
  statistical analysis 624, 628
  supervised learning 626
  white-box attack strategy 630
anti-honeypot techniques 144
anti-pattern 533
application binary interface (ABI) 531
application programming interface (API) 222
applications layer 303
APT. *see* Advanced Persistent Threats (APT)
area-dividing random route mutation (ARRM) approach 585
AS-GANT 235
AS identification, impact analysis, and mitigation strategies 301–303
AS-RNT 235
asymmetric cryptography 512–513
ATRIUM 444, 447
attackers 123
attacker stepping stones 274
attack exploration space 427, 428
attack graphs (AGs) 61–63, 77, 219, 335–336
attack model 224–225
attack sophistication *vs.* intruder technical knowledge 297–298
attack surface analysis 252–253
  attacks due to devices 253–255
  attacks on network 254–256
  attacks on service 256–257
attack surfaces (ASs) 301
attack trees (ATs) 219, 335–336
attribute-based access control (ABAC) model 509
augmented ICT infrastructure 187

authentication 258–259, 261, 457
authentication delay 464
authorization architecture, for IoT
centralized architecture 510–511
decentralized architecture 511–512
authorization model, in IoT
attribute-based access control model 509
capability-based access control model 509–510
role-based access control model 508–509
automata theory
automaton 534, 535
deterministic 536
finite-state machine (FSM) 543–548
linear-bounded automata 536
non-deterministic 536
state 534
turing machine 535
automated reasoning engines 200–201
average consensus protocol 282

b
backdoor attack 320
battery draining attacks 320
Bayesian attack graphs 62
Bayesian Nash equilibrium (BNE) 31, 34–35
beamforming techniques 80
behavioral monitoring 187–189
behavior authentication module 313
behavior-based detection 250
belief 34, 43, 49
Bellman–Ford algorithm 274, 292
Bellman optimality equation 113
biased min-consensus protocol 282–283
in dynamical network and shortest path 283–284
binary state space
equilibrium concept 37–38
equilibrium results 38–41
game-theoretic model 35–37
bitstream downloading channel 412
bitstream generation 415
bitstream reverse engineering 429
black-box attack strategy 630
Blind DDoS attack 585
blockchain 507
asymmetric cryptography 512–513
consensus algorithm 513
digital signature 513
extend 514
generate block 514
hash function 513
peer-to-peer (P2P) network 513
verify and consensus 514
work process 513
blockchain-based security solutions 641
blockchain-enabled decentralized CapAC (BlendCAC) 516–517
advantages 523
challenges 523–524
decentralized authorization 523
edge computing-driven intelligence 523
experimental results 521
experiment and evaluation 520–522
feasibility 520–521
fine granularity 523
general cost incurred 521
lightweight design 523
operation and communication modes
authorization validation 519
capability propagation 519
registration 517, 519
smart contract deployment 519
prototype design 520
system architecture 517, 518
token data caching solution 522
bluejacking 320
bluesnarfing attack 320

Bluetooth 2
Bluetooth-flow-characteristics (BFC) 321–322
Bluetooth Low Energy (BLE) 175
Bluetooth protocol
  "anding" of heuristic probabilities 322
  backdoor attack 320
  battery draining attacks 318, 320
  behavior analysis module 322
  bluejacking 320
  bluesnarfing attack 320
  feature set 320–321
  host controller interface (HCI) protocol frame 322
  IDS architecture 322–323
  Jelinek-Mercer Smoothing method 322
  observation-flow 321–322
  pairing attack 320
  piconet 319
  positioning and tracking attack 320
  precision for various classifiers 324–325
  recall for various classifiers 324, 326
  replay attack 320
  RoC area for various classifiers 324, 326
  scanning for Bluetooth addresses and surveillance 320
  state machine 319
  test bed 322–323
  unique n-grams 324–325
BoardPUF 446
brute-force approach 480

C

camouflage 481, 499
capability-based access control (CapAC) model 509–510
capability-based context-aware access control (CCAAC) model 510
Carrier Sense Multiple Access/Collision Avoidance (CSMA/CA) 304
causality 207–208
cybersecurity deception concept 156
centralized access control architecture 510–511
channel coefficients 93, 100
channel state information (CSI) 93
cheap-talk signaling game 125
cheap-talk signaling game with evidence 36
Chomsky hierarchy 536
Cisco Visual Networking Index 1
client honeypot 142
cloud-based speech to text (STT) conversion services 300
cloud computing 485
cognitive IoT (C-IoT) 82
cognitive radio and spectrum allocation 111
cognitive radio networks (CRNs) 80
collaborative multi-attribute moving target defense 645
common-off-the-shelf (COTS) Wi-Fi routers 191
Common Vulnerabilities and Exposures (CVE)/National Vulnerability Database (NVD) 231
Common Vulnerability Scoring System (CVSS) 64, 168, 231, 274
communication layer 302
compilers
  domain-specific language (DSL) 544–546
  linker 543
  loader 543
compress-and forward (CF) relaying 86
computational game theory (CGT) 111
computer-aided design (CAD) 408

computer architecture
  memory management unit (MMU) 544
  paging 544
  pipelining 549
  speculative execution 530, 549
  translation lookaside buffer (TLB) 544
confidentiality, integrity, and availability (CIA) triad 572
conjugate-prior method 50
Conservative detector 39, 40
constrained access 259
Constrained Application Protocol (CoAP) 570
continuous state space
  deceivability 43–44
  equilibrium concept 45–46
  equilibrium results 46–48
  game-theoretic model 42–43
  knowledge acquisition 44
control-flow attestation (C-FLAT) 443–444, 447
control theory model 148–149
Convolution Neural Network (CNN) 624
cooperation for enhancing secrecy in CRNs 89–91
cooperative communication techniques 88
cooperative IoT security 488, 489
cooperative jamming 81, 88
cooperative relaying 86
cooperative spectrum sharing and trust management
  beamforming techniques 80
  cognitive radio networks 80
  common-model paradigm 80
  cooperative jamming 81
  database control spectrum sharing mechanisms 79
  drawback 80
  energy constraint 79
  game-theoretic spectrum leasing model
    reputation-based game-theoretic model 96–102
    system model 92–96
  information-theoretic secrecy-based techniques 81
  limited spectrum mobility 79
  multiple-input multiple-out techniques 80
  nonvolatile memory 81
  overview to physical layer secrecy 83–84
    cooperation for enhancing secrecy in CRNs 89–91
    information-theoretic secrecy 85–89
  problem statement 81–83
  security paradigm 80
  Stackelberg game 81, 91–92
  TV white spaces 80
core BLE platform 176–177
correlation power analysis (CPA) attacks 410–411
  dynamic masking plus error deflection method 417–418
  power distribution network noise-based countermeasure 418
counterfeiting chip 408
cross-layer security
  assumptions of operation 265–269
  feature vectors comparison 269
  in-network profiling and decision 264–265
  motivation and benefits 263–264
  offline learning and mining 269
cryptographic methods 80
cryptographic system 148
curse of dimensionality 113
cyberattacks 332

cyber characterization 191
cyber deception 59, 60, 62, 64, 225–226, 244, 636–638
  adaptive strategic cyber defense for APT
    case study 51–52
    equilibrium concept 50–51
    multistage dynamic Bayesian game model 48–50
  advantages 331
  anti-honeypot techniques 144
  binary state space
    equilibrium concept 37–38
    equilibrium results 38–41
    game-theoretic model 35–37
  client honeypot 142
  cognitive biases 146
  computer defenses 143
  continuous state space
    deceivability 43–44
    equilibrium concept 45–46
    equilibrium results 46–48
    game-theoretic model 42–43
    knowledge acquisition 44
  cyber defense 146
  cyber infrastructure components 143–144
  cyber kill chain model 145
  deceivees 30
  defenders 29–30
  design and implementation of 147
  digital conflict 143
  dynamic security model (*see* dynamic deception model)
  game-theoretic models 30, 155–157
    signaling games 151–152
    Stackelberg game 152–153
  game theory in security 32–33
    perfect Bayesian Nash equilibrium 34–35
    signaling game model 33–34
  intelligence driven security model 145
  interdisciplinary teamwork 643
  kill chain strategy 142
  lifecycle of 142–143
  methods of decision process 147–151
  MTD and 219–221
  Nash equilibrium 30
  open challenges and opportunities 158
  organizational biases 146
  periodic health checks 147
  phishing 29, 144
  related work 31
  research goal and contributions 333
  security breach-prevention mechanism 145
  security systems 141–142
  server honeypot 142
  taxonomy for
    honey-x 155
    mixing 154–155
    moving target defense 154
    obfuscation 154
    perturbation 154
    schematic 153
  taxonomy of information protection mechanisms 144–145
cyber deception modeled as a first-level hypergame 64–67
cyber infrastructure components 144
cyber kill chain 7
Cybermoat 483–484
cyber-physical games, equilibrium analysis of
  FlipIt game 396–397
  zero-sum game 394–395
cyber-physical Stackelberg game model 392–393
cyber-physical systems (CPS) 435, 447
  fog computing 163

Industrial Internet of Things 163
Internet of Things 163
IoT-based system architecture 164
multi-modeling approach 169–170
assessment of physical vulnerability 178–181
OpenMETA multi-modeling environment 170–175
smart home sensor 175–177
system-level impact analysis 182
open architecture 163–164
physicality 164
research directions 183
scale 164
security risks 357
vulnerabilities in IoT
different layers 165
effects of inter-layer interactions 167–168
effects of physicality 168–169
open architecture and IoT security 166–167
security risks 166
cyber-physical systems (CPSs) 17
cyber security 111
Cyberspace 29
cyber-threads, timeline of 297–298
cycle of length 275

d

DARPA 445
database control spectrum sharing mechanisms 79
data confidentiality delay 464
data-dependent approach 190
data distribution service 571
data exfiltration attacks 224, 227
data falsification attack 616, 617
data integrity delay 464
DDoS Resistant Multicast (DRM) 595
deceivability 43–44
deceivees 30
deceivee's sequential rationality 45
deceiver's sequential rationality 45
decentralized access control architecture 511–512
decentralized authorization, BlendCAC 523
deception 111. *see also* cyber deception
deception-based security
attackers 123
cheap-talk signaling game 125
dynamic honeypot 126
game-theoretic deception model 124, 126–130
high-interaction honeypot 125
honeypot 125
low-interaction honeypot 125
medium-interaction honeypot 125
Nash equilibrium 124
simulation results 135–138
in a tree-based network structure 132–135
using *N* computing devices 130–132
deception game 31
deception security systems 141–142
deception server 482
Deception Toolkit (DTK) 148
deceptive routing in relay networks 484–485
decision process for cyber deception 147–151
decode-and-forward (DF) relaying 86
decoding algorithms 367–368
decoys 481, 483–484, 500
decoy system 218
decoy system as defensive cyber deception 225–226
deep Q-Learning 113
deep reinforcement learning 113
defenders 29–30

defender's action
  cost function 341
  end-to-end packet delay 341–342
  POMDP model 337
  utility function 340–341
defender's belief update algorithm 342
defender's perceived game 68–69
defender's strategies 65
Defense Advanced Research Projects Agency (DARPA) 59
defense cost 234
defense model
  decoy system as defensive cyber deception 225–226
  network topology shuffling-based MTD 226–227
Dempster–Shafer Theory (DST) 619
denial of service (DoS) 412
dependency graph, for cyber defense system 334
destination nodes 283
detector 36
device attestation 435
  attacker capabilities 437–438
  definition 437
  objective 437–438
  scheme requirements 438
    liveness of measure 439
    trustworthiness of prover 439
    unforgeability of measure 439
  static and dynamic 440
  verifier 436
device layer level security
  authentication 258–259
  constrained access 259
  improved operational realities 260
  split architecture 259–260
differential electromagnetic analysis (DEMA) attacks 410
differential power analysis (DPA) 410
digital economy governance, IoT
  applying resilience to 378–379
  centralized governance approach 377
  current status 375
  laissez-faire approach 376
  pre-emptive strategy 376–377
  stewardship model 377
digital signatures 573
directed edge 275
directed graph 275, 276
directed multigraph 276
directed weighted Graph 275
direct-sequence spread spectrum (DSSS) 304
distributed capability-based access control (DCapAC) mechanism 511
distributed denial of service (DDoS) attacks 59, 254, 457
distributed protocol 281
distributed SDN networks, technical challenges of 498–499
Distributed Sensor Network project 59
DNP3 459, 461
DNS-based Authentication of Named Entities (DANE) 261–262
Dolus system 485–486
domain-specific language (DSL) 544–546
double hopping approach 586
Double Hopping Communication (DHC) 586
dynamic address translation, RDS 482
dynamical model, of 3D-printing system 387–389
dynamic data 565, 576–577
dynamic deception model
  attack graphs 335–336
  attack trees 335–336
  deploying fake nodes/networks 336
  experimental results 346–352
  exploit dependency graph 335–336

POMDP model
defender's action 337
defender's belief update 339–340
defender's observation 339
state space 336–337
threat model 337–338
schematic illustration 335
dynamic detection of malicious flows, RDS 482
dynamic device attestation 440
dynamic flit de-permutation (DFDeP) 420
dynamic flit permutation (DFP) technique 420
dynamic graph 281
dynamic honeypot 126
dynamic network 565, 577
dynamic network-identity 577
dynamic network-infrastructure 577
dynamic network-path 577
dynamic network traffic configuration 577
dynamic permutation 416, 420
dynamic platform 565, 577
dynamic runtime environment 565, 577
dynamic software 565, 577

e

ECC decoder (DeECC) 420
edge computing-driven intelligence, BlendCAC 523
e-governance 186
eHealth domain 222
electromagnetic analysis attacks 410
ELF-based access control (ELFbac) 531, 541–542, 642
code *vs.* data 542–543
design 543
implementation 543–544
Spectre Variant 1 549–551
SSH Roaming Bug 547–549
using Mithril 544–546
elliptic curve cryptography (ECC)-based authentication 509
emulation-based decoys 226
encryption 457
end-devices layer 302
end-to-end delay 453
end-to-end packet delay 463–464
end users/applications layer 303
energy delivery systems (EDS) 453, 454
bulk generations 456
customers 456
distribution 456
markets 456
operations 456
QoS metrics for
authentication delay 464
data confidentiality delay 464
data integrity delay 464
end-to-end packet delay 463–464
throughput 464–465
total end-to-end packet delay 464
and SCADA communication network 456–458
service providers 456
system resilience 454
transmission domain 456
equilibrium analysis of cyber-physical games
FlipIt game 396–397
zero-sum game 394–395
error control code (ECC)-based error deflection mechanism 417
error deflection method 417
error ratio 474
Ethernet 2
event-condition-action (ECA) 197
evidence 35, 44
evolution/dynamic games 158

Executable and Linkable Format (ELF) 542–543
Executable and Linkable Format (ELF)-based access control (ELFbac) 22
exploit complexity score 276
exploit dependency graph 335–336
Extreme Studentized Deviate (ESD) 620

f

FairAccess 516
fault analysis (FA) attacks 417
federated capability-based access control model (FedCAC) 511–512, 517
federated capability-based delegation model (FCDM) 517
Field-Programmable Gate Array (FPGA) 410, 412–415, 418, 422, 424–429
fine granularity *vs.* minimization 262–263
fingerprinting 187–189
finite-state machine (FSM) 543–548
first-hand reputation 97–98
first-in first-out (FIFO) 419, 420
first-level hypergame
  cyber deception modeled as 64–67
  equilibrium of a 67–69
Fisher–Yates shuffle algorithm 596
fixed shuffling (FS) 234
FlipIt game 391–392
flow_score database 312
flow security 261
fog/edge computing 517
formal language theory 532
  alphabet 534
  Chomsky hierarchy 535, 536
  context-free 536
  decidable 536
  equivalence problem 536
  expressiveness 535, 536
  grammar 534, 535
  language 534–536
  recognizable 535
  recursive 536
  recursively enumerable 535
  regular 535
  string 534
  symbols 534
FPGA-Oriented Moving Target Defense (FOMTD) method 422, 424, 429
frequency-hopping spread spectrum (FHSS) 304
FS-GANT 235
FS-RNT 235
full OS-based decoys 226
Fuzzy C-Means and Artificial Neural Network (FCM–ANN) method 620

g

GA-based network shuffling optimization 229–230
game followers 91
game leaders 91
games 91
game-theoretic deception model 126–130
game-theoretic models 30, 62, 124
  cyber deception 155–157
    signaling games 151–152
    Stackelberg game 152–153
game-theoretic reputation-based mechanism 82
game theory 111–118, 485, 637
game theory in security 32–33
  perfect Bayesian Nash equilibrium 34–35
  signaling game model 33–34
GANT 235

Gaussian wiretap model 86
genetic algorithm (GA) 218, 219, 229, 231, 234–236, 239–244, 454
GitHub 256
Global Environment for Network Innovations (GENI) 481, 492–496
governance
 description 374–375
 digital economy, IoT
  applying resilience to 378–379
  centralized governance approach 377
  current status 375
  laissez-faire approach 376
  pre-emptive strategy 376–377
  stewardship model 377
 formal 375
 informal 375
graph-based temporal modeling 199
graphical security model (GSM) 218, 219, 221–222, 227, 230, 231, 244

h

halting problem 535–536
Hammer 540, 541, 555–556
Hamming distance 363–366
Hamming weight 366
hardware attestation 441, 446
hardware security threats 407
 hardware Trojan 411–412, 418–422
 from multiple stages of IC design flow 408, 409
 side-channel analysis attack 408, 410–411, 416–418
hardware tampering 408
hardware Trojan (HT)
 combinational 411
 countermeasures, high-level overview of 419
 destructive method 419
 detection 419
 dynamic flit permutation technique 420
 hit rate 426–429
 mitigation method 419
 network-on-chip bandwidth depletion 419
 nondestructive methods 418–419
 router architecture 420, 421
 sequential 411
 structure of 411
 taxonomy 412
 traffic hotspot migration and bandwidth depletion induced by 422, 423
Harmonic Mean to Arithmetic Mean ratio (HM–AM ratio) 619–620, 628
Hash-based Message Authentication Code (HMAC) function 461
hierarchical attack representation model (HARM) 219
hierarchical density based spatial clustering of applications with a noise algorithm (HDBSCAN) 193
hierarchical time-series feature extraction (HTSF) method 620
high-interaction honeypot 125
hit rate 426–429
honeypot 125, 220
honeypot placement, RDS 482
honey-x deception method 155
horizontal attack surface 256
host-centric approach 578, 579, 582
host controller interface (HCI) protocol frame 322
hot-swappable submodule assembling technique 426
HT countermeasures (HTC) 419
HTTP Secure (HTTPs) 570
human-CPS (H-CPS) systems 165

hybrid-IoT 516
hybrid methodology 190
hypergame 60–62, 64–68, 74, 75
hyper-game-based cyber deception strategy 637
hypergame-based defense strategy
  Advanced Persistent Threat attacks 62
  attack graphs 62
  cyber-deception strategy 60
  equilibrium of a first-level hypergame 67–69
  experiments 69–75
  game-theoretic models 62
  honeypots 60
  hypergames 61–62
  Insider Threat 62
  IoBT 59
  IoT devices 59
  meta-game models 62
  spread of an attack 62–63
    cyber deception modeled as a first-level hypergame 64–67
    time-varying attack graph 63–64
Hypertext Transfer Protocol (HTTP) 570

i

identification 187–189
IEEE 802.11. *see* Wi-Fi protocol
IEEE 802.11a 304
IEEE 802.11ac 305
IEEE 802.11b 304
IEEE 802.11g 304
IEEE 802.11n 304–305
IIoT-technology-enabled Energy Delivery Systems 641
Industrial Internet of Things (IIoT) 453
industrial IoT (IIoT)
  future challenges
    integrated cyber and physical MTD measures 602, 644
    MTD-aware SDN deployment 602, 644
  overview 567
  SDN-based MTD 600–601
  security challenges 573
Infer 558
Information Aggregation Services (IAS) 506
information and communication technologies (ICTs) 186
Information Centric Network (ICN) 499–501
information technology (IT) services 186
information-theoretic secrecy 85–89
inner Vickrey auction 81
Insider Threat 62
Integrated Circuit (IC) 408
integrated proactive defense mechanisms
  attack model 224–225
  comparing schemes 234–235
  defense model
    decoy system as defensive cyber deception 225–226
    network topology shuffling-based MTD 226–227
  experimental setup 231–233
  GA-based network shuffling optimization 229–230
  GSM for the IoT MTD 230–231
  metrics 233–234
  MTD and cyber deception techniques 219–221
  network model 223–224
  network topology shuffling with decoy nodes
    adaptive shuffling of a network topology 228–229

initial deployment of nodes 227–228
selection of a network topology to shuffle 228
node model 224
overview 219
research goal and contributions 218–219
SDN technology 222–223
security failure conditions 227
security models and metrics 221–222
simulation results 238–243
simulation steps and parameter details 235–238
Intellectual property (IP) piracy 408, 415
Intelligent Electronic Device (IED) 453
intent 529–530
interface-assisted secret communication 88
Internal Block Diagrams (IDBs) 171
Internet Engineering Task Force (IETF) 587
Internet of Battlefield Things (IoBT) 9, 59–63, 65, 124, 635
future research challenges
local intelligence 602, 644
SDN Control Channel 601–602, 644
SDN-based MTD 599–600
Internet of Things (IoT) 79, 273, 299, 371, 386, 435, 479, 505, 530, 615, 635
adaptation and deception in adversarial cyber operations 10
anomaly behavior analysis 15–16
application 407
authorization architecture 510–512
authorization model 508–510
blockchain-based security solutions 515–517
book roadmap 5–6
building blocks 568–569
coding theoretic view of secure state reconstruction 17
collaborative-aware services 506
communication requirements 569–571
in critical infrastructures 563
cyber-physical vulnerability analysis 12
data analytics 645
decentralized access control 21–22
deception for cyber adversaries 11
device attestation (*see* device attestation)
digital economy governance
applying resilience to 378–379
centralized governance approach 377
current status 375
laissez-faire approach 376
pre-emptive strategy 376–377
stewardship model 377
dynamic cyber deception 16–17
game-theoretic analysis of cyber deception 7–8
game-theoretic model for deception-based security 10–11
game-theoretic modeling of trust management 9–10
governance for the Internet of Things 18
hyper-game-based defense strategy 8–9
identity-related services 506
information aggregation services 506
integrated proactive defense mechanisms 13–14
intent as a security design primitive 22–23
IoT device attestation 20

Internet of Things (IoT) (*cont'd*)
leverage SDN for cyber security deception 21
moving target defense against hardware threats 19
moving target defense mechanisms 23–24
MTD-Based IPv6 for 589–592
OSI layer 571
overview 1–2, 4–5, 566–567
pervasive networked computing capabilities 273
polymorphic advanced threats 14
privacy threats 574
resilience-based approach 374
resilient system 373–374
risk assessment 372–373
robust outlier detector 24
secure and resilient control of IoT-based 3D printers 19
security and privacy challenges and opportunities 2–4
security and privacy issues 572–574
security risk assessment 273
security threats
hardware 407
hardware Trojan 411–412, 418–422
from multiple stages of IC design flow 408, 409
side-channel analysis attack 408, 410–411, 416–418
software (*see* software security threats)
smart cities 12–13
smart moving target defense 602, 645
software-defined networking for cyber resilience 20–21
stepping stone attack 15
transport layer protocols 570–571
ubiquitous services 506
Internet Protocol version 4 (IPv4) 570
Internet Protocol version 6 (IPv6) 570
interquartile range (IQR) 624
in-the-wild analysis 257
intrusion detection system (IDS) 218, 250, 336, 339, 340
intrusion detection system for Bluetooth protocol 318
intrusion detection system for Wi-Fi protocol 303
intrusion detection system rules 316–317
intrusion prevention systems (IPSs) 218
IoT architecture 299–300
IoT aware Smart Healthcare System architecture 345–346
CIA-triad 345
confidentiality 345
integrity 345
low-power wireless personal area network 345–346
with network decoys 348
RFID-enhanced wireless sensor network 345
IoT-based 3D-printer. *see also* 3D-printing system
APT-type attacker 390–391
architecture of 386
attack model of 390–391
cyber-physical Stackelberg game model 392–393
discrete-time Markov jump system 389
dynamical model of 387–389
FlipIt game 391–392
game-theoretic approach 387
numerical experiments, to evaluate defense mechanism
control inputs 400
move-cost ratio 402, 403

system parameters 397, 398
tracking performance 398–401
transition matrices 398
IoTChecker 206
IoT federation 201–209
IoTFuzzer 207
IoTHound 192–196
IoT sensor design 175–176
IoT sensor node 175–176
IoT Sentinel 198
IoT threat modeling methodology 301–303
IPv6 over Low-Power Wireless Personal Area Networks (6LoWPAN) 2
IT/OT networks 332

j

Jelinek-Mercer Smoothing method 322

k

knowledge acquisition 44
KNOX 442

l

LangSec parsers 205
language-theoretic security (LangSec) 22, 531–533, 642
Chomsky hierarchy 536
equivalence problem 536
input recognition 537–539
into system design and development 539–540
theory of computation 534
undecidability 536
leader–follower strategy 283
learning and rational play 112
levarage SDN-based cyber-deception platform
camouflage 481
cooperative IoT security 488, 489
Cybermoat 483–484
deceptive routing in relay networks 484–485
decoys 481, 483–484
Dolus system 485–486
game theory 485
GENI experiment
attack alert time 493, 494
attack-information sharing time 495, 496
experimental topology 492
flow-installation time on OpenFlow switches 493–495
total time 494, 496
hardware experiments 496–498
misrepresentation 481–482
ONOS 457
probabilistic logic 484
protocol design
attack-information message handling 490–491
initialization 489
normal packet handling 490
registration 489–490
reconnaissance deception system 482
threat model 487–488
lifecycle of cyber deception 142–143
limitations of game theory 117–118
linear time-invariant (LTI) system 359
secure state reconstruction 359
Hamming distance 365–366
Hamming weight 366
minimum Hamming distance of affine code 366–367
sparse strong observability 364–365
strong observability 364–365
logging and profiling 262
Long Range Wide Area Network (LoRAWAN) 2
loss of confidentiality 224, 227

loss of system integrity 224
low-interaction honeypot 125
6LoWPAN 570
low-powered wireless personal area networks (LPWPANs) 220

m

Macaroons 511
machine learning 621
Markov decision processes (MDP) 113
measure of the device 436
median absolute deviation (MAD) 620, 628
medium-interaction honeypot 125
message authentication codes 573
message queuing telemetry transport (MQTT) 552
messages authentication code (MAC) 459
M-estimator 624
meta-game models 62
Micro-Moving Target IPv6 Defense (μ−MT6D) 591–592
mimetic deception 158, 642
min-consensus protocol 282
min-plus algebra 286–292
Mirai botnet attack 572
Mirai distributed denial of service (DDoS) attack 193
misrepresentation 481–482
Mithril 544–546, 556–557
mixed-strategy equilibrium 33
mixing 154–155
mobile ad-hoc networks (MANET) 487, 498–499
  MTD-based identity virtualization for 592–595
model-based systems engineering (MBSE) approach 169
modeling smart cities 196–201
Mondrian 542
Monte Carlo simulations 274
Monte-Carlo tree search (MCTS) 343
moving proximity base station defense (MPBSD) 595
moving target defense (MTD) 154, 218–222, 225–230, 234–239, 242–244, 639, 641
moving target defense (MTD)-based design 18
moving target defense (MTD) techniques 422, 564–566, 576–580
  collaborative multi-attribute, future challenges 602–603
  feasibility 586–588
  games, future challenges 602
  SDN-based 580–586
Moving Target IPv6 Defense (MT6D) 589–590
MTD and cyber deception techniques 219–221
MTD-based DDoS Resistant Multicast (DRM) 595
MTD-based identity virtualization 592–595
MTD based network infrastructure randomization 595–596
MTD based network traffic configuration randomization 596
MTD based on identity randomization 589–595
MTD based on non-identity randomization 595–596
MTTC 234
MTTSF 233
multi-domain system model hierarchy 171
multiple-input multiple-out (MIMO) techniques 80
multiprotocol label switching (MPLS) 585
multistage dynamic Bayesian game model 48–50

n

Named Data Networks (NDN) 481, 499–501
Narrow Band IoT (NB-IoT) 2
Nash equilibrium (NE) 30, 33, 112, 124, 396
Nash games 33
National Institute of Standards and Technology (NIST) 456
National Vulnerability Database (NVD) 167, 274
N computing devices 130–132
netlist 412, 426
network-aware notification 262
network-centric approach 578, 582, 583
network interfaces (NIs) 419
network layer 251, 252
network-layer approach 190
network layer level 260–261
  authentication 261
  flow security 261
  network-aware notification 262
  profiling and logging 261
  secure naming and resolution 261–262
network model 223–224
network-on-chip (NoC) 419
network topology-based shuffling (NTS) 244
network topology shuffling 218
network topology shuffling-based MTD (NTS-MTD) 218, 226–227
network topology shuffling with decoy nodes
  adaptive shuffling of a network topology 228–229
  initial deployment of nodes 227–228
  selection of a network topology to shuffle 228
neurodynamic programming 113
n-grams 311–312
node model 224
noise forwarding (NF) 88
non-dominated sorting genetic algorithm (NSGA-II) 454
non-operation-critical services 466, 472–473
nonvolatile memory (NVM) 81
Norbert Weiner's system theory 148
number of attack paths toward decoy targets 233

o

obfuscation-based cybersecurity 154
off-line POMDP solver 342
one-way delay (OWD) 453
online defense algorithm
  description 342
  POMCP 342
    action selection procedure 343
    belief nodes 343
    belief state updates 343
    optimal policy 345
    optimum action 343
    schematic illustration 343, 344
ONOS 457
open architecture and IoT security 166–167
OpenFlow (OF) 222
OpenFlow protocol 459
OpenFlow switch 480, 493–495, 575
OpenMETA multi-modeling environment 170–175
OpenPLC 447
Open Syringe Pump 447
Open Virtual Switch (OVS) software 492, 493
operation-critical service 465, 470–472
organizationally unique identifier (OUI) 193
orthogonal frequency-division multiplexing (OFDM) 304
outlier detection method 619

p

Packet Header Randomization (PHEAR) technique 583, 584
pairing attack 320
Parametric Exploration Tool (PET) 172
Pareto front 466, 467, 471–474
Pareto-optimal front 474
parsers
  combinator 540
  recognition 532
  sanitization 532
  shotgun 533
  validation 532
partially observable Markov decision process (POMDP) model
  defender's action 337
  defender's belief update 339–340
  defender's observation 339
  state space 336–337
  threat model 337–338
partially observable Monte-Carlo planning (POMCP) algorithm 17, 342
  action selection procedure 343
  belief nodes 343
  belief state updates 343
  optimal policy 345
  optimum action 343
  schematic illustration 343, 344
passive identification methods 190
Path Hopping SDN Network Defense (PH-SND) 586
pattern analysis module 624
payload circuit, hardware Trojan 411, 412
PCB design 176–177
perfect Bayesian Nash equilibrium 31, 34–35, 37, 45
Performance Measurement System (PeMS)
perturbation 154
phasor measurement unit (PMU) 459
phishing 29
physical attack 408
physical characterization 191–192
physical-layer based approach 190
physical layer secrecy, overview to 83
  cooperation for enhancing secrecy in CRNs 89–91
  information secrecy *vs.* cryptography methods 84
  information-theoretic secrecy 85–89
  vulnerability of key storage mechanism 84
physical unclonable functions (PUF) 441, 446
pipelining 549
Place&Route stages 426
playbook 459
Policy Rule 509
POMCP algorithm 640
POMDPy software package 347
pooling equilibrium 152
positioning and tracking attack 320
Postel's robustness principle 539
power analysis attacks 410–411
power distribution network (PDN) noise-based countermeasure 418
predator–prey games 115
primary users (PUs) 80
principal component analysis (PCA) 617
principles
  least expressiveness 538
  least privilege 538, 541
  parser equivalence 538
  Postel's 539
printed circuit board (PCB) 173–174
privacy threats 574
privilege separation, intra-process memory isolation 541

proactive defense mechanism 218, 219, 221, 227, 230, 231, 233, 244
Probabilistic Suffix Tree (PST) 208
profiling and logging 261
property-right spectrum leasing mechanisms 90
protocol
  advanced message queuing protocol (AMQP) 552–557
  message queuing telemetry transport (MQTT) 552
  OpenSSH 547
  SSH 547
  supervisory control and data acquisition (SCADA) 552
prover (device) 436
pseudorandom replica selection 425
PSOC Core 176–177
PUF IC attestation 446
pure-strategy Nash equilibrium 33

q

Q-learning algorithms 113
quadrature amplitude modulation (QAM) 188
quality of service (QoS) 453
  collector 460
  metrics for EDS
    authentication delay 464
    data confidentiality delay 464
    data integrity delay 464
    end-to-end packet delay 463–464
    throughput 464–465
    total end-to-end packet delay 464
Quartus Chip Planner 415–416

r

radio frequency (RF) sensing 188
random route mutation (RRM) 585
Random Route Mutation Technique 334
rate-equivocation pair 85
"rational" best-response learning 117
real-world adversarial situations 114–117
received signal strength information (RSSI) 191
reconnaissance attacks 224
reconnaissance deception system (RDS) 482
reconnaissance phase 332
reconnaissance tools 334
recursive-PCA (R-PCA) algorithm 618–619
Regional Transmit Unit (RTU) 453
register-transfer level (RTL) 412
reinforcement learning 113
relay channel 86–87
reliable 92
replay attack 320
representational-state-transfer (REST)-based application programming interfaces (APIs) 197
reputation 97
reputation-based game-theoretic model 96–97
  definition of reputation 97
  first-hand reputation 97–98
  second-hand reputation 98
  solution of the proposed game 99–102
  utility of the game players 98–99
resilience-based approach 374
resilient IoT system 373–374
REST interface 576
reverse engineering 408
reverse engineering attack 408
risk-based/vulnerability-based security model 222
RNT 235
robust PCA (RPCA) 619
role-based access control (RBAC) model 508–509

route mutation, RDS 482
rules-based detection 250

S

safe and secure communications 204–205
Samsung KNOX 442
satisfiability modulo convex (SMC) programming 368
satisfiability modulo theory (SMT) solving 368
S-Box 417
Scalable Embedded Device Attestation (SEDA) 445
SDN 222–223
SDN-based collaborative dynamic network 586
SDN-based dynamic network-identity 582–584
SDN-based dynamic network infrastructure 585–586
SDN-based dynamic network-path 584–585
SDN-based IoT 219
SDN-based on-demand routing protocol (SDOV) 599
SDN-based switches for Smart Home 597
SDN-based vehicular network 598, 599
SDN-based *vs.* traditional network switches 574, 575
SDN Controller 482
SDN-enabled communication network 459
SDN-enabled Evolved Packet Core (EPC) 598
SDN-enabled IIoT-based EDS
  best effort service 473–474
  multi-objective optimization problem 465–469
  non-operation-critical services 465, 472–473
  operation-critical service 465, 470–472
  performance evaluation of optimized security settings 474
  security risk levels
    authentication risk level 462
    confidentiality level 461–463
    end-to-end integrity level 461
    maximum integrity level 462
    total 463
Seasonal Extreme Studentized deviate (S-ESD) 628
Seasonal-Hybrid-ESD (S-H-ESD) 620, 628
secondary users (SUs) 80
second-hand reputation 98
secrecy capacity 85
secrecy rate 85
secure, affordable defense service 219
secure identity-based capability (SICAP) system 510
secure naming and resolution 261–262
secure state reconstruction
  attack detection 360
  attack model 359–360
  classical coding theory 362
    affine code 363
    Hamming distance 363–364
    linear code 363
  decoding algorithms 367–368
  as error-correction problem 360–362
  linear time-invariant system 359
    Hamming distance 365–366
    Hamming weight 366
    minimum Hamming distance of affine code 366–367
    sparse strong observability 364–365
    strong observability 364–365

notation 358–359
system model 359
security (system) 537
security failure condition 1 (SFC1) 227
security failure condition 2 (SFC2) 227
Security Layer for Smart Home (SLaSH) 14, 249, 257, 639
security patrols 111
self-adaptive end-point hopping technique (SEHT) 586
selfish 80
self-play 114
self-reporting reputation methods 83
SELinux 542
separating equilibrium 152
separating in low states and pooling in high states (SLAPH) strategy 47
server honeypot 142
service layer 252, 303
service layer level
fine granularity *vs.* minimization 262–263
logging and profiling 262
service-oriented software-defined security (SO-SDSec) 459, 460
Session Keys 461
side-channel attacks (SCAs)
dynamic masking plus error deflection method 417–418
electromagnetic analysis attacks 410
power analysis attacks 410–411
power distribution network (PDN) noise-based countermeasure 418
success of 408
timing attacks 410
signaling game model 33–34
signaling games 33, 151–152
signaling game with binary state space 35
signaling game with continuous state space 42
Simple Agile RPL multiCAST (SARCAST) 595
simple electromagnetic analysis (SEMA) attacks 410
simple power analysis (SPA) attacks 410
SI wave analysis 178, 180
SI Wave PCB model 178
SI Wave tool 178, 180
SLAPH 53
slice position selection through user constraints file 424–425
SMART 442, 443, 447
smart cities
connecting and federating IoT technologies in 201–202
attack surface minimization 206
operation and maintenance 207–209
provenance 205–206
risk assessment and scenario prediction 203–204
safe and secure communications 204–205
security assessment 206–207
security requirements 202
sunset/disposal 210
unanticipated compositions 203
detection, identification, fingerprinting, and behavioral monitoring in 187–189
cyber characteristics 191
IoTHound 192–196
physical characteristics 191–192
state of the art 189–190
epilogue 210
modeling
automated reasoning engines 200–201
devices and interactions 199–200

smart cities (*cont'd*)
state of the art 196–198
temporal properties 200
origins 186–187
smart contract
auditability 515
automation 515
description 514
hierarchical structure 515
interoperability 515
Smart Grid AMI networks 596
Smart Grid initiative 567
smart home sensor 175–177
Smart speaker 300
sniffer module 313
software attestation 440–441
ATRIUM 444, 447
control-flow attestation 443–444, 447
Samsung KNOX 442, 447
SMART 442, 443, 447
Software-Defined Internet Exchange Points (SDXs) 485
software-defined networking (SDN) 218, 219, 222–225, 227–230, 244, 454, 455, 479, 565, 566
controller 459
distributed SDN architectures 486–487
in MANET 487
optimization model 460
shuffle protocol operations 582
shuffle technique 582
software-defined networking (SDN)-based IoT environment 218
Software Defined Wireless Sensor Network (SDWSN) 598
Software Guard eXtensions (SGX) 254
software security threats
countermeasure against
hardware Trojan hit rate 426–429
hot-swappable submodule assembling technique 426
moving target defense technique 422
pseudorandom replica selection 425
slice position selection through user constraints file 424–425
FPGA design software
Altera Quartus, attacks on 415–416
contaminated FPGA design suite 413
L-1 attacks 414, 427–428
L-2 attacks 414, 428–429
untrusted FPGA CAD tools 412
Xilinx ISE, attacks on 414–415
source nodes 283
sparse observability 364
sparse strong observability 364–365
spectrum scarcity 79
split architecture 259–260
SQLite 520
SSH Roaming Bug 547–549
Stackelberg 81
Stackelberg equilibrium solution 92
Stackelberg game 32, 91–92, 152–153
Stackelberg game model 392–393
Stackelberg Security Game (SSG) 153
state of a device 436
static device attestation 440
static leaders system 283
stepping-stone attacks 276, 280
stepping-stone dynamics
fixed topology case 284–285
switching topology case 285
stepping-stone path 276
cost of 276–277
subtractive manufacturing (SM) 385
supervised learning 626
Supervisory Control and Data Acquisition (SCADA) 453, 454, 552

communication network and EDS 456–458
master 459
slave 459
swarm attestation model 440, 445
system development life cycle (SDLC) 202
system model, device layer 250–251

t

TAP-based port and address hopping (TPAH) 578
taxonomy for cyber deception
honey-x 155
mixing 154–155
moving target defense 154
obfuscation 154
perturbation 154
schematic 153
taxonomy of information protection 145
Tennessee Eastman 51
theoretical computing
formal language theory 534
halting problem 535–536
theory of computation 534–536
Things component 2
3D-printing system
cyber-physical Stackelberg game model 392–393
discrete-time Markov jump system 389
dynamical model of 387–389
FlipIt game 391–392
threats in 386
zero-sum game framework 389–390
3D-printing technology. *see* additive manufacturing (AM)
three multi-stage multi-view stacking ensemble (TMSE) machine learning model 620
through-silicon vias (TSVs) 418
timeline of cyber-threads 297–298
time-series data 615, 616
time-varying attack graph 63–64
timing attacks 410
TinyOS environment 223
token data caching solution 522
tools
American Fuzzy Lop (AFL) 558
Hammer 540, 541, 555–556
Infer 558
Mithril 544–545, 556–557
Shodan 552
Topological Vulnerability Analysis (TVA) 347
traceroute function 334
Transmission Control Protocol (TCP) IP protocol stack 480
tree-based network structure 132–135
trigger circuit 411, 412
Turing machines 535
Turing-undecidable language 535
Turing-unrecognizable languages 535
2D and 3D analysis of circuit board 173–174
two-layered HARM 221
two-player zero-sum Markov game model 62

u

UCT algorithm 343
undirected graph 276
unmanned combat air vehicle (UCAV) engagement 111
Update Key 461
Urban Cybernetics 186
user constraints file 424–425
User Datagram Protocol (UDP) 570
utility 98

v

verifier 436
vertex/node 275

vertical attack surface 256–257
virtual local area networks (VLANs) 223
virtual network view generator, RDS 482
virtual personal assistant (VPA) system 207
vulnerabilities
  Android Master Key 532
  Heartbleed 532
  Mirai Botnet 530
  Shellshock 532
  Spectre 530
  SSH roaming 541
  zero-day 531, 551
vulnerability graphs 274, 277–279
vulnerability multigraph 276, 280
vulnerable host placement, RDS 482

W

WannaCry attack 297
Web of Things (WoTs) framework 508
weighted adjacency matrix 275
weird machine
  gadgets 533
  instructions 533
white-box attack strategy 630
wide area network (WAN)
  cyberattack types 456
  internal and external attacks 456–457
Wi-Fi 2
Wi-Fi frame structure 305
Wi-Fi header 305–306
Wi-Fi protocol 303–306
  data-link layer behavior of the 306–308
  experimental results and evaluation 314–318
  Wi-Fi IDS to detect attacks 308–314
Wi-Fi protocol state machine 306
wireless flows (Wflows) 312
wireless n-gram (W-gram) 309–310, 314
wireless sensor networks (WSNs) 220
wiretap model 85

X

Xilinx ISE, attacks on 414–415

Z

zero-sum game framework, 3D-printer 389–390
ZigBee protocol 2, 569

www.ingramcontent.com/pod-product-compliance
Lightning Source LLC
LaVergne TN
LVHW011534200926
848484LV00001B/262